FIELD INSTRUMENTATION
in
Geotechnical Engineering

FIELD INSTRUMENTATION
in
Geotechnical Engineering

A symposium organised by the
British Geotechnical Society
held 30*th May*–1*st June* 1973

LONDON
BUTTERWORTHS

THE BUTTERWORTH GROUP

ENGLAND
Butterworth & Co (Publishers) Ltd
London: 88 Kingsway, WC2B 6AB

AUSTRALIA
Butterworths Pty Ltd
Sydney: 586 Pacific Highway, NSW 2067
Melbourne: 343 Little Collins Street, 3000
Brisbane: 240 Queen Street, 4000

CANADA
Butterworth & Co (Canada) Ltd
Toronto: 14 Curity Avenue, 374

NEW ZEALAND
Butterworths of New Zealand Ltd
Wellington: 26–28 Waring Taylor Street, 1

SOUTH AFRICA
Butterworth & Co (South Africa) (Pty) Ltd
Durban: 152–154 Gale Street

First published in 1974

ISBN 0 408 70514 0

Printed and bound in Great Britain at The Pitman Press, Bath

Contents

Part 1: Papers

Part 2: Discussion

Preface

Improvements in design methods stem from knowledge of the behaviour of designed structures. Abnormal deformations indicate that the value of the factor of safety has fallen to one and in the early days of geotechnical engineering, failures were a powerful stimulus to development in design techniques. The types of measurements made to study behaviour depend on theories available at the time. Deformation measurements, particularly those to show settlement, were amongst the first field measurements and results were compared with predictions based on laboratory oedometer tests. The use of effective stress theory in stability analyses led to field measurements of pore and total pressures. During this period there was little emphasis on deformation measurements, although *ad hoc* observations were often made if failure appeared imminent. Interest in the relationship between pore pressure and strain led to strain measurements within the body of the soil. Recent developments in analytical methods that give the designer the power to predict movements, have aroused an increasing interest in field measurements of deformations. Field instruments, in monitoring behaviour, also give a continuous check on stability.

Instruments have been developed by numerous organisations and expertise in their installation and use is fragmentary. It was the purpose of this Symposium to gather together a comprehensive selection of this available knowledge with the aim of improving field instrumentation generally. The 37 papers chosen cover a wide range and several new instruments and techniques were introduced during the discussion.

The Organising Committee, consisting of A. D. M. Penman (Building Research Station), R. H. Jones (Nottingham University), P. A. Green (Scott Wilson Kirkpatrick and Partners), and P. R. Vaughan (Imperial College, London) wishes to thank the authors, general reporters, delegates to the Symposium, the exhibitors and guarantors for their support. It is grateful to the Institution of Civil Engineers for its initial work including the call for papers, to the Conference Department of the Institution of Electrical Engineers for handling the whole Symposium and to Butterworths for publication.

A. D. M. Penman
Chairman of the Organising Committee
and Technical Editor

THE MEASUREMENT OF SPATIAL DEFORMATIONS BY GEODETIC METHODS

V. Ashkenazi

Reader in Surveying, University of Nottingham.

REMOTE MEASUREMENT OF DEFORMATIONS

The remote observation and systematic monitoring of structural and spatial deformations by geodetic methods involves the measurement of horizontal and vertical theodolite angles and of distances to a series of ordinary and light reflecting targets. These mixed angle and range observations are usually processed on a computer, by well established statistical-geodetic methods, to provide not only the most probable absolute and instantaneous three-dimensional positions of the target points, but also an estimate of the accuracy of their determination which is essential for the correct interpretation of the results.

One can use the same statistical methods in reverse and predict optimum combinations of equipment and observational procedures which are necessary in order to achieve a certain degree of positional accuracy. Surveyors and geodesists express accuracy in terms of the root-mean-square or standard error, and consider it highly unlikely (1% chance) that the true error of a determination is larger than 2.5 times its standard error. The required standards of positional accuracy in the measurement of displacements may vary considerably, depending largely upon the magnitude of the displacements involved. Thus, for monitoring relatively large displacements, where one would be prepared to tolerate individual positional errors as large as 100mm, one would only need a method with an estimated average standard error of 40 to 50mm whereas, when dealing with small deformations, one would rather use a procedure leading to a positional accuracy of 5mm or better.

Special precautions or intricate refinements, such as instrument forced centring devices, would only be appropriate if their contribution to the overall accuracy of the measurement was commensurate with the ultimate positional standard error likely to be achieved by the adopted observational procedure. The use of the right procedure would speed up field observations and lead to substantial savings in manpower and equipment. It would also facilitate the correct interpretation of the results and give an adequate assessment of the true size of the actual displacements.

The following is a brief description of the rigorous statistical-geodetic method used for processing mixed angle and range observations. This, in turn, is followed by a sample of statistical predictions which illustrates the way in which different angle and distance measuring instruments could be combined, over target ranges from 50m to 1500m, in order to achieve varying orders of positional accuracy. The paper also contains a section listing the observational accuracies which can reasonably be expected from currently available theodolites and lightwave distance measuring instruments, and a brief account of the results achieved in field experiments. A short summary of the relative strength of pure and mixed triangulation, trilateration and traversing, as geometrical layouts for survey control networks, concludes the paper.

COMPUTATION OF MIXED ANGLE AND RANGE OBSERVATIONS

The simultaneous adjustment of horizontal and vertical angles and distance observations, with standard errors of different sizes and physically different dimensions, is best done by the well known method of 'variation-of-coordinates', (Bomford (1971), Ashkenazi (1967)). The following is a brief illustration of the method in the case of a determination of the unknown coordinates, x, y and z, of a target point, from two horizontal grid bearings, α_1 and α_2, two zenith angles, β_1 and β_2, and two distances, ℓ_1 and ℓ_2, measured from two stations with known coordinates, $x_1 y_1 z_1$ and $x_2 y_2 z_2$ respectively.

Notation

x_o, y_o, z_o	:	approximate values of the coordinates x, y, z
dx, dy, dz	:	unknown corrections to x_o, y_o, z_o
$x = [x_j]$	:	vector of the unknowns, dx, dy, dz
α_1, α_2	:	horizontal grid bearings (clockwise from the x direction)
β_1, β_2	:	zenith angles (downwards from the observer's zenith)
ℓ_1, ℓ_2	:	3-d distances from stations to target
$d\alpha$, $d\beta$, $d\ell$	:	small changes in α, β and ℓ

$A = [a_{ij}]$: coefficient-matrix of the observation equations

$b = [b_i]$: vector of the right-hand-side terms

$W = [w_{ii}]$: dimensional weight matrix

$v = [v_i]$: vector of residuals, corrections to observed quantities

$\sigma_x, \sigma_y, \sigma_z$: positional standard errors in x, y, z

σ : overall positional standard error, $(\sigma_x^2+\sigma_y^2+\sigma_z^2)^{\frac{1}{2}}$

ρ : 206265 (seconds-per-radian)

$\sigma_\alpha, \sigma_\ell$: observational standard error in bearing, in distance

Computational Procedure

The procedure consists of the following computational steps:

1 Compute, by any means, the approximate position of the target.

2 Calculate and set-up the elements of matrix A from the following differential formulae:

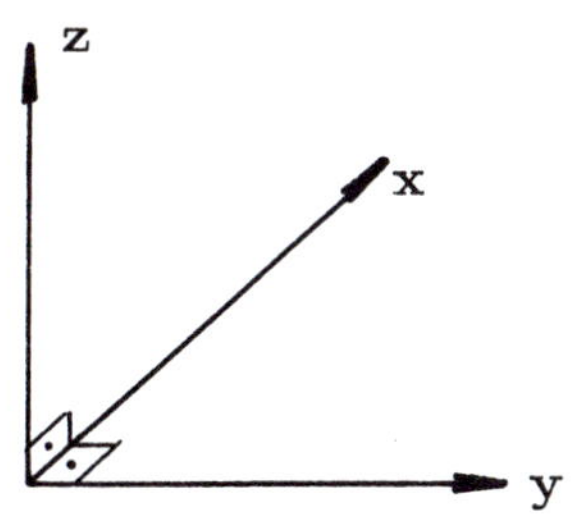

Figure 1 Reference system

	dx	dy	dz
$d\alpha_1$	a_{11}	a_{12}	a_{13}
$d\alpha_2$	a_{21}	a_{22}	a_{23}
$d\beta_1$	a_{31}	a_{32}	a_{33}
$d\beta_2$	a_{41}	a_{42}	a_{43}
$d\ell_1$	a_{51}	a_{52}	a_{53}
$d\ell_2$	a_{61}	a_{62}	a_{63}

Matrix A

$$d\alpha^{sec} = \left(-\frac{\rho\sin\alpha}{\ell\sin\beta}\right)dx + \left(\frac{\rho\cos\alpha}{\ell\sin\beta}\right)dy + (0)\,dz$$

$$\left.\begin{aligned} d\beta^{sec} &= \left(\frac{\rho\cos\alpha\ \cos\beta}{\ell}\right)dx + \left(\frac{\rho\sin\alpha\ \cos\beta}{\ell}\right)dy \\ &\quad + \left(-\frac{\rho\sin\beta}{\ell}\right)dz \end{aligned}\right\}$$

$$d\ell^{m} = (\cos\alpha\ \sin\beta)dx + (\sin\alpha\ \sin\beta)dy + (\cos\beta)dz$$

3 Compute the elements of vector b of the observation equations,

$$Ax = b + v\ ,$$

by subtracting, from the observed values of the six measured quantities, the corresponding values computed from the approximate coordinates of the target, x_o, y_o and z_o.

4 Establish the diagonal weight matrix W, with the reciprocals of the estimated mean-square-errors of the observations as diagonal elements.

5 Form and solve the (normal) equations

$$A^T WA\ x = A^T Wb$$

and obtain the final coordinates, x, y, z, of the target.

6 Invert the matrix $(A^T WA)$ to obtain, along the diagonal of the inverse matrix, estimates of the three positional accuracies σ_x^2, σ_y^2 and σ_z^2.

Step 4 above is essential for the adjustment of mixed observations with different physical dimensions. When the angular and linear units used in the computation are seconds-of-arc and metres, individual weights should be expressed in units of $1/\text{sec}^2$ and $1/\text{m}^2$. Thus, if angle and distance observations have respective average standard errors of, say, 1 sec and 1mm then the corresponding weights would be $1(1/\text{sec}^2)$ and $10^6(1/\text{m}^2)$. Failure to weigh the observation equations accordingly would effectively result in completely eliminating the contribution of the distance observations. Slight variations in the numerical values of the weights, on the other hand, would not affect the results significantly.

Some range measuring instruments (e.g. Geodimeter) are sometimes used in conjunction with an observation procedure leading to the measurement, to a very high degree of accuracy, of the differential distance (i.e. linear displacement) between two series of observations. One could deal with such a measurement by adding the 'observed differential distance' to the initial distance, as computed from the final adjusted coordinates of the target after the first series of observations, and by treating the resulting 'absolute distance' as an observed quantity and incorporating it in the second series of observations.

Usually, the reference system x, y, z is determined by the horizontal projection of the line joining any two of the observations pillars (y), the horizontal direction perpendicular to it (x) and the vertical direction perpendicular to both (z). The top of one of the observation pillars could conveniently be defined as the origin with suitable false coordinates. But, where a control network consisting of several observation stations exists, the reference system is based upon the false coordinates of one of the stations in the network or of a distant station, and the arbitrarily fixed horizontal bearing of a long line, preferably to a distant reference target.

Simple observational or computational devices are used to make all observations consistent with one another. For instance, a distance measured by a lightwave instrument may require a correction to change its observed value to what it would have been if it were measured from the same height as the trunnion axis of the theodolite. But, the correction would only be applied if the accuracy of the measurement warrented it.

STATISTICAL PREDICTION OF ACCURACIES

It can be shown (Ashkenazi (1970)) that the diagonal elements of the inverse of the normal matrix are statistical estimates of σ_x^2, σ_y^2 and σ_z^2, the positional mean-square-errors. Since the normal matrix, A^TWA, is only a function of the approximate geometry of the layout (matrix A) and of the estimates of the observational standard errors or weights (matrix W), it is possible to compute estimates of σ_x, σ_y and σ_z for any given geometry of the stations-target layout and any conceivable combination of observational accuracies. Conversely, one can obtain several practical combinations of angle and distance observational accuracies (corresponding to specific equipment and observational procedures) which would lead, for a given geometrical layout, to a pre-determined accuracy in the positioning of the targets.

The following is a simple example based upon the planimetric (two-dimensional) layout illustrated in figure 2.

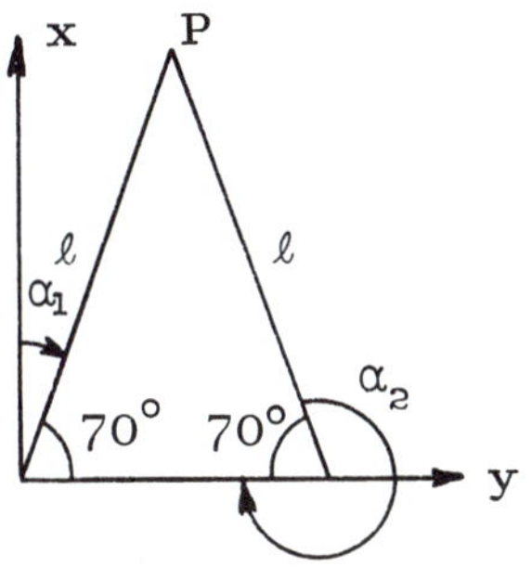

Figure 2 Layout geometry

	dx	dy
$d\alpha_1$	-47.03	+129.2
$d\alpha_2$	+47.03	+129.2
$d\ell_1$	+0.9397	+0.3420
$d\ell_2$	+0.9397	-0.3420

Matrix A (ℓ = 1500m)

If target P is determined either from only two angles or from only two horizontal distances (with no redundant observations), then its calculated position is not affected by the estimates of the observational accuracies. However, assuming that the two observations are free from blunders, the accuracy of the determination of P is a function of the estimated observational standard errors. Thus, a pure angular determination with an average standard error $\sigma_\alpha = 1$ sec would lead, for $\ell = 1500$m, to $\sigma_x = 15.0$mm, $\sigma_y = 5.5$mm and an overall positional accuracy of $\sigma = 16.0$mm, whereas a pure linear determination from two distances of about 1500m, with an average $\sigma_\ell = 1$mm, would result in $\sigma_x = 0.75$mm, $\sigma_y = 2.1$mm and $\sigma = 2.2$mm. A change in the range of P from the two observing stations, from 1500m to a smaller value, would lead to a proportionate decrease in the corresponding positional σ's, in the case of a pure angular determination, but it would not affect the accuracy of a determination by pure trilateration.

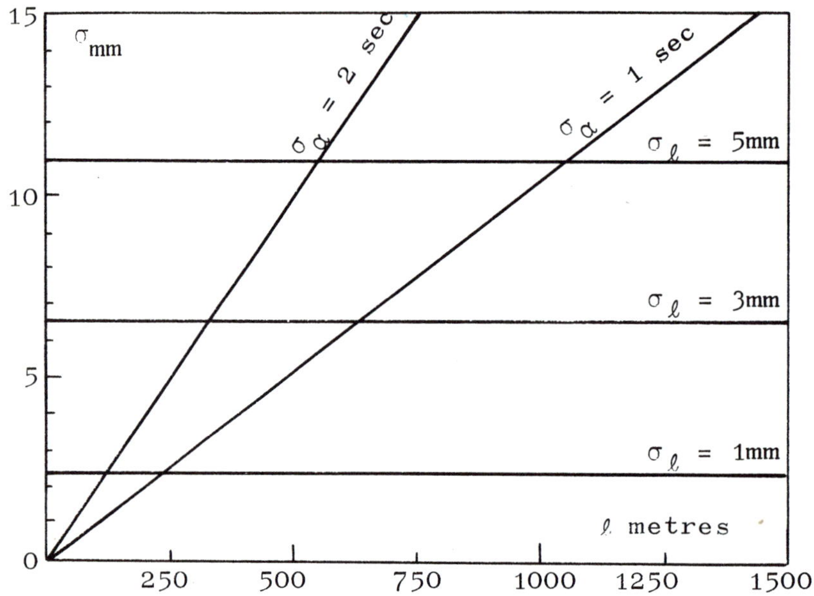

Figure 3 Pure angle and pure distance determinations

Figure 3 illustrates the variation in the estimated values of the overall positional standard error, σ, with change in the target range ℓ, in pure trilateration, for a number of specified observational accuracies. Table 1, on the other hand, lists the overall positional accuracies, σ in mm, resulting from mixed triangulation-trilateration observations, for a series of practical combinations of observational accuracies and for varying target ranges.

Both figure 3 and table 1 refer to the geometric layout in figure 2, but the relative magnitude of the results would not alter appreciably with small changes in the general shape of the layout. The combinations of σ_α and σ_ℓ used in table 1 were chosen realistically to correspond to currently (1972) available standard one-second-reading theodolites and distance measuring instruments. It is not a difficult computational task to produce a new table 1 for any other combination of angle and range measuring instruments and any

ℓ(m) / σ_α σ_ℓ	2 sec 2 mm	2 sec 5 mm	2 sec 10 mm	5 sec 2 mm	5 sec 5 mm	5 sec 10 mm
50	0.9	1.0	1.1	1.6	2.3	2.5
100	1.4	1.9	2.1	2.2	3.5	4.6
200	2.0	3.1	3.8	3.1	4.9	7.0
500	3.1	4.9	7.0	4.1	7.8	11.0
1000	3.9	7.0	9.9	4.3	9.7	15.6
1500	4.2	8.4	12.1	4.4	10.4	18.1

Table 1 Positional standard error in mixed observations

other conceivable elementary shape of layout.

Figure 4 illustrates the overall positional accuracies which are likely to be achieved, again for the geometrical layout in figure 2, over a variable range, for a very typical (in 1972) high accuracy combination of σ_α = 2 sec and σ_ℓ = 5mm.

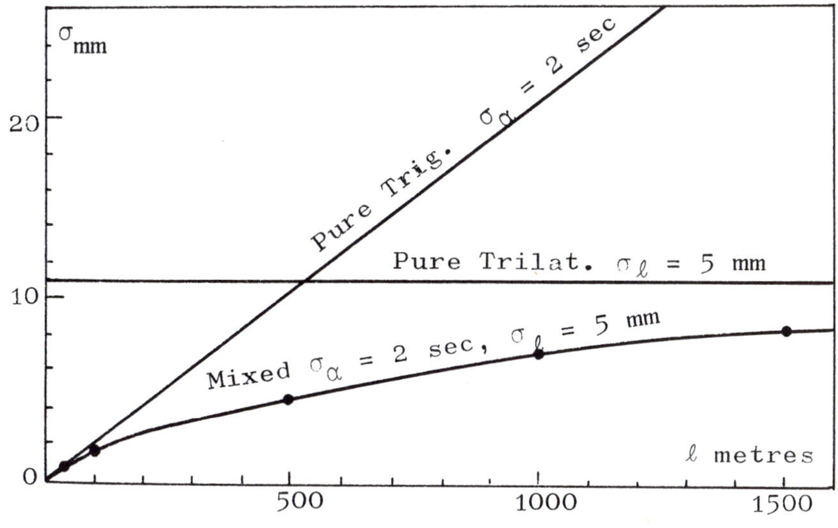

Figure 4 Positional accuracies by pure and mixed observations

The general pattern in figure 4 is typical, for any reasonable combination of angle and distance measuring instruments and for geometrical layouts roughly similar to that of figure 2. It seems, therefore, reasonable to conclude that for short target ranges, up to a few hundred metres, angular measurements alone are sufficient whereas, for very long target ranges beyond 1 km, distance measurements alone are essential. Combined angle-range observations seem particularly advantageous for middle target ranges between 400 and 1000 metres.

The conclusions above apply only to planimetric (two-dimensional) determinations of position. When vertical displacements are also involved, pure trilateration determinations lead to ill-conditioned (unstable) geometrical layouts and should not be contemplated. In general, a combination of distances and vertical angles would be sufficient for the spatial (three-dimensional) determination of a target. However, once a theodolite is set-up, centred, levelled and is pointing to a target, the extra effort involved in taking the horizontal angular reading, in addition to the vertical (or zenith) reading, is negligible and very worthwhile.

OBSERVATIONAL ACCURACIES FOR CURRENT EQUIPMENT

Theodolite Angles

It is generally accepted that the mean values of two or three complete rounds of horizontal angle readings, taken with one-second-reading theodolites, centred upon stable pillars, are likely to have an average standard error ranging from 2 sec to 4 sec, depending upon the quality and adjustment of the instruments used, the expertise and care taken by the observer and the prevailing atmospheric conditions. A further decrease in the average standard error towards the 1 sec mark, which approaches the accuracy of primary triangulation, would require a very substantial increase in the number of rounds of readings and could only be recommended under very special circumstances.

Vertical angles, on the other hand, would normally average observational standard errors of about 7 sec to 10 sec (equivalent to 1/3 to 1/2 of a division of a 20 sec vertical circle bubble), reaching 2 sec to 5 sec when great care is exercised in levelling the split prism bubble, just before taking the reading.

Distances

Estimating the observational standard errors of distances measured by the great variety of visible or infra-red light-wave instruments, which are currently (1972) available, is a far more complex task. Values quoted by the manufacturers refer loosely to accuracy, precision, repeatability, resolution, root-mean-square-error and mean-square-error. To complicate matters further, the values quoted are almost always made up of a constant component and a component which is proportional to the distance measured and expressed in p.p.m. or in mm per km. However, for ranges up to 1 km, the constant component tends to be overwhelmingly dominant and could be considered safely to represent the overall accuracy.

Subject to the instruments being in good calibration (both for instrumental constant and for all frequencies), repeated tests by various observers seem to indicate that the overall standard errors of distances measured by currently available instruments vary within the range from 2 or 3mm to 10 or 20mm, with the great majority of the instruments between 5 and 10mm. These values exclude instrument centring errors, which may vary from 0.1mm (for highly sophisticated instrument centring methods) to 0.5mm (for optical plummets in good adjustment), and atmospheric corrections which usually amount to a few p.p.m.

As mentioned earlier, some of the lightwave instruments may be used in a differential displacement technique, based on a single frequency, which may lead to a differential distance (linear displacement) accuracy of 1 to 2mm. This was confirmed in field tests (Maughan (1972)) with a Geodimeter Model 6A, over a range varying from 20 to 750m. The technique is applicable to small displacements and fails when the displacements involved are more than a few centimetres.

EXPERIMENTAL RESULTS

A series of tests were conducted at Nottingham to measure the amount of spatial displacement, by the methods outlined in this paper, of light reflecting targets which were designed to move in three mutually perpendicular directions by precise pre-determined amounts. The differences between the observed displacements and the corresponding pre-set actual values were taken as an indication of the accuracy of the technique used. The contributory effect of the relative movement of the observation and reference points was practically eliminated from the observations, by using forced centring on pillars and by taking the pre- and post-displacement observations in quick succession.

Pure triangulation observations, involving horizontal and vertical theodolite angles, were made to two targets from a range of about 50m. The discrepancies between the observed and actual displacements, resulting from six sets of observations are listed in table 2, (Maughan (1972)).

Table 2 seems to confirm the theoretical expectation that pure triangulation, from a very close range of about 50m, could lead to accuracies of about 1mm in every direction. This corresponds approximately to an angular displacement of 4 seconds and is well within the capabilities of an average surveyor equipped with a one-second-reading theodolite. Considering the thickness of the cross hairs of some of the theodolites used in these tests and the rather unrefined target marks adopted, it is thought that one would probably achieve the same order of linear displacement accuracies for a target range extending up to 100 metres.

Set	Lower Target			Upper Target		
	Δx	Δy	Δz	Δx	Δy	Δz
1	-0.80	0.21	0.95	-0.54	-0.03	0.29
2	0.66	-0.07	0.59	-0.90	-1.39	-0.63
3	1.04	0.18	0.38	2.35	-0.23	0.57
4	0.56	-3.61	0.41	-4.97	-2.10	1.47
5	-0.14	0.18	0.87	-0.54	0.08	0.31
6	0.31	-0.15	-0.97	-0.89	-1.39	-0.71
mean	0.58	0.73	0.70	1.70	0.87	0.66
r.m.s.	0.66	1.48	0.74	2.32	1.17	0.77

Table 2 Observational errors in mm

The inclusion of absolute linear measurements by Geodimeter to light reflecting targets (consisting of truck reflectors marked by cross-hairs) did little to improve upon the accuracies already achieved by pure triangulation. Unlike differential distances measured by displacement techniques, leading to accuracies of 1 to 2mm, these absolute distances had an estimated average standard error of 5 to 7mm which obviously was far too large to contribute significantly at the target range of 50m.

SURVEY CONTROL NETWORKS

Throughout the preceding sections of this paper, it has been assumed that the two or three reference stations, from which the observations were made, are perfectly stable. This is a safe assumption only in low accuracy work or when all the station pillars are built upon very stable foundations (e.g. rock).

But, in general, the precise monitoring of changes in the absolute coordinates of target points must include independent observations made to determine changes in the positions of the stations with respect to the common and arbitrarily held fixed reference system. The latter is usually defined by the coordinates of one station (thought to have a relatively higher degree of stability) and the bearing of one line joining that station to an equally stable (and preferably distant) reference target.

Computer simulated optimisation tests were carried out at Nottingham (Ashkenazi, Parker (1972)) to find the relative strength of various geometrical configurations, for a small engineering survey control network, made up of about 15 stations and with side lengths averaging 500 metres. The following are some of the conclusions based upon the results of these tests.

1) Triangulation, strengthened in scale by a small number of measured distances, is the strongest geometrical layout for almost any practical combination of angular and linear observational accuracies corresponding to theodolites and lightwave distance measuring instruments which are currently (1972) available or are being developed.

2) A scheme of traverses generally constitutes a good alternative and may even be superior to a mixed triangulation-trilateration network when the relative accuracy of the measured distances is much higher than that of the observed angles (e.g. 1mm and 5 sec).

3) Pure trilateration, even with a substantial number of redundant measurements, usually constitutes a weak network, particularly in terms of orientation control. It could be considered as a substitute for a mixed triangulation-trilateration network or for traversing only in low precision displacement work or when the network distances could all be measured to a very high degree of accuracy (e.g. 1mm).

ACKNOWLEDGEMENTS

The experimental work leading to this paper was carried out in the Department of Civil Engineering of the University of Nottingham, under the direction of Professor R.C. Coates. The writer is grateful to Dr. R. Wood for reading the manuscript and for many useful suggestions.

REFERENCES

Ashkenazi, V. (1967) Solution and Error Analysis of Large Geodetic Networks - Part 1: Direct Methods. Survey Review, 146.

Ashkenazi, V. (1970) Adjustment of Control Networks for Precise Engineering Surveys. Chartered Surveyor, 7, January.

Ashkenazi, V. and Parker, D. (1972) Optimisation of Engineering Survey Control Networks. Proc. F.I.G. Conf., Tel-Aviv.

Bomford, G. (1971) Geodesy 3rd ed. Clarendon Press, Oxford.

Kobold, F. (1961) Measurement of Displacement and Deformation by Geodetic Methods. Proc. ASCE Surv. and Mapp. Div., July.

Maughan, J.D. (1972) Measurement of Structural Deformations. B.Sc. thesis, Nottingham.

Penman, A.D.M. and Charles, J.A. (1971) Measuring Movements of Engineering Structures. B.R.S. paper CP32/71.

Reynolds, J.D. and Dearinger, J.A. (1970) Measuring Building Movement by Precise Survey. Proc. ASCE Surv. and Mapp. Div., April.

Robbins, A.R. (1972) Geodetic Measurements for Crustal Movements.

POSTSCRIPT: DEVIATION OF THE VERTICAL

Throughout the paper it has been assumed that the directions of the vertical at the various observation stations are parallel to one another and to the vertical at the station of origin, i.e. the z-direction of the coordinate system. Neglecting the generally unknown and negligible effect of irregular changes of mass distribution, the angle between the directions of the vertical at two (not too distant) stations is approximately 32.4 seconds per kilometre distance between the stations.

To counter this, strictly, one has to correct all observed horizontal theodolite directions, α, and all zenith angles, β by C''_α and C''_β respectively.

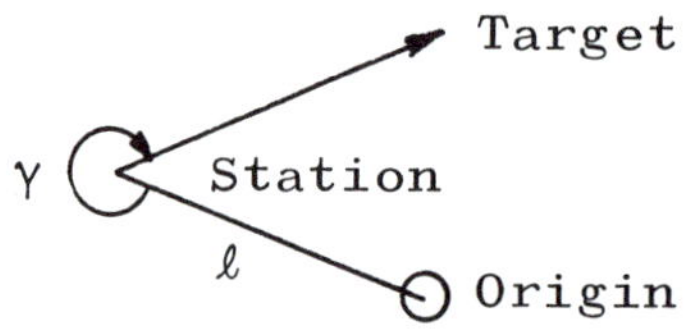

$$C''_\alpha = +\xi \sin\gamma \cot\beta$$
$$C''_\beta = -\xi \cos\gamma$$
$$\xi'' = \rho\ell/R = 32.4\ell \text{ (Km)}$$

where

ℓ = approximate distance of station from origin,
R = mean radius of the earth,
γ = clockwise angle at station, between origin and target,
ρ = 206000

However, since the corrections to be applied to the periodical values of the absolute coordinates of a target point are almost identical, there is no change in the amounts of deformation and, consequently, this extra refinement is unnecessary.

THE DEVELOPMENT AND USE OF TRIAL EMBANKMENTS

A.W. Bishop
Professor of Soil Mechanics, Imperial College, London

P.A. Green
Senior Engineer, Scott Wilson Kirkpatrick & Partners

INTRODUCTION

The increasing scale of earthworks in the past few decades and the growing range of reliable field instrumentation now available have led many engineers to reappraise the design procedures most likely to lead to safe and economical construction on a clearly foreseeable time scale. In this reappraisal increased emphasis has been given to instrumented trial embankments and it is the purpose of this paper to examine the part which is, or should be, played by such tests in design and in research.

In designing highway or water-retaining embankments the engineer will encounter a variety of problems relating to the nature of the foundation or of the fill (or to their interaction) and it is possible to distinguish at least seven different approaches which may be adopted:

(1) The use of empirical knowledge which has evolved from experience in a particular locality or with the use of particular foundation strata or fills.

(2) The use of index properties and a detailed geological description.

(3) The use of laboratory testing to determine the appropriate engineering properties coupled with quantitative methods of analysis.

(4) The quantitative analysis of the observed performance of similar structures (back-analysis).

(5) The use of model testing.

(6) The use of insitu field testing.

(7) The use of test sections, either full-scale or sufficiently close to full-scale to simulate realistically the principal variables.

It will be seen that these various approaches are in general complementary and not mutually exclusive. In deciding which approaches to adopt the engineer will be influenced by the scale and cost of the project, by the economic consequences of inadequate performance (in terms of variation of the contract or subsequent maintenance) and by the magnitude of the risk to life and property resulting from complete failure. In all but the most insignificant projects the basic approach is likely to be (3) - the use of laboratory tests and quantitative analysis - coupled with (2) to ensure that the samples are representative, relevant to the particular problem of stability, deformation or settlement, and of an adequate size[1] not to misrepresent the structure, or fabric, of the soil. The simpler insitu tests (6) are also likely to be included, in particular the constant-head permeability test as a check on field permeability and hence on the field value of coefficient of consolidation, and the vane test[2] if the end of construction stability of an embankment on soft clay is under consideration.

However, the engineer will sooner or later encounter one of a number of problems which are not readily amenable to standard procedures of testing and analysis or involve risks which call for a double check on the design. These problems relate to strength, both undrained and drained; rate of consolidation, both primary and secondary, as affecting both strength and settlement; and the magnitude of both compressibility and deformation modulus. These problems may be classified broadly into four categories:

(A) Problems of sampling and testing (for example, see Bishop (1966), Bjerrum (1967), Bjerrum and Kenney (1967), Rowe (1968), Bishop and Al-Dhahir (1969), Bishop (1971a), Marsland (1971) and Rowe (1972)). These problems occur in a range of soils, including

---

[1] So far as this is practicable. In other cases this data alone may indicate the need for a trial embankment.

[2] The status of the vane test as a direct measurement of insitu undrained shear strength has to be modified as a result of its correlation with the performance of trial embankments such as those reported by Eide and Holmberg (1972), Pilot (1972) and others. The present position is discussed by Bjerrum (1972), who suggests the use of substantial correction factors correlated with the plasticity index of the clay.

very sensitive soils, particularly of low strength; fissured overconsolidated soils of relatively high plasticity, such as the London Clay, Lias Clay, etc., in which both strength and rate of pore-pressure dissipation are influenced by sample size; all soils in which anistropy is significant; normally or lightly overconsolidated soils where consolidation is to occur under relatively small stress increment ratios close to the preconsolidation load; and very heterogeneous soils having an erratic structure.

(B) Problems of analysis (for example, see Bishop (1967), Kaufman and Weaver (1967), Bishop (1971 b) and Bishop (1972)). These problems include progressive failure, involving a soil of high brittleness index or the interaction of two or more soils of different stress-strain characteristics which are in general time-dependent; stress-strain and consolidation rate analysis in three dimensions; and analysis of settlement and rate of consolidation in heterogeneous strata.

(C) The physical properties of new types of fill or fill compacted using new techniques (for example, see McKechnie Thomson and Rodin (1972), Scott Wilson et al (1966) and Guilford and Chan (1969)). These types of fill and techniques include fly-ash (pulverised fuel ash), colliery spoil, till placed at very high moisture contents and fill placed under water.

(D) The consequences of using various geotechnical processes (for example, see Wilkes (1972), Murray (1971) and Dallard (1971)). Such consequences include the smear and remoulding produced in forming various types of vertical sand drain and their effect on subsequent performance; effects of preloading etc.

When faced with these problems the engineer may adopt a very conservative design procedure or he may resort to the use of special sampling and laboratory techniques and to the use of very large samples. He may also use large size insitu tests. All three are expensive and all three involve the further step of analysis. The engineer may therefore turn with some justification, and perhaps with relief, to the idea of an instrumented trial embankment, the last of the seven approaches listed. However, the use of a trial embankment does not permit the problem of analysis to be evaded, unless the foundation strata are so uniform (or so uniformly heterogeneous on a scale comparable with the size of the trial embankment) that a satisfactory prototype can be extended over the full length of the project without further reference to the foundation conditions. More generally the approach will be that exemplified by Eide and Holmberg (1972) in their study of the soft Bangkok clay which varied regularly along nearly 60 km of highway. They concluded that 'generally the standard vane test must be calibrated by means of large-scale field tests before being used as a basis for design', and found that the calibration factor (using the circular arc analysis) for their particular site was 1.61 for a bank with berms and 1.46 without, due to the greater influence of the lower shear strength in the horizontal direction in the former case. It is the need to make a realistic analysis of the mechanisms of failure or deformation and their time dependence which accounts for the scale of the instrumentation in a number of the trial embankments under discussion.

In justifying to himself, his clients or employers, the cost and expenditure of time involved the engineer will have to explain why the back-analysis of similar structures (4) is inadequate for the purpose and what could be gained from model tests (5). Too often 'similar' earth structures are not sufficiently similar geologically, the instrumentation or observations were insufficient for the project in hand, or the delays in reporting and publication (if not intentional secrecy) render the information effectively inaccessible. It is probably in the field of earth- and rock-fill dam design that the profession has benefitted most from the back-analysis of the published performance data of well-instrumented structures. Yet here the amount of data relating to the state of limiting equilibrium, either during construction, use or drawdown is, perhaps fortunately, very limited. The increasing use of strain-softening fills and foundations clearly justifies, for example, such trial embankments as those for the Atchafalaya levees (Kaufman and Weaver, 1967) and for the Empingham Dam (unpublished reports and Wade 1972).

The value of model tests lies primarily in being a replacement for analysis in intractable cases[(3)]. With embankments for highways and dams this in general means centrifugal testing in all cases except for seepage, and this raises the question of the stresses and strains within the model embankment unless it is constructed under the appropriately scaled-up gravitational acceleration. A second, and perhaps more important, difficulty arises in relation to the classes of problem represented by the trial embankments referred to

---

[(3)] Bolton, English, Hird and Schofield (1973) refer to cases where 'the three dimensional geometry of the boundaries create a problem of analysis that is outside the present bounds of theoretical solution', and indicate that 'much of the value of centrifugal testing lies in its complete independence from mathematical or computer models'.

in Table I. Just as the major and minor geological features must be realistically represented in a quantitative analysis, so must they be realistically represented in any model which is to replace the analysis. Problems in class (A) involving heterogeneous soils and soils with a large-scale macrostructure varying with depth are not readily modelled, and a single model test on a 'representative' block may be very misleading[4], and a great deal more costly than a number of large-scale strength or consolidation tests on samples distributed over the relevant zones.

A further difficulty arises in attempting to model progressive failure in strain-softening soils (a problem in class B) since the post-peak loss in strength is related to displacement on a surface or in a thin zone and not to general shear strain. True modelling of progressive failure would appear not to be possible, and Palmer and Rice (1972) have raised a further theoretical objection.

There are thus a considerable number of engineering situations when a trial bank can be justified in principle. How often can it be justified in practice? How complex should the instrumentation be? Before discussing these questions, it is relevant to review the development and use of trial banks in the four decades since the 1st International Conference on Soil Mechanics and Foundation Engineering in 1936, and to examine the scope of their instrumentation and their costs relative to the engineering works concerned. An estimate of their benefits is perhaps too subjective to tabulate!

DEVELOPMENT AND APPLICATION OF TRIAL EMBANKMENTS

One of the first 'modern' applications of an instrumented trial embankment to a design problem is given by Ringeling (1936) in the Proceedings of the 1st International Conference. This case concerned the possibility of constructing a highway embankment over peat adjacent to an existing tramway embankment and the instrumentation included settlement points and hydraulic piezometers.

Following the 1939-45 war there was a rapid development in instrumentation, both in Europe and the United States. Perhaps the foremost organisation in this field was the U.S. Bureau of Reclamation, and it is appropriate that one of the next well-documented trial embankments was constructed by the U.S.B.R. as part of their design study for the Willard Dam,

[4] Even on a full-scale trial embankment Lambe et al (1971) obtained settlements of 0.94 m, 1.34 m and 1.74 m at three adjacent stations at approx. 30 m intervals, the compressible foundation being about 27 m in depth and principally of Boston Blue Clay. The elapsed time was 13 years.

which encloses an arm of the Great Salt Lake in Utah. The preliminary site investigation was reported in 1951, the instrumented trial section constructed between June and September 1955, and the 23 km long dam constructed in three stages between 1958 and 1964 (Walker, 1967).

The Northeast Test Embankment near Interstate Highway I-95 (near Boston) was constructed in 1957, but was not reported on by Lambe et al until 1971, when 13 years of valuable records became available. Hydraulic piezometers added in 1967 show substantial excess pore pressures remaining after 10 years, corresponding to an estimated degree of consolidation of from 0.62 to 0.71.

Clearly there are many similar valuable records still unpublished or relatively inaccessible, and Table I can be considered as little more than representative. It is, however, useful to classify the cases into trial embankments built for highways, for flood levees, for dams and for research. Before considering the cases separately it is worth noting that most trial embankments are multipurpose, though with varying emphasis, and that most trial embankments for design purposes relate to production embankments which are very long in relation to their height. Often the trial section can be incorporated subsequently in the permanent works as a measure of economy.

Trial banks for design purposes are usually built to investigate one or more of the following problems:

(a) Settlement, deformation and pore-pressure characteristics of compressible foundations, including time dependence.

(b) End of construction stability.

(c) Construction methods to accelerate, reduce or elimate post-construction settlements.

(d) The behaviour of the embankment fill material.

TRIAL EMBANKMENTS FOR HIGHWAYS

A selected list of recorded uses of trial embankments for highways is given in Table I(a). It will be noted that, numerically, this forms the major category of trial embankments in Table I and reflects the rapid growth of high-quality roads in the last two decades, many of them across soft, compressible strata.

It will be noted in Table I(a)that all but one of the trials investigated the foundation condition rather than the

Table I : Recorded Uses of Trial Embankments

Locality	Predominant foundation soils or fill material (as appropriate)	Principal aspects investigated: Foundations	Fill material	Stability	Failure condition	Deformation	Sand drains	Stability analysis	Back-analysis[1]	Reference
(a) Highways										
France (3 sites)	Organic clays	X				X	X			Bourges et al (1969)
Bangkok (3 sites)	Soft marine clay	X		X	X	X		X		Cox (1971)
Bangkok	Soft marine clays	X		X	X			X		Eide & Holmberg (1972)
Valencia, Spain[2]	Organic clays	X				X				Justo (1969)
Boston, Mass.	Boston blue clay	X				X				Lambe et al (1971)
Portland, Maine	Organic clay	X		X	X	X	X	X		Ladd et al (1969)
Portsmouth, N.H.	Sensitive marine clay	X		X	X	X		X		Ladd (1972)
Westmorland (M6 M'way)	Boulder clay		X			X				McLaren (1968)
Avonmouth (M5 M'way)	Soft clay & peat	X				X				Murray (1971)
France (5 sites)	Soft clays, organic clays and peat	X		X	X			X	X	Pilot (1972)
Mexico City	Soft volcanic clay	X				X				Rico et al (1969)
Netherlands	Peat	X				X				Ringeling (1936)
California	Soft bay mud	X		X	X	X	X	X		Weber (1966)
King's Lynn	Peat and Soft organic clays	X		X	X	X	X	X		Wilkes (1972)
(b) Levees										
Quebeć	Soft lacustrine clay	X		X	X			X		Dascal et al (1972)[3]
Japan[4] (2 sites)	Sandy clay		X	X	X					Fukuoka & Yamamura (1965)
Louisiana	Organic clays	X		X	X	X		X		Kaufman & Weaver (1967)
Mexico	Soft volcanic clay	X				X				Ortiz (1967)
Launceston, Tasmania	Soft organic clays	X		X	X	X		X	X	Parry (1968)
Bangkok (AIT)	Soft marine clays	X		X		X		X		Scott Wilson et al (1970)
San Francisco	Soft bay mud & debris fill	X	X	X		X				Taylor & Buchignani (1972)
(c) Dams										
Hong Kong (Plover Cove)	Decomposed granite		X	X		X				Guilford & Chan (1969)
Volta R., Ghana	Clay core		X	X						Scott (1958)
Empingham, Rutland	Lias clay foundation and fill	X	X	X	X	X		X		Wade (1972)
Willard, Utah	Organic clays	X		X		X				Walker (1967)
Backwater Dam, Dundee	Boulder clay		X			X				Wilkinson et al (1970)
Grafham Water, Huntingdon	Boulder clay & Oxford Clay	X	X	X				X		Hammond and Winder (1967)
(d) Research										
Gloucester, Ottawa	Soft silty clay	X		X		X		X		Buzuzuk & Leonards (1972)
Ska-Edeby, Sweden	Varved clay	X				X	X			Holtz & Broms (1972); Holtz & Lindskog (1972)
Bangkok[5]	Soft marine clays	X				X				Moh et al (1972)
Mastemyr, Oslo	Norwegian quick clay	X		X	X	X		X		Clausen (1970)
Drammen, Norway	Peat & soft clay	X				X				Engesgaar (1970)

Notes: (1) Refers to back-analysis of production embankments.

(2) This project included a diversion channel with a road on the adjacent embankments.

(3) This reference deals only with stability consideration, a second trial embankment and a future paper will examine deformation behaviour.

(4) These embankments were built primarily to investigate the effects of seepage and rainfall on their stability.

(5) This embankment is the same as that referred to by Scott Wilson et al (1970) under levees; it had a duel design/research role.

behaviour of the fill material. In addition all but two considered the problem of deformation or settlement, which reflects the fact that even small deformations of a highway embankment after the construction of the pavement can cause serious embarrassment and expensive maintenance.

Many trials have been directed towards obtaining realistic values for the rates and amounts of consolidation settlement. The early trial by Ringeling (1936) and those reported by Bourges et al (1969) and Rico et al (1969) all had this as one of their principal objectives. In Great Britain similar studies have been reported by Murray (1971) as part of the investigation for the M5 Motorway near Bristol and by Wilkes (1972) for a road at King's Lynn.

Apart from its square shape in plan (most trial embankments are several times longer than their width) the M5 trial embankment is a typical example of a trial to determine more precisely the settlement characteristics of a soft clay foundation. The foundation (Fig.1) is a recent alluvial

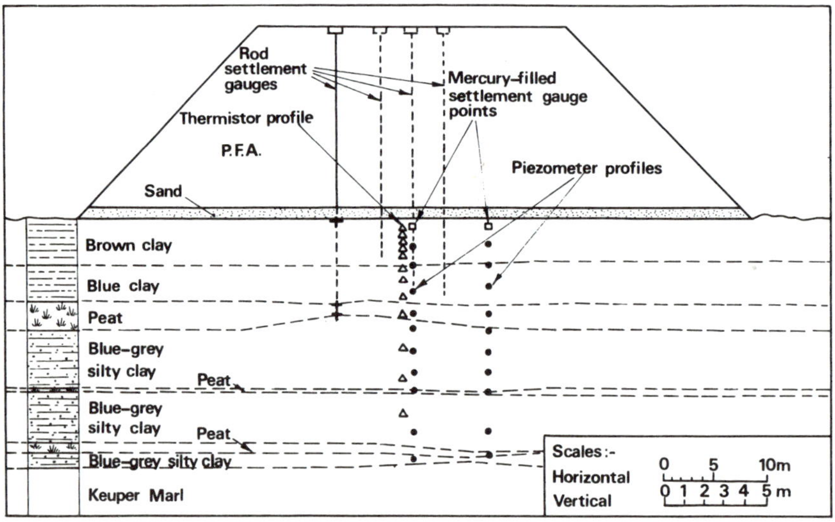

Fig. 1 Trial embankment for M5 Motorway
Simplified geological section and layout of instrumentation (after Murray, 1971)

deposit, probably formed within the last 7000 years (Hawkins, 1972) consisting of clays, silty clays and peats and intersected by small creeks. The instrumentation is simple but adequate to permit reasonable correlations to be obtained with theoretical predictions based on laboratory tests and field measurements of permeability (for example, Fig.2). Indeed, a theoretical analysis would in general be necessary for converting pore-pressure dissipation rates from a square trial embankment with three-dimensional drainage to a long highway embankment where drainage approximates to plane-strain conditions.

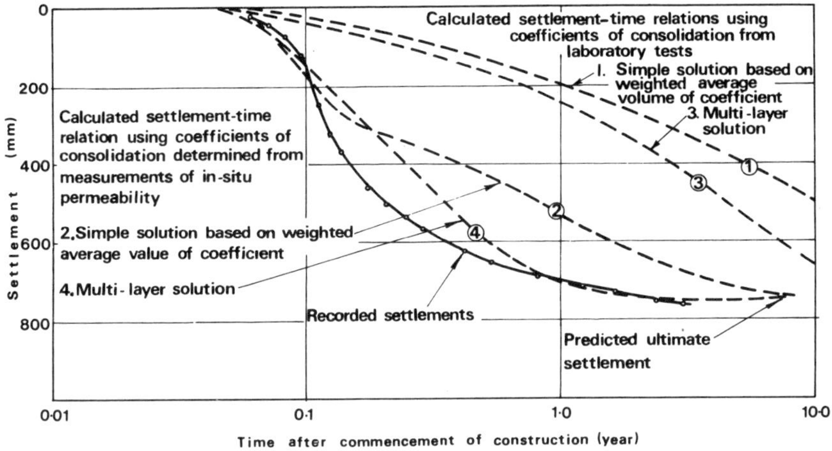

Fig. 2 Trial embankment for M5 Motorway Measured settlement beneath centre of embankment compared with the calculated settlement – time relations using coefficient of consolidation for the mean effective stress from both laboratory and insitu tests (after Murray, 1971)

It will be seen from Table I(a) that in only half of the references is stability given as a major aspect of the trial and in these cases the embankment was taken to failure, though in most other cases the observations yield some data relevent to the factor of safety. End of construction failure, with a consequent permanent reduction in the drained strength on the rupture surface may be of less importance in roadworks than in water-retaining structures for which a higher long-term factor of safety is usually demanded, but is clearly to be avoided, if possible without general overdesign. The recent examples of highway trial embankments for stability problems have mainly examined the 'correct' value of undrained strength to be used in the stability analysis, or have been used to examine the methods of analysis themselves. Eide and Holmberg (1972), Ladd (1972) and Pilot (1972) all report failures which occurred at calculated factors of safety much greater than unity when insitu vane shear strengths were used with conventional methods of undrained stability analysis. Bjerrum (1972) comments on the implications of these results in terms of rate of testing, anisotropy and progressive failure.

An examination of construction method, in this case surcharging, was one of the purposes of the M5 trial embankment, but the other four cases listed in Table I(a) (Bourges et al, 1969; Ladd et al, 1969; Weber, 1966; Wilkes, 1972) relate to embankments where one of the main purposes was the examination of the effectiveness of sand drain installations. Generally such trials consist of two, more-or-less similar, instrumented embankments, one with and one without sand drains.

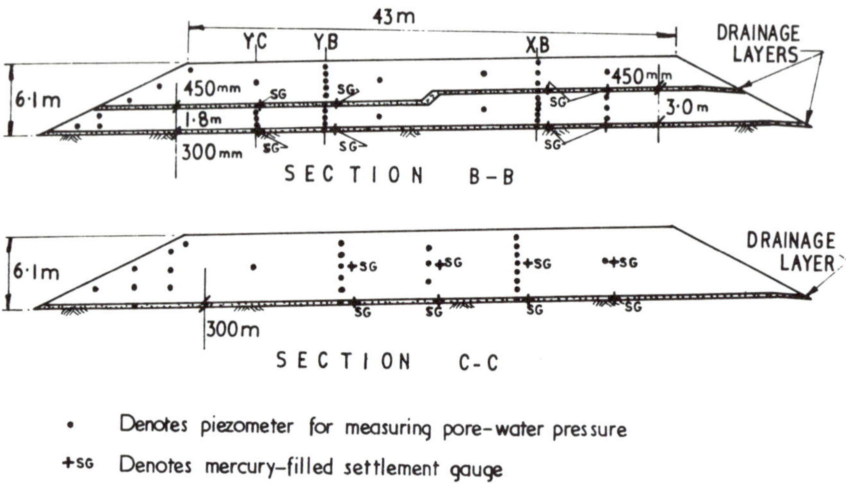

Fig. 3 Trial embankment for M6 Motorway
Cross sections showing positions of instrumentation and drainage layers

Fill properties formed the major study in only one instrumented trial embankment, that for the M6 Motorway (Scott Wilson et al, 1966; McLaren, 1968), although Rico et al (1969) mention two trials for the assessment of alternative fill materials. The M6 trial (Fig.3) consisted of a 6.1 m high embankment having three zones with drainage layers at different levels. The fill material was boulder clay (glacial till) of high stone content but of relatively low permeability and likely to be placed at a very high water content due to the prevailing high rainfall and the initially wet state of the clay.

The principal object of the trial was the determination of the full-scale one-dimensional compressibility and coefficient of consolidation of the very wet boulder clay, with the secondary objects of investigating field and laboratory control methods and the performance of plant in excavating, loading, dumping, crossing and compacting the material. Incremental analysis of the pore-pressure isochrones showed that the wet material had properties which departed drastically from those assumed in Terzaghi's theory of consolidation, which meant that the rate of consolidation increased by a factor of 3 as the effective stress increased

(Fig.4). Laboratory tests on a 102 mm dia. x 102 mm high triaxial specimen showed that this increase was associated with a 10-fold decrease in compressibility (Bishop and Al-Dhahir, 1969). Crest settlements were small, and after a few months became negligible on the sections with intermediate drainage layers.

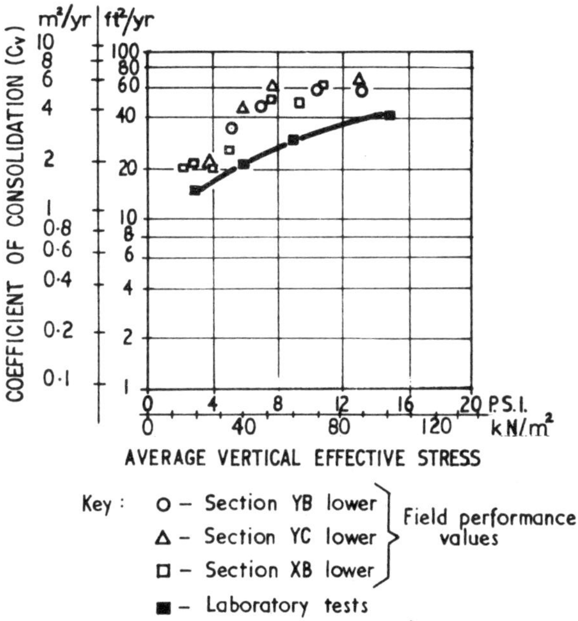

Fig. 4 M6 boulder clay fill
Comparison of values of Cv from laboratory dissipation tests on a recompacted sample with field performance values (after Bishop and Al-Dhahir, 1969)

TRIAL EMBANKMENTS FOR LEVEES

Flood levees or bunds comprise the second most numerous category of works for which instrumented trial embankments have been widely used. Like highway embankments, these structures are generally low earth banks many kilometres in length, usually sited on soft alluvial foundations, often in restricted locations adjacent to waterways and other engineering works. Though small differential settlements are not important, the maintenance of a fixed minimum crest level is critical, as is stability under flood conditions. However, maintenance work to keep up the crest level is clearly more acceptable than on a motorway.

It will be seen from Table I (b) that again all but one of the trials was concerned with foundation conditions and all but two considered deformations. All the trials took account of stability, but only about half investigated the failure condition. The Bangkok trial was not taken to failure for design purposes, but was failed after 11 months for a research investigation.

One of the most interesting examples is the trial embankment, comprising three test sections, for the Atchafalaya levees in Louisiana (Kaufman and Weaver, 1967). These were constructed in the years 1964 and 1965 to examine progressive failure in a deep alluvial foundation of soft clay and peat which showed an undrained brittleness index (Bishop 1971a) of the order of 50-70% and a peak strength which appeared to be strongly time-dependent. The instrumentation (Fig.5) is more elaborate than for a typical highway embankment, but has produced valuable data on the location of the zones of maximum shear. It is of interest to note that two of the three test sections moved laterally (at least 600 mm in one case) sufficiently to crack the fill forming the levee and were still moving slowly when reported. Failure, for the given conditions of brittleness index, time dependence and geometry, was not catastrophic, in contrast to the massive failure of the highway trial embankment on a sensitive marine clay at Portland, N.H. (Ladd, 1972) where the undrained brittleness index[5] was at least 90% throughout much of the foundation. Here the crest dropped 3 m in a bank height of 6.6 m.

TRIAL EMBANKMENTS FOR DAMS

It is necessary to distinguish between test fills and instrumented trial embankments. Carlyle (1968), Mann (1964), Reynaud and Rosset (1965), Zeller and Zeindler (1957) and many others have reported the use of test fills. These have usually been built to investigate the excavation, handling, placing, and compaction of the fill material, as well as to provide material for insitu tests and 'undisturbed' samples for laboratory testing. Though seldom instrumented, and therefore not falling within the scope of this paper, they are nevertheless of great practical value and also of importance at the design stage, since in the more plastic fills the pore pressure, strength and deformation characteristics of laboratory compacted fill are found to

---

[5] Using the relationship:

$$I_B \geqslant \left(1 - \frac{1}{S_t}\right) \times 100\% \quad \text{(Bishop, 1971a)}$$

Where I_B denotes undrained brittleness index

S_t denotes sensitivity

differ radically from those of fill compacted by full-scale equipment (Bishop and Vaughan, 1971).

Relatively few instrumented trial embankments have been constructed for dams and reservoir embankments. This probably reflects to some extent the greater selectivity of

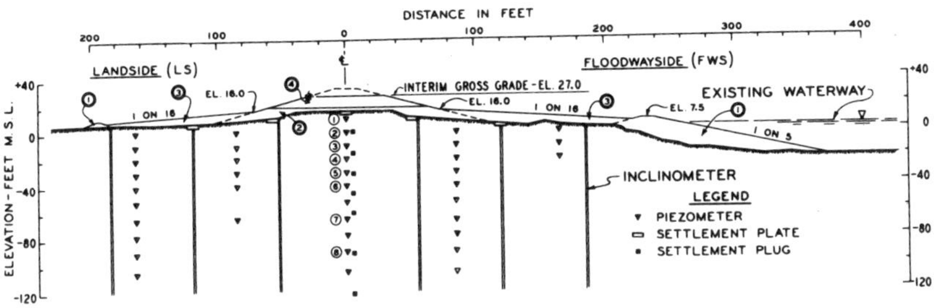

Fig. 5 Trial embankment for Atchafalaya levees
Test section 2 showing instrumentation
(after Kaufman and Weaver, 1967)

dam engineers with regard to what are considered to be suitable sites, and the greater amount of instrumented monitoring and control planned for the construction stage. While the 'design as you build' philosophy (for example, Li, 1967) is now becoming more acceptable to both clients and contractors, it is necessary to establish before letting the main contract that the goal can in fact be achieved without unreasonable cost or delay.

A <u>full-scale</u> test section may be justified only if the embankment is very long and the only example reported in Table I(c) is the Willard Dam (Walker, 1967) which is 23 km in length and 11 m in height - a length to height ratio of about 2000. Here the problem was both stability and settlement on a deep, soft heterogeneous foundation, and the decision on whether to build or not depended on the performance of the trial embankment.

At the Empingham Dam in Rutland the ratio of bank length to height is about 35. Here the foundation is of fissured Lias clay, of high plasticity, and badly sheared and distorted by valley bulging. The instrumented trial embankment, built largely within the final design profile in the early stages of construction to nearly 60% of the maximum dam height, is to confirm the design assumptions about the strength of the foundation by stressing a much greater area than is possible in any other form of field or laboratory test and by providing data on the role of progressive failure. By including the trial bank in the main contract and not constructing it earlier, sufficient confidence is assumed in the combination of tests on borehole samples, plate loading tests in boreholes and back-analysis of other dams to put forward acceptable upper and lower limits to the cross-section, which can then be refined on the basis of results from the trial embankments and other monitoring systems.

The partly submerged embankment dam at Plover Cove, Hong Kong (Guilford and Chan, 1969) has a length to height ratio of about 52. Here an instrumented trial embankment (Fig.6) was constructed to about 60% of the maximum height of the proposed embankment to investigate (i) the feasibility of placing rock fill directly into the bay mud underlying the site to produce a stable impermeable base for the dam (it was found not to be feasible), (ii) the maximum slope of dredged excavations in the bay mud, (iii) the pore-pressure dissipation, permeability and settlement behaviour of decomposed granite and sand fill deposited through water to form an embankment and (iv) the steepest slope that core-type material would adopt when placed under water (it was found to be 1 vertical to 3 horizontal for the equipment used in the trial).

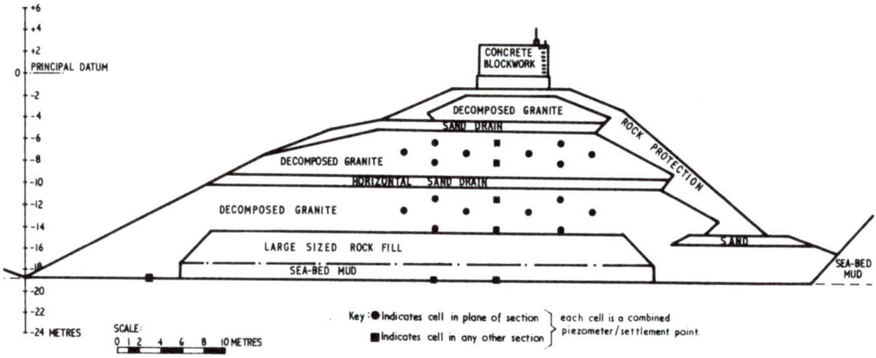

Fig. 6 Trial embankment for Plover Cove Dam East-West section showing positions of piezometer/settlement cells

In Fig. 7 the height of the trial embankment (h) is plotted as a percentage of the height of the actual structure (H) against the length to height ratio (L/H) of the actual dam, the values of h also being plotted. It is clearly practical and economical to construct a full-scale test section when the L/H is of the order of 2000 as at the Willard Dam. However, the working range of shear stresses (and it is these that are relevant to undrained foundation failure) can be achieved with lower embankment heights by using steeper slopes, though a bank of less than half height may not critically stress the relevant zone of the foundation. This depends on the rate of increase in strength with depth (and the strength of the fill), which permitted the use of an h/H of about 54% at Empingham. Even with this height a fairly elaborate analysis of observed strains is necessary to assess the relative importance of progressive failure in the trial bank and in the full-scale embankment, associated with the different length of potential slip path in the foundation and in the rolled clay fill in each case. In contrast, a dessicated crust of significant thickness may necessitate the use of an h/H value of 100% if stability problems are involved.

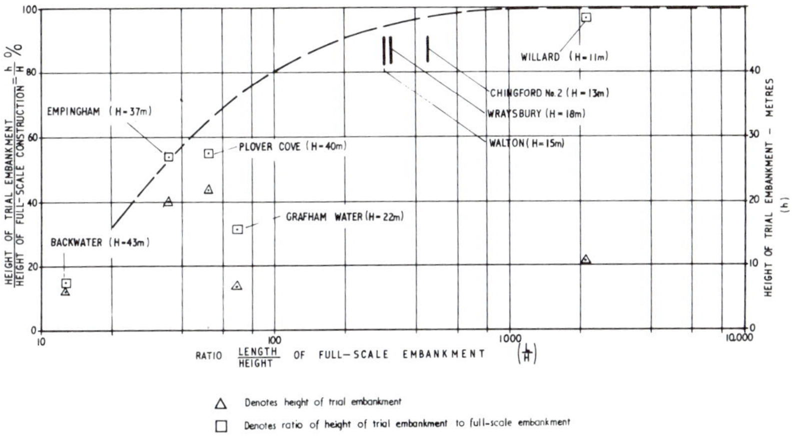

Fig. 7 Correlation between height of trial embankment and length to height ratio of dam

The trial embankments at Plover Cove, Backwater and Volta River[(6)] were all concerned with fill material, and their h/H values reflect not only decreasing L/H values, but also the relative amount of information about probable behaviour of the fill available in the literature, based on the observed and published performance of other full-scale structures. Even where the working range of normal stress cannot be approached in a trial embankment for practical or economic reasons, the behaviour of the fill in the low stress range and the negative pore pressure in the fill as placed serves as a valuable check on the realism of the laboratory programme. The trial bank also provides 'undistributed' samples for testing in the high stress range.

---

(6) Omitted from Fig. 7 due to absence of precise data.

interest to note that the large pond-type reservoirs Chingford No. 2 (Cooling and Golder, 1942), Walton , 1948) and Wraysbury, which have undergone major s in design on the basis of additional data about tion strengths emerging after the initial contract had let, lie in the length to height bracket which would suggest in retrospect that an instrumented trial section, possibly of full height, would have been a justifiable and economical design measure.

TRIAL EMBANKMENTS FOR RESEARCH INVESTIGATIONS

In the commissioning of some trial embankments the object is not the design of a particular structure, but a general advance in knowledge which may be relevant not only to future embankments, but to the behaviour of foundations and fills in general. Some examples are given in Table I(d) The back analysis of production structures should, and in some cases does, provide much of this data, particularly where time is the most important factor (as in secondary consolidation, creep, and the dissipation of pore pressures in thick strata) and cannot be scaled.

The time scale and continuity of observations may require the support of a research organisation and the type of instrumentation must be chosen with long-term durability and stability in view. Where the loads are relatively small and the embankment itself inexpensive, the relatively high security necessary to prevent damage to the instruments may favour a trial embankment rather than a more public working structure.

INSTRUMENTATION

The types and numbers of instruments which have actually been installed in some of the trial embankments under discussion are summarised in Table II. It can be seen from this table that the range is large, with the total number of items for any one trial ranging from 14 at the Volta River to 336 at King's Lynn! A comparison of Tables I and II shows no obvious correlation between the instrumentation and the principal aspects investigated. This may reflect different philosophical or theoretical approaches to the function of the trial embankment, the availability of funds, an uneasy feeling that if one only instruments for the predicted, the unpredicted may elude observation; and the knowledge on the other hand that a decision cannot be made until all the results have been analysed and too many results may overwhelm the office.

The danger of depending too heavily on only one instrumented section, particularly with regard to deformation, is highlighted by the results (quoted earlier) of the three adjacent settlement plates on Lambe et al's Northeast Test Embankment,

Table II : Instruments Used in Trial Embankments

Locality	Total no. of instruments used														Reference
	Piezometers	Type of piezometers (1)	Surface monuments (2)	Settlement plates (3)	Hydraulic or mercury settlement points	Horizontal settlement tubes (4)	Borehole extensometers	Inclinometers (5)	Horizontal movement/ strain gauges	Lateral movement pipes & boards	Lateral movement and alignment stakes	Failure stakes (6)	Slip-surface indicators	Earth pressure cells	
(a) Highways															
Bangkok	Few		14	-	-	-	-	-	-	-	-	-	13	-	Eide & Holmberg (1972)
Valencia, Spain	12	H	-	12	-	-	-	10	-	-	-	-	-	-	Justo (1969)
Boston, Mass.	4	H	44	5	-	-	-	-	-	-	-	-	-	-	Lambe et al (1971)
Portland, Maine	4	VW	21	-	-	-	-		-	-	30	-	-	-	Ladd et al (1969)
	9	H						5							
Portsmouth, N.H.	12	VW	-	8	-	-	-	4	-	-	18	-	-	-	Ladd (1972)
	Few	S													
Avonmouth (M5)	27	H	12	6	4	-	-	-	-	10	-	-	-	-	Murray (1971)
Westmorland (M6 M'way)	65	H	31	2	16	-	-	-	-	16	-	-	-	-	Scott Wilson et al (1966)
	2	S													
Mexico City (2 émbankments)	36	P	60	14	-	-	-	14	-	-	-	-	-	-	Rico et al (1969)
	24	S													
King's Lynn	132	H	126	12	16	-	8	20	-	18	-	-	4	-	Wilkes (1972)
(b) Levees															
Quebec	17	P	20	6	-	-	-	-	-	-	-	-	-	4	Dascal et al (1972)
Mexico	18	P	21	-	-	-	-	16	-	-	-	-	-	-	Ortiz (1967)
Launceston, Tasmania	6	S	12	-	-	-	-	-	-	-	-	-	-	-	Parry (1968)
Bangkok (AIT)	52	H	19	39	-	-	-	-	-	-	23	18	-	-	Scott Wilson et al (1970); Moh et al (1972)
	(+ few others)														
(c) Dams															
Hong Kong (Plover Cove)	30	H	-	-	30	-	-	-	-	-	-	-	-	-	Guilford & Chan (1969)
Willard, Utah	72	H?	99	5	-	-	-	-	-	-	-	-	-	-	Leonards & Narain (1963)
Volta R., Ghana	12	H	-	-	-	-	-	-	-	-	-	-	-	2	Scott (1958)
Empingham, Rutland	16	H	20	-	-	4	-	4	24	-	-	-	-	-	Authors
(d) Research															
Gloucester, Ottawa	24	S	-	15	-	1	-	5	25	-	-	-	-	13 (3 types)	Buzuzuk & Leonards (1972)
Mastemyr, Oslo	21	H	5	23	-	-	-	4	-	-	-	-	-	-	Clausen (1970)
	19	VW													

Notes: (1) H = Hydraulic; VW = Vibrating wire; P = Pneumatic; S = Standpipe

(2) To measure lateral and/or vertical movement on surface of the fill or ground adjacent to the fill.

(3) Includes plates and similar types of equipment placed in the fill, at the bottom of the fill or in the foundation strata.

(4) These tubes are now often combined with plates for the measurement of horizontal strain; where appropriate, such plates have been recorded in the horizontal movement/strain gauge column.

(5) Usually inclinometers are also used as vertical extensometers.

(6) Failure stakes have similar role to lateral movement stakes.

where the vertical movement differed by a factor of almost two over a distance of 60 m.

The instrumentation may be directed towards two distinct purposes. The first is the observation of overall performance. For the settlement of a highway embankment this could mean surface monuments and a limited number of settlement plates at the base of the fill (for example, Lambe et al 1971). For the stability of an embankment this could mean surface monuments (or photogrammetric observations), together with some form of slip-surface indicator (easily sheared stakes or tubes inserted in the probable shear zone)- for example Eide and Holmberg (1972).

For an understanding of the mechanisms involved, the settlement problem calls for piezometers (added by Lambe et al, 1971 at a later stage) and included in all the Table II(a) embankments. These should preferably be hydraulic to permit insitu measurement of permeability and allow the removal of gas in organic soils. Inclinometers should be added to observe the contribution due to lateral yield and borehole extensometers to observe the vertical distribution of foundation settlement. For a more refined analysis the distribution of normal and shear stress and displacement across the base of the fill should be observed, but this has only been seriously attempted in the research project reported by Bozuzuk and Leonards (1972).

The stability problem calls for a rather more detailed picture of the internal displacement pattern and in particular of the growth of zones of large shear strain which may precede rupture. As the maximum movement of the slope indicator near the toe of the 6.4m embankment reported by Ladd (1972) was less than 37 mm when massive failure occurred, these observations need to be taken with a high degree of accuracy (and any early warning system based on observed displacements needs to be treated with reserve in the case of brittle soils). The observation of significant shear strains in relatively thin zones remains one of the more intractable problems of field instrumentation, as does the measurement of post-rupture displacements on a slip surface as it begins to propagate, possibly long before overall shear failure occurs.

To obtain the complete strain pattern in plane strain, the compressive strains in two directions at right-angles, together with the associated shear strain, should be observed. This can be done with a system of vertical slope indicators, if they are used as, or associated with, borehole extensometers as well. As usually spaced (for example Fig. 5) slope indicators alone would not detect adequately significant variations in horizontal compression or extension, and horizontal displacement tubes at and above ground level give valuable additional data. At Empingham both systems were adopted since a study of progressive failure as instability is approached formed a major purpose of the trial embankment.

To obtain the complete stress pattern is not yet a practical reality. The pore-pressure component is easy enough to observe, but total stress cells can at present only be installed meaningfully at or above the contact between the fill and the foundation. The number of normal stress cells required to define the state of stress at a point is generally 5 if plane strain is assumed and the analysis is complicated by cell-action factors. So far only two trials have used them to check the state of stress (or a component of it) at the fill-foundation contact (Table II). At Empingham total stress measurements by hydraulic fracture have been made.

The location of instruments within the cross section should be the result of a careful theoretical analysis based on an adequate site investigation. However, the principle of 'design as you build' should also apply to the instrumentation of trial embankments, and provision should be made for rapid additions to the installation if the observations indicate unforeseen developments.

COSTS

Some information on costs is given in Table III. Information on costs is rather hard to come by, perhaps because engineers are unwilling to admit to themselves or to their clients what the actual costs are. Why this should be so is not apparent, since in all but the smallest earthworks recorded in Table III the cost of the trial embankment is less than 1% of the civil engineering construction costs, and 3% or less of the embankment or earthworks construction costs. If the contractors likely to tender for the earthworks visit the site of the trials (as with the M6 trial embankment) or see them on film, the additional knowledge of the feasibility of the scheme and in particular of the handling of fill materials should reduce the rates by at least this amount - a figure of up to 30% has been suggested. This is quite apart from the design advantages obtained.

Just as the economic advantages of good design are difficult to evaluate, so also are those of the inclusion of a trial embankment in the design process, due to lack of a 'control' for purposes of comparison. There are five considerations, all of which have strong financial implications:

(a) Completion of the embankment to full height is more certain (there are at least two major dams in the UK which have been terminated short of the maximum design height in the period under review).

Table III : Costs of Trial Embankments

Locality	Approx. cost[1] (£) for :		Cost of trial embankment as % of:		Reference
	Trial embankment complete[2]	Instruments	Embankment or earthworks construction costs	Civil engineering construction costs	
Avonmouth (M5 M'way)	25,000	10,000	3.2	0.75	Dallard (1971)
Westmorland (M6 M'way)	40,000	2,500	0.5	0.12	Authors
Bangkok (AIT)	5,400	2,000	6.4	4.10	Authors
Hong Kong (Plover Cove)	90,000	-	1.4	0.75	Authors
Empingham, Rutland	32,000[3]	5,000[3]	0.8	0.16	Authors

Notes: (1) At time of trial embankment construction.

(2) This cost does not usually include labour costs for installing or reading instruments, or analysing the records obtained; nor does it usually include costs of any of the ground investigation at the site of the trial or the subsequent laboratory testing.

(3) Estimated extra cost resulting directly from having a trial embankment.

(b) Changes in design at the construction stage which fall outside those envisaged in the contract (which many involve acquisition of additional land) can be minimised.

(c) Completion on time is more likely.

(d) Maintenance is reduced.

(e) The margin of safety under operating conditions is likely to be more clearly defined and the interpretation of monitoring observations is made more certain.

CONCLUSIONS

While every published record of an instrumented trial bank or working structure may reduce the need for another trial embankment under similar circumstances, the near future is likely to see an increase in their effective and economic use under a wide range of site conditions. The variability along the centre-line of the long embankments which most clearly justify their use must not be overlooked; nor must the fact that short-term stability may not be the most critical case in deep soft strata where the rate of loss of undrained strength with time may exceed the gain in strength due to consolidation (for example, Kaufman and Weaver, 1967), or where lateral redistribution of pore pressure may initially lower the factor of safety with time (for example

Ward et al 1955).

ACKNOWLEDGEMENTS

The authors are grateful to Binnie and Partners, Scott Wilson Kirkpatrick and Partners, T.C. Hawksley and Dr P.R. Vaughan for the use of unpublished information.

REFERENCES

Bishop, A.W. (1948) Some factors involved in the design of a large earth dam in the Thames Valley.
Proc. 2nd ICSMFE, Rotterdam, Vol. 2 pp 13-18.

Bishop, A.W. (1966) The strength of soils as engineering materials.
Sixth Rankine Lecture, Geotechnique, Vol. 16 pp 89-130.

Bishop, A.W. (1967) Progressive failure - with special reference to the mechanism causing it.
Proc. Geotech, Conf., Oslo, Vol. 2 pp 142-150.

Bishop, A.W. and Al-Dhahir, Z.A. (1969) Some comparisons between laboratory tests, in situ tests and full scale performance, with special reference to permeability and coefficient of consolidation.
Proc. Conf. on In Situ Investigations in Soils and Rocks, pp 251-264, British Geotechnical Society, London.

Bishop, A.W. (1971 a) Shear strength parameters for undisturbed and remoulded soil specimens.
Roscoe Memorial Symposium, Cambridge, pp 3-58.

Bishop, A.W. (1971 b) The influence of progressive failure on the choice of the method of stability analysis.
Geotechnique, Vol. 21 pp 168-172.

Bishop, A.W. and Vaughan, P.R. (1971) Interim Report No. 3 to the Water Resources Board.

Bishop, A.W. (1972) The stability of tips and spoil heaps.
Regional Conf. of Engng Group of Geol. Soc., Bristol (Quarterly Jour. Engng Geol., in print).

Bjerrum, L (1967) Engineering geology of Norwegian normally-consolidated marine clays as related to settlements of buildings.
Seventh Rankine Lecture, Geotechnique, Vol. 17 pp 81-118.

Bjerrum, L. and Kenney, T.C. (1967) Effect of structure on the shear behaviour of normally consolidated quick clays.
Proc. Geotech. Conf., Oslo, Vol. 2 pp 19-27.

Bjerrum, L. (1972) Embankments on soft ground.
ASCE, Proc. Speciality Conf.on Performance of Earth and Earth-Supported Structures, Purdue, Vol.2 pp 1-54.
Bolton, M.D., English, R.J., Hird, C.C. and Schofield, A.N. (1973) Ground displacements in contrifugal models. Preprint for 8th ICSMFE, Moscow.
Bourges, F., Chiappa, J., Legrand, J. and Paute, J.L. (1969) Etude de tassement des vases supportant des remblais. (Settlement of organic soft clays under embankments) Proc. 7th ICSMFE, Mexico, Vol. 2 pp 35-43.
Bozuzuk, M. and Leonards, G.A. (1972) The Gloucester test fill. ASCE, Proc. of Speciality Conf. on Performance of Earth and Earth-Supported Structures, Purdue, Vol. 1 pp 299-317.
Carlyle, W.J. (1968) The River Towy Scheme for the West Glamorgan Water Board.
Water and Water Engineering, pp 371-382.
Clausen, C-J.F. (1970) Resultater av et belastningsforsøk på Mastemyr i Oslo.
NGI Publication No. 84, pp 29-40 (English summary p 70)
Cooling, L.F. and Golder, H.Q. (1942) The analysis of the failure of an earth dam during construction.
Jour. ICE, Vol. 19 pp 38-55.
Cox, J.B. (1971) Engineering characteristics of the soils along the Thon Buri-Pak Tho Highway, Thailand.
Proc. 4th Asian Regional Conf. SMFE, Bangkok, Vol. 1 pp 249-255
Dallard, N.J. (1971) Design and construction of embankments on an alluvial plain.
Paper 7400, Proc. ICE, Vol. 49 pp 157-170.
Dascal, O., Tournier, J.P., Tavenas, F. and La Rochelle, P, (1972) Failure of a test embankment on sensitive clay. ASCE, Proc. Speciality Conf. on Performance of Earth and Earth-Supported Structures, Purdue, Vol. 1 pp 129-158.
Eide, O. and Holmberg, S. (1972) Test fills to failure on the soft Bangkok Clay.
ASCE, Proc. Speciality Conf. on Performance of Earth and Earth-Supported Structures, Purdue, Vol. 1 pp 159-180.
Engesgaar, H. (1970) Restulatater av to belastningsforsøk på Sundland i Drammen.
NGI Publication No. 84, pp 41-47 (English summary pp 70-71)
Fukuoka, M., and Yamamura, K. (1965) Full-scale model tests of slope failure of river embankments.
Proc. 6th ICSMFE, Montreal, Vol. II pp 467-471.
Guilford, C.M. and Chan, H.C. (1969) Some soils aspects of the Plover Cove marine dam.
Proc. 7th ICSMFE, Mexico, Vol. 2 pp 291-299.
Hammond, T.G. and Winder, A.J.H. (1967) Problems affecting the design and construction of the Great Ouse water supply scheme.
Jour. Inst. Water Engs, Vol. 21 pp 15-66.

Hawkins, A.B. (1972) Discussion on paper by Dallard, 1971.
Proc. ICE, Vol. 51 pp 715-716.
Holtz, R.D. and Broms, B. (1972) Long-term loading tests at Ska-Edeby, Sweden.
ASCE, Proc. Speciality Conf. on Performance of Earth and Earth-Supported Structures, Purdue, Vol. 1 pp 435-464.
Holtz, R.D. and Lindskog, G. (1972) Soil movements below a test embankment.
ASCE, Proc. Speciality Conf. on Performance of Earth and Earth-Supported Structures, Purdue, Vol. 1 pp 273-284.
Justo, J.L. (1969) Instrumentation of a new channel in soft ground.
Proc. 7th ICSMFE, Mexico, Vol. 2 pp 599-607.
Kaufman, R.I., and Weaver, F.J. (1967) Stability of Atchafalaya Levees.
ASCE, Jour. Soil Mech. and Found.Div.,Vol.93 No.SM4 pp 157-176.
Ladd, C.C., Aldrich, H.P. Jnr. and Johnson, E.G. (1969) Embankment failure on organic clay.
Proc. 7th ISCMFE, Mexico, Vol. 2 pp 627-234.
Ladd, C.C. (1972) Test embankment on sensitive clay.
ASCE, Proc. Speciality Conf. on Performance of Earth and Earth-Supported Structures, Purdue, Vol. 1 pp 101-128.
Lambe,T.W., D'Appolonia, D.J., Karlsrud, K. and Kirby, R.C. (1971) The performance of the foundation under a high embankment.
MIT Department of Civil Engineering, Soils Publication 277, Research Report R 71-22.
Leonards, G.A. and Narain, J. (1963) Flexibility of clay and cracking of earth dams.
ASCE Journal of the Soil Mech. and Found. Div., pp 47-98.
Li, C.Y. (1967) A voice of dissent, earth dams - design as you build.
Civil Engineering, ASCE, Jan. 1967 pp 70-71.
Mann, A.P. (1964) Underwater insepction of test fill in Gatun Lake.
Civil Engineering, ASCE, May 1964 pp 31-33.
Marsland, A. (1971) The shear strength of stiff fissured clays.
Roscoe Memorial Symposium, Cambridge, pp 59-68.
McKecknie Thomson, G. and Rodin, S. (1972) Colliery spoil tips - after Aberfan.
Publication of ICE, London.
McLaren, D. (1968) M6 trial embankment at Killington.
Ministry of Transport, RRL Report LR 238.
Moh, Z.C., Brand, E.W., and Nelson, J.D. (1972) Pore pressures under a bund on soft fissured clay.
ASCE, Proc. Speciality Conf. on Performance of Earth and Earth-Supported Structures, Purdue, Vol. 1 pp 243-272.
Murray, R.T. (1971) Embankments constructed on soft foundations: settlement at Avonmouth.
Dept. of the Environment, TRRL Report LR 443.
Ortiz, I.S. (1967) Zumpango test embankment.
ASCE, Jour. Soil Mech. and Found. Div., Vol. 93 pp 199-209

Palmer, A.C. and Rice, J.R. (1972) The growth of slip surfaces in the progressive failure of overconsolidated clay.
Presented at the 13th Int. Congress on Theoretical and Applied Mechanics, Moscow.

Parry, R.G.H. (1968) Field and laboratory behaviour of a lightly over-consolidated clay.
Geotechnique, Vol. 18 pp 151-171.

Pilot, G. (1972) Study of five embankment failures on soft soils.
ASCE, Proc. Speciality Conf. on Performance of Earth and Earth-Supported Structures, Purdue, Vol. 1 pp 81-100.

Reynaud, L.E. and Rosset, F. (1965) Essais de chantier pour la mise en oeuvre des terres au barrage du Mont-Cenis. (Construction tests on the soils used in the Mont-Cenis dam)
Proc. 6th ISCMFE, Montreal, Vol. 2 pp 532-536.

Rico, A., Moreno, G., and Garcia, G. (1969) Test embankments on Texcoco Lake.
Proc. 7th ICSMFE, Mexico, Vol. 2 pp 669-676.

Ringeling, J.C.N. (1936) Measuring groundwater pressures in a layer of peat, caused by an imposed load.
Proc. 1st ICSMFE, Cambridge Mass., Vol. 1 pp 106-111.

Rowe, P.W. (1968) The influence of geological features of clay deposits on the design and performance of sand drains.
Paper 7058S, Proc. ICE.

Rowe, P.W. (1972) The relevance of soil fabric to site investigation practice.
Twelfth Rankine Lecture, Geotechnique, Vol. 22 pp 195-300.

Scott Wilson Kirkpatrick & Partners (1966) Report on ground investigations; trial embankment.
(Report reprinted January 1968).

Scott Wilson Kirkpatrick & Partners (1970) AIT Bangkok: Notes on the analysis of the trial bund results and the design of the perimeter bund.
Internal Report KBCJ/BTP/68080

Scott, P.A. (1958) Observations on an experimental clay embankment to assist the design of the impermeable layer of a large rockfill dam.
Trans. 6th Congress on Large Dams. Vol. III pp 237-255.

Taylor, H.T. and Buchagnani, A.L. (1972) Field test of debris fill over soft soil.
ASCE, Proc. Speciality Conf. on Performance of Earth and Earth-Supported Structures, Purdue, Vol. 1 pp 395-414.

Wade, S. (1972) Trial bank saves fill at Empingham.
New Civil Engineer (ICE London), 5th October, 1972, pp 22-23.

Walker, F.C. (1967) Willard Dam - behaviour of a compressible foundation.
Jour. ASCE, Vol. 92 SM4 pp 177-198.

Ward, W.H., Penman, A.D.M., and Gibson, R.E. (1955) Stability of a bank on a thin peat layer.
Geotechnique, Vol. 5 pp 154-163.

Weber, W.G. Jr. (1966) Experimental sand drain fill at Napa River.
Highway Research Record, No. 133 pp 23-44.

Wilkes, P.F. (1972) An induced failure at a trial embankment at King's Lynn, Norfolk, England.
ASCE, Proc. Speciality Conf. on Performance of Earth and Earth-Supported Structures, Purdue, Vol. 1 pp 65-79.

Wilkinson W.B., Barden L., and Rocke, G. (1970) An assessment of insitu and laboratory tests in predicting the pore pressure in an earth dam.
Proc. Conf. on Insitu Investigations in Soils and Rocks, pp 277-284, British Geotechnical Society.

Zeller and Zeindler (1957) Test fills with coarse shell materials for Goschenalp Dam, Switzerland.
Proc. 4th ICSMFE, London, Vol. II pp 405-409.

THE MEASUREMENT OF IN SITU STRESS AND STRAIN IN SOILS

S.F. Brown

Lecturer in Civil Engineering, University of Nottingham

INTRODUCTION

The general problem of in situ stress and strain measurement has several main subdivisions.

1 Determination of free field stresses and strains, i.e. effects remote from boundaries.
2 Determination of boundary stresses.
3 Determination of transient effects, e.g. stresses beneath pavements.
4 Determination of long term changes in stress and strain, e.g. within earth dams.

This paper is primarily concerned with the problems of free field measurements which are more complex than those associated with the measurement of boundary stresses. The experimental work which forms the background to the paper has been concerned with the measurement of transient effects. However, the basic design considerations for both stress and strain transducers are similar to those involved in long term measurements. A significant difference between the two problems is the need for transducer elements in the latter case which remain stable over long periods.

Design considerations for stress and strain cells to be used in the measurement of free field stresses have been well documented in the literature. A useful review was presented by Selig (1964) and more recently papers by Morgan and Gerrard (1968), Dunn and Billam (1966), Thomas and Ward (1967) and Brown (1971) have described the development of particular instruments while reviewing the general design problem. The vast majority of effort has been devoted to pressure cells, relatively few strain measuring devices having been described.

The purpose of this present paper is to briefly summarise the design considerations for both types of instrument, describe some recent developments on theoretical predictions of pressure cell performance and assess the potential of some new instruments.

BASIC DESIGN CONSIDERATIONS

The introduction of a measuring instrument into a soil mass disturbs the stress and strain distribution. The overall aim in designing such instruments is to obtain a measure of the stress or strain which would have existed at the location of the instrument had it not been present. This can best be achieved by designing an instrument which has a minimum disturbing effect though, as in most design problems, this involves certain compromises because of conflicting considerations.

Pressure Cells

The majority of free field earth pressure cells are of the diaphragm type such as the one shown in Fig. 1 which has been developed at Nottingham. Various sensing elements have been

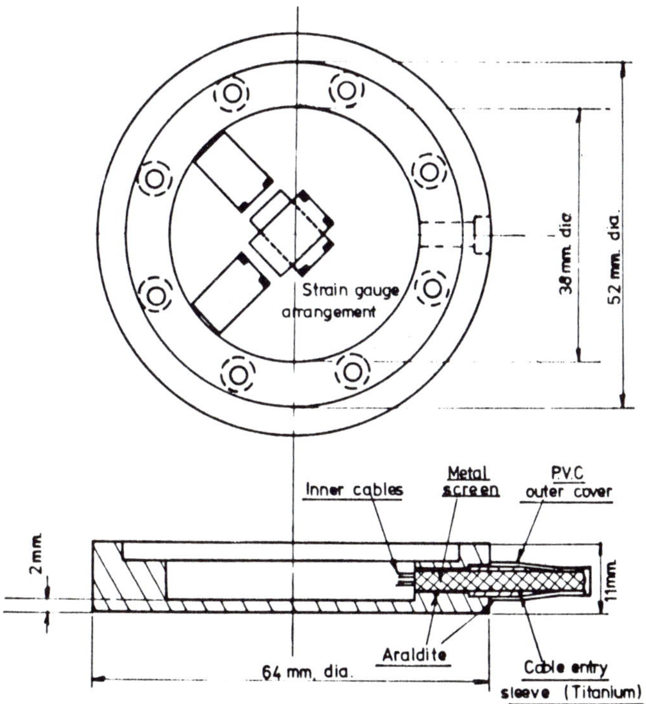

Figure 1 The Nottingham earth pressure cell

used to measure the deflection of the diaphragm and these include bonded strain gauges of various types, vibrating wire gauges, piezoelectric crystals and recently a small L.V.D.T. displacement transducer. The strain or deflection measurements made on the sensitive diaphragm are correlated with pressures applied to the diaphragm during a simple calibration test normally carried out in a pressure chamber.

The presence of the cell in a soil mass causes a redistribution of stress over the diaphragm as shown in Fig. 2 so that in general the instrument will either over- or under-register the field stress.

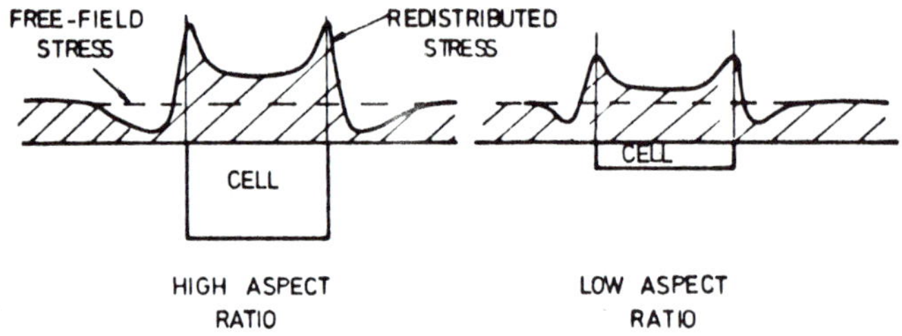

Figure 2 Redistribution of stress around a pressure cell

The two factors which chiefly affect this error in registration are the cell geometry and the stiffness of its diaphragm relative to that of the soil. Fig. 3 illustrates

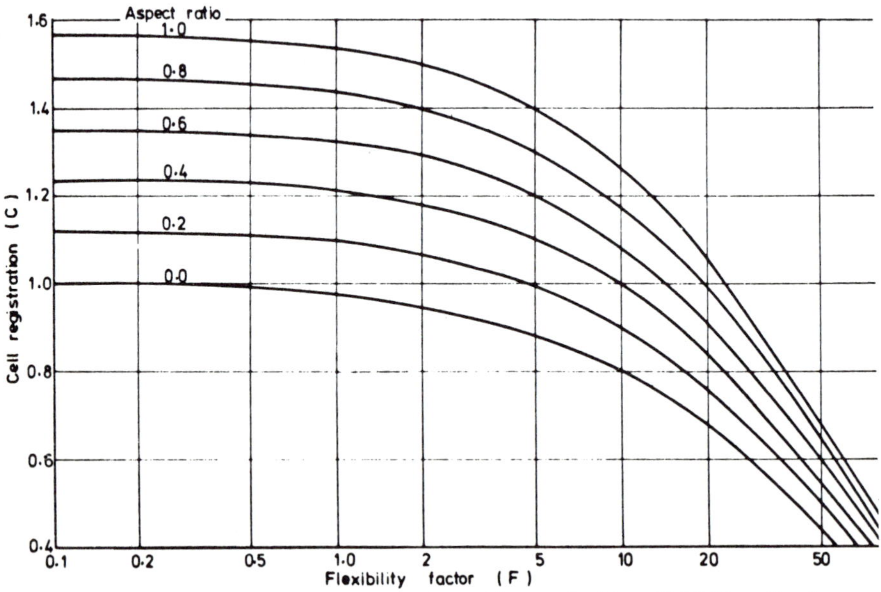

Figure 3 Influence of aspect ratio and flexibility factor on pressure cell registration

the effect of variations in these two variables on the cell registration, which is defined as

$$C = \frac{\text{stress indicated by the cell}}{\text{true or field stress}}$$

The variables are expressed as:

aspect ratio = cell thickness/cell diameter

flexibility factor, F = soil stiffness/diaphragm stiffness

$$= E_s d^3 / E_c t^3 \qquad (1)$$

where E_s, E_c = Young's modulus of soil and cell material respectively.

t, d = diaphragm thickness and diameter respectively.

The curves in Fig. 3 represent a somewhat simplified approach the extension of which is discussed in the next section dealing with theoretical predictions of performance.

Figs. 2 and 3 indicate that pressure cell design should aim to produce an instrument which is flat and stiff relative to the soil, in order to achieve a registration as close as possible to unity, while operating on the horizontal portions of the curves where cell registration is relatively unaffected by changes in soil modulus.

There are certain other detailed design considerations which were highlighted by the experimental work of Peattie and Sparrow (1954). The active diaphragm should not occupy more than 45% of the total area of the cell face in order that a reasonably uniform stress distribution occurs across it. Fig. 2 shows the high stress at the edge of the cell which is confined to the annular ring around the diaphragm if this limitation is observed.

This consideration results in the cell having a relatively rigid annular ring which is of importance in designing against cross-sensitivity. This refers to the need to design a cell which is only sensitive to stresses normal to the diaphragm. Brown and Pell (1967) and Brown (1969) have discussed this problem and some further comments follow in a later section in the light of recent considerations of this phenomenon. It is clearly desirable, however, that the pressure cell should be stiff in the diametrical direction so that cross-stresses do not influence the output from the transducer element.

The overall size of the cell should be related to soil particle size and the anticipated stress gradient in which it is to be placed. While larger soil particles can be kept away from the diaphragm during installation, the soil in contact with the instrument should not differ greatly from that in the surroundings otherwise further redistributions of stress are likely. Clearly, the diaphragm diameter must be much larger than the "particle size" and a minimum ratio of 50 to 1 has been suggested by Kallstenius and Bergau (1956).

The cell diameter should be small enough for any stress variation over the diaphragm due to an in situ stress gradient to be minimal. If this is not the case a complex stress redistribution is likely to result. While certain in situ effects can be quantified by appropriate laboratory calibrations, this particular one would be difficult to deal with accurately.

The particular transducer element to be used will depend very much on the in situ problem. For dynamic measurements adequate frequency response is required, and either strain gauges or piezoelectric crystals may be used. These latter in fact cannot be used for static applications where vibrating wire gauges, L.V.D.T.'s or strain gauges are normally used.

The requirement that a stiff diaphragm should be used conflicts with the need for an adequate electrical signal from the transducer, except in the case of the piezoelectric gauge. All the other transducers require an adequate deflection of the diaphragm to produce a measurable output. A compromise is generally possible however.

Two other design considerations are the need to provide effective waterproofing and resistance to corrosion.

Strain Cells

A strain measuring device is required to move with the soil in which it is placed without reinforcing it in any way. In order to obtain a measure of strain, the relative movement of two fixed points at either end of a gauge length needs to be monitored.

There have been relatively few strain cells described in the literature. At Nottingham an instrument was developed to measure small strains beneath road pavements (Sparrow and Tory (1966), Brown and Pell (1967)) and this has recently been improved by changing the transducer element from a strain gauged cantilever to an L.V.D.T. as shown in Fig. 4. The resulting instrument is similar in conception to that used by the T.R.R.L. (Potter (1969)).

The problem with such instruments is that there is a positive connection between the two end plates required to transmit the movement of one relative to the other. This requires a shear break to bridge the gap between the uniform deformation of the

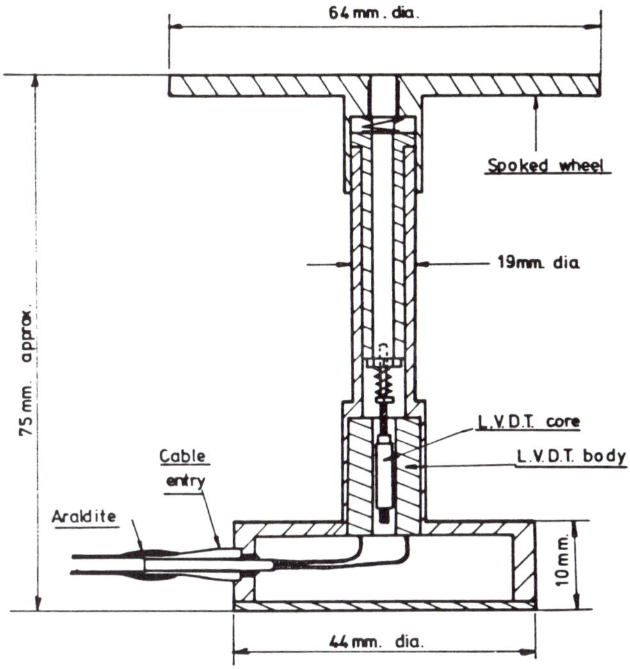

Figure 4 The Nottingham strain cell

soil and the concentrated relative movement of the two end plates. Brown and Pell (1967) have discussed this in presenting calibration test data on the earlier Nottingham instrument.

A potentially superior idea which overcomes the problem of interconnected end plates is the use of strain coils (Selig and Grangaard (1970)). A pair of coaxial wire wound coils are placed at either end of the gauge length and an oscillating current in one induces a signal in the other, the magnitude of which is a function of the gauge length. The application of this idea originally put forward by Truesdale and Anderson (1964) has taken some time to develop. Morgan and Gerrard (1968) have discussed the use of this technique.

THEORETICAL PREDICTIONS OF PRESSURE CELL PERFORMANCE

Early work by Taylor (1947) and Monfore (1950) was extended by Tory and Sparrow (1967) for a cell with a flexible diaphragm in a uniaxial stress field. Fig. 3, used in the previous section, is taken from the results of Tory and Sparrow. In practice, a pressure cell is used in a three dimensional stress field and the influence of stresses perpendicular to the direction of measurement is clearly of importance.

This problem has recently received attention from Fossberg (1970) who used finite element analysis to determine cell registration for a variety of conditions. He showed in effect that cell registration depends on the applied stress ratio as well as the flexibility factor.

Though Fossberg's results related to the particular geometry of his cell it is possible to apply the results to other cells of similar geometry. This has been done for the Nottingham cell in Fig. 5 by shifting the curves vertically

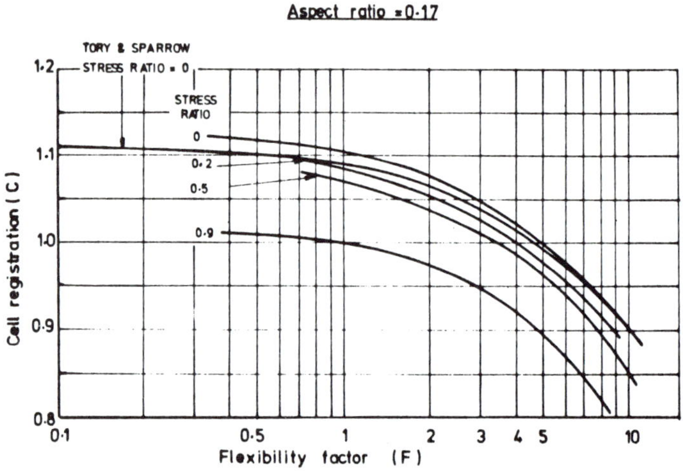

Figure 5 Theoretical cell registrations for the Nottingham cell

to allow for the slight difference in aspect ratio (0.13 compared with 0.17), the shift being based on the results of Tory and Sparrow (1967). This can be justified by noting that Fossberg's results compared well with Tory and Sparrow's for the special case considered by the latter. When a pressure cell is used in a particular soil both the stress ratio and the flexibility factor can change as the stress conditions change. The magnitude of the resulting changes in cell registration has been determined for two soils which have been used for pressure cell calibration tests at Nottingham, earlier results of which were described by Brown (1971). The soils were Keuper marl and a well graded crushed rock of 10 mm maximum particle size (Meldon dust). The non-linear stress-strain relationships for each soil were determined during pressure and strain cell calibration tests under various stress conditions and are summarised in Fig. 6.

Typical pressure cell calibration tests involved a variation of deviator stress from 0 to 125 kPa in six increments under confining stresses of 70 and 140 kPa. Hence the variation of stress ratio is known and by reference to Fig. 6 so too is the change in soil modulus. For the Nottingham cell the

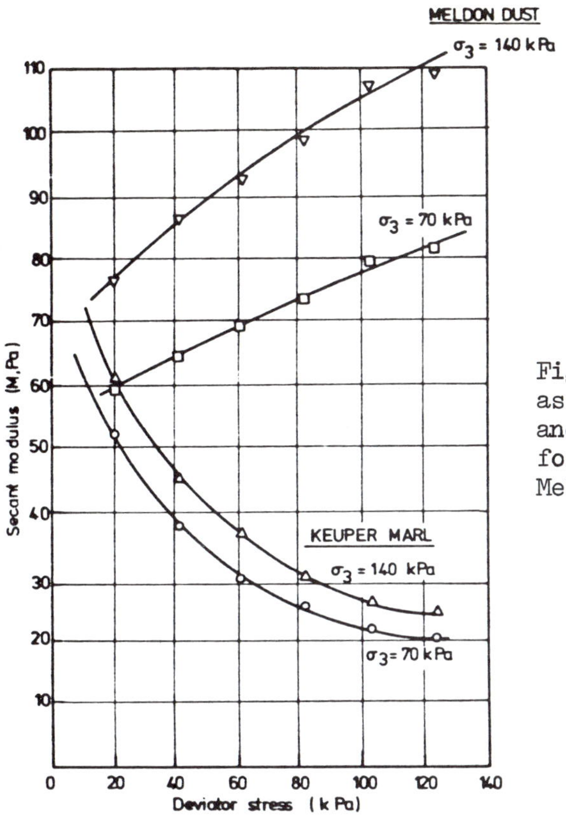

Figure 6 Secant moduli as a function of deviator and confining stresses for Keuper marl and Meldon dust

flexibility factor from equation (1) is: $F = 4.06 \times 10^{-4} E_s$ and using this relationship changes in F may be specified.

Fig. 7 shows the theoretical changes in cell registration for each test. This figure is a replot of Fig. 5 with the predicted cell registration paths superimposed. The same information could have been plotted on Fig. 5 but with less clarity.

The main conclusion to be drawn from Fig. 7 is that the change in cell registration is not very great for practical purposes over the stress range considered, the maximum being 6% for the Keuper marl tests. In practice, even under controlled laboratory conditions it is difficult to obtain repeatable results to within 10% of one another. This implies that practical factors, particularly placement technique, introduce errors which are likely to swamp those considered in this analysis. However, the procedure may well produce significant results for other soil-cell combinations.

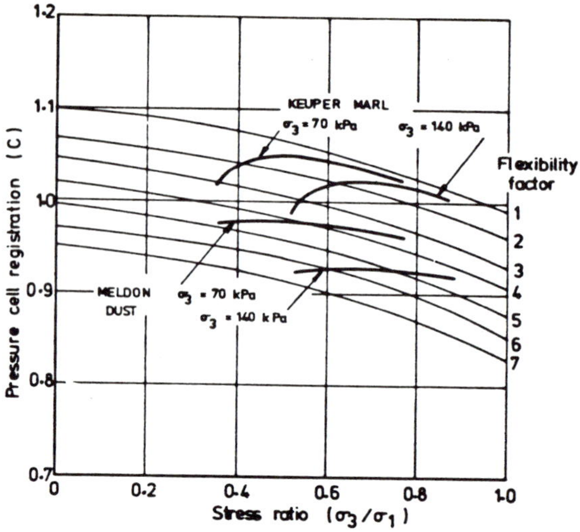

Figure 7 Theoretical variation of cell registration for the laboratory calibration tests

LABORATORY CALIBRATION TESTS

The foregoing theoretical analysis has indicated that a pressure cell will in general not exactly register the field stress though with well designed instruments the error may not be very large. Experience with controlled calibration tests, however, (Brown (1971), Morgan and Gerard (1968)) has indicated larger errors than those predicted by theory. These errors are due perhaps to the difficulty in precisely modelling the in situ conditions, particularly those resulting from placement techniques.

It is thus clearly desirable to calibrate pressure cells, and also strain cells, under controlled laboratory conditions which reproduce as closely as possible, the field situation. This implies attention to moisture content and compaction of the soil generally and in particular around the instrument, following the same installation technique as on site and subjecting the cell to the range of stresses in all directions, which are anticipated and doing this at the appropriate rate of loading.

Two types of test have been used for pressure cells, both employing large diameter (225 mm to 950 mm) cylindrical samples with the cell installed centrally. The samples have been either tested in a rigid cylindrical container reproducing Ko conditions (Peattie and Sparrow (1954), Williams and Brown (1971)) or as triaxial samples (Buck (1963), Brown (1971) Morgan and Gerard (1968)) when the vertical and lateral stresses can be varied in different combinations.

The second of these techniques has been used by the Author for both pressure and strain cells the detailed techniques having been described previously (Brown and Pell (1967), Brown (1971)). The tests referred to in discussing theoretical predictions of performance involved four to five samples of each of the soils and the average calibration figures are summarised in Table 1 together with the mean theoretical predictions taken from Fig. 7. Reasonable

Soil type		Cell Registration	
		σ_3 = 70 kPa	σ_3 = 140 kPa
Keuper Marl	Exp. Theo.	0.97 1.04	0.96 1.01
Meldon Dust	Exp. Theo.	1.10 0.98	0.94 0.93

Table 1 Comparison of experimental and theoretical pressure cell registrations

agreement is shown between the two sets of figures, although the tests in Meldon dust at a cell pressure of 70 kPa are more at variance with theory than the others.

This work has shown in general that theory may be used as a guide to trends in cell performance, both under static and dynamic conditions. Typical calibration curves obtained by Brown (1971) for the two soils showed that a straight line calibration was justified over the range of stresses investigated and for this cell the same conclusion was reached over a wider stress range by Williams and Brown (1971). Brown (1971) discussed the slight curvature noted in these curves indicating that these reflected changes in soil modulus. They do not, however, reflect the combined influences of stress ratio and soil modulus as indicated theoretically in Fig. 7.

A significant stress ratio effect has, however, been noted experimentally by Morgan and Gerrard (1968) for their 25 mm diameter cell as it was earlier by Buck (1963), using a similar instrument. This difference seems to be a function of the particular instrument and may be explained in terms of the mechanical cross-sensitivity effect described by Brown and Pell (1967).

There would hence seem to be two reasons for the influence of lateral stress on cell output:

1 A mechanical effect due to the action of the cross stress on the instrument influencing the transducer element.

2 A modification of the stress distribution on the diaphragm as demonstrated theoretically by Fossberg (1970).

The presence of a significant mechanical effect can be checked for quite simply by a bench test. In the Author's opinion it is this which contributes most to the overall stress ratio effect.

Calibration tests analogous to those described above have been used for strain cells. With the instrument installed at the centre of a large triaxial sample, the main problem is to obtain a measure of the "true strain" which forms the basis for the calibration. This requires direct "on sample" deformation measurements remote from the influence of end platens. Various techniques have been used for this including travelling microscopes focussed on targets (Brown and Pell (1967)) and recently for dynamic work the use of "strain collars" of the type described by Snaith and Brown (1972) for bituminous materials. Strain cells such as the Nottingham one (Fig. 4) having a positive connection between the end plates may be subject to a zero error caused by friction between the soil and the central tube. The magnitude of this error depends on installation details and values of about 100 microstrain were reported by Brown and Pell (1967).

The main object of calibration tests on strain cells is to establish the correct gauge length for the instrument. Experience with the Nottingham cell has shown that this is about 5% less than the overall length of the instrument in its mid-travel position.

One of the most useful aspects of calibration tests such as those described above is the development of successful installation techniques, which can be somewhat difficult, particularly for strain cells.

RECENT DEVELOPMENTS

Pressure Cells

A new instrument has been developed by the Transport and Road Research Laboratory. It is a diaphragm cell using a miniature L.V.D.T. as the transducer element and has the potential for long term and transient stress measurement. A number of these cells have been installed in full scale road experiments and others are being used for the determination of stresses round buried pipes.

A miniature pressure cell has been developed for the U.S. Federal Highway Administration (Walter et al (1971)) which has several potential advantages over existing instruments. The sensing element is a piezoresistive gauge which is suspended in an oil medium between two thin stainless steel discs. It has a very low aspect ratio and because it is filled with oil a high stiffness. Both these factors have been shown to be desirable design features. It is suitable for both static and dynamic applications.

Strain Cells

The commercial availability of strain coils with their associated electronics has renewed interest in this concept and several agencies are currently using this equipment.

A strain cell with no positive connection between its ends is potentially attractive and wider experience with this equipment will reveal whether this potential is realised. Selig and Grangaard (1970) have described the various applications of this equipment which may be used for both static and dynamic problems. Tests by Morgan (1971) have shown that the strain coils can interfere with strain distribution significantly at high strain levels and that this is a function of the surface finish given to the coils. However, calibration tests of the type described in this paper would provide data which allowed for this. Such tests are in progress at Nottingham and preliminary results indicate linear behaviour under static and dynamic conditions over the strain range 0 to 8,000 microstrain with a resolution down to 50 microstrain.

CONCLUDING DISCUSSION

In reviewing recent developments of theoretical predictions of pressure cell performance together with experimental investigations three main factors were shown to influence cell registration:-

1 Flexibility factor
2 Aspect ratio
3 Applied stress ratio

In addition to giving attention to these factors in design, care should be taken to ensure that mechanical cross-sensitivity is minimised.

The importance of calibration tests which reproduce the anticipated in situ conditions is emphasized. Such tests allow information to be obtained on cell registration and the development of satisfactory installation techniques.

Strain cells with a positive connection between their end plates present important difficulties if small strains are to be measured. The wider use of separated strain coils is thus regarded as an important development.

In designing in situ instrumentation it is important to prevent future problems arising from corrosion or poor waterproofing.

Several recent developments show good potential particularly for solving the problems of obtaining long term measurements.

ACKNOWLEDGEMENTS

The experimental work referred to in this paper was carried out by Mr. B.V. Brodrick under a Science Research Council grant. Facilities were provided by Professor R.C. Coates, Head of the Department of Civil Engineering, University of Nottingham. The cooperation of the Structural Properties Section of the Transport and Road Research Laboratory was appreciated in dealing with piezoelectric pressure cells.

REFERENCES

Brown, S.F. (1969) Discussion on earth pressure cells. Proc. Conf. on In situ Investigations in Soils and Rocks, B.G.S., London : 243-245.

Brown, S.F. (1971) The performance of earth pressure cells for use in road research. Civil Eng. and Pub. Wks Rev., London, 66, 775 : 160-165.

Brown, S.F. and Pell, P.S. (1967) Subgrade stress and deformation under dynamic load. Journ. S. Mech. and Found. Div., ASCE, 93, SM1 : 17-46.

Buck, G.F. (1963) An investigation of earth pressure cell response and the stress distribution in test pavement subgrades. Ph.D. thesis, Univ. of Birmingham.

Dunn, C.S. and Billam, J. (1966) A study of the response characteristics of a new miniature earth pressure cell. Civil Eng. and Pub. Wks Rev., London, 61, 715 : 181-186.

Fossberg, P.E. (1970) Load deformation characteristics of three-layer pavements containing cement-stabilised base. Ph.D. thesis, Univ. of California.

Kallstenius, T. and Bergau, W. (1956) Investigations of soil pressure measuring by means of cells. Royal Swedish Geot. Inst. Proc. 12.

Monfore, G.E. (1950) An analysis of the stress distribution in and near stress gauges embedded in elastic solids. Struct. Res. Lab. Report No. SP-26, U.S. Dept. of Interior, Bureau of Reclamation.

Morgan, J.R. (1971) Discussion on strain coils. Stress-strain behaviour of soils. Cambridge : 495-496.

Morgan, J.R. and Gerrard, C.M. (1968) Free field measurement of stresses and strains in soils. Proc. 4th Conf. Aust. Road Res. Board, 4, 2 : 1743-1760.

Peattie, K.R. and Sparrow, R.W. (1954) The fundamental action of earth pressure cells. Journ. Mech. and Phys. of Solids, 2 : 141-155.

Potter, J.F. (1969) Gauge for measuring dynamic and long term strain in soil. Road Research Lab. Rep. LR251.

Selig, E.T. (1964) A review of stress and strain measurement in soil. Proc. Symp. on Soil-Structure Interaction, Univ. of Arizona : 172-186.

Selig, E.T. and Grangaard, O.H. (1970) A new technique for soil strain measurement. Mat. Res. and Standards, 10, 10 : 19-36.

Snaith, M.S. and Brown, S.F. (1972) Electro-hydraulic servo-controlled equipment for the dynamic testing of bituminous materials. RILEM Int. Symp. on the Deformation and Rupture of Solids Subjected to Multiaxial Stresses, Cannes, 3 : 139-154.

Sparrow, R.W. and Tory, A.C. (1966) Behaviour of a soil mass under dynamic loading. Journ. S. Mech. and Found. Div., ASCE, 92, SM3 : 59-93.

Taylor, D.W. (1947) Pressure distribution theories, earth pressure cell investigations and pressure distribution data. Soil Mech. Fact Finding Survey, W.E.S. Vicksburg, Miss.

Thomas, H.S.H. and Ward, W.H. (1969) The design, construction and performance of a vibrating-wire earth pressure cell. Geotechnique, London, 19, 1 : 39-51.

Tory, A.C. and Sparrow, R.W. (1967) The influence of diaphragm flexibility on the performance of an earth pressure cell. Journ. Scientific Inst. 44 : 781-785.

Truesdale, W.B. and Anderson, M.E. (1964) A new device for soil strain measurement. Proc. Symp. on Soil-Structure Interaction, Univ. of Arizona : 129-137.

Walter, D., Kriebel, A.R. and Kaplan, K. (1971) URS Report 785-6.

Williams, A.A.B. and Brown, S.F. (1971) Proc. 5th Reg. Conf. for Africa on S. Mech. and Found. Eng., Luanda, 1 : 5-15 to 5-21.

IN-SITU MEASUREMENTS OF EARTH PRESSURE AND ANCHOR FORCES FOR A DIAPHRAGM RETAINING WALL

J. Bundred

Greater London Council, Dept. of Planning and Transportation

LOCATION

Wandsworth Bridge Southern Approach provides direct dual three-lane access to and from Wandsworth Bridge on the south side of the River Thames towards Tooting. A large part of the route is in a cutting between retaining walls, passing under the East Hill and North Side Wandsworth Common Road's bridges; this underpass rises southwards from the North Side Common Bridge to ground level on Wandsworth Common and the diaphragm wall is situated here on the east side of the underpass (Fig.1).

Diaphragm
Some 143 m length of wall was constructed as R.C. diaphragm by casting successive 9.14 m long by 0.79 m thick panels in a bentonite-supported trench; 82 m length of deeper panels, rising to 8.8 m maximum height for panel 1 at the wall's northern end are anchored at higher level and 61 m length of shallower panels at the wall's southern end are vertical cantilevers, all being propped at road level. The latter varies as the road slab rises southwards out of the underpass onto Wandsworth Common.

Panel 1 is instrumented as shown on the wall cross-section of Fig.2 in order to obtain data on the performance of wall and anchors as a structural unit and as a check on design assumptions; also to attempt an assessment of conventional site investigation procedure and soils laboratory testing and the validity of design parameters used for fissured clay. The instrumentation project was initiated in February 1968 with the casting of wall panel 1 including earth cells and slope indicators; comprehensive measurements commenced at the end of July 1968 when the wall anchors were positioned in the ground ready for stressing.

GEOLOGY AND SITE INVESTIGATIONS

By wall panel 1 the strata consist of made ground overlying

Taplow gravel over London clay more than 30 m thick, the made ground approximately 1 m thick being a sandy clay with ash and brick fragments. The Taplow is comprised of dense sand and gravel, thickness varying from 1 m to 2 m under Wandsworth Common. In 1966, the London clay was proved by boring to a depth of 20 m below existing ground level and the major part of the underpass cutting is within stiff grey silty over-consolidated fissured London clay. As is common beneath alluvial gravel, natural weathering to brown clay has been confined to a thin zone in the clay-gravel interface region. Above this the level of the Water Table is present in the gravel about 2.5 m below ground surface. Claystone bands and isolated clay stones occur at various levels within the height of the diaphragm wall. In 1968 and 1969, the 1966 information was augmented by tests on soil samples obtained near wall panel 1 or carried out 'in situ' in the underpass excavation.

Stiff clay

Soil Properties
London clay properties relevant to the wall design, as determined by standard laboratory tests on 38 mm and 102 mm undisturbed and other samples and 'in-situ' shear box tests on 0.6 m square undisturbed clay blocks are summarised in Table 1, shear strength being expressed in both total and effective strength parameters. Shear strength values, consolidation and compressibility co-efficients, bulk density, moisture contents, index properties and sulphate content are normal for fissured London clay.

Wall Design Criteria Peak strength parameters $C^1 = 8.95$ kN/m^2, $\phi^1 = 23\frac{1}{4}^o$ and residual strength parameters $C_r{}^1 = 1.65$ kN/m^2, $\phi^1_r = 16^o$ were adopted as design criteria.

DETAILS FOR WALL PANEL 1

Panel 1, being typical of all others, is restrained by 8 prestressed 18 m long inclined earth anchors acting 2 m below the wall's crest, at 1.15 m centres along the wall and stressed in Aug. 1968 before underpass excavation proper commenced. Each anchor was originally post-tensioned to 150 kN working load. The wall is very stiff and it was supposed that measured earth pressures might lie finally between 'at rest' values and the fully 'active' state.

Wall Design
A major uncertainty in the wall design being the possible magnitude of the earth pressures emphasised the desirability of measuring 'in situ' pressures, etc. in this case. Preliminary design assumptions for immediate post-construction conditions were for horizontal ground load factors against the wall of K = 0.28 for terrace gravels and K = 1.00 for clay, plus water pressure, with a 0.305 m thick R.C. road slab assumed to act as a low level stiff prop after completion of the underpass. Underpass Excavation took place in stages as indicated in Figs. 2 and 5. The underpass excavation programme is important in its relation to the wall's performance as measured.

Soil Properties for L.C.	Imperial Units	S.I. Units	
Bulk density	115–125 lb /ft^3	1850–2000kg/m^3	Laboratory Tests
Natural moisture content	22–29%	22–29%	
Plastic limit	24–30%	24–30%	
Liquid limit	73–81%	73–81%	
Undrained shear strength	1400 (at 10ft) 6800 (at 65ft) lbf/ft^2	67,000–325,000 N/m	
Drained shear strengths C^1	187 lbf/ft^2	8953 N/m^2	In-situ shear box
ϕ^1	23½°	23½°	
C^1_r	34 lb/ft^2	1650 N/m^2	
ϕ^1_r	16°	16°	
Coefficient of compressibility (M_v)	0.004–0.0170 ft^2/ton	0.000037–0.00016 m^2/kN	Lab. test

TABLE 1

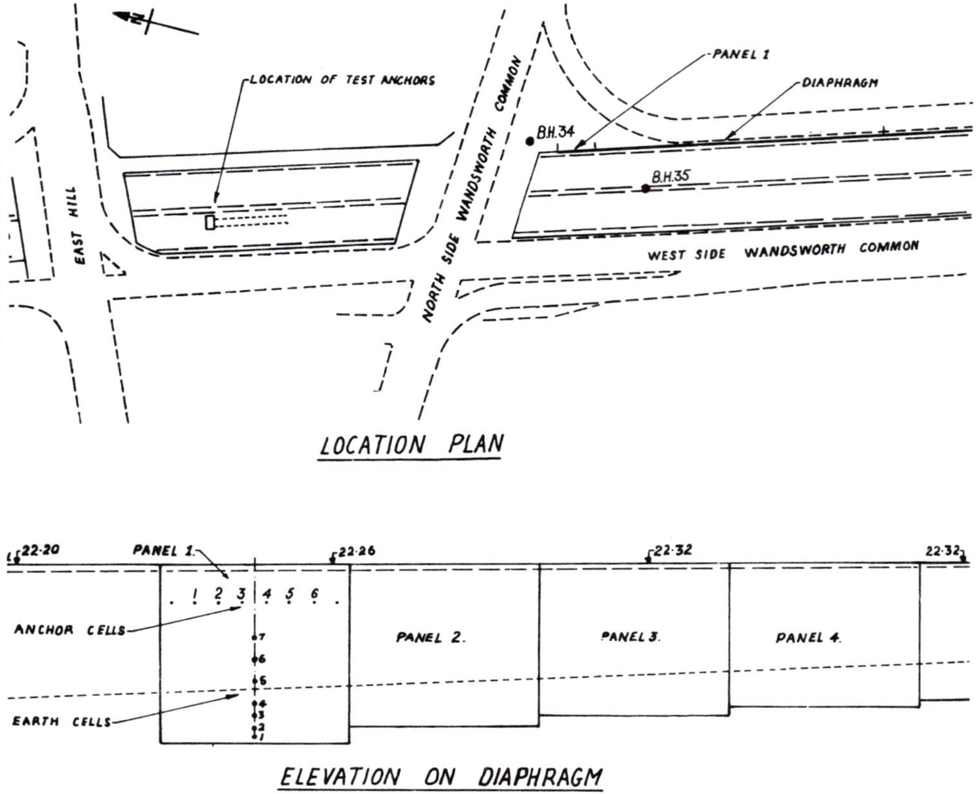

Figure 1 Location plan and elevation on diaphragm

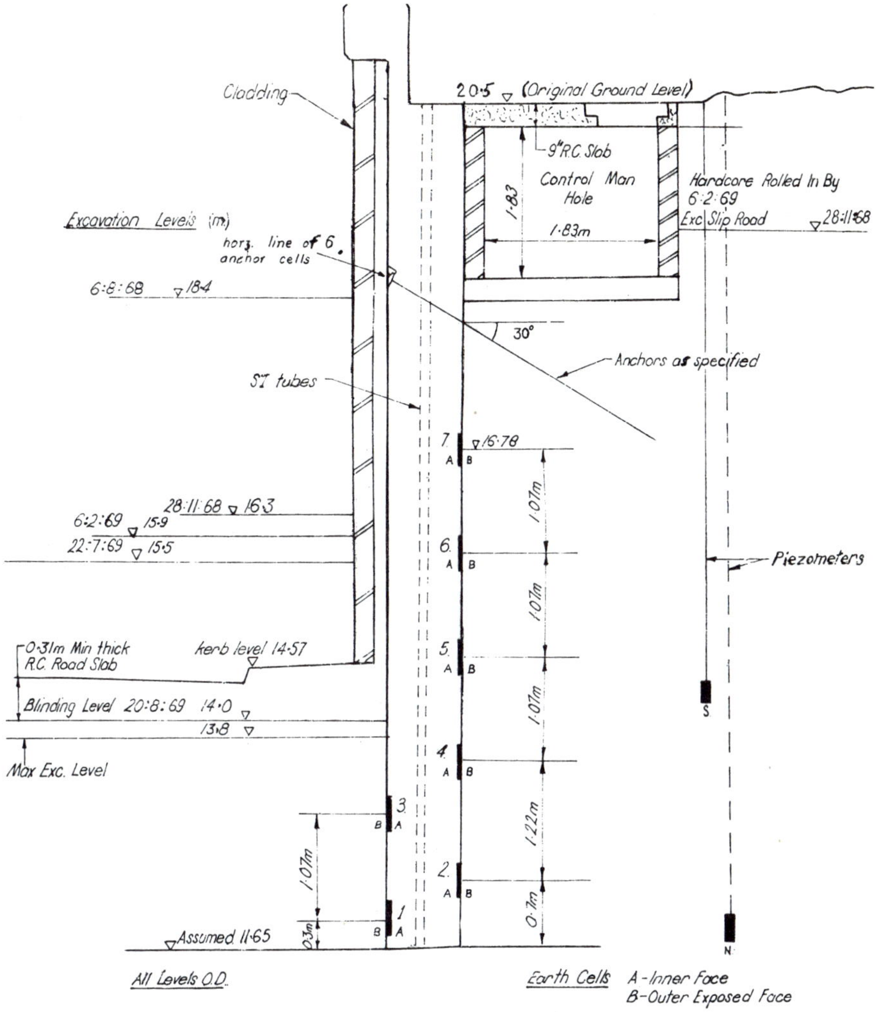

Figure 2 Instrumentation showing earth cell positions, etc.

Earth Anchor Tests	Test No.1	Test No.2	Test No.3
Number of wires	12	12	1
Load at failure(kN)	674	770	71.5
Load per wire (N)	56,167	64,167	71,500
Cross-sectional area per wire (square mm)	50.3	50.3	50.3
Stress at failure (kN/m^2)	1.1	1.3	1.4
Total nett extension movement (gauge deflection - mm)	87.6	106.7	113.0
Age at testing (days)	5	7	34

TABLE 2

EARTH ANCHORS

The anchors adopted have a 12 m free length resin protected and 6 m fixed end grouted, using the Soletanche patent 'Tube à Manchette' with multi-strand wire rope and dependent upon a regrouting technique to provide their load capacity. For a permanent anchor in clay soil, Littlejohn (1970) has suggested that C^1 and ϕ^1 are the relevant shear strength parameters. Each hole was wash-bored, then cleaned out with weak grout and compressed air, the whole operation taking approximately five-six hours. Casing was necessary only through the gravel layers overlying the clay. The working anchor cables made up from six 8 mm wires were placed in the holes and could be grouted with sulphate resisting cement in two or more stages dependent on local ground conditions.

Grouting and Stressing Data

For each anchor 1422 kg of grout was provided to achieve an anchor block of 800 kN capacity. This was usually accomplished in two main grouting stages. The pressure required in the second stage varied on average from 1378 to 2068 kN/m^2 and the grout pump was capable of reaching 6890 kN/m^2. Seven days after completion of grouting each anchor was stressed to working load and wedged. The Specification required all anchors to be tested to working load and 10% to 2 WL; ie, 300 kN, and to be held at this load for 24 hours. Any doubt as to the load capacity of an individual anchor resulted in further grouting being carried out to increase the anchor's capacity to the required value.

Earth Anchor Tests Before commencing the main anchor installation work, tests to destruction were carried out on two 12 wire trial anchors sited between East Hill and North Side Wandsworth Common (Fig.1). These failed through some of the wires pulling out of the 6 m anchor block and later single wires which had not failed in bond were pulled out individually. The results of these tests are summarized in Table 2 where test No.3 represents a summary of the average results from the tests on single wires. It was apparent from load-deflection curves, which start tangential to the theoretical load-extension curve of a cable 12 m effective free length, that the bond had failed progressively down the anchor block until the wires pulled out. The corrected measured extensions obtained from test results combined with a typical stress/strain curve for the wire used forms the basis for deducing the stress/strain characteristics and extensions of the wall anchor wires under working loads. Thus elastic extension of the anchors at their working load of 150 kN will be small (about 4 mm) but it cannot be assumed that this and similar variables will have negligible effect on the wall movement as measured, eg., the tests showed also that movement of the whole anchor held during a 36 hour test at constant 2 x Working Load (300 kN) was very small. At the working load of 150 kN, relaxation of anchor load capacity due to ground consolidation

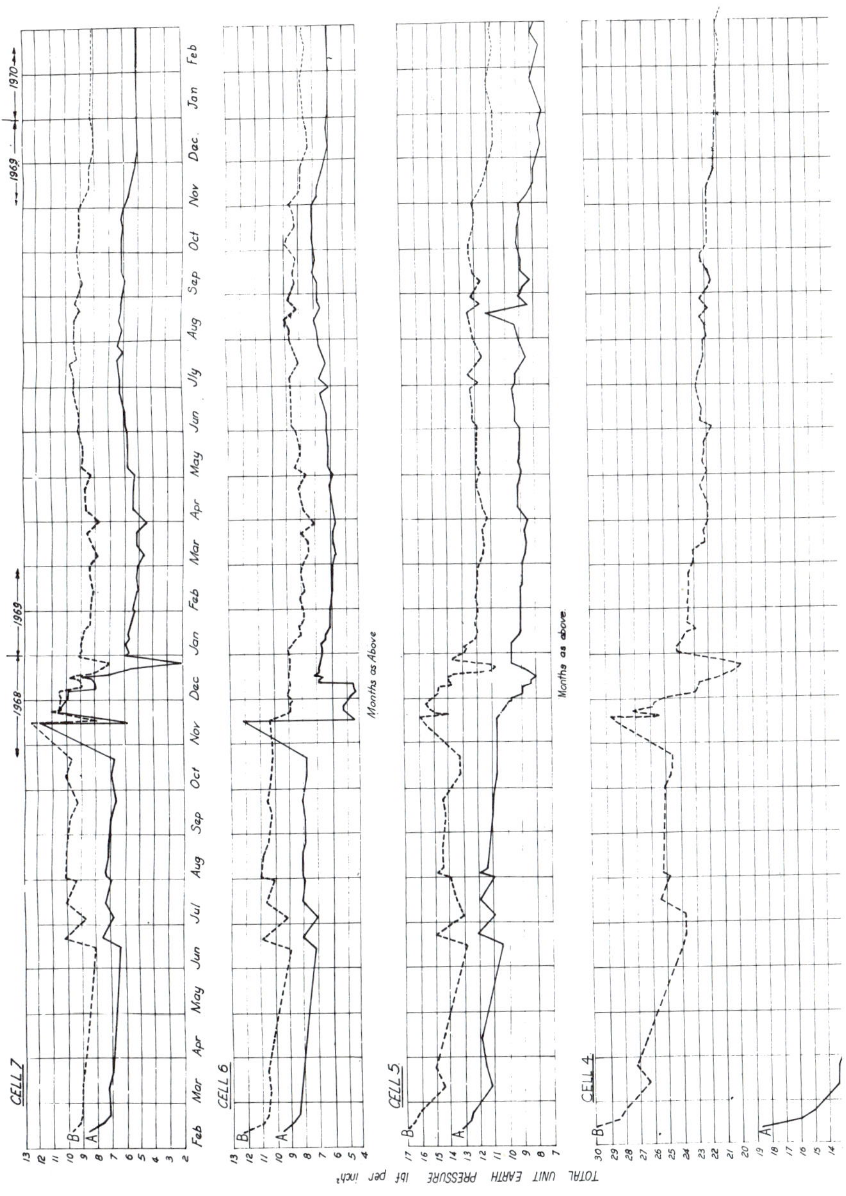
CELL 7
CELL 6
CELL 5
CELL 4
A
B
1968
1969
1970
Feb Mar Apr May Jun Jul Aug Sep Oct Nov Dec Jan Feb Mar Apr May Jun Jly Aug Sep Oct Nov Dec Jan Feb
Months as Above
Months as above
TOTAL UNIT EARTH PRESSURE lbf per inch²

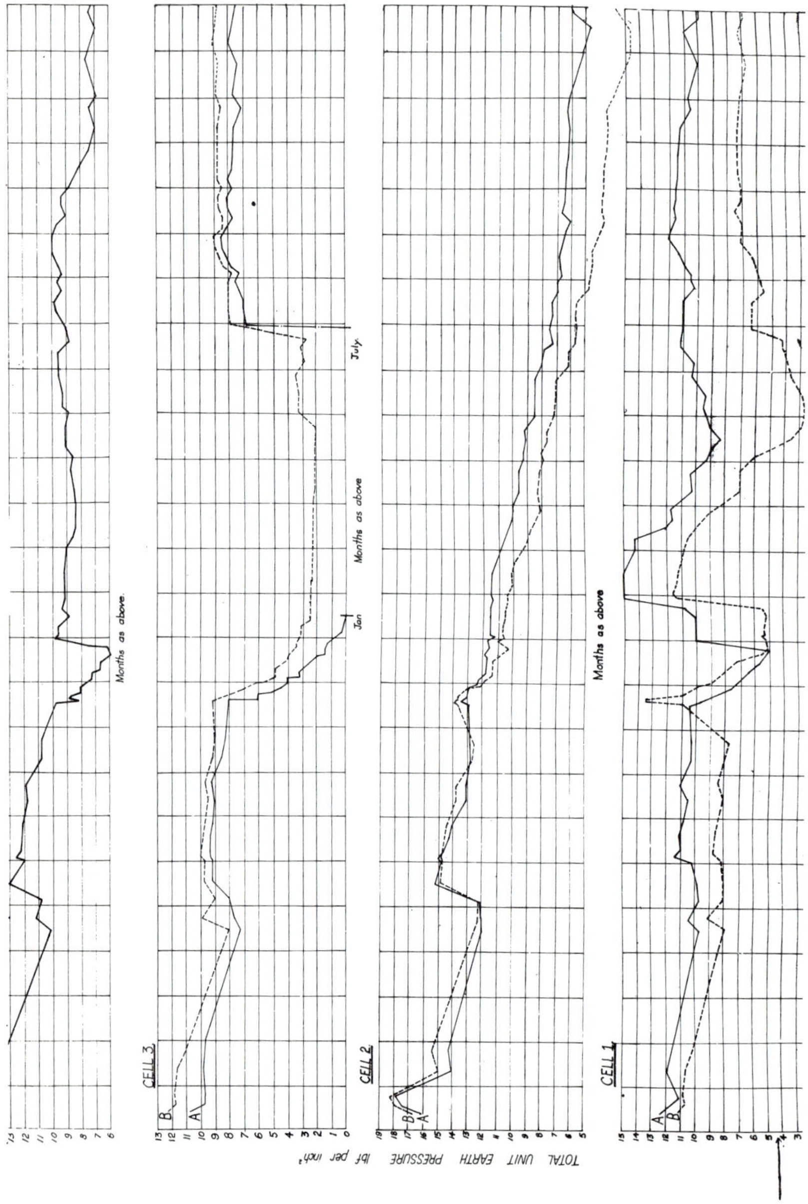

Figure 3 Record of earth pressure cells' readings

should thus be small; ie., anchor yield was expected to be less than 8 mm in the long term. No provision was made to measure anchor yield (or creep of wires) on wall panel 1.

FIELD MEASUREMENTS PLANNED

Instruments were installed in or near to panel 1 to obtain the following, the form of instrumentation being such that readings would be maintained for several years without any necessity to disturb the underpass cladding or the slip roads, services, etc. in use -

(a) The active and passive earth pressures acting.
(b) Records of variation in ground water and selected pore water pressures using piezometers.
(c) Measurement of the tensile force acting in 6 anchors.
(d) Data for a study of long term earth pressures, anchor loads and relative wall movements/deformation.

The measurements also provided an opportunity to gain experience of the types of instrument adopted.

Instrumentation

Details of the instrumentation are summarised in Appendix 1 with layout as shown in Figs 1, and 2. Summaries of some readings obtained with detailed data are given in Fig.3 and Table 3. A precedent had been established for the methods of placing, reading, interpreting results and evaluating performance of the earth cells by those employed in a trial quay wall at Seaforth, Liverpool. Ref. should be made to Uff (1969) for more details.

Earth Cells - Observations. The 7 total earth pressure cells were laboratory 'zeroed' on 2 Feb. 1968, 'zeroed' on site on 7 Feb., pressure checked in bentonite in situ on 9 Feb. and pressure checked after concrete was poured on the same day. Within one week of casting panel 1, earth pressure readings commenced on 14 Feb. and continued to 27 Mar. to observe how the earth cells settled down. Pressures were initially high but fell to become steady after one month and then continued to decrease slowly for the next three months. Excavation in front of the wall then proceeded to 2 m depth to allow installation of the earth anchors in July 1968. Thereafter from 20 June regular fortnightly instrument readings were taken; it was observed that the cells behaved satisfactorily and their results are considered to be reliable, with significant changes being recorded later at times of anchoring, major excavation, etc. Pressure values are obtained by reading off laboratory derived frequency calibration graphs for the two faces of each cell. Some sample pressure measurements are plotted against time in Fig.3; in general, higher pressures are recorded on the outer (B) faces of the cells than on the inner (A) faces. Before the panel 1 trial there had been two cases where the same type of earth pressure cell was used respectively in Boulder and London clays (Uff 1969 and Armishaw 1970). In both cases the same pattern of readings was obtained - the B faces giving pressures higher than the A. It appears that higher readings on the B face is characteristic of the BRS type modified cell when cast into concrete. At times this has caused difficulty in interpretation of the correct pressure value to record for an individual cell; eg, cell 4 shows a large difference in A and B readings.

Anchor Cells - Observations The 6 load cells were 'zeroed' on their anchorages on 31 July 1968 by representatives of the installers and GLC. Installation was completed on 1 Aug. and the installer's separate anchor load readings agree reasonably with GLC comparator readings of 2 Aug. Subsequent fortnightly readings after mid-Aug. showed little change until Oct. 1968 when definite load increases began to occur. Load values interpreted from laboratory calibration graphs are tabulated as in Table 3.

PERFORMANCE OF CELLS

The design of the total earth pressure cell being such that its reading is dependent on the pressure distribution on the cell diaphragm, the measured high value of pressure recorded on B face of cell 4 could be due to a point load, probably from the presence of a clay stone acting near the centre of the cell. A tentative correction was devised in keeping with the performance of the other cells to bring this high B face value in line with the B values of the remainder. Typical sets of pressure readings were plotted against cell positions along with the average of the two faces of all 'active' cells except cell 4. The sum of the deviations of the B values from A face values was obtained and then the average of these deviations was added to the A face value of cell 4 to obtain a corrected B value for this cell. Uff showed that individual cell wire readings may differ widely while averages agree reasonably with predicted pressures. Trial jacking of cells into a soil can produce widely differing readings on the soil diaphragm B faces which may be due to random isolated point contacts. The inner A face readings show much less scatter and the average in the case of all the cells under trial agreed well with the applied pressure. Therefore all results presented here for active earth pressure calculations and diagrams are based on the average of A and B readings, the latter having first been corrected for cell 4. There were no problems with the anchor cells in interpreting load values.

Earth Cells and Anchor Cells - Interpretation of Results
Earth cell readings taken on 2 Aug. 1968, after completion of all panel 1 anchor stressing, were compared to those taken on 31 July just before the anchors were post-tensioned. They showed a small increase of 3.5-7 kN/m^2 on passive earth cells 1 and 3 but little effect on the corresponding active earth cells except above future underpass road level where a similar increase was recorded. Thereafter fortnightly readings remained reasonably constant until 25 Oct. when the instruments'

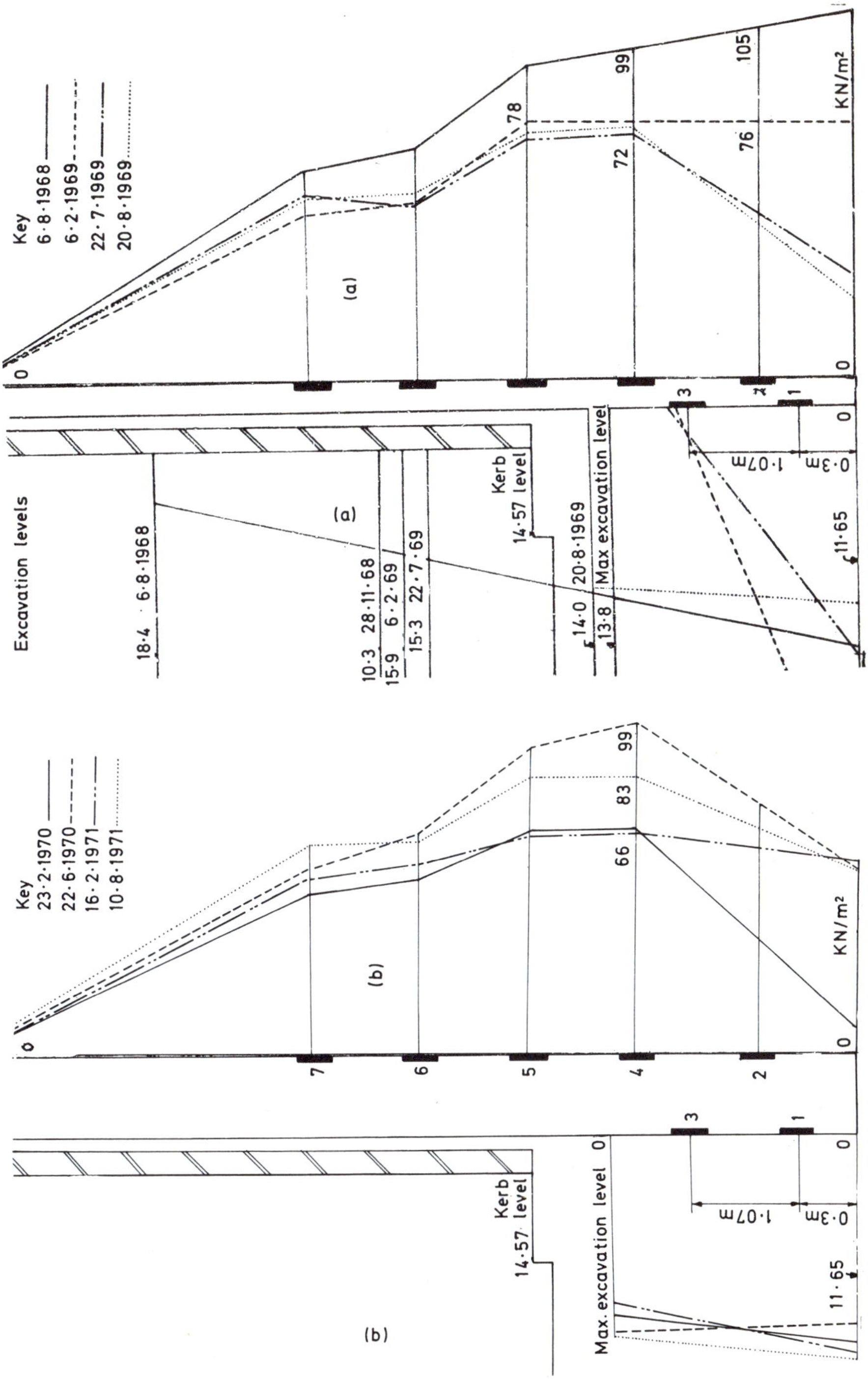

Figure 4 Total earth pressure distributions;(a) before, (b) after underpass road slab completion

control manhole was flooded due to a burst water main on site and reliable readings took some months to re-establish. The changes of earth pressure distribution recorded throughout the various stages of underpass excavation were plotted for the cross section of the diaphragm wall (Fig.4). All pressure diagrams obtained during the period Aug. to Nov. 1968 revealed a maximum unit pressure 105 kN/m^2 on the backface near the wall base. After initial losses the early anchor cell readings recorded a meaned horizontal load of 83.0 kN/m run of wall (Fig. 5) compared to the design applied load of 108.8 kN/m run. The stress loss was immediate after anchoring and apart from some possible creep there was no material change subsequently until further excavation began in steps to the +15.9 m level, the final step being carried out in the weeks from 6 Jan. 1969, by which time a definitive upward trend in anchor loads had been established. The maximum unit active earth pressure on 6 Feb. 1969 (excavation level +15.9 m above O.D.) was 78 kN/m^2 on the wall backface, with anchor loads averaging 119.5 kN per anchor. By April the earth pressure had decreased to a backface unit maximum of 72 kN/m^2 and anchor loads had increased to an average 125 kN per anchor. Regular readings were continued at fortnightly intervals throughout 1969. The maximum unit earth pressure value recorded on 29 Aug. one day after the final stage of underpass excavation to +13.8 m O.D. level was 76.6 kN/m^2 with the mean anchor load increased to 130.0 kN.

The earth and anchor cells seem to have behaved satisfactorily up to the present. Summarised results for all measurements are presented in Fig.5 - such as the total horiz. earth pressure computed directly from 'active' cell readings, 'Pa'. The anchor forces have varied within narrow limits during their upward trend to the present, as may be inferred from their mean load plot, T_m. However fluctuations in the water table can have an effect on Pa and T values; the relevant ground and groundwater levels should be considered when comparing the Pa, T values and unit 'total earth cell' pressures for different dates.

WALL MOVEMENT MEASUREMENTS

Slope indicator casings 1 and 2 were cast in panel 1 on 9 Feb. 1968 and initial zero datum readings were established on 23 Feb. Readings taken on 30 July just prior to anchor stressing showed that the wall top at No.2 position had moved outward (ie, westward into the 2 m cut) relatively by a maximum of 11.4 mm (Fig.5). One week after anchor stressing was complete, readings at positions 1 and 2 showed that the wall had moved back 3.8 mm and there was then a permanent outward maximum relative deflection at No.2 top of 7.6 mm. Readings taken on 14 Feb. 1969 after the underpass excavation to the +15.9 m O.D. level, showed a marked increase in relative outward movement of the wall top to 11 mm. The next readings taken on 30 April showed little change so that the wall deflections can be presumed to have settled down by this date, three months after the previous

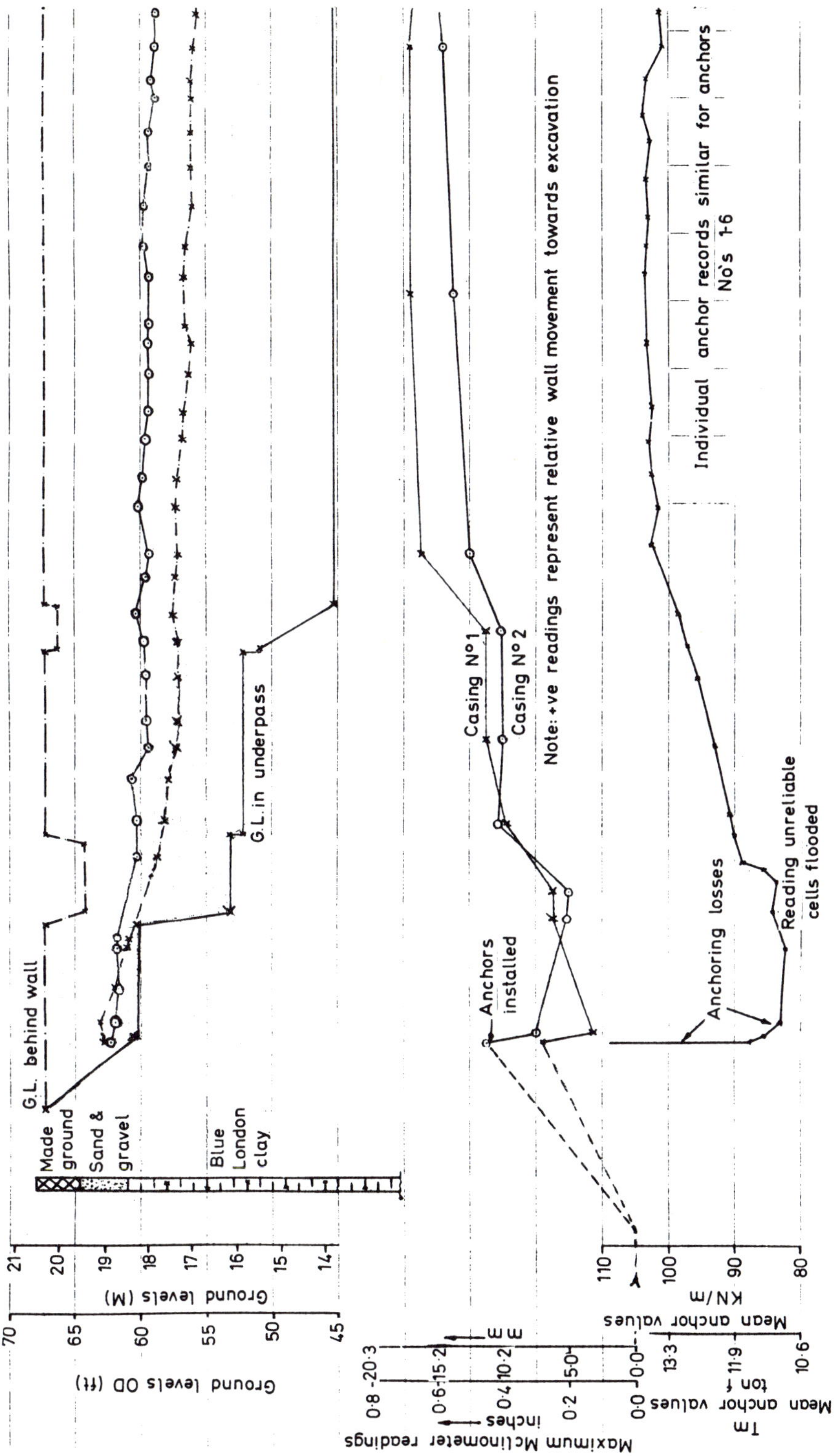

G.L. behind wall
Made ground
Sand & gravel
Blue London clay
G.L. in underpass
Ground levels OD (ft)
Ground levels (M)
Maximum Mclinometer readings
inches
mm
Mean anchor values
KN/m
Mean anchor values
ton f
TM
Anchors installed
Casing N°1
Casing N°2
Note: +ve readings represent relative wall movement towards excavation
Anchoring losses
Reading unreliable cells flooded
Individual anchor records similar for anchors No's 1-6

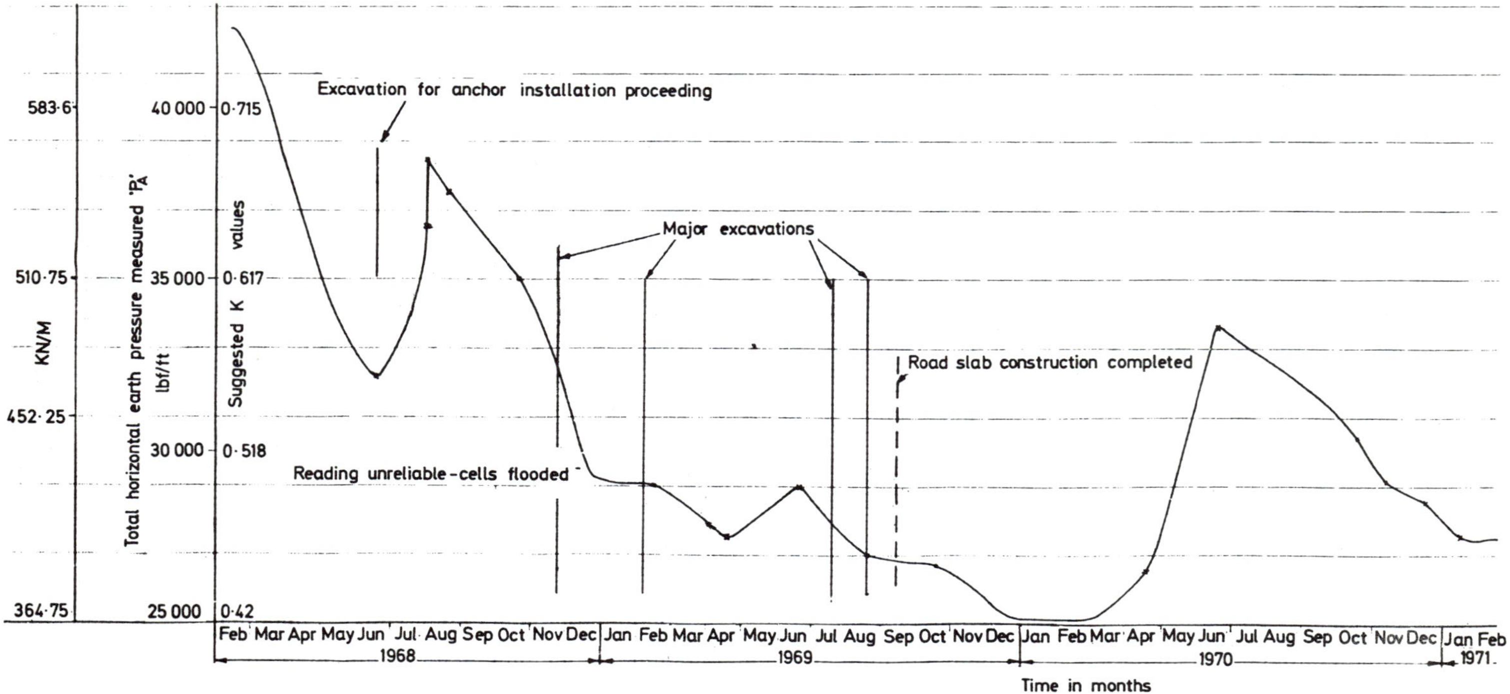

Figure 5 Summarised results of field measurements

excavation stage. Deformation measurements were continued at quarterly intervals into 1970 and latterly are taken at half yearly intervals.

Slope Indicator Readings as plotted - Fig.5

These are presented as maximum horizontal deviations of the casings from the vertical as a function of time. The maximum deviation has not altered materially since mid-1970 when the relative outward movement plotted for the wall top was 15.2 mm, some four months after the completed underpass came into traffic use. Readings are plotted so that maximum deviation occurs at the top of the wall; direct survey measurements indicate that the S.I. readings are near absolute values and may probably be read as such.

ANALYSIS OF RESULTS

The active pressure distribution retains its initial shapes despite significant changes in the total active pressure throughout the construction period and afterwards. The distribution appears to lie somewhere between triangular and parabolic but further information concerning the validity of the consistently high readings of pressure cell No.4 and the earth pressure in the vicinity of the anchors is needed to enable a more positive conclusion to be drawn. The passive pressure distribution can, unlike the active pressure, be separated into distinct patterns coinciding with variations in external conditions. An initially even distribution over the two passive cells prior to any excavation or anchoring gradually changed during underpass excavations, due probably to local excavation conditions near to cell 3. This situation was reversed during the three months following the construction of the road slab when the pressures reverted to the initial state. There has been little change in either the total passive pressure or the pattern of distribution since then and calculations would seem to indicate that the full passive pressure has been mobilised.

Theoretical Formulae

An effective stress analysis proposed by Bjerrum and Kirkedam (1958) can be utilised to obtain an estimate of active earth pressures when an appropriate value of C^1 is used. Bell's formula is also appropriate and values for C^1 are suggested by Henkel (1957) among others for London clay. Although the diaphragm wall is near impervious, the change in drainage pattern brought about by construction and a gravel syphon at the northern extremity of the anchored section are likely to cause deterioration of the clay.

CONCLUSIONS

From the results so far, effective stress analysis gives a good estimate of active earth pressure using $C^1 = 9$ kN/m^2 and $\phi^1 = 23\frac{1}{4}^{\circ}$. The pressures measured are consistent with a coefficient of fully active earth pressure K_A confirming the value 0.43 quoted by Packshaw (1946). Although it is not yet clear whether a consistent long-term trend has been established for the 'active' earth pressure, K = 0.67 after anchoring is the estimated upper limit of the co-efficient so far obtained. The 1970-72 results indicate a probable long-term seasonal variation in earth pressure which seems to be related to possible swelling and shrinkage of the retained clay.

Wall Movements - Relation to Mobilised Earth Pressures

The wall deviations plotted cannot be used as a wholly reliable quantitative assessment of the wall movement necessary for active pressure mobilisation although it is felt that 12 mm top movement would be enough to induce a fully active state. The deviations with time graph shows however some mirrored resemblance to the time-variations in 'active' earth pressure during construction and the resulting change in K value. Major excavations in the underpass in mid-Nov. 1968 and July/Aug. 1969 both result in active pressure reduction, the value of K falling towards the fully active value K_A. The relationship is not a simple one; periods of little or no relative movement between Feb. and April 1969 and also Nov. 1969 to March 1970 correspond to subsequent significant rises in pressure. This could be due to a 'tightening up' process in the clay wedge following a certain amount of stress release brought about by the previous wall movement. If this is so, one might expect the active pressure to remain fairly constant (apart from the effects of seasonal factors) from Sept. 1970 onwards in the short term, this pressure being distributed for horizontal equilibrium between the anchors, the road slab reaction and passive pressure, K_p value 2.7 approximately.

Experimental Accuracy - Comments It has been assumed that all readings are correct and no assessment of accuracy is given apart from a purely qualitative one in the case of the S.I. readings. The greatest inconsistency is in the recorded values of earth pressure at Cell 4 which have been continuously high due to the high reading on this cell's 'active' side; several rows of cells would have helped interpretation here. Other factors to be considered are (a) the possible change in drainage pattern of the clay and the resulting changes in the effective cohesion C^1 (b) the contribution of vertical wall movements if any, (c) anchor creep and (d) the value of the angle of wall friction. Of these, factors (b) and (c) could have been evaluated with further instrumentation, eg, several earth pressure cells placed in the base of the wall. A further significant contribution could have been made had absolute wall movement been precisely measured, this being an important omission in confirming instrumentation results. It must be emphasised that to obtain accurate information on the relationship between earth pressure, wall movement/deformation, mobilised shear strength of a clay and anchor loads is difficult under construction conditions; the present study was therefore somewhat limited in scope. It has been a useful trial of field instrumentation carried out during the everday realities of a civil engineering contract. Some similar more carefully controlled studies will be necessary to establish reliability of the Wandsworth results and confirm their validity.

ACKNOWLEDGEMENTS

The work described was undertaken by the Greater London Council to whom the Author is indebted for permission to publish this paper. The GLC is not responsible either for the statements made or for the opinions expressed. Suppliers and installers of instruments were: Soil Instruments, Ltd for the earth pressure cells; Deakin Instrumentation, Ltd for the anchor load cells; Nuttall's Geotechnical Services for the slope indicators.

References

Armishaw, J. (1970) Private Communication.
British Standards Institution (1951) CP.2002 - Earth Retaining Structures.
Bjerrum, L. and Kirkedam, R. (1960) An effective stress analysis for stiff fissured clays. Norwegian Geotechnical Institute, Pub. No. 33.
Bjerrum, L. and Kirkedam, R. (1958) Some notes on earth pressures in stiff fissured clay. Proc. cont. on Earth Pressure Problems, Brussels, Vol.1: 15-27 and Norwegian Geotechnical Pub. No. 33.
Henkel, D.J. (1957) Investigations into two long term failures in London clay at Wood Green and Northolt. Proc. 4th Int. Conf. S.M.F.E. London, Vol.2: 315-320.
Littlejohn, G.S. (1970) Proc. Symposium on Ground Engineering, London, I.C.E.
Packshaw, S. (1946) Earth pressure and earth resistance, published in 'A Century of Soil Mechanics', 1969 - London, I.C.E.
Thomas H.S.H. and Ward W.H. (1969) Geotech., Lond. 19, 1:39-51
Uff, J.F. (1969) In-situ measurements of earth pressure for a quay wall at Seaforth, Liverpool. Proc. Conf. on In situ Testing of Soils, London, I.C.E. - Paper 18.

APPENDIX 1

Equipment installed for instrumentation of Panel 1, etc.

7 B.R.S. type modified total earth pressure cells (Thomas and Ward 1969) range 0-700 kN/m^2, each reading independently on two opposite faces, laboratory calibrated by hydraulic loading and kept continuously in a dry nitrogen atmosphere; 5 measuring 'active' pressures and 2 for measuring 'passive' pressures. 6 load cells attached under the anchor wedge cones, one to each of Nos 2 to 7 anchorages of the 8 inclined anchors installed.

Terminal units and B.R.S. type vibrating wire comparator, reading in $\frac{(\text{frequency})^2}{1000}$ units and accurate to $\frac{2f^2}{1000}$, installed in control manhole and used for reading both earth pressure cells and anchor load cells (Table 3).

4 slope indicator tubes cast in at positions along the centre

line of the wall - 2 No. in panel 1, north and south of manhole and 2 No.76 metres southwards beyond the tied section of wall.
2 piezometers adjacent to instruments behind wall back face at depths of 2/3 x wall height and full wall height below G.L. on the south and north side respectively of the manhole. 1 piezometer 30 metres east of wall in Wandsworth Common at 9 m depth on centre line of instruments.
Conventional survey methods were used to check measure vertical and horizontal movements of the wall top.

Anchor Load Cell No.		1	2	3	4	5	6
Date							
31.7.68	Mean Zero as $(\text{frequency})^2 \times 10^{-3} = fo^2$	1394	1375	1390	1394	1389	1380
19.8.68	Mean of $(\text{frequency})^2 \times 10^{-3} = f^2$	1185	1193	1185	1246	1203	1179
	Anchor force in tonf=T	11.46	10.79	13.32	8.58	10.28	11.67
25.10.68	f^2	1183	1194	1187	1248	1205	1184
	T	11.57	10.73	13.19	8.46	10.77	11.38
28.11.68	f^2	1173	1182	1183	1247	1206	1185
	T	12.12	11.44	13.45	8.52	10.11	11.32
20.12.68	f^2	1176	1186	1186	1249	1210	1189
	T	11.96	11.20	13.26	8.41	9.89	11.09
21.2.69	f^2	1157	1167	1168	1231	1190	1171
	T	13.00	12.33	14.43	9.45	11.00	12.13
23.4.69	f^2	1153	1164	1163	1226	1187	1167
	T	13.20	12.51	14.75	9.74	11.16	12.36
25.6.69	f^2	1146	1158	1159	1222	1183	1163
	T	13.62	12.86	15.01	9.97	11.38	12.60
22.7.69	f^2	1145	1156	1154	1216	1177	1154
	T	13.66	12.98	15.34	10.32	11.71	13.12
21.8.69	f^2	1143	1152	1151	1214	1174	1155
	T	13.77	13.22	15.53	10.44	11.88	13.06
23.10.69	f^2	1127	1139	1143	1207	1168	1149
	T	14.66	13.99	16.05	10.84	12.21	13.41
27.11.69	f^2	1125	1141	1139	1209	1173	1153
	T	14.76	13.87	16.31	10.73	11.93	13.18
22.12.69	f^2	1122	1137	1138	1206	1173	1152
	T	14.92	14.11	16.37	10.90	11.93	13.23
27.1.70	f^2	1120	1136	1137	1207	1173	1153
	T	15.03	14.17	16.44	10.84	11.93	13.18
23.2.70	f^2	1122	1139	1137	1206	1173	1152
	T	14.92	13.99	16.44	10.90	11.93	13.23

TABLE 3

Figure 6 View looking north on panel 1 with road slab completed, showing earth cell and anchor positions

THE MEASUREMENT OF GROUND DISPLACEMENT AROUND DEEP EXCAVATIONS

by J B Burland, Head, Geotechnics Division
and J F A Moore, Principal Scientific Officer, Geotechnics Division, Building Research Station

INTRODUCTION

During the last few years the Building Research Establishment has been engaged in a study of the deformations of the ground beneath and around excavations. The purpose of this work is to provide data on the magnitude and extent of movements from a number of sites with varying ground conditions and construction methods. The provision of such data is fundamental to gaining a better understanding of ground behaviour and to the determination of realistic parameters for future designs. This work has created a need for simple reliable and sensitive measuring equipment for use under rigorous field conditions.

The paper describes a variety of pieces of equipment and measurement techniques which have been used in the study of ground movements around excavations. Many of these can be used for routine monitoring of movement to check on the design and safety of a project. Examples of actual measurements made with each piece of equipment are given.

TYPES OF MOVEMENT ASSOCIATED WITH EXCAVATIONS

During the process of excavation vertical and horizontal pressure is reduced in the surrounding ground. This gives rise to predominently upward movements in the base of the excavation and inward and vertical movements (both up and down) outside the excavation.

To illustrate the type of movement that can occur Fig 1 shows predicted and observed short term displacements associated with an open excavation in London Clay (Cole and Burland, 1972). Two important features are first that significant movements extend a considerable distance from the excavation and second that the horizontal movements can be much larger

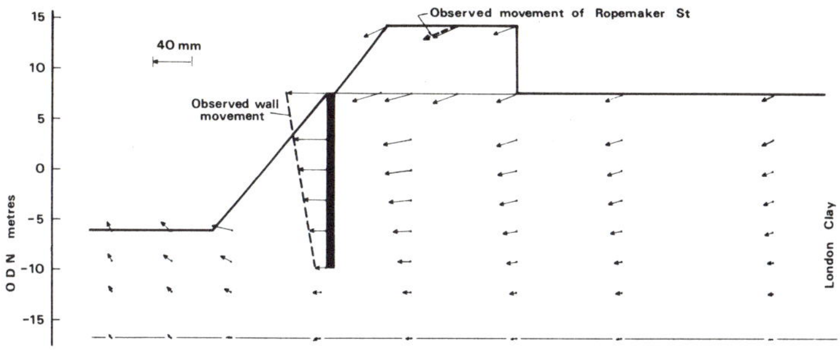

Fig 1 Section through excavation showing vector displacements

than the vertical movements.

It is apparent from the foregoing example that, when planning observations of movement around a proposed excavation, it is necessary to establish stable reference points at a distance of at least three times the depth of the excavation away from its edge. It is also evident that any comprehensive set of measurements may have to include not only the measurement of horizontal and vertical displacements of surface points but also of points within the ground. Instruments for measuring internal movements (eg inclinometers) only yield relative displacements and these will usually have to be tied in to the surface reference points.

Experience at the Building Research Station has shown that movements which are of significance to most buildings need to be measured to an accuracy of about 1 mm. However, there are often very real advantages from a research point of view in measuring relative movements between certain points to ten times this accuracy. In order to achieve such accuracies attention needs to be given to the stability of datums and measuring points and the next section will be concerned with this.

Before entering a detailed discussion of various pieces of equipment it is important to emphasise the features that are essential to most of them. They should be simple, reliable and easy to install and use. Perhaps the most important requirement is that the installations should be robust and not prone to either accidental or wilful damage. The success of site measurements depends on attention to detail in the design and construction of the equipment and on the ability to maintain it under very hostile and arduous conditions. Time and energy spent in achieving a dependable installation are seldom wasted as a significant long term expense is that of staff time for observing and processing data.

DESIGN OF SURVEY REFERENCE POINTS

The use of precise surveying techniques to achieve accuracies of 1 mm or better requires detailed attention to the design of reference points, which include datum points which are required to be stable for a relatively long period and movement points which are required to follow ground displacements. In outlining the use of the precise level to determine relative vertical position Cheney (1973) has described a vertical reference point or levelling datum and a movement point (levelling station) for use on structures. In this section a number of reference points are described which have been used to provide accuracy of lateral as well as vertical position.

Datum pillar. The most effective base for surveying instruments is provided by a rigid column of convenient height around which there is freedom of movement for the observers without risk of disturbance to instruments. Although a datum pillar may be founded most easily on a rock outcrop it will usually be necessary to found on some sub-surface firm stratum.

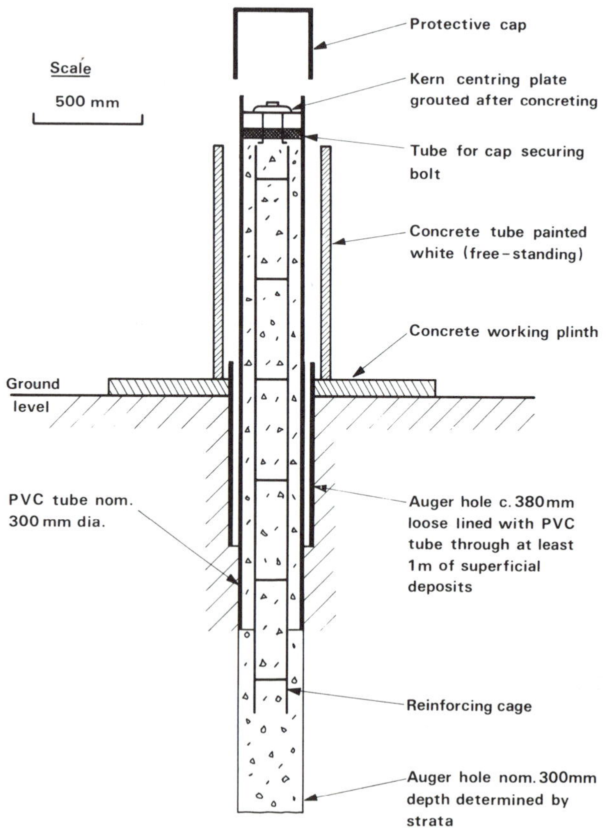

Fig 2 Datum pillar

The pillar should be isolated from the immediate surface strata to avoid local moisture/temperature effects. A compromise depth of isolation shown in Fig 2 which has been found reasonable is about 1 metre. When there is a greater thickness of strata considered too weak for founding the advantages of isolating the pillar from them to a greater depth must be balanced against possible instability of a long thin column. The column above ground should be protected from direct sunlight by an independent tube.

Precise interchangeability of instruments on the pillar requires the installation of a permanent plate which provides forced centering for any item of equipment. The autocentering system manufactured by Kern is particularly convenient, giving a re-centering accuracy of 0.03 mm. As well as catering for a wide range of standard instruments (theodolites, targets, optical plumments and distance measuring equipment) it is easy to fit other special equipment to the standard centering plates eg taping heads, special targets.

Sub-surface reference point. When it is impracticable to provide a permanent reference point above the surface eg in built-up areas, a datum may be established several metres below the surface to achieve stability away from surface effects and on a firm stratum. Lateral position is referred vertically from datum to surface by optical plumbing from a tripod, which is probably more desirable than a long unsupported pillar but less convenient, while a permanent levelling rod provides a vertical datum.

The arrangement of Fig 3 which has been used to 7 m depth comprises a survey plate grouted to the bottom of a loose-lined auger hole. The plumbing point is a 1 mm diameter hole filled with black wax at the top of a white hemisphere, illuminated as necessary. The levelling rod rests on a separate spherical seating. This same design has been used as a movement point (eg when ground level was subsequently to be reduced) with a suitable diameter/depth ratio of the hole to allow for the full range of anticipated movement.

Surface movement point. Although the sub-surface movement point can be used to within 1 m of the surface the alternative design of Fig 4 may be used if it is impracticable to establish the necessary tripods. It consists of a reference rod isolated from local moisture and temperature effects on top of which is screwed a kinematic support (fitting A) just below ground level which accepts the spherical feet of removable targets. The levelling target has a simple domed surface whereas the intersection target has a rod, painted matt black, with a conical end containing a central drill hole. The latter target may be intersected directly with a theodolite or plumbed from a tripod.

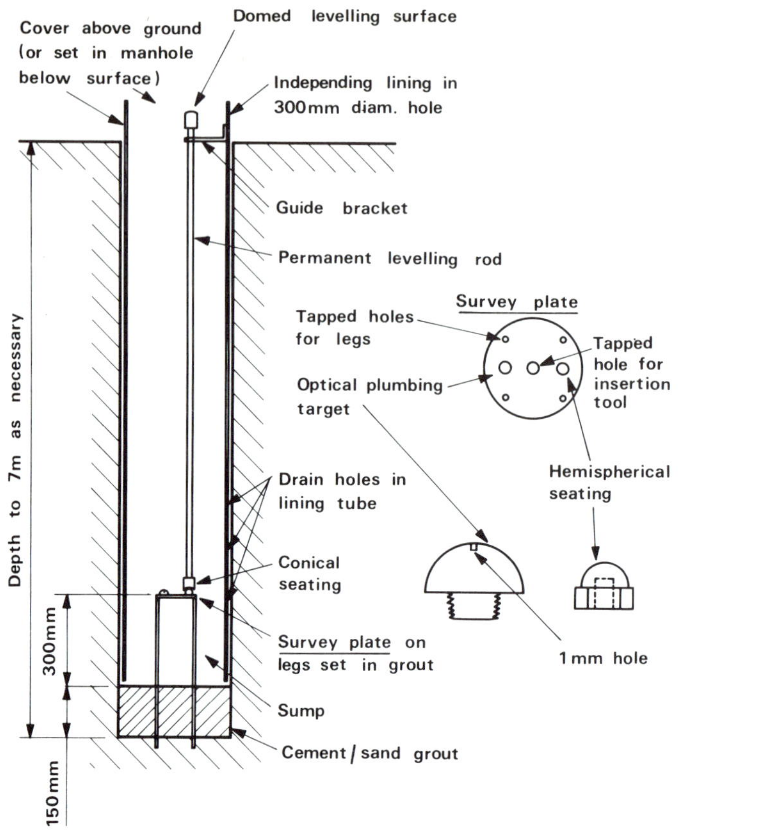

Fig 3 Sub-surface reference point

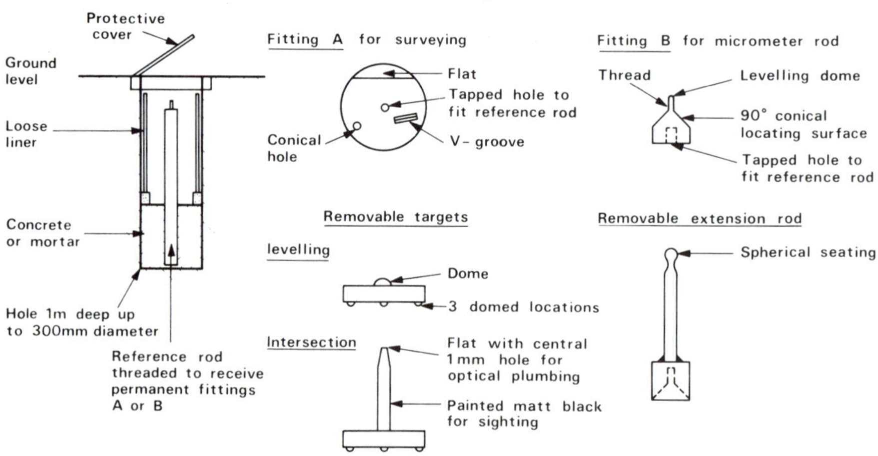

Fig 4 Surface movement point

A tilt of 1 in 1000 in the ground, and the movement point, would cause an error of 1 mm over the combined length of the reference rod and target, but the lateral movements associated with such a tilt would probably render the error insignificant.

Taping plug. The taping plug shown in Fig 5 has been designed so that the plan position of the BRS levelling station (Cheney, 1973) may be determined by distance and angle measurement, the angle being observed to the tip of the levelling plug.

Wall alignment. Horizontal displacements normal to a retaining wall may be measured by establishing a theodolite line of sight along the wall to intersect a number of scales (Burland and Cole, 1972) or micrometer targets which move with the wall.

Typical movement. Figure 6 shows the lateral movement of the top of a 25 m high clay bank determined by triangulation at intervals over a period of 24 months relative to datum pillars (as in Fig 2) some 200 m distant. Movement was caused by successive stages of excavation. The movement point of Fig 4 was successfully immune to the passage of contractors' heavy plant.

INCLINOMETERS

An inclinometer is used for measuring the inclination of a borehole at various depths from which its profile may be calculated. The inclinometer used by BRS is the MkII version manufactured by Soil Instruments Ltd and is described by Green (1973).

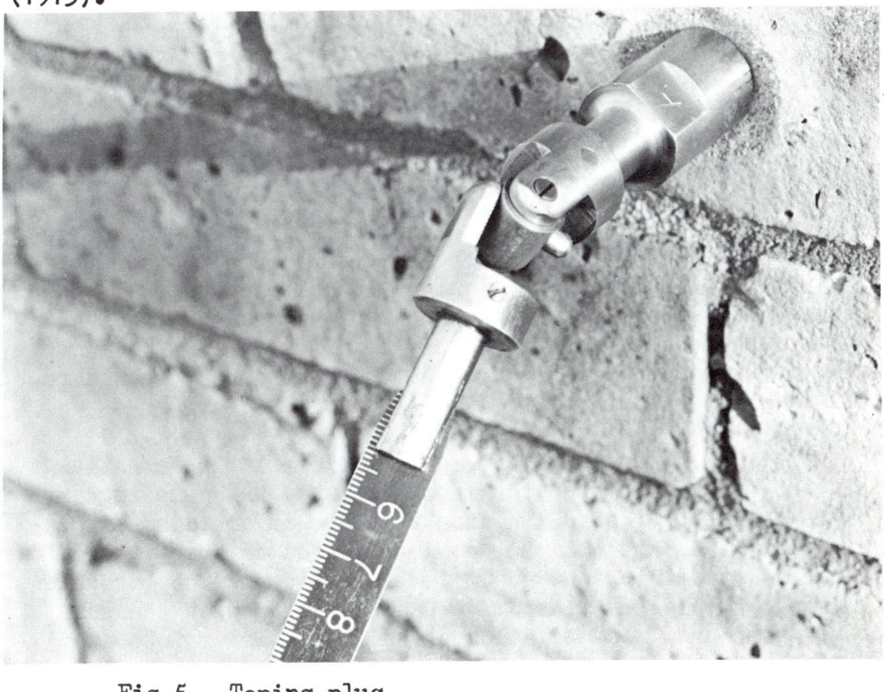

Fig 5 Taping plug

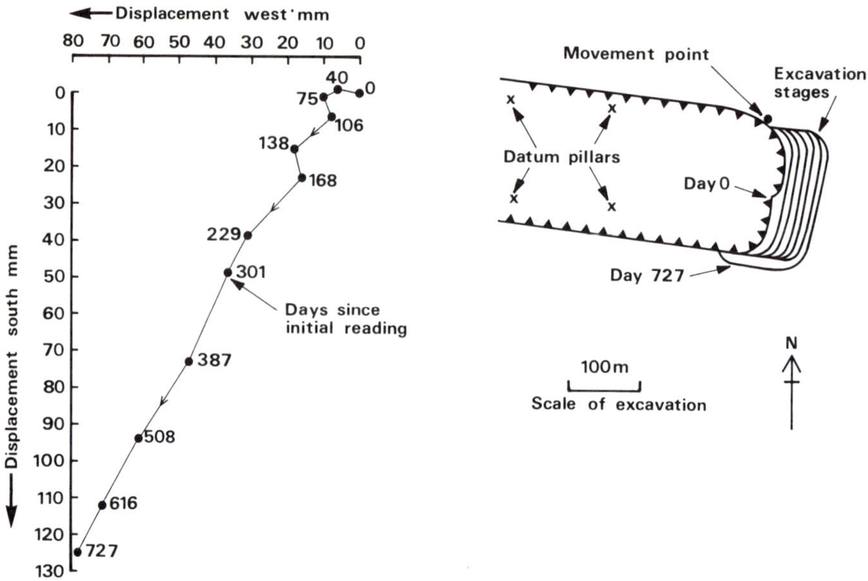

Fig 6 Lateral displacement of a steep bank by precise surveying

Operation. An accurate method of determining the profile of the borehole is to take readings at intervals of depth equal to the wheel base of the instrument. The greater the effective length the fewer the number of readings and the greater the accuracy. The metre long instrument used by BRS has proved to be very satisfactory.

The most accurate results are obtained by: cleaning the inclinometer tube with a stiff brush to remove grit from the tracks; by positioning the instrument to an accuracy of 1 mm in depth by using a steel tape or wire; leaving the instrument switched on and suspended in the hole for 20 minutes before taking readings; taking readings on opposite faces of the inclinometer tube. The inclination at each depth is then given by half the numerical sum of the two readings which eliminates the need to determine zero readings. Half the sum of each set of two readings, which is the zero error, should remain nearly constant provided no zero drift occurs.

Installation. For long-term installations it is essential that the exterior of the inclinometer tubes and couplings are coated to prevent corrosion when the tube is embedded in concrete. The tube should be completely grouted over its full depth by introducing the grout to the bottom of the borehole. The grout should be of about the same stiffness as the surrounding ground. Installation of an inclinometer tube in a diaphragm retaining wall may be carried out by wiring it to the steel reinforcing cage before or while it is lowered into

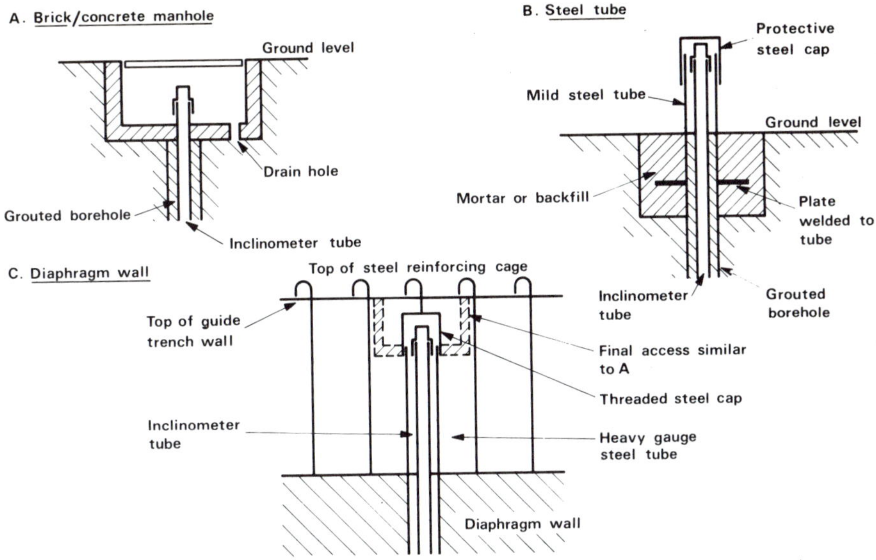

Fig 7 Termination of inclinometer tubes

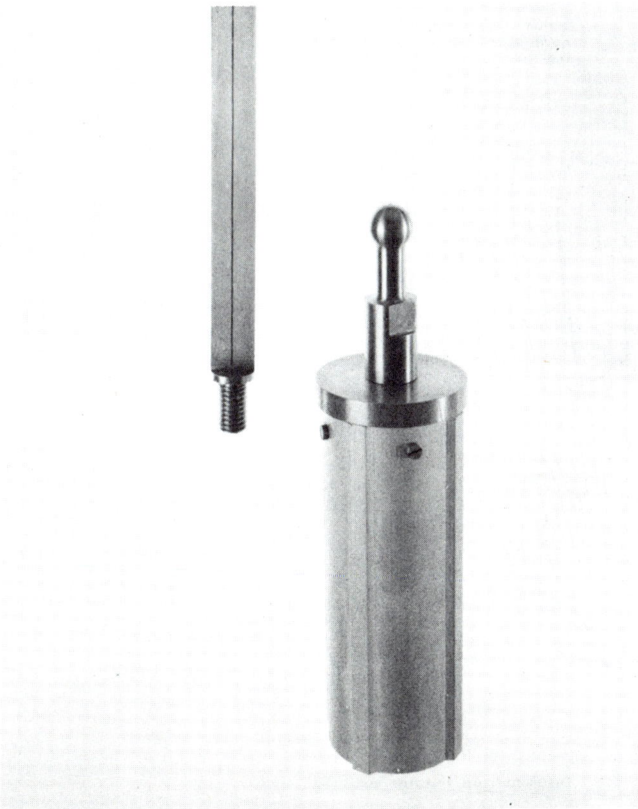

Fig 8 Inclinometer tube targets

the trench.

The top of an inclinometer installation is very vulnerable and Fig 7 shows three methods of protection which have been used successfully for various conditions. Figure 8 shows simple survey points for levelling and intersection which can be mounted on the top of the inclinometer tube.

<u>Results</u>. When frequent sets of measurements are made, as is usually the case during monitoring of an excavation, data accumulates very quickly and the observations are best processed on a digital computer. When results are required rapidly a data link between the site and a computer is worth considering.

A convenient method of recording results is to plot relative horizontal displacement for various depths against time. The record should include information relevant to the instrument and the progress of the job. This basic plot can be used to prepare vertical profiles of horizontal displacement corresponding to various times and events.

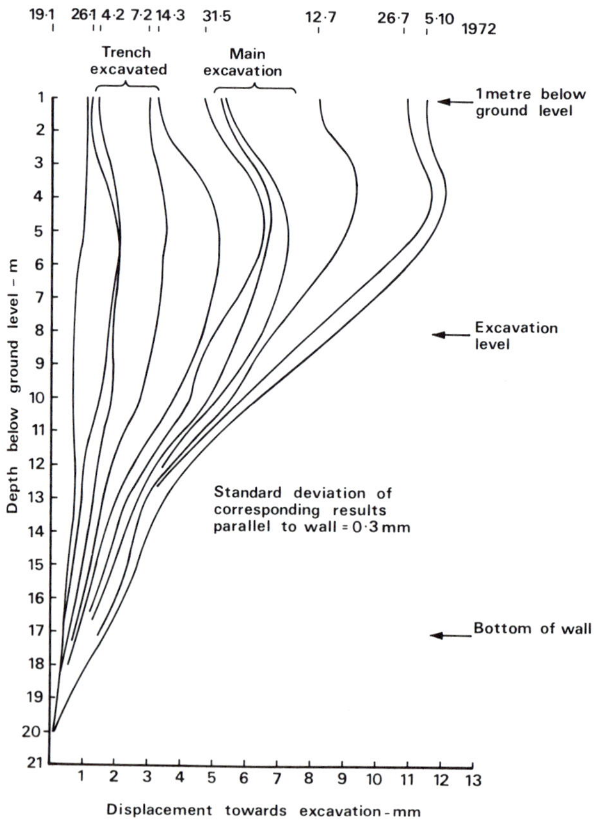

Fig 9 Displacement behind a diaphragm wall due to excavation

Figure 9 shows some typical results for an inclinometer tube located one metre behind a cantilever diaphragm wall for an 8 m deep road cutting in London Clay. Corresponding measurements made parallel to the cutting, where the movement would be expected to be small, had a spread of only 3.2 mm, the standard deviation of the mean being 0.3 mm.

MAGNET EXTENSOMETER

Recently Burland, Moore and Smith (1972) described a precise borehole extensometer which consists of circular magnets embedded in the ground, which act as markers, and reed switch sensors moving in a central access tube which locate the positions of the magnets. The system can be used in horizontal and inclined boreholes as well as in vertical boreholes.

Mounting the magnets. The method of mounting the magnets in the borehole depends on the type of ground and the size of the borehole. Four types of magnet holder which have been used in stiff clays and soft rocks are shown in Fig 10. Types A and B are installed by means of a special insertion tool in holes of about 100 mm and 250 mm in diameter respectively. Type B may also be used in smaller holes subject to overbreak. Both these types require a flush jointed access tube to allow the holders to slide along it. It is usually advisable to grout the hole after installing the magnets.

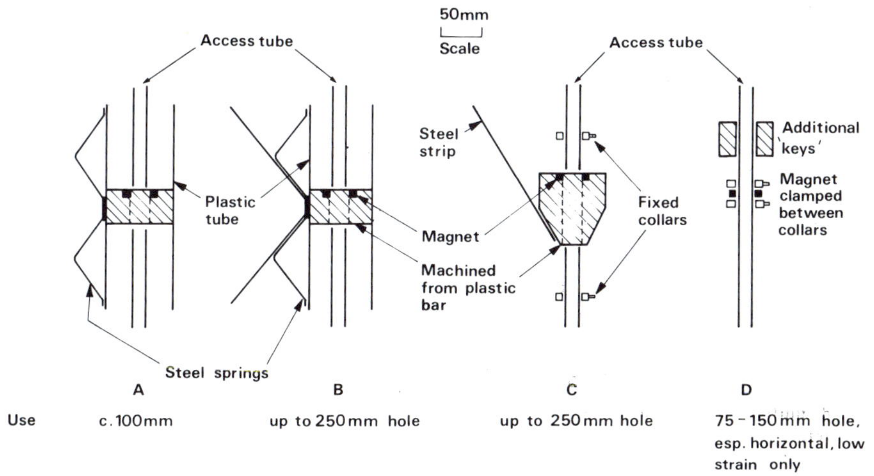

Fig 10 Magnet holders for medium and stiff clays

Type C requires no insertion tool and has been used successfully in boreholes which are to be grouted. For shallow holes the holders are pushed over a flush-jointed access tube to the required depth. In holes greater than about 20 m in depth the holders are assembled in sequence between fixed collars about 400 mm apart on the access tube as it is installed. The borehole is grouted and the access tube, which should be heavily greased, is immediately pulled back about 200 mm so that each magnet holder is clear of the collars.

Type D has been used where the ground is relatively strong and only small strains are anticipated, as for example in soft rock. The magnet is clamped to the guide tube which is grouted in the borehole. Additional keys are attached to the tube to ensure that it strains with the ground. The grout used with all types should have about the same stiffness as the surrounding ground.

The top of the access tube should be finished off in a concrete plinth with a ring magnet embedded in it. By levelling to the plinth the movements of the magnetic markers can be related to surface movements. If the bottom magnet is located in a firm stratum the installation can be used as a deep datum.

Sensing systems. Each magnet can be located to an accuracy of about 1 mm by lowering a reed switch sensor on the end of a graduated steel tape down the access tube. Accuracies of about 0.05 mm can be achieved by installing a permanent datum rod or tape in the access tube with a reed switch mounted opposite each magnet. The small axial movements of the datum required to sense each magnet are measured by a precise measuring head (see Burland et al 1972).

Application. Magnet extensometers have been widely used for measuring vertical movements outside excavations. They have also been installed in horizontal boreholes drilled normal to the excavation face. They are particularly well suited for

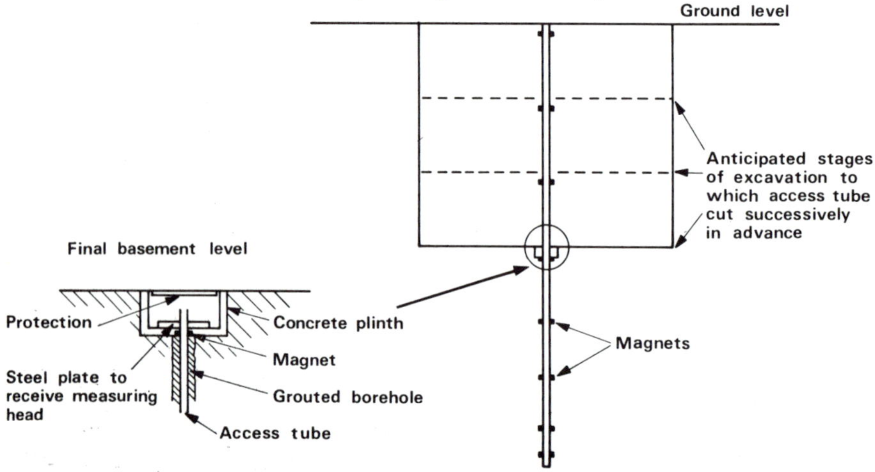

Fig 11 Magnet extensometer for heave measurements during basement excavation

the measurement of heave at various depths beneath excavations as the upper magnets can be located in ground which is to be excavated while the lower magnets are safe from disturbance (see Fig 11). During excavation the access tube, whose plan position should be known, is sealed at the required level with an expanding plug and cut internally just above this level. When final basement level is reached the top of the access tube is finished off as shown in Fig 11.

Many deep basements have to be designed against base failure due to hydraulic uplift. In such situations magnet extensometers have been used in conjunction with piezometers to monitor the stability of the base.

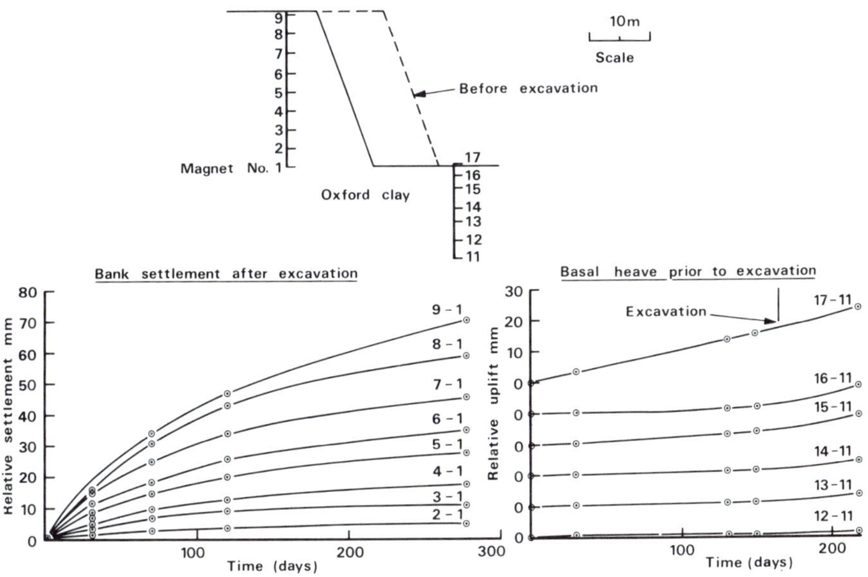

Fig 12 Vertical displacement by a steep bank

Figure 12 shows some results from two magnet extensometers installed alongside and in the base of a 25 m deep brick pit in Oxford Clay. It is evident that heave is taking place in the base, particularly near the surface, and settlement is occurring over the full length of the borehole outside the excavation.

MICROMETER ROD

The rigorous disciplines of precise surveying methods often make them unsuitable to the physical and temporal constraints of a construction site. However, displacements around an excavation may occur under approximately plane strain conditions eg behind a long diaphragm wall, and it may be sufficient to

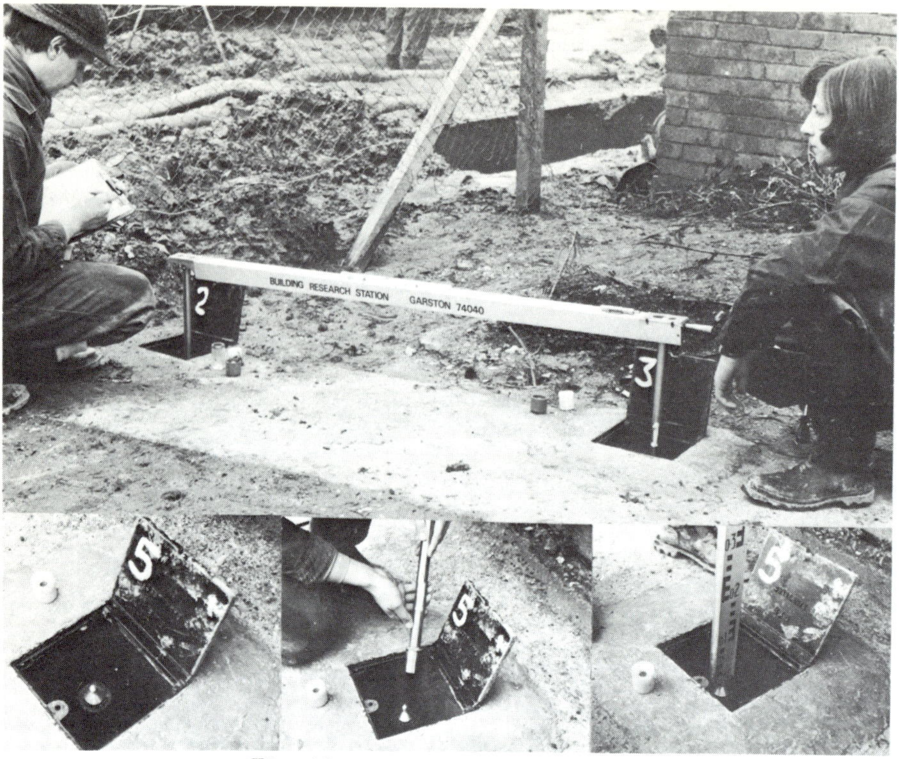

Fig 13 Micrometer rod

measure surface lateral displacement normal to the edge of the excavation.

The surface movement point (Fig 4) with fitting B (a conical location) accepts an extension rod with a spherical seating. A micrometer rod with alternative fixed locations at one end and a micrometer reading to 0.01 mm at the other is used to measure between spherical surfaces (Fig 13). Unless the rod is made from invar temperature corrections to length must be made. To maintain an accuracy better than 0.1 mm, correction may also be necessary for differential vertical movements, by levelling directly on to the movement points, and for tilt of the reference/extension rod, by using a clinometer. A measuring length of 3 m (using aluminium) is convenient for most work, although 1.5 m is useful in high strain regions. Lengths of 6 m may be measured using a temporary intermediate change point.

Figure 14 shows the horizontal strain behind a diaphragm wall (the same as in Fig 9) due to construction of the wall, subsequent excavation and time dependent deformation. Quite large strains occurred near the wall, with an unexpected increase about 7 m back. The results suggest that a tension crack may have formed between points 5 and 6. The consistency of observations during a year is indicated by the minor variations more than 10 m back.

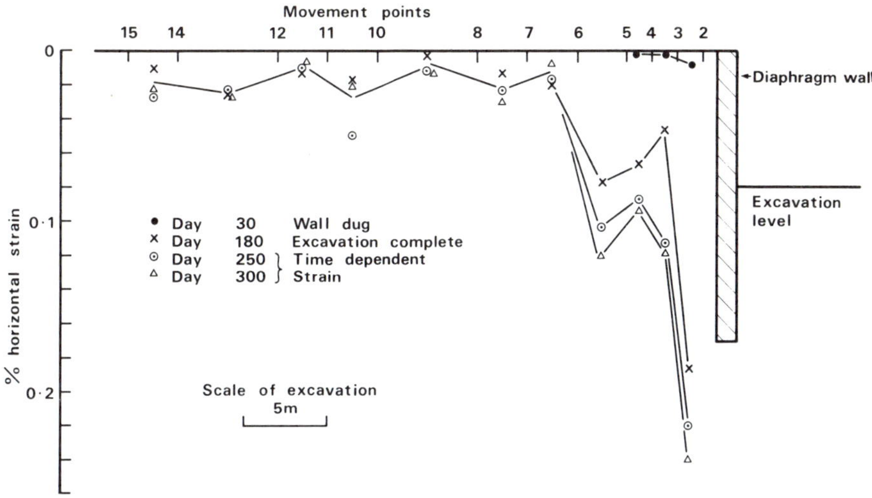

Fig 14 Horizontal strain due to excavation

CONCLUSIONS

All the instruments and measurement techniques described have been used in the study of ground movements around excavations both for research purposes and for routine monitoring of movement to check on design and safety. In developing the equipment emphasis has been placed on simplicity, reliability and ease of installation and use. The success of any field installation depends on its adequate protection from accidental or wilful damage and the methods described in the paper have been developed as a result of experience gained on a number of construction sites.

ACKNOWLEDGEMENTS

The observations plotted in Fig 6 were made by the Central Survey Unit of the Property Services Agency, Department of the Environment. The authors are grateful to their many colleagues at BRS who have assisted with the work described in the paper. The paper is published by permission of the Director of the Building Research Establishment.

REFERENCES

Burland, J B, Moore, J F A and Smith, P D K (1972): A simple and precise borehole extensometer. Geotechnique, vol 22, pp 174-177.

Cheney, J E (1973): Techniques and equipment using the surveyors level for accurate measurement of building movement. BGS Symp on Field Instrumentation, London.

Cole, K W and Burland, J B (1972): Observation of retaining wall movements associated with a large excavation. Proc 5th European Conf SM & FE, Madrid, April 1972, vol 1 pp 445-453.

Green, G E (1973): Principles and performance of two inclinometers for measuring horizontal ground movements. BGS Symp on Field Instrumentation, London.

TECHNIQUES AND EQUIPMENT USING THE SURVEYORS LEVEL FOR ACCURATE MEASUREMENT OF BUILDING MOVEMENT

by John E Cheney

Senior Scientific Officer, Building Research Station

SUMMARY

The Building Research Station (BRS) has used precise levelling techniques for over 20 years. Commercial equipment used by BRS and specialist items designed at BRS, and now commercially available, are detailed. Techniques for site installation and the precautions necessary to record settlement observations to an accuracy better than 0.5 mm are covered. The method of presentation of results is also discussed and illustrated by graphs from a BRS installation at Windsor.

The paper is addressed to engineers and surveyors who have not been previously involved in work of this nature but who may wish to embark on the monitoring of structures with which they are concerned.

INTRODUCTION

There are two main reasons for taking settlement observations: first the observations can yield important information about the ground; and second very little is known about the way actual buildings perform, what differential settlements occur, and what distortions are tolerable for various types of construction and finish. There are also situations, where some degree of damage has been observed, where prompt installation of a system to monitor further movements can prove very useful in ascertaining the cause of damage or potential failure.

Equipment and techniques used by BRS are described in sufficient detail for those who may wish to make settlement observations on buildings with which they are concerned. The BRS levelling station for setting into buildings is described and its installation detailed. The requirements for datums, or bench marks, follows. Installation procedure is given for one type of BRS datum which, like the levelling stations, is now commercially available. The use of the precision level and staff follows with a description of the recommended equipment and with details of procedures to be observed in order to obtain accurate results.

The method of presentation of results is illustrated by observations taken over the last 13 years on a terrace of cottages at Windsor. In this case damage suspected of being caused by settlement of one wall was shown to be due to differential uplift following the removal of trees from the clay site before building commenced. Movements are still taking place 20 years after construction.

SCOPE FOR INSTRUMENTATION

It is important that settlement observations be made both during construction and for a period of years afterwards. Although some civil engineers and building firms may well be interested in carrying out the work themselves, it is anticipated that the continuity will be more readily available from organisations such as local authorities and civil engineering consultants or polytechnics and technical colleges working either from direct interest or in liaison with others.

If the valuable time of first-class personnel is put into settlement observations it is not only essential to use first class equipment but also to choose suitable structures and publish the results of levelling together with other information concerning the site and structure so enhancing the value of the work. Concerning the type of structure chosen the following may serve as a guide:

1. Large structures in which a great deal of money has already been invested in the form of comprehensive and detailed soil investigations.
2. Buildings that are particularly simple in structure and plan as this makes for ease in interpreting the results. Similarly buildings for which a simple uniform ground condition exists.
3. Buildings founded in soil strata of which there is only limited experience.
4. Buildings in which there are local high concentrations of load where differential settlement might be troublesome.
5. Structures that are subjected to large fluctuations in load such as bridge piers etc.
6. Structures where movement damage has already taken place and where there is reason to suspect that the movement is continuing and may lead to some measure of failure.
7. Existing buildings that may be adversely affected by proposed large works nearby.

Information additional to the levelling records should include as much of the following as possible, although it is emphasized that the absence of some of the information should not preclude publication.

1. A detailed profile of the ground underlying the structure to a depth of $1\frac{1}{2}$ times its width, or to a firm stratum, whichever is the lesser. A detailed description of the soil structure and fabric should be given and the presence of sand layers noted.
2. A description of the sampling equipment and methods.
3. Laboratory results, including Atterberg limits, grain size distribution, shear strength and the results of oedometer tests presented as curves.
4. Details of the structure and foundations including plans, cross sections, loads and construction sequence.
5. The value of any settlement study will be greatly enhanced if a record is kept of the performance of the structure and finishes of the building. In this way it is possible to obtain quantitative information about tolerable foundation movements. Perhaps the simplest method of recording the performance of the finishes is by high-grade photography.

LEVELLING STATIONS

It is considered that levelling stations should be purpose-made units built into the structure. They should be designed not only to enable accurate levelling to be undertaken, but also to satisfy a number of other requirements. They should not be liable to accidental damage during construction or to damage by vandals. They should be capable of being installed either during or after construction and should be designed so that they may be easily adapted to permit continuity of observations when cladding or finishes are added to the basic structure. Levelling stations should also be non-corroding, capable of lasting at least 20 years, and acceptable to client and architect. The station described below has been used for many years and seems to satisfy these requirements.

BRS levelling station

Description: The BRS levelling station as shown in Fig 1 consists of a stainless steel socket, 65 mm long and 22 mm diameter, which is set in a hole in the structure, and a removable levelling plug of about the same dimensions on which the base of a staff is held when levels are taken. When not in use the socket is sealed from dirt by a protective Perspex bung. The face of the bung is flush with the structural finish and the Perspex takes on the colour of the finish, being barely visible at a few metres distance. This latter feature has proved useful in overcoming aesthetic objections sometimes voiced by an architect on behalf of his client when a settlement installation is proposed.

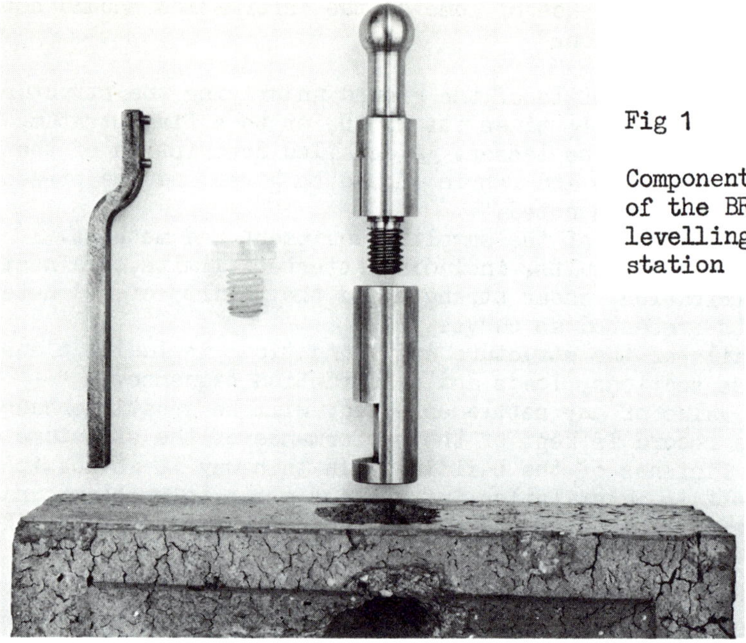

Fig 1

Components of the BRS levelling station

The levelling station is designed to ensure that the plug will position with repeatable accuracy. The thread is a loose fit and is used only to pull the plug into the socket. The plug is positioned radially to an accuracy of 0.03 mm by a precise machined spigot and socket turned concentrically with the threads on the levelling plug and socket. As the plug is finally tightened into the socket their mating faces pull the two components coaxial. It is only necessary to tighten the plug firmly by hand. Spanner flats are provided for emergency use only.

The levelling stations are obtainable from Tusroke-Beaver Ltd, 673 Dunstable Road, Luton, Beds LU4 ODR, Telephone: Luton 57731, and may be ordered as below. The prices quoted are as at December 1972 for small quantities.

BRS type A373/52 levelling system:

Socket	£ 1.73	
Perspex bung	0.64	one required per socket plus 10 per cent spares
Wrench for bung	0.95	two recommended per site
Levelling plug	3.45	two recommended per site

<u>Installation in prepared holes</u>: A satisfactory height for level stations is about 600 mm from the ground. An approximately horizontal hole, 32 to 38 mm diameter, should be drilled to a depth of 75 to 80 mm at the selected position. It is

generally more satisfactory to drill into a brick even if this entails breaking into the frog rather than attempting to grout the socket between bricks where it is likely to become loose due to inter-brick movement.

The socket should be de-greased with a suitable solvent and a strong mortar mix of 1:2.5 cement:sand used for fixing. A plasticiser additive will assist the work and accelerators may be used. The hole should be well wetted and if a frog has been broken into it should be at least partly plugged with mortar. Using a loosely inserted bung as a hand-hold the socket is positioned in the hole so that the front face of it is at least 6 mm recessed in the wall. It is important to ensure that the key-way towards the rear of the socket is filled with mortar. Mortar is then tamped firmly all around the socket using a suitable tool, such a thin steel bar. When the hole is filled to within about 2 mm of the socket face, the bung is carefully unscrewed and the mortar wet-finger finished so that it is visibly clear of the socket mating face as shown in Fig 2

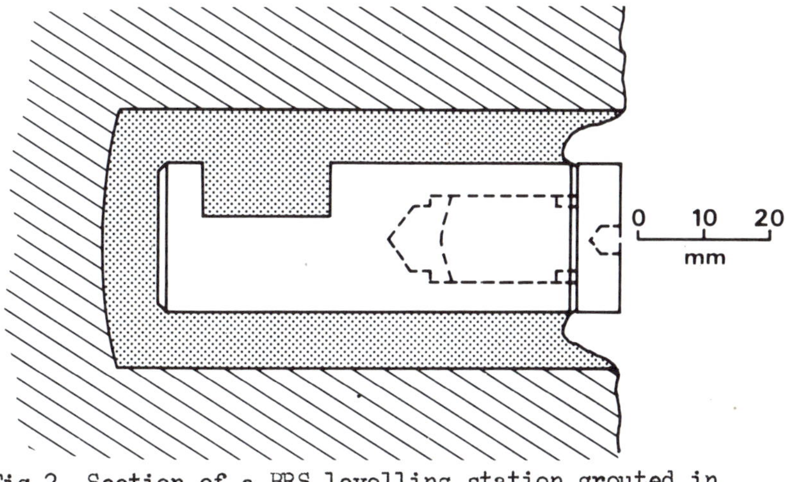

Fig 2 Section of a BRS levelling station grouted in a prepared hole

The socket needs only be set approximately horizontal as the ball-shaped end of the levelling plug on which the staff rests will accommodate considerable off-level installation of the socket. Indeed, for special applications a levelling station may be installed at any angle up to the vertical, though a rubber washer under the bung should be incorporated if cement wash is anticipated.

There is no objection, in principle, to the use of hard-setting epoxy resin for installing the socket, but BRS has not used this technique. It is considered that as wet bricks and concrete are likely to be encountered there is no advantage to be gained by the use of resin as against the traditional cement mortar techniques which are readily available on all construction sites.

Casting in concrete: The levelling socket may be cast into concrete by attaching it to the inside of shuttering before the concrete is poured. The female thread in the socket is 0.5 inch Whitworth so that the socket may be held to the casting face of the shuttering by a bolt of suitable length, passing through a 14 mm diameter hole drilled in the shuttering. In order to prevent grout leakage to the socket or through the shuttering it is essential to use a washer between the socket face and the shuttering. This washer should be cut from 6 mm thick reinforced hard rubber sheet (rubber insert), be a close fit on the 0.5 inch bolt and about 30 mm external diameter. The bolt is removed before the shuttering is struck, subsequent removal of the rubber washer forming a recess to accommodate the protective bung.

Extension of levelling stations: If cladding is to be added, which would obscure the socket or foul the staff, an extension piece (socket extension) may be permanently fixed to a socket to permit continued use of the levelling plug. The socket extension has a male end exactly as the levelling plug and a female end similar to that on the socket. The originally installed socket will not be exactly horizontal so levels taken from an extended station will differ from those taken on the original. To provide continuity of readings this difference must be determined before the cladding is fixed by taking a close level sight on a staff using the original socket immediately followed by a sight using the socket extension temporarily fitted. The extension may then be removed until it is permanently installed after the cladding has been fixed.

For permanent installation a hole of suitable size will have to be provided in the cladding through which the bung may be removed from the original socket enabling the socket extension to be screwed hard home using an epoxy resin on its male thread. The bung may then be re-used for protection of the extension face. Socket extensions should be ordered from Tusroke-Beaver Ltd quoting the distance between the original and final building surface. This firm can also supply tools for removing the bung from the original socket through the hole in the cladding, and also for tightening the socket extension. This extension system is only practicable for lengths up to about 120 mm.

LEVELLING DATUMS (or bench marks)

The provision of deep datum points is undoubtedly very expensive and is not always necessary. The important thing is that any datum point should be reliable. In many cases it is sufficient to use nearby structures. This is particularly so where they have been in existence for a number of years and where they are founded below the depth of seasonal influence which is about 1.5 m where no trees are involved. The involvement of trees is defined as the presence on clay soil of individual trees within a distance equal to their own height and in the case of groups or rows of trees within a distance of 1½ times their own height; or that trees have been felled within the above distance limits during the previous 20 years or so. Where it is decided that an existing structure may be used, a datum may conveniently take the form of a BRS levelling station on that structure. To check reliability two or three such stations should be established.

Where trees are involved or no suitable existing structure can be used, it will be necessary to install one, or preferably two datums in the ground. Where no large excavations are involved and the new building is not too large, a simple sleeved datum may be installed at a depth of about 6m. A BRS datum of this depth is described below and could be adapted for datums between 2 m, which should be regarded as an absolute minimum for any settlement work, and about 10 m depth.

If the proposed structure is very large, or where there are deep excavations, piles or other factors involved that could affect ground water levels or otherwise produce movements in the ground in the vicinity of the structure, it will be necessary to install datums at a considerable depth. Sleeved rod systems at greater depths than about 10 m become expensive to install and a datum based on the reed-switch magnet-sensing system developed at BRS is preferred. This system is described by Burland, J B and Moore, J F A (1973).

A BRS 6 m datum

Description: An economical 6 m datum constructed from an inner datum rod made from 10 mm bore galvanised tube surrounded by a guard tube of 25 mm bore together with specially manufactured end pieces may be purchased from Tusroke-Beaver Ltd by ordering a 'BRS 6 m levelling datum complete'. Additionally required are: one 600 mm length of 100 mm diameter salt-glazed drain pipe, and one 140 mm circular sealing plate with locking screws (Broads No 460A). The datum is shown in Fig 3.

Installation: A pit about 600 mm deep and about 900 mm square is excavated. In the centre of the pit a hole 150 to 300 mm diameter is then bored to a depth of 5.5 m ± 0.1 m measured from the proposed finished ground level (ground surface). The datum rod is then liberally greased with a soft waterproof lithium grease (eg Shell Retinax A) along its whole length and the datum rod and guard tube assembled with all attachments, except the protective cap, firmly screwed on. The datum rod will protrude about 150 mm from the top of the

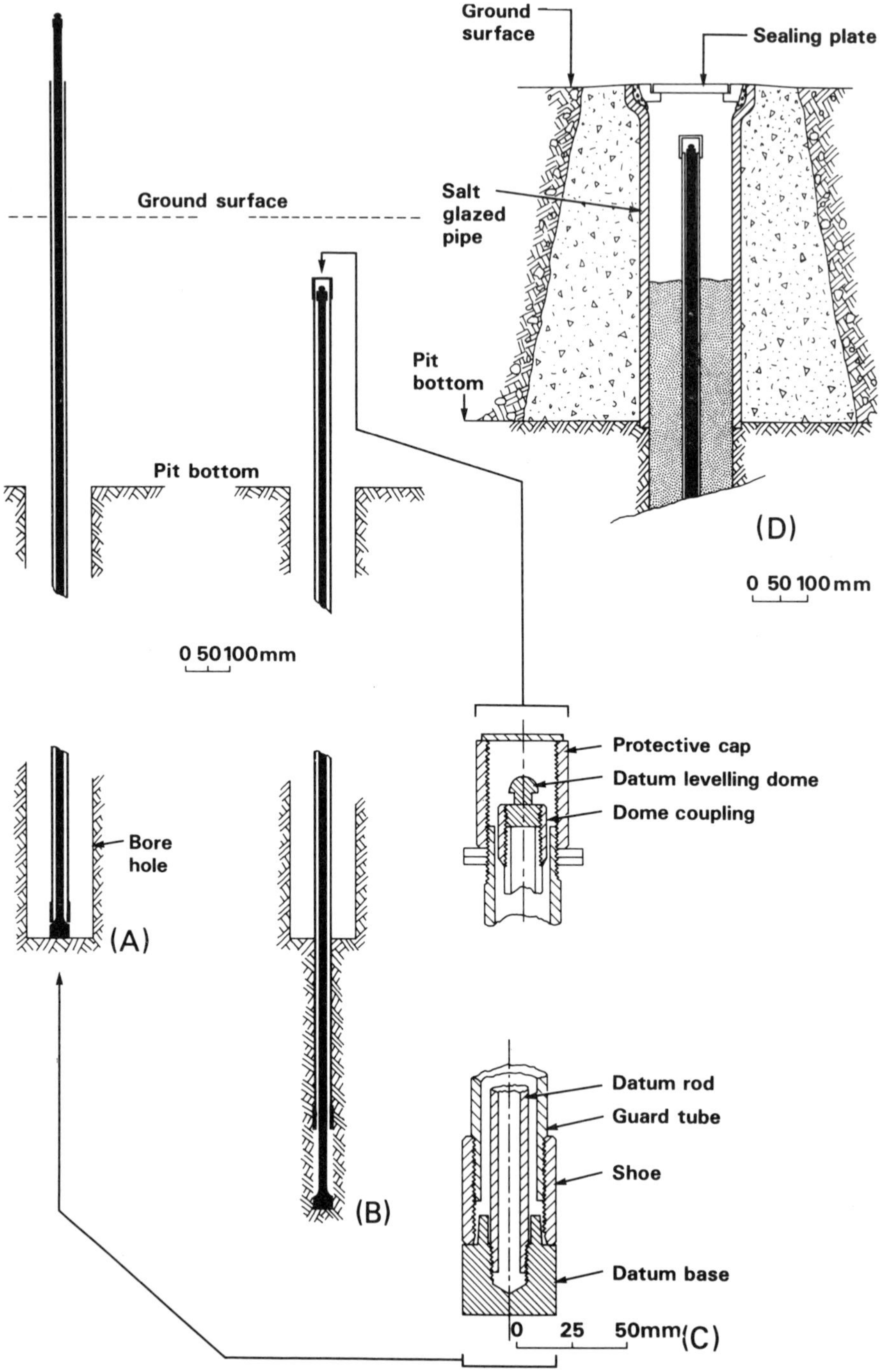

Fig 3 Installation of BRS 6 m datum

(A) Before driving (B) Installed position
(C) Component detail (D) Site protection for the installed datum

guard tube to which it must be firmly strapped to prevent the rod falling out as the whole assembly is positioned in the borehole. With the datum base resting on the bottom of the borehole and the strapping removed, the datum levelling dome will be about 520 mm above ground surface (see Fig 3A). The actual height should be noted.

A double longscrew of 250 mm length is screwed on to the end of the guard tube to act as a dolly for a sledge or for jacking. The whole assembly is then driven into the ground a distance equal to the noted height, plus 140 mm, to an accuracy of 10 mm. It is convenient to mark the driving distance on the guard tube up from a marker timber placed across the top of the borehole. It will be noted that the guard tube pushes the datum base which in turn draws the datum rod down. After driving, the longscrew is removed and the guard tube carefully raised until the dome coupling has half entered the guard tube. This final installed position of the datum is shown in Fig 3B and Fig 3C upper. The lifting of the guard tube may be accomplished using a clockwise rotation together with upward pull imparted by stillson wrenches attached to the guard tube at 180° to each other. It is very important to check that the dome coupling does not catch on the top of the guard tube while the latter is being raised. A postcard tubed in the annulus between the two tubes for the critical time of lifting can be helpful.

The protective cap should be loosely fitted before backfilling of the borehole. In order to prevent any possibility of grouting the guard tube to the datum rod, sand should be used for the lowest 0.5 m followed by a further 0.5 m of concrete or mortar. Backfilling can then continue with any convenient material provided that the top 1 m of borehole fill is clay which should also extend about half way up the inside of the salt-glazed pipe.

The top of the datum installation should be finished as in Fig 3D which presents a standard building component generally acceptable and of little interest to vandals. It will be noted that the method of finishing also ensures that if surface loads such as vehicle wheels depress the outer protection, no movement will be imparted to the datum system until such depression results in the sealing plate touching the protective cap. Even at this stage the surveyor can be assured that the datum itself is undisturbed if he observes that the protective cap itself does not touch the datum levelling dome.

If it is necessary to use a borehole of diameter greater than 300 mm arrangements must be made to ensure that the installed position of the datum base, Fig 3B, is such that its final depth below the bottom of the borehole is at least twice the diameter of that hole.

THE LEVELLING INSTRUMENT AND PROCEDURE

The precision level used by BRS is the Wild N3. It is probably the most accurate level commercially available and has proved very satisfactory for all the Station's precise geotechnical work.

The N3 level is used with an invar staff graduated at 10 mm intervals. It is not necessary for the observer to estimate the fraction of the staff interval as an internal optical micrometer is built into the level. This micrometer is used to displace the line of sight vertically and parallel to itself and is calibrated over a range of exactly one staff interval. The surveyor visually straddles the staff graduation line with the wedge-shaped lines of the telescope reticule by turning the micrometer knob. The micrometer scale is then read giving the fraction of staff interval directly in mm and 0.1 mm and, by estimation to 0.01 mm. This figure is then added to the multiple of 10 mm as indicated on the staff graduation seen in the telescope thus giving the true staff reading.

The levelling equipment may be obtained from Wild Heerbrugg (UK) Ltd, Revenge Road, Lordswood, Chatham, Kent, ME5 8TE, Telephone: Chatham (STD 0634) 64471. The equipment required, prices current at December 1972, is as follows:

1 - Wild N3 Precise Level (10 mm)	Code	212289	£316.80
1 - 1.8 m Invar Staff (GWL 182)	"	196724	77.20
1 - Special foot for GWL staff	"	236951	8.65
1 - Tripod type GST			33.80
2 - Ground plate Code 197000 @ £6.70			13.40
		Total	£449.85

Wild Ltd can also supply a transport box for the staff.

The author has noted that closing errors obtained by a observer using the N3 level, but employing techniques sufficient for normal civil engineering work are frequently about 2 to 5 mm, but that the same person will quickly attain accuracies of 0.2 to 0.3 mm when the procedures essential to precise levelling are adhered to.

A careful study of the excellent instruction book, supplied with the N3 level will be amply repaid. It is frequently not possible in settlement work to equalise back and fore sights, therefore it is essential that the horizontality of line of sight (collimation) of the instrument be maintained at a high accuracy. It is to be noted that, anticipating disturbance during transit, the manufacturer only sets the collimation to within 5 second of arc (1 mm in 40 m) leaving the final setting to the user. This setting must be done, using the simple techniques described in the booklet, on receipt of the level, at 3 monthly intervals and whenever the level has been subjected to rough handling. The user should aim at setting the collimation to 0.5 second of arc (0.1 mm in 40 m). The cross error of the instrument should also be checked on receipt and each year. The bubble of the staff must be both set and used on only one of its positions, the lower one being generally satisfactory. The level and tripod should be left in the temperature of use for 15 minutes before work commences and the level should be protected from direct sunlight throughout the operation.

The following procedure differs slightly from the Wild recommendation for focusing, but the author has found it useful for operators having uncorrected astigmatism. De-focus the telescope. Against a light background focus the eye piece to the reticule concentrating on attaining equal clarity of the two wedge lines. Off-set this focus and re-set it. Focus the telescope to a staff interval and then moving the eye up and down slightly, re-focus the telescope as necessary until no apparent relative movement takes place between reticule and staff. Regardless of the apparent absolute clarity of the staff there is now no parallax, and the staff is correctly focused in the plane of the reticule. It is now permissible to re-focus the eye piece to obtain maximum comparative sharpness of the reticule wedge and staff interval. The dioptric reading of the eye piece should now be noted as a constant for that observer. The observer should record the staff reading to 0.01 mm using the vernier about 6 times to obtain the best repeatable reading.

The work of the staff man in maintaining the staff bubble central for long periods of time is fatiguing. A routine of commands from the level operator of 'bubble' and 'relax' only calling for 'bubble' when the micrometer is being used will relieve fatigue and result in the staff man being able to concentrate his utmost effort to keep the bubble central at the important time. A bipod leaning frame is a useful accessory.

It is recommended that levelling sights be no longer than 15 m and if a change point has to be used it be a well-trodden ground plate. An effort should be made to equalise back and fore sights although it is recognised that site conditions will frequently not allow this, nor can intermediate station sights always be equal to the fore and back sight. Levelling should be done in over-cast weather with a slight breeze if possible.

It is always better to take fewer high quality rounds than many of indifferent accuracy. The observer should aim at the highest accuracy he can attain but certainly a round that produces closing errors greater than 0.5 mm should be repeated. If the level and staff are in good adjustment and the observer finds that two rounds will not produce his usual accuracy the

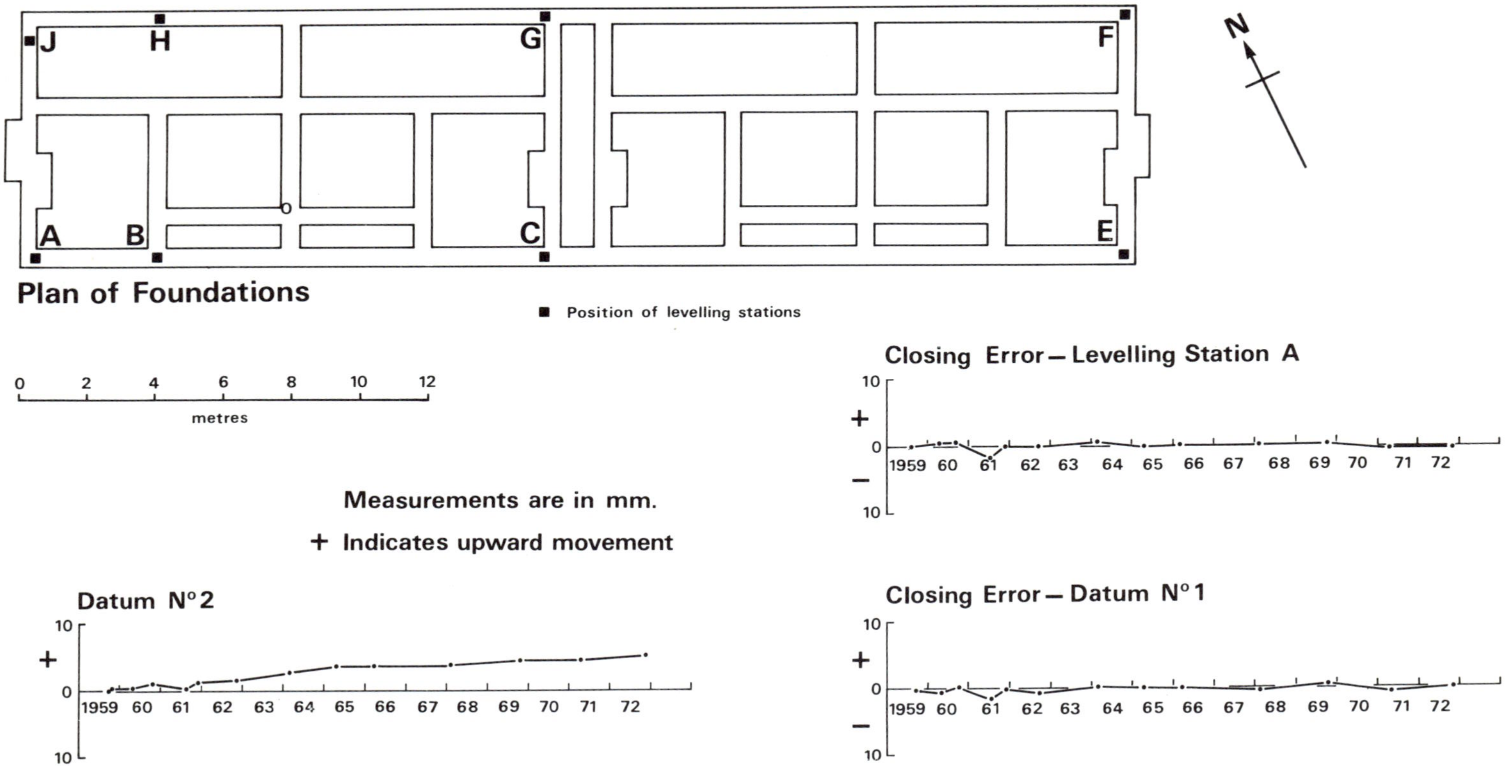

Fig 4 Levelling records of Queen's Gate cottages, Windsor. Upper: changes in level with time based on main datum (No 1); Lower: foundation plan, changes in level of datum No 2 based on main datum, levelling closure errors for main datum and for station A.

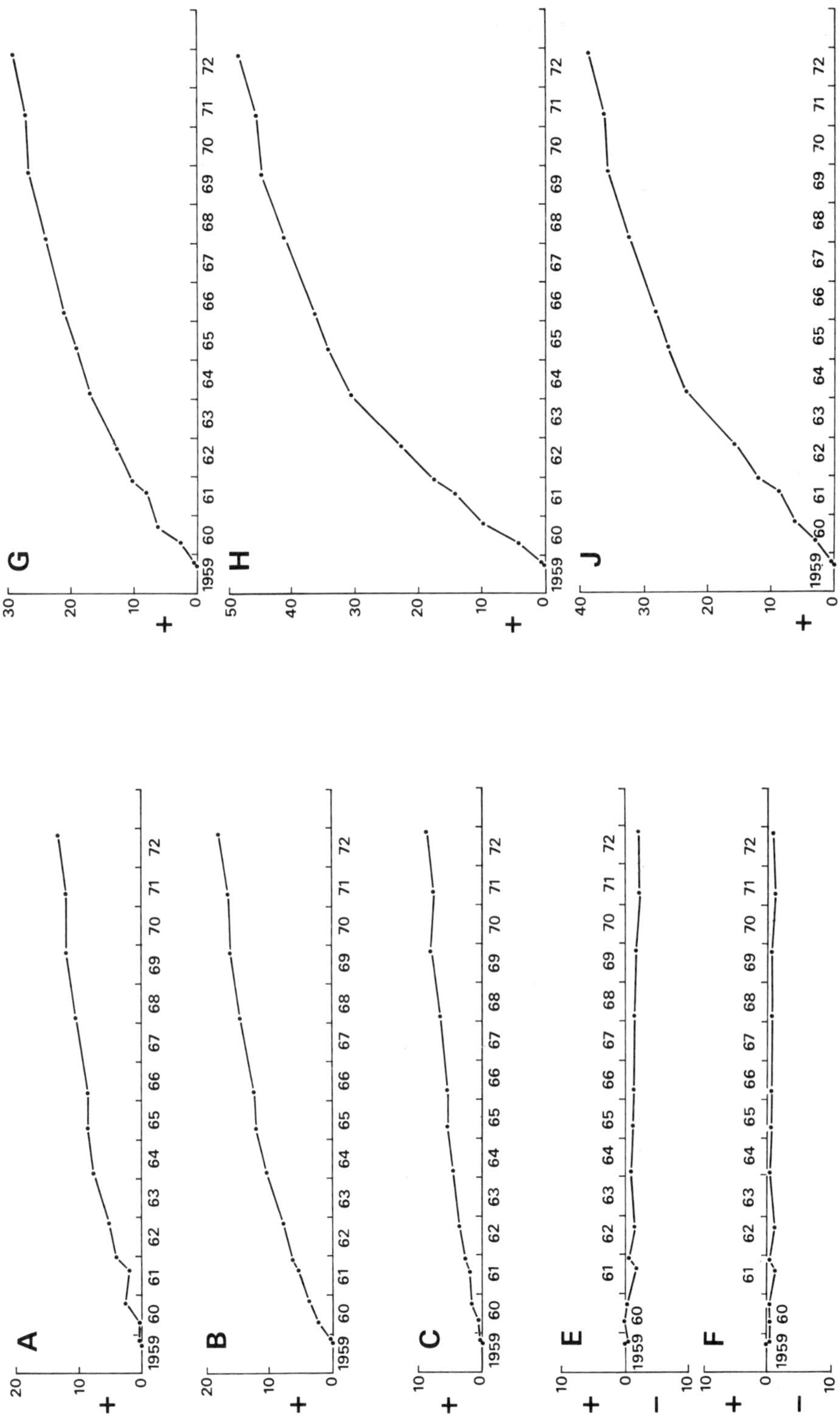

A
B
C
E
F
G
H
J
+
−
1959
60
61
62
63
64
65
66
67
68
69
70
71
72
0
10
20
30
40
50

fault is probably due to atmospheric refraction which can produce considerable errors in certain conditions. Rather than persevere too long he should accept the best round and re-level another day.

THE LEVELLING SEQUENCE AND PRESENTATION OF RESULTS

The datum judged to be most reliable should be called the main datum (D1) and its level designated 0.00 mm or, as necessary to avoid negative levels, 1000.00 mm etc.

Levelling should start at D1, proceeding to the nearest levelling station (a) on the structure and then round the structure, where possible, levelling each station lettered alphabetically (omitting the letter D) closing to A and then to D1. Where-ever possible, levelling stations should be utilized for change points and subsidiary datums (D2, etc) levelled by intermediate sights.

To establish the initial level of the stations and of the subsidiary datums, based on D1, several rounds of levelling should be taken and levels computed on D1 as 0.00 mm (or 1000.00 mm etc) for each round. Closing errors should not be distributed, thus for each round there will be two level values for A and a second for D1. The round of levels giving the smallest closing errors for D1 and for A should be selected and used for the first entries on a result table which should be prepared showing two columns for each levelling station. The left column is for successive levels taken throughout the investigation and the right is to indicate movements since commencement of the work; the first right hand entry for each station will thus read 0.00 mm. Settlement should be indicated by negative figures. D2, D3 etc should be recorded in the same way. The table should also provide a single column entry 'closure D1' where the second D1 level is recorded and a two column entry 'closure A' where the second 'A' reading and the 'A' difference is noted. Other columns should indicate date, time, temperature, wind, sun, rain, observer, staff man and progress of construction. The table should record all rounds of levelling taken, the best only of each day being plotted on graphs which should be prepared and kept up to date. A suitable form is shown in Fig 4. A convenient vertical scale is 2x natural, the time scale may be 2 mm per week during construction followed by a total re-plot and continuation, at 1 mm per month, for long term work.

The frequency of levelling will be dictated by the job. As a guide weekly readings might be suitable during construction followed by monthly for a year, lengthening the interval to twice and then once a year for long term studies.

It will be seen that the presentation shows as much information as possible and does not conceal errors. In the example shown D2 is lifting relatively to D1 but the coice of main datum is confirmed by the knowledge that trees have been removed near D2. The kinks in 1961 are due to a poor closing probably caused by a mis-reading between C and E; it is known that the observer on that date read B as the first station. Generally closing errors are shown as satisfactory.

ACKNOWLEDGEMENTS

The author is pleased to acknowledge the invaluable help of Mr D Burford who has carried out the majority of the observations and maintained the records of the Windsor site and also of staffman, Miss J Nicholls. The kindness and interest of the consultant architect for that site, Mr R Tatchell, and of the Crown Estate Office in allowing our installation is also gratefully acknowledged.

The author also wishes to acknowledge the early stimulation of his interest in precise levelling especially in relation to effects of vegetation and seasonal movements by Dr W H Ward, and also the encouragement to publish this paper given by his Division Head, Dr J B Burland. The paper is published by permission of the Director of the Building Research Establishment.

REFERENCE

Burland, J B and Moore, J F A (1973): The measurement of ground displacement around deep excavations. Proc Brit Geotech Soc Symp on Field Instrumentation. London.

A PRACTICAL METHOD FOR FIELD CONTROL OF LOADING

D. A. COOK* and T. S. INGOLD**

Bath University* Freeman Fox and Partners**

INTRODUCTION

1. In 1957 Bjerrum and Overland reported the practice of a Norwegian firm who maintained their oil tanks topped up with water during construction. Fabrication was undertaken from a floating platform inside the tank, so the test load coincided with the building and completion. When in, 1952, the erection method on this same site was amended to separate the two activities, the water test followed at a much more rapid rate and provoked a bearing failure. It was appreciated at that time nevertheless, that an improvement to bearing strength was obtainable by a gradually applied load. In the Petroleum industry it was then common practice, when erecting tanks on poor foundation material, to restrict the rate of the water test. Subsequently the widespread use of field instrumentation provided scope for continuous assessment; and several case histories are now recorded in the literature of their successful employment in controlling loading operations - Gibson and Marsland (1961), Darragh (1964), Lambe (1967), Penman (1968). More frequently they find their application in estuarine regions where the laminated nature of the recent alluvial clays offers rates of consolidation within acceptable time limits.

2. A controlled loading operation relies on the results of instrument measurements and subsequently their interpretation. In this respect the measurement of settlements and pore water pressures have developed to a high degree of refinement; and Engineers entering this sphere of activity for the first time may predict with reasonable accuracy what contractural commitment is involved. It is less easy however to forecast the time involvement in sorting and interpretation of what can amount to a very large quantity of data - and it is frequently underestimated. Yet, where conditions are marginal the effectiveness of the venture focusses on the speed at which a

constantly changing factor of safety may be determined. Consequently a system of pre-analysis must be undertaken, which is sufficiently comprehensive to cover all reasonable conditions likely to arise; and which can then be readily applied. The following deals with a method of setting out the pre-analysis for a controlled loading operation which can be used by the site personnel to assess stability at intermediate stages of construction, at the same time as the complementary instrumentation is being recorded.

METHOD OF PRE-ANALYSIS

3. The use of pore pressure measurements for estimating stability during construction demands an effective stress analysis of the form developed by Professor Bishop (1955). Now that computer methods can quickly deal with the numerical

Circle	R_u	F
1	0.25	1.9
	0.50	1.5
	0.75	1.1
2	0.50	1.7
	0.75	1.1
3	0.50	2.1
	0.75	1.3
4	0.75	1.5

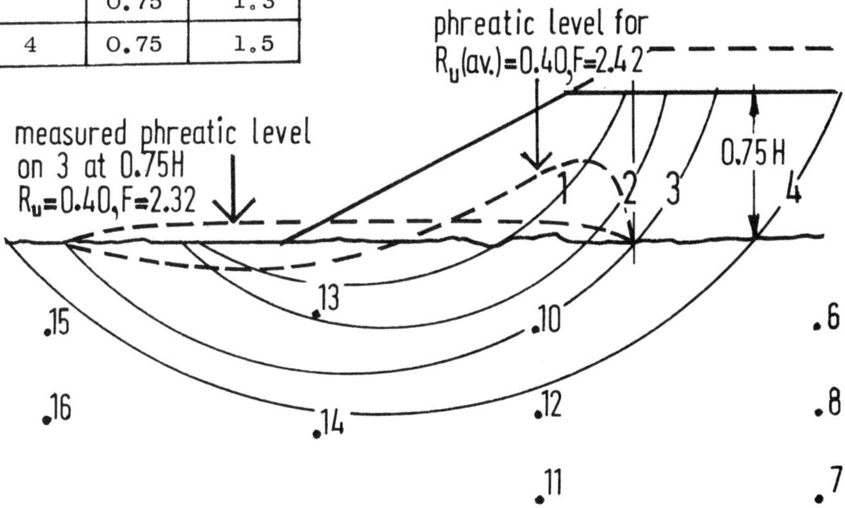

Figure 1 A 8 diversion embankment (Erskine Bridge)
Summary of stability analysis at three quarters full height

work, the better benefit from this facility can be obtained by its fuller employment before work starts on site, to give the widest prediction of the anticipated range of foundation performance. An example of what is inferred by this is given in Figure 1, in the context of the A8 diversion on the south approach to Erskine Bridge. The junction coincided with an isolated pocket of soft silty clay in an area where the main surface deposits were boulder clays. The decision to instrument the section of bank was made, in this case, in favour of incurring a positive delay in the earthworks programme.

4. The setting out of the pre-analysis follows on from the procedure where computer programmes, investigating circular arc failure surfaces, will calculate factors of safety referenced to grid points of each circle centre combined with a fixed pore water pressure ratio and tangent level. Provided the grid is positioned correctly then the zone offering the lowest safety factor is identified. The process repeats for different pore pressure ratios, tangent levels and if necessary, a displaced grid. Because the field pore pressure distribution is largely indeterminate, the pre-analysis must assume average pore pressure ratios. Also, where the subsoil conditions are reasonably uniform, the tendency lies towards higher values of F with increasing depth. Figure 1 indicates this, by the increasing factors of safety as calculated between circles at tangent levels 1 to 4.

5. However, where there is a drainage blanket across the foundation base, a steep excess pore water pressure gradient naturally develops in the vertical direction towards this drainage outlet. Typical profiles showing this are given in Figure 2, from field records in the Severn Estuary and at Erskine. The purpose of the drain blanket is to create exactly this condition; and it resembles a normal one-dimensional consolidation test. Under these conditions it is not possible to predict at what level the worst pore water pressure condition will occur with respect to stability; because lower values of F, at relatively shallow depth may offer greater stability than at a deeper level, where the excess pore pressure (or pore pressure ratio) remains high.

6. The format of figure 1 drawn to enlarged scale, with the range of analytical solutions included, also provides the basis on which a complementary stability condition may be checked in the field. This is done by transposing the measured pore pressure variations around the path of the potential slip surface (this will usually require interpolation) and then hand plotting the values as a stand pipe head measurement.

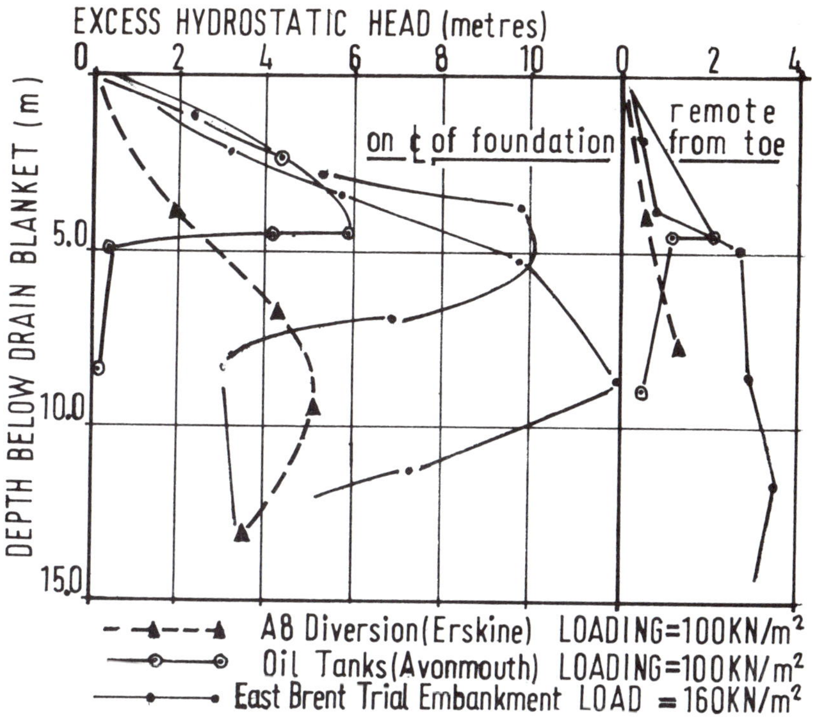

Figure 2 Typical observed variations of excess pore pressures plotted against depth below drain blanket level

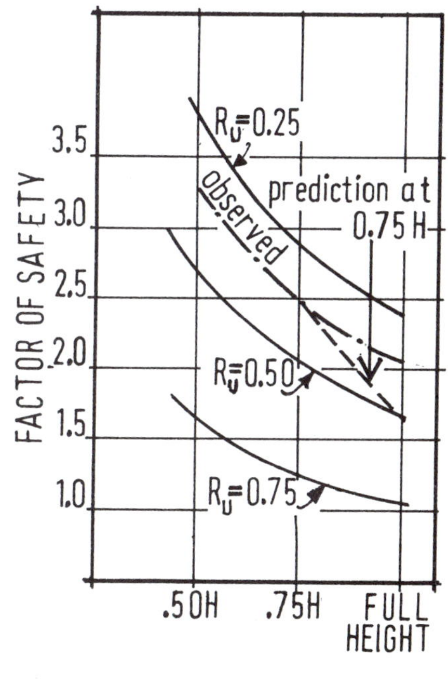

Figure 3 A 8 Diversion Embankment (Erskine). Summary of stability analysis over full height range for tangent level,3

In this text the average pore pressure ratio is defined and measured from the diagram as:-

$$R_u = \frac{\text{Weight of water within the area contained by the phreatic line and the slip surface}}{\text{Gross weight of soil and water vertically above that part of the slip surface within the foundation}}$$

The measured standpipe head variation for the most critical circle at level 3 is shown in Figure 1 for the bank at three-quarters full height. Graph relations similar to that shown in Figure 3 can also be made up to correlate the full range of the pre-analysis at each selected tangent level.

7. That the method does not give completely accurate values of F is evident from the differences between actual variation of water pressure and assumption of a constant ratio. The divergency between the actual and average is also shown on Figure 1. In the Authors' experience it is not normal practice either, to measure pore pressures within road works embankments up to heights of 26 ft. (8.00m). In the foregoing calculations the assumption is that pore pressures are zero above the drain blanket. There is the question also of what strength parameters should be assigned to the embankment fill material. In the case of wide road embankments on soft alluvial clays where failure would occur mainly through the foundation material, it is also argued that the final development of tension cracks minimises the contribution from the embankment and should be neglected. In the case cited, a re-analysis at level 3 using the actual field data gave a value F = 2.32 (vis-a-vis F = 2.42 for R_u av. = 0.40).

8. Perhaps of more significance than absolute accuracy however is the provision of a basis for predicting performance at a particular level. It is not intended to minimise the importance of achieving the best accuracy obtainable. What is emphasised here is the need to know in advance if a marginal situation is likely to arise, so that whatever extra attention is necessary can then be paid to the progress. In this respect the analysis of field results at intermediate stages are of more importance than that at full height; when it is either superfluous, or too late. Frequently road works and oil tanks are built on recent alluvial soils with undrained shear strengths below the zone of desiccation in the region of 500 lb/sq.ft (25 kN/m^2). Where field control is provided for - and sometimes where it isn't - bearing values around 1.00 - 1.25T/ft (107 - 135 kN/m^2) are employed at Factors of Safety approaching unity measured on undrained strength. These deposits, however, very frequently possess a light degree of pre-consolidation which may be attributed to delayed decompression (Bjerrum 1967) or desiccation during deposition (Hawkins, Cook 1972). This contributes to a pore pressure response during the initial loading stages which is

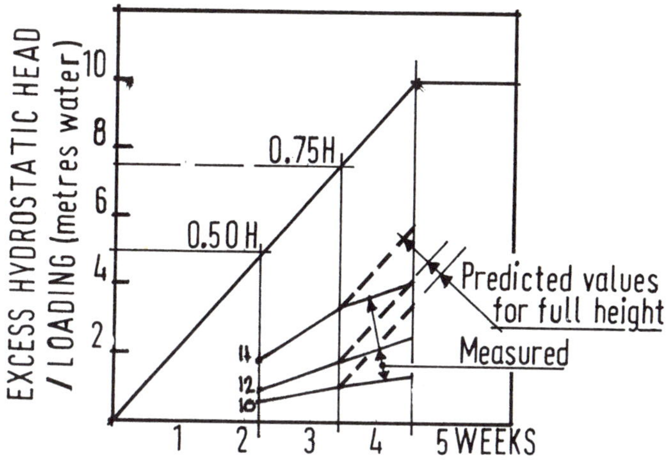

Figure 4(a) A 8 Diversion Embankment (Erskine)
Records of excess pore pressures at tips 10, 11 and 12 and foreward predictions from 0.75H to full height

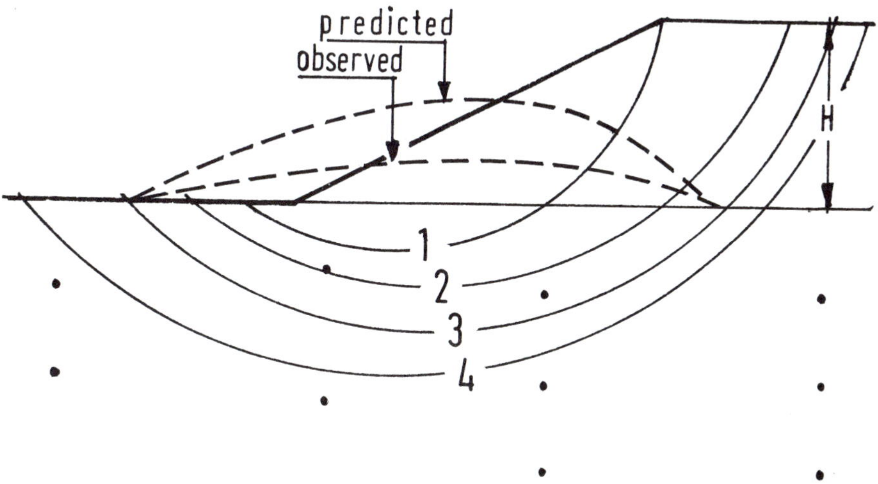

Figure 4(b) Comparison of actual and predicted excess pore pressures at full height for tangent level 3

comparatively low. In this context it may be in the region of 50% of the applied vertical pressure. At a later stage however the pore pressure rise will probably accelerate to a rate equal to the vertical pressure increase or slightly greater. Consequently a forward prediction of pore pressure should make this assumption and not a continuation of what has been observed up to that time.

9. The A 8 (Erskine Bridge) example is continued in figures 4(a) and 4(b). In Figure 4(a), 3 piezometer records are compared with the bank loading; but a prediction of the full height situation was made at 0.75H by extrapolating the measurements up to that point, parallel to the loading curve. The predicted, vis-a-vis the actual final factors of safety and pore pressure ratios for tangent level 3 are compared in Figure 4(b). In the event it can be seen the result was conservative. This case history was selected however because of the completeness of the data and not because of any 'cliff hanging' characteristics.

OIL STORAGE TANKS AT AVONMOUTH

10. The same method of control as that described was used for a pair of Oil Tanks erected by Messrs. Philblack at Avonmouth during 1969. The alluvial subsoil contained a crustal zone of desiccation to 2.00 m depth followed by soft to firm interlaminated silts and silty clays to Keuper Marl at 10.00m depth. Around 3.50m depth the succession is punctuated by a peat horizon, between 0 - 1.0 m thick. The site is about two miles from the first M5 Trial embankment at Avonmouth reported by Dallard (1971) and Murray (1971). Expectedly the subsoil at the two sites is very similar; with the single qualification that at Philblack the materials below the peat band show a distinct classification within the silt/fine sand range. The tanks were water loaded to 100 kN/m in 36 days, and the final factor of safety reduced to 1.35 at an average R_u value of 0.36. The layout, pre-analysis summary for full loading and typical pore water pressure records are given in Figures 5, 6 and 7.

11. The experience of the Oil Tank loading contained points of interest worthy of discussion, because they highlight some problems of a general nature. The first concerns the results obtained from 18 piezometer tips, the final readings for which are marked up on the layout in Figure 5. For better comparisons the half section only is drawn so that points in an equivalent radial position are grouped together. Because of the presence of the peat and the known laminated soil structure a ring of sand drains centred at 3.50 m, was installed on the periphery of the drainage blanket. But as some doubt existed as to their efficiency 6 tips were left

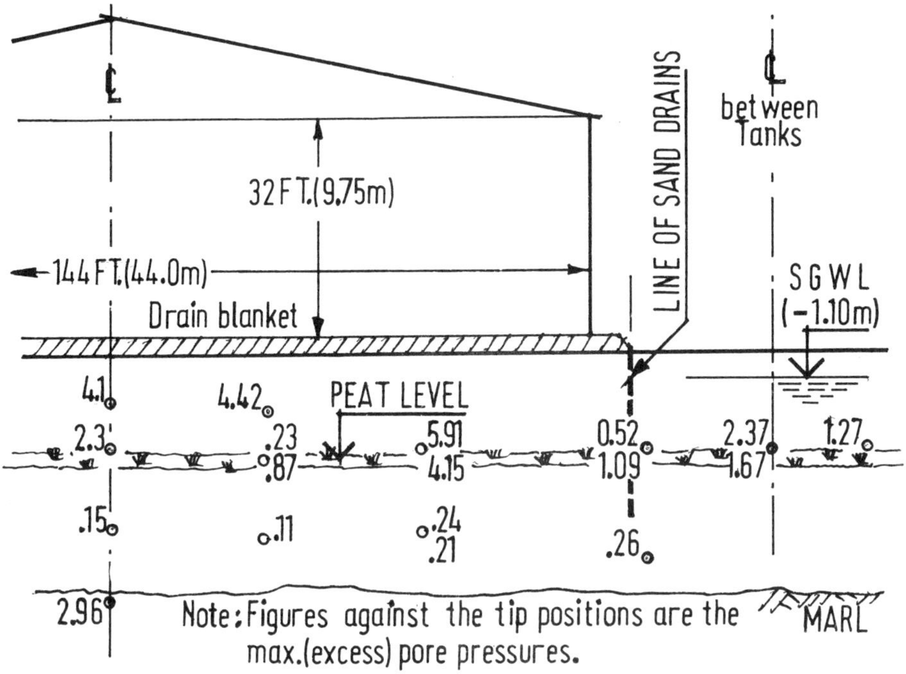

Figure 5 Philblack (Avonmouth) Section through Oil Tank and Foundation with max. recorded (excess) pore pressures.

Circle	R	F
1	0.3 0.6	1.39 0.92
2	0.3 0.6	1.53 1.00
3	0.3 0.6	1.60 1.10
4	0.3	2.61

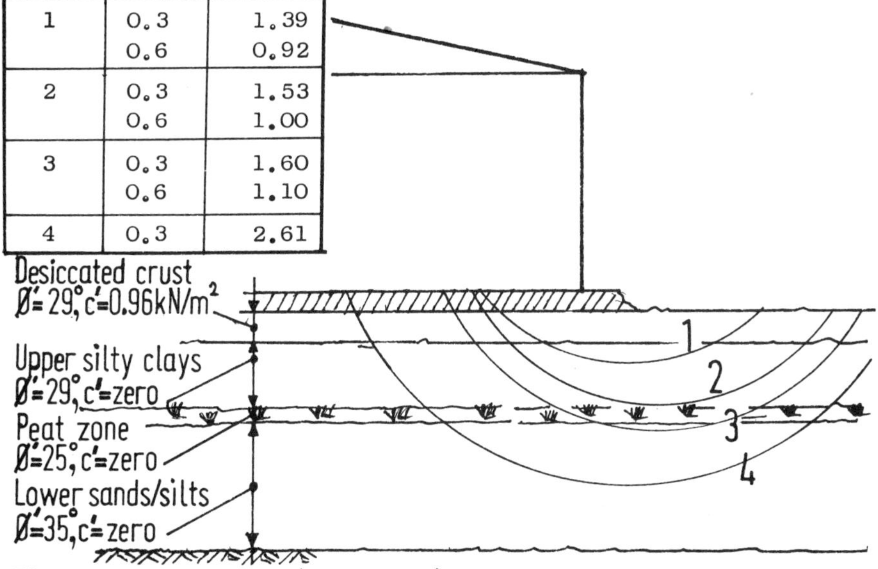

Figure 6 Philblack (Avonmouth) Summary of Soil Properties and Pre-analysis for full load condition

outside the curtain. Readings from these varied between 0.52 - 2.37 m (excess head). The higher value occurred midway between the tanks as the second was being loaded. It raised the question as to what were the relative contributions from, a partial working only of the curtain of sand drains, and undrained shear distortion within this central zone. From either cause the interaction of the two tanks appeared to give an unfavourable result.

12. Except for a single tip sealed into the Marl the response of all those below the peat was negligible; and served to confirm the sand/silt classification below this level. Within the peat and above it the pore pressure measurements gave a much wider scatter of results than had been anticipated. This variability was attributed to the natural character of alluvium in the Severn Estuary. Whilst in general, the silt/clay sequence regularly provides high ratios of horizontal to vertical permeability, Hawkins (1972) pointed out that the accretion process also left behind abrupt changes in the consistency, within quite short distances. Laminations cannot always be relied on to offer a continuous drainage path therefore. Nevertheless, at the time, the results raised some doubts as to the correct performance of the instrumentation. For the stability analysis the higher reading at each point was used always, without averaging of any sort.

13. However in this context it is worth refering to Professor Peck's comment (1968), that if you put in instruments you have an obligation to accept what they tell you. The exercise is pointless if you reject those that do not 'appear' to be right. The truth of the foregoing was impressed on the first Author in 1967 during an embankment construction. In this instance a number of the settlement gauges on the centre line showed positive values after 5ft - 6ft of fill had been placed. This was in reverse to what was expected so the results were treated with a measure of disbelief. This mistrust persisted through the latter stages of construction when the same gauges reflected settlements of about one foot within a few days. Consequently it was not appreciated they were forecasting the beginnings of a slip. In fact they were reading correctly in the first instance; the degree of overconsolidation within the alluvium was sufficient to cause a slight dilatency under the initial loading. For largely the same reasons the initial piezometric readings frequently give a negative rise (Figure 7), which should not be attributed to their "settling down".

14. Lastly, the requirement becomes more and more demanding to forecast the time rates of consolidation, to the stage where stability can be maintained. In other words, 'what is a minimum or maximum loading period in any given instance?'

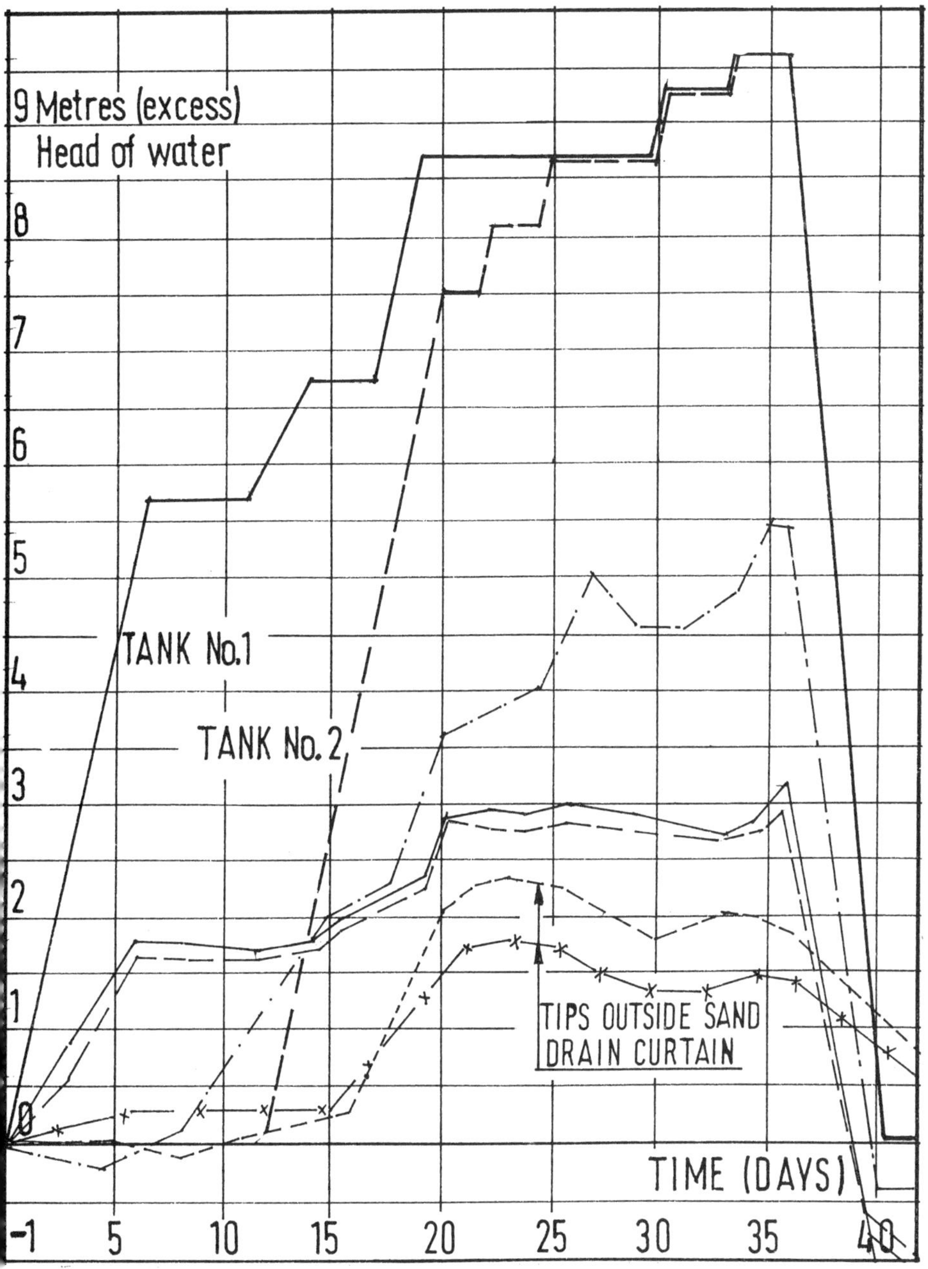

Figure 7 Philblack (Avonmouth) Pore water pressure response versus loading for 5 typical Tip points

Where earthworks are involved the problem is probably not quite so acute, because the seasonal nature of the job incorporates periods when a surcharge, or consolidation effect can be planned without much overall loss in time. Tanks on the other hand suffer no such restrictions, nor do Oil Companies run to loose schedules. This brings home the fact that the favourable aspects of the method from a foundation cost point of view might not outweigh the consequences of an unexpected delay; if for instance the completion had to dovetail with the date of a Tanker discharging its cargo.

CONCLUSIONS

15. Computer solutions for the effective stress analysis of stability problems provides the means for a much more intensive study of foundation performance. For controlled loading operations the greater benefit derives where a comprehensive pre-analysis is made before work starts. It is shown how this can be set out in a format which is usuable by site personnel; and it also provides the means by which a forward prediction of stability may be made at an intermediate stage of loading.

16. On recent alluvial deposits, the presence of a slight degree of pre-consolidation is frequently reflected in the response of the piezometers and settlement gauges during the initial loading stages. Variability of the soil structure however, can cause equally wide variations in the recorded pore pressures to a degree which might lead to doubt as to the correct performance of that instrumentation.

17. The difficulty of making an accurate forecast of the necessary consolidation time required to maintain stability in any instance, must be very fully appreciated by an Employer if the consequence of a delay is likely to be serious.

ACKNOWLEDGEMENTS

The information contained in this paper is presented by permission of G. F. Norris, FICE, MInstHE, Chief Roads Engineer, Scottish Development Department; and of Philblack (Avonmouth) Limited. The authors wish to thank Freeman Fox and Partners and G. C. Mander and Partners (Bristol) for allowing records taken during the work to be reproduced herein. Thanks are also due to Dr. C. P. Wroth for his valuable interest and encouragement during the period of the job at Avonmouth.

REFERENCES

Bishop AW 1955 "The use of the Slip Circle method in the stability analysis of earth slopes" Geotechnique 5 pp 7-17

Bjerrum L and A Overland 1957 "Foundation failure of an oil tank in Fredrikstad, Norway" Proc 4th Int Conf Soil Mech Found Engng Vol.1 p 287-290

Bjerrum L 1967 "Engineering Geology of Norwegian normally consolidated marine clays as related to settlement of buildings" Geotechnique 17 No.2 p 81-118

Dallard NJ 1971 "Construction of embankments on alluvial flats" Proc Inst C.E. Paper No.7400

Hawkins AB and DA Cook 1972 Contributions to discussion on I.C.E. Paper No.7400 (Dallard 1971)

Darragh RD 1964 "Controlled water tests to preload tank foundations" Proc ASCE Conf on Design of foundations for control of settlement (Evanslon, Illinois)

Gibson RE and A Marsland 1961 "Pore water pressure observations in a saturated alluvial deposit beneath a loaded oil tank" Proc Conf Pore pressure and suction in soils Butterworths, London p 112-118

Lambe TW 1968 "The behaviour of foundations during construction" Proc ASCE (SM and FE Div) Vol 94 SM1

Murray RT 1971 "Embankments constructed on soft alluvial foundations; settlement study at Avonmouth" Road Research Laboratory Report No.LR419

Peck RB 1969 "Advantages and limitations of the observational method in applied soil mechanics" Geotechnique 19 No.2 pp 171-187

Penman AD and DM Watson 1968 "Foundations for Storage Tanks on reclaimed land at Teesmouth" Building Research Station, Paper No.29/68 reprinted from Proc Inst Civ Engrs, vol 37 (May 1967), pp 19-42.

HORIZONTAL INCLINOMETERS FOR THE MEASUREMENT OF VERTICAL DISPLACEMENT IN THE SOIL AROUND EXPERIMENTAL FOUNDATIONS

R W Cooke and G Price

Building Research Station

SUMMARY

The development of three inclinometers based on electrolytic liquid levels is described. The inclinometers are used to measure the inclination to the horizontal of the soil in the vicinity of experimental foundations and are normally operated from a trench or shaft outside the soil zone affected by the foundation. Used with commercial AC strain indicators or with AC data logging equipment they are capable of resolving to 1 second of arc. Vertical displacements can be computed with an accuracy of the order of ± 0.02 mm. Tests to ascertain the reliability of the inclinometers in field conditions are described and some observations close to an instrumented friction pile loaded in compression and uplift are presented.

INTRODUCTION

Deep foundation design is evolving from a state in which settlements are presumed to be within acceptable limits if the factor of safety against failure is adequate towards methods which permit prediction of the settlement. To further this approach it is necessary to know how foundation loads are transmitted to the supporting soil and to observe the soil strains and displacements around loaded foundations. This paper describes the development and operation of horizontal inclinometers for measuring the vertical movements of the soil in the vicinity of foundations as the foundations are installed and loaded. The transducer around which the inclinometers have been designed is an electrolytic liquid level. It is used to record changes in the inclination to the horizontal at various locations along small holes bored from a trench outside the soil zone affected by the foundation. Two of the instruments (A and B) have been employed successfully to determine the soil displacements during the installation and test loading of an instrumented steel tube pile 168 mm diameter and

3.5 m long in London Clay (Cooke, R W and Price, G, 1973). Large numbers of a simpler version (C) have been prepared for the simultaneous recording of the movements around a group of three similar piles at the same site. The technique is capable of considerable development and there are no reasons why similar inclinometers should not be operated from a lined shaft or between two or more large diameter bored piles.

THE ELECTROLYTIC LEVELS

The levels consist of short lengths of glass tubing into which an electrolyte is sealed. Electrodes near each end and at the centre of the tube are partially immersed in the electrolyte and the resistance between the central electrode and the end ones varies as the tube is tilted in the vertical plane containing the tube. By connecting the levels to an AC strain indicator or in an AC bridge circuit with a precision voltmeter very small changes in inclination to the horizontal can be measured.

Levels are now commercially available having a variety of sensitivities and operating ranges but at the time instruments A and B were being developed it was only possible to obtain them with high sensitivities and limited ranges. The level chosen, which was supplied by Hamlin Inc of Wisconsin, USA, has a linear range of approximately ± 30 minutes of arc and a sensitivity on a portable AC strain gauge bridge of about 100 instrument divisions per second. For most of the observations it was found convenient to reduce this by shunting the output with high stability resistors to 1 or 2 instrument division per second. The stability of individual levels on a heavy floor in controlled laboratory conditions was found to be within ± 10 seconds during a period of 2 weeks.

<u>The effects of temperature</u>
Electrolytic levels are affected by the ambient temperature and the manufacturer's literature suggests that the sensitivity increases by approximately 7 per cent for an increase in temperature from 5°C to 25°C. Since the temperature changes in the ground were expected to be small this was not considered important, but calibration tests were carried out at 10, 12.5 and 15°C. The variation at both 10°C and 15°C about the calibration factor at 12.5°C was found to be less than 0.5 per cent.

Whilst the effect of temperature variation on the level calibration was negligible it was evident that temperature changes, and particularly temperature gradients along the level tube due to air movement, affected the zero readings. To examine this a level was mounted longitudinally between two small aluminium water tanks, the gap between each end of the level and the vertical sides of the tanks being bridged with transformer oil. Thermocouples were soldered to each of the level electrodes and the whole assembly insulated from external air movement with polystyrene. Both tanks were filled with water at a temperature 10°C above room temperature and allowed to cool naturally so that the very small temperature gradient along the level tube remained constant. The variation in the electrical output corresponded to approximately 16 second of arc per °C change in temperature.

By raising the initial water temperature in one of the tanks it was possible to apply a temperature gradient along the level tube. In tests with temperature differentials between electrodes of up to 1°C the zero drift corresponded to approximately 6 minutes of arc per °C temperature difference. It must be born in mind however, that this was an extremely severe test of the level characteristic. In practice the levels are mounted in a brass or copper casing and installed in a lined borehole in the soil, the open end of the borehole being closed with cotton wool. Observations of temperature in the boreholes in which the levels were to be used showed that over a period of three weeks during the Autumn the temperature did not vary outside the range of 12-13°C. Under these conditions, when stability has been attained, the temperature gradient along individual level tubes may well be less than 0.01°C.

The effects of both overall and differential temperature changes on the electrical output of the levels should be eliminated in a test of two levels clamped rigidly together and connected in electrical opposition. This test, which has the added advantage of eliminating the minute changes in inclination caused by people walking on the floor nearby, showed that the zero drift in ideal conditions was less than 7 seconds over a period of 3 days.

The effect of inclination to the vertical

To examine the errors likely to be introduced if the levels were not set up with the plane containing the electrodes vertical, calibrations were made using a level mounted with its longitudinal axis on the axis of a precise vertical circle clamped to a calibration beam. By this means the plane through the electrodes could be inclined at known angles to the vertical while the level was tilted longitudinally in its normal operating mode. The tests showed that the calibration factor changed by less than 4 per cent for inclinations either side of the vertical of up to 20°. Since the levels can be located in the boreholes to within ± 5° of the vertical these errors are not significant.

Varying the inclination to the vertical of the electrode plane while maintaining the device horizontally in the longitudinal direction appeared to have a marked effect on the zero.

Rotation of 5° out of the vertical plane caused a zero change equivalent to 4 minutes of tilt in the operating direction, the relationship between rotation and zero change being sensibly linear. As long as adequate range is available in the recorder this is unimportant in practice unless the inclination of the plane through the inclinometer electrodes changes during a test.

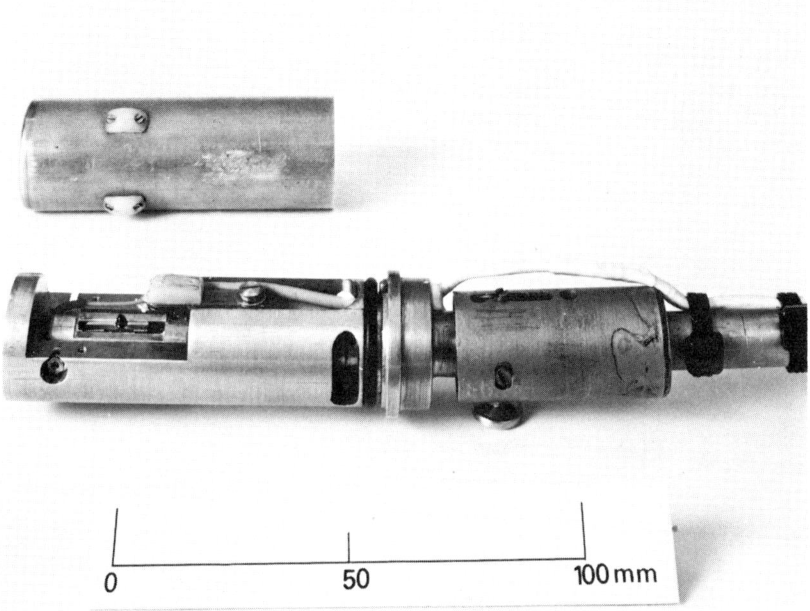

Fig 1 Inclinometer A with the casing removed.

THE INCLINOMETERS

Instrument A

Because of the limited operating range of the electrolytic levels available the inclinometers in which they were to be used had to include provision for extending the range mechanically. It was anticipated that the holes in which the instruments were to be installed could be bored and lined with PVC tubing to within $\pm 2^o$ of the horizontal and that the soil movements due to foundation loading would be predominantly downward towards the foundation. The inclinometers were therefore designed to operate over a range from 6° downward to 3° upward. In inclinometer A, Fig 1, this was achieved by mounting one end of the level on a horizontal pivot and supporting the opposite end on a sliding wedge with a face angle of 14° to the horizontal. The thick end of the wedge was extended to form a nut and rotation of a lead screw terminating in the nut caused the wedge to slide smoothly between guides formed in the brass body of the instrument. The inclination of the level, which was held in contact with the wedge by a small leaf spring at the pivot end, could therefore be varied over the required range by rotating the lead screw. Another spring attached to the wedge made contact with the top of the casing to prevent the wedge from lifting as the lead screw was adjusted. The lead screw was operated remotely while the instrument was in the borehole by a 1.6 m long drive shaft connected to it through a miniature universal coupling. A casing tube containing the drive shaft terminated in a spherical seating concentric with the coupling at the end of the inclinometer body. By this means handling movements of the casing tube, while the drive was rotated to adjust the wedge, were isolated from the inclinometer. The main components of the inclinometer are shown in the section drawing in Fig 2.

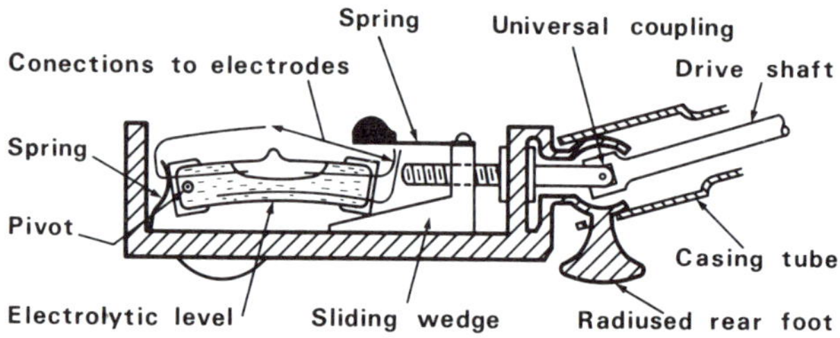

Fig 2 Longitudinal section through inclinometer A showing the main components.

Controlled rotation of the drive shaft was provided by a micrometer mounted axially at the end of the casing tube. So that longitudinal movements of the micrometer stem were not transmitted to the shaft a pin driven diametrically through a bush clamped to the micrometer stem located in a fork at the end of the shaft. The position of the wedge was thus precisely determined by the micrometer setting and movements of the micrometer housing and casing tube, other than in the longitudinal direction, did not disturb the inclinometer.

The inclinometer was supported on two curved nylon feet near the front of a length of brass tube forming the instrument casing and on a single rear foot. This foot was radius to its distance from the centre of the universal coupling and attached to the spherical seating vertically beneath the coupling centre. A soft leaf spring mounted on top of the casing at the centroid of the three feet applied some pressure to the feet to improve stability.

The electrical connections to the level were carefully water-proofed and brought out through the end of the inclinometer body, where they were sealed with Silcoset. An O-ring between the inclinometer body and its casing prevented moisture from penetrating to the moving and electrical parts. Graduations on the drive casing tube enabled the position of the inclinometer in the borehole to be determined to within ± 0.5 mm, and a small spirit level attached to the micrometer housing perpendicular to the axis permitted alignment of the inclinometer in the vertical plane.

Operation

With instrument A two modes of operation were possible. For the large movements which occurred as a pile was jacked into the ground it was convenient to survey each borehole using the micrometer to level the instrument and the electrical output on the strain indicator as a null detector. In this mode the sensitivity was reduced to 1.5 micrometer divisions per minute of arc but readings could be taken rapidly at a large number of stations along the boreholes. At this sensitivity, when the instrument was moved and replaced at a station to an accuracy of ± 0.5 mm, imperfections in the surface of the borehole lining had little effect on the results. Vertical displacements could be determined to an accuracy better than ± 0.1 mm.

For measurements of soil displacement around a pile loaded only to the design load this method was insufficiently sensitive. At design load the settlement of a pile is normally less than 1 per cent of its diameter and soil movements are discernable to at least 10 diameters from it. Inclinometers for studying soil displacements around a pile under these conditions must resolve to a few per cent of an angle of 1×10^{-3} radians (240 seconds). Using the second mode of operation a sensitivity of ± 1 second could be obtained provided the instrument was not moved between successive readings. In this mode the inclinometer was initially levelled with the micrometer and readings were then made with the strain indicator. At loads up to 50 per cent of the ultimate in short term tests in London Clay movements of the experimental pile and the corresponding soil movements were found to be elastic and recoverable. Surveys of the boreholes could therefore be made by loading and unloading the pile with the inclinometer at each station in turn. Examples of the data obtained in this manner are given in the final section of the paper.

Instrument B

Experience with inclinometer A showed that while it was extremely useful for most situations where the deformations of the borehole were small, because of the length and rigidity of the drive shaft it was not suitable for observations beneath the pile tip where the changes of curvature were large. Inclinometer B (Fig 3) dispensed with the remote micrometer control chosen for inclinometer A and the rear end of the level was raised or lowered by a cam coupled directly to a low speed electric motor built into the instrument body. The end of the level was retained in contact with the edge of the cam by means of a leaf spring and very fine control of the cam position could be obtained by an inching switch in the power supply to the motor. To allow for pronounced changes of curvature in the borehole lining tube it was necessary for the connection to the instrument to be flexible while in order to adjust the instrument for verticality it needed adequate torsional rigidity. A steel measuring tape was found to be satisfactory for this purpose and at the same time to provide the means of accurate location of the inclinometer in the lining tube. The instrument was pushed to the end of the tube with a flexible wooden dowel and drawn back to each new position after the appropriate readings had been taken. Verticality within the required limits of $\pm 5^{\circ}$ was obtained by sighting on a line scribed at the end of the instrument. A miniature light bulb mounted on the instrument casing provided the necessary illumination.

Fig 3 Inclinometer B showing the motor driven cam for levelling the electrolytic bubble tube.

Operation

Instrument B was not designed for use in the electrical null mode and no method of measuring the rotation of the cam was provided, although in further development this could be included. The instrument was extremely effective for observations made electrically in the same manner as instrument A. Some of these taken beneath the pile tip are included in the final section of the paper.

Instrument C

A major criticism of inclinometers A and B was that because of the impossibility of replacing the instruments exactly in a position from which they had been moved, the soil displacements near foundations under small loads could only be recorded when the inclinometers were left at each station for a complete loading run. This was a laborious process and it became clear that simpler, cheaper devices were necessary. These could be made available in sufficient quantities for trains of them to be installed in the boreholes and left untouched for the duration of the test. Fortunately electrolytic levels had by this time become available with acceptable sensitivities and larger operating ranges. Inclinometers based on these levels, if correctly constructed and installed, would not require mechanical resetting and the main limitation was likely to be imposed by the range of the data logger.

Fig 4   Part of a train of inclinometers C at close and double spacing.

Inclinometer C therefore consisted of a wide-range level carefully aligned in a 28 mm diameter 75 mm length of copper tubing and held in position by Silcoset silicon rubber. A horizontal copper plate soldered into each tube before the level was installed provided a space through which the leads from other inclinometers in the train could pass. Brass end plates screwed to the lower portion of the tube containing the level formed moulds for the liquid Silcoset during pouring and bulkheads by means of which the inclinometer units could be connected together. Trains of units were connected through universal couplings by lengths of brass rod located in the end plates. Each inclinometer was supported on a brass foot vertically beneath the coupling. The foot was machined to the radius about the coupling centre so that movements applied to one unit had minimal effect on adjacent ones. Inclinometer units at both ends of a train had two additional feet soldered to the casing so that they were supported at three points to give the required stability. A soft leaf spring fitted to the top of each inclinometer pressed on the soffit of the borehole lining and ensured good contact at the feet. To economise in the number of inclinometers used the brass rods between the couplings and the units were increased in length at positions in the trains where the soil displacements were expected to be small. Figure 4 shows part of a train of inclinometers at close and double spacing and the modified leaf spring used in the latter case to locate the pressure over the feet. This prevented bending in the rods which might have occurred at large spacings if all the springs had been placed centrally over the inclinometer units.

The effect of environment

The effect of temperature variation on the behaviour of the electrolytic levels has been reported earlier in this paper. In the case of inclinometer C, in which the levels were bedded in Silcoset instead of in sprung contact with machined metal surfaces (instruments A and B), it was desirable to study further the effects of environmental changes.

Calibration tests on a train of 13 levels were carried out under cover at the field test site on occasions when the air temperature was 10, 12.5 and 15°C. There was no obvious relationship between calibration factor and temperature, the overall variation being less than 1 per cent. Zero drift was also unrelated to temperature and corresponded to less than ± 40 seconds over this temperature range.

To check the stability of the inclinometers under field conditions observations were made at 2-hourly intervals for 7 days of trains installed in two lined boreholes 2.4 m long at depths of 0.5 m and 1.8 m. The drift of units in the lower hole remote from the trench face during this period corresponded to less than 30 seconds of arc. In the upper hole some of the units drifted appreciably more, probably due to temperature and moisture changes in the soil and the trench timbers, but in the majority of cases the drift was less than ± 100 seconds. A displacement of 1 mm over a radius of 2 m is equivalent to 100 seconds and soil movements of this magnitude due to moisture changes are common within about 1 m of the ground surface. In all cases the drift over a few hours during daytime was less than 20 seconds suggesting that the stability of the inclinometers was adequate for short term tests.

Interference between the inclinometer units

The interference between inclinometers was measured by taking readings on every unit of a train resting on a horizontal beam before and after lifting each foot in turn by means of an 0.2 mm feeler gauge. The process of mannually lifting each inclinometer to insert the feeler gauge introduced errors that would not occur in the field but the response of units either side of the two directly affected was generally less than 0.5 per cent of the applied inclination and in all cases less than 2 per cent.

CALIBRATION

For calibration, the inclinometers were strapped to an aluminium box beam 2 m long, supported at one end on a roller and at the other end on a ball, resting on ground steel plates fixed to the platten of a large testing machine. Each end of the beam was raised in turn on high precision slip gauges. In the case of inclinometer C it was convenient to calibrate completed trains as they were made up for installation in boreholes.

Inclinometer A read with a Peekel Strain Indicator had a calibration factor of 1.5 micrometer divisions per minute of arc in the electrical null mode. The calibration factors of inclinometers A and B were between 1.5 and 50 Peekel instrument divisions per second in the direct reading mode, depending on the values of the external shunt resistances included in the circuit. Inclinometers C were connected to a high speed AC data logger and the amplifier was adjusted to give either 1 or 0.5 bits per second, depending on the range of movement required.

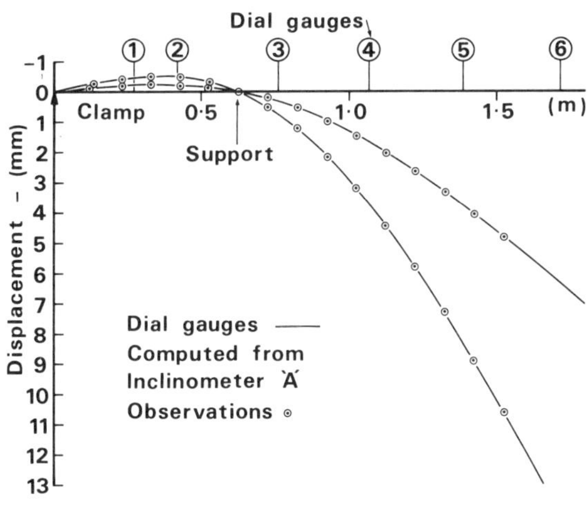

Fig 5 Displacements of a copper tube measured by dial gauges and inclinometer A.

ASSESSMENT OF THE INCLINOMETERS

To determine the vertical movements in the soil by means of

these inclinometers it is necessary to convert readings of slope to displacement over known horizontal distances and to sum them from a point where the movement is measured by some other means, or is known to be zero.

The feasibility of this approach was investigated in the laboratory by measuring the displacement from the horizontal of a 38 mm ID copper tube both indirectly with the inclinometers and directly with dial gauges. The tube, which was 2 m long, was clamped and supported at points near one end and deformed by applying known displacements close to the other end. Fig 5 shows the agreement between the tube displacements measured by dial gauges and values computed from inclinometer observations in two tests with instrument A using the electrical null mode of operation.

Similar tests were made with a train of inclinometers C linked together in the final form in which they were to be installed in a borehole and connected to the data logger. Figure 6 shows the agreement with dial gauge readings obtained in tests with five different deformations applied to the tube. In all cases the differences between inclinometer and dial gauge observations were less than 2 per cent of the mean displacement.

INSTALLATION AT THE SITE

The inclinometers were designed to be used in 38 mm bore spirally reinforced hose which would easily deform to follow

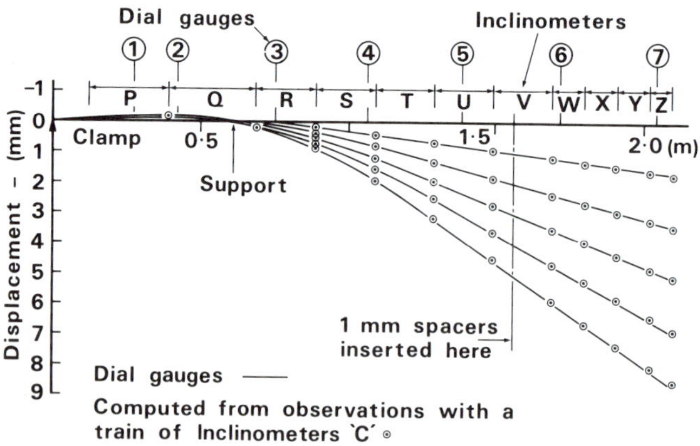

Fig 6 Displacements of the copper tube measured by dial gauges and a train of eleven inclinometers C.

local soil movements. Griflec Surcoflex braided transparent PVC hose type HM110 was found to be satisfactory for this purpose. The hose was installed in horizontal boreholes drilled from the face of a well-strutted trench which has been excavated some months beforehand. A 50 mm diameter flight auger, modified to fit standard drilling rods, was used to form the boreholes in London Clay and a special drilling guide was devised so that the holes would slope upward at 1° to the horizontal in order to drain any water which might collect. It was necessary to keep the hose straight for insertion in the holes and mandrels were formed by making three saw cuts at 120° along 38 mm OD, thick-walled steel tubing which could be expanded to fill the hose and later withdraw. Both compressed air within a length of bicycle wheel inner tubing and a rod with conical wedges were satisfactory for expanding the mandrel walls. Screw type jacks acting on the mandrel head applied the pressure necessary to install the hose, which was a tight sliding fit in the clay borehole. To minimise movement at the mouth of the hose it was plastered into the slightly larger hole previously cut in the trench timbering before the mandrel was withdrawn and the hose sawn off flush. After

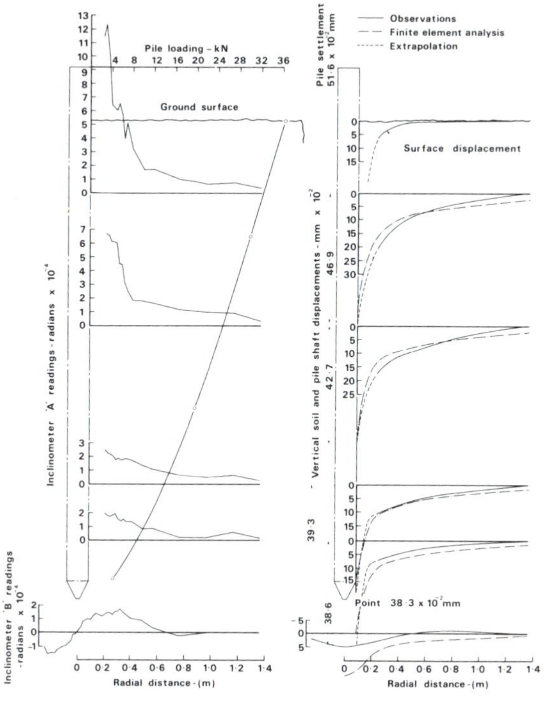

Fig 7 Inclinations and computed displacements in the soil around a loaded friction pile compared with the results of a finite element analysis.

the inclinometers had been located in the hose cotton wool was loosely packed at the mouth to minimise air movements.

THE EFFECT OF BOREHOLES ON CLAY BEHAVIOUR

In order to have some assessment of the effect of boring holes and lining them with hose on the behaviour of the clay close to a foundation, simple bending tests were made on a 300 mm cube of clay with and without a 50 mm lined hole formed along the axis of bending. The specimen was supported at two opposite edges and loaded centrally over the full width of the top. Inclinations were observed at points on the top surface either side of the loading plate. The test showed that installation of the hole and lining modified the load:inclination curves for the clay block by less than 5 per cent. It seems reasonable to presume that the effect of holes in the much larger clay masses stressed in field investigations will be substantially less than this.

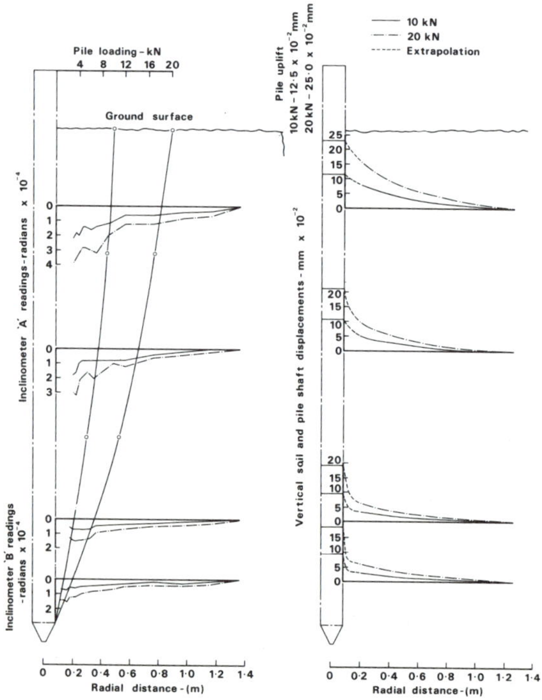

Fig 8 Inclinations and computed displacements in uplift tests of the pile.

FIELD EXPERIENCE

Trains of inclinometer C have at present only been used in the field to observe their stability and the movement of unloaded ground. Inclinometers A and B have been used in compression and uplift tests of an instrumented pile 168 mm diameter, jacked 3.5 m into London Clay. In the compression tests measurements were made with inclinometer A in four boreholes from a trench 10 pile diameters from the shaft to within 100 mm of the shaft surface. Inclinometer B was used in a borehole extending 250 mm beyond the pile tip. Figure 7 shows the inclinations observed and the computed soil displacements when the applied load was 36 kN (0.57 x Ultimate Load) and the settlement of the pile head was 0.5 mm. The soil displacements have been extrapolated to the computed pile shaft settlements at corresponding levels and compared with a finite element analysis using elastic parameters for the soil given by tests at the site.

The pile was subsequently loaded to failure by the Constant Rate of Penetration Method and two months later subjected to uplift loads of 10 kN and 20 kN (0.2 and 0.4 Ultimate Uplift Load). Figure 8 shows the inclinations and displacements obtained using inclinometer A in the two upper boreholes and inclinometer B in the two lower ones.

CONCLUSIONS

The inclinometers described in this paper are capable of measuring inclinations in the ground to an accuracy of a few seconds of arc. They have been used in short term loading tests of an instrumented pile without significantly affecting the behaviour of the soil. Sufficient observations have been made for complete soil displacement profiles under working load conditions to be determined with an accuracy of the order of ± 0.02 mm. The instruments are cheap to build and simple to install and operate, once a suitable access trench or shaft has been provided.

ACKNOWLEDGEMENTS

The authors gratefully acknowledge the assistance of their colleagues in constructing and assessing the inclinometers and in providing the logging facilities. The work described forms part of the Research Programme of the Building Research Establishment and the paper is published by permission of the Director.

REFERENCE

Cooke, R W and Price, G (1973). Strains and displacements around friction piles. Proc 8th Int Conf Soil Mech & Found Eng, Moscow 1973.

PERFORMANCE CHARACTERISTICS OF THE SLOPE INDICATOR SERIES 200-B INCLINOMETER

D. H. Cornforth

Managing Director, Nuttall Geotechnical Services Ltd., Colnbrook, Bucks.

USE OF INCLINOMETERS

Inclinometers are used to observe horizontal movements below ground. The method of observation is as follows. A purpose-made casing is installed vertically within the ground or structure. The lower end of the casing usually penetrates below the estimated level of movement and is used as the reference point for movements at higher levels. The inclinometer is lowered down the casing and the inclination of the casing is measured at short intervals between the top and bottom of the casing. If y represents the horizontal direction and x represents the vertical direction, then the inclination of the casing measures $^{dy}/dx$. By summing up the inclinations from the base of the casing (the reference point), the initial deflected profile of the casing can be obtained. Subsequent sets of readings will show whether or not the casing has moved from the initial position. In most practical applications, it is the differences from the initial set of readings which are computed and plotted, rather than the actual shapes of the casing.

The practical problems associated with the use of inclinometers may be subdivided into three groups:

(a) the measurement of lateral movements of soils or structural members. Examples include the use of inclinometers in earth dams during construction, movements of soil or retaining walls towards excavations, horizontal movements of laterally-loaded piles, deflected form of tied-back excavations and flexible sheet-pile walls, etc.

(b) observation of landslides to determine the depth to the slide interface. This information is especially useful for stability analysis and for selection of sampling levels. The absolute values of the horizontal movements are of secondary importance, unlike the problems in group (a).

(c) monitoring the ground to provide early warning against slope failure. The inclinometer casings are installed in stable ground and are observed during and after construction. In this role, inclinometers lessen the risk and may allow the engineer more freedom to make a balanced engineering design.

PERFORMANCE OF INCLINOMETERS

There is an increasing number of inclinometers becoming available, and the engineer usually has to make his selection from technical specifications and the competing claims made by sales-oriented organizations. The inherent dangers of this procedure have been described in several recent papers, notably by Peck (1970). The technical specification provides data about the resolution capabilities (sensitivity) of the instrument under workshop test conditions. It does not give any guarantee of long-term reliability, which is largely dependent on the principle of measurement, nor does it give assurance about ruggedness, which is the ability of the instrument to withstand the hazards of field use. It is the three R's of instrumentation - resolution, reliability and ruggedness - which control the overall performance of an instrument and relates directly to the requirements of the practising engineer.

The present paper provides some factual evidence about the field performance of the Slope Indicator Series 200-B inclinometer, based on case histories accumulated by the author's company over the past five years. It answers two questions which are of most concern to the practising engineer: What is the accuracy of the instrument under field conditions? What is the maximum horizontal deflection that can be measured by the instrument? The first question is the more important because it is impossible to judge whether an observed deformation is significant without a detailed knowledge of the instrument's accuracy. The question is answered by an appraisal of the performance of the instrument at three sites where inclinometers performed a monitoring role (group (c), referred to earlier) in essentially stationary ground. Sufficient sets of readings were obtained from each site to allow the results to be treated statistically. The analysis shows the spread of results that can be expected as normal deviation about the mean within a stationary casing.

Figure 1 Slope indicator in field use

Spread of readings

There are two methods of presenting the spread of results obtained from inclinometer readings. The first method is to compute the deviations from the mean, and in this case the spread (i.e. twice the deviation from the mean) is given in absolute units of inches or millimetres. This method is of practical use in relating the accuracy of the instrument to observations of horizontal movements. The second method is to express the spread in terms of angular displacement (in radians). In this case, the deviation from the mean is divided by the distance from the base of the casing; thus, a deviation that increases linearly from the base of the casing has a constant value of angular displacement. It will be shown later that it is not a linear relationship, so that any quoted values of spread must be qualified by the distance to which it relates.

Slope indicator instrument

The slope indicator was designed by S.D. Wilson and first marketed in 1954; the present series (200-B) has been in worldwide use since 1965. The slope indicator is part of an inclinometer system comprising the torpedo instrument, connecting cable, readout box, and grooved casing (Fig. 1). The principle of measurement has been fully described in several earlier papers (e.g. Wilson & Hancock, 1965; Wilson, 1967, Green, 1969) There is also a summary in the manufacturers handbook (Slope Indicator Co.) and in promotional literature. For these reasons, no further details will be given.

Previous work

Gould and Dunnicliff (1971) and Dunnicliff (1971) summarize the current state-of-art on field instruments for measuring deformation. Green (1969) has performed a laboratory study on the slope indicator. Meese (1970) analysed field results obtained from slope indicator casings installed in the concrete cylinder piles of the Seattle Freeway. Both studies indicated an accuracy of the order of $\pm 2 \times 10^{-4}$ radians, but reference should be made to the original papers for further details.

Manufacturers specification for the 200-B model

The potentiometer in the readout box has a scale range of 1000 units. There are approximately 35 dial units for 1° angular displacement, thus making 1 unit (the minimum resolution or sensitivity) equivalent to about 1.7 minutes of arc or 5×10^{-4} radians. The maximum range of angular displacement is thus $\pm 14^{\circ}$ from the vertical. The manufacturer uses conservative values in promotional literature, quoting a sensitivity of only 3 minutes of arc and a range of $\pm 12^{\circ}$ from the vertical.

Field problems

The problems that can arise in the field measurements are too numerous to be listed in detail. Briefly, they can be subdivided into three categories:

(i) faulty installation of casings, which can include excessive casing twist (usually a fault in manufacture), unsatisfactory backfilling (poor workmanship or insufficient room between inclinometer casing and borehole), damage on withdrawal of borehole casing, backfill entering the inclinometer casing at the joints, no allowance for vertical movements (sometimes but not always necessary), corrosion in certain soils (especially made ground and alkaline soils), etc.

(ii) faulty behaviour of the instrument under field conditions which can include damage by transportation (especially at the site itself), errors caused by dirt and water at the connections, instrument susceptibility to changes in temperature and humidity, malfunctions not apparent at the time of observations, water vapour within the instrument housing, etc.

(ii) poorly skilled observers, which can cause transcribing errors, faulty handling of the equipment before, during and after the field work, inaccurate computations, etc.

Items (i) and (iii) are common but avoidable problems. Errors listed in item (ii) depend to some extent on the measuring principle, design and manufacturing standards of the inclinometer; these govern the longer

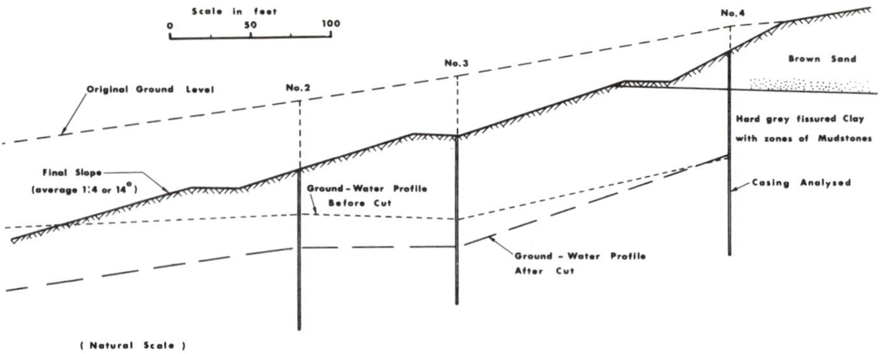

Figure 2 Slope section, Site A

Figure 3 (right) Observations of upper casing, Site A

term reliability and ruggedness for field use.

Field accuracy of the slope indicator

To check the field performance of the slope indicator, statistical analyses have been made on the results obtained from three different sites. In all cases, sets of readings have been taken over a comparatively long period and in sufficient number to permit a statistical treatment. The slope indicator casings were in essentially stationary ground during the observation period.

At Site A, four slope indicator casings were installed to forewarn against landslide movements during the excavation of a deep cut (Fig. 2). The casings were installed in 8-inch (200 mm) dia. boreholes and backfilled with tamped gravel. The sets of readings taken on the casing at the top of the slope have been analysed.

It is a matter of interpretation whether or not the casing has moved at all, but the observations (Fig. 3) show an apparent relative movement of about 0.2 inch (5mm) near the top, which could result from the lower part of the

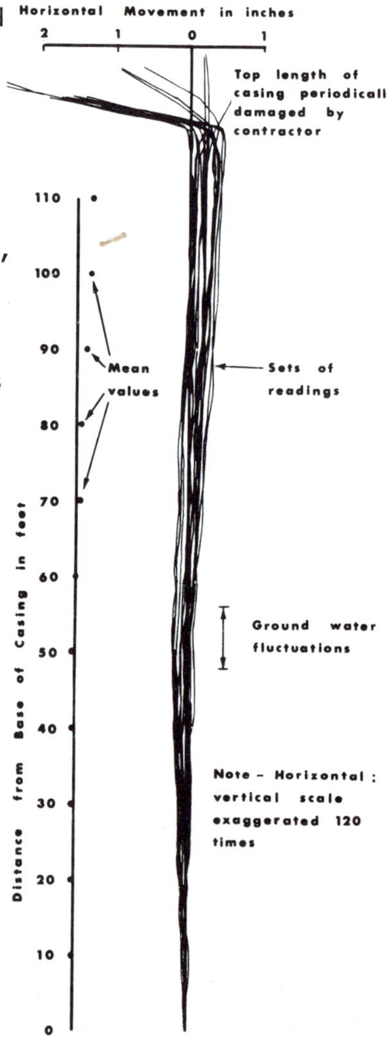

casing moving downslope relative to the top. The casing was installed shortly after excavation had begun and construction had moved a considerable distance away from the casing within three weeks. Sets of readings were taken at one-week intervals for the first eight weeks and thereafter at increasing intervals over a period of almost three years. A total of 23 sets of readings were taken by 6 different observers. Weather conditions varied from ground frost to warm summer, and ground water levels within the casing varied over a range of 8 feet (3m). The instrument mechanism was not serviced during the period of observations. None of the sets of readings has been rejected as unsatisfactory or taken again for any reason.

The casing was made up from 10-foot (3m) lengths with short (1-inch) gaps allowed at the joints for vertical displacements. Readings were taken at 2-foot intervals in each groove of the casing at 1, 3, 5, etc. levels. The only corrections made to the readings were at deeper levels where the instrument straddled the joint due to the cumulative effect of the gaps. These locations were easy to detect because the reading summation on opposite grooves was outside the normal range; these individual readings were rejected and replaced by the average value of the adjacent readings immediately above and below the joint.

The analysis was made on the 19 sets of readings taken four weeks or more after installation (Fig. 3). Any lateral movement due to the excavation is likely to have been completed at this stage and this is generally confirmed by the fact that the average results for the first ten sets of readings agree very closely to those for the last nine sets of readings.

The apparent horizontal deflections of the casing at 10-foot intervals, starting from the base of the casing, were listed for each set of readings. The mean deflection at each depth was calculated (Fig. 3) and deviations from the mean value obtained for each set of readings. The scatter about the mean was roughly binomial in distribution, and it is thus possible to estimate a standard deviation as the range embracing 13 out of the 19 individual readings (68%).

The results have been plotted in terms of three criteria of deviation on Fig. 4. The maximum deviation represents the total spread of readings at each depth (the spread being twice the deviation from the mean). This criterion is useful as an indication of the possible spread in results before a 'significant' movement has occurred. However, it is partially dependent on the number of readings taken and will tend to increase with increasing numbers of readings. It is also susceptible to fluke readings (e.g. the 1 in 1000 extremity) being included in the statistical sample and thus affecting the result. The second criterion is the 90%

limits and this is probably more acceptable statistically for setting the outer boundaries of spread. The third criterion is the standard deviation referred to earlier.

At Sites B and C, several slope indicator casings were installed in stiff fissured clay slopes to observe the ground in, and adjacent to, landslide areas. The method of casing installation was similar to the method at Site A, except that the casings were butt-jointed. The casings were read at 1-foot intervals.

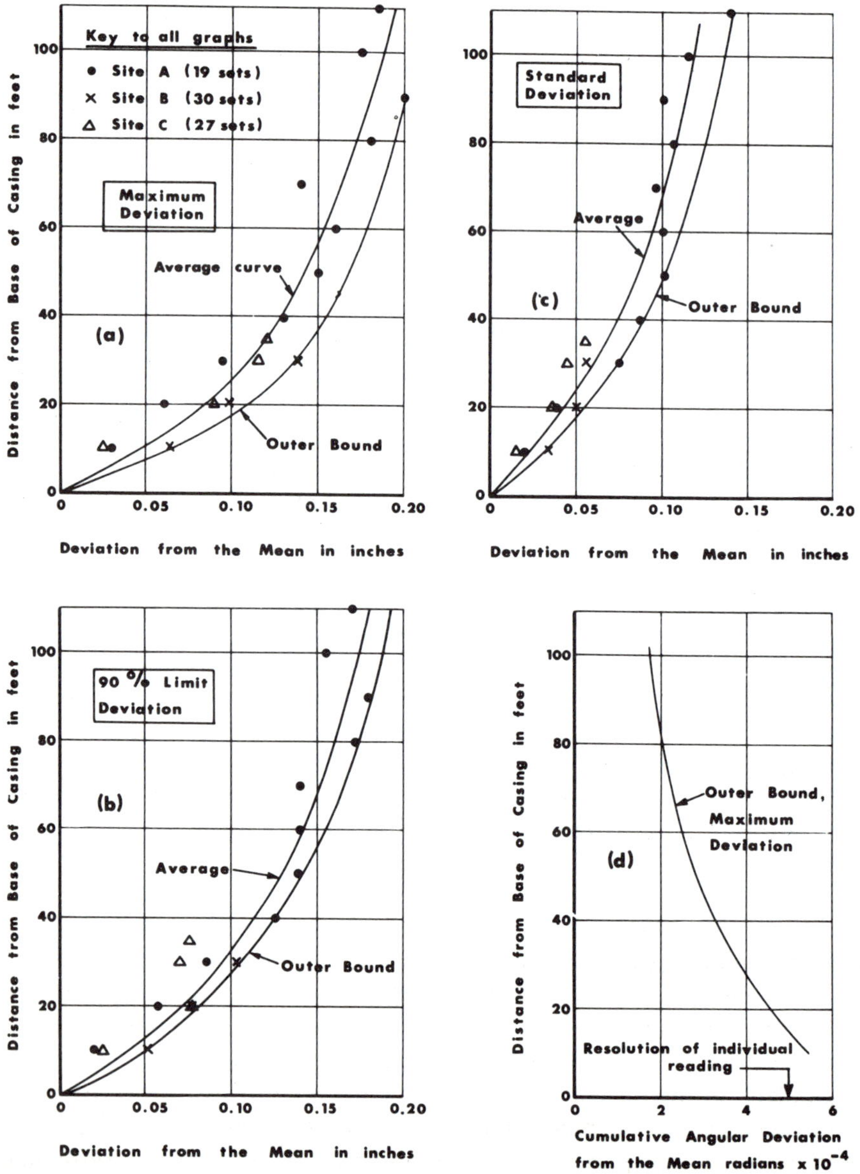

Figure 4 Computed deviations from the mean at three sites

Casings installed in stable ground were selected for analysis. At both sites, the mean values of deflection differed by less than 0.06 inch (1.5mm) from the initial set of readings. The other relevant details are:

	Site B	Site C
Length of casings	35 feet (10.7m)	40 feet (12.2m)
Period of observations	seven months	nine months
Sets of data obtained	30	27
Number of different observers	7	6

The data has been analysed by the methods described previously. Two casings were studied at Site B; the average results have been entered on the summary graphs, Fig. 4.

The results of the analyses from the three sites are reasonably consistent. In addition to the average curves, an estimated 'outer bound' curve has been drawn using the same shape as the average but passing through the extremes of deviation for each criterion. All the curves are non-linear showing that the spread of results increases at a slower rate than the increasing distance from the base of the casing. This result can be expected when the individual errors of reading are random rather than systematic.

The trend of the maximum deviation from the mean is approximately as follows:

Distance from base of casing		Maximum deviation from the mean	
20 ft.	6 m	0.10 in.	2.5 mm
50 ft.	15 m	0.15 in.	3.8 mm
100 ft.	30 m	0.20 in.	5.0 mm

The standard deviation from the mean (Fig. 4(b)), within which two-thirds of the readings should lie, is about two-thirds of the maximum deviations given above.

As mentioned earlier, the resolution of individual readings is 5×10^{-4} radians. The maximum deviation from the mean attains this value in the lower 20 feet of a casing but the figure decreases to less than 2×10^{-4} radians at more than 80 feet from the base (Fig. 4d). These figures are 2 to 5 times more accurate than the conservative value of 10×10^{-4} radians quoted in the manufacturers specification.

To sum up, the graphs of Fig. 4 show the normal range of deviation that can be expected from inclinometer readings taken in stationary casings. When these deviations are exceeded, it is reasonable to infer that a

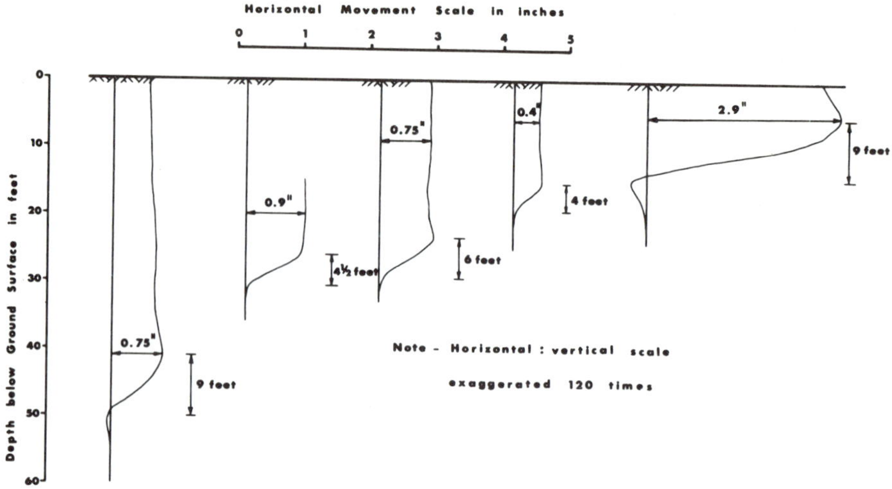

Figure 5 Typical examples of slope indicator observations in landslides.

lateral movement has probably occurred. In such cases, further sets of readings are usually taken to confirm or extend the information.

Measurement of very small horizontal movements

The limitations of the inclinometer system preclude their use for measuring very small horizontal displacements, and extensometers are generally used for this purpose. However, it is possible that the overall accuracy of inclinometer measurements could be improved by taking numerous sets of readings and treating the results statistically. For example, by splitting up the readings into groups of 10 to 20 sets in chronological sequence, a comparison of the mean values from each group may show a significant trend that may not be apparent from the analysis of fewer sets of readings.

The manufacturer also produces the Digitilt, an inclinometer with a resolution capability 10 times more accurate than the Series 200-B slope indicator. A report on the field reliability of the Digitilt should become available in the future.

Measurement of large horizontal movements

The two factors that limit the measurement of lateral movement are the tilt range of the instrument, referred to earlier, and the curvature of the casing. If the curvature of the casing is gradual (e.g. lateral movements within an earth dam) displacements of more than 1 foot (300 mm) have been recorded without difficulty (e.g. Wilson & Marano, 1968). For studies of landslides, the displacement range is much smaller due to the sharp changes in inclination of the casing at the two extremities of the shear zone. Some typical examples of ground movements within unstable clay slopes are plotted on Fig. 5. Common features in perhaps a majority of such failures are (i) a well-defined shearing zone 4 feet (1.2m) to 9 feet (3m) thick (ii) the soil above the shear zone moves laterally without shear distortion, i.e. the casing remains approximately vertical during the movement. As a consequence, the maximum displacement that can be measured before the instrument jams in the casing is in the order of 3 to $4\frac{1}{2}$ inches (75 to 110mm) depending on the depth of the shear zone. However, with well-defined shear zones, relative displacements of only $\frac{1}{4}$ inch (6mm) are often sufficient to confirm the position of the shearing zone, as evidenced by the curves reproduced in Fig. 5.

References

Dunnicliff, C.J. (1971) Equipment for Field Deformation Measurements Proc. 4th Panamerican Conf. SMFE, San Juan. Vol. 2, p.319-332.

Gould, J.P. & Dunnicliff, C.J. (1971) Accuracy of Field Deformation Measurements. Proc. 4th Panamerican Conf. SMFE, San Juan. Vol. 1, p.313-366.

Green, G.E. (1969) Report on Tests on the Performance of the Wilson Slope Indicator and the Soil Instruments Inclinometer. Imperial College, London. Unpublished. Summary in Green's paper to this Symposium.

Meese, R.H. (1970) Interim Report on Performance Analysis of Cylinder-Beam Retaining Walls, Seattle Freeway, Seneca Street to Olive Way. University of Washington. Report submitted to Washington State Highway Commission.

Peck, R.B. (1970) Observations and Instrumentation: Some Elementary Considerations. Seminar on Field Observations in Foundation Design and Construction. New York Met Section, ASCE.

Slope Indicator Company. Instruction Manual. Available from 3668 Albion Place N., Seattle, Washington 98103, U.S.A.

Wilson, S.D. & Hancock, C.J. (1965) Instrumentation for Movements within Rockfill Dams. ASTM Spec. Tech. Publ. 392, p.115-130.

Wilson, S.D. (1967) Investigation of Embankment Performance. Journal ASCE. Vol. 93, SM4, p.135-156.

Wilson, S.D. & Marano, D. (1968) Performance of Muddy Run Embankment. Journal ASCE. July, p.859-882.

OBSERVATIONS AND ANALYSIS OF A FLOW SLIDE IN SAND FILL

D. H. Cornforth

Managing Director, Nuttall Geotechnical Services Ltd., Colnbrook

E. G. Worth

Principal Engineer, Cornell Howland Hayes & Merryfield, Corvallis, Oregon, U.S.A.

W. L. Wright

Vice-President, Shannon & Wilson, Inc., Portland, Oregon, U.S.A.

INTRODUCTION

The site was at Portland, Oregon (north western United States), where Terminal 2 on the waterfront of the Willamette River was being reconstructed as a deep fresh water terminal (Fig. 1). The river bottom in the area was dredged to remove a cover of brown organic silt, after which sand was pumped onto the site to form a fill area approximately 1000 feet square. The contractor then had to deepen the river channel and dress the sides of the fill to a 1 : 2 (vertical : horizontal) design slope before constructing an anchored sheet pile bulkhead part way up the slope. The design height of the slope was 68 feet, with only about 20 feet being above river level during the construction period.

The soils problem arose during the deepening of the river channel adjacent to the fill slope. The technique employed by the marine contractor was to dig out the sand from the base of the slope with a clamshell and allow the sand higher up the slope to run into the excavation ('glory hole' excavation). Within a short period of time, four separate failures occurred at different localities in the slopes of the fill.

The four slope failures were broadly similar in their characteristics. The upper scarp of the unstable slope was either just behind the top edge of the fill or within the slope above the water line. Hydrographic surveys showed that the sand had slumped to a flat slope of about 1 : 6 (vertical : horizontal) below the water.

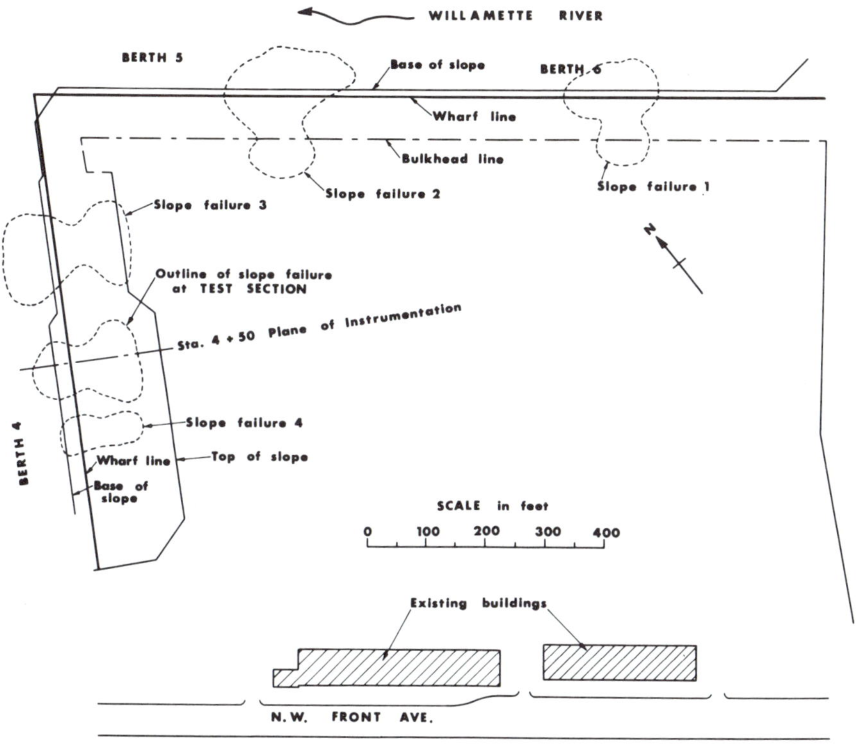

Figure 1 Site plan

Several boreholes were put down adjacent to the slope failures, using a lorry-mounted rotary boring rig with mud flush. SPT's were performed and 3-inch dia. thin-wall shelby tube samples were recovered by hydraulic jacking. Soil properties measured in the upper 30 feet of the sand in the fill slope adjacent to slope failure were:

Average SPT blow count	-	3
Relative dry density	-	20 to 40%
Static angle of repose	-	34 degrees

Gradation curves are shown on Fig. 2.

There were two theories of failure. First, that the slides were occurring through very weak organic silt layers immediately beneath the sand fill or, second, that the slides were caused by liquefaction in the sand fill itself. Neither theory seemed to be entirely satisfactory, but it was reasoned that, because of the random spacing of the four slope failures that had occurred, other parts of the slope may also have a marginal factor of safety against failure. To learn more about the nature of the problem, the decision was taken to instrument a section of the slope and try to induce a slope failure by the excavation techniques previously used by the contractor.

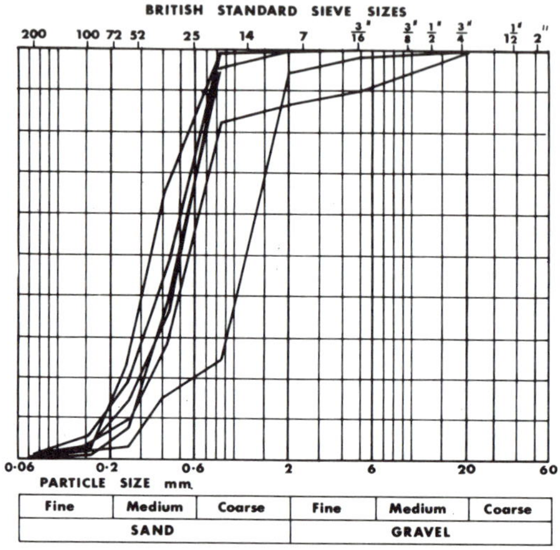

Figure 2   Gradation curves of sand fill

TEST SECTION

The selected site for the test was approximately mid-way between the two slope failures in the slip berth (Fig. 1) and close to where the largest slide had occurred. In this part of the site, the original hydraulic fill had been underbuilt and the contractor had pushed sand over the edge of the slope with a bulldozer to make up the required section. Above the river level, the fill was compacted in 6- to 8-inch lifts by a 5 ton vibratory roller. The interface between the hydraulically placed and the bulldozed fills is shown by a broken line on Fig. 3.

Two boreholes were put down at the test section (Fig. 3). The SPT blow count profile at the mid-height of the slope had very low N values, being less than 2 blows/foot in the upper 12 feet and rising to a maximum value of 12 blows/foot deeper within the sand fill. A weak organic silt layer, 6 feet thick, was encountered immediately below the sand fill so that the test section was ideal for checking the conflicting theories of failure.

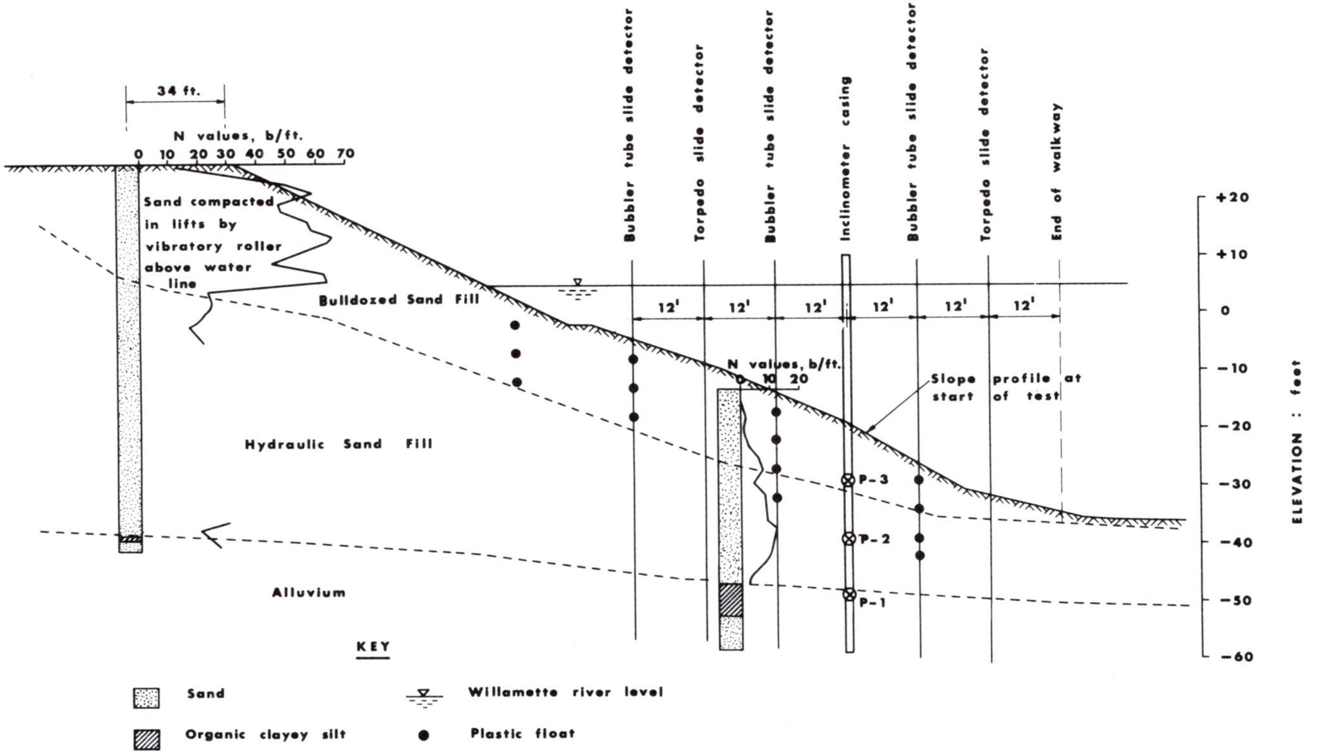

Figure 3 Instrumentation layout at test section

The borehole at the top of the slope had higher SPT blow counts, typically around 50 blows/foot, which implies that compaction had significantly increased the relative density. There was a decrease in the penetration resistance within the hydraulic fill below the compacted sand.

INSTRUMENTATION

The instruments were installed in essentially a plane section of the slope. The layout at the start of the test is shown on Fig. 3 and consisted of:

- one inclinometer casing, the lower end being embedded about 10 feet into the river alluvium below the sand fill. The inclinometer was a Slope Indicator Series 200-B instrument, which was manually lowered and raised within a 3-inch diameter grooved aluminium casing (Wilson & Hancock, 1965; Cornforth, 1973).

- three pneumatic piezometers installed at the same position within the slope section as the inclinometer casing but offset 10 feet from it. The porous tips (P-1, P-2, P-3) were set at three different levels, the lowest tip being located at the top of the river alluvium and probably within the weak organic silt. Rapid response piezometers were chosen so that any surge in pore water pressure resulting from sand liquefaction could be monitored. The pneumatic piezometers were Pore Pressure Indicators; the tip section is illustrated on Fig. 4. The operating principle is that the pore-water pressure deflects the instrument diaphragm, thus opening the spring-loaded valve. Air pumped through the piezometer tubes circulates freely until the pressure is sufficient to return the diaphragm to the equilibrium position and thus close the valve. The air pressure on the outlet side then remains stationary and is equal to the pore-water pressure at the tip.

- groups of three to four plastic floats were installed in the sand fill at four separate locations in a plane about 20 feet offset from the line of pipes. The floats within a group were spaced at about 5-foot intervals. Each float was marked for identification of its position, and was attached to a wood float on the water surface so that it could be recovered.

- slide detector devices, which provide a simple means of measuring the position of the slide interface after failure. The bubbler tube device consisted of a thin glass tube protected by a 2-inch diameter metal pipe. The air pressure required to continuously pass bubbles into the ground was measured. If bending of the pipe at the slide interface breaks the glass tube, the required air pressure decreases. The second type of slide detector was a pipe with a wire-supported torpedo at the lower end of the pipe. After failure, the torpedo is raised until it cannot pass the bends in the pipe caused by ground movement. Similarly, a second torpedo is lowered from the top of the pipe to confirm the position of the shear zone. Three bubbler tube and two torpedo slide detectors were installed in the test section, each pipe being embedded about 10 feet into the alluvium below the sand fill.

The instrument readings were monitored from a floating pontoon moored above the test section (Fig. 5). Inclinometer readings were taken at 10 to 30-minute intervals and piezometric pressures were observed more frequently. The bubbler tube devices were operated continuously. Measurements of the slope surface below the water were taken hourly from the walkway tied alongside the pontoon.

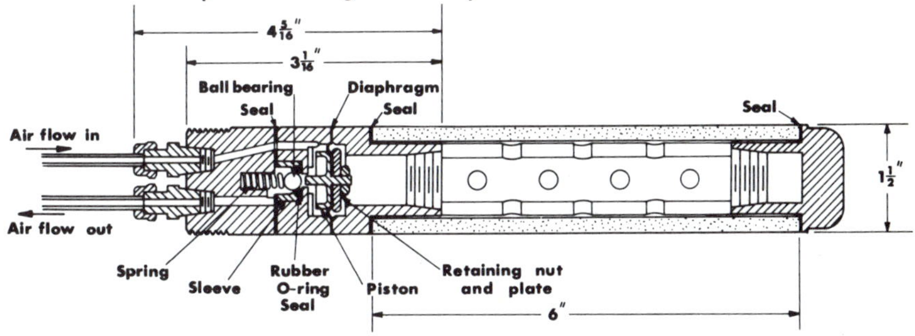

Figure 4 Pneumatic piezometer tip

OBSERVATIONS OF SLOPE FAILURE

Excavation at the foot of the test slope began at 2:20 p.m. on January 17, 1967. The floating dredge excavated the sand with a clamshell bucket to form a trench approximately 120 feet long parallel to the shoreline. During the experiment, the design depth of excavation was only exceeded by 3 feet.

The dredge initially deepened the foot of the slope with little measurable effect on the sand further up the slope. Short periods of bubbling in local areas at and immediately upslope of the dredge indicated that surface ravelling and slope readjustments were occurring. The first slippage of consequence occurred at 6 p.m. when bubbles appeared at the water surface for a period of several minutes and the dredge operator reported that the toe level had risen. Measurements showed that the slope below elev. -20 feet was standing at an angle of 34 degrees and the slope at higher levels had been relatively unaffected

Figure 5 Instrumentation section prior to test

Figure 6 Failure scarp and silt scum on river immediately after failure

by the excavation. The inclinometer casing was deflected at the top, but the lateral deformation did not extend more than 1 to 2 feet below the slope surface.

A second minor slope movement occurred shortly before 9 p.m. Water turbulence and bubbling were seen as far upslope as the uppermost metal pipe. The lower part of the slope had an overall slope angle slightly more than 30 degrees and slope inclinations of up to 33 degrees were measured between adjacent pipes. The inclinometer casing was deflected to a depth of about 3 feet below the slope surface. Two plastic floats and one piezometer tip had been undermined by the excavations at this stage. The other two piezometers had registered no change in pressure.

Dredging was halted for the day at 9 p.m. The slope movements which had occurred during this first day were consistent with surface wave movements of sand (i.e. ravelling) that can be expected when dredging temporarily oversteepens the lower part of the slope. Observations were continued throughout the night but there were no significant changes.

The dredging was resumed at 8:37 a.m. the next morning, and failure of the slope occurred about one hour later. At the start of the failure, violent agitation and bubbling were observed in the river near the shoreline. A thick scum of brown aerated silt formed on the water surface. Dredging was stopped immediately. Within about 5 minutes, at 9:55 a.m., the subaqueous movements had progressed back up the slope to the waterline, and the slope above the water started to break away in thin cupped slices as the submerged sand below it was flowing away. The near-vertical scarp gradually enlarged itself back into the slope as the failure progressed. It eventually stopped about 4 feet in front of the top edge of the slope (at the centre) and formed a concave scarp extending to a width of about 150 feet at the waterline (elev. +6 feet). The main slide activity occurred over a period of about 20 minutes, although minor sloughing continued until approximately 10:30 a.m. The west side of the scarp, and the thick scum of silt on the river, is shown in Fig. 6.

The observed failure of the test section was identical in its characteristics with observations made by the dredging crew when one of the earlier failures was witnessed. The final shape of the slope was similar to all the previous slope failures.

The slope conditions immediately before failure are shown on Fig. 7(a) and after failure are shown on Fig. 7(b). The ground level at the foot of the slope had risen by about 18 feet. The boundary between the sand fill bulldozed over the side of the slope and the underlying hydraulically placed sand fill is plotted on the two drawings. It probably represents the approximate boundary between very loose sand and sand of higher relative density (see SPT blow count profile, Fig. 3). The final slope of the ground after failure is roughly parallel to this boundary and 7 feet above it, as shown on Fig. 7(b).

The three plastic floats from position 'A' and the upper plastic float from position 'B' came to the surface during the slope failure. Float A-1 was located just above the bulldozed/hydraulic sand fill boundary, thus showing that liquefaction had occurred to at least this depth. All the other plastic floats remained buried and were recovered by water jetting. The estimated position of these floats after failure are given

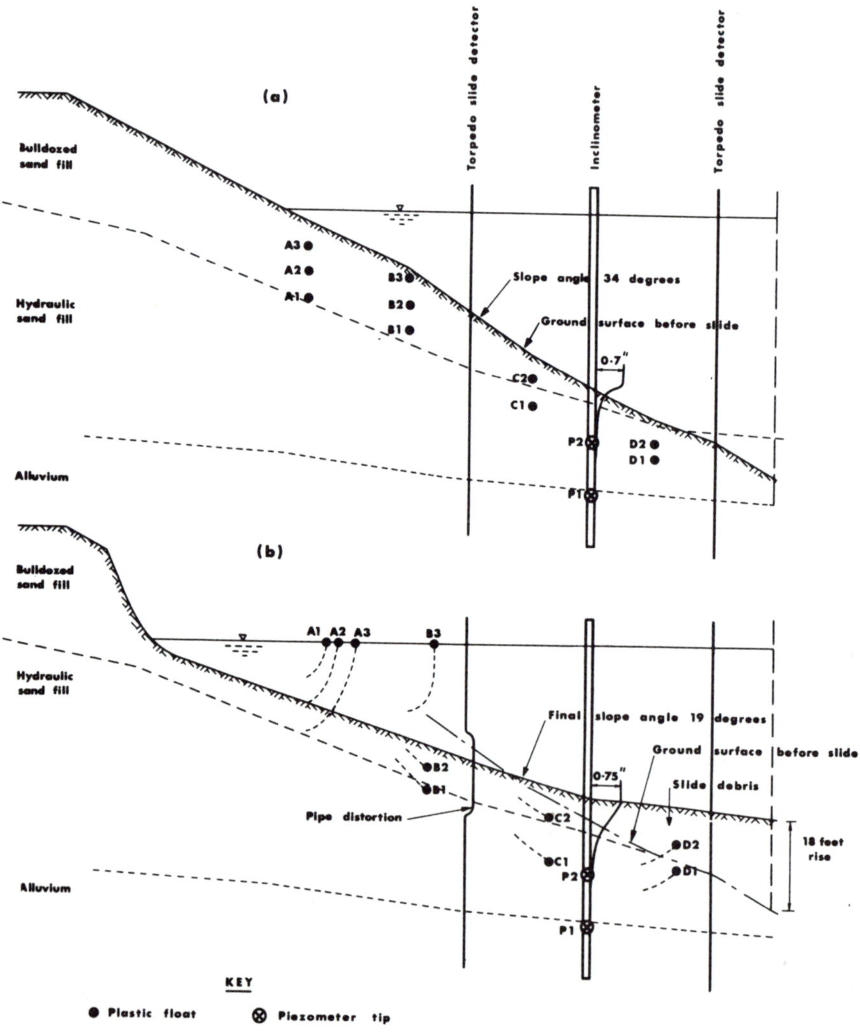

Figure 7 Slope conditions (a) before failure (b) after failure

on Fig. 7(b) with broken lines showing the movement from the original position. However, these estimates are considered to be unreliable because it is extremely difficult to accurately determine the relative movements of floats embedded in the slope 30 to 60 feet below the water surface using the methods employed by the contractor. There is also an obvious anomaly in the final position of float D2, which is shown to be embedded in the slide debris. It is very improbable that any float could come out of the slope and not reach the water surface, and the evidence is contrary to the information obtained from the adjacent piezometers and inclinometer. For these reasons, the estimated movements of the embedded floats can be ignored.

The inclinometer casing showed virtually no change in lateral deflection during the failure, despite an increase in the ground level of 5 feet at this location. It shows that the flow of sand exerted very little pressure on the casing. The maximum deflection of only $\frac{3}{4}$ inch had occurred on the previous day. Ground movement was no deeper than a few feet below the slope surface prior to failure, and probably terminated at or around the bulldozed/hydraulic sand fill boundary (Fig. 7(b)). There was no movement at the fill/alluvium boundary.

The two piezometers did not register any change in pore-water pressure during the failure. If the sand at this locality did not liquefy, any deep-seated movement at the sand fill/alluvium boundary would have been observed by the inclinometer. Since this did not occur, it is concluded that the weak silt alluvium played no part in the failure. The thick scum of silt seen on the river after failure is thought to be caused by entrapped air within the bulldozed fill which was released during liquefaction and carried out the silt fraction with it.

Finally, the uppermost torpedo slide detector was distorted at two different levels (Fig. 7(b)). The distortion at the top of the tube, which prevented the torpedo from being lowered any further down the tube, occurred at a level corresponding to the approximate surface of the fill slope before failure. The lower distortion of the tube, which prevented the torpedo at the bottom of the tube from being raised any higher, occurred near the bulldozed/hydraulic sand fill boundary.

The lower torpedo slide detector tube did not distort by a sufficient amount to block the movement of the torpedoes. The bubbler tube slide detectors did not operate satisfactorily during the experiment.

A hydrographic survey of the site after the various slope failures indicated that the sand had flowed out through a narrow opening near the mid-height of the slope. Because of the small scale, the survey did not define this effect as clearly at the test section as at the adjoining slope failure, where the fan-shaped cone at the bottom of the slope and the concave-shaped upper scarp are clearly shown (Fig.8).

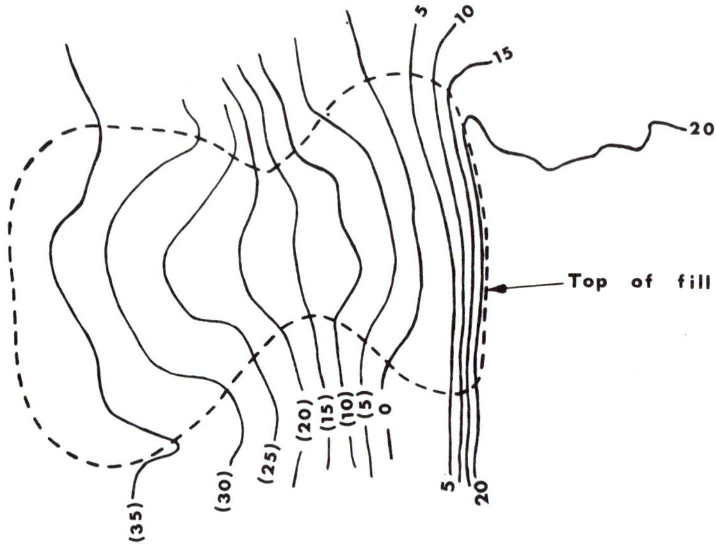

Contours in feet, negative values bracketed

Survey 10 weeks after failure

River level approx. elev. + 6 feet

Figure 8 Ground contours at slope failure 3

To sum up the main points obtained from the field observations:

(i) failure of the slope was due to liquefaction of the sand fill and did not involve the weak alluvium beneath the fill.

(ii) liquefaction was initiated at a position near the mid-height of the slope, causing the sand to flow out through a neck and fan over the slope below.

(iii) movements of sand within the slope probably occurred to a depth at or near the boundary between the very loose bulldozed sand fill and the underlying hydraulic sand fill.

LIQUEFACTION ANALYSIS

A simple form of stability analysis has been performed, using the observed data as the model (Fig. 9). It is assumed that a shear failure occurs along a plane surface BC downslope from a liquefaction face AB; in the present case, the plane BC is given by the boundary separating the bulldozed and hydraulic fill. This gives a wedge of soil, roughly triangular in shape, which participates in the slope movement.

A static wedge analysis is then carried out to compute the factor of safety of the wedge with the liquefaction force acting on it. By taking a series of sections, Fig. 9b, a graph can be obtained of the factors of safety at several points in the slope, Fig. 9c. It is of particular interest that the minimum factor of safety of 1.04 was computed at section X, since observations showed that liquefaction probably was initiated at this part of the slope. The low factors of safety near the top of the slope result from the potentially higher liquefaction pressures that can be generated in partly submerged slopes. In practice, it seems unlikely that liquefaction would occur in these regions except under unusual circumstances such as earthquakes or rapid drawdown.

The computed factors of safety against failure by liquefaction forces at the start of the test are also plotted on Fig. 9c. At the critical section, the original factor of safety is 1.31.

Analyses have also been made on the section with the assumption that

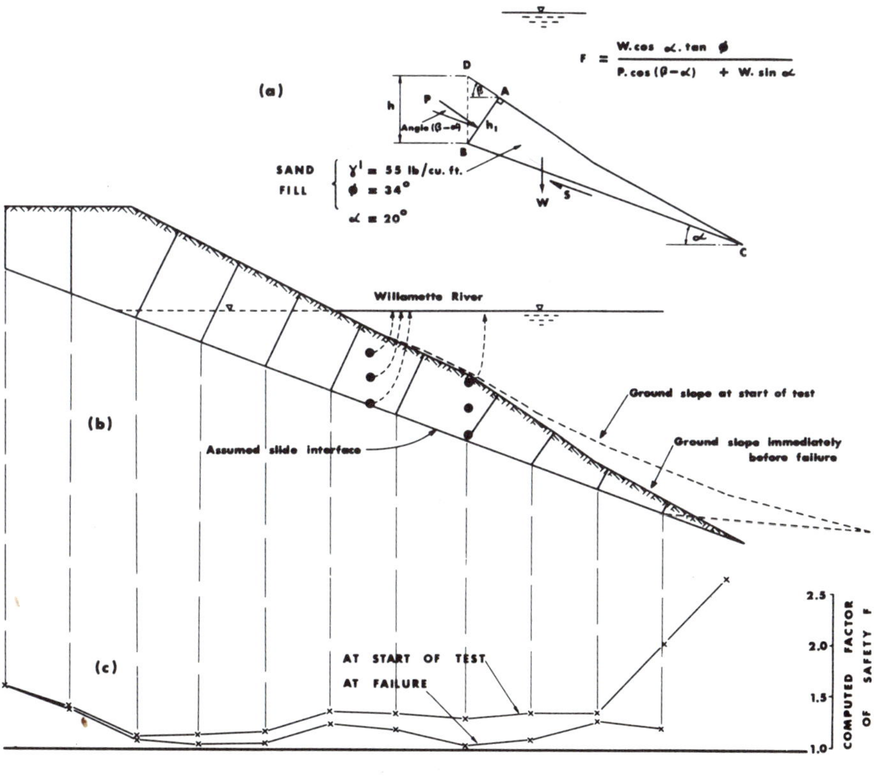

Figure 9 Wedge analysis of slope (inclined liquefaction face)

liquefaction occurs behind a vertical face BD instead of behind a face normal to the outer slope. This analysis is less acceptable mathematically, but it is interesting to note that the computed factors of safety are almost identical. This suggests that the results of the analyses are not particularly sensitive to the assumed inclination of the liquefaction face. The statically determinate liquefaction face, which would allow the three forces acting on the wedge to pass through a common point, would be inclined in a direction somewhere between the two cases considered.

Finally, an approximate theoretical analysis has been made for a straight-side wedge section subjected to liquefaction pressure upslope (see Appendix A). For a fully submerged slope, this analysis gives the factor of safety F as

$$F = \frac{\tan \emptyset}{\tan \beta} \qquad \ldots\ldots. (2)$$

where β = outer slope angle
$\emptyset$ = friction angle of sand within the slope.

There are two interesting aspects to this result. First, the liquefaction analysis is not dependent on the height of the slope or on the inclination of the shear interface below the plane of liquefaction. Second, equation (2) is identical to the result of the conventional static analysis of a soil element within an infinite slope. The latter analysis studies the statics of a shear interface parallel to the outer slope and ignores the contribution to stability which must be provided by the passive resistance to break-out in the lower part of the slope. Assuming that the soil conditions are uniform, the more critical condition must be near the surface where such 'end effects' are small, thus accounting for the observed tendency for sand slopes to readjust by shallow wave movements when they are undercut.

The liquefaction analysis, like the infinite slope analysis for static conditions, indicates that stability becomes critical as the outer slope angle β approaches the value of the friction angle $\emptyset$. However, the extent of a liquefaction-type failure (i.e. its depth of penetration into the slope) once this critical condition has been reached depends only on the ability of the sand to generate transient liquefaction pore-water pressures. This in turn depends on the relative density of the sand and the external triggering mechanism that can induce liquefaction pore-water pressures.

The mechanism that triggers off liquefaction at a point in the slope is not fully understood. Two possibilities are suggested. First, a shallow wave movement of sand going down the slope may cause a suction action on the underlying sand. If sand is drawn out by the suction, liquefaction pore-water pressures could be induced within the slope below. A second explanation is that steepening the lower part of the slope reduces the lateral support stress. The resultant lateral strain is sufficient to generate liquefaction presssures in very loose sand. Once the liquefaction pressure is generated, by whatever triggering mechanism, the wedge of sand below the point of liquefaction is pushed downslope, allowing the upslope sand to flow out through the liquefaction zone.

This type of failure probably can occur in any sand deposit if the sand near the outer edge of the slope is loose enough to be susceptible to liquefaction. Although the bulldozed sand/hydraulic fill sand boundary has been cited extensively in the present analyses, it has significance only as the approximate demarcation between very loose and denser sand at the test section. At least two of the other failures at Terminal 2 occurred in areas where no fill had been bulldozed over the edge of the slope, but the SPT blow counts were very low.

Flow slides in the banks of the Mississippi River appear to have similar failure characteristics (Banks & Strohm, 1965). The failures usually occur in point bar deposits, where the river scours and oversteepens the lower part of the banks.

REMEDIAL MEASURES

The remedial measure chosen for Terminal 2 was to densify the sand by driving timber piles into the slopes using a McKiernan-Terry 11B3 underwater hammer. Piles were driven 30 to 40 feet into the sand on a grid pattern with centre to centre spacing of 4 to 6 feet. Approximately 3000 piles were driven to stabilize the sand slope in front of the bulkhead, and 2000 behind the bulkhead. The apparent improvement in density, as indicated from SPT borings before and after densification, was quite erratic, due partly to the wide spacing and limited number of the comparative boreholes. A statistical analysis of the SPT results indicates an average improvement of about 10 percentage points in relative density (Schroeder & Byington, 1972).

Other remedial measures which were considered but rejected on grounds of economy, reliability, or design objections, included excavation of loose sand and replacement with sandy gravel, flattening the design slope, vibroflotation, explosives, chemical grout, wellpoints, tamping, and a pinch pile wall. Underwater vibration methods have since been used successfully by others.

SUMMARY AND CONCLUSIONS

1. The failure of the sand fill slope occurred in two distinct stages. In the first stage of the test, the lower part of the slope responded to the dredging operation by surface ravelling, the slope angle being maintained at, or slightly below, the angle of repose of 34 degrees. As excavation proceeded, this steeper slope angle extended further and further upslope until a stage was reached when a liquefaction pore-water pressure was induced somewhere near the mid-height of the slope. It is then postulated that the rise in pore-water pressure was sufficient to cause a shear failure in the wedge of sand below it, thus allowing the liquefied sand to flow out of the slope into the river. The backward extent of the slope failure was limited by the presence of compacted sand at the top of the slope.

2. This explanation of the failure mechanism is consistent with the measurements of the field instrumentation. It is doubtful whether such a dynamic mechanism could have been revealed by any other means. The key factor in the analysis is that the instruments defined the approximate position and extent of the liquefaction that initiated the flow slide.

3. The 'glory hole' technique of dressing the slope was responsible for initiating the flow slide, and the larger slope failures occurred where sand had been bulldozed over the top edge of the slope to make up the required section. It suggests that particular care should be taken on riverfront fill areas to prevent (i) loose sand, containing entrapped air, being placed over the edge of a fill without compaction (ii) temporarily oversteepening the sand fill slope during the channel dredging and slope dressing operations. Consequently, it seems preferable to initially overbuild the hydraulically-placed fill and then dress back to the design profile, thus removing the loose sand at the edge of the fill.

Acknowledgements
The authors wish to thank the Port of Portland (former owner: Commission of Public Docks) and the Director of Engineering: A.M. Eschbach, for permission to publish the results of this study. Consultants to the Commission were Cornell, Howland, Hayes & Merryfield (Corvallis, Oregon) who were assisted by Shannon & Wilson (Portland, Oregon) on soil mechanics studies for the project.

References

Banks, D.C. & Strohm, W.E. (1965) Methods of preventing flow slides. U.S. Army, Corps of Engineers Report 12-16, Vicksburg, Mississippi.

Cornforth, D.H. (1973) Performance characteristics of the Slope Indicator inclinometer. Symposium on Field Instrumentation.

Schroeder, W.L. & Byington, M.L. (1972) Experiences with compaction of hydraulic fills. 10th Idaho Engineering Geology and Soils Engineering Symposium, Moscow, Idaho.

Wilson, S.D. & Hancock, C.W. (1965) Instrumentation for movements within rockfill dams. ASTM Special Technical Publication 392.

Appendix A Theoretical wedge analysis

Refer to Fig. 9(a) for terminology except that outer slope AC is linear and inclined at angle β to the horizontal. Liquefaction force P is applied normal to plane AB. The analysis is approximate because the ine of action of P does not precisely intersect the resultant of the other two forces W and S. However, as stated in the main text, slight variations in the inclination of plane AB have little practical effect on the computed factors of safety and thus the small error in the statics of the problem is not considered to be significant to the evaluation of the analysis.

Liquefaction force $P = \frac{1}{2}\gamma' h h_1$

$= \frac{1}{2}\gamma' h_1^2 \sec\beta$

Submerged weight of soil wedge $W = \frac{1}{2}\gamma' h_1^2 \cot(\beta - \alpha)$

Shear force $S = W\cos\alpha . \tan\phi$

$$\therefore \text{Factor of safety } F = \frac{S}{P.\cos(\beta-\alpha) + W.\sin\alpha}$$

Removing the common multiplier $\frac{1}{2}\gamma' h_1^2$, the equation becomes

$$F = \frac{\cot(\beta-\alpha)\cos\alpha . \tan\phi}{\sec\beta . \cos(\beta-\alpha) + \cot(\beta-\alpha).\sin\alpha}$$

$$= \frac{\cos\alpha . \tan\phi}{\sec\beta . \sin(\beta-\alpha) + \sin\alpha}$$

$$= \frac{\cos\alpha . \cos\beta . \tan\phi}{\sin\beta . \cos\alpha - \cos\beta . \sin\alpha + \sin\alpha . \cos\beta}$$

$$\therefore \quad F = \frac{\tan\phi}{\tan\beta}$$

FIELD LOADING TESTS AT CANVEY ISLAND

P. J. George — Dames & Moore

R. H. G. Parry — Cambridge University

SYNOPSIS

Field loading tests are being carried out at Canvey Island to provide information for the design of oil storage tanks. The site consists of about 8 m of soft clay overlying dense sand. Two circular embankments were constructed of 30 m diameter and with a planned height of 8.6 m. Sandwicks were placed under one bank only, to determine their value in accelerating pore pressure dissipation. Instrumentation in the soft clay underlying the embankments includes three separate methods of measuring settlements (the results of which are compared) together with inclinometer tubes and peizometer points.

The writers' experience regarding contract problems, usefulness of instrumentation and an evaluation of cost and benefit are also recorded.

INTRODUCTION

It is proposed to construct the Thames Oil Refinery on an undeveloped soft soil site, adjacent to the river Thames at Canvey Island, Essex. The refinery will have a total storage capacity of 1,255,000 m^3, for crude oil and petroleum product (120,000 barrels per day).

If local practice were to be followed, high expenditure would be incurred in piling to support the tanks. However, it should be possible to construct limited height ground supported tanks, providing reasonably long water pre-loading programmes are scheduled. Economic and production considerations put a limitation on the maximum period of preloading, and thus emphasised the need for confident foundation recommendations, if such a scheme is to be adopted.

A preliminary site investigation had shown that the upper 7

to 10 m of soil were soft, highly compressible and slow draining, but uniform in thickness and characteristics.

To study the feasibility of the scheme, a comprehensive site investigation was put in hand, which included the construction of two comparative embankment load tests.

This paper describes the application of field instrumentation to these trial fills to establish basic design criteria, and to design a full scale tank monitoring scheme. Commercially manufactured instrumentation systems were used throughout.

The embankments were constructed to study the effect of vertical drains as a method of ground treatment. Embankment No. 1 was constructed to a maximum height of 7.3 m without ground treatment. Embankment No. 2 was taken to 8.6 m when a slip occurred. The ground under this embankment was treated with 8.0 m long, 60 mm diameter sandwicks (Dastidar and Gupta 1969) installed at 1.5 m and 2.0 m spacing.

The embankments were constructed as circular frusta having base diameters of 30 m, and were separated by 40 m. Circular loading was adopted to simplify analysis (axisymmetric conditions) and simulate tank foundations. In addition a considerable saving in fill material was achieved over equivalent width loads adopting square or rectangular constructions.

If ground treatment were to be recommended as a result of the trials the possibility of using Kjellman paper drains was to be considered as an alternative to sandwicks.

In view of the proposed refinery production programme it was decided that the crude tanks (6 No. 70 m diameter by 22 m high) should be pile supported. The reduction in benefit resulting from this is discussed below under Cost Benefit Evaluation of Field Trials. The remaining tankage holding the refined products would consist of either 20 piled tanks, or 34 earth supported tanks, having capacities ranging from 10,000 to 50,000 m^3, depending on the results of the embankment load tests.

COST BENEFIT EVALUATION OF FIELD TRIALS

The following cost benefit analysis is based on current tank construction contractors' prices used on similar projects in Europe. The sums quoted, therefore, bear no relationship to the subject project, except insofar as they indicate the magnitude of benefit derived from the engineering studies under discussion.

Regarding steel erection costs, attention is drawn to the old rule: the taller the tank the cheaper it is. This tends to dwindle the benefits of constructing shorter earth supported tanks. However, the sums involved are small compared with the difference in foundation costs, as illustrated in Tables 1 and 2.

Table 1

TANK CONTENT	TOTAL STORAGE CAPACITY (m^3)	COSTS (£)			
		Tank Steel & Erection	Piled Foundation	Sandwicks & Tank Pad	Kjellman Paper Drains & Tank Pad
Crude Oil	500,000	1,500,000[1]	850,000	-	-
" "	"	1,800,000[2]	-	450,000	260,000[5]
Petroleum Products	725,000	2,400,000[3]	1,300,000	-	-
" "	"	2,700,000[4]	-	690,000	410,000[5]

Notes:
1. Six tanks 21.9 m high, on piles
2. Twenty tanks 12.2 m high, earth supported
3. Most economic use of product tankage assuming maximum allowable height of 21.9 m (approx. 20 No. tanks)
4. Thirty four No. tanks 12.2 m high, earth supported
5. The cost of sandwicks and Kjellman paper drains is based on a spacing of 1.0 m.

Table 2

SCHEME		COSTS (£.000)			
		Tank Steel & Erection	Piled Foundation	Earth Supported Foundation	Total Cost
A.	All tanks piled	3,900	2,150	-	6,150
B.	All tanks earth supported with sandwicks	4,500	-	1,140	5,640
C.	All earth supported no ground treatment	4,500	-	270	4,770
D.	Crude tanks piled/product tanks sandwicked	4,200	850	690	5,740
E.	Crude tanks piled/product tanks earth supported no ground treatment	4,200	850	170	5,220
F.	All tanks earth supported with Kjellman drains	4,500	-	670	5,170
G.	Crude tanks piled/product tanks Kjellman drained.	4,200	850	410	5,460

A range of saving from £1,380,000 to 410,000 can be shown by comparing the actual construction cost of various methods of treatment with the all-piled case (Scheme A). However, since the proposed production schedule requires the immediate use of the crude tanks, it was decided that they should be pile supported in any case. This reduces the savings accordingly, and in particular brings the maximum end of the range to £930,000.

Estimated costs of providing those savings are as follows:

1. Test embankment cost including engineering supervision, fill, plant hire, instrumentation and contract labour:

less salvagable instrumentation and fill from embankments £22,000

2. Installation of instrumentation at tank sites ...£15,000
(*7,000)

3. Supervision of installation and monitoring of tanks £33,000
(*9,000)

4. Estimated costs of more sophisticated engineering, field and laboratory studies associated with earth supported tank design, over that for pile design analysis £15,000

5. Possible re-levelling costs assuming all 34 tanks need to be re-levelled once (Estimated cost of re-levelling one 50 m diameter tank £5,500) £190,000
(*109,000)

6. Interest on capital investment in product tankage for six month non-productive period during water testing programme £130,000
(*88,000)

Total cost: £405,000
(*213,000)

Note: *Additional amount assuming crude tanks are also earth supported.

Therefore, assuming the more costly form of ground treatment is chosen (sandwicks) a nominal saving of £5,000 is possible. More likely, however, a Kjellman paper drain system would be adopted saving some £285,000; or in the no ground treatment alternative, £525,000. Similarly it can be estimated that a maximum saving of £762,000 could be achieved if production schedules allowed all tankage to be earth supported, without ground treatment.

Expense No. 6 above emphasizes the value of careful monitoring and pre-engineering, that is a saving in time during tank loading can realise a substantial saving in finance.

Further saving could be achieved if temporary P.V.C. tank bases were to be utilized during water testing. This possibility is at present being studied.

It should also be noted that for any set of tanks, the total volume of crude and product, if spilled, must be retained within fire walls, having a statutory minimum height. Therefore land costs associated with the tank farm are virtually constant for a given refinery capacity at a given location regardless of the foundation solution.

SITE CONDITIONS

The site of the proposed refinery is situated in relatively flat farmland, protected from flooding by dykes along its southern boundary. The surface elevations at the site range from 0.5 m to 2 m O.D. The sea protection dyke at present has a crest elevation of 6 to 7 m O.D., but is to be heightened to in excess of 7 m, as part of the Thames flood protection scheme. The site is drained by a large number of natural and excavated channels which discharge via tide gates into the Thames.

The test site was chosen on uniform soft soil strata of typical thickness and close to a plentiful and inexpensive source of fill. The subsurface conditions of the test site were investigated by sampling soil from test pits, boreholes (including one 410 mm diameter borehole for 254 mm piston sampling);Delft continuous sampling; Dutch Deep Sounding; and in-situ vane testing. Constant head field permeability tests (Gibson 1963, 1966 and 1970 and Wilkinson 1968) were also performed at the site.

In general, Table 3 summarizes the subsurface conditions.

Fig 1 shows the soil state below embankment No. 1

Table 3

SOIL TYPE	CONSISTENCY	AVERAGE DEPTH TO TOP OF LAYER (m.)	APPROXIMATE LAYER THICKNESS	
			Tank Site General (m.)	Embankment Site Particular (m.)
Brown-grey silty clay	"Crust" soft to stiff, slightly compressible	0	1.0 - 2.5	1.5 - 2.0
Grey silty clay, organic matter and peat	Soft, very compressible	2	5.0 - 7.0	6.5
Silty fine sand, and clayey and sandy silt	Medium dense, stiff; slightly to moderately compressible	8	7.0 -10.0	7.5 - 9.0
Silty and sandy clay; clayey silt	Medium stiff; slightly compressible	16	1.0	0.0 - 0.5
Sandy gravel (Thames Ballast)	Dense, incompressible	17	8.0 -10.0	6.0 - 9.0
London Clay	Very stiff, very slightly compressible	26		Greater than 20

FIELD AND LABORATORY TESTING

The following tests were performed on the soft soil strata:

1. Consolidation tests on samples 75 mm diameter, 19 mm thick.

2. 254 mm diameter consolidometer tests.

3. Consolidated undrained triaxial compression tests with pore pressure measurement on samples, 65 mm diameter, 152 mm high.

4. In-situ constant head permeability tests.

Computed permeabilities from the 254 mm laboratory consolidation tests are shown to be in the range 0.5 to 24 x 10^{-8} cm per second for vertical permeability and 1.5 to 36 x 10^{-8} cm per second for horizontal permeability. Field permeability tests gave a range of 2.0 to 4.2 x 10^{-8} cm per second.

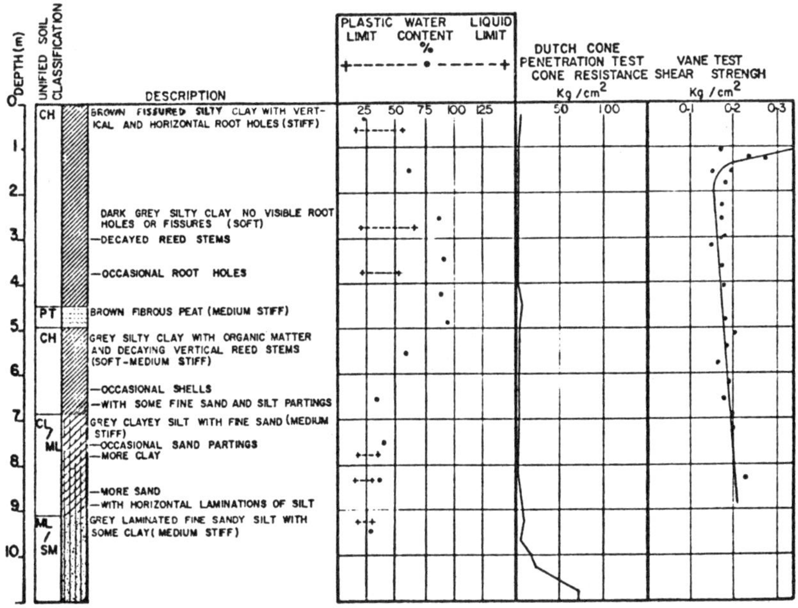

Fig 1 Subsurface conditions: Embankment No 1

SITE PREPARATION AND FILLING

The proposal to construct the embankment load test was approved by mid-February 1972. Although cost estimates for instrumentation and filling had already been obtained, installation did not proceed until late March. Instrument installation was completed at embankment site No. 1 on 22nd April, whence filling proceeded.

The sandwick treatment of embankment site No. 2 was carried out between 17th April and 1st May and the remaining instrumentation was completed by 4th May ready to begin filling on 8th May. This indicates the minimum time for preparing such a programme.

INSTRUMENTATION - TEST EMBANKMENTS

Most of the instrumentation under each embankment was positioned in order to compare directly an instrument associated with one embankment with its duplicate below the other. Table 4 lists instruments installed beneath each embankment, and an average cost in the ground for one instrument or measurement point. The cost includes instruments and materials, and subcontractor charges and technical supervision involved in installation.

Table 4

INSTRUMENT	NUMBER OF INSTRUMENTS UNDER EMBANKMENT			COST OF INSTRUMENT OR MEASUREMENT POINT (£)
	No. 1	No. 2	Total	
Piezometer	12	9	21	110
Hydraulic settlement Gauge Points	4	3	7	120
Borehole Settlement Gauge Points	11	-	11	30
Mole Settlement Access Tubes	3	3	6	30/5m of profile (15)
Inclinometer Access Tubes	4	3	7	180/10m inclinometer tube (60)*
Heave Markers	18	18	36	0.75

Note: *reducing to ( ) in full monitoring programme, e.g., for mole settlement access tube, £30 per 5m of profile reducing to £15 in full monitoring programme.

Fig 2 below shows the location of each of the instruments.

The individual instruments are indicated as follows:

- initial figure indicates embankment number
- letters indicate type of instrument as follows:
  P - piezometer, I - inclinometer, H - hydraulic settlement gauge, B - borehole settlement gauge
- last figure indicates number of particular instrument or measurement point, e.g. 1P7 indicates piezometer No. 7 in embankment No. 1.

Settlement measurement

Three independent systems of measuring settlement beneath the embankment were used. The hydraulic settlement gauges and the mole settlement profiling system were installed so that a comparison could be made between two methods of remote settlement measurement, which might be used beneath tanks. The borehole settlement gauge was included to give settlement of individual layers of substrata. A comparative plot of settlement versus time below the centre of the tanks for the three methods of measurement is shown in Fig 4. Sensibly identical results were given by the hydraulic gauge and the surface borehole settlement gauge; the mole gave somewhat higher settlements.

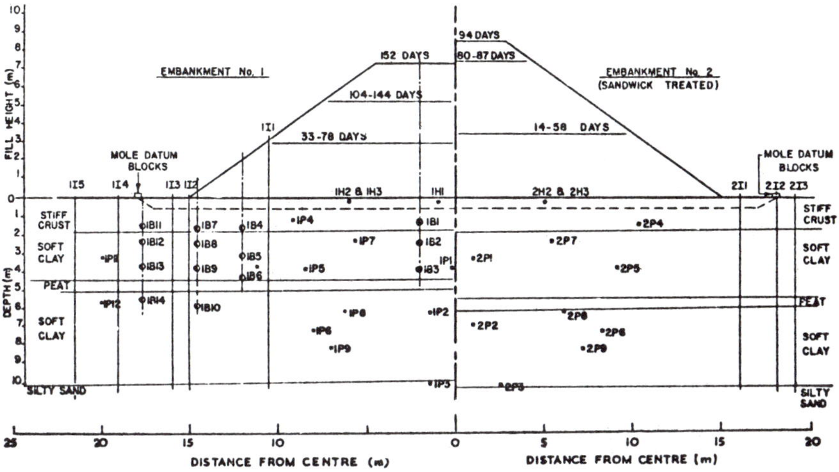

Fig 2 Location of instruments below each embankment

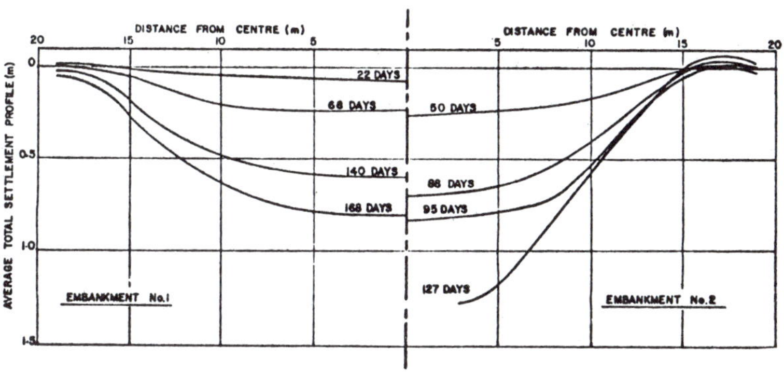

Fig 3 Typical settlement profiles derived from mole

Hydraulic Settlement Gauges A modified version of the Building Research Station pattern overflow type was used. The modification was necessary since the gauge house installation was located at a level above the cell weir, making it necessary to apply a partial vacuum to the ends of the stand pipe to raise the equilibrium level. Despite the agreement in Fig 4, this method proved generally to be unreliable insofar as readings were difficult to repeat and it appeared to be susceptible to operator factors. Moreover, operation was tedious and time-consuming.

Provision and installation costs associated with this system were high and even had the system functioned reliably it is doubtful that it would have been used for tank monitoring.

Mole Settlement Gauge The mole settlement gauge was potentially the most useful item of instrumentation included in the embankment load test, as it can give a complete profile of settlement immediately beneath a loaded area. This is particularly important when considering its application in storage tank loading programmes, where tank bottom shape is of critical importance: it is normally overstress of base plates, which heralds tank failure. Hitherto it has only been possible to measure settlement of tank shells, and isolated points below tanks using systems such as the hydraulic or mercury settlement gauges. Profiling of the base plates can normally be achieved only by levelling within an empty tank, or by using diver inspection during water testing.

The basic design of the mole is due to Bergdahl and Broms (1967) and a primitive version was manufactured by the writers for tank loading tests at Tilbury (Parry 1970).

The commercially manufactured system used in the present tests consists of a flexible access tube, which is placed in a trench below the loaded area, and a probe (the mole) which is inserted into the access tube. The instrument measures settlement at any number of points across the profile, relative to a concrete datum block. The probe contains a bladder connected to an electrical transducer by a nylon tube filled with liquid (water). The transducer is located in a recorder box which rests on the datum block. Change in negative pressure in the transducer is converted to an electrical signal, and is indicated on a direct reading meter as a difference in elevation between datum and probe.

Three flexible access tubes disposed at equal horizontal angles were placed under each embankment in the base of sand-filled trenches approximately 0.6 m deep. The tubes crossed at the centre and were concreted at both ends into 0.5 m square 0.3 m thick datum pads. It can be seen in Fig 4 that settlements recorded by the mole were larger than recorded by the other two measuring devices.

It was found that small fluctuations in temperature caused the system to give errors of up to 70 percent during the early stages of loading, when the settlements were less than 0.2 m. It was necessary to lay the tube out on the ground before insertion, exposing it to atmospheric conditions. It proved to be very sensitive to the sun's heat, and to reduce errors, a policy of night time operation was adopted. An obvious improvement would be to use a probe fluid with a smaller coefficient of expansion.

Mole readings could be obtained from all access tubes below

embankment No. 2 until the slip occurred. After the slip, one of the access tubes which passed through the slip zone kinked up 0.3 m above ground level in the heave zone between the toe of the bank and the datum block. From the other end it could only be penetrated 19 m to the centre of the bank. Another tube passing through the slip zone could be penetrated for a distance of 1 m from the heave zone end, and 20 m from the other end. The datum block in the heave zone in this case moved out radially by about 50 mm. The third access tube, which appeared not to pass through the slip zone, could be probed 19 m from either end to the centre of the bank, but not beyond this point. The centre settlement at this time was about 1.25 m.

Even so, the system has continued to give further information with respect to the continuous settlement of the embankment, while all other methods have failed. Fig 3 shows typical settlement profiles derived from use of the mole for both embankments, and can be compared with fill loads shown in Fig 2.

Borehole Settlement Gauge The system is described by Burland Moore and Smith (1972) and consists of magnetic rings set in a borehole at the levels where settlements are to be measured. The magnetic rings were set in short lengths of rigid PVC cylinder which were spring-loaded against the side of the borehole and surrounded a central PCV access tube. A probe containing reed switches was lowered down the access tube to determine the depth of each magnetic ring. An audible signal indicates reed switch actuation. The borehole was backfilled with cement/bentonite grout after placing the units.

This equipment was installed only below embankment No. 1 to obtain additional information regarding settlement of various strata: its design would obviously prohibit its use beneath tank foundations.

This system of settlement measurement was certainly the simplest and quickest to operate and potentially the most accurate of the three. However, the design of the system adopted was not considered ideal for the soft clay conditions.

Problems encountered were:

i) When settlements at the surface reached about 0.5 m, the large movements and downdrag of the fill caused the PVC access tube to distort, preventing probe entry. Thus the readings for the instruments shown in Fig 4 terminated before the fill had reached full height; and

ii) Certain of the sprung PVC magnetic ring units

appeared to jam against the probe access tube, and slip relative to the soft sides of the borehole.

These limitations might be overcome by using telescopic access tube and magnetic units with mechanical anchors, which penetrate further into the borehole wall.

Pore Water Pressure Measurement - Piezometers

All piezometers were of the type described by Wilkes (1970); basically a simple Casagrande piezometer tip designed to be pushed into soft and loose soils at the base of a borehole. A few of the piezometers (1P1, 1P4, 1P6 and 1P8) were placed in sand cells at the bottom of boreholes. All boreholes were sealed with bentonite/cement grout. The piezometers were each connected to a double line mercury manometer pressure measurement system. All piezometers were installed below the water table.

The in-situ constant head permeability tests referred to above were performed at three piezometers installed midway between the two embankments at depths of 3.4, 5.0 and 7.0 m.

It was found that the system required frequent de-airing, in particular during the early stages of loading, when piezometric levels were below the header tank level, which established at 3.32 m O.D. However, once the pore pressures had increased to a level corresponding to the header tank, little or no de-airing was necessary. Prior to this pressure being attained, de-airing was necessary after about every tenth reading.

A general rule regarding the need for de-airing was enforced such that: i) manometers were de-aired if air bubbles could be detected visually; and ii) piezometers were de-aired if the difference between the two manometer readings was in excess of 10 mm.

Fig 4 shows changes in pore pressure at 1P1, 1P2, 2P1, 2P2, and also height of fill, both plotted against time in days. Plots of excess pore pressure against fill height are shown in Fig 5.

The ground pressure is not a linear function of fill height, because the amount of fill placed per unit of height decreases. The ratio $du/d\sigma_1$ of excess pore pressure to increase in major principal stress (from elastic stress distribution) of these piezometer tips for fill heights of 3 m and 6 m is given below:

Piezometer	1P1		1P2		2P1		2P2	
Fill Height (m)	3	6	3	6	3	6	3	6
$du/d\sigma_1$	.60	.83	.46	.75	.62	1.0	.46	.73

It can be seen that the ratio $du/d\sigma_1$ increases with fill height as expected for a lightly overconsolidated clay.

The piezometer system was simple to operate but time consuming and appeared to give consistent and reliable data.

The large strains associated with the slip in embankment No. 2 broke the piezometer leads and put them out of action. The maximum settlement at this time was 1.0 m. Fig 5 shows that no clear indication of the imminent slip in embankment No. 2 was recorded by the piezometer system.

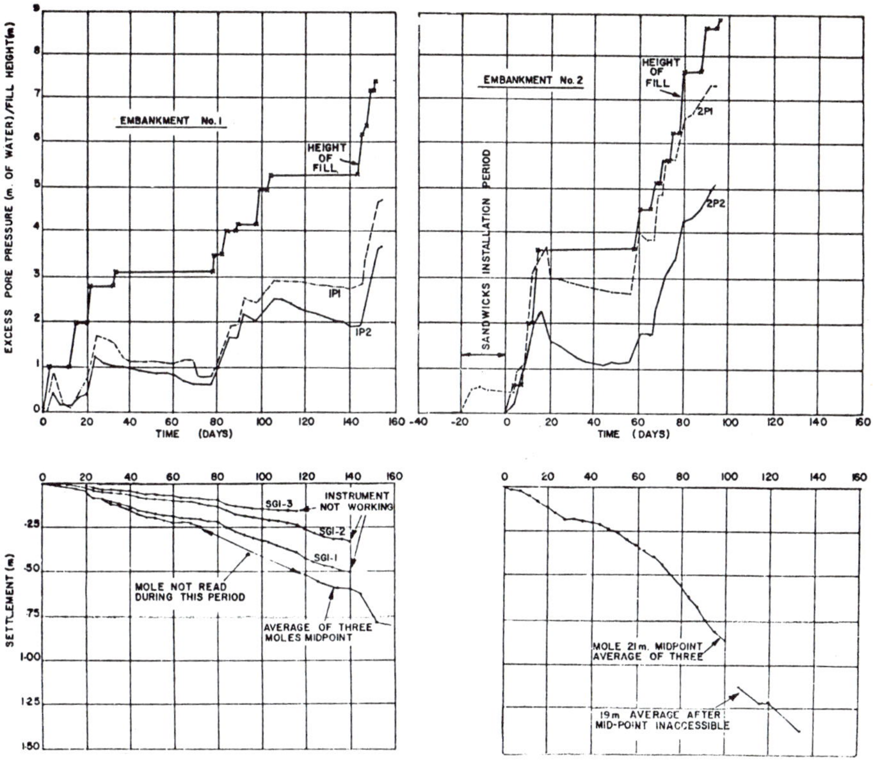

Fig 4 Fill Height, Settlement and Excess Pore Pressure versus Time

Lateral Earth Deformation Measurement - Inclinometer

A pendulum type inclinometer with digital direct reading unit was employed to measure the lateral deformation of substrata. The torpedo was 0.7 m in overall length and ran on four keyways inside 40 mm internal diameter aluminium access tube, grouted into 0.15 m diameter boreholes. The results from this instrument were considered adequate and its operation straightforward, but there were periods when the instrument was out of operation due to faults, arising in some measure from operator inexperience.

Following the slip in embankment No. 2 the inclinometer torpedo became jammed inside tube 2I1 and it was necessary to pull out that access tube to free the instrument. The tube had sheared at a depth of 2.5 m below ground; its distorted shape compared closely to that plotted by the instrument. Fig 6 shows a plot of lateral movement of 2I1 versus depth.

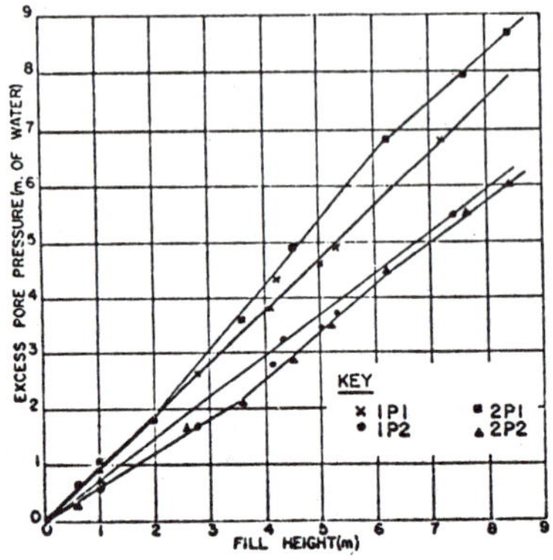

Fig 5 Excess pore pressure versus fill height

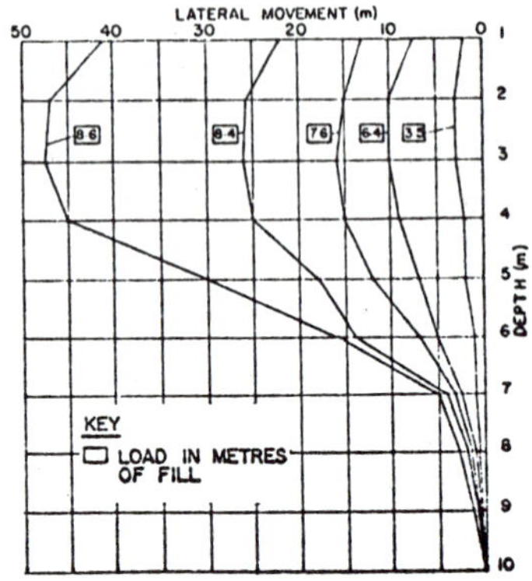

Fig 6 Lateral movements of inclinometer 2I1

Instrument Reading and Reduction Times

Table 5 indicates the time required by a technician for data collection and reduction for each instrument. Reduction times are based on hand methods: computerization would reduce them by 70 percent.

Table 5

INSTRUMENT	DATA COLLECTION TIME (Min per instrument)	REDUCTION TIME (Min per instrument)
Inclinometer (per 10 m length, one reading each metre)	20	15
Moles (40 m length, reading each 2 m)	30	30
Hydraulic Settlement gauge (one point)	30	5
Borehole settlement gauge (4 points per borehole)	5	5
Tank Settlement lugs (16 per tank)	30	20
Piezometer	1 (15*)	5

Note: * Time to de-air one piezometer

INSTRUMENTA ION - TANK MONITORING PROGRAMME

It is proposed that the earth supported tanks will be monitored using the following instrumentation scheme:

a) All tanks to have shell settlement lugs welded at eight points around the circumference;

b) 50 percent of tanks to be instrumented with four inclinometer access tubes at the tank quarter points and with two orthogonally placed mole access tubes passing below the tank pad; and

c) At least two tanks to be instrumented with eight piezometers.

CONCLUSIONS

The commercial instrumentation used for the trial loading tests has, in general, proved satisfactory in producing results of the required accuracy. A particular advantage of the embankment test has been the opportunity to experiment with various systems, particularly settlement measurement, to enable suitable instruments and location of instruments, to be chosen for monitoring tanks. As a result of these trials some instruments originally intended to be used for this purpose have been found to be either unnecessary or unsuitable and the trials have unquestionably led to a saving of time and money.

ACKNOWLEDGEMENTS

The study reported here was commissioned by Occidental Refineries Ltd. The direction of Mr R. H. Stubbings of that firm is especially acknowledged.

REFERENCE LIST

Bergdahl, U. and Broms, B.B. (1967) New methods of measuring in-situ settlements. Journal of S.M. A.S.C.E.

Burland, J.B., Moore, J.F.A. and Smith, P.D.K. (1972) A simple and precise borehole extensometer. Geotechnique Vol 22 No 1.

Dastidar, A.S., Gupta, S. (1969) Application of sandwicks in housing project. Seventh International Conference of the International Society of Soil Mechanics and Foundation Engineering, Mexico.

Gibson, R.E. (1963) An analysis of system flexibility and its effect on time-lag in pore water pressure measurement. Vol 13 No 1.

Gibson, R.E. (1966) A note on the constant head tests to measure soil permeability in-situ. Geotechnique Vol 16 No 3.

Gibson, R.E. (1970) An extension to the theory of the constant head permeability test. Geotechnique Vol 20 No 2.

Parry, R.H.G. (1970) Discussion: In-situ Investigation in Soils and Rock Conference Proceedings. British Geotechnical Society.

Wilkes, P.F. (1970) A note on the installation of piezometers in small diameter boreholes. Geotechnique Vol 20 No 3.

Wilkinson, W.B. (1968) Constant head in-situ permeability tests in clay strata. Geotechnique Vol 18 No 2.

PRINCIPLES AND PERFORMANCE OF TWO INCLINOMETERS FOR MEASURING HORIZONTAL GROUND MOVEMENTS

G E Green

Lecturer in Soil Mechanics, Imperial College, London.

INTRODUCTION

Field observations of both the magnitude and rate of subsurface ground movements are required in connection with, for example, natural slope stability, earth and rockfill dam construction and deep excavations (Wilson, 1967, 1969, 1970). Precise surface surveys (Penman, 1971; Gould & Dunnicliff, 1971) will often be necessary but insufficient. Vertical movements may be determined by a variety of methods such as settlement tubes with cross-arms or electrically detectable metal plates or magnet rings (USBR, 1968; Penman, 1971; Burland *et al*, 1972) or by a water level (Bishop & Vaughan, 1962) or by rods or tensioned wires (Gould & Dunnicliff, 1971). None of these methods readily provide a continuous vertical profile of horizontal movements.

An inclinometer measures tilt angles, small or large, usually by means of a gravity-operated pendulum which generates a transmittable electrical signal. They range from instruments designed to operate in an unlined oil-well hole with limited accuracy to very sensitive earthtide meters designed to resolve angles as small as 1×10^{-8} radians (Simon *et al*, 1968). For many civil engineering purposes a range of about $\pm 20^{\circ}$ and a sensitivity of $0.01 - 0.05^{\circ}$ is suitable. Such an inclinometer consists of a torpedo containing a pendulum and fitted with guide wheels or skids, which is lowered on an electrical cable down a grooved casing to control orientation (Ward, 1956; Wilson, 1959; Bromwell *et al*, 1971). The inclination with respect to gravity is measured at incremental depths and the vertical profile obtained by integration. Sets of readings over a period of time enable both the magnitude and rate of horizontal movement to be calculated.

DESIGN, INSTALLATION AND OPERATION FACTORS AFFECTING ACCURACY

The overall accuracy and reliability of the displacement profile is dependent on the angular measuring device used, the readout system, the quality of the casing and its installation, observational techniques and method of data analysis.

Angular measuring system

The pendulum in the torpedo which measures its tilt with respect to gravity may incorporate a rotary potentiometer, an LVDT, a bonded resistance strain gauge, a vibrating wire gauge or an accelerometer. A low-hysteresis elastic suspension is preferable since bearings may stick due to dirt or wear. If a rotary potentiometer is used the moving contact should be disengaged until the pendulum reaches equilibrium to eliminate friction, and then engaged. Brinelling of this contact can occur. Bonded resistance strain gauges must be properly temperature-compensated and drift-free. Magnetic or oil damping is advantageous to stabilise the pendulum rapidly. It is desirable to lock the pendulum during storage and transit to prevent overstress or shock damage. The tilt sensing device must be housed in a robust torpedo, sealed to prevent the ingress of water, particularly at the cable entry gland. A pair of fixed wheels (or skids) on the torpedo define a gauge length and a second opposed pair of sprung wheels (or skids) locate the torpedo in the casing. Clearances should be as large as possible and the torpedo should be streamlined so that it does not easily jam. The torpedo is connected to the surface by an electrical cable which should be flexible, waterproof, robust and preferably be permanently and accurately graduated. It must not stretch under load or with time and the outer waterproof sheath must not creep over the inner core or errors will result. Alternatively the cable may be unmarked and a steel tape or separate graduated inextensible cable can be used, although this is much less convenient. An alternative is a metering reel with an unmarked cable but this appears to possess no direct check that the depth setting is correct and the cable must still be inextensible and the sheath creep-free.

The readout unit may be a simple direct reading voltmeter, a manual null-balance bridge circuit, a digital voltmeter or an automatic null-balance bridge circuit with digital display together with an integrating facility to sum incremental displacements. Digital units may in addition allow recording on magnetic or paper tape suitable for direct computer input.

Casing

Metal or plastic tube with four grooves, is coupled together on site and backfilled in a bore-hole. The tube must resist external earth pressures and yet flex as the ground deforms. The grooves must be smooth and free from twist and the casing should be corrosion resistant. Extruded aluminium alloy tube (Fig 1 b,d) can be made with a random twist consistently better than 1°/10 ft length and may be coated in epoxy paint to resist alkaline attack if cement/bentonite grouts are used. Plastic tube (Fig 1 a) is corrosion resistant but cannot be consistently extruded without excessive twist. If the grooves are machined rather than extruded (Fig 1 c) this is better but warping may still occur during storage. Unfortunately twists are not generally random so that measurement and selection from a batch of tube is expensive. A survey can be made after installation and the twist allowed for in the data analysis although this is inconvenient.

Extruded metal tube is readily joined by riveted telescopic couplings of similar shaped cross-section, with little loss in alignment. Plastic tube may be joined by cemented plastic couplings or by crimped metal sleeves. Experience shows that joint alignment is less reliable with plastic tube. Uncemented joints should be sealed to prevent the backfill entering the casing. Telescopic metal couplings allow large vertical movements (e.g. in dams) to occur as the rivets can shear whilst the torpedo will track across the couplings.

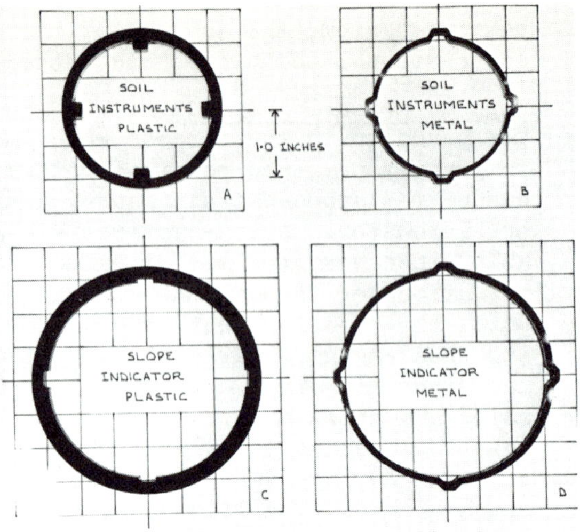

Fig 1 Inclinometer casings

Installation

The tube is assembled and lowered into the borehole. It should be sealed and filled with clean water. The torpedo should be inserted to check tracking before the tube is backfilled. Poor backfill will result in initial scatter in the data as it settles. Grouting via a tremie pipe is the most positive method. Present experience suggests that where long-term measurements with epoxy-coated metal casing are required in aggressive conditions it should in addition be coated with a thick bitumastic or rubber paint. Adequate protection from damage by construction traffic and vandals is essential.

Observations

The same instrument should be used during an entire project preferably by the same observer who also analyses the data. It must be used, cared for, cleaned and stored properly. A simple calibration frame is desirable for daily checks. It is essential that the operator is able to verify the reliability of the data before leaving the site. Ingress of moisture is usually indicated by drift or inexplicable readings.

The reading interval is a compromise between accuracy and excessive time and quantity of data. An interval equal to the wheel-base of the torpedo gives an exact mathematical analysis of the profile. Closer intervals will give greater accuracy. It is common practice to use a reading interval 1.5 to 2.0 times the wheel base and although errors are introduced they are small and usually acceptable. Errors can arise if the torpedo is not located at the correct depths in the tube. This

can be caused by stretching of the cable, inaccurate graduations or metering errors. It can be particularly serious where large vertical movements must be accounted for or where the casing is extended during dam construction. In all cases the intervals should be chosen so that readings with a wheel in or very near a joint are avoided.
Since the inclinometer will not normally read zero with the torpedo precisely vertical it is necessary to eliminate face errors by taking readings 180° apart. This enables the face error (a) to be eliminated and (b) to be determined for each pair of readings. Ordinarily the face error should be constant and this provides an important field check against faulty readings, poor techniques or transcribing errors.
Normally measurements are made on each pair of casing grooves, i.e. N-S and E-W. In some cases little or no movement will be anticipated in one direction if the casing is initially orientated correctly, e.g. in the plane of a diaphram wall. Readings in this direction indicating zero movement provide a valuable check on the instrument.

Data analysis

Having measured the inclination, θ, at a particular depth, the absolute horizontal displacement from the vertical, $\delta H \simeq L\theta$, where L = the reading interval, so that the total displacement from the vertical $\Sigma\delta H = L\,\Sigma\,\theta$. Thus the change in position is $L\,\Sigma\,\delta\theta$. The error in assuming that the horizontal movement is equal to the arc length $L\,\delta\theta$ increases rapidly if the casing is initially non-vertical by an angle α° and overestimates the deformation (Table 1). For inclined casings (say $\alpha > 20°$) a more refined analysis is required to exclude these errors.

α degrees	Percent error in displacement		
	θ = 1°	θ = 2°	θ = 5°
0	0.005	0.020	0.13
5	0.46	0.56	0.89
10	1.7	1.9	2.5
20	6.8	7.1	8.3
30	16.3	16.7	18.6

Table 1 Displacement errors with non-vertical casing

If the instrument indicates angles these will be recorded manually and/or on tape. If displacements are indicated and integrated by the instrument these will similarly be recorded manually and/or on tape. As noted earlier face errors must be eliminated by averaging two sets of readings 180° apart. The face error should also be calculated in the field to check on the reliability of the data. Although automatic integration of displacements appears attractive it may be difficult to check the face error at each depth increment since the calculations are more tedious. Unless done at the time the readings are taken much time and effort may subsequently be wasted.

The field data should be scrutinized and some judgement and experience is needed to screen out errors while useful information is not discarded. The screened data may then be reduced manually, which is tedious on a large job, or by computer.

PERFORMANCE TESTS ON TWO INCLINOMETERS

A series of controlled laboratory tests were conducted on two inclinometers systems (Green, 1969);
(a) The 200 B Series Wilson Slope Indicator. Two instruments were tested, SINCO 200 B/59 owned by Imperial College and SINCO 200 B/37 loaned by Nuttall Geotechnical Services.
(b) The Mark 1 Soil Instruments Inclinometer. Two instruments were tested SIL MK 1/A with nylon skids and SIL MK 1/B with stainless steel skids, both loaned by Soil Instruments Ltd. The readout unit for SIL MK 1/A was also used with the SIL MK 1/B torpedo and cable as a second readout unit was unavailable. Calibrations were performed to study their stability with time and temperature and full-scale tests were carried out with 80 ft of casing set up in a stairwell. For each system the casing was surveyed by each instrument and by precise offsets from a pair of fixed plumb-lines (Fig 2).

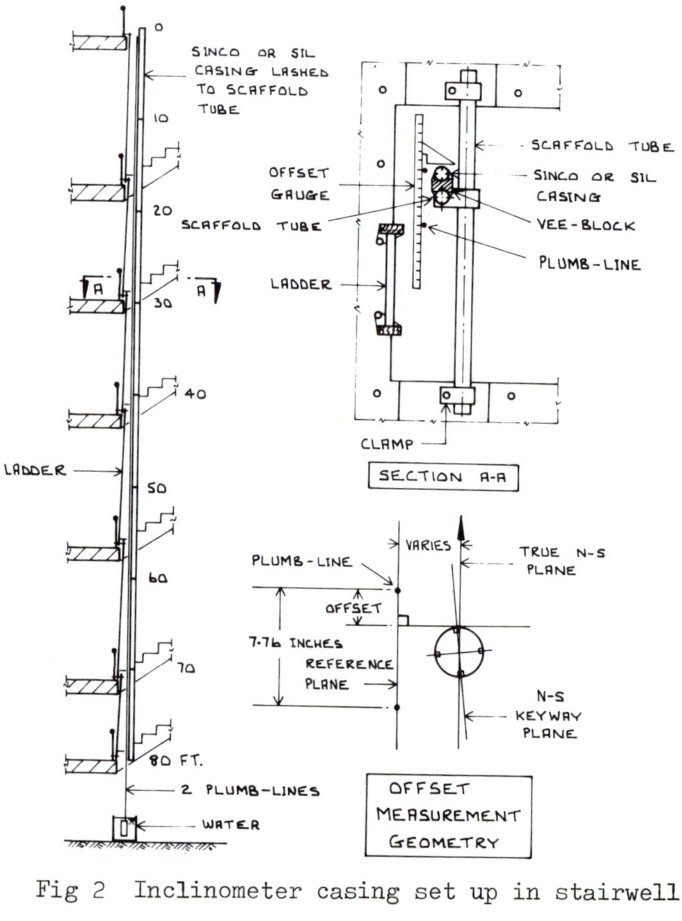

Fig 2 Inclinometer casing set up in stairwell

Wilson 200 B Series Slope Indicator

This operates by means of a magnetically-damped pendulum-actuated rotary potentiometer. A pendulum-mounted contact, closed by a relay in the torpedo, divides a curved fixed resistance element into two portions directly proportional to the angle between the pendulum and the torpedo. Since the contact is normally 'open' the only friction on the pendulum is in the bearings which support it and, provided they do not become contaminated, the friction forces are negligible. The pendulum is free to swing at all times and is protected by rubber stops. The torpedo is guided in the casing by a pair of fixed stainless steel wheels 12 in apart together with a second opposed pair of sprung wheels, each wheel running on sealed bearings.

The control cable has a Bowden wire core and is covered in neoprene rubber with yellow vulcanised bands every foot (replaced every 5 ft by red bands). The readout unit is a dry-cell battery-powered manual null-balance Wheatstone bridge circuit with a mechanical digital display. No battery check facilities are provided. No provision is made to simply adjust the zero or scale span as this is not usually necessary and a calibration frame is not supplied. The instrument has a range $\pm 15^\circ$ and a sensitivity of 0.03°.

Extruded aluminium alloy casing (Fig 1 d) with an epoxy paint finish was butted together in 10 ft lengths joined with telescopic couplings held by self-tapping screws.

Soil Instruments MK 1 Inclinometer

This operates by means of a beryllium-copper leaf spring with two pairs of bonded resistance strain gauges back-to-back in a full bridge circuit. The tapered brass weight is oil-damped and is free to swing at all times. The torpedo is guided in the casing by fixed skids 12 in apart and is held against the guide by two pairs of sprung runners 120° apart.

The PVC sheathed control cable has a stranded steel wire core and is marked every 2 ft by white printed bands protected by transparent sleeves. The readout unit is a dry-cell battery powered voltmeter reading in two switched ranges $\pm 5^\circ$ and $\pm 25^\circ$, graduated to 0.1° and 0.5° and by estimation to 0.01° and 0.05° respectively. The $\pm 5^\circ$ scale was used in the present tests. A battery check is provided together with adjustments for zero and scale spans and a calibration frame was available.

Extruded PVC casing (Fig 1 a) in 10 ft lengths was joined with cemented couplings with the aid of an alignment mandrel.

Calibrations

Angular scale span and zero stability Each inclinometer was tested at 20°C on an angular calibration frame with a range $\pm 5^\circ$ in 1° increments and accurate to $\pm 0.01^\circ$ in any 1° increment. All tests showed no significant variation in span calibration with time, Fig 3, SINCO 200 B/37 and SIL MK 1/A. The SINCO inclinometers each showed a very linear calibration, no hysteresis and a calibration factor within 0.3% of the desired true value. This error was neglected in computing displacements. The SIL inclinometers each showed a non-linear calibration with no hysteresis. (This was confirmed in tests on a third SIL MK 1 inclinometer.) The deviation from the average straight line is as much as 5% but since it is not easy to account for non-linearity in the analysis, measured average values were used in computing displacements.

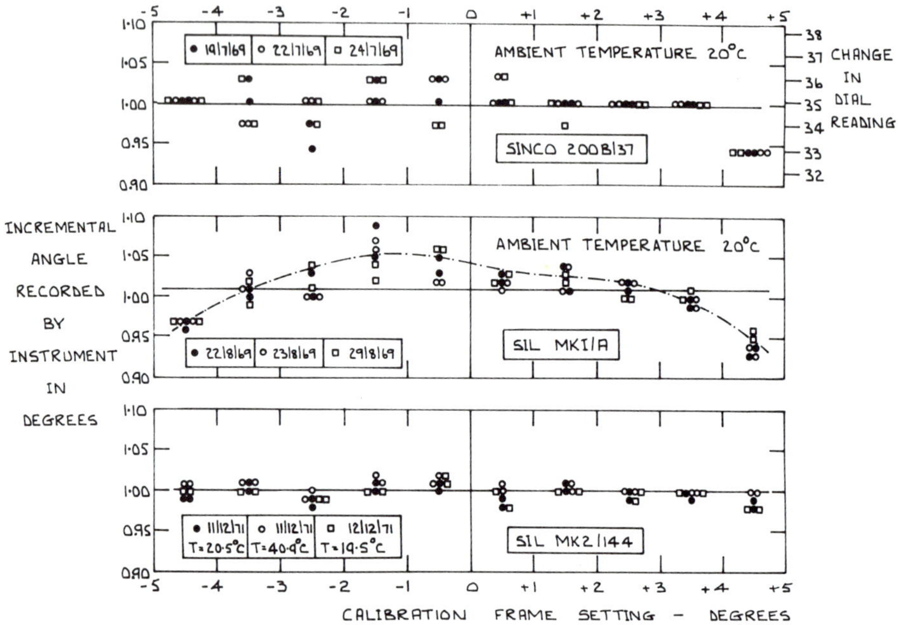

Fig 3 Angular calibrations of inclinometers

The effect of varying ambient temperature was investigated by fixing a short length of casing to the calibration frame into which the torpedo was inserted. Water was steadily circulated from a thermostatically-controlled heated water-bath. Water at laboratory temperature (20°C) was circulated for 1 hour after which the heater was switched on and the temperature raised in increments to 40°C. One hour elapsed after each temperature adjustment to allow a steady state to be reached before angular calibration. Calibrations at each elevated temperature and on the following day at 20°C clearly showed no measurable temperature effect either on the span or the absolute values for SINCO 200 B/59. Similar tests on SIL MK 1/A torpedo (Fig 4) show a steady drift with increasing temperature but which is reversible - the solid points in Fig 4. The average temperature drift rate of an absolute reading is 32 secs/°C which is excessive. This was confirmed by independent tests at BRS (Burland 1969). The effect on the scale span is much less, 2 secs/°C/degree, which is insignificant provided the torpedo temperature is constant during a set of readings 180° apart throughout a particular casing. If the down-hole SIL torpedo temperature varied linearly by 10°C from top to bottom of the hole this would give an apparent displacement of 1.0 in over 100 ft, an unacceptable error but in most cases an unlikely situation.

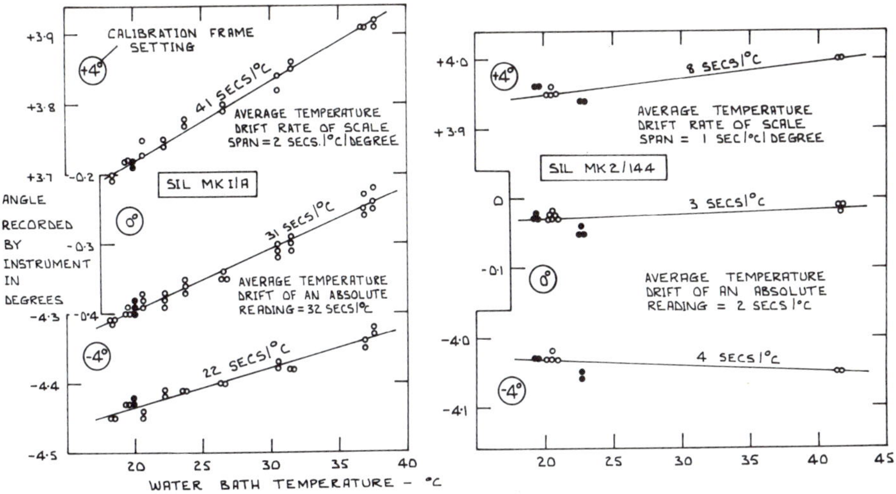

Fig 4 Temperature calibrations of SIL inclinometers

The temperature equalisation rate (Fig 5) was obtained for SIL MK 1/A after calibration at 40°C by quickly transferring the circulating pump input into water at 20°C, and recording the change in reading with time with the torpedo fixed. About $\frac{1}{4}$ to $\frac{1}{2}$ hour elapsed to reach 90% equalisation compared to tests at BRS in air which required about 1 hour for 90% equalisation. Thus the SIL torpedo should preferably be used in a water-filled casing and left down-hole for about $\frac{1}{2}$ hour before taking readings. Transfer between keyways and from one casing to another must be as rapid as possible.

Fig 5 Temperature equalisation rate tests on SIL inclinometers

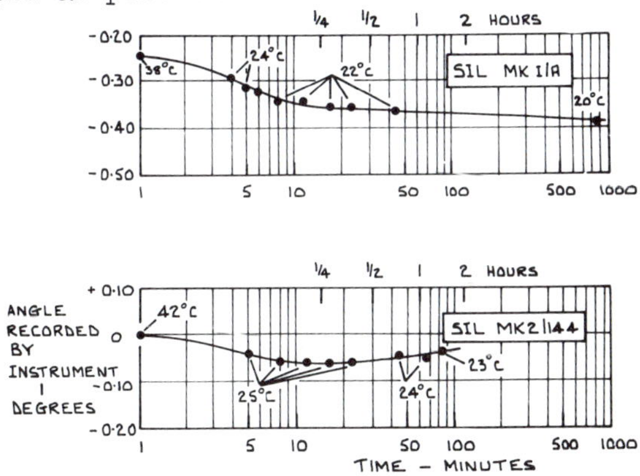

Cable markings The cable markings were checked by freely suspending the torpedo on 80 ft of cable adjacent to a steel tape in a stairwell. The markers on SINCO 200 B/37 (Nuttall's hire instrument) were within $\pm \frac{1}{4}$ in of the true position except adjacent to the torpedo which was set $\frac{1}{2}$ in low. The SINCO 200 B/59 cable (new and unused) was within $\pm \frac{1}{8}$ in . The errors in SIL MK 1/A (SIL's hire instrument) were generally within + $\frac{1}{2}$ in

but locally errors of up to $1\frac{1}{2}$ in in 10 ft occurred due to slip of the markers. The errors in SIL MK 1/B cable (new and unused) were $\pm\frac{1}{2}$ in with a cumulative error of + $1\frac{1}{2}$ in over 80 ft probably due to the cable being stretched when the torpedo was stuck in the casing during a test run. The SINCO cable is considered adequate whereas the SIL cable is unreliable due to slip stretch and being insufficiently robust.
No account was taken of marking errors in subsequent tests.

Casing Casing twist and joint alignment errors may be serious if cumulative. No twist could be detected in any of the 10 ft lengths of extruded aluminium alloy SINCO casing. Inspection of the joints indicated that an error of 2° per joint could occur. In spite of this no twist was detected in the 80 ft assembled length freely hung in the stairwell although no special precautions were taken during erection. The total twist measured on the freely suspended SIL casing was 18° RH. This was assumed to be uniform and was removed for the full scale surveys by clamping each end in proper alignment. After these tests the SIL casing was cut up and the individual twists were measured (Table 2).

Twist in casing			Twist at joints		
Casing	RH	LH	Joint	RH	LH
0-10	3	-			
10-20	3	-	10	-	1
20-30	4	-	20	-	2
30-40	1	-	30	-	1
40-50	3	-	40	-	2
50-60	3	-	50	-	0
60-70	3	-	60	-	0
70-80	3	-	70	-	1
Total = 23° RH			Total = 7°LH		
Cumulative twist = 23-7 = 16°RH					

Table 2 Twist in SIL plastic casing

The consistent LH twist at the joints is a result of operator technique and could be avoided. Cumulative twist in the casing cannot be avoided. Fortunately the assumption of overall uniformity was reasonable here but 2.9° per ft average twist consistently RH is undesirable in the field.
The torpedo must be free running and accurately located in the guides and must track across couplings. The SINCO system is good in this respect whereas the SIL system is much poorer due to friction on the skids, low quality extrusion and cement exuding inwards at the joints. The two pairs of sprung runners set 120° apart on the SIL torpedo may align the torpedo improperly if their spring stiffnesses are different or if dirt gets jammed in the housings. Their travel and clearances are relatively limited inside the casing.

Casing displacement tests
An 80 ft length of each casing was set up in turn in a stair-

well being rigidly clamped to a vertical scaffold tube (Fig 2). Horizontal displacements could be imposed in one plane (N-S), these being surveyed by 'before' and 'after' runs with each instrument both 'up' and 'down' the casing.

The true displacements were obtained from offset measurements at 2 ft intervals in the N-S plane from a pair of damped plumb-lines. Two plumb-lines were used together with a special gauge graduated to 0.1 in (read by interpolation to 0.01 in) to eliminate parallax errors. Fixed ladders gave access for measurement. Before taking a reading the operator attached himself by a chest harness to the ladder so freeing both hands. At each position readings were taken against each plumb-line so that the difference, 7.76 in, provided a check against gross reading or transcribing errors. It is considered that the error in displacement at any level determined from the plumb-line offsets is within ± 0.02 in.

For each of the three casing positions sets of readings were taken with each SINCO inclinometer, one working 'up' and one 'down', in each case at 2 ft intervals, 1,3,5 ..... 77,79 ft. The face errors were very consistent throughout. The measured displacements (Fig 6) are in good agreement with the true displacements. In Case I the error in a single run is less than ± 0.2 in over 80 ft and if an average of an 'up' and 'down' run were taken, less than ± 0.1 in. In Case II where the displacement is larger and the casing sinusoidal the error is greater, ± 0.6 in, and a number of important features emerge. The errors are not cumulative but are self-cancelling. They are similar for each instrument and for 'up' or 'down' readings and the peak errors occur at the joints. Taking an average of 'up' and 'down' readings would not reduce the error significantly in this case. Even so, these errors would be acceptable in most field situations since consistency is of far greater importance than absolute accuracy.

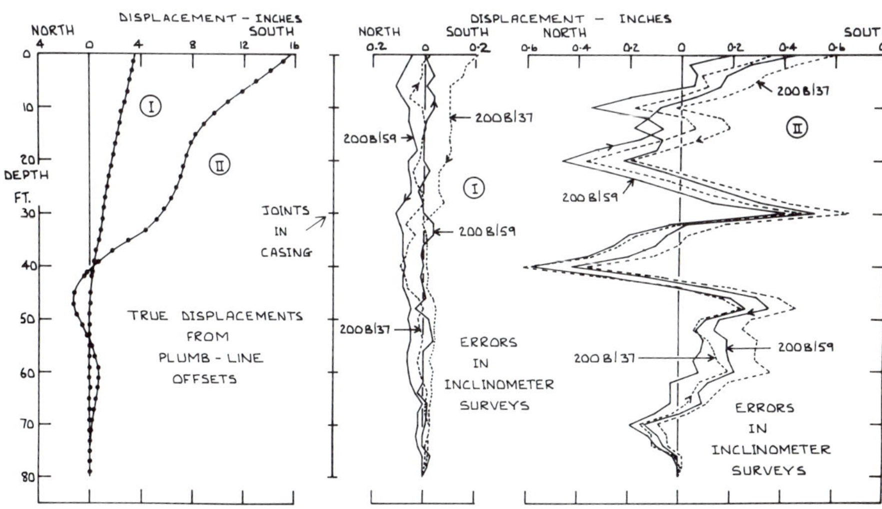

Fig 6 Slope Indicator 200B: actual and measured displacements

Similar tests with each SIL inclinometer also show good agreement (Fig 7) between true and measured displacements (see also Murray & Irwin, 1970). In Case I the error in a single run is less than ± 0.15 in over 80 ft and if an average of an 'up' and 'down' run were taken, less than ± 0.1 in. In Case II the error is greater, -0.1 to +0.5 in, but is less than that obtained in the Case II SINCO tests and the pattern is quite different. The errors are cumulative in the upper half of the casing and the 'up' and 'down' runs for each SIL instrument are very similar. Taking an average of 'up' and 'down' readings for either instrument would reduce the error to less than +0.3 in.

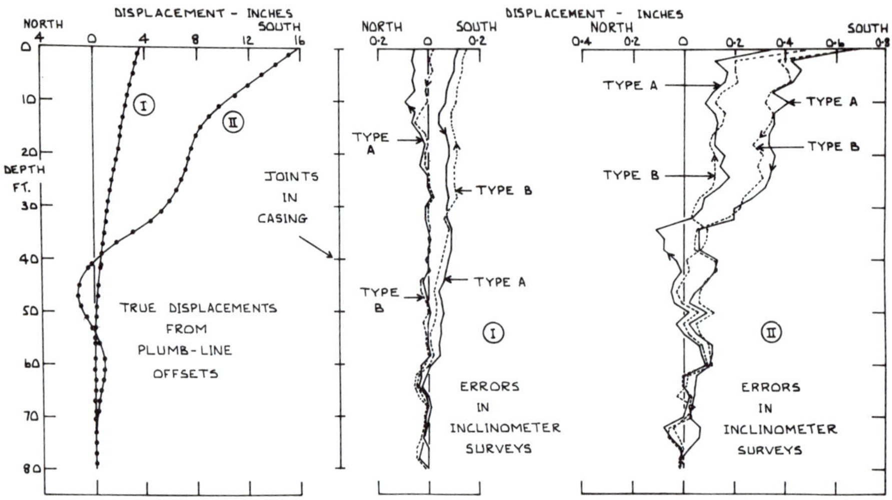

Fig 7 SIL inclinometer MK1: actual and measured displacements

The difference between the SINCO and SIL Case II error patterns is probably due to differences in the assembled casing stiffness. The metal SINCO casing is relatively rigid and the joints which are held by self-tapping screws are flexible. The deformed SINCO casing becomes dog-legged with little bending in each 10 ft length. In contrast the plastic SIL casing is more flexible and the cemented couplings are much stiffer so that the casing follows a smoother curve when deformed.
Face errors were less consistent in the SIL tests and the SIL MK 1/B inclinometer exhibited a sudden 1° shift in face error between two surveys although the variation in face error during any particular survey was constant, which is what really matters. However sudden zero shifts in the field are undesirable since they confuse the observer and can lead to large errors. Inconsistent face errors occurred in data from three SIL inclinometer casings at Balderhead Dam in 1968 and the deformations were difficult to reconcile.

DESIGN OF SOIL INSTRUMENTS MK 2 INCLINOMETER

The above tests indicate that the SINCO 200 B and the SIL MK 1 inclinometers can each measure horizontal ground movements fairly accurately, and indeed in these tests the SIL MK 1 inclinometer was marginally more accurate. Field experience with the SINCO 200 B instrument indicates that it is simple, robust, reliable and relatively trouble-free whereas more limited experience with the SIL MK 1 instrument shows that it is a poorer instrument. Most of the difficulties arise from moisture ingress, temperature effects, zero shifts, unsuitable casing and guidance system and poorly marked cable.

Many of the difficulties are overcome in the SIL MK 2 inclinometer (Fig 8) with digital readout, which is designed to operate in extruded aluminium alloy casing (Fig 1b) with an epoxy coating to reduce alkaline corrosion. This has now replaced the MK 1 system. Tests indicate that twist is generally less than 1°/3m length. Temperature-matched strain gauges are bonded to a tapered spring-steel leaf which is encastred by soldering. An oil-damped, high-density metal tapered bob-weight gives a high sensitivity and range. The sealed torpedo is guided by a pair of fixed wheels and is located by a second pair of sprung wheels. A transit locking float inside the torpedo is engaged on the bob-weight by inverting the torpedo. The torpedo may than be transported or stored in a horizontal or inverted position with comparative safety. The automatic digital display readout unit indicates angles over $\pm$ 20° with a resolution of 0.01^{o} and reads and integrates displacements over $\pm$ 10 m with a resolution of 1 mm and computing to 0.1 mm. It operates from a rechargeable battery and has data output facilities. Facilities are available for measuring vertical settlements using a reed relay (Burland et al, 1972). At present an adequately marked cable is unavailable, and is being developed.

Angular Span and Zero Stability

Angular calibrations at 20°C (Fig 3) on SIL MK 2/144 indicate a hysteresis-free, linear response over a $\pm$ 5° range and is typical of three instruments tested. Elevated temperatures have a much smaller effect than on the SIL MK 1 inclinometer (Fig 4). The average temperature drift of an absolute reading is reduced to 2 secs/°C and the average scale span drift rate is lower, 1 sec/°C/degree. The solid points in Fig 4 again demonstrate reversibility. The temperature equalisation time (Fig 5) is about $\frac{1}{4}$ hour but since the angular shift is 1/15th of that measured on the SIL MK 1 inclinometer this is less of a problem.

No full-scale controlled laboratory survey has been done with the SIL MK 2 inclinometer but it may be expected to perform at least as well as the MK 1 instrument and probably significantly better. Its field performance has been very satisfactory so far and it has proved to be more robust and reliable. Well documented records obtained at a number of sites indicate excellent repeatability. Sixteen sets of readings over a 10 month period in a 22 m casing just behind a diaphram wall indicate an integrated displacement at the surface parallel to the wall having a range of 3.3 mm with a standard deviation of 0.3 mm (Burland & Moore, 1973).

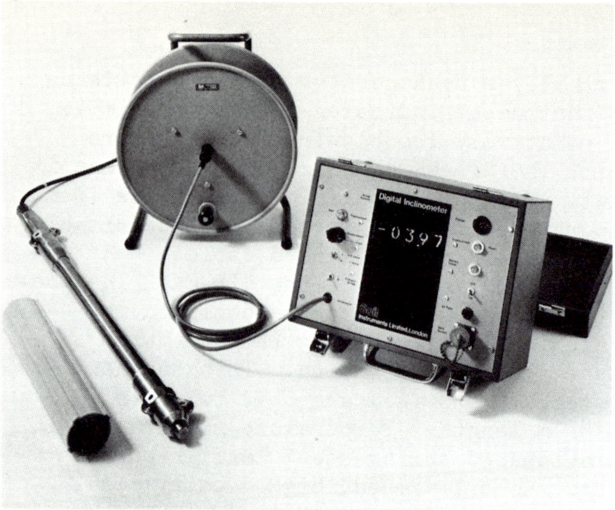

Fig 8 Soil Instruments inclinometer MK 2

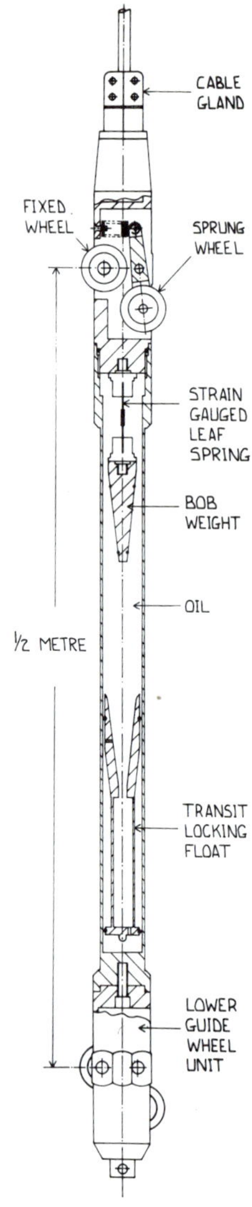

CONCLUSIONS

Vertical profiles of horizontal movements may be measured with inclinometers of various types. Accuracy is important but reliability, robustness and simplicity in use are of far greater importance.
Reliable data will only be obtained even with a perfect instrument if it is properly installed and read by careful and skillful persons. The data must be processed immediately and not allowed to accumulate unreduced, undigested, and uninterpreted.
Tests on two inclinometers, the SINCO 200 B and the SIL MK 1 show a similar performance under ideal conditions with an accuracy which is acceptable for many engineering purposes and which could even so, be improved by decreasing the reading interval and by averaging repeated readings. Whereas the SINCO 200 B inclinometer is a robust and reliable instrument the SIL MK 1 was not and was redesigned to eliminate the defects, resulting in the SIL MK 2. Experience to date with the new inclinometer appears very satisfactory.

ACKNOWLEDGEMENTS

Tests on the 200 B Series Slope Indicator and the Soil Instruments MK 1 inclinometer were financed by a number of firms both in Britain and in North America to whom thanks are due. Equipment for these tests was kindly loaned by Nuttall Geotechnical Services and Soil Instruments Ltd. The development of the SIL MK 2 inclinometer was carried out by the author in conjunction with Soil Instruments Ltd.

REFERENCES

Bishop, A W and P R Vaughan (1962) Selset reservoir: design and performance of the embankment, Proc. I.C.E., 21, 305-346.

Bromwell, L G, C R Ryan and W E Toth (1971). Recording inclinometer for measuring soil movement, Proc. 4th Pan Amer. Conf. Soil Mech., Puerto Rico, 2, 333-343.

Burland, J B (1969) Personal communication.

Burland, J B, J F A Moore and P D K Smith. (1972) A simple and precise borehole extensometer, Geotechnique, 22, 174-177.

Burland, J B and J F A Moore (1973). The measurement of ground displacement around deep excavations, Symp. Field Instrumentation, British Geotechnical Society, Butterworths.

Gould, J P and C J Dunnicliff (1971). Accuracy of field deformation measurements, Proc. 4th Pan Amer. Conf. Soil Mech., Puerto Rico, 1, 313-366.

Green, G E (1969). Report on tests on the performance of the Wilson Slope Indicator and the Soil Instruments Inclinometer (unpublished report).

Murray, R T and M J Irwin (1970) The performance of an inclinometer for measuring lateral movements in soil, Road Research Laboratory Report LR 332.

Penman A D M (1971) Instrumentation for embankment dams subjected to rapid drawdown, Building Research Station, Current Paper CP1/72 Jan.

Simon, I, A G Emslie, P F Strong and R K McConnell (1968) Sensitive tiltmeter utilizing a diamagnetic suspension, Rev. Scientific Instru., 39, No. 11, pp. 1666-1671.

U S Bureau of Reclamation (1968) Earth manual, First Edn., Revised, U S Gov. Printing Office.

Ward, W H (1956) A remote reading inclinometer, Bldg. Res. Note E697.

Wilson, S D (1959) Application of the principles of soil mechanics to open-pit mining, Quart. Colorado School of Mines, Symp. on Rock Mechs. Golden, Colo., 54, No. 3

Wilson, S D (1967) Investigation of embankment performance, Proc. Am. Soc. Civ. Engrs., 93, SM4, 135-156.

Wilson, S D (1969) Field instrumentation for excavations, Proc. 18th Annual Soils Conf. Kansas, Univ. of Kansas.

Wilson, S D (1970) Observational data on ground movements related to slope instability, 6th Terzaghi Lecture, Proc. Am. Soc. Civ. Engrs., 96, SM5, 1519-1544.

FOUNDATION INSTRUMENTATION FOR THE NATIONAL WESTMINSTER BANK TOWER

E.T. Haws

Director, Engineering & Resources Consultants, (Ercon) U.K.

D.C. Lippard

Partner, Pell Frischmann & Partners, London

R. Tabb

Senior Engineer, Ercon Australia

J.B. Burland

Head of Geotechnics Division, Building Research Station

INTRODUCTION

Construction of the new Head Office for the National Westminster Bank is proceeding in the City of London. The 185m high tower block will be the highest office building in London and the Owners, Architects, Consulting Engineers, CIRIA, and BRS, are jointly participating in a major programme of instrumentation.

A sectional elevation of the tower and foundation is shown in Fig. 1.

The main foundation analyses involved Finite Element methods for prediction of raft/piled soil contact pressures and raft concrete stresses. The selected scheme was for a total of 375 bored piles, 1.2m diameter and 26.5m deep, of which 348 were belled to 2.14m dia, all capped with a polygonal reinforced concrete raft of 54m equivalent dia and up to 4.5m total thickness. A proposal for instrumentation of the main building piled foundations was approved by CIRIA, with additional support from BRS in the form of a deep settlement gauge. The CIRIA sponsored scheme was primarily to investigate the short and long term loads taken respectively by the raft/clay contact face and the pile group for what is probably the highest foundation bearing pressure of any office building yet constructed in London. The first and major part of the proposed instrumentation, consisting of 8000 kN photoelastic pile load cells, twenty Gloetzl total pressure cells and the BRS settlement gauge, was installed early in 1972.

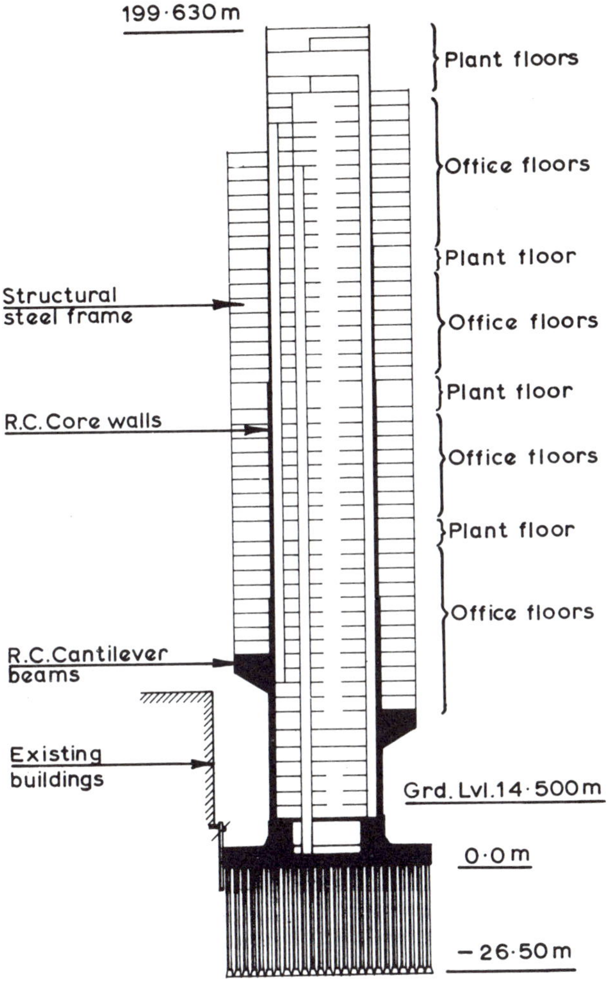

Figure 1. Sectional elevation of tower and foundation

In analysing and designing the anchored bored pile retaining walls, which were to surround the foundation excavation, the prime criterion was adequate support for the ten storey structures directly adjacent to the 12m deep excavation. Whilst the method of analysis used, together with appropriate material parameters would predict upper bound values for lateral movements, wall stresses and anchor loads, it was thought advisable that suitable instrumentation should be installed to allow monitoring of the wall and adjoining structures during excavation. A total of eight annular vibrating wire ground anchor load cells, five pile deflection tubes and two extensometer lines were installed during the progress of bored pile wall and ground anchor contracts.

By the end of 1972, the construction of the reinforced concrete foundation will be partly complete representing only initial loading to the foundation instruments. The casting of the massive foundations will, however, effectively prop the multi-anchored bored pile wall between excavation level, 0.0m RL and +4.5m RL eliminating the requirement for the lower two rows of ground anchors. It is hoped that load measurements of the then redundant ground anchors will continue, but at less frequent intervals, for the next four years making a total duration of records of some 5 years.

BORED PILE WALL INSTRUMENTATION

Optical deflection tubes, designed by J. Mowlem & Co. Ltd., were cast within the reinforcement cage of five individual concrete piles forming part of the bored pile wall, located as shown on Fig. 2.

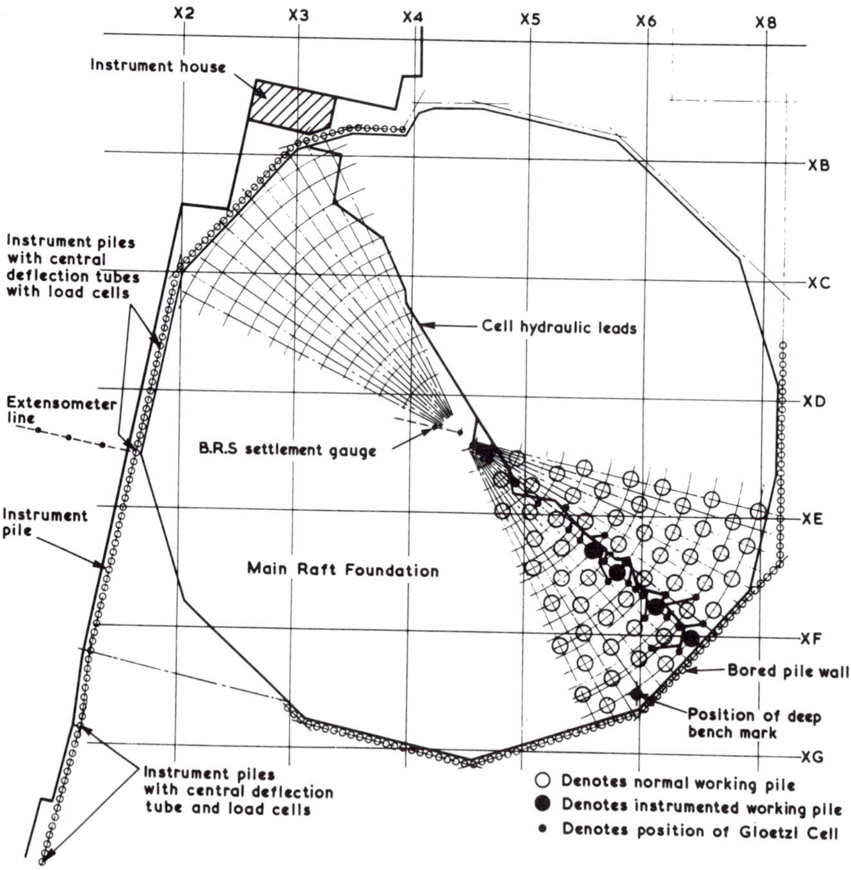

Figure 2 Plan on foundations and retaining structures showing arrangement of Instrumentation

The steel tubes consist of 100mm x 100mm square sections extending the full depth of the pile, having a 12m free height and extending 5m below excavation level for the maximum retained height. Within the box section, a series of brass pins are welded to the walls at 2.25m vertical centres. The plan arrangement of pins is staggered thus enabling unobscured sighting of each when illuminated by a recoverable device lowered down the tube. Angular measurements to each pin are taken with a Watts Microptic theodolite mounted in an inverted position at the pile head. Sightings made onto surrounding structures and the extensometer end point locate the instrument over the 'displaced' pile centre. A maximum set-up error of $\pm\frac{1}{2}$mm is estimated assuming the measurement of lateral movement by the extensometer is correct, a topic further discussed later. Repeatability of observed relative pin displacement of .02mm for the upper pin position and 0.1mm for the base pin has been achieved.

Extensometer Line

An extensometer line has been installed normal to the bored pile wall within the basement of the Owner's adjoining property, the position of which is shown on Fig.2. The line is adjacent to the deepest excavation and consists of a total of 17 bays of 3m, and constructed by screwing dome-headed pins into plates set in the basement floor. A datum bar with dial gauges is used to determine changes in bay lengths. The remote end pin, located some 50m from the face of the bored pile wall, is assumed not to move laterally towards the excavation. The summation of individual bay movements is used to determine the lateral movement of the pin fixed to the head of a pile forming part of the wall. The average of five temperature-corrected measurements is taken as representing a bay length.

Ground Anchor Load Cells

A total of eight annular vibrating wire load cells were installed to monitor the behaviour of the 550kN working load ground anchors. Six were installed at the section of deepest excavation corresponding to two deflection tubes; the remaining two were allocated to the single anchored wall also close to deflection tubes. Each cell was sandwiched between bearing plates with machined faces, typical plate dimensions being 600mm x 300mm x 57mm thick stiffened with two longitudinal bars (57mm x 57mm). Calibration tests indicated variation in cell response with irregular bearing plates, and so final machine work to $\pm$0.02mm tolerance was performed only after completion of welding and burning. The test load for each cell was 800kN, some 250kN greater than expected working load requirements.

During installation, the ground anchor load loss on lock-off was observed as up to 10%; this factor was used as an initial over-stress value.

Design Criteria

Adjacent to the existing ten storey structures, design criteria for the retaining bored pile wall in respect of both stresses and lateral displacement are severe. Finite Element analysis was considered but this was rejected since modelling of the system would have required too much time-consuming development work. It was therefore decided to use an analytical approach based on the solution of a beam on elastic foundations. The main problem lay in evaluation of soil reaction spring constants for use in the analysis, involving broad assumptions which could lead to inaccuracies. Use was therefore made of the computer to model a stage by stage excavation sequence using upper and lower bounds of soil reaction springs. The analysis modelled anchor forces as loaded springs.

For the net lateral pressure imposed on the wall at any excavation stage, consideration was given to initial Ko conditions, with lateral stress release following wall displacement towards the excavation. Even with a cut off for stress release at Ka conditions, the method could lead to serious under estimation of intermediate lateral stresses. It was therefore decided to assume a constant state of lateral stress throughout wall deflection, with a coefficient for London Clay of 0.6 lying between Ko and Ka conditions. For the overlying Thames gravel the range between Ko and Ka conditions is smaller and for conservatism the Ko condition was chosen, the value of the coefficient being predicted by the method of Jaky (1944).

In the sequential modelling of the wall behaviour, account was taken of net lateral loading with deductions made for restraint on the 'passive' face. Spring constants were selected after study of Terzaghi (1965) and Broms (1971). For gravel, two sets of spring values of different orders of magnitude were taken and for the clay three sets based upon low, medium and high modulus values. These five different spring parameter assumptions were combined with three trial loading distributions, triangular, trapezoidal and parabolic in form, to give a wide spectrum of design investigation. The lower spring values result in higher lateral displacement predictions, the trapezoidal pressure diagram in higher anchor force predictions, the parabolic pressure diagram in higher wall moment of resistance requirements and the triangular distribution in the requirement of a 5m toe-in below excavation level. A typical bending moment diagram and displacement prediction is given in Fig.3. The design adopted resulted from the envelope of these requirements. This conservative approach was felt necessary for the safety of the high adjoining structures and for maintaining uninterrupted progress on the main foundations.

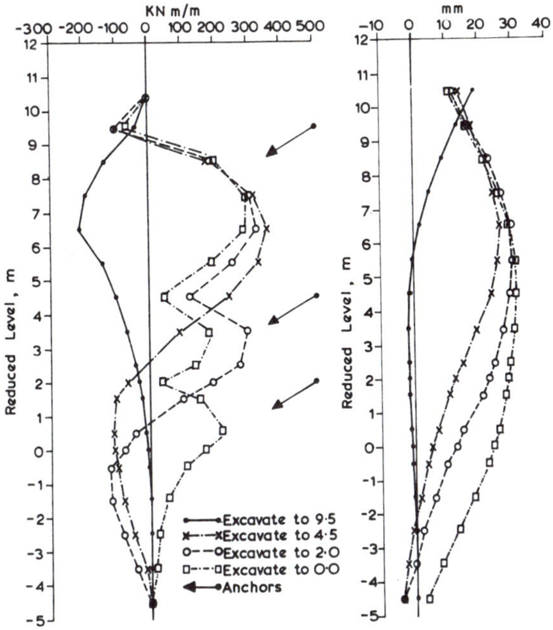

Figure 3 Typical bending moment diagram and deflected form predicted for deep section by analysis using a trapezoidal loading diagram.

RESULTS OF WALL INSTRUMENTATION

Extensometer Line

Readings of the extensometer line showed the pile head movements to be less than 3.5mm during excavation. The results have fluctuated, both during excavation and since. However, the difference between successive readings was often less than 1mm and it is considered probable that such changes were mainly due to climatic factors.

There is an indication from the detailed extensometer results that some local compression of the restraint material has occurred close to the retaining wall. This is probably due to the anchor stressing which resulted in high local lateral forces close to the top of the wall. In general, the performance of the extensometer line system was satisfactory.

Ground Anchor Cells

The annular load cells used to monitor ground anchor loadings have performed satisfactorily. Results (Fig.4) show that the anchors have generally lost up to 10% of their initial stressing load over the 1 year monitoring period with two approaching 15%. This loss of pre-stress cannot be associated with wall deflection causing reduction in the free anchor length and can only be the result of creep within the anchorage system. At this stage the creep is attributed to the London Clay itself, the loading/time relationship often being logarithmic, (Fig.4). The anchors constructed with an under-reamed shaft generally show a better long term response than the straight shafted pressure grouted anchors.

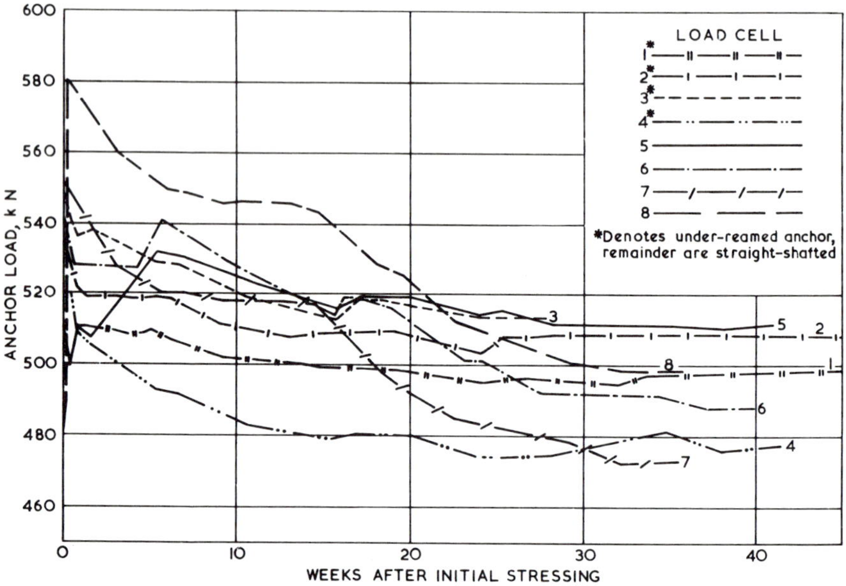

Figure 4 Ground anchor load / time curves

Deflection Tube Readings
The readings have not revealed the predicted bowed shape of the wall (Fig.3) but have shown almost linear lateral deflection with depth (Fig.5) and indications are that most of this movement has occurred prior to the installation and stressing of the ground anchors. This also means the computer prediction of changes in anchor load due to subsequent wall movement have not been realised in practice and resulting total measured movements are lower than predicted. Internal corrosion of the deflection tubes has not been a problem over the two year reading period. The somewhat vulnerable brass observation pins have not suffered damage from lowering the reading light source. The deflection tube has proved almost ideal for measurements under site conditions whilst still providing a high degree of resolution.

CIRIA INSTRUMENTATION

Following excavation for the raft foundation construction, a protective blinding slab was laid over the area with hydraulic and electrical connections leading to boxouts for pressure cells and temperature sensors.

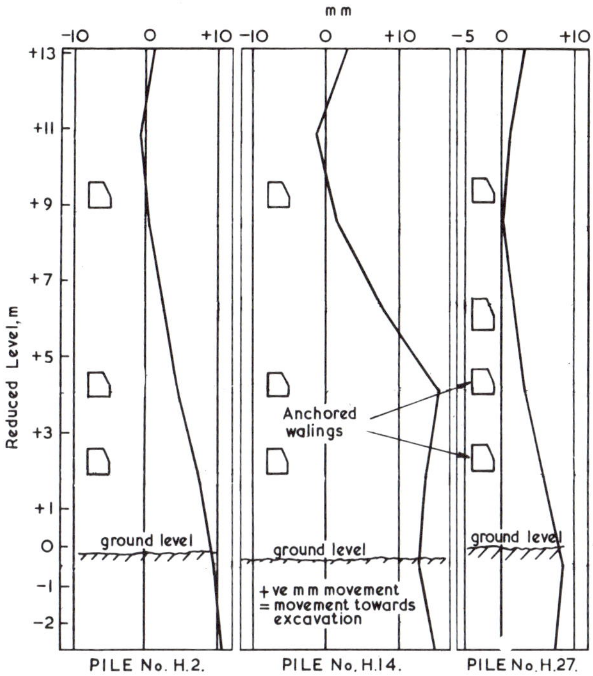

Figure 5 End of excavation deflected form of deep section deflection tubes

Gloetzl Cells

After completion of the main piling a total of twenty cells were installed at the raft/clay interface. The cell pressure pad is 170mm diameter by 4.5mm thick and is formed from an upper and lower steel plate connected at their perimeters. The space between the plates, some 1mm deep, is oil filled. The diaphragm/pressure chamber is connected to the pressure pad by a single metal tube (Fig.6). The principle is that of

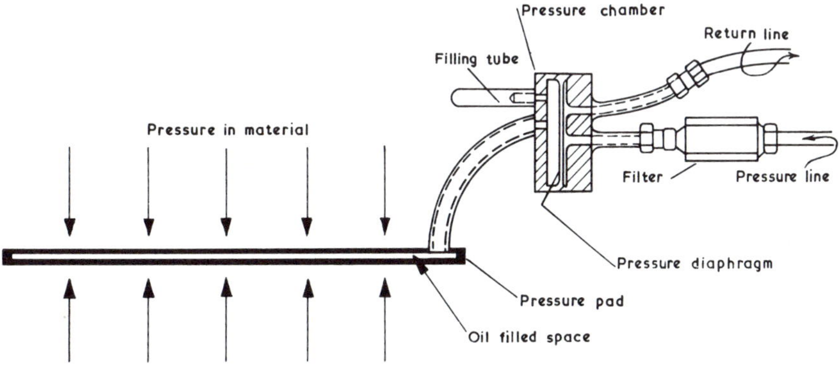

Figure 6 Gloetzl cell operating principle

measurement of diaphragm opening pressure required to maintain flow through the pressure chamber against a back pressure from the pressure pad.

Problems exist when predicting interface stresses by use of a buried, relatively rigid, inclusion and a Finite Element analysis was carried out. Several moduli combinations were investigated representing possible installation configurations under both undrained and drained clay conditions and with both short and long term concrete parameters. Selection of a suitable cell modulus was difficult, and in the absence of measured load displacement characteristics, bounds to the problem were studied. A first investigation assuming a 'concrete modulus' cell in a concrete surround against a clay interface indicated a predictable action factor of unity and it is inferred that a 'clay modulus' cell would behave similarly. With this check datum a 'steel modulus' cell was investigated under both undrained and drained clay conditions. The results indicated that under both conditions, the cell would possibly over read by some 13%, with the existence of high edge stresses. Surrounding the model steel cell in a low modulus backfill gave little improvement in action factor. The analysis is not conclusive but indicates the possible order of errors. CIRIA are considering a supplementary laboratory investigation into total pressure cell action factors but in the meantime it was decided to place the cells at the raft/clay interface and backfill with structural concrete. The 20 Gloetzl cells are located on a single radius of the foundation raft, on which the pile load cells are also located, so that predictions of load distribution, centre to perimeter and raft/clay to piles, can all be assessed. (Fig.2).

The thin protective layer of concrete left in the boxouts was carefully removed and the relatively undisturbed clay surface was cleaned and levelled. Prior to casting the concrete backfill the hydraulic and electrical leads were checked and connected to the cells laid in position. The electrical leads serve temperature sensors fixed to five cells to register the heat of hydration during concreting. The pressure reading system was changed from hand pump to steady rate electric pump, selected for long term measurement because the recorded Gloetzl Cell pressure is a function of the pumping rate.

The concreting of the foundation raft was commenced in November 1972 and Gloetzl cell readings were started.

<u>B.R.S. Settlement Gauge</u>

The values of Young's modulus, E, for the ground used in the design of the foundations are based on a limited amount of data obtained by site investigation and by back analysing the observed settlements of other large buildings in the London area. The bearing pressures are larger than used previously and conservative values for E for the London Clay and Woolwich and Reading beds were adopted. It is desirable to check on these values and the best method is to measure the compression of the ground at various depths beneath the foundations. The Building Research Station have developed a borehole extensometer which permits such measurements to be made simply and with considerable precision.

The extensometer has been described in detail by Burland, Moore and Smith (1972) and Burland and Moore (1973). It consists essentially of a number of permanent ring magnets housed in plastic holders which are fixed to the walls of the borehole at various depths. The magnets are threaded on to a small diameter PVC tube which runs down the centre of the borehole. The relative movements between the magnets can be detected by means of reed switches, which are used as sensors, mounted opposite each magnet on a datum bar which is permanently suspended inside the PVC tube.

The magnet extensometer was installed after the blinding slab for the raft had been cast. A 150mm dia. hole was bored to a depth of 64m below blinding level so as to penetrate the chalk bedrock by some 7.5m. An admixture was used with the drill fluid to ensure stability of the hole where it passed through the Thanet sands which overly the chalk and are about 14m thick at this location. A total of 10 magnets were installed in the borehole, the lowest being located in the chalk to act as a datum. Four magnets were installed in the Woolwich and Reading beds and the remainder at various levels in the London Clay. Subsequent to installing the magnets and central PVC tube the borehole was filled with a cement/pfa/bentonite grout designed to have approximately the same stiffness as the surrounding ground.

Pile Load Cells

A total of four photoelastic pile load cells have been installed along the same radial line used for positioning of the Gloetzl pressure cells. Photoelastic devices were preferred to electrical types on the grounds of long term stability at reasonable cost. To measure pile loads effectively it is necessary to provide a load cell which transmits the whole pile load and which maintains the structural integrity of the pile with similar load displacement characteristics to those of a non-instrumented pile.

The load cell is 1.2m dia and is constructed from 55mm thick top and bottom plates separated by a total of eight photoelastic load columns. Each of the photoelastic load columns is 110mm dia and 300mm high constructed from EN24 steel and has a working load capacity of 1000 kN for an approximate four fringe response. The 32mm dia photoelastic cylinder is located by an adjustable wedge mechanism in a 40mm hole, machined through the column, with a 0.5 fringe pre-load. The calibration curves are non linear due to the load displacement response of the glass plug in diametral compression but the curves for each of the eight load columns within a cell are practically identical, allowing a single calibration curve to be used for the assembled load cell. It is expected that loading resolution should be better than ±0.05 of a fringe, a little greater than 1% of the designed working load. Much of the design and field testing of this type of cell construction has been done by J.A. Hooper (1972). His work has included a laboratory long term sustained load test which showed a linear change in fringe order at maximum working load of less than 2% in 1 year. The photoelastic load columns are relatively insensitive to temperature changes and such changes should be extremely small, some 8m below basement level. There are, however, two thermal effects to be considered in interpretation of measured loading. The first applies to the true load imposed on the complete cell by differential temperature effects within the pile concrete and load cell and this can be likened to thermal loading of structures. The second temperature effect applies within the load gauge itself and is caused by the differing temperature coefficients for the glass and steel. Although comparatively insensitive to temperature changes, significant variations in fringe order have to be accounted for in practice. A temperature sensor will be located within the load cell to enable fringe corrections to be made based on observed fringe/temperature correction curves. Temperature variations in practice should be less than 5°C requiring a load correction of just over 1% of the full range. The load column detail is shown in Fig.7.

The pile load cell has been provided with a 'backup' measuring system for use if the primary photoelastic system becomes unreadable in any way. The backup system works on the principle of measuring by recoverable transducer the load cell compression at two diametrically opposed positions. Load interpretation will be made on the basis of the load displacement calibration performed on the local columns. Two datum bars fabricated from EN24 steel are positioned closely adjacent to diametrically opposed load columns. The transducer, manufactured by Sangamo Weston, ref. DC/DC DR ± 5mm, is located in an annular brass mounting and lowered down the 50mm dia backup measuring tube to seat on three stainless steel domes. The reading arm of the transducer is equipped with a captive return spring and domed head to ensure proper seating against the datum bar. The resolution of applied concentric axial load should be better than 3%. A temperature sensor is attached to the annular transducer mounting.

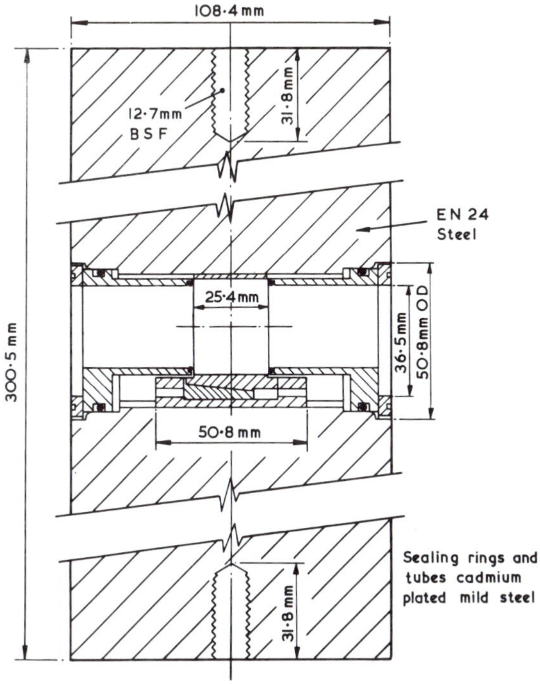

Figure 7 Section through
1000 kN photo elastic load column

The load cell top and bottom plates were surface finished and annealed and the eight calibrated load columns and two datum bars were bolted and screwed in position with great care to avoid gaps at their ends. To protect the metal surfaces of the load cells, observation tubes and backup tubes, a coating of epoxy paint was applied. The cell assembly was completed by covering the perimeter with neoprene sheeting.

The complete cell weighed almost 1000 kg and full calibration would have required loading to 8000 kN. A similar 1m dia pile load cell having a 5000 kN capacity and financed by CIRIA had been tested previously at the BRS. At any point on the loading curve, the total applied loading could be predicted from the summation of the readings from individual load columns based on their calibration. It was therefore decided not to undertake another complete calibration.

The Finite Element analysis of the pile, raft and soil system predicted dishing of the raft under load. This imposes an angular rotation to the heads of the piles, fully fixed to the underside of the raft foundation. Predictions of pile bending moment with depth were produced based on imposed restraints and to ensure the instrumented piles are typical of the group positioned at that radius the load cells are positioned at an elevation of zero tension, the cell top being some 3.4m below the pile head and 7.9m below the top of the completed raft.

Piling progress exceeded programme and in the instrumented piles the concrete was curtailed at -3.9m RL, with a lining above. To prevent transference of load past the cell the toe of the lining was set into the London Clay and it was coated internally with a bitumastic material. The adhesion transference through the lining to the clay has been limited thereby to an estimated 1% of that for untreated casing. The load cell was lowered until firmly seated on a still-fluid mortar layer on the scabbled primary concrete, and a check was made for level and bedding. The annulus between the cell and lining was grouted except for a 12mm insert of soft bitumen at the top for prevention of load transference. The reinforcement cage and measurement tubes were positioned and the remaining section of the shaft concreted. (Fig 8)

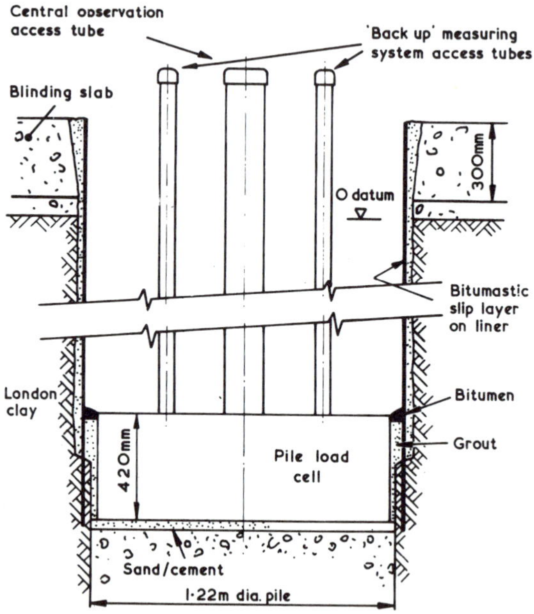

Figure 8 Sectional elevation through pile shaft showing cell and permanent lining prior to concreting above cell

Photoelastic load column observations require that polarized light be passed through the glass cylinder. A recoverable 12V 100W tungsten halogen light source and polariser is provided on a doubly hinged rod. Fringe observations are made at the surface through an endoscope with an analyser mounted over the eyepiece. Trials have been undertaken to check the performance of the backup measuring system and good repeatability was obtained.

Deep Datum

In 1965 a deep datum was anchored into the Thanet Sand. It consists of an oilfilled 50mm dia protective casing with an internal 25mm dia tube anchored at the base, and its position in association with the BRS gauge is such as to allow raft tilt to be recorded. Static and dynamic measurements will be recorded by a DC/DC ±50mm displacement transducer. Measurements on twelve levelling pins will also be used to observe raft dishing.

CONCLUSIONS

The bored pile wall instrumentation confirms the satisfactory performance of the support system during excavation and back analysis by Finite Element will assess the pressures operating. The main foundation instrumentation will provide information in both short and long term regarding ground deformations to depth and load distributions areally and as divided between raft/clay interface and the piles. This may confirm the feasibility of even more heavily loaded foundations on London Clay with possible design economies. Instrumentation results will be presented later.

ACKNOWLEDGEMENTS

The authors acknowledge assistance given by colleagues, by project staff of J. Mowlem & Co. Ltd., Soil Mechanics Ltd., Economic Foundations Ltd., and by Dr. J.A. Hooper. Thanks are due to Principals of Ercon, Pell Frischmann & Partners, the Director of the Building Research Station and CIRIA for permission to publish this paper.

REFERENCES

Broms, B.B. and Ingelson, I. (1971): "Earth pressure against the Abutments of a Rigid Frame Bridge". Geotechnique 1971 pp 15-28.

Burland, J.B., Moore, J.F.A. and Smith, P.D.K. (1972): "A simple and precise borehole extensometer". Geotechnique, March 1972 pp 174-177.

Burland, J.B. and Moore, J.F.A. (1973): "The measurement of ground displacement around deep excavations". Proc Symp on Field Instrumentation, Brit Geotech Soc, London.

Hooper, J.A. (1972): "The theory and design of photoelastic load gauges incorporating glass element transducers". Int J Rock Mech Min Sci Vol 9 pp 363-401.

Jaky, J (1944): "The Coefficient of earth pressure at rest". J Society of Hungarian Architects and Engineers pp 353-358.

Terzaghi K. (1955): "Evaluation of coefficients of subgrade reaction". Geotechnique Vol 5 1955 pp 297-326.

INSTRUMENTS DEVELOPED BY THE TRRL FOR STUDYING THE BEHAVIOUR OF EARTHWORKS

M.J. Irwin

Senior Scientific Officer, Transport and Road Research Laboratory, Department of the Environment

SYNOPSIS

A description is given of instruments which have been developed by the Earthworks and Foundations Division of the Transport and Road Research Laboratory for measuring settlement, porewater pressure and total pressures in soil masses. In addition a recently developed apparatus for recording automatically the data from in-situ permeability tests is described.

The instruments have been employed at a number of full-scale investigations of the settlement and stability of road embankments and the performance of the gauges in these studies is briefly described.

INTRODUCTION

The advance of theoretical soil mechanics in recent years has created a great need for reliable field data from full-scale studies. In addition to the requirement for field observations for purposes of research, measurements may also be necessary in connection with the control of construction to avoid the occurrence of instability in foundations and slopes. As a result new instruments and techniques are being developed continually by engineers and research workers.

During the last decade, the Earthworks and Foundations Division of the Transport and Road Research Laboratory have devised a range of instruments to assist in the research being carried out to improve the design and construction of new roads. This paper describes many of these instruments which have been developed for measuring stresses, strains and permeability in soil.

INSTRUMENTS TO MEASURE SETTLEMENT

The mercury-filled settlement gauge, developed by the Laboratory for measuring the settlement of the ground surface beneath road embankments (Irwin 1967) was introduced to overcome the many serious disadvantages which existed with both the rod-type and water-level gauges.

In its simplest form, the mercury-filled gauge consists of two units: a cell placed on the surface of the soil where a measurement of settlement is required and an indicator unit situated some distance away. The two units are connected by plastic tubing buried in a shallow trench. Changes in the level of the cell relative to the indicator unit are determined by recording the change in air pressure required to hold a column of mercury in one of the plastic tubes in a fixed position. A more sophisticated version of the gauge, the multi-point gauge, is also available and is capable of measuring the settlement at several points along the same run of tubing.

Mercury-filled settlement gauges were first used during the construction of the Staines by-pass in 1960. Since that time both single-point and multi-point gauges have been used successfully at a considerable number of sites both in the United Kingdom and abroad. Fig. 1 shows the records of settlement from a full-scale investigation carried out on the Oxford Southern By-pass Extension (Murray 1972). The results demonstrate the very satisfactory performance of the gauge over a period of about 10 years. In general settlements can be measured to an accuracy of better than ± 2 mm.

The primary application of the mercury-filled settlement gauge has been for measuring the settlement of the ground surface in situations where the difference in elevation between the cell and the indicator unit are relatively small (i.e. up to about 3 m). In situations which would necessitate large differences in elevation between these two components, such as the settlement of points within high earth banks or deep foundations, the large gas pressures required to operate the gauge introduce many problems. A new type of settlement gauge has therefore been devised by the Laboratory for use in such situations.

A schematic diagram of the apparatus is shown in Fig. 2. In principle the measurements of settlement are obtained by monitoring the small variations in pressure resulting from changes in elevation of the cell. This assumes a constant volume of water in the system. Although a settlement gauge employing a constant volume of water has been described by Peignaud (1965) the gauge developed by the Laboratory differs very considerably in the method of operation.

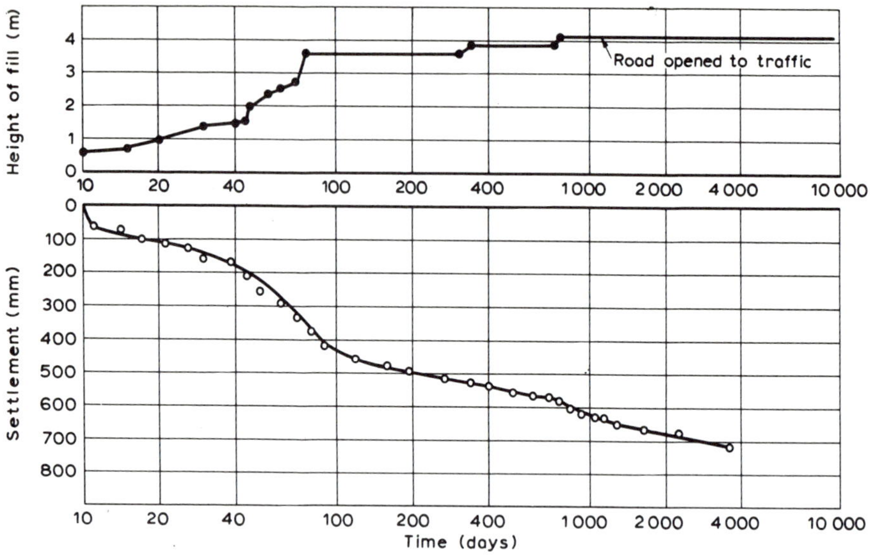

Figure 1. Settlement of subsoil at
Oxford Southern By-Pass Extension

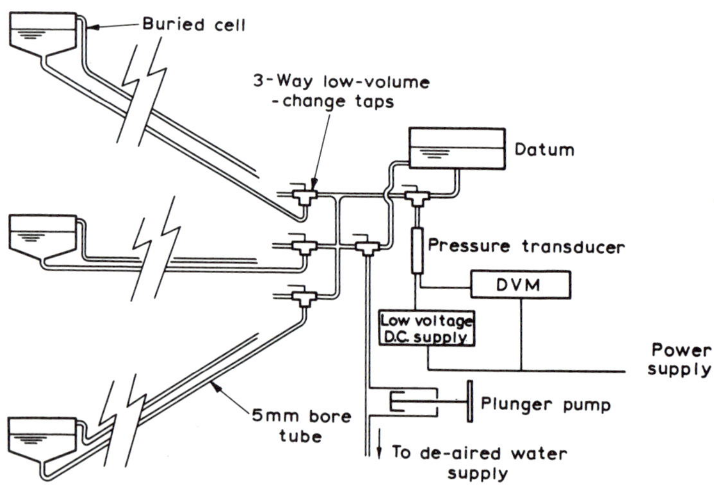

Figure 2. Schematic diagram of the water-level gauge

The pressure changes can be recorded using a standard manometer system but it is preferable to employ an electrical pressure transducer. The volume of water in the system must remain constant. A plunger pump assembly of the type normally associated with de-airing operations in triaxial apparatus provides a simple and effective method of introducing a specific quantity of water into the system.

Calculations have shown that a variation in temperature of 10°C produces a change in elevation of the surface of the water in the cell of only about 0.07 mm. However, variations in the density of the water resulting from this temperature change have a much greater influence on the measurements. Results of a typical calculation involving a difference in elevation between the cell and the transducer of 6.5 m indicate an error in reading of 1.0 mm per degree Centigrade. Proportionally greater errors will occur if the differences in elevation of the cell and transducer are increased. Thus if the full potential accuracy of the system is to be achieved it will be necessary to make corrections for the effect of the temperature. Generally this will be necessary only when the elevations of the cell and transducer differ by more than about 3 m.

Laboratory studies have indicated that, provided a precise calibration of the transducer is carried out for the working range of pressure likely to be involved, an accuracy of ± 1 mm can be achieved. However, under field conditions an accuracy of the order of ± 2 mm is more likely as shown in Fig. 3. At this site the cell was positioned about 3 m above the transducer and no corrections for temperature have been made.

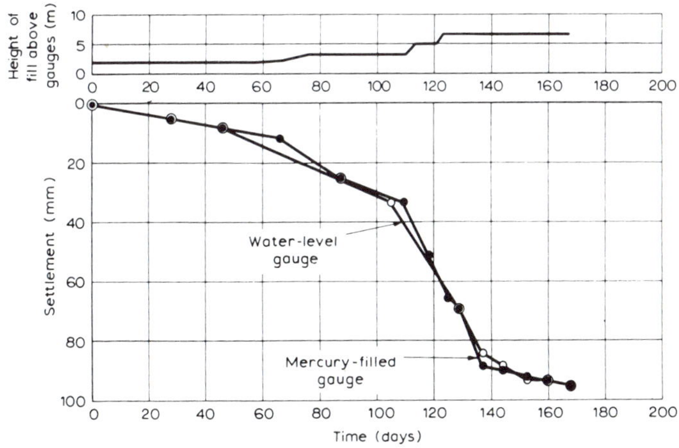

Figure 3. Settlement record for a water-level gauge and a mercury-filled gauge embedded together in concrete at M40, Denham

APPARATUS FOR THE MEASUREMENT OF POREWATER PRESSURE

In the construction of major earthwork projects, large numbers of piezometers are often installed and considerable staff effort is required to make the readings. To overcome this problem the Laboratory has developed apparatus for the automatic recording of porewater pressures in saturated soils (Fig. 4.)

Figure 4. Automatic porewater pressure recording apparatus

In the apparatus, solenoid valves and a single electrical pressure transducer replace the manometers and Bourdon gauges normally employed as indicator units in the conventional twin-tube hydraulic system. Each piezometer is connected in sequence to the pressure transducer through a solenoid valve. As the plunger of the valve is fully immersed in water, the volume changes which occur when the valve is actuated are negligible. Permanent records of the measurements are obtained using a chart recorder or, when a mains electricity supply is available, by collecting the information on punched paper tape for direct analysis by computer.

Ideally, the pressure transducer which is employed with the system should have a pressure range which corresponds closely with the range of pressures anticipated at a specific scheme. However, a transducer having a range of 0 - 170 kN/m^2 has been found to have a wide application in monitoring the porewater pressures in the subsoil beneath road embankments. The maximum volume change resulting from the deflection of the internal diaphragm of the transducer is less than 0.36 mm^3 and thus the response time for the system is relatively small when compared with that involving a mercury manometer or Bourdon gauge. In soils having a permeability coefficient of 1 x 10^{-7} mm/sec, the response time required for the system to reach an equilibrium condition is less than 10 seconds.

The apparatus and earlier prototypes (Margason, Irwin and Huxley, 1966) have been used on a number of full-scale investigations carried out at road embankment sites and the continuous records which have been obtained have proved to be of considerable value in the interpretation of the results.

MEASUREMENTS OF EARTH PRESSURE

It is extremely difficult to make accurate measurements of total stress in soil because the presence of the cell in the soil mass can have a considerable influence on the stresses in the vicinity of the cell. Ideally the cell should have the same elastic properties as the surrounding soil. However, the elastic moduli of soils vary considerably and in addition the moduli change with further application of stress. In practice, therefore, the results obtained with earth pressure cells are subject to some error. To minimise the magnitude of the error, described as the cell action factor, two major criteria have been established for designing earth-pressure cells. Recommendations were given that the ratio of the thickness to the diameter shall not exceed 0.2 and the ratio of the diameter of the cell to the deflection of the diaphragm must be 2000 or greater (US Waterways Experimental Station (1944)).

In 1960 the Laboratory had a requirement for earth-pressure measurements and, although several types of earth-pressure cells were available, they did not always conform to the established design criteria or, alternatively, the measuring systems were incapable of long-term measurements with a performance free from drift and temperature effects. A new type of cell for measuring static or slowly changing total stresses in earth masses was, therefore, developed (Fig. 5).

The cell was designed to measure any magnitude of earth pressure likely to be encountered in earthworks construction and, by incorporating a simple pressure-sensitive device not previously used in this application, some of the limitations of earlier pressure cells were overcome.

A schematic diagram of the cell is shown in Fig. 6. The body of the cell is 140 mm in diameter and 6 mm thick. Pressure applied to the upper surface of the cell body is transmitted through a thin metal diaphragm to a fluid-filled chamber beneath. The chamber is connected by small-bore copper tubing to a Bourdon tube housed in a pressure cylinder of 50 mm diameter by 100 mm in length. To avoid the cylinder

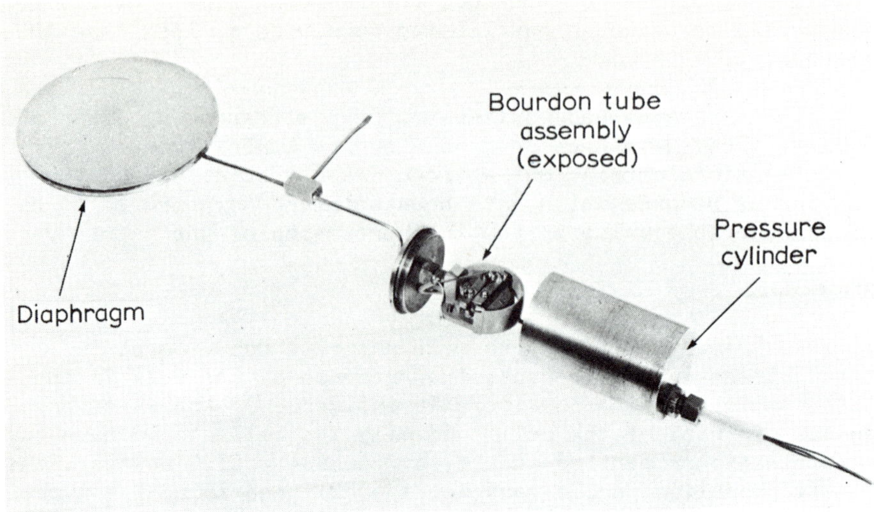

Figure 5. Earth pressure cell

influencing the stresses surrounding the cell body, the two units are separated by a distance of about 200 mm. The pressure cylinder is connected by nylon tubing to a pressure indicator unit located clear of the earthworks. In addition to allowing the passage of gas, the nylon tubing contains three electrical wires associated with the operating system.

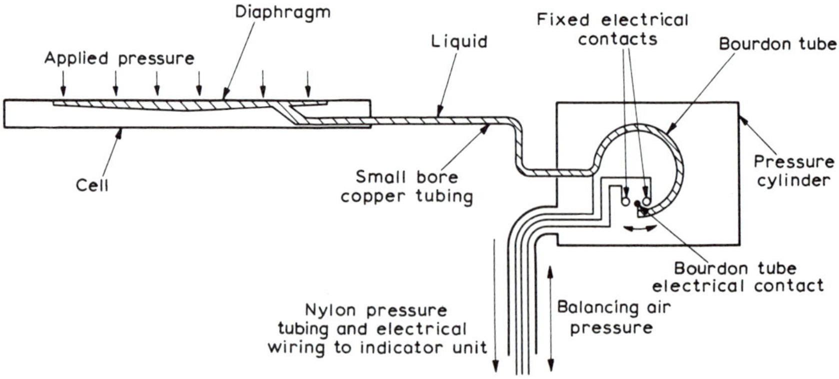

Figure 6. Diagrammatic sketch showing principle of operation of earth pressure cell

Increasing or decreasing pressure on the diaphragm of the cell produces a deflection of the free end of the Bourdon tube from its unstrained position. If the deflection is greater than 0.04 mm in either direction a closure of electrical contacts energises solenoid valves which adjust the air pressure in the pressure cylinder to restore the Bourdon tube to the unstrained position. The balancing air pressure is a direct indication of the pressure applied to the cell diaphragm.

The cell, if calibrated in a fluid-pressure chamber through a complete cycle of increasing and decreasing pressure over the range from zero to 200 kN/m^2, gives an indicated pressure to an accuracy of within $\pm$ 1.4 kN/m^2 of the applied pressure. The major part of this inaccuracy results from the need to have a finite gap between the electrical contacts. When an electrical contact attached to the Bourdon tube moves across the finite gap between the fixed electrical contacts, the maximum volume change which can occur in the system is about 3 mm^3 irrespective of the applied stress. Thus, assuming this volume change occurs when an increment of stress is applied, the "modulus of elasticity" for a cell having an active face of 127 mm diameter will vary from about 67 MN/m^2 to a value of 5200 MN/m^2 at the maximum working stress. In general, however, it would be expected that the moving contact would normally lie at about the mid-point of the finite gap and the maximum value of the "modulus of elasticity" would then be as great as 10,400 MN/m^2.

Over the range of temperature likely to be encountered in field measurements the calibration in a fluid-pressure chamber is unaffected by variations in temperature.

The cells have been employed in full-scale studies carried out on road schemes at Tickton in Yorkshire and Over in Gloucestershire. The long-term stability of the cell is demonstrated by records from the site at Tickton (Fig. 7) where records have been obtained over a period of about 4 years. Measurements of the temperature of the cell are also shown in this figure. At both sites (Fig. 7 and Fig. 8) comparisons of the recorded and calculated pressures indicate that generally the cells have over-registered by between 20 and 25 per cent. The over-registration of the cells is probably associated with their stiffness in relation to that of the surrounding soil which at both sites were soft clays (E values of the order of from 3 to 5 MN/m^2). In stiffer soils it would be expected that the errors would be somewhat smaller.

During the early stages of construction at Over when the height of fill above the cells was only about 4 m, considerable disparities were obtained between the measured and calculated stresses. These differences are believed to be

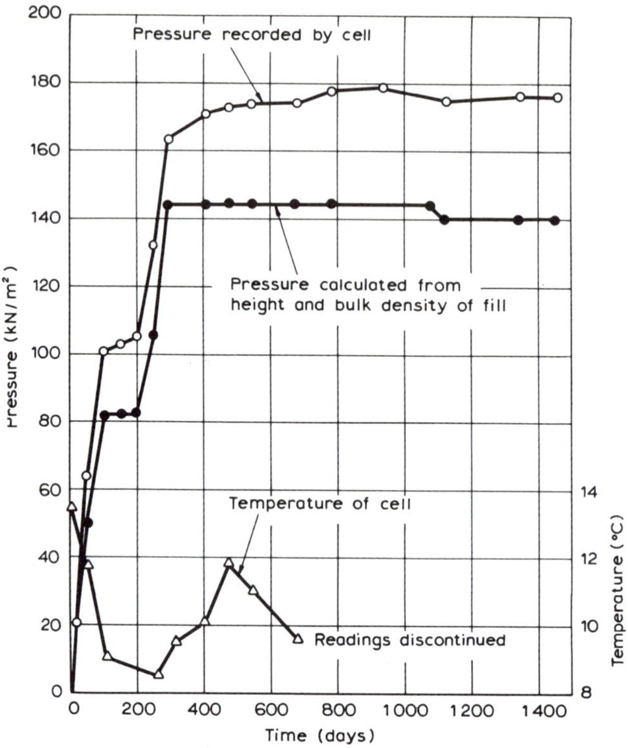

Figure 7. Measurement of earth pressure beneath a road embankment at Tickton, Yorkshire

mainly the result of compaction plant inducing stresses which were temporarily "locked-in".

A modified version of the cell has recently been constructed in which the Bourdon tube assembly is replaced by an electrical pressure transducer. For a pressure range of 0 - 170 kN/m^2, the transducer has a maximum volume change of only 0.36 mm^3 (about an eighth of the volume change occuring with the Bourdon tube system). Thus the effective modulus of elasticity using the pressure transducer is about 38,500 MN/m^2. The design has a particular application in the measurement of earth pressures behind retaining walls where the diaphragm of the cell is installed flush with the surface of the wall. The use of a stiff cell is likely to reduce the problem of under-registration which would otherwise occur because of the very great stiffness of the wall.

AUTOMATIC DATA RECORDING FOR IN-SITU PERMEABILITY

Recent research into the behaviour of road embankments founded on compressible soils has indicated that predictions of the

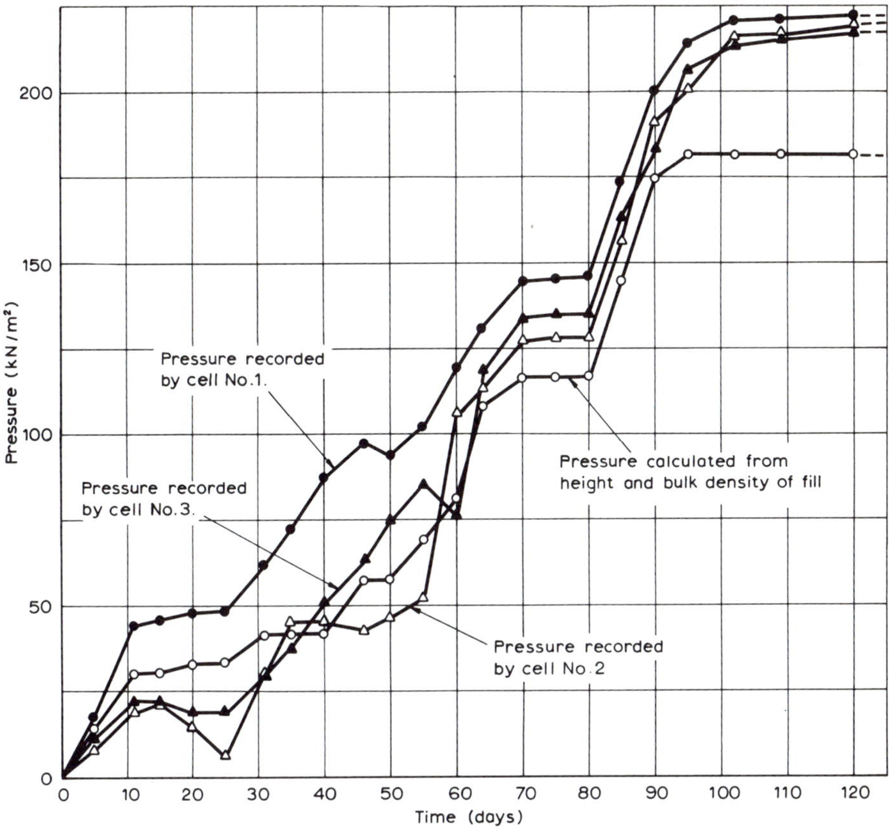

Figure 8. Measurement of earth pressure beneath a road embankment at Over, Gloucestershire

settlement based on the compressibility parameters from conventional laboratory consolidation tests are frequently inaccurate, particularly with regard to the rate of settlement. Much closer agreement with the predicted and measured relations between settlement and time were obtained at a number of sites, however, using the values of in-situ permeability in the settlement calculations (Murray 1971, 1972). To enable measurements of in-situ permeability to be more easily made and also to overcome certain disadvantages inherent in manually operated apparatus, automatic recording apparatus has been developed for monitoring the flow of water through a piezometer system. The apparatus has been designed for battery operation enabling it to be used at those sites where a mains supply is not available.

The apparatus (Fig. 9) consists essentially of a twin piston-and-cylinder assembly connected through a gear train to an electric motor. The gearing is arranged to produce directionally opposed travel of the pistons. The displacement of each piston is limited to about 20 cm^3 by the actuation of a micro-switch which, by means of a control unit, reverses the motor and hence the direction of travel.

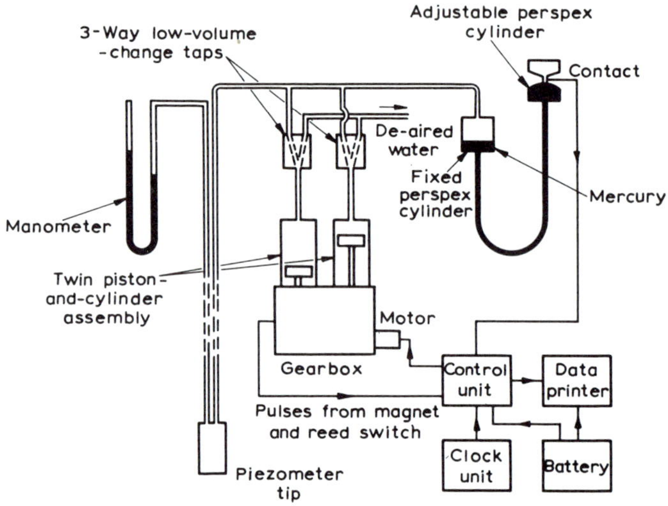

Figure 9. Schematic diagram of the automatic permeability recording apparatus

A constant pressure or suction is maintained in the system using a mercury-filled U-tube assembly comprising both an adjustable and a fixed perspex cylinder. A change in volume in the system alters the level of the mercury in the adjustable perspex cylinder. The variation in level interrups the continuity of an electric circuit causing the pistons to move and adjust the quantity of water within the system until the original level of mercury is restored.

Each revolution of a selected gear in the gear train, equivalent to a volume change of 0.01 cm^3, generates a pulse by means of a reed switch and magnet combination. A counter incorporated in a data printer unit is actuated by the pulse and provides a cumulative count of volume change. Permanent records of volume change are obtained at predetermined intervals of time by means of a pulse from a timer unit which energises the print mechanism and produces a print-out of the elapsed time and the total volume of water flowing through the system.

The accuracy of the system was determined by weighing the quantity of water discharged at various intervals of time. The errors were generally not greater than 0.02 cm^3 even at a rate of flow considerably in excess of that normally encountered in field measurement (2 cm^3/min).

The apparatus has been used very successfully on a number of full-scale studies at road embankment sites where measurements have been recorded over a period of several days.

CONCLUSIONS

It is now commonplace to instrument earthwork sites with a range of equipment for recording deformation, changes in porewater pressure and total stress in soil masses. The cost of the equipment together with the effort involved in obtaining the measurements can be considerable. It is very important if full benefit is to be obtained from this expenditure and effort that the instruments should perform satisfactorily over several years under the adverse field conditions normally experienced. Furthermore, because of the large quantity of data involved, it is desirable to employ modern data-logging equipment to enable the records to be directly analysed by computer.

In the past, insufficient attention has been given to these aspects and as a result field instrumentation has not always been very satisfactory. Instruments have now been developed which overcome many of these earlier problems and in the future it will be possible to carry out more reliable field studies for both research and engineering control.

ACKNOWLEDGEMENTS

This paper was prepared in the Earthworks and Foundations Division of the Highways Department.

The author also wishes to acknowledge helpful comments and assistance from the following members of the staff of the Laboratory:

W A Lewis, G Margason, R T Murray, J E Cross and M D Ryley.

This paper is presented by permission of the Director of Transport and Road Research Laboratory.

REFERENCES

IRWIN, M J. (1967) A mercury-filled gauge for measuring the settlement of foundations. Ministry of Transport, RRL Report LR 62, Road Research Laboratory, Crowthorne.

MARGASON, G, M J IRWIN and M A HUXLEY. (1968) The effect of tides on subsoil porewater pressure at a site on the proposed Shoreham by-pass. Ministry of Transport, RRL Report LR 195. Road Research Laboratory, Crowthorne.

MURRAY, R T. (1971) Embankments constructed on soft foundations: settlement study at Avonmouth. Department of the Environment, RRL Report LR 419, Transport and Road Research Laboratory, Crowthorne.

MURRAY, R T. (1973) Embankments constructed on soft foundations: settlement studies near Oxford. Department of the Environment, TRRL Report LR 538. Transport and Road Research Laboratory, Crowthorne.

PEIGNAUD, M. (1965) La measure du tassement des ouvrages. Mise au point d'un nouvel appareil. Bul. liaison Labo. P. et Ch. No. 11, 4.1-4.12. Laboratoire Central des Ponts et Chaussees, Paris.

US WATERWAYS EXPERIMENT STATION. (1944) Soil pressure cell investigation. Technical Memorandum No. 210-1. Waterways Experiment Station, Vicksburg, Miss.

FIELD MEASUREMENTS OF EARTH PRESSURES AGAINST MOTORWAY RETAINING WALLS AND BRIDGE ABUTMENTS

C. J. F. Jones

Senior Section Engineer, West Riding County Council Sub-Unit of N.E.R.C.U.

INTRODUCTION

The West Riding County Council is responsible for maintaining and repairing some 1600 bridges which were built before the introduction of the Ministry of Transport standard loading in 1922. The majority of these structures were constructed in either brick or masonry and in a number of them bulging and cracking has been found in the spandrels and wing walls. Occasionally the trouble is due to deterioration of materials, but the bulging and cracking at mid height of the walls which make up the majority of cases cannot be attributed to this. The vehicular traffic passing on or adjacent to these structures is thought to have contributed to their deterioration.

These commonly observed faults in retaining walls and bridge abutments are not readily explained by the classical earth pressure theories. In 1966 the West Riding County Council in conjunction with the Ministry of Transport, decided to investigate, over a number of years, the lateral pressures acting on a large retaining wall on the M.1 Motorway back-filled with conditioned hopper ash. (See Figure 1) Details of the investigation and the initial results were reported by Sims et al (1970).

In 1970, in view of the success of the initial M.1 experiments the West Riding County Council, now acting as part of the North Eastern Road Construction Unit, in conjunction with the Department of the Environment, decided to augment the investigation and to obtain additional qualitative and quantitative data of the earth pressures generated by other filling materials and various standard methods of placing the fill. Tests on Hunsworth Dyers Retaining Wall on the Hartshead-Gildersome section of M.62 and the North Ings Interchange Bridges on the Hatfield-Thorne Section of the M.18, both of which started construction in the summer of 1970, were started.

Details of the instrumentation used on the M.62 and M.18 Motorways are described in this paper together with examples of the earth pressures recorded.

Figure 1

HUNSWORTH DYERS RETAINING WALLS

Description of the Test Wall and Object of Investigation

The M.62 Motorway constructed during 1970-72 passes close to the works of Hunsworth Dyers Ltd. near Chain Bar Interchange. To restrict the land requirements of the motorway near the Dyers main water tanks the motorway embankment was supported by a reinforced concrete cantilever retaining wall. This wall is 160 feet long and has a maximum height of 39 feet. It is divided into panels each 16 feet long and was backfilled with the fill that was being used to construct the neighbouring embankment. The wall stands on sidelong ground and was designed in accordance with standard practice, to withstand an equivalent fluid pressure of 30 lb./cu.ft. plus an H.B. surcharge.

Preliminary boreholes indicated a difficult foundation situation, in which the uphill end of the wall would be founded on hard fissured sandstone rock, while the downhill end would be founded in soft weathered shales. These conditions were verified during the construction of the wall. In view of the complex foundation conditions it was decided to support the uphill sections of the wall on spread footings and the downhill sections on insitu concrete piles. The fissures in the sandstone were grouted after the wall had been built.

It was decided to instrument this wall to obtain quantitative and qualitative data of the earth pressures acting on a retaining wall and to study the behaviour of the wall when subjected to these loads. The investigation was also designed to check the range and validity of the conclusions drawn from the previous retaining wall test. Two adjacent panels near the midpoint of the wall, one supported on spread footings the other on piles, were chosen for the investigation.

The backfill for Hunsworth Dyers wall was grey silty mudstone, sometimes clayey with occasional siltstone and sandstone fractions. It was the same as that used for the adjacent motorway embankments and was classified as a Class B fill. Laboratory tests, prior to placing, suggested that the material had a maximum dry density of 133 lb/cu. ft. at an optimum moisture content of 8.3 per cent with 10 per cent air voids. All the material laid behind the wall was excavated by TS14 or TS24 excavators and levelled in approximately 6 inch layers by D6 or D8 bulldozers in accordance with the Specification. The fill was compacted by Hyster Grid, Sheeps' foot roller or DW21 vibrating roller as the conditions dictated. Compaction immediately adjacent to the wall was achieved using a Bomag 90 vibrating roller. Plate bearing and sand bottle tests of the backfill were conducted during the whole of the filling period.

Instrumentation and Testing

Lateral and vertical pressures in the backfill were measured using Building Research Station Total Earth Pressure cells. Thirty-two gauges were used in the investigation laid horizontally and vertically at five levels in two rows behind the two panels. Five gauges were set into the backface of the wall flush with the surface in one panel with a sixth similarly fixed in the other panel as a control.

Internal concrete strains near both faces of the wall stem were measured using Road Research Laboratory type RRL/SS vibrating wire strain gauges, which were concreted into the wall in two rows at the same levels as the pressure cells. In addition RRL/SS vibrating wire gauges were concreted into the top of the piles and guide tubes to measure the inclination of the wall stems were concreted into the wall stems of both panels. Sufficient concrete cubes were made for long term testing while shrinkage beams and creep beams (20" x 4" x 4") were cast using concrete from the wall and piles in accordance with the recommendations of Tyler (1969). Details of the positions and types of gauges used in investigations are shown in figures 2 and 3.

A temporary tower was built in front of the wall to house the gauges and gaugehouse during the construction period. When the motorway embankment was completed a permanent instrument house was built behind the motorway verge between the panels.

After an initial period to allow the gauges to stabilise the zero readings of the total pressure cells were established. Regular readings of the internal strain gauges commenced in December 1970 and the earth pressures were measured once the filling had started. The inclination of the wall was measured before an after filling. It is planned to continue measurements for several years. The motorway past the wall was opened to traffic in September 1972.

NORTH INGS INTERCHANGE INVESTIGATION

Description of North Ings Interchange Bridges

To conclude the earth pressure investigation on standard motorway earth retaining structures, it was decided to investigate the earth pressures acting on a typical motorway bridge. The object of the investigation, as at Grange Mill Lane and Hunsworth Dyers, was to obtain both quantitative and qualitative data of the earth pressures acting on abutments and wing walls under various standard backfilling conditions. In 1970 the construction of the Hatfield-Thorne section of the Rotherham-Goole Motorway (M.18) was started and consequently it was decided to instrument the bridges on the North Ings Interchange.

The North Ings Interchange overbridges connect the M.18 Motorway with the Scunthorpe Spur on Thorne By-Pass. The bridges are right two span standard overbridges using standard beams and reinforced concrete abutments and wing-walls. The South Bridge was backfilled with Bunter sand, taken from a borrow pit on site, and the North Bridge abutments were backfilled with a well graded made up limestone and bunter sandstone fill. Figure 4 shows the arrangement of the backfill wedges behind the abutments, together with details of the positions of the total earth pressure cells used to measure the lateral and vertical earth pressures.

The abutments were designed in accordance with standard practice to withstand an equivalent fluid pressure of

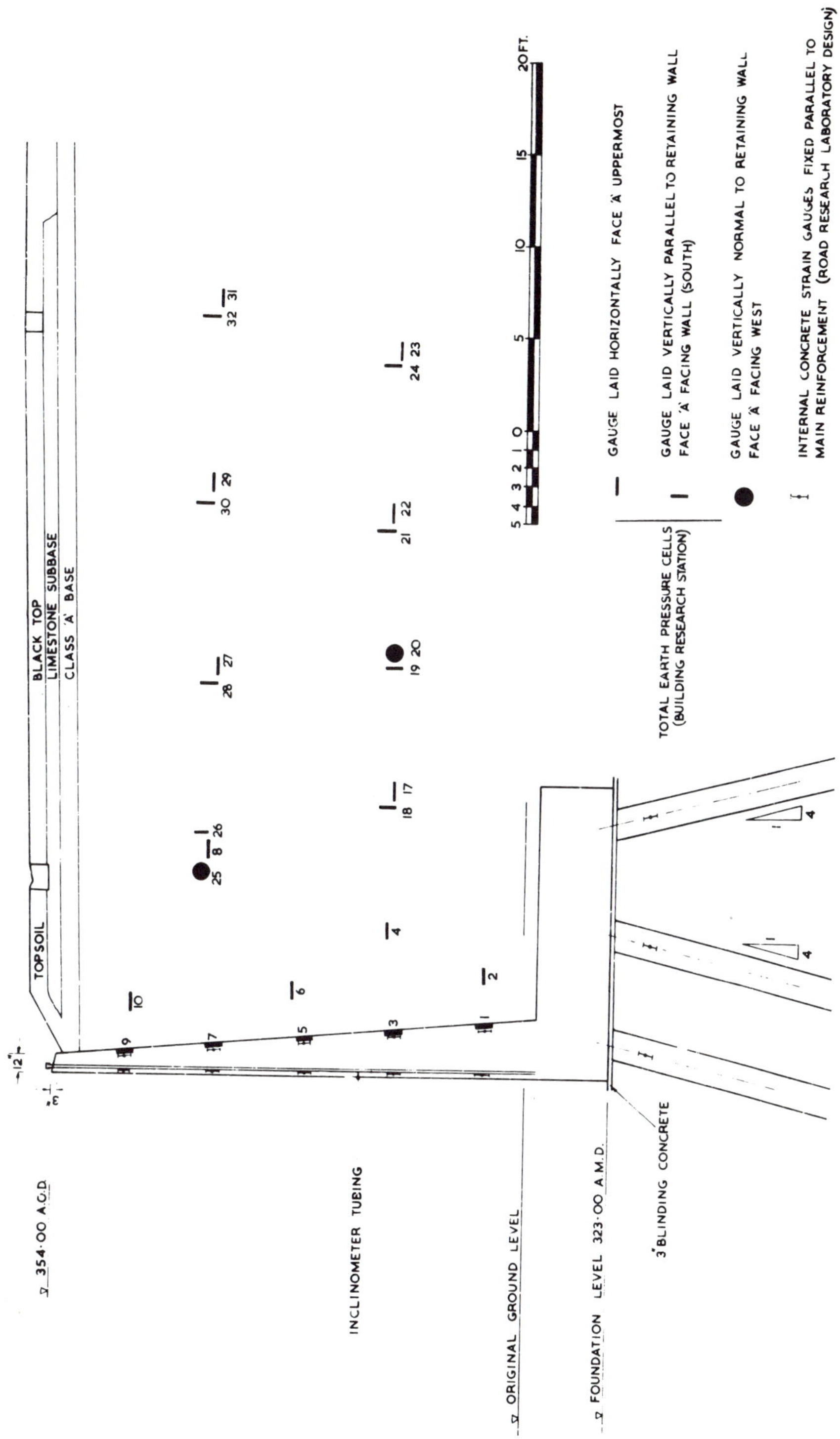
BLACK TOP
LIMESTONE SUBBASE
CLASS 'A' BASE
TOPSOIL
12"
354·00 A.O.D
INCLINOMETER TUBING
ORIGINAL GROUND LEVEL
FOUNDATION LEVEL 323·00 A.O.D.
3" BLINDING CONCRETE
5 4 3 2 1 0 5 10 15 20FT.
TOTAL EARTH PRESSURE CELLS (BUILDING RESEARCH STATION)
GAUGE LAID HORIZONTALLY FACE 'A' UPPERMOST
GAUGE LAID VERTICALLY PARALLEL TO RETAINING WALL FACE 'A' FACING WALL (SOUTH)
GAUGE LAID VERTICALLY NORMAL TO RETAINING WALL FACE 'A' FACING WEST
INTERNAL CONCRETE STRAIN GAUGES FIXED PARALLEL TO MAIN REINFORCEMENT (ROAD RESEARCH LABORATORY DESIGN)

Figure 2

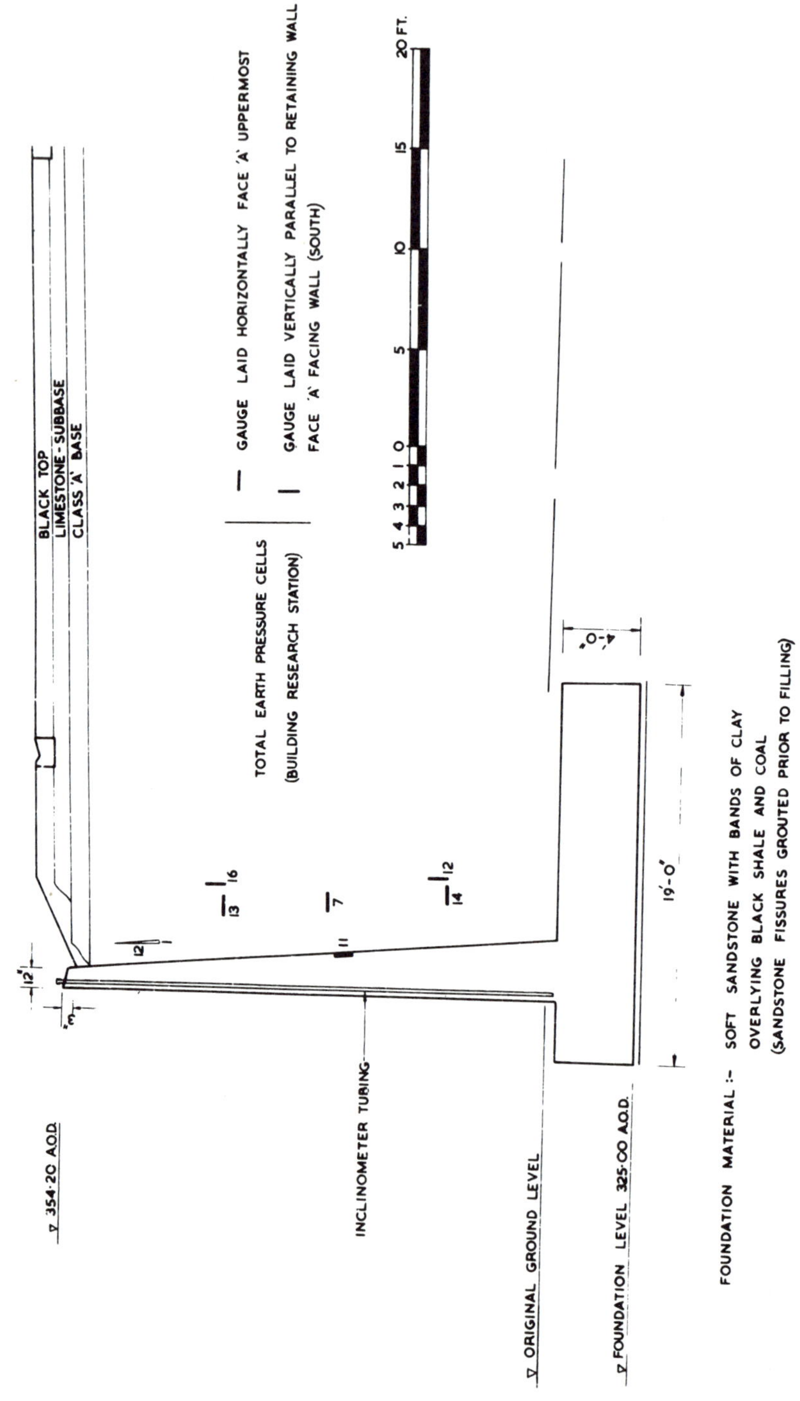
▽ 354·20 A.O.D
BLACK TOP
LIMESTONE - SUBBASE
CLASS 'A' BASE
12"
3"
INCLINOMETER TUBING
▽ ORIGINAL GROUND LEVEL
▽ FOUNDATION LEVEL 325·00 A.O.D.
19'-0"
4'-0"
12
11
13
16
7
14
12
TOTAL EARTH PRESSURE CELLS
(BUILDING RESEARCH STATION)
GAUGE LAID HORIZONTALLY FACE 'A' UPPERMOST
GAUGE LAID VERTICALLY PARALLEL TO RETAINING WALL
FACE 'A' FACING WALL (SOUTH)
5 4 3 2 1 0 5 10 15 20 FT.
FOUNDATION MATERIAL :- SOFT SANDSTONE WITH BANDS OF CLAY
OVERLYING BLACK SHALE AND COAL
(SANDSTONE FISSURES GROUTED PRIOR TO FILLING)

Figure 3

SECTION C-C

WEST ABUTMENT

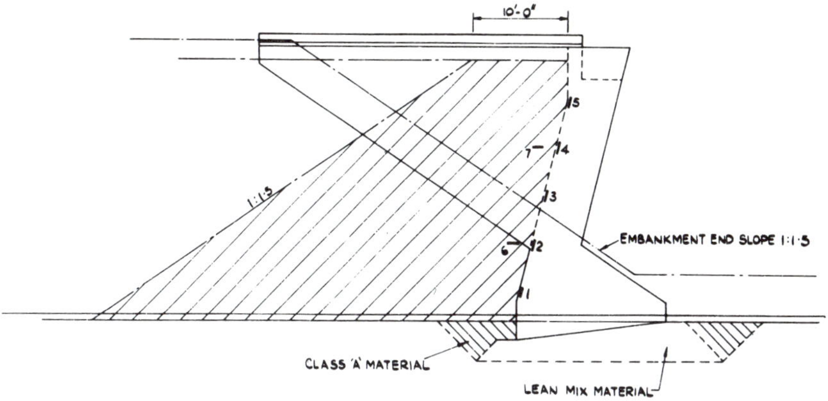

SECTION B-B

EAST ABUTMENT

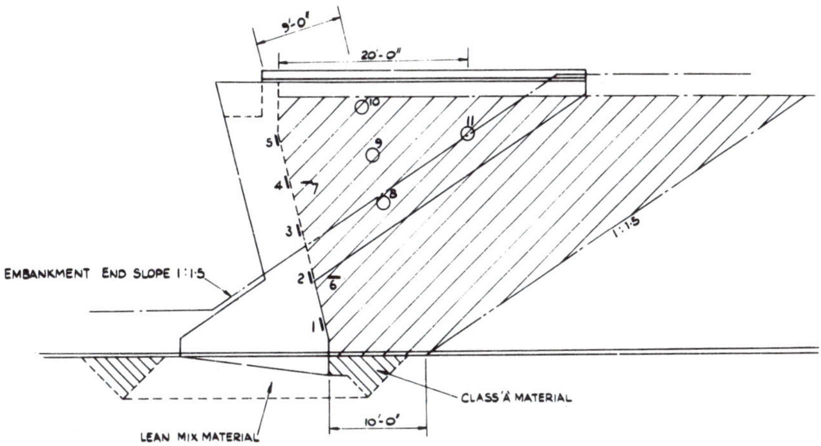

BRIDGE	ABUT	FILL	METHOD
NORTH INGS SOUTH	WEST	BUNTER SAND	SECTION B-B
NORTH INGS SOUTH	EAST	BUNTER SAND	SECTION C-C
NORTH INGS NORTH	WEST	CLASS 'A'	SECTION B-B
NORTH INGS NORTH	EAST	CLASS 'A'	SECTION C-C

Figure 4

30 lb./cu.ft. with an H.B. surcharge. Except for minor details all four abutments are identical and as a result, these bridges offer a unique opportunity to study the effects, under identical traffic conditions, of different backfill materials and backfill shapes, on the lateral pressures.

Instrumentation and Testing

As with the previous investigations lateral and vertical earth pressures in the backfill were measured using Building Research Station Total Earth Pressure Cells. Seven gauges were used behind each abutment, five being set flush with the backface of the abutments midway between the wingwalls, while two were laid horizontally in the fill, 2 feet away from the abutments. In addition four cells were set flush with the inside face of the north eastern wingwall to measure the lateral pressure acting on the wing walls. Figure 5 shows the position of the cells in the backface of the abutments.

Internal concrete strains near both front and back faces of the abutments were measured using Building Research Station vibrating wire strain gauges.

Figure 5

All cable leads from the gauges were terminated in four manholes constructed next to the individual abutments. Records of the pressures and strains were started as soon as the filling commenced. These were repeated at regular intervals and it is planned to continue the tests for several years. In June 1972 the motorway was opened to traffic.

EARTH PRESSURE CELL

The Building Research Station Earth Pressure Cell developed by Thomas and Ward (1969) was used to record the earth pressure at all the sites. The main reason for choosing this cell was that it had been specifically designed for reliable long term measurements. The cell was used in two forms, as a basic cell when laid in the fill, and as a boundary cell to measure pressures against a structure. Figure 6 shows the basic cell. The boundary cell is adapted from the earth pressure cell by encasing it in a thick circular trough $\frac{1}{4}$ in. deeper than the cell and $\frac{1}{2}$ in. larger in diameter. The gap between the trough and the cell is fitted with a cold-setting rubber compound. The boundary cell is mounted in the structure so that the uncovered diaphragm of the cell, the edge of the trough and the rubber annulus are all flush with the surface.

Installation Procedures

Grange Mill Lane Retaining Wall M.1 At the time of the first investigation boundary cells had not been developed. Following the recommendations of the Building Research Station the cells were placed in the fuel ash backfilling at a distance approximately 2 ft. back from the wall in a shallow depression that had been excavated out of the compacted fill. The cells measuring the vertical pressures were laid (Face A uppermost) on a prepared horizontal surface trimmed as plane as possible with a steel straight edge. Fuel ash was then recompacted over the cell to the original level by hand using a standard rammer. The cells measuring the horizontal pressures were held on edge and the fuel ash filling compacted in layers on both sides.
(From the experience gained at Grange Mill Lane the author suggests that better results could be obtained by forming artificial quicksand conditions in the pulverised backfill and "floating" the cells into position).

Hunsworth Dyers Retaining Wall M.62 At this site boundary cells were available to measure the pressure on the retaining wall together with the basic cells for measuring the earth pressures. The boundary cells were installed in the wall by attaching the casing to shutter immediately prior to concreting. The cell was held in place by being trapped between the back face reinforcement and the shutter. The cabling was concreted into the wall, emerging at the parapet.

North Ings Interchange Bridges M.18 Both basic earth pressure cells and boundary cells were installed at North Ings Interchange. The basic cells were installed using the same procedure as at Grange Mill Lane and Hunsworth Dyers. The installation of the boundary cells differed from the method adopted on the M.62 in that the Contractor 'boxed out' the back of the wall during the pour leaving a recess for the cell to be placed at a later date. The boundary cell was concreted into the wall during backfilling care being taken to ensure a sufficient time lapse for the concrete surrounding the gauge to set. The cabling in this method was exposed and was ducted to the gauge houses through P.V.C. pipes specially provided for the purpose. The exposed diaphragm of the cells in the wall were treated with bitumastic paint in accordance with the treatment given to the backfaces of abutments.

TYPICAL RESULTS

A very large number of earth pressures measurements have been recorded on the M.1, M.18 and M.62 motorways, both during and after construction; the readings are being continued. It is intended to publish details of the results when the investigation has been concluded, however, typical values are shown below.

Grange Mill Lane Retaining Wall M.1
The initial results of the M.1 investigation was published in 1971, these results showed that the lateral pressures were increasing and developing a parabolic distribution. Figure 7 shows the lateral pressures recorded at approximately six monthly intervals during the last 4½ years. An equilibrium condition appears to have been achieved.

Hunsworth Dyers Retaining Wall M.62
Figure 8 shows the lateral earth pressures against Hunsworth Dyers retaining wall as three separate occasions. It can be seen that an initial high lateral pressure produced during compaction fell off during the final stages of the construction of the wall. However, once construction of the motorway was complete the lateral pressures began to increase becoming parabolic in distribution, similar to that recorded on the M.1 Motorway.

North Ings Interchange Bridges M.18 Figure 9 shows the lateral earth pressures against the middle of the north bridge abutments immediately prior to the opening of the motorway. The containing action of the abutment wingwalls is apparent. The main conclusion that can be deduced from these measurements is that the lateral earth pressures acting on a standard bridge abutment are three dimensional in character, and depend upon the shape and position of the wing walls.

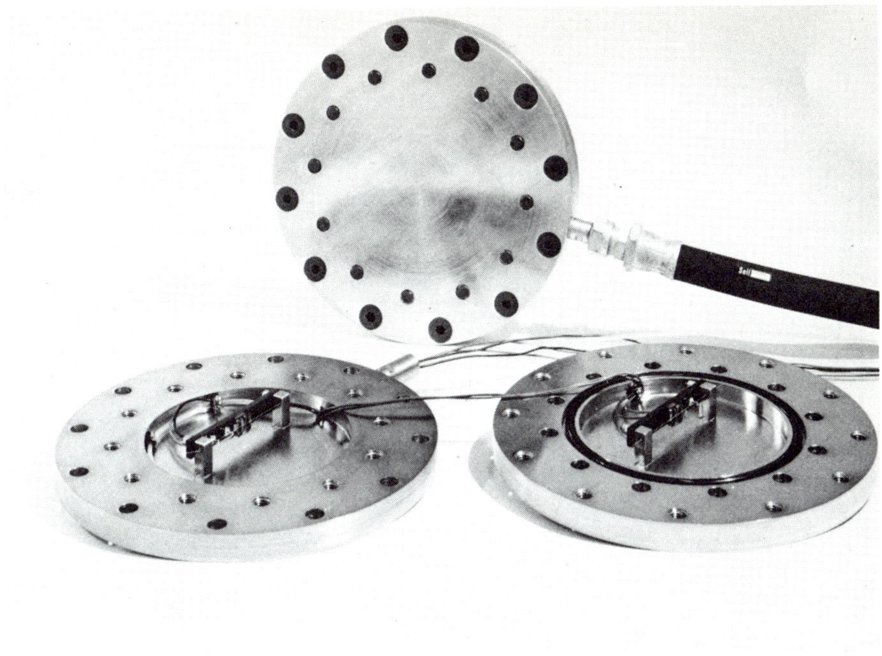

Figure 6

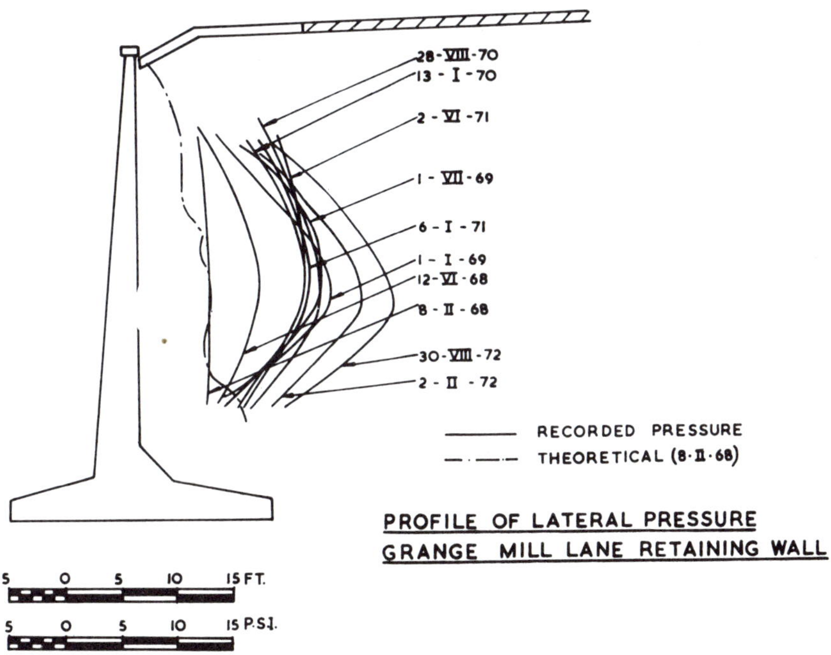

Figure 7

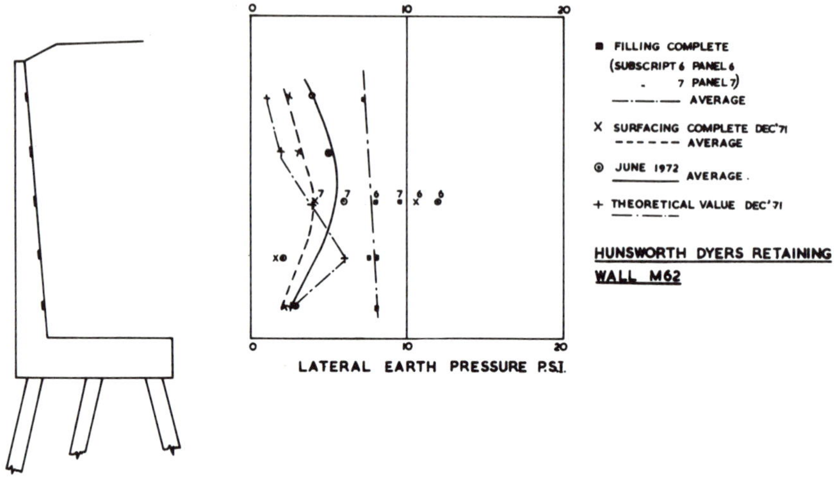

Figure 8

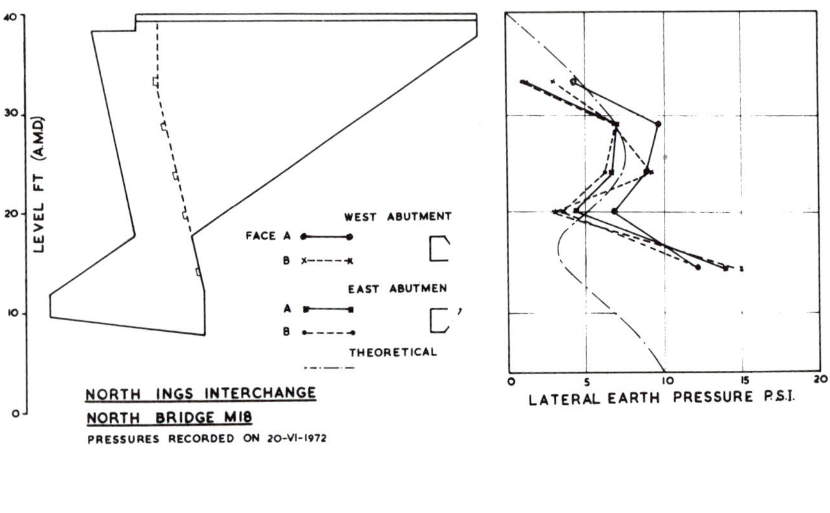

Figure 9

Theoretical Considerations

The results of the investigations show that the lateral pressures are greater than those expected and that the distribution is different from that assumed during design. The intensity of the lateral pressures and their distribution appear to be dependent on both season and on traffic intensity. The assumptions of the plastic earth pressure theories appear unjustified and the results of the tests suggested that an elastic approach to the problem was possible. A theoretical investigation based upon finite elements was started. The results of two dimensional analyses of the conditions at Grange Mill Lane on the M.1 and at Hunsworth Dyers retaining wall on the M.62 are shown plotted on figures 7 and 8. They give close agreement with the lateral pressures recorded in the immediate post construction period. Three dimensional finite element analyses of the conditions at North Ings Interchange bridges are being studied with the University of Leeds. The results of an initial analysis are shown plotted in figure 9.

Acknowledgements

The Author wishes to thank Mr. J. A. M. MacKenzie, F.I.C.E., Director, North Eastern Road Construction Unit, and Mr. J. A. Gaffney, B.Sc., F.I.C.E., the County Engineer and Surveyor of the West Riding County Council for permission to publish this paper, together with Mr. F. A. Sims, B.Sc., M.I.C.E., the Superintending Engineer Bridges, under whose direction the investigation has been undertaken. The Author would also like to acknowledge the work and assistance given to him by his colleagues in the Highways and Bridges Sections of the Department.

References

Sims, F. A., Forrester, G. R., and Jones, C. J. F. (1970) 'Lateral Pressures on Retaining Walls'. Jour. Inst. High. Eng. 6:19-30.

Thomas, H. S. H., and Ward, W. H. (1969) 'The design, construction and performance of a vibrating wire earth pressure cell'. Geotechnique <u>19</u> 1:39-51.

Tyler, R. G. (1969) 'Determining stress in concrete structures'. Civil Eng. June and July.

FIELD INSTRUMENTATION WITHIN A CIVIL ENGINEERING CONTRACT

M.F. Kennard

Partner, Rofe, Kennard & Lapworth

Introduction

This paper is based on the author's experience on several earth dam and embankment projects constructed in the last 20 years in the British Isles, and designed by Consulting Engineers. His personal views may not necessarily be applicable to all other types of projects where instrumentation is involved. Earth dam projects often involve large programmes of instrumentation with many types of instruments, which are required to be installed and observed from the short term, for constructional behaviour, to the long term for performance data of the structure. This may lead to a different approach than, for example, for the instrumentation for a motorway trial bank or for the stability of a slope of a deep excavation.

In a large earth dam project several parties can be involved in the instrumentation, including the Consulting Engineers, specialist consultants from universities, research establishment personnel, instrument manufacturers and suppliers, and the main contractor.

The contractual procedure must enable the following requirements to be achieved.

1. Reliable types of instruments to be available to be installed as required
2. Allowance for new developments in instrument design that become available, to be introduced
3. Installation to be carried out correctly and connected to recording equipment quickly
4. Testing of installation can be carried out
5. Layout of instrumentation to be flexible and to permit variation or additional instruments to be placed as work proceeds and results obtained
6. Initial readings to be reliable and soil samples and information to be obtained

7. All information to be recorded and made available to the Engineer quickly
8. Spares and replacement of consumable items to be available
9. Ancilliary equipment and fittings to be reliable
10. Cost to be reasonable

Some of these points may be obvious, but the author could quote examples,where individually,they have not always been achieved.

So that examples of instrumentation can be related to the observations, the end product of the exercise, the references given in this paper enable actual observations to be studied in their correct context. Most of the examples quoted are inevitably drawn from the author's experience.

Contractual methods

One procedure that is not recommended is for the instrumentation layout and details to be completely designed, specified and billed in individual items for the main Contractor to price. This makes it difficult to vary the design as the work proceeds and more information becomes available, and will inevitably lead to contractual arguments on variations, rates and possibly claims. It could also lead to the tenderer basing his rates on the use of an outside sub-contractor who is not acceptable to the Engineer.

In general, there are three methods of arrangment for the instrumentation on a large Contract, namely :-

Nominated Sub-Contract A firm, probably with specialist experience, is invited to tender for the supply and installation of the apparatus. This would sometimes include for maintenance and the taking of observations. The requirements and details of the apparatus would sometimes be specified in detail, but in some cases be left to the tenderer to detail. The tenderer cannot know all the design assumptions and problems that the Engineer has in mind.

Discussions with the specialist sub-contractor should be held before tendering to evaluate their expertise, and to agree on the type, reliability and range of the instruments to be offered.

This method does not involve the main Contractor, except for co-operation and providing facilities when the work is being undertaken in the field. The method can be useful for special requirements within other more general arrangements.

Nominated Supplier This method is similar to the previous example, but is limited to the supply of the apparatus. The design of the apparatus could be to the supplier's design or the Engineer's design, or a combination of both. It is unusual on a large project for the whole of the equipment to be required at a certain date and the order or contract should allow for additional items to be ordered and supplied later, as the work proceeds. Inspection of the instruments, before delivery to the site, is most desirable in all arrangements, but especially in this case where the supplier's responsibilities are limited to supply.

In this method the installation, and observations, can be carried out by the Contractor or the Resident Engineer's staff, or by a combination of their staffs.

A danger with the whole of the apparatus being specified and supplied by one firm, is that it could be difficult to make use of developments in instruments made by other suppliers, or research laboratories, that occur during the course of the contract.

Provisional Sum A more flexible method is to include a Provisional Sum in the Bill of Quantities for the supply and installation of the apparatus, and measured items for trenches, instrument houses, and other similar items of work. In this way, the main Contractor is not involved at the tender stage in very small items, which may be outside his experience, and the Engineer is free to make arrangements for the supply and installation of the instruments as the work develops. It is easy to arrange for a combination of different firms and specialists to suit particular requirements. If one firm is found not be be satisfactory, then changes could be made. This would not be so easy if the main Contractor was committed to a particular sub-contractor.

Where the client (when no consulting engineers are employed) or the consulting engineer has sufficient specialist knowledge and experience, the author considers this method to be the most satisfactory in use. Recent projects where the work has been carried out on these lines include Wraysbury Reservoir and Cow Green Dam, amongst many others.

The method also has the advantage that it is not necessary to specify all the apparatus in the Specification for the main Contract, other than a short description of the instruments required so that the Contractor knows what may be involved.

tegories of instrumentation

is desirable when planning the instrumentation for an tual project and the consequential contractual arrangements divide the work into three general categories :-

1. Instrumentation for constructional control, where

rapid feed back of information is required. Examples include deformation and pore pressure measurements where failure is being approached at critical stages of construction.

2. Instrumentation for constructional control where only a slow feed back of information is required. Examples include pore pressure measurements during early stages of construction where the observed rate of consolidation may control the need for further drainage measures or the extent of weight berms.
3. Instrumentation for semi-research aspects of design. It may be decided to instrument certain features of the structure to find out more about their behaviour during construction, without knowing whether the information will be immediately relevant. The information may not be immediately relevant but could well assist future designs. Slow feed back of information may be sufficient. Examples in this category have included the installation of strain gauges and pressure cells in a clay core.

The best contractual arrangement probably depends on the general category in which the instrumentation lies, and the personnel involved. In a particular case, the semi-research category could be in the hands of a research establishment, and the more routine constructional control categories in the hands of a specialist sub-contractor. Early work, such as for a trial bank, might also lead to a different arrangement than for the main constructional control instrumentation.

Examples

As an example of the kind of development that can take place, where a flexible contractual arrangement exists, mention can be made of Selset Dam (Bishop & Vaughan 1962). It was not possible to determine all the instrumentation requirements prior to construction, as the foundation conditions which were revealed led to the design and installation of sand drains, and the performance of the boulder clay fill needed to be observed so that the spacing of vertical and horizontal drainage blankets could be determined. The subsequent decision on spacing of blankets led to a different layout of instruments and the need for further instrument houses.

Discussions between Professor A.W. Bishop (Imperial College), Dr. A.D.M. Penman (Building Research Station) and the Consulting Engineers, led to various developments in instrument design and techniques, including the installation of water level settlement gauges; fine porous stone piezometer tips in preference to the BRS design which was based on the USBR design; an attempt at an internal displacement or strain gauge; and grouting of boreholes with a bentonite/cement grout above piezometer tips installed between sand drains. At this site, and at that period of time (1955-1960), it would have been more difficult, more time-consuming, and more costly to have carried out all this work through a more rigid nominated sub-contract.

Similarly at the Balderhead Dam, the decision to install pressure cells of the BRS vibrating-wire type (Thomas & Ward, 1969), which required special manufacture (by the BRS) and experimental installation techniques, necessitated an open method, which enabled BRS personnel, specialist Consultants and Resident Engineer's staff, with daywork assistance from the main Contractor, to install and operate the cells. Since then it has been possible to carry out such work on a more formal nominated supplier basis, but on one occasion (at Mangla Dam) it has been reported that the unsatisfactory manufacture of this type of cell by a supplier led to a number being unreliable (Binnie et al, 1967).

The satisfactory performance of the pressure cells at Balderhead Dam, was followed by the use of similar cells at a junction between the concrete dam and earth dam at Cow Green Dam (Vaughan & Kennard, 1972). Here the cells were manufactured by a specialist supplier, and the experience at other sites of these particular cells were considered by the Consulting Engineers before orders were placed.

In the last decade, specialist instrument suppliers have become established, so that certain instruments and techniques, which were developed on a semi-research basis, are now standard, Consequently the arrangements made on one job, may not be the most suitable for the next.

Another example, where the Consulting Engineer was able to carry out necessary instrumentation quickly and efficiently without the need for relying on outside suppliers or sub-contractors, was at Fiddlers Ferry Power Station Ash Lagoons (Al-Dhahir, Kennard & Morgenstern, 1970). Movement on a soft foundation necessitated a quick appreciation of the extent of the area where high pore pressures were occurring. It was possible for the Consulting Engineer's site representative to arrange for the driving of M.S. tubing with a loose pointed shoe through the fill and into the foundation, to observe the water levels and to plot the area of high pore pressure in a reasonably short time. Where all the instrumentation on a site was put into the hands of an outside firm, it would not necessarily be possible to arrange for such non-standard work to be carried out quickly and at modest cost.

One type of instrument that has caused the author concern is the inclinometer, as installed in the core and foundations of earth dams. The installation and observation techniques have proved to be much more difficult, time-consuming and expensive than would appear from a simple description of the apparatus. There are examples where the grooved tubing has been installed with an unknown orientation, where the tubing has a twist, (a disadvantage of plastic tubing), and where telescopic joints have caused a narrowing of the tube. The weather conditions at upland dam sites are usually more demanding than in the laboratory tests, and considerable improvisation has been necessary to obtain satisfactory results at the time when critical conditions existed in an earth dam foundation. Field experience has led manufacturers developing and improving many aspects of this type of instrument, but after site work at the cost of the clients who have obtained such instruments. Considerable recent developments are overcoming these problems and future installations will reap the benefit of other's teething problems. The difficulties of installing vertical tubing and maintaining them in the midst of heavy constructional traffic will remain with this type of instrument. The author's experience has shown that sometimes it is necessary for the supply of an instrument by a specialist manufacturer to be backed up by an advisory service, possibly with some form of guarantee.

If the supply and installation of inclinometers were left to a main Contractor, serious problems and, possibly, lack of useful results could arise. The use of an inclinometer may require surveying and levelling to be carried out when observations are made. This involves more than direct reading of the instrument by an observer and assistants, and this can usually only be undertaken by the Resident Engineer's staff.

The cheapest type of instrument, or cheapest type of component is not always the most satisfactory. Proved reliability elsewhere, or improvements based on laboratory and field use should be considered of prime importance. A supplier may sometimes offer a reduced price, for an alternative component which becomes available on the market, but is improved. An example is the use of Nylon 6 in place of Nylon 11 for the leads of hydraulic type piezometers. Because of the lower price for Nylon 6 tubing, it was supplied for use at Llyn Celyn and Derwent Dams, whereas earlier tests on Nylon tubing at Selset Dam had shown Nylon 11 to be satisfactory. At these two dams where Nylon 6 was used it was found that the tubes were becoming blocked, and flushing with a solvent, such as methyl alcohol was necessary. At Derwent Dam, 10 tubes were lost because of this (Buchanan, 1970). Although this type of incident could happen with any contractual arrangement, it is more likely to occur when the expertise of the specialist engineers involved is not fully taken into account and the matter of instrumentation left solely to a commercial supplier.

Causes for concern

There are two general aspects of instrumentation that can give cause for concern (Kennard, 1971). These are the escalation in the numbers of instruments now being installed, and whether the instruments themselves could have any effect on the safety of the structure. Both these aspects are related to the methods adopted for ordering, installing and reading instruments for a large contract.

If the numbers and types of instruments to be installed are the result of collecting ideas from a number of advisers and specialists within and outside the engineer's firm or office, then escalation of instrumentation is inevitable. The problem involved in installing them and then the physical effort required in reading and plotting the results utilises competent personnel whose time could be better spent on other supervisory matters. It is very probable that the results would not be plotted and assessed as quickly as may be necessary.

There are dangers in installing instruments that could affect the safety of the structure. For example, piezometer leads in a trench in a clay core if poorly backfilled or filled with the wrong material, could act as a drain through the core, and the drainage would cause lower pore pressures which on analysis would indicate a higher factor of safety occurring than actually exists. In another case, standpipes in a foundation if imperfectly sealed could lead to leakage downwards to an underlying aquifer, or if the water pressure in the aquifer is artesian, could lead to a flow upwards into the fill. There could be a possibility that an outside agency, not familiar with all the design problems and assumptions, could, perhaps inadvertently, undertake work that would affect the safety of the structure. This is especially so when the ground water conditions will be different after impounding from those at the time when the instrumentation was installed. It is therefore important that the contractual methods of supply and installation do not lead to such imperfections. Supervision by experienced personnel, especially by the Resident Engineer's staff, is essential and they should have sound knowledge of the design requirements.

Observations

The author would not be happy to see an item in a Bill of Quantities for the main contract to cover taking readings of pressure cells, settlement gauges, inclinometers, etc. by the Contractor on a priced rate basis.

It is considered that this work should be carried out by the Resident Engineer's staff in all cases. It is recognised that probably there are projects where the Contractor, or a sub-contractor, has undertaken this work satisfactorily, but in general, there could well be defects with this method.

Casagrande (1965) quotes a case on the Great Salt Lake railroad fill where an overworked surveyor kept results of heave pipes in his notebook, instead of getting them plotted and examined promptly by others. Whilst the records were in his pocket, the embankment failed! Such an occurence is more likely to happen when the observer is not directly responsible to the Resident Engineer or Consulting Engineer.

The taking and recording of observations on the site and the transmission of all relevant data to the Engineer can involve considerable numbers of site staff. It is most desirable that provision is made early enough for adequate staff to be provided on site and in the office, and for the client to know that such staff will be necessary. Sufficient experience must be available to enable personnel to be trained on the site to undertake the more routine work of installation and observation. In the planning of the work, provision must be made for the cost, accommodation and travel of the staff required.

Conclusion

In the brief review of "Instrumentation for earth and rockfill dams", Penman (1969) stated

> "......we now have reliable instruments which will measure pore pressures, total pressures, and vertical and horizontal strains. All have horizontal connexions to avoid disruption to construction and for a cost of 1 - 2% of the cost of the dam, provide the designer with information on its behaviour against which he can check his design assumptions, and if necessary modify the design during construction to effect further economies or prevent excessive deformations."

This development has been achieved through numbers of engineers, who have incorporated instruments in earth dams in Great Britain in the last 20 years. By following a flexible system of dealing with the instrumentation, rather than a rigid system, such as set out in the "Earth Manual" (published by the U.S. Bureau of Reclamation), the present more comprehensive position has been reached. Nevertheless, problems have arisen due to poor and inadequately developed equipment, poor installation and maintenance, and failure to read and interpret the instruments. Whereas inadequate planning and control, inadequate and unsuitable contractual arrangements and inexperienced personnel have been causes of these problems, the author considers that as the considerable knowledge and expertise of the recent decade or so is utilised in the future, these problems will be overcome. This will be of considerable benefit to the design, construction and performance of civil engineering structures involving geotechnical aspects.

Acknowledgment

The author acknowledges the benefit of many discussions with Dr. P.R. Vaughan in leading to the views expressed in this paper.

REFERENCES

Al-Dhahir, Z.A., Kennard, M.F. and Morgenstern, N.R. (1970) Observations and pore pressures beneath the ash lagoon embankments at Fiddler's Ferry Power Station. Conf. In-Situ Investigations in Soils and Rocks, 1970: 265-276.

Binnie, G.M. et al, (1967) Engineering of Mangla Proc. Inst. Civ. Engrs. 1967, 38: 343-544

Bishop, A.W. & Vaughan, P.R. (1962) Selset reservoir: design and performance of the embankment. Proc. Inst. Civ. Engrs. 1962, 21: 305-346

Buchanan, N. (1970) Derwent Dam - construction Proc. Inst. Civ. Engrs. 1970, 45: 401-422

Casagrande, A. (1965) Role of the "Calculated Risk" in Earthwork and Foundation Engineering J. Am. Soc. Civ. Engrs. 1965, SM.4: 91: 1-40

Kennard, M.F. (1971) Discussion on Application of Instrumentation to Earth Dams, by Rofe, B.H. & Tye, P.F. J.Inst.Wat.Engrs. 1971,25:153

Penman, A.D.M. (1969) Instrumentation for earth and rockfill dams BNCOLD News & Views, 1969, 5:7-10

Thomas, H.S.H. & Ward, W.H. (1969) The design, construction and performance of a vibrating wire earth pressure cell. Geotechnique, 1969, 19: 39-51

Vaughan, P.R. & Kennard, M.F. (1972) Earth Pressures at a junction between an Embankment Dam and a Concrete Dam: Proc. 5th European Conf. on Soil Mech. & Found. Eng. Madrid, 1972: 215-221

EXPERIENCES WITH INSTRUMENTATION FOR EMBANKMENT DAM PERFORMANCE

A. L. Little

Geotechnical Consultant, Binnie and Partners

The experiences described in this paper all relate to dams constructed under the supervision of Binnie and Partners from about 1951 to date.

Hydraulic piezometers

The first post-war installation of embankment dam instruments for Binnie and Partners was made by Mr. A. D. M. Penman of the Building Research Establishment in 1951 at the 135 ft high Daer dam in southern Scotland. Twelve piezometer tips, all on the downstream side of the concrete core wall were connected to a gauge manifold with small bore polythene tubes. The tips were of the flat disc type with 2 inch diameter porous stones. Only two Bourdon gauges were provided initially and the various tips were connected to the gauges by manipulating cocks at the manifold. Although no appreciable pore pressures developed, the long response time of the tips meant that the manifold arrangement was unsatisfactory and the system was changed so that each tip had an individual gauge.

The very low pore water pressures were checked by drilling in standpipes when the dam had been constructed to about half height in 1952. Water tipped into the standpipes flowed away, leaving little or no residue, confirming the low pore pressures measured by the piezometers.

Once the manifold was replaced, the installation functioned satisfactorily but, because the porous stones in the tips were coarse grained and the polythene tubes, although waterproof, were permeable to air, de-airing had to be done frequently (Penman, 1961).

The next installation for Binnie and Partners, also made with the assistance of Mr. Penman, was at the Usk dam, beginning in 1951. After experience at Daer, each tip was provided with an individual gauge, although all tips were connected to common "bus bars" in the gauge house. The de-airing equipment was essentially that still used in many of the smaller installations in UK. Indeed, it is a tribute to the designers of this early apparatus that similar installations, differing only in points of detail, have continued to be provided down to the present day. Concrete block gauge houses were built on the downstream stone toe of the dam. A good deal of the trouble was experienced with freezing of the installation during the winter. Although paraffin heaters in the gauge houses helped, the plastic tubes crossed to the first gauge house in the open air from the embankment and froze up when the air temperature was particularly low. The zero readings of some gauges were affected but the most startling effect of freezing was an apparent increase in pore pressure which created at least one scare. For a second gauge house which was built at the site, the crossing was made in the fill, although freezing still affected the results occasionally.

Very high pore pressures were measured in the Usk fill in the 1951/52 season and horizontal drainage mattresses were introduced to prevent failure (Sheppard and Aylen, 1957). Fifteen additional standpipes were driven into the fill in 1952 and readings from these confirmed the high pore pressures. Nine of these standpipes were converted to gauge reading in 1953.

The Usk installation was fully justified by events and the piezometer readings proved invaluable for the redesign of the embankment and checking that its subsequent performance was satisfactory.

Because of the critical condition of the embankment, the piezometers continued to be read for over 10 years. At the end of that period one BRE type tip and 4 standpipes had ceased to function. The remainder were disconnected from their gauges and the tubes capped for use in the future if they should be required.

In 1955, 24 piezometers were installed in a small embankment (c 30 ft high) being constructed from over-consolidated London clay at Hanningfield in Essex. Although the tips were of the coarse stone type, negative pore pressures down to 3 psi were measured which gradually increased in about $3\frac{1}{2}$ years to small positive pore pressures in the downstream shoulder and to about reservoir level in the upstream shoulder. Negative pore pressures were also measured in another embankment of the same height, built of overconsolidated clay (probably glacially redeposited Forest Marble) at Foxcote, near Buckingham (Little, 1958).

At Hanningfield, because of the particular circumstances of the site, it was not possible to provide a gauge house at the outer toe of the dam and the gauges were accommodated in a dry well adjoining a pump sump. The well was very damp and electrolytic action developed between the brass cases of the gauges and aluminium spacers inside the gauges. It was surprising that this trouble had not been experienced previously, for example in the hot, humid atmosphere of boiler houses, but the gauge manufacturer refused to take any interest in the problem. The gauges were eventually replaced with another make containing no dissimilar metals in contact.

In 1958, with the construction of the Ayer Itam dam, Penang, it was realised that to de-air the 50 piezometers installed, some of them 800 ft from the gauge house and 150 ft above it, would require considerable time and effort with the hand apparatus then current. Accordingly, a design was evolved using a submersible pump to circulate the de-aired water and flow control by a rotameter. The design was successful and is still in use but later developments have been simplified (Little and Vail, 1961). At Shek Pik, Hong Kong, it was found necessary to insert a cooling coil (adapted from a local air conditioner) in the circulating water to prevent an excessive rise in the operating temperature of the submersible pump. At Mangla, with very long leads and high heads, non-submersible Mono pumps were used successfully. A minor problem at Penang was created by the supply of flat disc type tips with no water passage behind the porous stone (Penman, 1956). This was discovered when it proved impossible to circulate de-aired water, fortunately before installation. Shortly after the Penang installation, Mr. Penman drew attention to the permeability of polythene tubes to air. After that, all installations were made with nylon or, latterly, polythene coated nylon, to reduce permeability to both air and water.

At Penang, both upstream and downstream foundation piezometers recorded erratically before impounding; the downstream piezometers continued to record erratically after impounding. It was concluded that the coarse stone tip piezometers were attempting to record pore air pressures in the partially saturated foundations, rather than pore water pressures (Little, 1965). Since that time, most piezometer installations have been made with fine grained high air entry value porous stone tips (Bishop, Kennard and Penman, 1961)

In October, 1964, during a statutory inspection of the Deep Hayes dam under the Reservoirs (Safety Provisions) Act, it became necessary to find the seepage line in the downstream slope. Some steel pipes with retractable points were driven to form piezometers resembling those described by Parry (1971); the seepage line was found to be reassuringly low, although the ground at the toe of the dam was wet. The device has subsequently been used at a number of other locations.

A bubbler type of piezometer was used for some locations at Mangla where gauges could not be accommodated at a sufficiently low level and the ground was initially dry (Little, 1968). A second tube is introduced into a standpipe terminating in a normal Casagrande pot. Air forced down this tube returns to the surface where it is detected in a bubbler as soon as the air pressure is equal to the pressure of the column of water standing above the Casagrande pot. This installation worked well at Mangla and at Withens Clough dam where it was used for otherwise inaccessible upstream piezometer positions.

Difficulty was experienced with the nylon leads used at Tryweryn dam (Llyn Celyn), near Bala, N. Wales in 1962. Leads, which were water filled during storage because of the absorption of water by the nylon, were found to contain a white sediment. Attempts to clear leads to de-air piezometer tips already installed led to their complete blockage. After a few experiments it was decided to leave the remaining leads alone and many of the tips have continued to give consistent readings. Consultation with a nylon chemist disclosed that the nylon in use was type 6, based on caprolactum; this is water soluble but as the water stored in the tubes became nylon saturated, it began to deposit excess nylon at inconvenient places. The type of nylon was changed to type 11 and no further trouble of this kind has been experienced at any site since.

Other tubing hazards spring from manufacturing defects. Variations in wall thickness cause weak spots and constrictions; occasional complete blocks are not unknown. It is difficult to test the tubing in lengths of up to a mile. A proposal to force a small steel ball by air pressure along the tubing to test its bore met with an unenthusiastic response from the manufacturer although at one site the ball was pulled through a complete coil of tubing with a magnet.

Rodents (and insects in tropical areas) have been caught eating polythene tubing in ducts. The cure for this seems to be to cover the tubing with sand or earth.

Gauge houses buried in the embankment slopes or shafts are less difficult to insulate against extremes of either heat or cold. However, condensation and the need to heat the colder gauge houses produces an unpleasantly warm humid atmosphere. Although this had not been found to affect the instruments unduly, it may affect the efficiency of the staff taking the readings.

At Arlington dam, Sussex, gauge shafts were provided from the bottom of the rather deep excavation required to remove soft alluvium from the foundations. As a matter of convenience, the Contractor elected to complete the shafts almost to their full height of 30 ft before backfilling the excavation. The consequent difficulty of access to the shafts made them unpopular with the Resident Engineer's staff who had to take the readings in them during construction.

Electrical piezometers

Fourteen electrical piezometers were installed in the foundations of the raised Tittesworth dam in 1960. Most of these still continue to give readings. The next installation of electrical piezometers made by Binnie and Partners was at Diddington in 1962 in the foundations and fill of a road diversion embankment about 30 ft high constructed of the heavily overconsolidated Oxford clay. Because of doubts expressed about the ability of electrical piezometers to record correctly in partially saturated fills with strong negative pore pressures, the electrical piezometers were paired with hydraulic piezometers fitted with fine grained high air entry value porous stones. Air entry values were about 1.5 psi for the electrical piezometer tips and about 30 psi for the hydraulic piezometer tips.

In the foundations, both sets of piezometers, electrical and hydraulic, were consistent and recorded similar results. However, in the partially saturated fill, there was no agreement between the reading of a hydraulic piezometer and the corresponding electrical piezometer. All the hydraulic piezometers showed negative pore pressures, sometimes very strongly negative. With the exception of one, the electrical piezometers recorded positive or small negative pore pressures. The exception at first recorded a strongly negative pore pressure which quickly became less negative; finally the piezometer indicated a small negative value; three out of the six electrical piezometers recorded atmospheric pressure after 11 months (Little, 1964). The pore pressures inside a moderately wet compacted fill of overconsolidated clay would be expected to be negative; it is, in any case, difficult to see how they could be at atmospheric pressure. It was concluded, therefore, that the hydraulic piezometers were correctly recording whereas the electrical piezometers were recording incorrectly or, in some cases, the pore air pressures.

An objection to electrical piezometers which has been raised before (Bishop, 1961) is that they cannot be de-aired. This objection applies of course to other types of piezometer which transmit piezometric pressure from a diaphragm, by oil for example. Not only can air accumulate up against the diaphragm, affecting the reading, but the datum to which the diaphragm works may also be affected. As a result of experience, electrical and other piezometers which cannot be de-aired are now used only in those positions where they will be permanently saturated.

Some difficulty was experienced at Mangla with electrical leads breaking under the large settlements which were taking place.

Pneumatic piezometers

Experience with pneumatic piezometers has so far been confined to a single installation in a trial bank being built in the Wash. Conditions at the site are particularly severe both for equipment and for operators who have to work at the top of an exposed scaffolding tower to be above high tide level.

Twenty pairs of hydraulic and pneumatic piezometers were installed in the sand fill and nine pairs plus eight individual pneumatic piezometers in the foundations. In the foundations, eight pneumatic tips no longer read. This is probably due to leaks in leads or at couplings. Some of the pneumatic tips gave piezometric levels 0.45 m higher than the hydraulic tips for readings of about 3 m of water. The action of the pneumatic piezometer depends on the closing of a valve when the applied air pressure equals the pore water pressure. Because of the long leads being used, initial charging of the system with air was at a "fast flow rate" until the valve closed; the pressure was then relieved slightly and the reading taken at the "test flow rate". The operator found that, depending on the amount of pressure relief which he gave before taking the reading, he obtained differences of up to one metre head of water in a reading of 10 metres. Even on short lead piezometers, using the test flow rate only, differences of 0.3 m were obtained for consecutive readings at short intervals. Water leaked into some of the pneumatic leads and it was found impossible under site conditions to remove it.

It would be wrong to condemn pneumatic piezometers on the basis of this single unfavourable experience. It is hoped to have another installation under better conditions elsewhere. One of the most serious drawbacks of the hydraulic piezometer is the need to keep the gauge house at or near the level of the tips; a satisfactory pneumatic piezometer would overcome this drawback and could be used at many places where the normal hydraulic piezometer is useless.

Total pressure cells

The first installation of total pressure cells on a Binnie and Partners' job were made at Mangla where 40 BRE design cells were installed in the fill of the main dam in 8 groups of 5 cells each. The cells are of the vibrating wire strain gauge type and are purged with dry nitrogen to prevent corrosion and keep the interior of the cells at atmospheric pressure. From the first this was a troublesome feature and the performance of the cells was, on the whole, disappointing. Most of the cells in the first group to be installed, quickly went out of action. Later cells were dismantled on arrival at the site and were found to be corroded to various degrees; some even contained measurable amounts of water. It seems likely that condensation on the journey accounted for the condition of the cells, since they were purged with dry nitrogen before leaving the factory. Each cell had a different calibration curve and this made interpretation of the readings a laborious business. In a later installation of cells at dams in Singapore and Arlington, Sussex, the dry nitrogen feature was eliminated but it is doubtful if any more of this type of cell will be installed.

At Brianne, Gloetzl cells were installed in groups of five to measure total pressures. The cells are oil filled and connected through a manifold to a bourdon gauge.

The Gloetzl total pressure cells have behaved well and have shown the development of strong arching action in the core of the dam and principal stress rotation in the shells (Carlyle, 1973).

Slope and settlement indicators

On earlier jobs, the USBR type of internal settlement gauge was used, with mechanical indication at cross arm positions. This was found to be a very satisfactory form of equipment except at one site where there was a tendency to lose torpedoes down the tubes; this was mainly due to a failure of the mechanism retracting the engaging pawls and, in consequence, there was no alternative to abandoning the torpedoes below the lowest cross arm. Fortunately, the extension length of tube into original ground forms a convenient sump for the reception of discarded objects.

A need to measure horizontal deflections within dams called for a more sophisticated instrument and at several sites, the deflection indicator manufactured by the Slope Indicator Company has been installed. Civil engineering sites do not provide ideal surroundings for sensitive electrical apparatus and from time to time, the torpedoes of this equipment have given trouble and have had to be returned to the USA for service. Nevertheless, useful records have been obtained and readings are still being taken at Mangla, for example where horizontal deflections up to 1.0 metre have been measured. A successful installation was made below water level at Hong Kong (Dunnicliffe, 1968). The instrument has been modified to read settlements as well as deflections. When the deflections become large, it is difficult to coax the torpedo round the rather sharp curves.

A simplified and, it is hoped, a more robust form of instrument has been purchased for use at the Singapore dam previously mentioned. In this instrument, settlement is measured by the magnetic effect on a reed switch of steel rings outside the plastic casing.

Seepage measurement

This is a most important parameter of dam performance. So far, installations have been of a comparatively simple type, measuring the seepage over brass or latterly stainless steel 90° V-notches. The installation is necessarily below ground level and it is common for the chamber to be finished at ground level with a manhole access. This tends to discourage frequent readings; it would be desirable to have a small house covering the site, perhaps in conjunction with the gauge house for piezometers and other instruments.

Conclusions

In conclusion, it might be said that the most successful instruments are simple and robust. The hydraulic piezometer is the outstanding example of a highly successful instrument which has survived almost in its original form since about 1938. Civil engineering sites are not sympathetic to elaborate types of equipment and are particularly inhospitable to those using any form of electrical energy.

Gauge houses need to be strong, well insulated and, in cold climates, kept above freezing point in the winter. The buried type of gauge house is easy to insulate but may bring problems of condensation. Accessibility should be good to encourage systematic reading of important instruments.

Acknowledgements

The Author is grateful to Messrs. Binnie & Partners for permission to publish the paper and to members of their staff, in particular to Mr. G. C. G. Beavan for assistance in its preparation.

REFERENCES

Bishop, A. W. (1961) Discussion on Session 2: Earth dams, Embankments and Foundations. Pore Pressure and Suction in Soils, Butterworth, London, 133.

Bishop, Kennard and Penman (1961) Pore Pressure Observations at Selset Dam. Pore Pressure and Suction in Soils, Butterworth, London, 91.

Carlyle, W. J. (1973) The Design and Performance of the Core of Brianne Dam. Eleventh ICOLD, Madrid, Q42, R26. (In the Press).

Dunnicliff, C. J. (1968) Instrumentation of the Plover Cove Main Dam. Geotechnique, London, 18, 3:282.

Little, A. L. (1958) Compaction and Pore Water Pressure Measurements on Some Recent Earth Dams. Sixth ICOLD, Q22, R42, 3, 205.

Little, A. L. (1964) Discussion on Q29. Eighth ICOLD, Edinburgh, 5, 292.

Little, A. L. (1965) Piezometer Observations During and After Construction of the Ayer Itam Dam. Proc. Fifth Symp. Civ. and Hydr. Engrg. Dept., Indian Inst. of Science, Bangalore, 1, paper A15.

Little, A. L. (1968) Mangla Instrumentation. Water Power, London, 20, 7:257.

Little, A. L. and Vail, A. J. (1961) Some Developments in the Measurement of Pore Pressure. Pore Pressure and Suction in Soils, Butterworth, London, 75.

Parry, R. H. G. (1971) A Simple Driven Piezometer. Geotechnique, London, 21, 2:163.

Penman, A. D. M. (1956) A Field Piezometer Apparatus. Geotechnique, London, 6, 2:57.

Penman, A. D. M. (1961) Introduction to Session 2: Earth Dams, Embankments and Foundations. Pore Pressure and Suction in Soils, Butterworth, London, 120.

Sheppard, G.A.R. and Aylen (1957) The Usk Scheme for the Water Supply of Swansea. Proc. Inst. Civ. Engrs., London, 7, 246.

THE DETECTION OF LEAKAGE THROUGH THE CORE OF AN EXISTING DAM

H. T. Lovenbury

Rofe, Kennard and Lapworth

INTRODUCTION

During repairs to the clay core of Balderhead Dam in 1968 a grid of standpipe piezometers was placed in the upstream shoulder to give advance warning of any future leakage through the core. The installation was based on the principle that any local depression of piezometric elevation (relative to reservoir level) arising in the shale fill from flow through the core would immediately be registered by adjoining piezometers. Its practicability was proved when in 1970 a small leak was detected and located in an area of the dam which had previously been left untreated, enabling remedial action to be taken quickly before serious erosion could develop. This paper describes the design and installation of the piezometer grid and presents the observations which led to the discovery of the 1970 leak.

PRINCIPLE OF DETECTION SYSTEM

The development of cracking in the Balderhead Dam core between 1966-67 and the methods adopted for its repair have been fully described elsewhere (Vaughan et al.,1970). Briefly, following first impounding in 1966, both the flow and turbidity of the main underdrain rose; subsequently localised settlements occurred along the crest and in 1967 two sinkholes formed. Exploratory drilling and testing revealed that considerable erosion had taken place within the boulder clay core at a number of positions; and it was later deduced that the cracking had been initiated by hydraulic fracture of the core under almost full reservoir pressure. Under the low flow conditions in the cracks the coarser material had segregated, leaving the fines to be washed out through the filter.

Re-examination of the piezometer observations (Vaughan et al., 1970 - Fig. 4) showed that two Bishop-type piezometers

(D4 and U7) in the relatively free draining upstream shale fill had exhibited increasing depressions of piezometric elevation during the period that leakage was taking place through the core. The maximum depressions reached, relative to reservoir level, were 3.0m at piezometer D4 and 2.1m at piezometer U7, whilst the overall increase in underdrain flow was approximately 30 l/s. When the leakage was halted in April 1967 on lowering the reservoir by 9.2m, the piezometric elevations returned to reservoir level. Piezometer D4, at a position 15.5m below crest level and 2.4m from the upstream face of the core, was at a distance approximately 18m from a zone of damage, extending from 9m to 24m depth beneath the crest. Piezometer U7, at a depth of 22.2m, was located 13.4m from the core and approximately 15m from the damaged zone.

These observations indicated that the velocity head loss (or cone of depression) due to seepage flow through the core could be registered by nearby piezometers in the upstream fill. It was recognised that this fact could be applied to devise an advance warning system capable of detecting any future trouble before severe damage resulted. Provided that the piezometers were spaced in the fill at sufficiently close intervals over the length of the dam, the system would also enable the position of the leak to be defined.

Monitoring would be conducted by recording the difference between the piezometric elevation and the reservoir level at regular intervals. Any consistent local increase in the piezometric difference would be indicative of some core leakage. Surveillance would be mainly focussed on the periods during which the reservoir was at or close to top water level when leakage would be most likely to develop. Furthermore, if the system was sufficiently sensitive it would allow leakage to be identified ahead of any measurable increase occurring in underdrain flow.

DESIGN AND PERFORMANCE OF PIEZOMETER INSTALLATION

Packer tests performed in two pilot holes indicated that the horizontal permeability of the upstream shale fill lay in the range 2 x 10^{-5} to 5 x 10^{-7} m/s. In selecting the horizontal spacing between piezometers, it was considered that the minimum identifiable piezometric response to leakage flow should be taken as 0.3m. Approximate calculations, based on radial flow theory, suggested that an equilibrium depression of this order would be produced at a range of 15m by a flow as small as 0.15 l/s. Thus the piezometer spacing was tentatively fixed at 30m.

So that the piezometric elevations could be determined to this accuracy and for ease of installation, it was decided to form the grid with standpipe piezometers. These could be placed close to the upstream face of the core in raking boreholes drilled from the dam crest. They also had the

,tage of being simple, direct reading and easily serviceable. At any section several piezometers would be needed at different depths to define the elevation at which leakage was taking place. And so to avoid the difficulty of satisfactorily sealing separate elements in a single borehole or alternatively having to drill a number of holes at one position, a multi-unit piezometer was developed. It was hoped that the use of this type of piezometer would enable the installation to be completed before the reservoir refilled.

This piezometer (Vaughan, 1969 - Fig. 7) can accommodate up to four elements in a single 75mm diameter borehole. Each element is annular in shape, consisting of a 57mm diameter external sleeve of porous polythene (supported by perforated pvc tubing) and a 25mm O.D. inner tube, joined at each end to a pvc bulkhead. A 9.5mm O.D. nylon standpipe is connected to each element. The elements are separated by flush-coupled spacer tubes, with the standpipes passing inside the inner tubes of the higher elements to the top of the borehole. In this installation the elements were 4.6m long and spaced at 9.2m centres. The piezometers were placed in NX-cased boreholes at an upstream rake of 5 to 1. The longest piezometer stem installed was 41m.

The units were cemented together as the stem was lowered down the hole by means of a wire attached to its base. The casing was raised to the top of the lowest element, allowing the surrounding shale to close up around the element and prevent the stem lifting on withdrawal of the remaining casing. The piezometer was then sealed in a sleeve of bentonite/cement grout injected down the inside duct of the stem and released through ports drilled in the spacer tube above the lowest element. The permeability of the grout was approximately two orders of magnitude less than that of the surrounding shale fill. However, when readings are taken only at daily or longer intervals, Vaughan (1969) has shown that the influence of the grout on the response time is insignificant.

The water levels in the standpipes are measured directly using a dipmeter with a special narrow probe. As monitoring is concentrated on the final 3m of reservoir filling (i.e. 3 - 6m distance below standpipe level), the use of such small bore standpipes was considered acceptable. Readings at greater depth, however, may be subject to error due to the difficulty of lowering the probe down the nylon tubing and to the presence of entrapped air.

To check the response of the piezometers located at a spacing of 30m, a trial 4-unit piezometer with 3m long elements was installed at Ch.401 and pumping-in influence tests were carried out from holes drilled at distances of 7.5m and 15m from it. The tests were performed at depths of 18.3m and 32.3m using a water flow rate of approx. 1.5 l/s. At a distance of 7.5m a piezometric rise of 0.20 - 0.30m developed

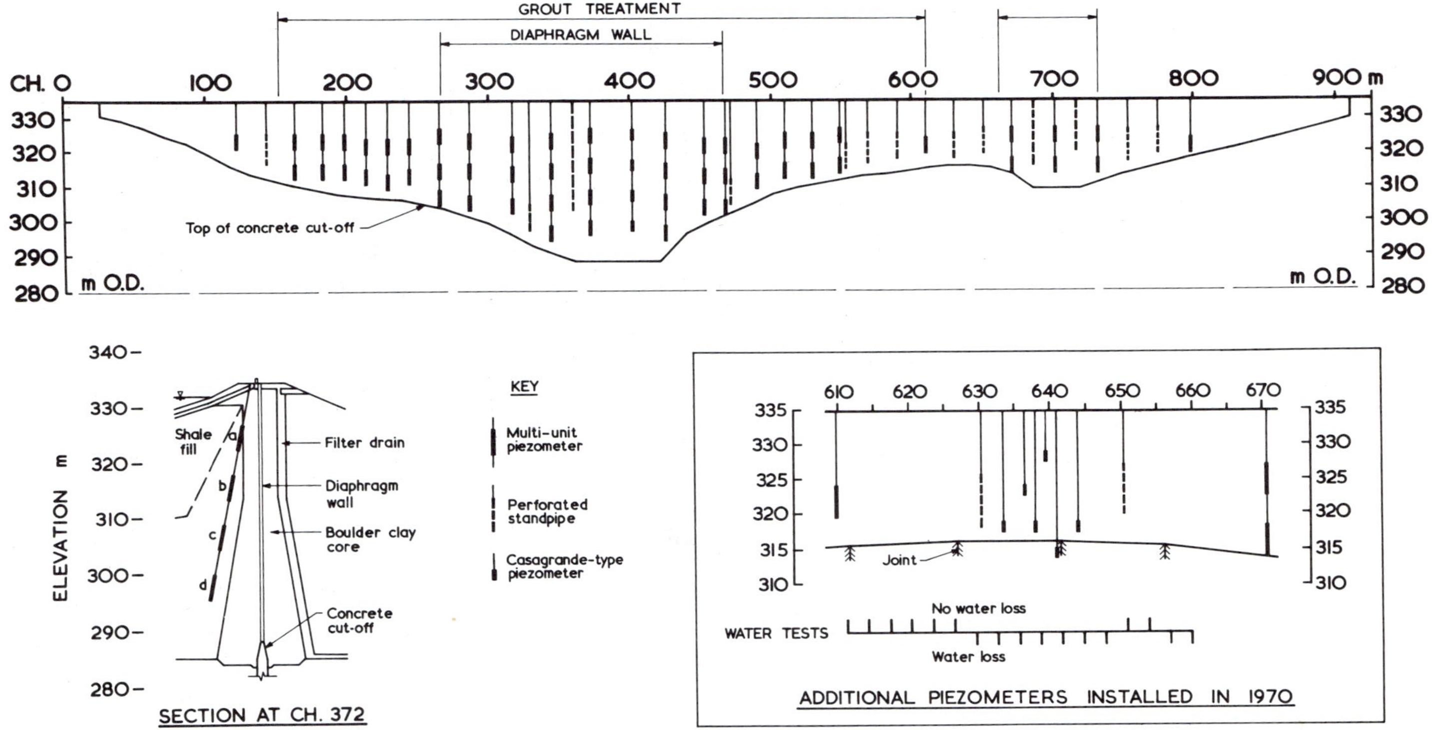

Figure 1. Layout of piezometer installation in upstream shale fill.

at the same level in the fill after 2 hours whilst in the same period the response from the test at 15m distance was negligible. Some difficulty was experienced in interpreting the test results due to the presence of air pockets in the fill resulting from the use of air circulation when drilling the holes. However, as other tests carried out from perforated standpipes at Chs. 360, 686 and 716 also showed a similar lack of response at 30m distance, it was decided to reduce the longitudinal spacing of the piezometers to approximately 15 - 23m intervals. This was achieved by supplementing the multi-unit piezometers with a number of perforated standpipes. The perforated sections were 7.6 - 9.1m in length and protected from ingress of shale by terylene gauze bound around them. Three additional standpipes were installed at Chs. 329, 470 and 550 to replace elements later found to be blocked.

Since the elements of the trial piezometer also showed little systematic vertical response to the induced flow, the length of the piezometer elements was extended to 4.6m. It was feared that, due to the stratification of the shale fill, a leak might remain undetected if it was confined within impermeable layers between two element positions. The layout of the installation, comprising a total of 69 operational piezometers extending across a dam length of 670m, is shown on Figure 1.

Falling-head response tests are carried out at intervals to check the performance of the piezometers. For the majority of the piezometers 90% equalisation is reached within one hour. In some cases the response is instantaneous. However, several piezometers, notably 671(a) and standpipes 631 and 651, exhibit slower response times so that the recorded piezometric elevations lag behind the reservoir level at times of rapid rise and fall. As a consequence any apparent depression of piezometric elevation would need to be confirmed by obtaining consistent readings over a period of at least two weeks under constant reservoir level. Observations were initially taken at 2 - 3 day intervals but, with increased familiarity of the system's performance, are now only taken weekly. In general, the readings obtained are within 100mm of the reservoir level.

DETECTION OF 1970 LEAK

In March 1970, following re-filling of the reservoir, a gradual decrease of piezometric elevation was detected by the two perforated standpipes at Chs. 631 and 651. The observations are presented on Figure 2 together with, for comparison, those recorded between January-May 1969 during the previous impounding. It has already been noted that these two standpipes have slow response times and their tendency to lag during previous sharp rises of reservoir level may clearly be seen. The records of the adjoining

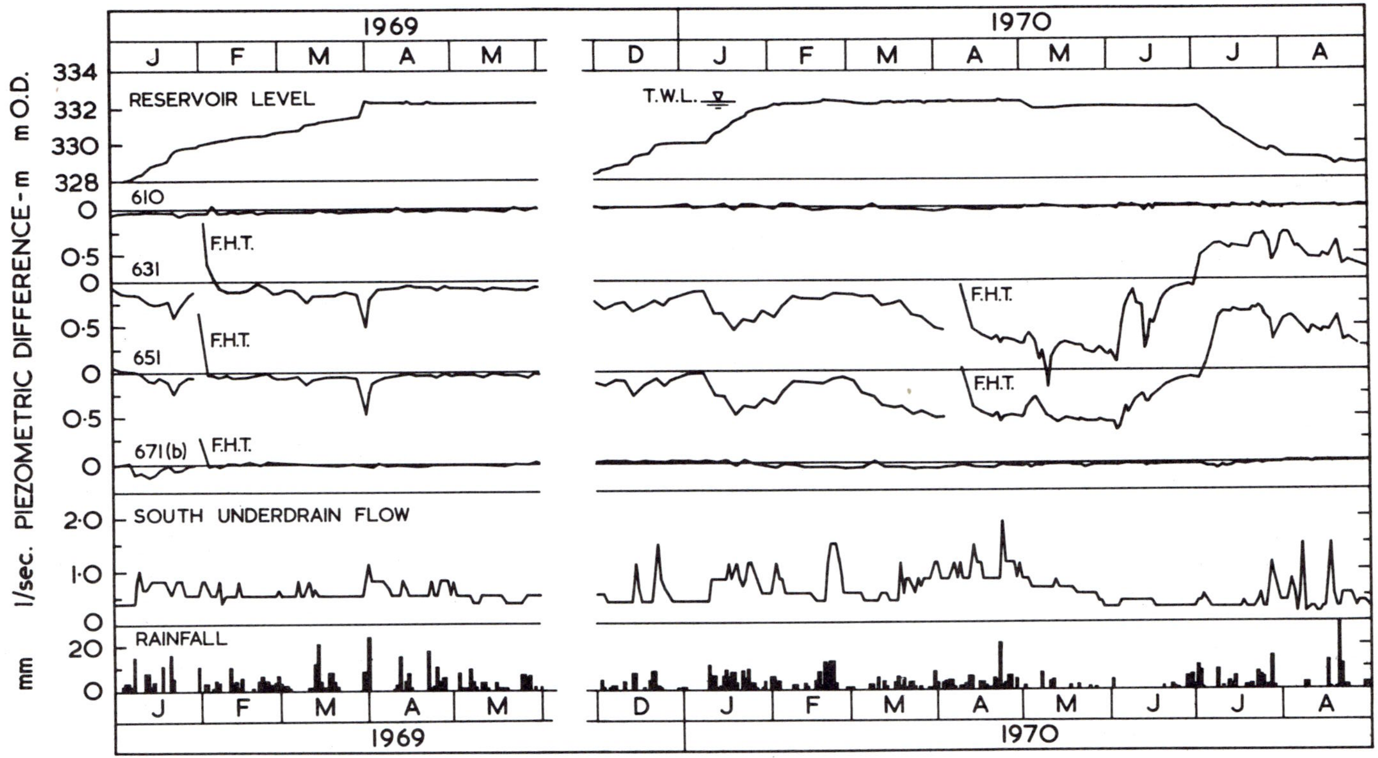

Figure 2. Upstream shale piezometer and south underdrain observations.

piezometers 610 and 671(b) showed no change.

Falling-head response tests (denoted by F.H.T. on Figure 2) were immediately carried out to check the readings. By mid-April the piezometric depressions had stabilised at 0.64m at Ch.631 and 0.49m at Ch.651. The development of renewed leakage under full reservoir pressure was confirmed when close examination of the south underdrain flow records (Figure 2) revealed that the "base flow" had increased by approximately 0.36 l/s since the beginning of March. This corresponded to a rise in head over the broad-crested weir of 1.6mm. No change in turbidity was identified, implying that the erosion process was not far advanced.

Five additional standpipe piezometers were quickly installed at different depths in the shale fill between Ch.631 - 651 (see Figure 1) in an attempt to define the zone(s) of leakage more precisely. A sixth piezometer (Ch.641) was placed in the shale foundation just beneath the top of the concrete cut-off spearhead, almost opposite a joint. The piezometers were of the Casagrande-type placed in 1.5m long sand cylinders. So as to maintain piezometric equilibrium and the level of leakage flow, the reservoir was kept at top water level throughout the investigation. These observations (Figure 3) indicated that leakage was taking place at the base of the clay core, with the centre of piezometric depression around Ch.641 (Table 1).

Chainage	Piezometric depression - m
631	0.73
634	0.82
638	1.52
(641	4.36)
644	1.52
651	0.55

Table 1. Piezometric depressions on May 18th, 1970

This position was within the area where the dam rests on the ridge between the main and south valleys and is only 18.6m high. The section was left untreated in 1968 when test holes drilled to the base of the core showed no appreciable water loss.

Three 26m long raking holes were then drilled through the clay core into the foundation at Chs. 642,639 and 645 so that they passed upstream of the spearhead. The condition of the core was examined by carrying out water tests at 3m intervals to a depth of 12m and at 1.5m intervals thereafter. Each hole was sleeve grouted and a tube-à-manchette inserted. The number of holes was then increased so that the complete length Ch.610 - 662 could be grouted at 3m centres. From the observed water losses during drilling, the results of the water tests and the recorded sleeve grout tests, as well as the responses of the piezometers to this drilling (Fig.3),

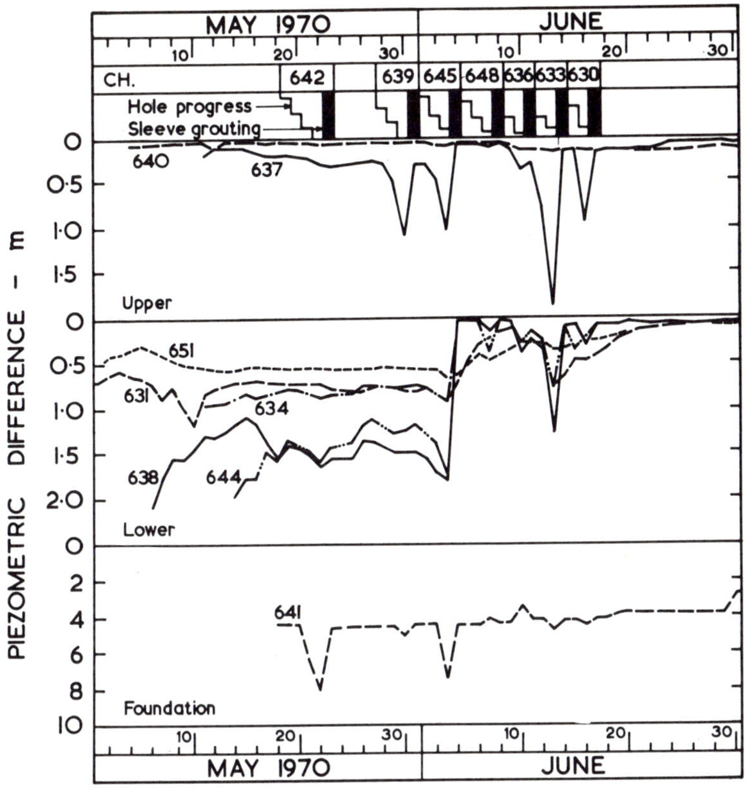

Figure 3. Observations of additional piezometers installed in 1970.

it was concluded that there were two areas of damage. One was centred on Ch.645, where the damage extended from a depth of approx. 13.7m to the base of the core, and the other, less severe, was at Ch.633. All the piezometers (Figure 3) showed a response to the drilling of both these holes; and, on sleeve grouting, instantly recovered from the previous depression of piezometric elevation. (The only exceptions were the slow responding standpipes 631 and 651 and piezometer 641 in the foundation). This behaviour suggested that the gravity sleeve grouting alone was successful in sealing the core locally. By the end of June, after completing all the grout holes, every piezometer (except Ch.641) was again recording full reservoir head and the underdrain flow had substantially decreased. The permanently low elevation registered by piezometer 641 in the soft weathered shale foundation, where the permeability may be higher than that of the compacted shale fill, probably arises from seepage under the dam.

As the available evidence indicated that the damage had been caused by hydraulic fracture, the grout injection treatment used in 1968 was again adopted in order to raise the vertical stress in the core. The marked increases in underdrain flow recorded in July and August (Figure 2) were due to the filter being flushed during grouting. Four test holes were drilled in the clay core after grouting had been completed and these exhibited no water loss.

CONCLUDING REMARKS

The piezometer installation at Balderhead Dam demonstrates the feasibility of instrumenting an existing embankment dam to complement seepage measurements in monitoring its condition. The example proves that confined leakage can be successfully detected at an early stage of development, thus preventing the occurrence of more severe damage. Since the position of leakage is also determined this obviates the need for extensive investigation. The installation, which is still observed regularly, is simple direct-reading and requires little maintenance.

There were two special factors contributing to the successful operation of the system which may not ensure its practicability elsewhere. Firstly, the value of permeability of the upstream shale fill allowed the formation of a sufficiently large piezometric drawdown for the leak to be detectable at a distance of 10m. Secondly, the dam owner has continued to meet the expense of the surveillance. Even so, such a system could possibly be of value, in suitable circumstances, for providing warning of trouble in old dams, especially when concern is felt about their internal condition.

ACKNOWLEDGMENTS

This paper is presented by permission of the Tees Valley and Cleveland Water Board. The piezometers were designed by Dr. P.R. Vaughan, manufactured by Soil Instruments Ltd. and installed by Soil Mechanics Ltd. The observations were taken by Mr. G.W.B. Turnbull. Joint Consulting Engineers for the remedial works and subsequent surveillance were Rofe, Kennard and Lapworth and Binnie and Partners.

REFERENCES

Vaughan, P.R. (1969) A Note on Sealing Piezometers in Boreholes. Geotechnique, 19, 3:405-413.

Vaughan, P.R., D.J. Kluth, M.W. Leonard and H.H.M. Pradoura (1970) Cracking and Erosion of the Rolled Clay Core of Balderhead Dam and the Remedial Works Adopted for its Repair. Trans. 10th Int. Congr. Large Dams, Montreal, 1, 73-93.

INSTRUMENTATION OF SOME EMBANKMENT DAMS IN SOUTHERN AFRICA

D.C.R. Mackellar
Senior Engineer, Ninham Shand and Partners, Cape Town
D.J. Nunn
Soils Mechanics Engineer, Min. of Water Development, Salisbury
P.J.N. Pells
Senior Research Officer, Geomechanics Div. N.M.E.R.I. Pretoria

INTRODUCTION

This paper describes the instrumentation of five embankment dams. Bridle Drift, Idas Valley, Raubenheimer and Xonxa dams are in South Africa and Lesapi Dam is in Rhodesia. They were constructed between the years 1967 and 1972. Only brief descriptions of the dams themselves are given as the authors have concentrated on the instrumentation used and on the value of the resulting data in evaluating the performance of the structures.

It must be mentioned at the outset that the scope of the instrumentation in these dams is rather limited by European standards. However, one advantage of this relative paucity of instrumentation is that the resulting data are not overwhelming and can be more readily analysed.

INSTRUMENTATION USED

Piezometer Systems

In the dry climatic conditions of Southern Africa construction pore pressures in embankment materials are very often negative. It became apparent at an early stage that the measurement of these negative pore pressures over long periods requires an extremely well constructed piezometer system, and the following points on the installation have been noted with this aspect in mind.

The piezometer installations at the five dams have all been based on the B.R.S. Imperial College System using Bishop type high air entry ceramic tips and twin hydraulic leads.

The placing of the ceramic tip presents the first problem in the installation of a reliable system of pore pressure measurement. In a hot climate, the excavation of the area around the planned position for the tip allow the fill to dry out locally and create capillary tensions that would not otherwise have been present. With construction forging on it was not always possible to plan the installations at such times to minimise this effect, and initial measurements might be inaccurate.

At Bridle Drift, Idas Valley, Raubenheimer and Lesapi a conical shaped solid mandrel was driven in to form the hole for the tip. This obviously causes severe over compaction of a zone immediately surrounding the tip location and the pore pressures measured subsequently must have been affected to some degree. The permeability of the fill must also have been altered locally. At Xonxa, however, a specially tapered auger was eventually substituted and the authors consider that this represents a definite improvement.

For the measurement of negative pore pressures it is important that polythene coated tubing be used. At Bridle Drift, the oldest installation, ¼" dia. nylon tubing was employed and a number of piezometers gave very erratic readings in the negative range. The system operated quite satisfactorily in recording positive pore pressures, but five of the twenty tips installed failed to give satisfactory readings during construction.

On the other dams 3/16" dia. polythene coated nylon tubes are used with greater success, but with these one minor problem was experienced. In stripping the polythene coating to make connections it is easy to nick the inner nylon tube accidently. Any subsequent bending of the tube may break the nylon at this point. This is particularly important to notice at the connection to the ceramic tip itself.

Inside the gauge house 1/4" or 3/16" dia. nylon tubing is preferred. At Bridle Drift use was made of 1/4" dia. copper tubing, but considerable trouble was experienced at the connections with the nylon piezometer tubes. Bourdon gauges have been used to record piezometer pressures in the four dams in the Republic of South Africa, as they are easy and quick to read and can be recalibrated if necessary. In Rhodesia the experience has been different. Gauge houses have been more humid and after a while gauges have become erratic or seized up completely. As a result a mercury manometer system has been installed at the Lesapi dam. Special care has been taken to provide ventilation for safety against mercury poisoning. No problems have been experienced in filling the manometer nor with the recording of data. The manometer system has the advantage of presenting the pressure in each tube simultaneously.

For the de-airing systems of all the projects except Bridle Drift use has been made of the apparatus produced commercially by Messrs. Soil Instruments Ltd. of London. At Bridle Drift a de-airing system on the same principles was built up in the CSIR Laboratories, Pretoria. A problem with the hydraulic leads generally has been the development of dust-like fungus in the tubing. It has been found that a liberal dose of sodium hypochlorite solution can destroy this growth without harm to the system.

The regular de-airing of a system to ensure continued accurate measurements presents little problem where pore pressures are positive. In cases where negative pore pressures develop in the fill during construction and remain for some time thereafter, the de-airing operation requires great care and often is very time consuming. This results in site staff de-airing piezometer leads less frequently than required to monitor the negative pore pressures properly. As a consequence there is a disappointing lack of field data on this aspect.

Settlement Gauges

At all the dams except Lesapi standard vertical settlement devices have been used. These consist of a vertical pipe of rigid PVC pipe passing through marker plates of aluminium or stainless steel buried at about 5m intervals in the embankment. This type of gauge is very simple and cheap. It has proved quite satisfactory in earthfill where the procedure is to bury the pipe while placing a lift and then excavate down to fit the extension. In rockfill this is not possible and the tube has to protrude above the level of the fill creating an obstruction to construction traffic.

The system consisting of horizontal PVC tubes combined with plates and a water level torpedo as used at Scammonden, Penman et al (1971) has the advantages of not obstructing construction traffic and of yielding displacement vectors and not merely settlements. It is hoped to make use of this system in the future.

Surface Monuments

These consist of survey targets suitable for level and triangulation observations, and although very simple are giving valuable information on the long-term behaviour of the embankment at Bridle Drift. The other dams have crest monuments only and detailed data have not been assembled.

Total Pressure Cells

B.R.S. vibrating wire gauges were installed at Lesapi to study the stress conditions in the very deep and narrow core trench on the right flank. This aspect is treated in a paper to be published by Prof. G. Blight (1973).

SUMMARY OF FIELD DATA

Bridle Drift Dam

The Bridle Drift Dam consists of a sloping core rockfill embankment 54 metres high, impounding 77,2 million cubic metres on the Buffalo River in the Eastern Cape. A typical section through the embankment is shown in Figure 1.

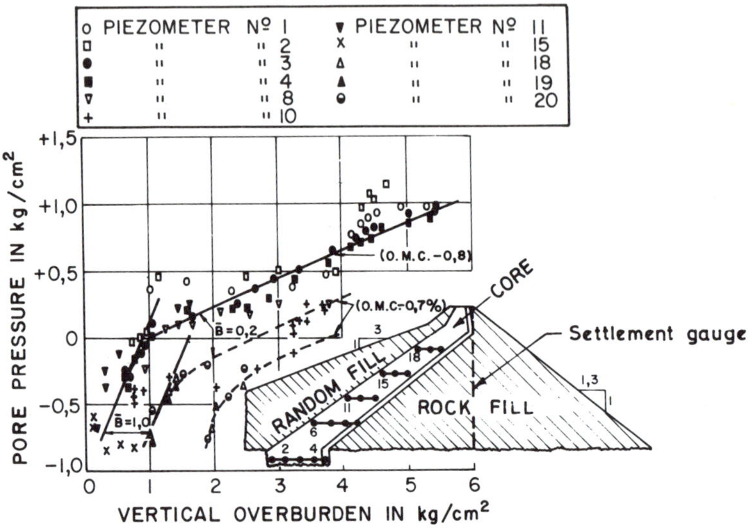

Figure 1 : Bridle Drift Dam; Construction Pore pressure After Blight (1970)

The instrumentation was designed to monitor the behaviour of the thin sloping core that was constructed very rapidly (4 months) after the downstream rockfill shell had been completed.

	Bridle Drift	Lesapi	Raubenheimer	Idas Valley	Xonxa
Liquid Limit %	39	-	20-39	35-45	35 (mean)
Plasticity Index %	23	19	14-24	11-18	12-18
Clay Fraction (<,002mm) %	30	-	25-40	20-30	10-20
O.M.C. (Proctor) %	16,1-20,5	17,1 (mean)	11-18	16-20	16 (mean)
M.D.D. (Proctor) gm/cm^3	1,68-1,82	1,77	1,77-1,96	1,67-1,78	1,67-1,80

Table I : Range of Properties of Fill Material

The properties of the decomposed shale used in the core are given in Table I. Material was placed at a moisture content

0,5% dry of Optimum and compacted to an average density of 101,5% of the proctor maximum.

The instrumentation consisted of a piezometer system with 20 tips (18 in one cross section), one settlement gauge in the downstream shell and surface monuments along the crest and on the slopes at the highest section.

A detailed description of the piezometer installation is given by Blight (1970). Figure 1 is reproduced directly from Prof. Blight's paper and shows the pore pressure build up with overburden pressure.

The interesting point regarding these data is the fact that, contrary to expections, the rate of pore pressure build up decreased with increasing overburden pressure. This was not due to partial drainage at the tips as it has been shown (Blight 1970) that with the very rapid construction of the core the soil surrounding the tips remained essentially undrained. Similar data were obtained at the Idas Valley and Lesapi dams. The probable reason for this behaviour is discussed below. The piezometers at Bridle Drift have recorded the development of the steady state seepage conditions that compare well with predictions based on a simple flow net.

The displacements of the settlement gauge plates have been recorded regularly as a check on embankment performance but they are not reproduced here. It must be appreciated that this type of gauge gives movements at each plate and the settlements measured are then cumulative compressions of all the layers of fill beneath. Variations in compressibility of the fill in the embankment are thus averaged and suppressed.

Figure 2 is a section through the embankment at the highest part and shows the movements of the surface beacons on this section for the first 1½ years after construction.

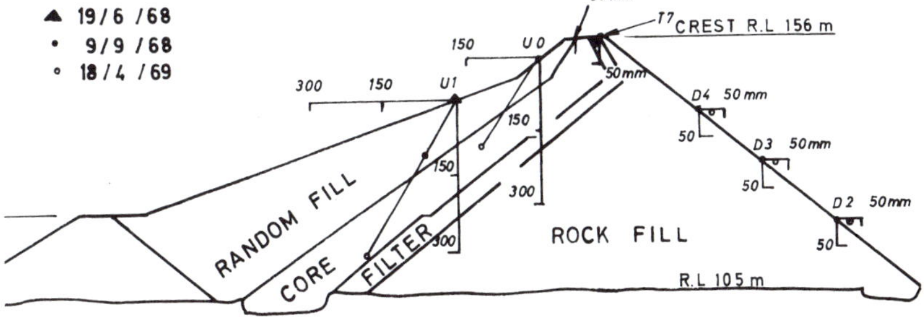

Figure 2 : Bridle Drift Dam : Movements of surface monuments

The relatively large deformations of the upstream shell became apparent shortly after completion following a rapid filling of the reservoir to within 12 metres of full supply level. A longitudinal crack appeared 2m back from the upstream edge of the crest. This crack enlarged and by July, 1969, when the reservoir had risen a further 5 metres, had a maximum width of 65mm and a relative vertical displacement of 300mm. At this stage the crest was regraded and when the reservoir finally filled, no further movement was apparent.

Longitudinal cracking along the crests of composite earth-rock dams is not unexpected. The cracks arise from excessive saturation settlement of the upstream shell and are not an indication of incipient slip failure.

Lesapi Dam

A typical cross-section of this 240m long and 43m high embankment dam is shown in Figure 3. The embankment, designed to pass floods over the fill during construction, had reached level 1342,64m and piezometers 1 to 5 had been installed when a flood was passed over the partially completed dam. Placing of earthfill was completed in January, 1971.

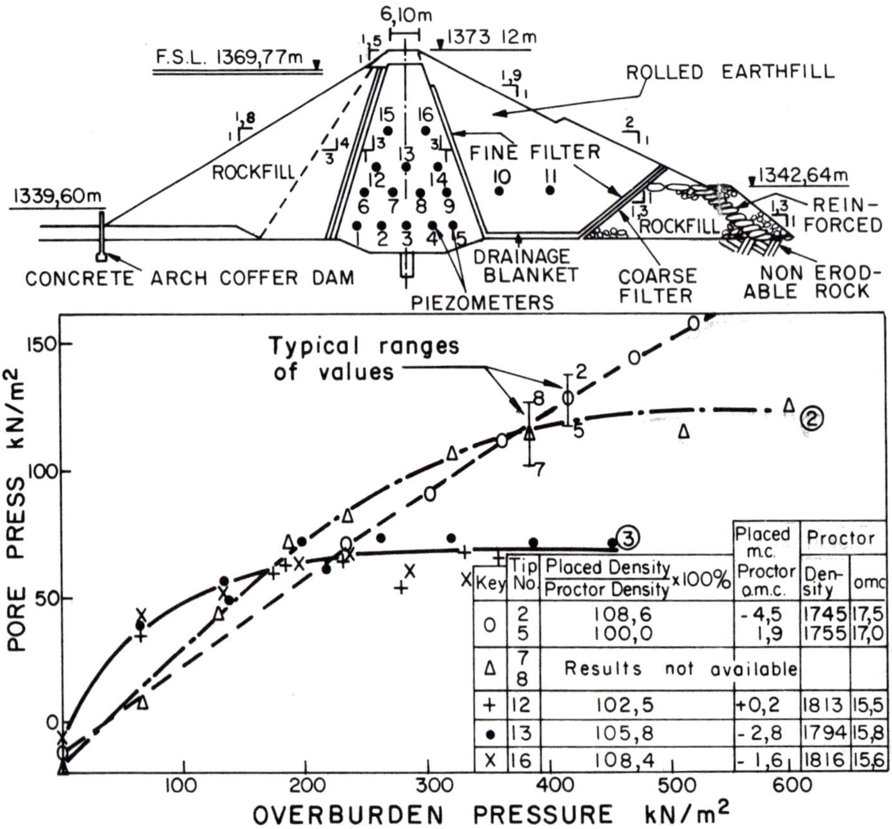

Key	Tip No.	Placed Density / Proctor Density x100%	Placed m.c. Proctor o.m.c.	Proctor Density	Proctor omc
O	2 5	108,6 100,0	-4,5 1,9	1745 1755	17,5 17,0
Δ	7 8	Results not available			
+	12	102,5	+0,2	1813	15,5
•	13	105,8	-2,8	1794	15,8
X	16	108,4	-1,6	1816	15,6

Figure 3: Lesapi Dam: Typical Cross Section and Construction pore pressures

The earthfill materials are residual soils, the core being derived from a granite/dolerite contact and the rolled earthfill from granite. Properties of the soils are shown in Table I, and Figure 3. The rockfill was clean, sound granite from the spillway excavation. All fill materials were compacted with 4 to 6 passes of a 10 tonne vibratory roller.

Instrumentation consisted of three total pressure cells and sixteen high air entry piezometers. The total pressure data are discussed elsewhere (Blight 1973).

Figure 3 gives the construction pore pressure data. It should be noted that the response of each layer of tips was different. The core material was continually changing in nature as the excavation in the borrow area moved through and away from the dolerite/granite contact. Material placed higher in the core is coarser and less plastic than that lower down. In any particular layer the material was similar although compaction density and moisture content varied. Nevertheless this fill did not develop the strong negative pore pressures as experienced during construction with shaley material of the other four dams.

During the first seven months of 1971 the reservoir filled partially to level 1362m. Over a period of 104 days from the 28th July, 1971, the reservoir level dropped some 15m. The piezometer data from tips 12 to 16 are given for this period in Figure 4. It is clear from this figure that significant drainage of pore pressures occured and that the design assumptions used, based on Bishop's method for instantaneous drawdown for impervious compressible soil were very conservative.

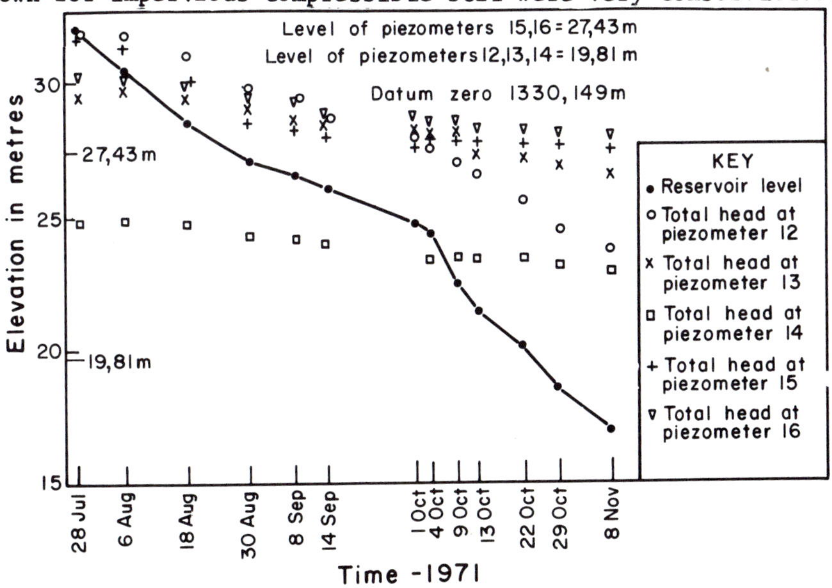

Figure 4: Lesapi Dam : Drawdown pore pressures

Figure 4 shows that at the end of drawdown the flow pattern in the core was nearly symmetrical with drainage in both upstream and downstream directions. However, a surprising phenomenon is that the pressures at tips 15 and 16 levelled off at a slightly positive value although the reservoir level dropped some 10m below the level of the tips. The reason for this is not clear as one certainly would have expected drainage of the core to lead to capillary tensions at these tips.

One final point of note with regard to the seepage pore pressures at Lesapi is that they indicate a definite anisotropy in the core permeability. If the piezometer readings of the 28th July, 1971, are considered it is clear that the phreatic surface was higher than that indicated by an isotropic flow net. It is appreciated that compacting cohesive soils with vibrating rollers leads to a horizontal permeability much greater than the vertical. By transforming the core section in the usual manner to fit the recorded data it appears that the anisotropy ratio is somewhat greater than 10.

Measurements of the surface beacons at Lesapi have shown no apparent movement and this has been ascribed to the good compaction achieved in the fill.

<u>Idas Valley Dam</u>

The embankment at Idas Valley is 36m high completed in 1967 and a cross section is shown in Figure 5.

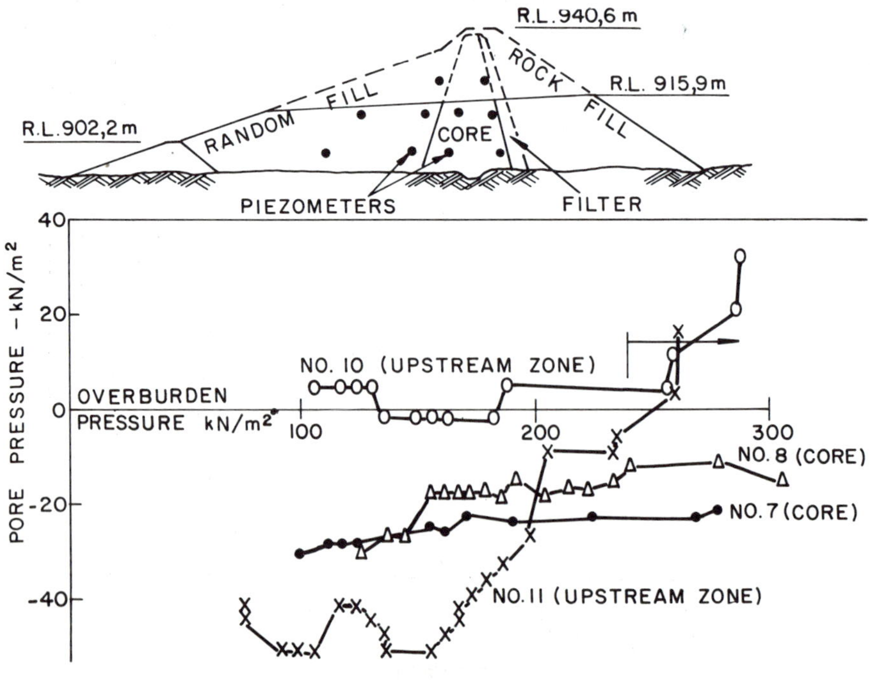

Figure 5: Idas Valley: Typical cross-section and Construction pore pressures

Three zones are indicated apart from filters and rip-rap, but as decomposed shale was used throughout, the embankment may be considered homogeneous. The properties of the shale used in the core are given in Table I. The material was placed at a moisture content between 0,5% and 1,0% dry of optimum and compacted to an average density of 103,5% of proctor maximum.

Instrumentation was limited, consisting of 7 hydraulic piezometers and a vertical settlement gauge with two surface beacons on the crest. Figure 13 gives the construction pore pressure data. Although the pore pressures were strongly negative they show a decrease in the rate of build up with increasing overburden pressure, as was the case at Bridle Drift and Lesapi.

Of the seven piezometers only one failed satisfactorily to record negative pore pressures. When properly de-aired the system was very sensitive and two tips still at a shallow depth recorded a decrease in pore pressure when 1m of overburden was removed during construction.

Readings taken a year after completion showed that a flow net had developed through the core much as predicted on the basis of flow net sketching.

Recordings of settlement in the embankment show that the consolidation of the lower 12m of fill was virtually complete by the end of construction whereas the upper portion of the embankment was still settling nine months after completion. This slow rate was ecorded even though the gauge is located adjacent to the chimney drain.

Raubenheimer Dam

A typical cross-section of the embankment 40m high is given in Figure 6 that also shows the location of the 11 piezometers and the internal settlement gauge. Construction was completed in 1970. Material for the core was derived from two sources. Initially a highly weathered residual clayey shale was available but the upper third is constructed of less weathered decomposed shale. The ranges of the properties of the fill are given in Table I the lower values being more applicable to the less weathered material. Placement of the core material was between 0,5% wet of optimum in winter and 2,5% dry of optimum in the summer months. Average compaction achieved was 101,4% of proctor maximum.

The construction pore pressure data are given in Figure 6. These are rather erratic. The results from a number of the tips showed a tendency for the pore pressures to decrease during the initial increase in overburden pressure, compared with the experience at the previous three dams. The reason for this behaviour is not clear. It could be the same phenomenon observed in the laboratory by Barden and Sides

(1970). However, some of the fill was placed over 3% dry of Proctor Optimum, and negative pore pressures well beyond the range of hydraulic piezometers must be expected initially. Water would have been lost from the system through the tips into the fill and increased the moisture content locally. This could have caused the apparent decrease in pore pressure during the initial increase in overburden.

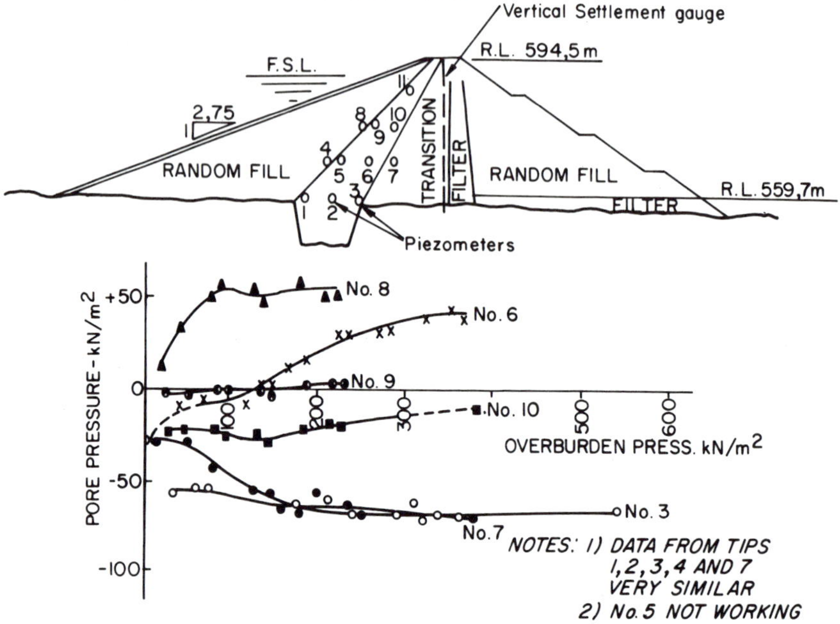

Figure 6 : Raubenheimer Dam : Typical cross-section and construction pore pressures

Xonxa Dam

This is a composite earth rockfill structure with a planned height of 50m and is still under construction. A feature of the design is the downstream reinforced rockfill shell provided to pass moderate floods over the embankment during construction without the need of elaborate diversion works. However, when the dam was at a height of about 25m floods passing over the structure caused severe washaways and the bulk of the instrumentation was lost. This is being replaced as construction proceeds.

Properties of the material used in the core are given in Table I.

Control of placement moisture was difficult in the very hot climate and the moisture content ranged from 2,5% to 1,0% dry of optimum. The average compaction achieved was about 97% of Proctor Maximum.

Nine hydraulic piezometers were installed prior to the washaway and the data are given in Figure 7. Much of the same trend is seen as noted for the Raubenheimer Dam. With one exception all the piezometers satisfactorily registered negative pore pressures. However, from the quantity of air that entered the system after initial de-airing it appears that very high suctions existed immediately after compaction.

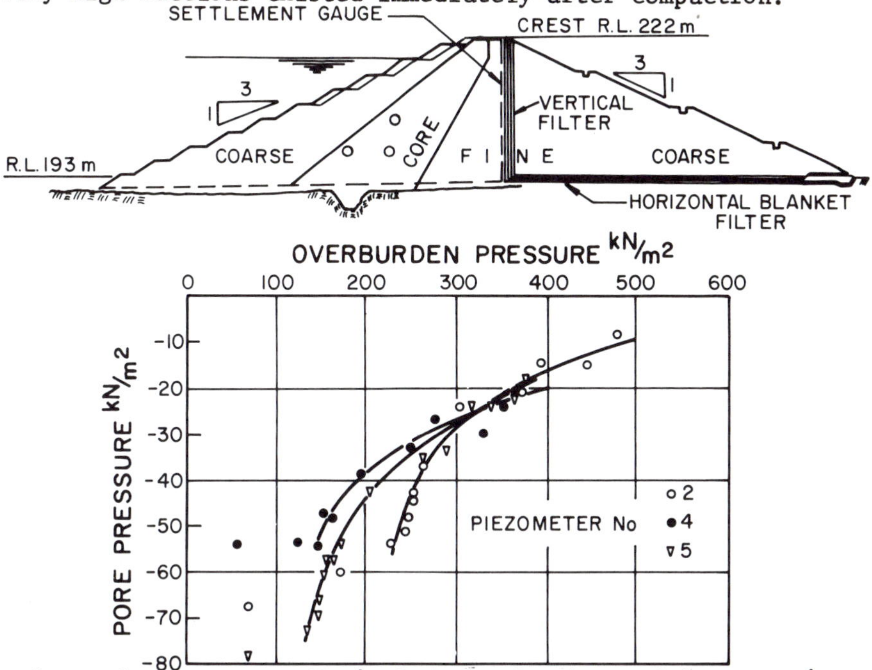

Figure 7: Xonxa Dam: Typical Cross-Section and Construction pore pressures

The increase in pressures recorded at the end by some of the tips was caused by the rising level of the reservoir during the rainy season.

INTERPRETATION OF PORE PRESSURE DATA DURING CONSTRUCTION AT BRIDLE DRIFT, LESAPI AND IDAS VALLEY

The trend of decreasing pore pressure ratio r_u with increasing overburden pressure under undrained conditions experienced at these three dams is noteworthy, although similar data have been recorded at other sites e.g. Manjirenji (Blight 1970). In European practice the initial rate of pore pressure increase usually indicates whether measures will be necessary to guard against "end of construction" instability. The experience in Southern Africa shows that the initial behaviour can be misleading.

A reasonable explanation for the trend shown by the field data can be given in terms of the stresses set up during compaction.

After heavy compaction at or below optimum moisture content, the material in a layer of cohesive fill is overconsolidated with horizontal effective stresses greater than the vertical, and frequently having negative pore pressures initially. Thus r_u will decrease steadily until there is a marked increase in the overburden pressure and the vertical effective stress exceeds the horizontal. The final value of r_u at the end of construction will, therefore, seldom approach the rather conservatively high figures often used in design assumptions. A detailed study of this phenomenon will be offered in a paper to be presented by Pells (1973).

CONCLUSIONS

The BRS Imperial College hydraulic piezometer system has been found to be very satisfactory in recording pore pressures in a number of embankment dams in Southern Africa, provided adequate care is taken with the installation. In the right conditions bourdon gauges are quite satisfactory but very accurate measurements are probably better recorded on mercury manometers.

Where negative pore pressures are encountered particular attention must be paid to placing the ceramic tip and to all leads and connections. De-airing the system regularly in these circumstances is very necessary to keep track of changes in pore pressures but can be a tedious process. This leads to less frequent attention by site staff and data on the negative pore pressure range are not recorded.

As expected piezometer systems can monitor seepages very effectively and have been used to check the anisotropy of a core compacted with a vibrating roller compared with the more homogeneous results achieved using sheepsfoot and rubber tyred rollers The piezometers at Lesapi dam have also indicated that significant drainage of an "impervious" material can take place under rapid drawdown conditions.

Internal settlements within earth and rockfill zones have been successfully recorded using a simple induction coil to check the level of gauge plates in a vertical standpipe. This information alone is not always sufficient for a thorough appraisal of the performance of an embankment and a method yielding displacement vectors would be preferable. Disruption of construction traffic on the embankment is a major disadvantage of the vertical device that would be avoided by a horizontal system.

In general, to be of practical use, instruments should be easy to read and requiring a minimum of maintenance. The number installed on any one project should not exceed that which the site staff can readily attend to, for overprovision of instruments can result in the whole system being inadequately serviced.

ACKNOWLEDGMENTS

The authors gratefully acknowledge permission from the following for the use made of project data on instrumentation.

Ciskeian Government Service	- Xonxa Dam
East London Municipality	- Bridle Drift Dam
Oudtshoorn Municipality	- Raubenheimer Dam
Stellenbosch Municipality	- Idas Valley Dam
Rhodesian Min. of Water Development	- Lesapi Dam.

REFERENCES

Barden, L. and Sides, G.R. (1970) Engineering Behaviour and Structure of Compacted Clay Proc. A.S.C.E. SM4, July 1970.

Blight G.E. (1970) Construction Pore Pressures in two Sloping-Core Rockfill Dams Trans. 10th I.C.O.L.D., Q36, R18, Montreal.

Blight G.E. (1973) Paper to be published Trans. 11th I.C.O.L.D. Madrid.

Pells P.J.N. (1973) Stress Ratio Effects on Construction Pore Pressure Paper to be published 8th Int. Conf. SM and FE, Moscow.

Penman A.D.M. Burland J.B. and Charles J.A. (1971) Observed and Predicted Deformations in a Large Embankment Dam during Construction Proc. I.C.E. Vol. 49 May.

MONITORING THE M5 MOTORWAY EMBANKMENTS ON SOFT GROUND IN SOMERSET

John M McKenna

Consulting Soils Engineer, Kingston Hill, Surrey

INTRODUCTION

For 35km, the M5 motorway crosses the Somerset Levels, see Figure 1, the second largest fenland in England, where the depth of alluvium is typically about 26m. Because of the poor ground conditions, the calculated factor of safety of the overbridge and motorway embankments was about 1.0, and large settlements were expected. The embankments were surcharged and allowed to settle for one year, in order to reduce the post construction settlement of the motorway to a design value of 150mm. To control the stability of the embankments during construction and to check the settlement and surcharge calculations, the vertical settlement, lateral ground movement and pore pressure build up and dissipation were monitored during construction and afterwards. The stability of the surcharged motorway embankments was in general monitored by observation of ground movements, and that of the side road embankments by ground movements and effective stress analyses. Although pulverised fuel ash (PFA) was used to build the side road embankments, stage construction and berms were necessary to build several to full height.

GEOLOGY AND SOIL PROPERTIES OF ALLUVIUM

The Somerset Levels, according to Green and Welch (1965), consist of Pleistocene to recent estuarine alluvial sediments which were laid down in two major depositions. The first, from 6300 to 5000 BC, due to a relatively rapid eustatic rise in sea level, deposited initially sandy, then grey estuarine clay with minor peat beds up to about Datum level. It was followed by a long period of peat accumulation over a wide area. This is the main peat bed and is generally found at Ordnance Datum. Another marine

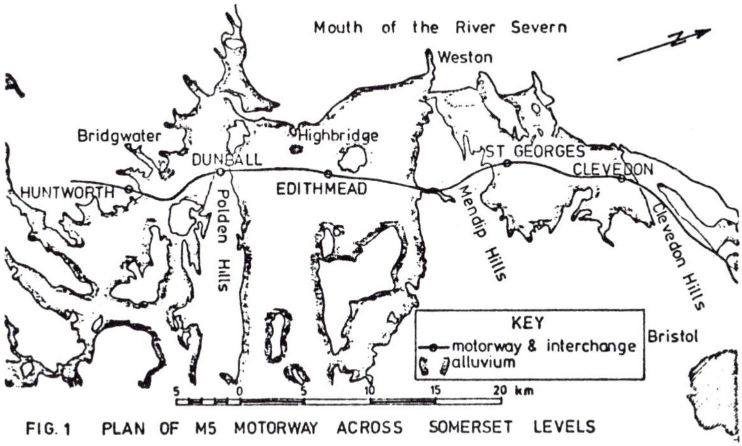

FIG. 1 PLAN OF M5 MOTORWAY ACROSS SOMERSET LEVELS

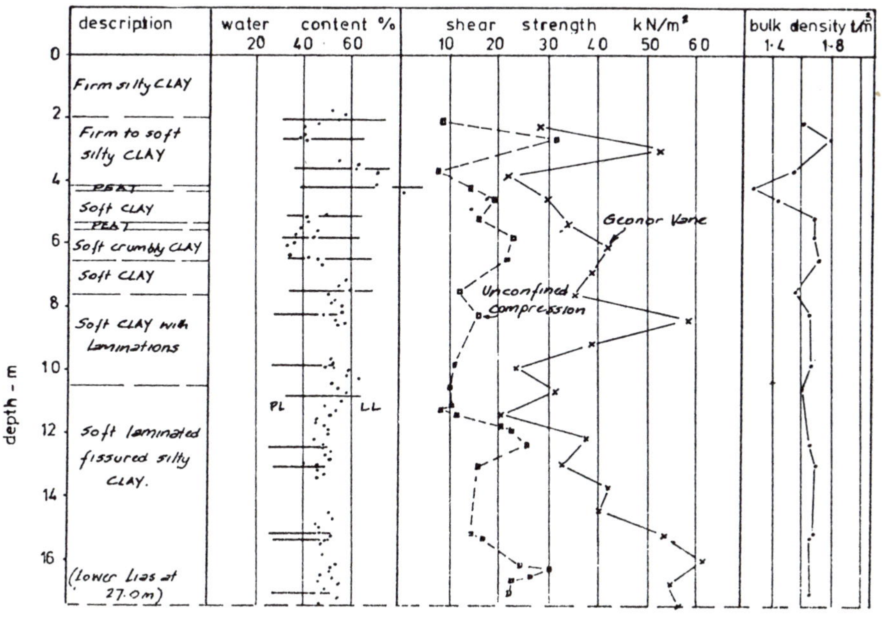

FIG. 2 WITHY ROAD SOIL PROFILE SUMMARY OF TEST DATA

incursion at about AD 250 (i.e. 1750 BP) deposited a blue grey estuarine clay about 4.3m thick over the peat with a flat upper surface 5.5 to 6.1m above OD, the top of which has since been desiccated.

Continuous piston samples were taken at selected sites along the motorway, and the samples were tested using standard laboratory techniques. For some lengths of the motorway the soft clays were continuous to bedrock. This is illustrated by Figure 2 which gives the soil properties at Withy Road, approximately half way across the Levels. The clays were lightly overconsolidated, with a quasi-preconsolidation pressure, pc, about 1.6 times the present effective overburden pressure, p_o. (McKenna 1971). This is in agreement with the findings of Bjerrum (1967). The shear strength was measured in situ using a Geonor vane. Selected piston samples were tested in unconfined compression without trimming. The unconfined strengths were in general lower than the in situ vane strengths.

MOTORWAY EMBANKMENT CONTROL

Settlement

A typical cross section of the motorway embankment is shown in Figure 3, the level of the pavement being chosen to reduce the risk of flooding.

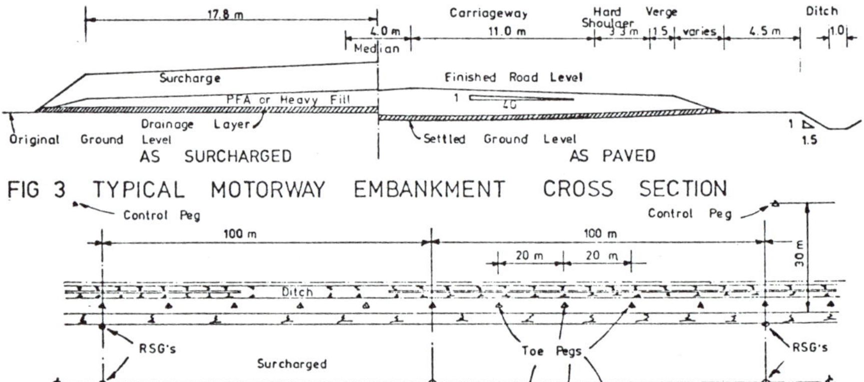

FIG 3 TYPICAL MOTORWAY EMBANKMENT CROSS SECTION

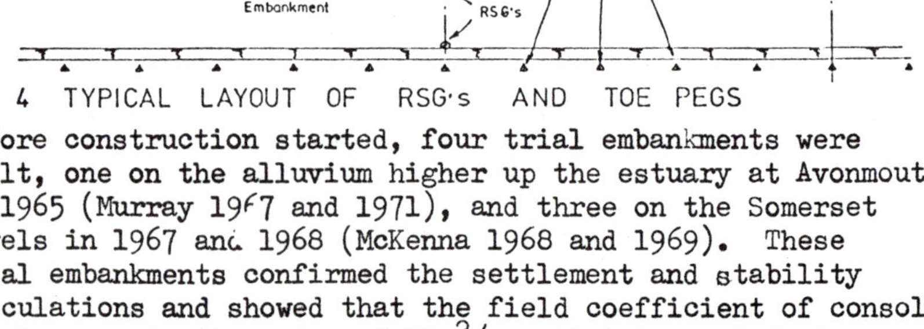

FIG 4 TYPICAL LAYOUT OF RSG's AND TOE PEGS

Before construction started, four trial embankments were built, one on the alluvium higher up the estuary at Avonmouth in 1965 (Murray 1967 and 1971), and three on the Somerset Levels in 1967 and 1968 (McKenna 1968 and 1969). These trial embankments confirmed the settlement and stability calculations and showed that the field coefficient of consolidation was in the order of $15m^2/yr$, that is, not fast enough for most of the settlement to take place during the construction period, but fast enough for surcharge to be of benefit. The embankments were therefore surcharged to limit the post-construction settlement to a design value of 150mm in the 50 years after the start of construction. The surcharge calculation required that the settlement at the end of the settlement period should be accurately calculated, for if the actual settlement were to be in excess of the design value, the actual amount of surcharge would be reduced by the amount of the excess settlement. Unless the excess settlement was due to more rapid consolidation than that assumed in the design, additional surcharge would be required to make up the deficit. The sooner these high settlement areas could be identified the better, for, with a relatively fixed twelve month settlement period, the more time the additional surcharge could be in place, the more effective it would be. To ensure accurate settlement records, rod settlement gauges were built on top of the original ground surface and the embankments constructed around them. Figure 4 shows a typical layout of rod settlement gauges. Settlement readings were taken every week during construction of the embankments and until the settlement pattern had become established. Thereafter, the readings were taken at longer intervals. Graphs of settlement against log time, in days since the start of construction, were prepared and updated throughout the settlement period. An example is shown in Figure 5.

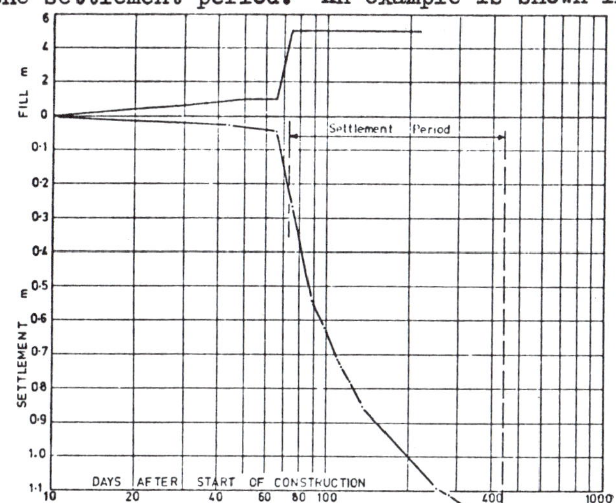

FIG. 5 EMBANKMENT SETTLEMENT vs LOG. TIME

By extrapolating these graphs, the probable settlements and rates of settlement at the end of the surcharge period were estimated. The probable net surcharge and the surcharge ratio were then calculated. A distinction has to be made between net surcharge and gross surcharge. Net surcharge is the difference in the embankment loading on the ground during the surcharge period and that after the pavement has been constructed. The gross surcharge is the loading due to the net surcharge plus the weight of the material which

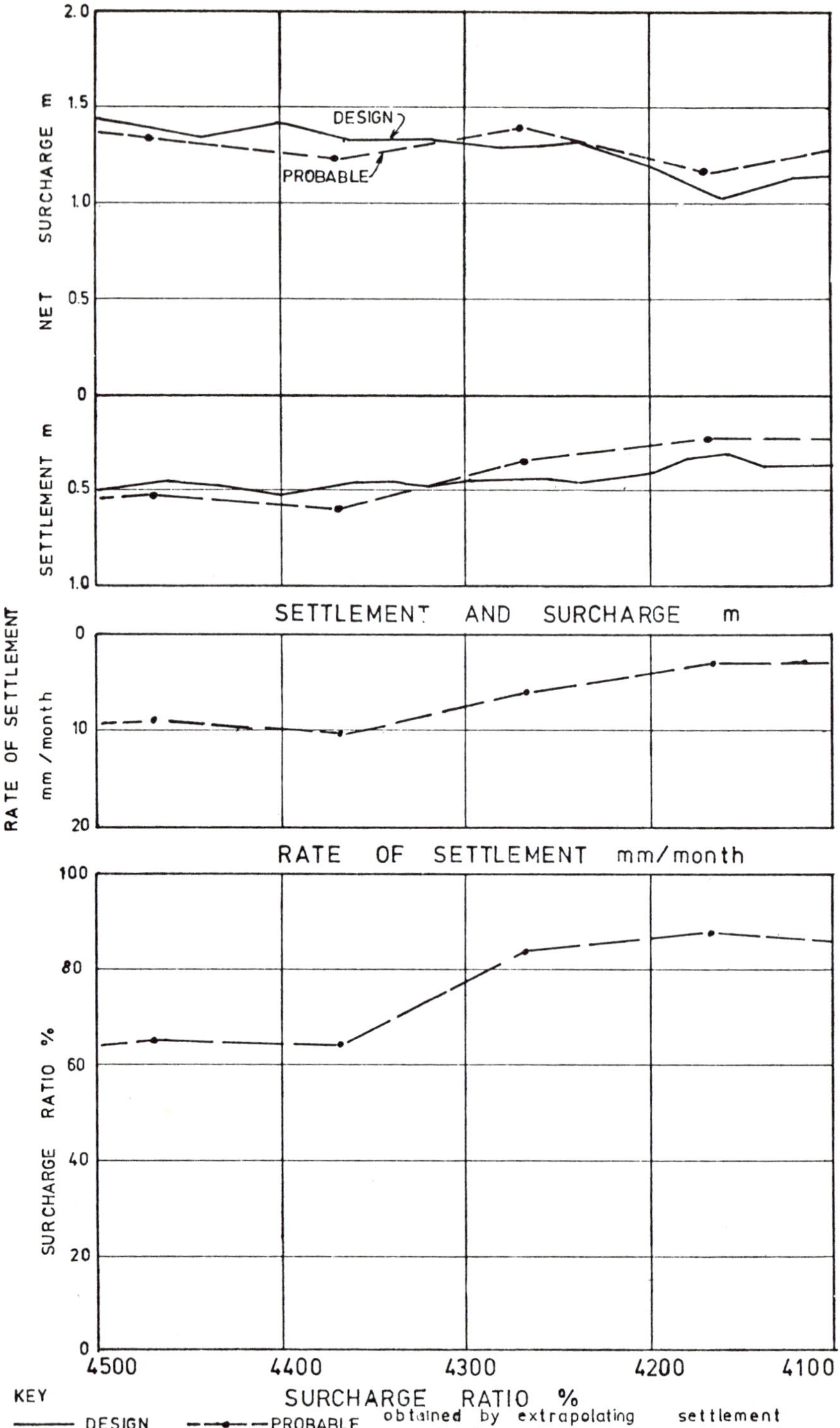

KEY
—— DESIGN —•— PROBABLE obtained by extrapolating settlement records to end of surcharge period

FIG. 6 SETTLEMENT AND SURCHARGE DRAWING

has to be excavated to make room for the pavement itself. If the embankment, surcharge and pavement are all of the same bulk density, then surcharge can be defined in terms of thickness of fill. The surcharge ratio is defined as the weight of the net surcharge divided by the weight of the finished embankment, allowance being made for submerged densities, if part of the embankment fill settles below the permanent water table.

Figure 6 shows typical settlement and surcharge results for a length of motorway. The rate of settlement at the end of the surcharge period was calculated for each rod settlement gauge and compared with the average rate for that length of motorway.

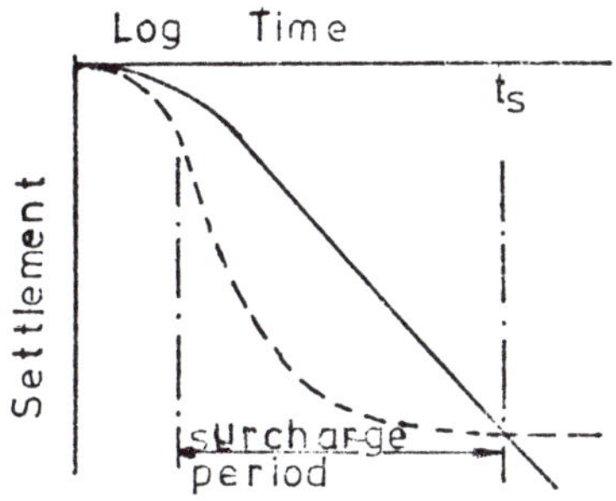

FIG 7 DIFFERENT RATES OF SETTLEMENT AT t_s

Figure 7 illustrates the need for this parameter to differentiate between areas where, although the settlements at the end of the surcharge period are the same, the post-construction settlement of one will be small, and compared with it, the other, disregarding the effect of surcharge, will be large. In those lengths where the rate of settlement at the end of the surcharge period was in excess of the average value of about 12mm/month, additional surcharge was placed, if this could be done without endangering the stability of the embankments; otherwise, the surcharge was left in place for as long as was practicable, if possible until the rate had slowed down to the average value.

Stability

The only practicable way of monitoring the stability of 35km of embankment was by observation of the lateral movement and heave of toe pegs placed along the toes of the embankment. Typical spacings are shown in Figure 4. The lateral movement of the pegs was checked as the embankments were constructed, and only if the pegs showed significant sideways movement was the vertical movement measured. Figure 8 shows results for an 1450m length of motorway. In addition, the rod settlement gauge readings at 100m intervals on centre line and at 200m intervals along the line of the hard shoulder were examined during construction for signs of acceleration of the rate of settlement which would indicate the onset of failure. The pore pressure build up was also examined for signs of increases in excess of the rate of loading.

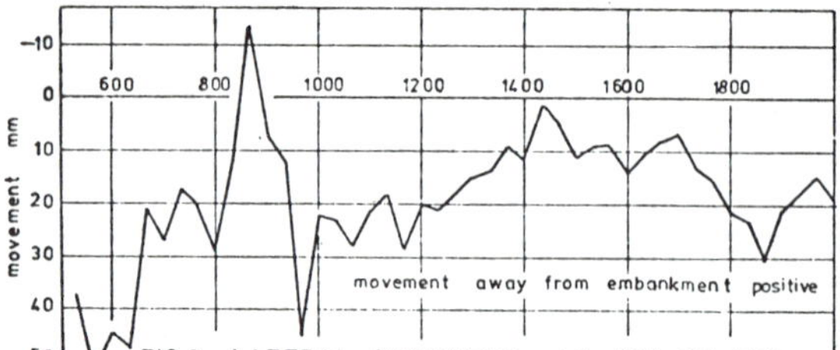

FIG 8 LATERAL MOVEMENT AT TOE OF EMB.

SIDE ROAD EMBANKMENT CONTROL

Stability

The maximum height of the approach embankments to the side road overbridges is about 8.5m, and PFA was used to build the higher parts of these embankments. Even allowing for the use of PFA, the stability of every embankment has to be closely monitored during construction. The rate of embanking was limited to 0.3m/day after 4m of fill had been placed. The embankments were instrumented with toe pegs, rod settlement gauges, Wilson slope indicators, horizontal settlement gauges and piezometers. During the construction of each embankment, the factor of safety was calculated in terms of effective stress from the results of a general stability analysis carried out by computer using the Little and Price programme (1958). In order to simplify the design charts to be used on site, the analysis was carried out in terms of the pore pressure dissipation under the centre of the embankment as a percentage of the embankment load. The pore pressures under the shoulder, in fact, control the overall stability, and to take account of this, two lateral pore pressure distributions were analysed. The two distribution curves A & B on Figure 9 were based on the actual pressures measured under two of the embankments.

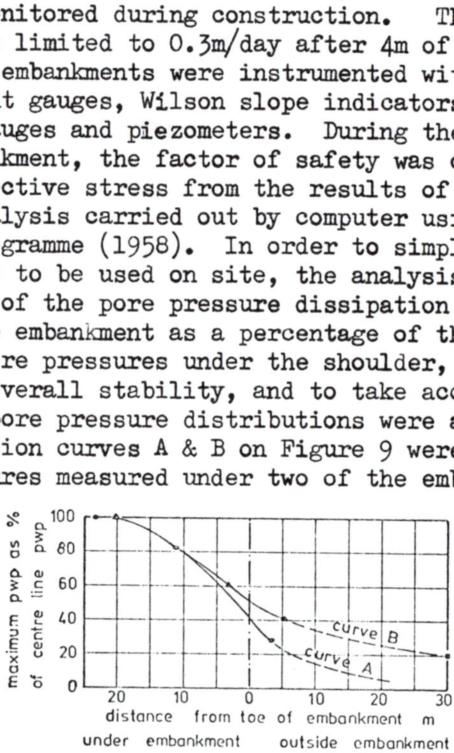

FIG 9 MAXIMUM PORE WATER PRESSURE DISTRIBUTION

The results of these analyses were used to prepare two design charts, the one based on pore pressure curve A being shown in Figure 10. The slip circle analysis assumed a soft clay

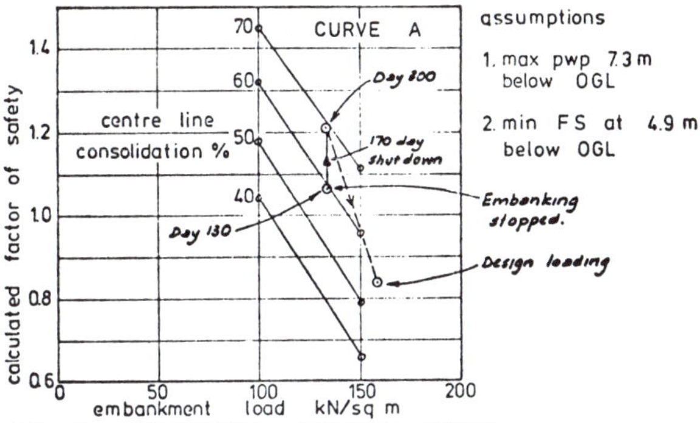

FIG 10 EMBANKMENT STABILITY CHART

foundation with a thickness of 14.6m, the maximum pore water pressure occurring at mid height, that is, 7.3m below ground level. Circles giving the minimum factor of safety were found to be tangential to a line 4.9m below ground level. The factor of safety of an embankment was calculated in the following manner.

1. On a cross section of the embankment the actual excess pore water pressures were plotted, and the percentage consolidation under the centre line to a depth of 7.3m was calculated.
2. Choosing the pore pressure distribution curve closest to the field conditions, the pore water pressures under the shoulder at each piezometer position were calculated and plotted on the cross section, and visually compared with the actual values. If the comparison was poor, then by a trial and error process the percentage consolidation on the centre line was altered, until the calculated values under the shoulder and the actual values agreed.
3. Knowing the embankment loading and the actual or assumed percentage consolidation on the centre line, the stability chart corresponding to the chosen pore water distribution curve was used to determine the factor of safety of the embankment.
4. The effect of placing further fill was investigated by extrapolating the graphs of pore water pressure against embankment load. From the extrapolated values, the factor of safety was recalculated.

Embanking was allowed to continue until the required height was achieved, or until the factor of safety fell to about

1.2. It was then stopped and the excess pore pressures were allowed to dissipate, until they had dropped sufficiently for another layer of fill to be placed safely. Where the rate of pore pressure dissipation was not fast enough for the embankment to be completed within the available time, wide berms were used to stabilise it, and to allow filling to continue to full height. Figure 10 shows the results for an embankment which required berms in this way.

As the alluvium can change appreciably in a short distance, it was considered unlikely that the piezometers would be positioned in the weakest ground. In addition, they had been placed under only one of the two embankments involved in every side road crossing. It was therefore necessary to monitor the ground movements using toe pegs and rod settlement gauges. Up to date graphs were kept of settlement vs embankment load. These were examined for evidence of the rate of settlement accelerating, an indication of impending failure. Figure 11 shows results for a typical side road embankment (Withy Road). This graph also gives the plots

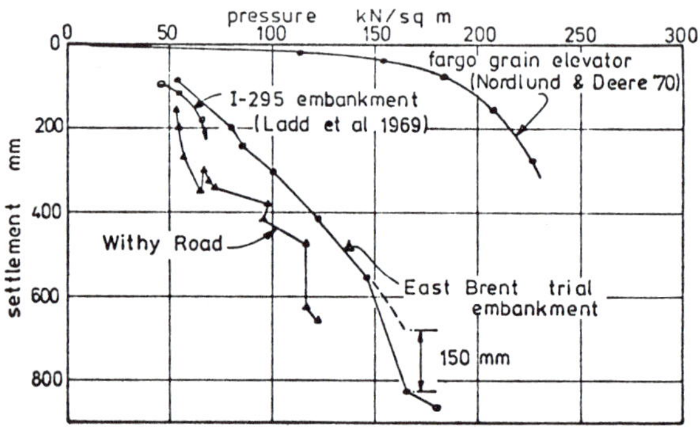

FIG 11 SETTLEMENT vs PRESSURE

for two trial embankments, both of which failed during construction, and a grain elevator which failed during first filling. The Fargo grain elevator, details of which are given by Nordlund and Deere (1970), was built on glacial lake deposits in North Dakota, and collapsed forty days after filling started in 1955. It is obvious that, had an up to date plot of settlement against load been kept, a clear indication of the impending failure would have been obtained. The Interstate I-295 trial embankment failed in 1967 during construction, when the calculated factor of safety based on field vane tests was greater than 2.0 (Ladd et al, 1969). Because of the high calculated factor of safety, the field records were not being plotted and studied at this stage in the construction, so there was no warning. In the case of the East Brent trial embankment which failed in 1967, 92 days after the start of construction, the sudden increase of 0.15m in the centre line settlement of part of the embankment adjacent to the slip indicated the onset of failure. Whereas the soft ground under the grain elevator and the two trial embankments was loaded at a relatively uniform rate, the graph for Withy Road shows the effect of loading in stages. It is important to take this effect into account when interpreting a graph of settlement against load. A reasonably high standard of levelling is required for this form of monitoring to be reliable.

At one of the side road embankment sites, Wilson slope indicators were installed. Figure 12 shows the results.

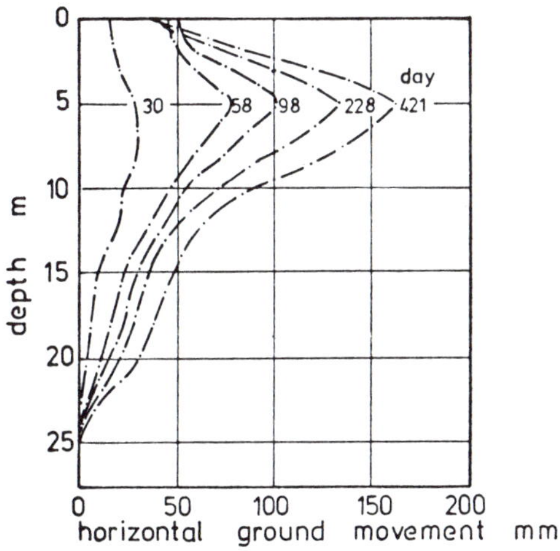

FIG 12 GROUND MOVEMENT AT TOE OF EMBANKMENT

The maximum horizontal movement during fill placing was about twice that measured at the surface. It is interesting to note, however, that during the settlement period, although the soft clay at depth continued to move outwards, the direction of movement at the surface after about day 100 was towards the embankment.

Settlement

Although PFA was used to construct the high side road embankments, it was found impossible for stability reasons to surcharge them without constructing wide berms. As this was considered to be uneconomic, the need for maintenance at the ends of the pile supported bridges was accepted, and approach slabs were not used. The graphs of settlement against log time were extrapolated to estimate the post construction settlements. Over the thicker deposits of compressible alluvium, the settlement at the end of 10 years was estimated to be about 0.4m, and allowance was made when trimming the final slops for the additional width of embankment required to maintain the design width after the post construction settlement had occurred.

Soil Instruments Ltd. were asked to provide the first horizontal strain and settlement gauge for the Clevedon Hills Contract, based on a method of measuring the complete settlement profile under an embankment described by Bozozuk in 1969. Six more were subsequently supplied for the Edithmead to Huntworth Contract. They were fitted with metal plates every 3.05m in order to measure the horizontal strain, but the experimental errors in the measuring system were of the same order as the changes in length between the plates, and therefore no meaningful results were obtained. They were of particular value, however, in recording the heave of the embankment when the surcharge was being stripped as the tubing was not liable to be damaged by the earth moving equipment, whereas the rod settlement gauges were extensively damaged by it.

PILE DEFLECTION CONTROL

In order to minimise the horizontal deflection of the piles driven near the toes of the embankments, driving was not permitted until towards the end of the one year settlement period. Then the records of the slope indicator were extrapolated to estimate the post construction deflection of the piles. Figure 13 shows the estimated post construction deflections of a pile based on the records of a slope indicator on the south bank of the River Huntspill.

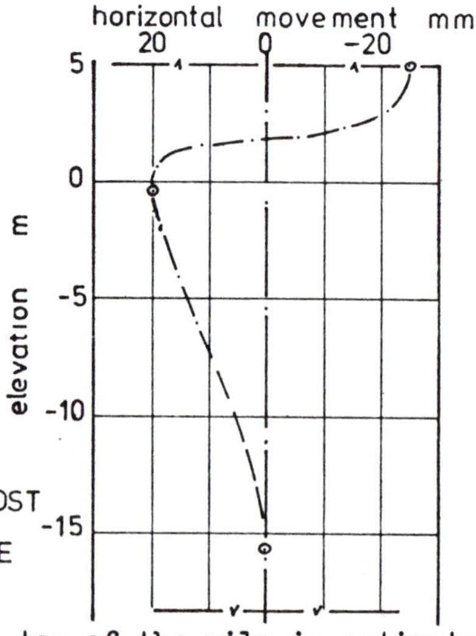

FIG 13 ESTIMATED POST CONSTRUCTION PILE DEFLECTION

The movement of the top of the pile is estimated to be away from the river, causing a lengthening of the span. The estimated post construction deflections were taken into account when checking the design of the piles.

CONCLUSIONS

The successful design and construction of a motorway across 35km of soft ground posed two major foundation problems. The first was to limit the post construction settlement of the motorway embankments to an acceptable amount (150mm). Surcharging, with a one year settlement period was adopted. It was realised, however, that surcharge design is about as demanding as any problem normally encountered in foundation engineering, and that construction supervision is an extension of the design phase. It was imperative, therefore, that the actual behaviour of the ground should be monitored during construction, in order that any lengths whose performance was poorer than the design assumptions should be identified, and the appropriate remedial measures taken. The second problem was the stability of the surcharged motorway embankments and of the side road embankments. Stage construction and berms were necessary in order to build some of them to full height. Again instrumentation to monitor the horizontal and vertical ground movements and the excess pore water pressures was essential in order to control the rate of construction. The paper presents typical results of the field measurements. A companion paper (McKenna and Roy, 1973) describes the practical problems associated with the instrumentation on the southern half of the motorway across the Somerset Levels.

ACKNOWLEDGEMENTS

The author was soil mechanics consultant to Messrs. Freeman Fox and Partners, who were the Engineers for Contracts 10B and 11, and to the South Western Road Construction Unit who were the Engineers for Contracts 12 and 13.

The paper is presented with the permission of the partners of Freeman Fox and Partners and the Director of the SWRCU.

The author wishes to thank in particular Mr W T F Austin and Mr D R Wolstenholme, the Freeman Fox Partner and Senior Engineer in charge of the northern half, and Mr P G Lyth, Mr A S Turner, Mr J B Bentley and Mr J C Billington who are respectively Director SWRCU, Chief Engineer, Superintending Engineer (Roads) and Superintending Engineer (Bridges) of the SWRCU Somerset Sub-Unit responsible for the southern half.

Many engineers were involved on site in taking all the readings required for the monitoring of four contract lengths of motorway, and the author gratefully acknowledges their enthusiasm and attention to detail. These include for Contract 10B, M G Lewis, P M Farmer, W T Bell; for Contract 11, K J James, P J Holland, I D Watt, C G Dean;

For Contract 12 and 13, R W Browning, A J Billinghurst, T Stamper, M Roy and B Heskins.

The Instrumentation Contractors were as follows:

M5 Motorway Contract		Contractor
No	Section	
10B	Clevedon Hills	Soil Mechanics Ltd.
11	Mendip Hills	Soil Instruments Ltd.
12 & 13	Edithmead to Huntworth	Soil Instruments Ltd.

REFERENCES

Bjerrum, L (1967) Engineering Geology of Norwegian normally consolidated marine clays as related to settlements of buildings. Geotechnique, 17, 2: 81-118.

Bozozuk, M (1969) A fluid settlement gauge. Canadian Geotechnical Journal, 6, 3:362-364.

Green, G W and F B A Welch (1965) Geology of the Country around Wells and Cheddar. HMSO, London. P 121.

Ladd, C.C, H P Aldrich and E G Johnson (1969) Embankment failure on organic soil. Proc 7th Int Conf Soil Mech. Mexico, 2:627-634.

Little, A L and V E Price (1958) The use of an electric computer for slope stability analysis. Geotechnique, 8, 3:113-120.

McKenna, J M (1968) Site investigation of a slip in the East Brent (Somerset) trial embankment. Soil Mechanics Ltd Report No 5101/1. Unpublished.

McKenna, J M (1969) Performance of the M5 Motorway trial embankments at Clevedon, Somerset. Soil Mechanics Ltd Report No 5190. Unpublished.

McKenna, J M (1971) Engineering properties of the Somerset alluvium. Paper presented at Bristol University, March 10.

McKenna, J M and M Roy (1973) Performance of the instrumentation used to monitor the M5 motorway embankments built on soft ground between Edithmead and Huntworth. BGS Symposium on Field Instrumentation.

Murray, R T (1967) An investigation of the stability and settlement of a trial embankment on the proposed extension of the Birmingham-Exeter Motorway (M5). Ministry of Transport RRL Technical Note TN 212.

Murray, R T (1971) Embankments constructed on soft foundations: settlement study at Avonmouth. Ministry of Transport RRL Report, LR 419.

Nordlund, R L and D U Deere (1970) Collapse of Fargo grain elevator. ASCE J Soil Mech, 96, SM2:585-607.

PERFORMANCE OF THE INSTRUMENTATION USED TO MONITOR THE M5 MOTORWAY EMBANKMENTS BUILT ON SOFT GROUND BETWEEN EDITHMEAD AND HUNTWORTH

John M McKenna[1] and Malcolm Roy[2]

[1]Consulting Soils Engineer, Kingston Hill, Surrey.
[2]Chief Materials Engineer, Somerset County Council, Taunton, formerly Section Engineer, M5 Contract 12, South Western RCU

INTRODUCTION

For 35km, the M5 motorway crosses the Somerset Levels where the depth of soft alluvium is typically about 26m. Because of the poor ground conditions, the stability of all the embankments was low, and large settlements were expected. To reduce the post construction settlement of the motorway, the embankments were surcharged and allowed to settle for one year. The side road embankments which could not be surcharged because of stability considerations, were built to the design height as rapidly as the strength of the ground permitted, and allowed to settle for up to a year before construction of the bridges started. This paper discusses the performance of the instrumentation used on the southern half of the 35km length. A general paper on the monitoring of the whole 35km appears elsewhere in the proceedings. (McKenna 1973).

EMBANKMENT INSTRUMENTATION

The object of the instrumentation was

1) to enable the actual vertical and horizontal movements of the foundations of the embankments to be measured in order to check the stability and the settlement and surcharge design calculations;

2) to measure the excess pore water pressure build up during construction, and the subsequent dissipation during the settlement period. From these records the factor of safety of the embankments was calculated in terms of effective stress, and the

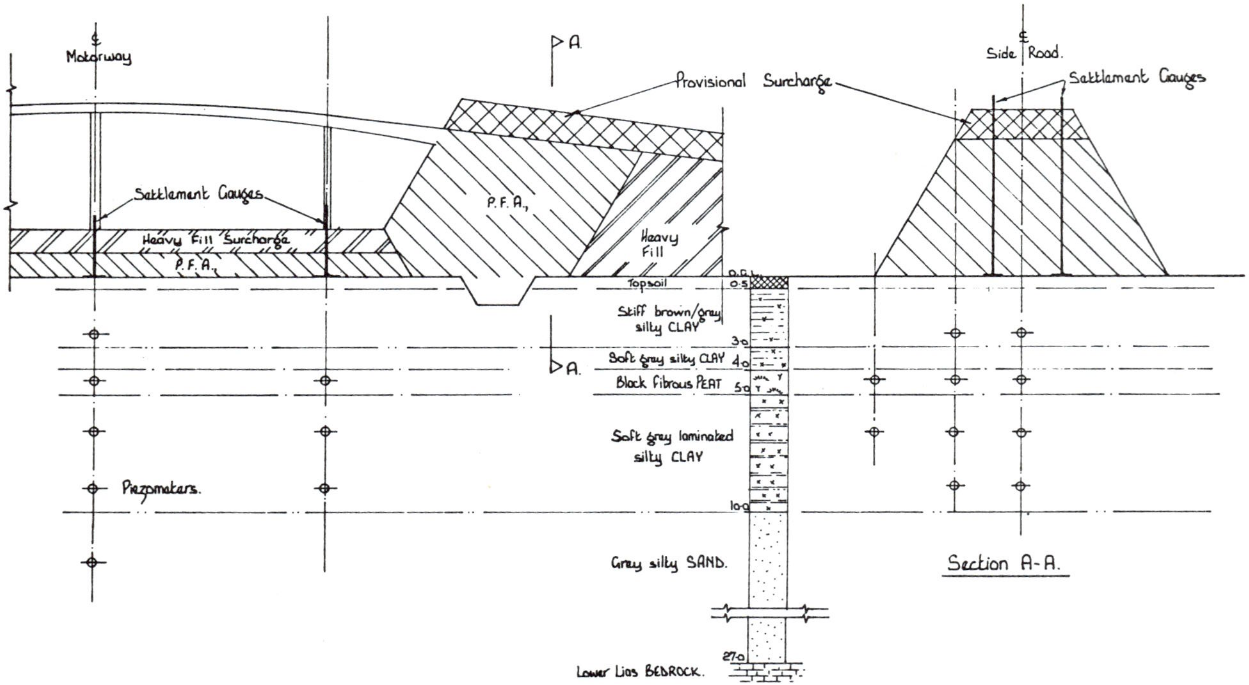

General Arrangement of Instrumentation.

Fig 1.

settlement calculations were checked.

Rod settlement gauges, inductive settlement gauges, piezometers, horizontal strain and settlement gauges, slope indicators, and toe pegs were used. See Fig. 1. The number of instruments used is given below:-

Rod settlement gauges	362
Perforated settlement gauges	76
Inductive settlement gauges	4
Piezometer huts	28
Piezometers	342
Horizontal strain and settlement gauges	6
Slope indicators	23
Toe pegs	1270

The costs referred to below are based on the tender submitted in January 1971.

ROD AND PERFORATED SETTLEMENT GAUGES

Details of the two types of gauge are shown in Figure 2. The rod settlement gauges were installed at 100m intervals on the motorway centre-line and at 200m intervals on the hard shoulder, offset from the centre-line by 17.3m. They were staggered in order to provide two gauges at every 100m along the motorway. They were originally designed with a concrete base to be cast in situ, but this was later changed to a 600mm square steel plate, 12mm thick, with the first section of the rod welded into a hole in the centre of the plate, so that the lower half was a spike which was driven into the topsoil to fix the gauge. The rods were sleeved off from the surrounding fill by 50mm diameter pitch fibre pipes in order to prevent any down-drag force due to compaction. To protect the gauge during filling, it was specified that a 750mm diameter dry stone wall should be built around the gauge. During construction, however, this was found to be unnecessary, and it was later omitted.

Two perforated settlement gauges were positioned over each piezometer group, allowing the standing water level in the fill to be measured. The outside of the 40mm diameter pipe was coated with bitumen to reduce the down-drag of the fill due to compaction.

All the gauges were extended by the main Contractor as required. The overall cost of each type of settlement gauge was about £35. The cost of the water dipper was £26.

The main objections to the rod settlement gauges are that instruments projecting above the fill interfere with the placing of the fill and are liable to be damaged by the plant.

It is difficult to be specific about how much the RSG's interfered with the construction of the embankments. The major effects were probably the additional compaction required around each gauge, and the difficulties the graders had in getting the windrows past them when maintaining the formation. To provide against damage by plant, a diamond shaped board 600x600mm was placed over each gauge as a warning. In spite of this, about 25% were damaged, and bent rods had to be replaced.

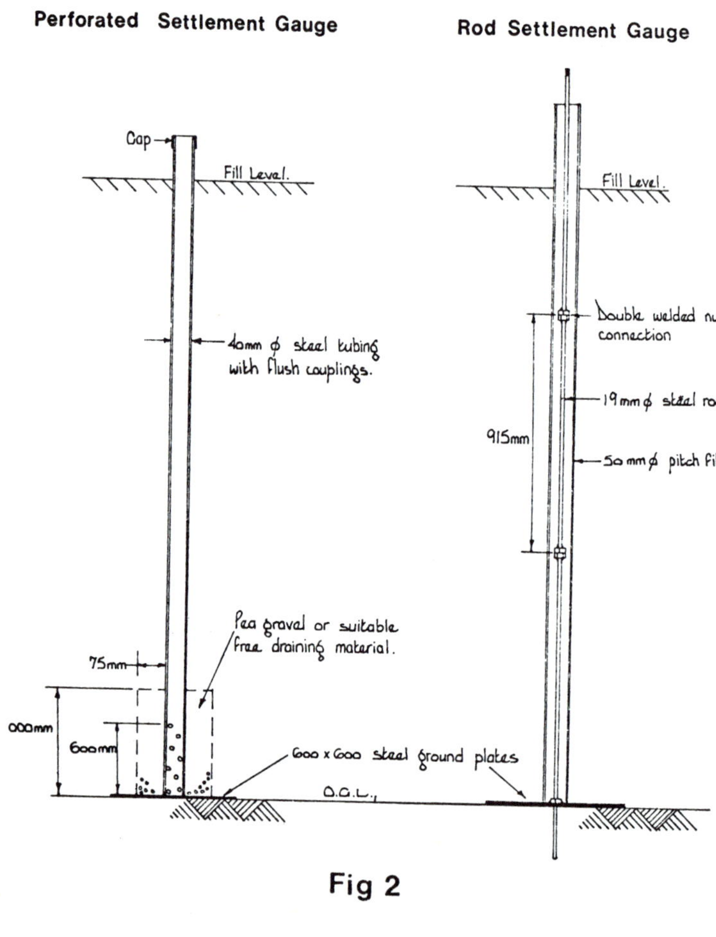

Fig 2

INDUCTIVE SETTLEMENT GAUGES

These were used to measure the settlement at different depths below original ground level. The gauge consists of 25mm and 38mm rigid pvc telescopic tubing installed in a 200mm diameter borehole to bedrock, details of which are given in Figure 3.

Inductive Settlement Gauge

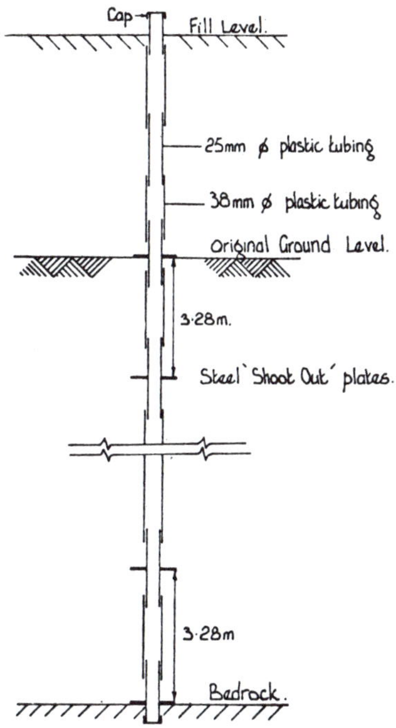

Fig 3

At 3m intervals, expanding plates were keyed into the sides of the borehole so that they were independent of the telescopic tubes. The hole was then backfilled with 3:1 bentonite:cement grout. Using an induction coil in a probe, the depth of each plate below a datum point, and hence the settelement, could be determined.

There is a basic difficulty in installing this type of instrument in sensitive soils, in that the expanding plates are supported by the soil on the perimeter of the borehole, and if this becomes disturbed when the borehole is sunk, it may not have sufficient strength to support the plate. In such a case, unless the grout is stiff enough to support it, the plate falls down the borehole until it takes up the slack in the tubing which has been used to pressurise the expanding mechanism.

At first, some trouble was experienced with the probe which was not watertight. The lead was found to need careful handling, as a kink could break the electric cable.

These gauges, being made of pvc, were easily damaged by earth moving plant, and, once damaged, were difficult to reinstate, as fill material could then fall down the tubing and lodge on one of the telescopic sections. The four gauges installed on this length of motorway were damaged beyond repair in this way early in the Contract.

The average cost of an inductive settlement gauge 24m long was about £380, and the cost of the probe was £112.

PIEZOMETERS

Groups of piezometers were placed under the motorway at about 600m intervals, on both banks of the three river crossings and under one approach embankment of each side road crossing. The first piezometer in each group was installed in a 1m long sand pocket at the bottom of a borehole from which continuous 54mm diameter piston samples had been taken. These samples were extruded in the site laboratory, cut into 150mm lengths and split open lengthways one half being used for moisture content and Atterberg Limit tests, the other being laid out on corrugated polythene trays to air dry. When partially dry, the samples were described in detail, in order to fix the depth of the piezometers for the rest of the group. The piezometers used had a low air entry value and a permeability of about 3×10^{-4} m/s. They were installed in 75mm diameter flight auger holes which were bored to 300mm above the required depth, pushed into position on the ends of detachable rods, and the holes were backfilled with a 3:1 bentonite:cement grout which was tremied in.

<u>Installation and De-airing</u>

Each piezometer was connected to a 12m length of twin tubing and installed dry as a construction expedient. The 12m length of lead provided a check on the depth of the piezometer below ground level. The leads were subsequently joined to leads from the instrument hut which had been filled with de-aired water. With one of the joints undone, water was forced under a low pressure down the other lead into the tip and up the return until it started to flow from the undone joint. With water weeping from each, the leads were joined and the complete circuit was de-aired using a fairly high positive pressure on one lead and a small vacuum on the other. Within a day or two, however, the mercury manometers gave out of balance readings, and it was realised that the initial de-airing pressures had generated a pressure at the tip in excess of the hydrostatic pressure, and that, as a result, some of the original air and water in the leads had been forced into the soil surrounding the tip. This problem arose because the Contractor, when using the standard formulae to calculate the de-airing pressures, did not take into account that the piezometer tip was below the instrument hut rather than above it, which is more usual. The de-airing pressures were then changed so that the pressure inside the tip equalled the hydrostatic pressure in the soil. De-airing using these pressures took longer.

As the settlements under the embankments were going to be in the order of 1 to 2m, it was desirable to use a grout back-fill for the piezometer holes which had similar compress-ibility and shear strength characteristics to the natural ground. If a strong grout were used, the column of backfill would act as a pile and force the piezometer down with the settlement of the embankment. Simple tests carried out on site indicated that the most suitable grout was probably the specified 3:1 bentonite:cement mix, with the mixing time kept to a minimum (15 minutes). A research project into the shear strength and compressibility characteristics of grout backfill would be to the benefit of the profession.

The leads from the tips were colour coded at 1m intervals from a point 2m within the embankment to the instrument hut. It would have been better to have colour coded the leads for the complete length, as several were cut along the centre-line of the motorway during drainage excavations. Without positive identification, pairing up the severed leads caused considerable problems. Only three piezometers, that is 1%, became unsatisfactory during the Contract.

Constant head in situ permeability tests were carried out on all piezometers prior to embankment loading, on completion of loading, and prior to surcharge removal, the applied heads being kept low. (Gibson 1963, 1966, 1970, Wilkinson 1968 and Bjerrum et al 1972). Because of the differences in level between the manometer unit, the header tank and the permeability test apparatus, the applied head was measured using the manometer on the piezometer board itself. The cost of making and backfilling a hole for a piezometer was approximately halved by the use of a flight auger instead of a shell and auger. The average cost of a piezometer in a flight augered hole was £125 and in a shell and auger hole £165. The additional cost of piston sampling, including the cost of the equipped site laboratory and store, was about £150 per hole.

HORIZONTAL STRAIN AND SETTLEMENT GAUGES

Six gauges were installed to measure the strain and settlement profile under selected embankments. Each gauge was a 48mm outside diameter plastic tube laid in a trench 600mm below original ground level. Metal plates, approximately 500x100mm, were threaded on to the tube at 3.05m centres. The trench was backfilled with sand or a fine granular fill and the ends located in concrete blocks with lockable covers. See Figure 4.

The settlement recording equipment consisted of a liquid line connected to a transducer housed in the cable drum, which measured the pressure changes directly in terms of settlement. The settlements were always read at the same points in the tube which were fixed relative to the metal plates along the tube. The plates were located using a probe containing a coil which formed the active arm of a parallel resonance bridge.

As the gauge was still being developed, several teething problems arose. The main ones on the settlement side were that the liquid line needed frequent de-airing and that it was subject to temperature changes. The horizontal strain measurements were of little value as the inaccuracies in the distance measurements were greater than the movements of the plates due to the embankment loading. The gauge has since been modified to facilitate easier de-airing, and a steel band is now used to measure the plate distances.

SLOPE INDICATORS

Slope indicators were installed at the three river crossings, on the northern approach to the Huntworth Viaduct, on one of the side road embankments and at four bridge sites to measure the lateral movements below ground. The extruded 80mm diameter aluminium tubing was installed in a 200mm diameter borehole sunk 0.3m into bedrock. The tubing was supplied in 3.05m lengths which were joined by external couplings. Gaps varying from 25 to 150mm were left between adjacent lengths of tube to allow for settlement. The annulus was backfilled with 3:1 bentonite:cement grout and the tube was capped with a steel pipe fitted with a lockable cover, the pipe being anchored in a concrete block.

The recording unit was powered by four 6 volt batteries. The aluminium body of the torpedo became scored during use, as it rubbed against the inside of the tubing, and slightly pitted at the junction with the stainless steel cap, due, it

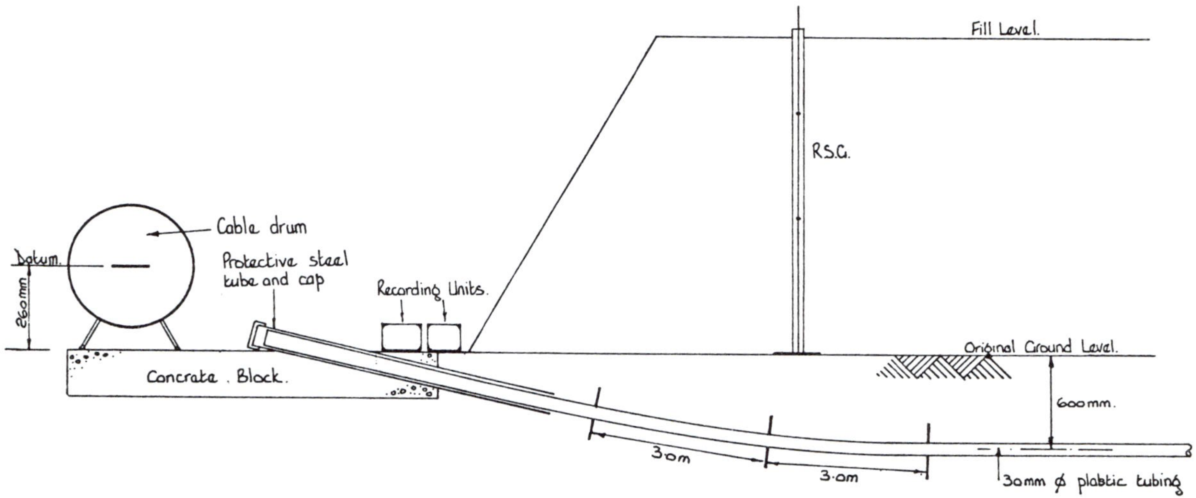

Horizontal Settlement Gauge.

Fig 4

is thought, to electrolitic action. There was some evidence of the aluminium tubing being attacked in the ground, frothing being noticed in three of the tubes.

As the cable between the recording unit and the torpedo requires careful handling, a cable drum was specified in the Contract.

The instrument has been used continuously for over two years on site, and during this time has only required one set of replacement batteries and the repair of two minor electrical connections. For these repairs, however, the instrument had to be sent off site. At the beginning, each tube was read twice and these readings confirmed that the instrument was accurate to 6mm in 30m. It was found that one pair of grooves 25m long took one hour to read in the field and another hour in the office to reduce the readings and plot the deflections. To facilitate the reduction of the readings, a table converting readings to deflections was prepared on site.

The cost of the slope indicator torpedo, cable, cable drum and recorder was about £2,000, and the cost of a tube installed in a 24m deep borehole about £360.

TOE PEGS

These were installed at 20m intervals along the toes of all embankments where potential failure conditions were thought to exist. This was for all side road embankments where the height of heavy fill exceeded 4m, and for the surcharged motorway embankments where the 2m deep drainage ditch was close to the toe of the embankment.

The pegs were controlled by survey lines 30m offset from the embankments. At each reading, the 30m offset was checked, and a theodolite used to determine the lateral movements of the pegs. Levels were also taken.

The pegs were 50x50x750mm timber stakes driven into the ground and surrounded by concrete. Errors arose because it was often two weeks after the stakes were driven before they were concreted in. To do this, the ground was cut away from around the top of the peg, and this caused some disturbance. Some of the stakes became loose in summer as the surface crust dried out, showing that where readings are required for several months during the summer, stakes longer than 750mm are advisable. Each toe peg cost £2.50.

SETTLEMENT PEGS

For measurement purposes, settlement pegs were placed on top of all the embankments at the start of the settlement period. These were spaced at 40m intervals along the centre-line and hard shoulder of the surcharged motorway and at 20m intervals along the verges of the side road embankments. They were levelled twice, at the beginning and at the end of the surcharge period.

STAFF

When embanking was in progress, seven men were employed on monitoring. These were a section engineer, an assistant engineer, a graduate engineer, a draughtsman, and three technical assistants. With several side road embankments under construction at the same time as the motorway embankment itself, continuous effort was required by these men to keep pace with the Contractor.

CONCLUSIONS

The instrumentation in general performed very well and the motorway and side roads were built without any failures. The construction of several side road embankments was stopped because of high pore water pressures in the foundations, and only started again when the effective stress analysis indicated that it would be safe. One embankment, where the pore pressures were not dissipating fast enough, required the addition of wide berms to allow it to be constructed to full height.

The only instruments which did not perform satisfactorily were the inductive settlement gauges, which became blocked with fill after being damaged by earth moving plant, and the horizontal settlement and strain gauges. These were in the course of development, and although the settlement gauge worked well, the horizontal strain measurements were not sufficiently accurate to allow the ground movements to be measured.

ACKNOWLEDGEMENTS

The paper is presented with the permission of Mr P G Lyth, the Director of the South Western Road Construction Unit, Engineer to the Contracts, and Mr A S Turner, Chief Engineer of the Somerset Sub-Unit of the SWRCU. The authors wish to thank them.

They also wish to thank, amongst many others, Mr B A Heskins and Mr A R Croall, for their assistance in taking the many readings.

REFERENCES

Bjerrum, L, J K T L Nash, R M Kennard and R E Gibson (1972) Hydraulic fracturing in field permeability testing. Geotechnique, 22, 2:319-332.

Gibson, R E (1963) An analysis of system flexibility and its effect on time lag in pore pressure measurements. Geotechnique, 13, 1:1-11.

Gibson, R E (1966) A note on the constant head test to measure soil permeability in situ. Geotechnique 16, 3:256-259.

Gibson, R E (1970) An extension to the theory of the constant head in situ permeability test. Geotechnique, 20, 2:193-197.

McKenna, J M (1973) Monitoring the M5 motorway embankments on soft ground in Somerset. BGS Symposium on Field Instrumentation.

Wilkinson, W B (1968) Constant head in situ permeability tests in clay strata. Geotechnique 18, 2:172-194.

INSTRUMENTATION OF FLOOD DEFENCE BANKS ALONG THE RIVER THAMES

by A Marsland

Geotechnics Division, Building Research Station

INTRODUCTION

During the next 5 to 6 years some 60 miles of flood defences along the Thames downstream of the proposed flood barrier are to be raised by up to 2 m in order to provide protection against river levels about 1 m above those reached during the last major flood in 1953. A large proportion of the existing defences are earth banks located on marshes underlain with soft, highly compressible clays and peats which have a total thickness of up to 15 m. The presence of these soft soils leads to considerable problems of stability and settlement. Slips have occurred on both the landward and riverward side of these banks since their reconstruction in 1953 (Marsland, A 1957(a) and (b)).

In addition to the problems involving large settlements and instability due to soft underlying soils, there is the additional complication of continuously changing pore water pressure in underlying pervious layers which are directly connected to the river. The pore water pressures in these layers can become the critical factor in the stability of the banks during flood conditions and cause violent failures (Marsland, A, 1961).

The programme of instrumentation for the new banks is in its early stages. Some instruments have been installed on the first length of trial bank and the first stage of construction comprising a loading berm on the landward side of the bank was completed in August 1972. Instrumentation of typical sections on other trial lengths and on the main construction are being planned. The paper outlines some of the problems associated with the instrumentation of flood defence banks, describes some of the equipment which has been installed and discusses ideas to be incorporated in the programme as it develops.

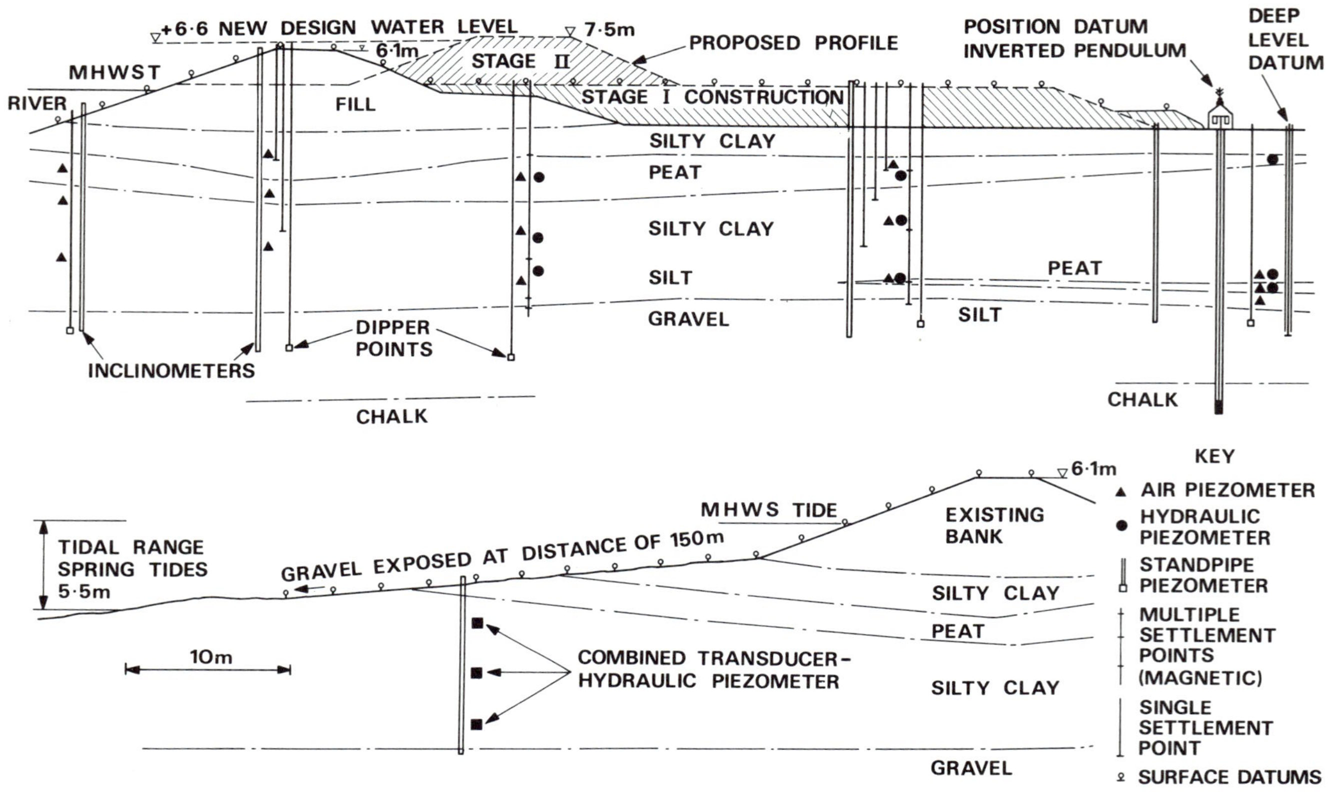

Fig 1 Cross-section of typical bank showing positions of instruments

DESCRIPTION OF A TYPICAL SITE AND ITS INSTRUMENTATION

A cross section through a typical bank is given in Fig 1. The underlying gravel layer outcrops in the deep channel which is 150 m from the front crest of the bank at this site but varies with location. Under normal conditions the river level changes with a twice daily cycle and has a range of up to 7 m.

Instrumentation is required to

i. monitor the movements of the riverward and landward slopes to provide a warning of excessive movements.

ii. monitor the settlement of the bank and the rate of consolidation of the soft soils under the bank during each stage of construction. These measurements will be used to decide when to proceed with the next stage of construction.

iii. provide pore water pressure data to check the assumptions made in the stability analyses, and to monitor unusual occurrences.

iv. provide pore water pressure data to enable predictions to be made of the uplift pressures which are likely to develop in underlying pervious layers when the river level reaches the design level.

The types and positions of instruments on a typical installation are shown in Fig 1. Both the type and location of instruments will be changed on existing and later installations if experience shows this to be necessary.

MEASUREMENT OF PORE WATER PRESSURES

Pore water pressure measurements are required to provide estimates of uplift water pressures in underlying pervious layers which have direct connection with the river, to monitor the rate of consolidation, to provide data to check the stability during construction and to observe the effects of cyclic tide loading on the riverside slopes. The continuously changing river and ground water levels, the need to measure pore pressures in ground on the riverside of the banks, and the possible presence of gas in the ground are important factors which must be considered when choosing a piezometer system.

Types of piezometer and their application

The type of piezometer used depends upon the permeability of the soil, the rate of change of the pore water pressures to be measured, its reliability and flexibility of operation in the particular application, the location of the piezometer in relation to the reading position, ease of reading, and its cost as compared with other types. From the point of view of application there are three basic types of piezometer. These are the vertical standpipe, the twin-tube closed circuit hydraulic piezometer, and piezometers with pressure transducers fixed directly to the porous filter tips.

Vertical standpipes comprising 12.5-20 mm diameter plastic tubes and fitted with porous ceramic tips, have been used extensively for measuring the changes in pore water pressure in layers of sand and gravel which have a direct connection with the river. Piezometric levels measured in the gravel at a site in the Dartford area are plotted against the tide level for a number of tides in Fig 2. Extrapolation of these measurements to the design tide level gives uplift pressures in the gravel which are sufficient to float the 6 to 10 m of peat and clay which overlie the gravel in the marsh.

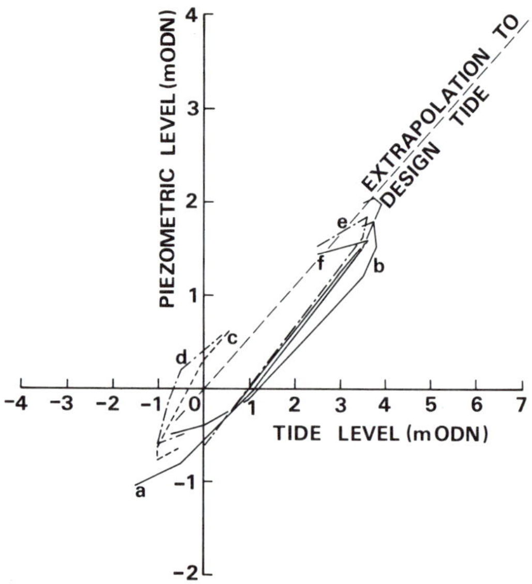

Fig 2 Variation of piezometric level in underlying gravel layers with tides

Most of the measurements have been made manually with an electrical dipper but the author has also taken continuous readings using a air back-pressure system which he developed during similar investigations along the Thames in 1953-1955 (Cooling, L F, 1962 Fig 10). Air was fed continuously through a 4 mm diameter tube, the lower end of which was fixed in the standpipe below the lowest anticipated water level. The flow of air was fixed at a very slow rate, just sufficient to cause a bubble of air to leak out of the lower end of the tube every few seconds. Since the quantity of air passing down the tube was extremely small, there was a negligible pressure loss along the tube and the pressure measured on the recorder was equal to the pressure head of the water in the standpipe above the level of the lower end of the air tube.

The advantages of the open standpipe piezometers are their simplicity, and their relatively easy installation and long life which makes them particularly useful for long-term observations at isolated sites. On the other hand they take a long time to reach equilibrium in soils with low permeability which makes them unsuitable for measuring pore water pressure changes in these soils caused by rapidly changing loads such as tidal loading.

Twin-tube closed system hydraulic piezometers (Penman, A D M, 1956) have been used extensively to measure pore water pressures in studies of various aspects of the stability of flood banks. They were used in 1953-1955 as an alternative system and as a check on measurements made in vertical standpipes installed in layers of fine sand or silt under flood banks (Marsland, A, 1957(a)). The measuring equipment was located in small insulated instrument boxes and huts and portable de-airing equipment was developed which could be carried by one man to isolated sites where access by vehicle was impossible in winter months. They were also used in investigations to measure the pore water pressures in clays, peats and clay-silts at a few sites where slips occurred on the riverward side of the banks. In these earlier applications ordinary polythene tube was used and placed over the banks to instrument positions on the landward side in trenches which were only a few feet deep. The water in the length of tube passing over the banks was often under a partial vacuum of up to half an atmosphere which increased the diffusion of air through the polythene and this made it necessary to de-air the system about once a week. At one site where slipping was occurring piezometric levels measured in the clays, peats, and clay-silts exceeded the water level in the river at all stages of the tide and a significant lag was observed during the fall of the tide. A pneumatic piezometer installed on the riverside of the bank below the high water level in the recent investigation has also given pressure heads above the river level. Since appreciable deformation was occurring when these observations were made it is possible that the pressures recorded were the actual pore pressures in the ground. The high values which have been measured could, however, have been due to methane gas which may be present in the organic soils, entering the porous filter tips. In this case the piezometer would have measured the pore gas pressure which is higher than the pore water pressure due to the surface tension forces developed in the menisci surrounding the gas bubbles. If gas is present then it is necessary to measure both the pore water and pore gas pressures in order to estimate the effective stress in the ground. (Bishop, A W, Kennard, M F and Vaughan, P R,

1964). The rate at which gas passes through the porous element can be reduced significantly by using very fine filter elements. Even so air will gradually enter by diffusion. Further measurements of pore water pressure will be made in the near future using a piezometer fitted with a transducer and facilities for de-airing in an attempt to overcome these difficulties (see later section).

The advantages of the hydraulic twin-tubed closed system piezometers are the relatively short time they take to come to equilibrium, they can be installed in areas subjected to inundation, cause comparatively little interference with construction, and can be recorded at central observation stations. Unlike the transducer type piezometer described in the next section, the fact that they can be de-aired means that provided they are fitted with fine porous elements they can be used for long-term observations in soils containing gas. They can also be used to measure the permeability of the less permeable soils in-situ, but great care is required in the installation if reliable measurements are to be made. Their main disadvantage is the need to prevent any but the lowest suctions from occurring in any part of the system. The flexibility of the long lengths of plastic tubing normally used in these installations could mean that they do not measure the correct pore water pressures in soils of low permeability where the load is continuously changing.

Piezometers fitted with pneumatic pressure transducers were developed in the USA (Warlam, A A and Thomas, E W, 1965) where they have been used reasonably successfully on a number of field projects since 1958. The principle of operation is to balance the pore water pressure which acts on one side of a flexible membrane by a measured air pressure on the other side. Pneumatic piezometers manufactured in the United Kingdom by Soil Instruments Ltd (see Fig 3) are being used on preliminary trials on the present project. Air from a pressure cylinder is supplied to the pneumatic pressure transducer at a controlled rate of about 0.5 litres/min. Initially this passes through the valve and back up the return tube to build up a pressure on the measuring gauge. When the air pressure acting on the flexible membrane becomes equal to the pore pressure in the ground it begins to deflect and closes the valve which prevents any further increase of pressure acting on the measuring gauge. Provided the deflection of the membrane does not produce any increase in the pore water pressure around the porous tip the pressure gauge records the pore water pressure in the ground. Having recorded the pressure the air supply is cut off and the air released from the measuring gauge and the transducer in readiness for the next reading. In practice the author re-applies the pressure to the piezometer which has just been measured to check that the system comes to equilibrium at the same pressure. If the readings differ the

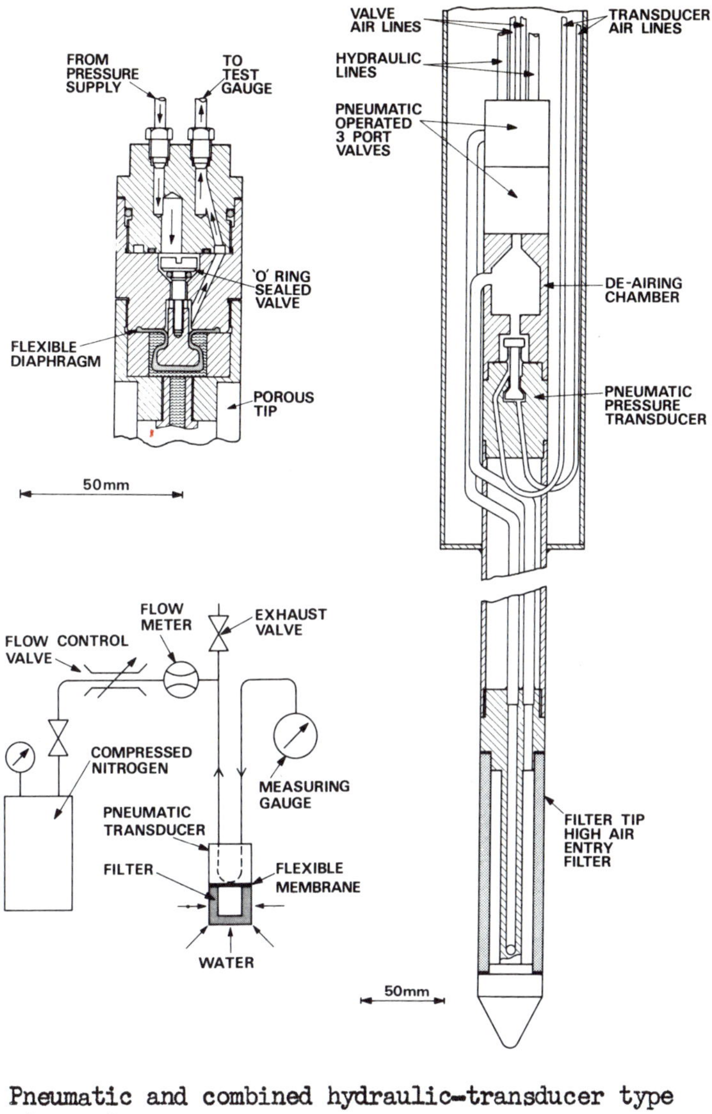

Fig 3 Pneumatic and combined hydraulic-transducer type piezometer

operation is repeated until at least two consecutive readings agree and the final reading is taken to be the measure of the pore water pressure in the ground. The first measurements taken on piezometers in soft alluvial clays often give pressures of up to 0.5 m water head greater than the final values. Smaller differences are measured in the more permeable and compressible peat and the differences between the initial and final readings in silts and sands are usually negligible. So far only a few comparisons have been made between measurements on pneumatic and twin-tube hydraulic piezometers over a limited period of a few months and so far the agreement has been reasonable. A number of the pneumatic piezometers have malfunctioned but some of the cases were probably due to insufficient care being taken during installation. At present it is too early to evaluate their overall and long-term performance.

Piezometers in which deflection of a stiff diaphragm is measured by a vibrating wire or electrical resistance gauge differ from the pneumatic piezometer in that the measurement can be made without moving the diaphragm.

Field trials are being carried out with vibrating wire piezometers of the type designed at the Building Research Station by Mr H S H Thomas, described by Cooling, L F (1962) and manufactured by Soil Instruments Ltd. They are provided with tubes for the circulation of dry nitrogen which is necessary to prevent condensation inside the gauge, Thomas, H S H and Ward, W H (1969).

Recent improvements in long-term stability of resistance wire type pressure transducers, particularly of the unbonded type, has made them a possible alternative for use in field applications. A fully recording system using piezometers incorporating these transducers in conjunction with a water-proofed battery operated magnetic tape data logger is being developed and measurements made with this equipment will be compared with those from other types of piezometer as soon as it is installed.

In theory the transducer type piezometers are ideal for measurements on the riverside of the banks because of their low volume change requirements and their freedom from hydraulic tubes. Unfortunately the presence of gas in the soil could preclude their use for long-term measurements, since it is necessary to de-gas the porous elements at regular intervals. Wherever possible transducer type piezometers are being provided with a long sleeve having a diameter slightly larger than the gauge and installing them by pushing them into the ground at the base of boreholes which are then filled with a thick soft bentonite grout so as to permit their easy retrieval for re-calibration in the event of unusual results being obtained.

Some pneumatic and electrical resistance type piezometers are being fitted with a duplicate hydraulic system which can be opened or closed by remotely operated valves situated near the tips. This arrangement will permit both de-gasing and re-calibration of the transducers in-situ. The hydraulic lines will be terminated as close to the piezometer as possible compatible with the requirements to permit de-airing and re-calibration to be carried out at low water. In this way possible errors in the hydraulic systems mentioned earlier in the paper will be eliminated.

Installation of piezometers

Installation of a piezometer within a sand filter will appreciably increase the rate of response of the piezometer. The best method of forming the sand filter is to pump a known quantity of sand with the necessary amount of water around the piezometer using a hand-operated diaphragm grout pump. Ideally, the best method of backfilling the borehole is to inject a grout having the same properties as the surrounding soil. It should be pumped through a pipe which extends down the borehole to a level just above the top of the sand filter. Various mixtures of bentonite, cement, sand and pulverised fuel ash have been used to grout in the piezometers, but lack of adequate data of the properties of grouts often means that the choice of grout is not as satisfactory as it could be. There is an urgent need for detailed study of suitable materials and methods of sealing piezometers and other instruments into boreholes in different types of ground. If accurate measurements of permeability of laminated clays and silts, clays or peats are required then smear caused by drilling of the borehole should be removed by careful water jetting prior to placing the sand or silt filter around the piezometer. When gas is present in the ground the placing of sand around a piezometer intended to measure pore water pressure is not recommended since it can act as a reservoir for gas collecting around the piezometer. Except for standpipe piezometers installed in sands and gravels it is preferable to fit fine filter elements to all types of piezometer. Piezometers of all types must be thoroughly de-aired before installation. The filter elements should be saturated by a vacuum technique or by boiling in previously de-aired water. The piezometer should be re-assembled under water and kept saturated until required for installation.

The zeros and calibration factors of vibrating wire, electrical and pneumatic piezometers should be checked on site over as wide a range as possible by taking readings at a number of definite levels as they are lowered into the water filled borehole. The temperatures of vibrating-wire and electrical resistance type piezometers should be taken together with readings on the piezometers at intervals after immersion in the water at the first level. The temperature of the water should also be measured and the piezometer kept at the first level until the readings have become constant.

MEASUREMENT OF SURFACE MOVEMENTS

Simple equipment can be used to make very useful measurements when movements are fairly large. In 1955 the author measured the movements of a bank at Longreach near Dartford while it was slowly slipping into the river by levelling on steel pins

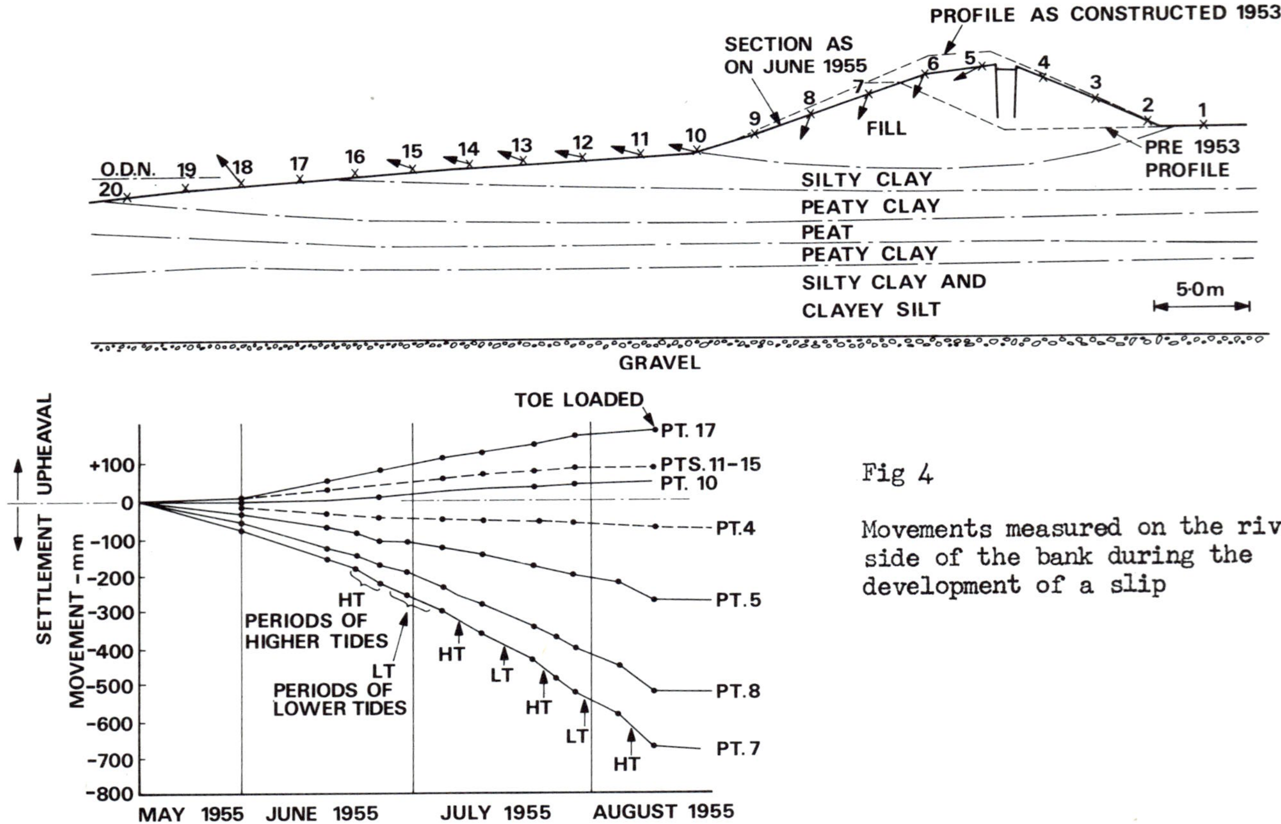

Fig 4

Movements measured on the river side of the bank during the development of a slip set in concrete blocks and measuring the horizontal movement using the levelling staff held horizontally in conjunction with a plumb bob. The direction of movements on the surface over a period of 3 months are shown by vectors on the cross-section in Fig 4. Changes in level at the various points across the cross-section show that the rate of movement was continuous and slowly accelerating and that there was a tendency for the rate of movement to increase during period of higher tides.

In order to study the movements of the banks prior to the onset of obvious failures more accurate measurements are required. Relative movements between specially designed measuring points (Fig 5), spaced 3 m apart over a cross-section of a bank are being measured using a micrometer rod similar to that developed by my colleagues (Burland, J B and Moore, J F A, 1973). The measuring points have been constructed to be reasonably water-proof and are fitted with a very flexible outer tube to accommodate the appreciable seasonal shrinkage and swelling which occurs on flood banks. All levels are referred to deep datums fixed in the gravel, but isolated from the overlying soils. At present lateral movements of the measureing points are only being made relative to one another. The distance between the banks and stable ground varies between 1 and 4 km. At a distance of 1 km the Mekometer Mk3, which is the most accurate instrument available will position a datum near the bank to within about 0.3 mm (Penman, A D M and Charles, J A, 1973). While it is intended to test this method as soon as the BRS Mekometer becomes readily available consideration is also being given to the use of inverted pendulums and optical plumbs for transferring horizontal movements at the surface to datums in the chalk 18.0 m below the marsh near to the banks. Some measurements made on the inverted pendulum installed at Scammonden Dam, which had a 30 m long wire, showed that it returned fairly quickly to its neutral position with an accuracy of about ± 0.3 mm. A trial installation using an inverted pendulum with an 18.0 m long wire is being installed behind the first trial bank to act both as a datum and to measure any movement of the marsh which occurs due to changes in the river level. It is anticipated that a positional accuracy of better than about 0.5 mm will be achieved. This is roughly comparable with what we would expect from an optical plumb system at this depth. Direct measurements will be made between the inverted pendulum and the surface measuring points using the micrometer measuring rod.

Precise level measurements between the deep level datum and the surface measuring points on the site of the first trial embankment (see Fig 1) have shown that the marsh behind the bank rises and falls with tides over a range of 0.5 mm during a normal tidal cycle. Much larger movements are anticipated at other sites where the thickness of the peat layers is much

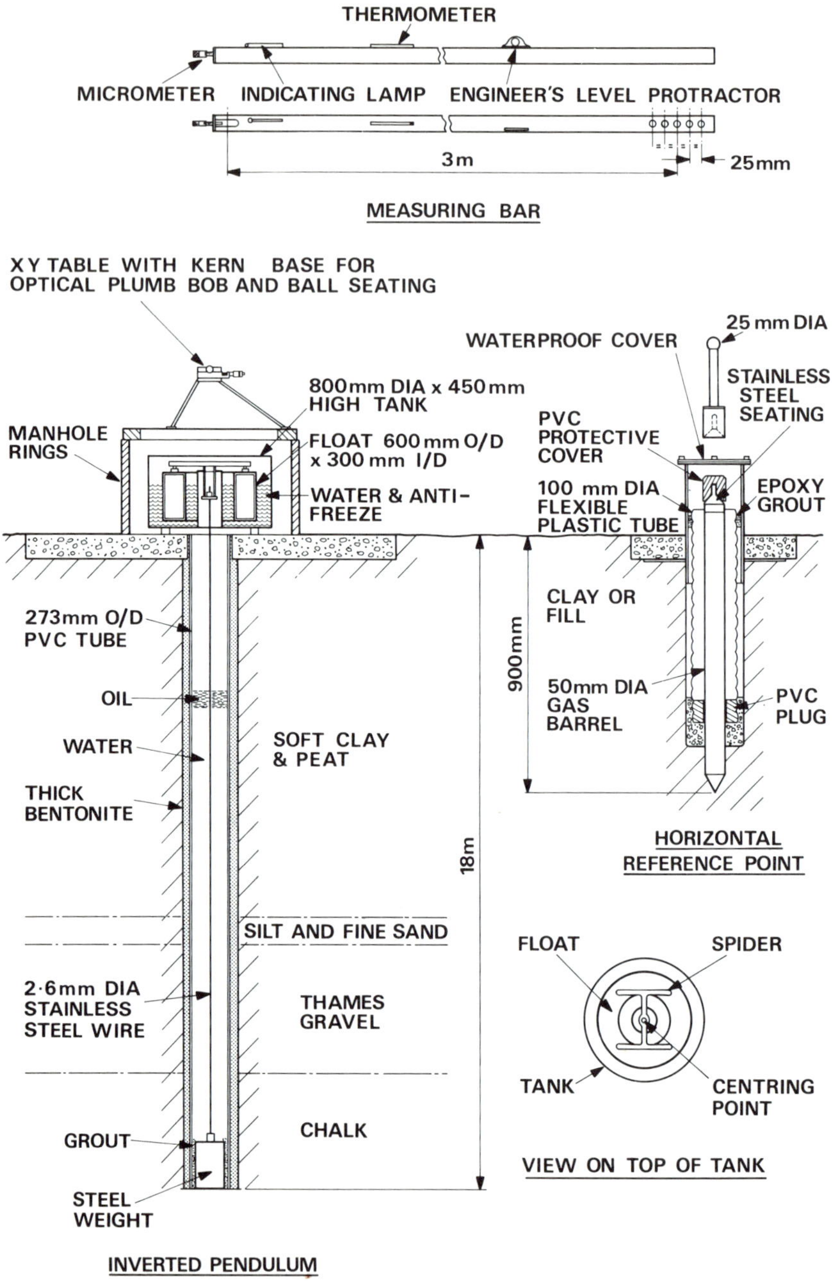

Fig 5 Inverted pendulum datum and surface measuring points

greater.

MEASUREMENT OF INTERNAL MOVEMENTS IN THE GROUND

Measurement of vertical movements

After considering the various types of instruments available for measuring settlement below ground level the author decided to use the simple measuring system shown in Fig 6(a) as the basic method for measuring vertical settlements in the first stage of the project. Each point consists of a hollow steel tip having a diameter of 100 mm and a length of 300 mm which is pushed into the ground at the bottom of a borehole. The surface area of the tip is sufficient to mobilize enough shear to carry a vertical load of at least 50 kg, which should be more than sufficient to prevent movement due to small accidental forces which may be applied during the levelling operations. The measuring tube was surrounded by a very flexible nylon and wire reinforced plastic tube and the remaining space outside the flexible tube was backfilled with a thick bentonite suspension. This arrangement permits the surrounding

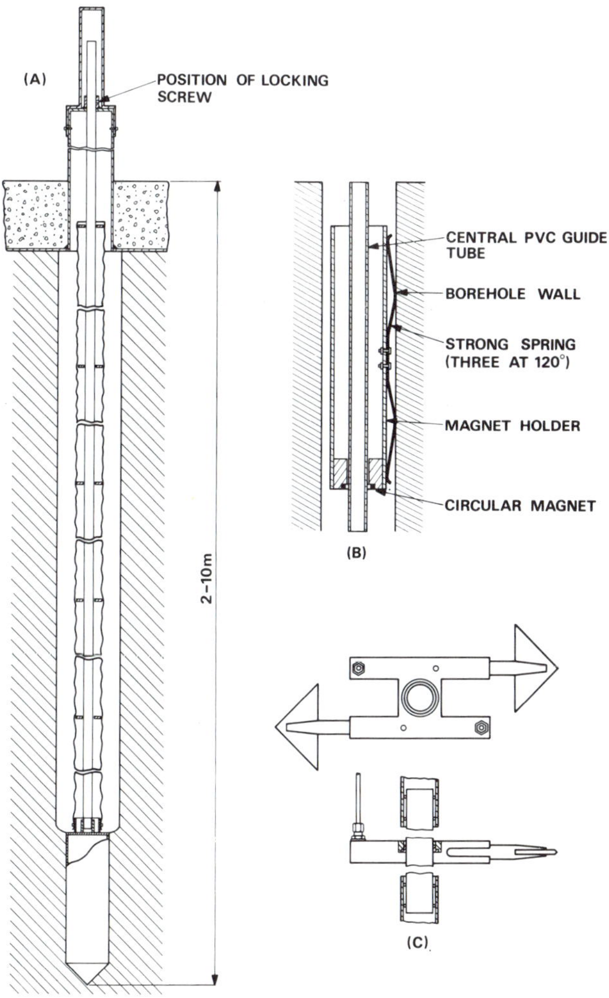

Fig 6 Deep settlement gauges

ground to move freely around the extension rods.

Two types of multi-point settlement measuring devices which are being considered as possible alternatives are shown in Fig 6(b) and (c). The levels of the magnets fixed to both types of devices are measured by probes fitted with reed switches (Burland, J B, Moore, J F A and Smith, P D K, 1972). Preliminary trials with the spring type of magnet holders (type B) in the soft alluvial soil were only partially successful, but it is intended to make further trials with this type and also type C which has been used elsewhere in soft soil. Since both these systems incorporate tubes which are much more rigid than the ground it is necessary to isolate them from the surrounding soils by filling the borehole with very deformable material free from particles large enough to cause jamming at the magnets and sliding joints. One possible solution is to fill the hole with a thick bentonite suspension prior to installation. A prototype of a new arrangement fitted with a very flexible outer tube (Fig 7) is being made. It is intended to pre-drill a hole the same diameter as the lower shoe and the magnet holders and to push the assembly into the hole by means of the central tube.

Lateral movements

Measurements of lateral movements are being made using the Mark II Inclinometer manufactured by Soil Instruments Ltd and described by Green, G E (1973). Because of the soft nature of the soil it was initially decided to take the measurements in plastic rather than aluminium tubes. Because of the difficulty in obtaining plastic tubes with straight grooves it was decided to use a plain plastic tube and to lower the inclinometer on rods to maintain the required orientation of the instrument. Measurements with this system proved to be rather slow and tedious. Aluminium tubes with longitudinal grooves have been installed to check the measurements made in the plastic tubes and also to measure the small initial movements more accurately. The procedures outlined in the paper by Burland, J B and Moore, J F A (1973) have been followed when taking measurements. Series of measurements can be taken much quicker in the grooved aluminium tubes and this has made it possible to follow horizontal movement of the bank as the tide rises and falls. These readings have shown that the front crest of the bank moves with an amplitude of about 15 mm during a tidal cycle. So far the inclinometer tubes have been installed in 150 mm diameter boreholes and completely grouted over their full depth by pumping grout through a pipe to the bottom of the borehole. A grout comprising 1 part by weight of cement to 10 parts P F A mixed in a 8 per cent bentonite-water suspension has been used. At present the bottom of the inclinometer tubes, which are located well into the gravel, are being used as datums for measuring movements at higher levels. The tops of the inclinometers are being integrated

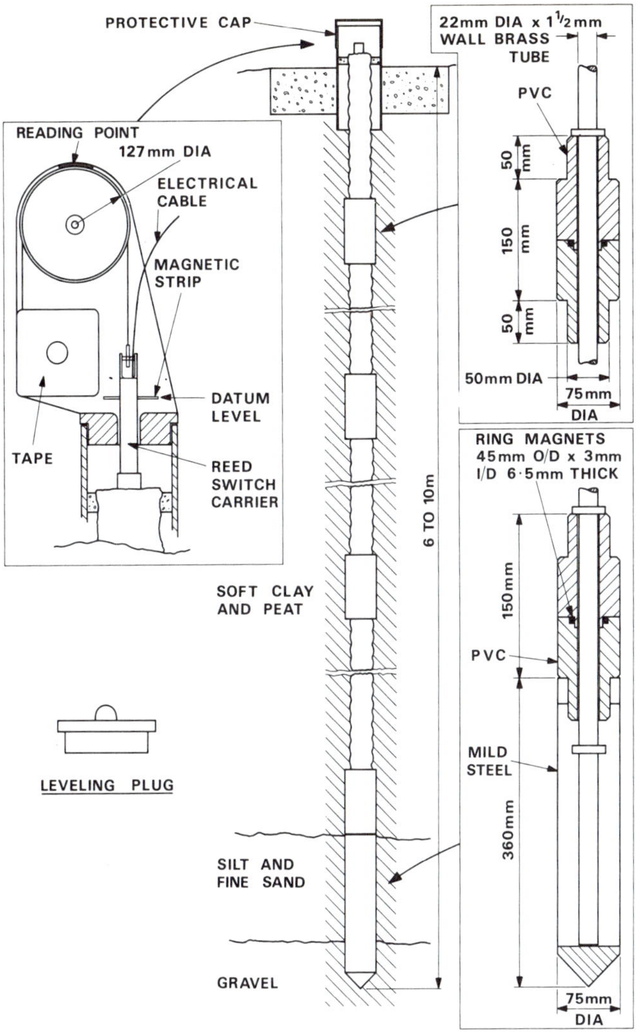

Fig 7 New prototype settlement system

with the system of surface measuring points which in turn will be related to the inverted pendulum datum and any other datums which are subsequently established.

CONCLUSIONS

Measurements of movements and pore water pressures in and under flood banks are complicated by the continuously varying river level which has a twice daily cyclic movement having a range of up to 7 m. Soft clays and peats below the banks make it necessary to pay special attention to the construction and installation of instruments so that they are compatible with the surrounding ground. The problems of instrumentation in these conditions emphasize the need to provide more checks by comparing measurements obtained with different types of instrument and with similar instruments installed in different ways. There is an urgent need for more detailed studies of grouts and other materials used to backfill boreholes around instruments. Much more attention needs to be given to the detailed design and method of installation of instruments to suit particular problems and soil conditions, since any one type of instrument is usually only satisfactory in a particular range of soils. Since the drilling and installing the instruments is a major item in the overall cost further consideration should be given to means of installing a number of instruments of the same or different types in a single borehole. This should not, however, be done at the expense of reducing the reliability of any of the individual instruments.

ACKNOWLEDGEMENTS

The author wishes to thank his colleagues Messrs B J Eason, J Dobbs and A P Butcher, R S Quarterman of the BRS Design Office, G G Brennan of the Kent River Authority and Soil Instruments Ltd for their assistance. The paper is published by permission of the Director of the Building Research Establishment.

REFERENCES

Bishop, A W, Kennard, M F and Vaughan, P R (1964): Developments in the measurement and interpretation of pore pressures in earth dams. Trans 8th Int Cong on Large Dams, Edinburgh 1964, vol 2, pp 47-72.

Burland, J B, Moore, J F A and Smith, P D K (1972): A simple and precise borehole extensometer. Geotechnique, March 1972, pp 174-177 (Also BRS Current Papers 11/72).

Burland, J B and Moore, J F A (1973): The measurement of ground displacement around deep excavations. Proc Brit Geotech Soc Symp on Field Instrumentation. London.

Cooling, L F (1955): The measurement of pore water pressures and the application in some engineering soil problems. Rilem Symp 18, Theme 1(b), Lisbon, Oct 1955.

Cooling, L F (1962): Field measurements in soil mechanics. Geotechnique, vol 12, June 1962, pp 77-103.

Green, G E (1973): Principles and performance of two inclinometers for measuring horizontal ground movements. Proc Brit Geotech Soc Symp on Field Instrumentation. London.

Marsland, A (1957(a)): The design and construction of earth flood banks. Jnl Inst Water Engrs, vol 11, no 3, pp 236–258. (Also Chapter 24, R B Thorn: River engineering and water conservation work, Butterworth).

Marsland, A (1957(b)): Discussion. Proc 4th Int Conf SM and FE, vol 3, p 257.

Marsland, A (1961): A study of a breach in an earth embankment caused by uplift pressure. Proc 5th Int Conf SM and FE, Paris 1961, vol 2, pp 663–668.

Penman, A D M (1956): A field piezometer apparatus. Geotechnique, vol 6, no 2, June 1956.

Penman, A D M and Charles, J A (1973): Measuring movements of embankment dams. Proc Brit Geotech Soc Symp on Field Instrumentation.

Thomas, H S H and Ward, W H (1969): The design, construction and performance of a vibrating-wire earth pressure cell. Geotechnique, vol 19, no 1, pp 39–51.

USBR (1963): Earth Manual. United States Dept of the Interior, Bureau of Reclamation, 783 pps.

Warlam, A A and Thomas, E W (1965): Measurement of hydrostatic uplift pressure on spillway weir with air piezometers. Symp on Instruments and Apparatus for Soil and Rock Mechanics. ASTM Spec Tech Publication No 392.

MEASUREMENT OF DISPLACEMENTS IN THE GROUND BELOW LOADED PLATES IN DEEP BOREHOLES

by A Marsland and B J Eason

Building Research Station

SUMMARY

Equipment has been developed for measuring displacement in the ground below large diameter plate loading tests. An essential feature of this equipment is that it can be quickly installed in the confined space at the bottom of a 900 mm borehole. Some typical measurements obtained with this equipment are given. They emphasize the need for measuring vertical movements at a number of points in the clay in addition to accurate measurement of the movement of the plate.

INTRODUCTION

Detailed studies are being made at the Building Research Station of deep in-situ loading tests on plates of various diameters up to 865 mm (Marsland, A, 1971(a), 1971(b), 1972). The equipment and test procedures used for these tests were designed to reduce the time between excavation and re-loading to a minimum in order to keep the expansion of the clay below the base of the borehole as small as possible. Deformation moduli determined from the load-settlement curves of loading tests on 865 mm diameter plates in 900 mm diameter boreholes in fissured London Clay were several times higher than those measured in standard triaxial tests on 38 and 125 mm diameter specimens (Marsland, A, 1971(c)). The values determined from the large in-situ tests were more compatible with displacements measured around foundations and excavations (Serota, S and Jennings, R A J, 1959, Cole, K W and Burland, J B, 1972, Ward, W H, 1971). Even though all possible measures were taken to remove disturbed material and to reduce the time that the clay was unloaded, some expansion of the clay below the base of the borehole was inevitable, mainly due to the rapid rate at which fissures open when stresses are reduced to near zero. As a result of this expansion the moduli determined from plate loading tests may be lower than the in-situ moduli corresponding to a loading condition not preceded by unloading.

In order to investigate the effects of bedding disturbance and expansion of the clay on the load-settlement behaviour of the large diameter plates it was necessary to measure movements in the clay below the plates. Other engineers have developed systems for measuring displacements in soils and rocks in the vicinity of applied loads and excavations (Wallace, G B, Slebir, E G and Anderson, F J, 1969), but these were unsuitable for rapid installation in the confined space at the bottom of a borehole. Prior to designing new equipment some preliminary measurements were made at a single point in the clay 400 mm below the plate using a displacement transducer to measure the relative displacement between the plate and a tube grouted into a small borehole. These measurements indicated that a substantial proportion of the settlement of the plate was due to compression of the clay in the 400 mm thickness of clay immediately below the plate. Improvements were made to this equipment and a more complex system was also developed to measure the movement in the clay at several depths below the plate. The essential requirement was for equipment which could be installed easily and quickly in the confined space at the bottom of a 900 mm diameter borehole. The present paper describes the equipment developed for this purpose and gives some typical results to indicate the importance of measuring the settlements at a number of positions in the clay below the plate as well as that of the plate itself.

DESCRIPTION OF EQUIPMENT AND THE METHOD OF INSTALLATION

Single point measuring system

This simply consisted of a length of aluminium tube pushed a short distance into the bottom of a 16 mm diameter hole bored to a depth of about 400 mm into the clay below the plate and coupled to a displacement transducer. Additional fixing was provided by a small quantity of quick-setting plaster which was poured into the hole through a separate tube prior to installing the measuring tube. The relative displacement between the plate and the top end of the measuring tube was measured by an AC induction type transducer mounted in a housing which also centralised the vertical measuring tube. The housing was installed in holes drilled through the existing loading plate located 250 mm on either side of the centre of the plate.

To install the equipment the 865 mm diameter loading plate was first lowered on to the prepared base of the borehole which was covered with a 8 mm thick layer of wet plaster. The wet plaster was prevented from entering the drilled holes in the plate by a plug of plasticine through which a thin sleeve was pushed into the underlying clay as soon as the plate was in position. This enabled the instrument hole to be drilled with a 16 mm diameter carpenters screw eye auger without waiting for the plaster to set. The auger was fitted with a long guide which fitted into the top part of the reamed hole in the test plate so that the hole could be drilled perpendicular to the plate with a fair degree of accuracy. When drilling was complete a small quantity of plaster was poured into the bottom of the borehole and the lower part of the transducer housing inserted into the top part of the hole in the loading plate. The aluminium settlement tube was then placed through this housing and tamped down to the correct level with the installation jig shown on the left hand side of Fig 1. The top part of the transducer housing with the transducer clamped at the correct level was then placed into position and the system was then ready for electrical connection to the AC strain meter used to measure the displacement of the transducer. The AC inductance transducers used in this system were accurate to 0.005 mm and had a travel of 25–0–25 mm (50 mm). The measurements made with this simple system showed that more

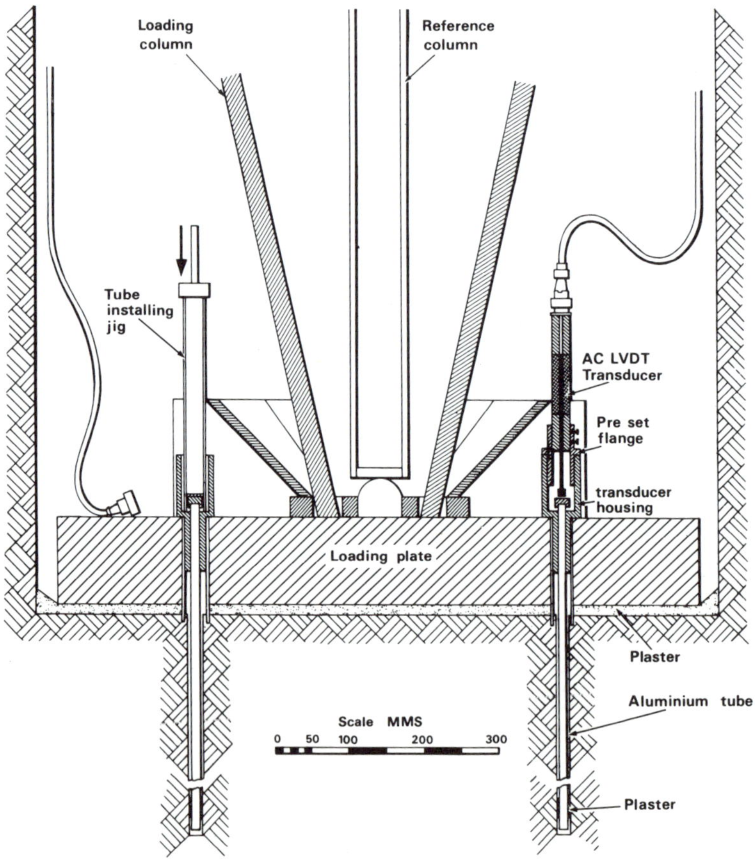

Fig 1 865 mm diameter loading plate with single-point measuring system.

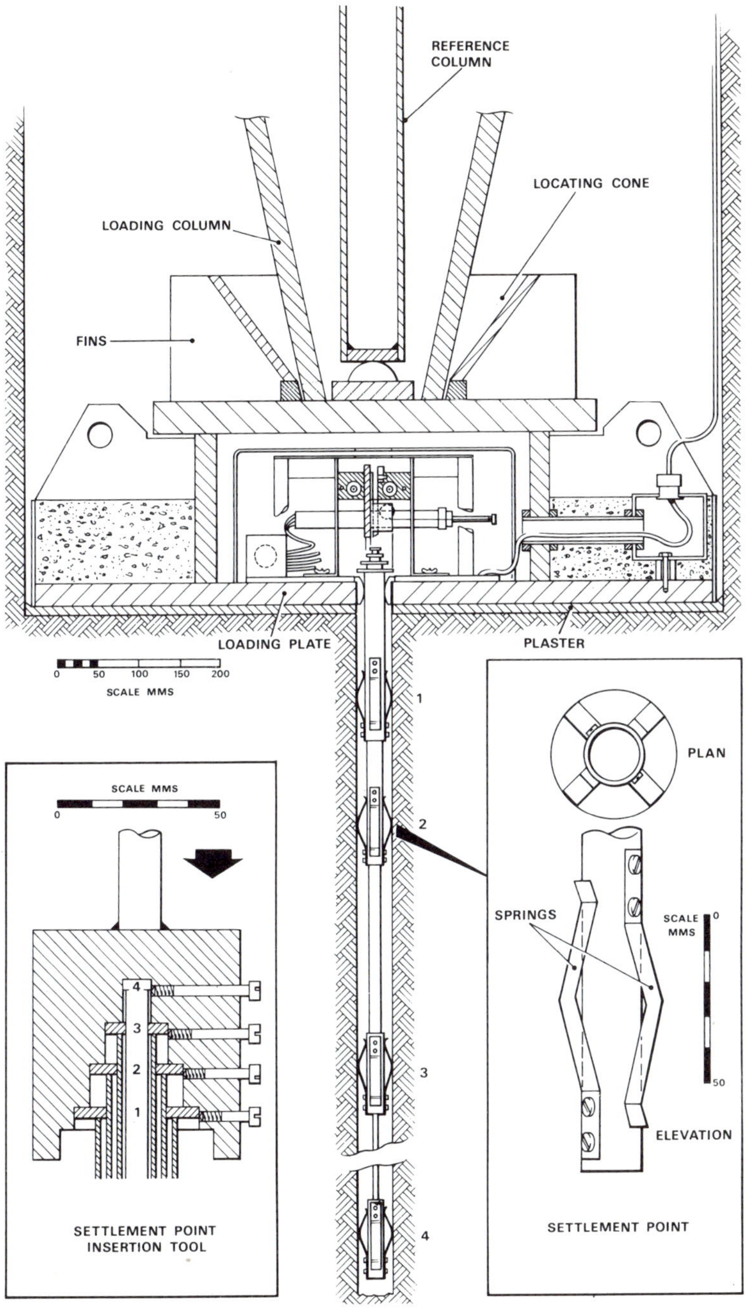

Fig 2 865 mm diameter loading plate with four-point measuring system.

than 50 per cent of the settlement of the plate was due to compression of the 400 mm thickness of clay immediately below the plate as against the theoretical value of 15 per cent for an isotropic homogeneous elastic material. When allowance was made for the excess compressibility of the clay immediately below the plate, moduli were obtained which were about twice those calculated from the load-settlement curves of the plate. While this simple system provides data which can be used for adjusting moduli to allow for the excessive compression of the clay immediately under the plate it does not provide any idea of the change in compressibility of the clay with depth below the plate.

Multi-point measurement system

The multi-point measurement system consists of two main parts, a settlement point assembly and a compact transducer assembly which fits into a central cavity provided in the loading plate, see Fig 2. The settlement point assembly consists of a set of four concentric tubes each of which are fitted with a circular reference platform at the top and four pre-formed steel strips at the base for gripping the clay in the side of the borehole. The four spring steel strips on each settlement point are equally spaced round the tube and fixed by the top and bottom alternately to balance any vertical component of their sideways movement as they grip the clay in the side of the borehole. The top of the settlement tubes pass through a double-coned bearing located in the hole in the loading plate in order to keep the top of the settlement point assembly central and at the same time prevent any bending of the settlement tubes due to uneven settlement of the plate during loading. This arrangement keeps any lateral movement of the concentric platforms at the top of the settlement tubes to within tolerable limits and permits reliable measurements of settlement to be taken for all loads almost up to failure. Location of the measuring system at the centre of the plate was also adopted to reduce the effect of tilt of the plate. The circular reference platforms at the top of the settlement tubes were also made as small as possible since the larger these platforms the greater the tilt component included in the displacement measured by the transducers.

All four settlement points were pushed into a pre-drilled 38 mm diameter borehole as one unit. The correct spacing of each individual unit was maintained by the insertion tool shown on the left hand side of Fig 2. The steel springs on each settlement point were set so as to just pass through a 50 mm diameter hole in a gauging tool so that the total force required to press the settlement point assembly into the hole was approximately 200 kN. This force was chosen to make it comparatively easy to install the assembly in the confined space at the bottom of a 900 mm diameter borehole, but still ensured that there was no slip of the measuring points when in use.

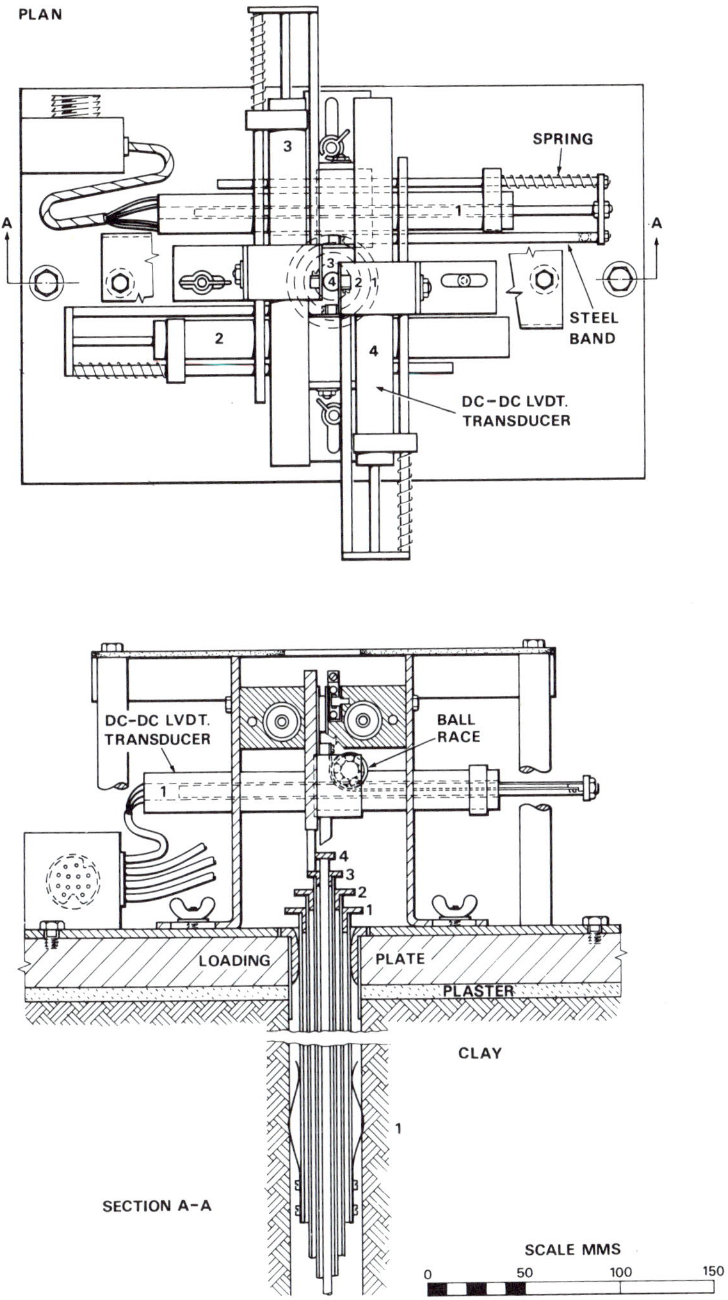

Fig 3 Details of multi-point measuring system.

When the measuring points had been inserted to the required depth the complete assembly was rotated clockwise through 45° which locked the protruding angles of the springs into the clay to provide an even greater grip. After removal of the insertion tool the transducer assembly was carefully lowered over the top of the protruding platforms. Adjustments made on a special rig in the laboratory ensured that the plungers which operate the transducers would automatically contact the correct reference platforms as the assembly was guided into positon without any need for further adjustment down the borehole. After a few measurements had been made to check the transducers the top portion of the loading plate was lowered into position so that erection of the loading and reference columns could proceed ready for the loading tests.

The transducers were mounted horizontally in order to keep the overall height of the loading plate as small as possible and thus reduce the effects of any eccentric loading from the loading column, see Fig 3. The vertical movement of the plungers resting on the circular platform at the top of the reference tubes are transferred to the horizontal transducers by spring-loaded steel tapes which pass over vertically mounted ballraces. The maximum downward force exerted by all the plungers acting together is about 50 kN which is less than one half of the force required to produce any measurable movement of the points in the ground. Since the overall mass of the transducer assembly exerts a downward force of about 120 kN it is not necessary to fix it down to the loading plate.

The displacement transducers used were DC to DC linear variable differential transformers (DC-DC L.V.D.T.) with a 25-0-25 mm travel. The large travel was provided to allow measurements to be made throughout a plate loading test to as near failure as possible. The transducers contain encapsulated oscillators and demodulators so that both the input and output are DC. This makes it possible to transmit both input and output through an unscreened multi-cord lead. It also eliminates the need to rigidly fasten the leads down which is necessary in the AC system to prevent errors due to changes in inductance between the leads. Transducer inputs were from a stabilized 6-volt supply which was frequently checked during the test and the output was fed via a manual switching unit to a digital volt-meter sensitive to ± 1 mV. The transducer assembly was calibrated as a complete unit by means of a large precision micrometer screw mounted beneath a series of stepped platforms identical with the circular reference platforms at the top end of the reference tubes.

One disadvantage of the DC-DC transducer is its relative high temperature sensitivity. Calibrations made in the -5°C, +5°C and 20°C constant temperature rooms showed that a temperature change of 1°C is equivalent to a movement of about 0.04 mm.

Fortunately temperature changes in the ground are usually small. Since all the individual transducers react to temperature in approximately the same way the effects on the differential movement between points measured by the transducers is very small. Even so these aspects require more attention, particularly for measurements made over longer periods and in very stiff soils. In future experiments, measurements will be made on a 'dummy' transducer and thermocouples attached to the assembly.

SOME TYPICAL MEASUREMENTS

Measurements obtained using the multi-point equipment are given for a test in London Clay which was made at a depth of 17.5 m below street level at a site in Central London. The test was made on a 865 mm diameter plate at the bottom of a 900 mm diameter borehole. The last 600 mm were drilled with a flat-bottomed bucket auger and additional 50 mm were removed by hand using a sharp spade and scraper. After removing the loose material with a hand brush the test plate was bedded on a layer of quick-setting plaster. Loading of the plate commenced 160 minutes after the drill reached its final level.

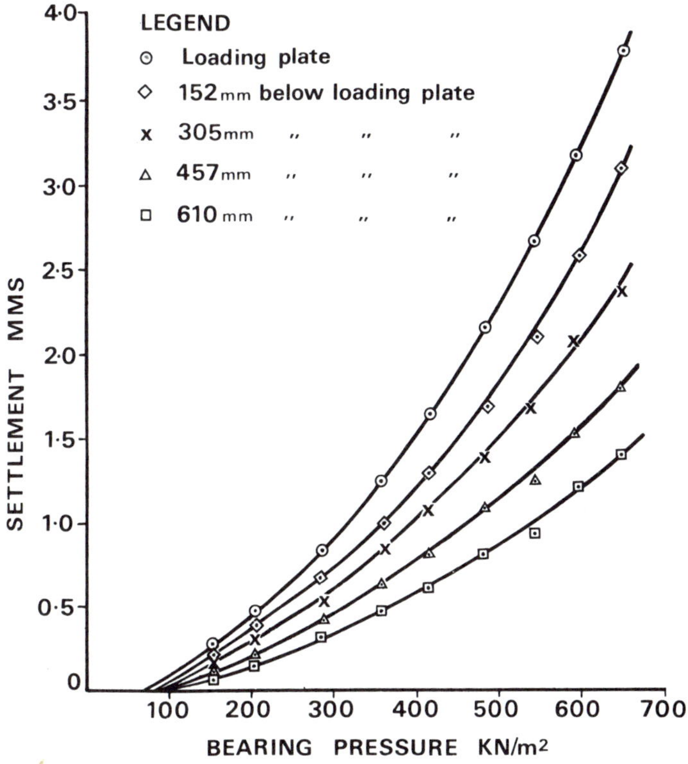

Fig 4 Vertical settlement of plate and measuring points during initial loading at a constant rate of penetration to half of the ultimate load.

In Fig 4 the settlements of the plate and the four points at depths of 152, 305, 457 and 610 mm below the plate are plotted against the average pressure applied to the plate. The pattern of these settlement measurements is very consistent indicating overall reliability of the measuring system. The settlement and heave curves for the plate and the point 610 mm below the plate are given for the first two load and unloading cycles carried out at a constant rate of penetration of 2.5 mm per minute in Fig 5. The interesting feature in this diagram is that a large proportion of the settlement of the plate during the first loading cycle was non-recoverable whereas nearly all the settlement which occurred at a point 610 mm below the plate was recoverable. The ratio of the measured settlement at each of the measuring points to that of the plate are plotted against depth in Fig 6. A curve of theoretical values obtained by finite element analysis for a loaded plate in a deep borehole within a uniform elastic medium is also included in this diagram. The measured values at a depth of 610 mm below the plate are only about 50 per cent of the theoretical

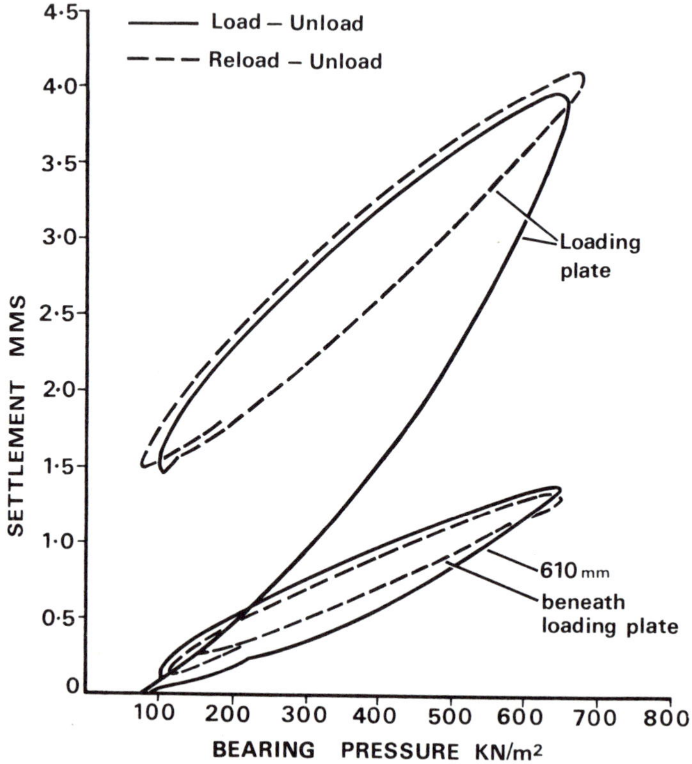

Fig 5 Vertical movement of loading plate and measuring point in the clay 610 mm below the plate during initial loading, unloading, and re-loading cycle at constant rate of penetration.

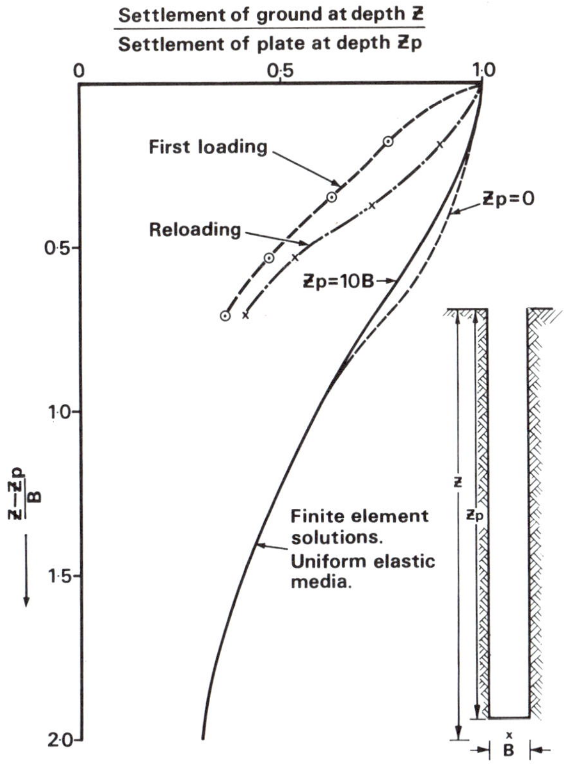

Fig 6   Comparison of measured settlements under the centre of the plate with theoretical values assuming a uniform elastic medium.

values, in spite of the considerable care that was taken to remove disturbed material and to bed the plate.

While detailed theoretical analysis of the results are outside the scope of the present paper it is useful to obtain an approximate idea of the effects of the high compressibility of the clay near the plate on the estimated values of the moduli. Average values of the moduli of the clay below each level can be obtained by using the measured values in conjunction with the theoretical curve in Fig 6 to give corrected estimates of the settlement of the plate which are then inserted into the expression:

$$E = \frac{q}{\rho} \frac{\pi}{4} \cdot B\,(1 - \nu^2)\,I_D$$

where E = modulus of clay
q = average pressure applied to the plate (Load/Area)
ρ = settlement of the plate
B = diameter of the plate
ν = Poisson's ratio of the clay
I_D = depth correction factor (see Burland, J B, 1969).

Values of E calculated in this way from the results of the present tests are given in Table 1.

TABLE 1 - Estimate of secant moduli for loads between zero and half ultimate

Depth of point below plate mm	Settlement of point mm	Theoretical $\frac{\rho \text{ point}}{\rho \text{ plate}}$	Corrected Settlement of plate mm	Average E below point MN/m^2
(0) plate	3.25	1.0	3.25	84
152	2.62	0.96	2.73	100
305	2.05	0.9	2.28	120
457	1.55	0.83	1.87	146
610	1.20	0.73	1.65	166

Even though these estimates are only approximate they indicate an appreciable increase in E when settlements at successively deeper levels are used. The moduli of the clay immediately below the plate is much lower than the lowest value given in Table 1 and more comparable with the values obtained in laboratory tests on specimens containing fissures. These values emphasize the need for measuring vertical movements at a number of points in the clay below the plate as well as making accurate measurements of the movement of the plate.

Another advantage of measuring settlement at a number of levels in the clay below the plate was observed during that part of the test where the load was held at selected values for a period of 15 minutes. The measurements taken during this stage of the test are shown in Fig 7. At the point 610 mm below the plate heave was measured under the two lowest loads while even under the lowest load the plate started to settle before it began to heave. Under the second load increment the settlement of the plate continued to increase with time. Had the plate not previously been loaded to about half the ultimate load, it is probable that recompression of the clay would have completely masked any evidence of heave of the plate even under the lowest load. Movements at points below the plate provide a more sensitive measure of the swelling pressure of the clay than do movements of the plate.

COMMENTS AND CONCLUSIONS

Equipment which can be quickly installed in the confined space at the bottom of a borehole has been developed to measure displacements in the ground under a loaded plate. This equipment has been used successfully to measure displacements in stiff clays at several sites. Measurements made in London Clay showed that 50 per cent of the settlement of the plate was due

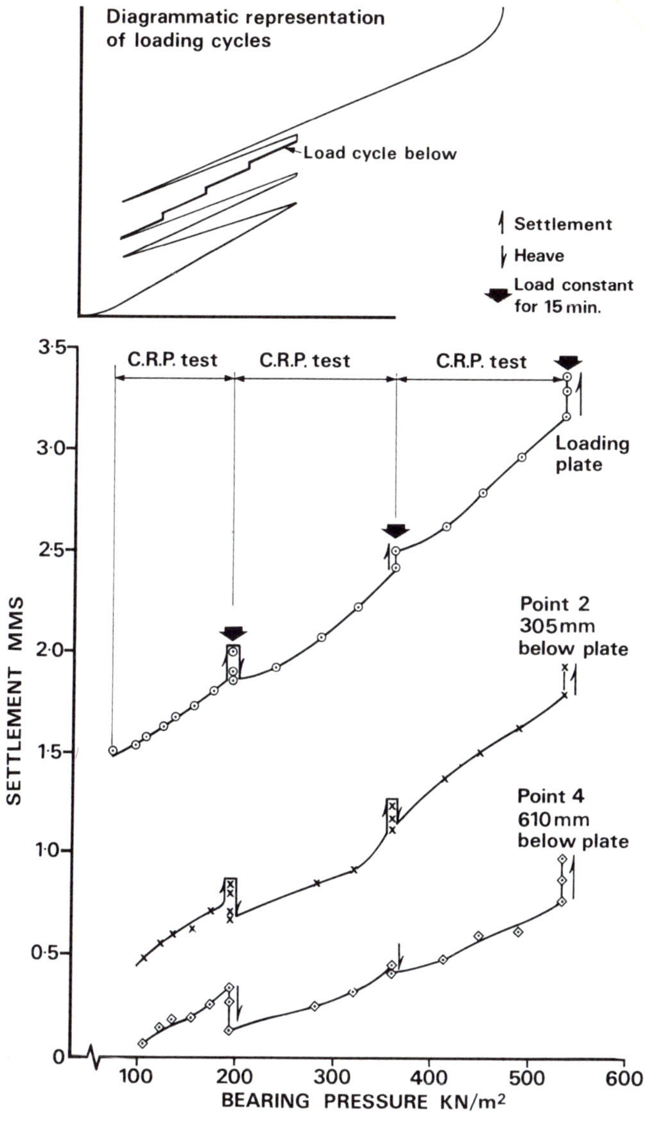

Fig 7 Movement of loading plate and measuring point at depths of 305 and 610 mm below the plate during incremental loading.

to compression of the clay within a depth equal to about half of the plate diameter immediately below the plate as against the theoretical value of 15 per cent for an isotropic homogeneous elastic material. It was considered that the high compressibility of the clay in this region was due mainly to the opening of fissures in the clay during excavation of the borehole. In a fissured clay, measurements of the movements below the plate provide a wide range of moduli which are dependent both on the amount that the stress is reduced and on the interval of time before re-loading takes place. As further measurements of movements under structures and around excavations become available it should be possible to choose the appropriate range of moduli for a particular problem from those obtained from in-situ plate tests in which movements are measured in the clay below the plate. There is obviously a need to make measurements at deeper levels below the plate. This can easily be done by spacing the lower measuring points further apart. If more points are required a larger diameter borehole and housing will probably be necessary. So far the use of the equipment described in this paper has been limited to stiff clays and it is therefore essential to check that the anchorage of the points into the ground are adequate in widely different soils.

ACKNOWLEDGEMENT

The authors wish to thank Mr G Brown for his assistance with the detailed design of the equipment. The work described in this paper forms part of the Research Programme of the Geotechnics Division of the Building Research Establishment and is published by permission of the Director.

REFERENCES

Marsland, A (1971(a)): The shear strength of stiff fissured clays. Proc Roscoe Memorial Symp, University Eng Dept, Cambridge, 29-31 March 1971. (Also BRS CP 21/71)

Marsland, A (1971(b)): The use of in-situ tests in a study of the effects of fissures on the properties of stiff fissured clays. Proc 1st Australia-New Zealand Conf on Geomechanics, Melbourne 1971. (Also BRS CP 1/73)

Marsland, A (1971(c)): Laboratory and in-situ measurements of the deformation moduli of London Clay. Paper presented to the Symp on the Interaction of Structure and Foundation, Midland Soil Mech & Found Soc, 1971.

Marsland, A (1972): In-situ plate tests in lined and unlined boreholes in highly fissured London Clay at Wraysbury near London Airport. Ground Engineering, vol 5, no 6, pp 24-31. (Also BRS CP 5/73)

Serota, S and Jennings, R A J (1959): The elastic heave of the bottom of excavations. Geotechnique, 9, pp 62-70.

Ward, W H (1971): Some field techniques for improving site investigation and engineering design. Proc Roscoe Memorial Symp, University Eng Dept, Cambridge, 29-31 March 1971.

Cole, K W and Burland, J B (1972): Observation of retaining wall movements associated with a large excavation. Proc 5th European Conf on Soil Mech & Found Eng, Madrid, April 1972, vol 1, pp 445-453. (Also BRS CP 8/72)

Wallace, G B, Slebir, E J and Anderson, F A (1969): In-situ methods for determining deformation modulus used by the Bureau of Reclamation. Symp on the Determination of the In-situ Modulus and Deformation of Rock. ASTM Spec Publication 477.

Burland, J B (1969): Reply to discussion. Proc Conf on Site Investigations in Soils and Rocks. Inst Civ Engrs, London, p 62.

ENVIRONMENTAL FACTORS AFFECTING THE INTERPRETATION OF GROUND MOVEMENT

I.K. Nixon, B.O. Skipp and B.F. Wareham

Soil Mechanics Limited

INTRODUCTION

When observing ground movement such as long term settlement, landslides or tunnel convergence, it is important always to remember that a number of extraneous factors peculiar to the environment may affect the interpretation of results, together with the more commonly recognised factors such as instrument performance and accuracy. The hostility of the geotechnical environment is often underestimated, especially when using sophisticated instruments. This hostility shows itself in two ways. Firstly, the measuring systems may have to contend with water, water vapour, corrosive atmospheres, noise and vibration and the rigours of a construction site, while secondly, the ground movements in which one is interested may have to be extracted from a background of unwanted information. The introduction of more sensitive and stable measuring equipment has ironically made observations more suceptible to environmental factors.

Measuring systems may be grouped as:

Direct measurements by	Mechanical Optical Hydraulic Electrical	means. System located at the surface or buried.
Indirect measurement by	Accoustic emission	surface or buried.

The movements of interest may be the direct effect of some known cause such as constructional loading or unloading, the duration of which is short compared with the period over which observations are to be made. On the other hand they may be caused by ill defined and even intermittent displacements which occur over a long period of time, for example when measuring hill creep or fault movement.

Environmental agencies may be listed as:

(a) Climatic effects.

(b) Groundwater and water content.

(c) Ground vibration (wind, traffic and other sources).

(d) Electrical noise.

(e) Inadvertent disturbance.

These agencies may affect the object under study, that is cause dimensional changes in the ground, or the measuring system, or both. The significance of each agency will depend upon the character of the measuring system.

CLIMATIC EFFECTS

If we are to consider systems for world wide application, the temperature range at ground surface may vary between -20°C and 110°C. In the U.K. the maximum daily range at any one place is around 40°C. The degree to which the surface mounted instruments are affected depends upon their thermal screening. When measuring the distance between two points by an extensometer or by electro-optical means, overal temperature changes in the ground may be significant (Gould & Dunnicliff 1971). Coefficients of linear expansion of rocks and clay soils range from 2 to 15 x 10^{-6} depending upon the porosity (Searle & Grimshaw, 1959).

Increase of temperature with depth can be in the region of 1°C per 30 metres of depth (Bullard & Niblett, 1951) but daily variations are known to penetrate to a nominal depth of 1 m and seasonal variations to a nominal depth of 20 m (King 1971). Short and long term near surface monitoring should be compared with surface temperatures on the basis of daily and seasonal penetrations to the above depths. The geothermal gradient is probably relevant when one is dealing with ground movements measured over depths in excess of 100 m and in the case of tunnels, especially such as deep mines where heat flux change due to ventilation is large, strains may be set up as thermal equilibrium is taking place and it may be necessary to separate them from the mechanical strains in the measured data.

Temperature fluctuations may also have a direct effect upon the movement measuring instrumentation itself. Where steel wires or rods are used, appropriate corrections may be made, as suggested by Gould & Dunnicliff (1971), by running a parallel wire of different thermal coefficient of expansion to a separate sensing element but anchored at the same point.

Hydraulic systems for settlement measurement should preferably have the reference hydraulic unit in the same temperature environment as the settlement sensing unit and gauges using water must be protected against freezing. Electrical transducers (LVDT, electrolevels) are usually temperature compensated to within $\pm 0.2\%$ over temperature range of -10° to $+30^{o}C$.

Mechanical extensometers are relatively simple to calibrate but require care. Frequent temperature calibrations must be made but, given a relatively stable temperature, very good results can be achieved. A line of brass studs has been measured by a mechanical extensometer with a dial gauge read-out in connection with the National Westminster Bank foundation excavation at Bishopsgate, London. The closure of the traverse which was in the basement of the adjacent property indicated accuracies of the order of 1 part in 140,000.

Precise settlement measuring systems often entail reference points anchored at depth in the ground. Generally, invar rod or wire is used (Skipp 1969; Meigh, Skipp & Hobbs 1973). Although below a few metres the ground temperatures are steady the introduction of a hole may, by convection processes, disturb the thermal equilibrium and give rise to small unwanted thermal strains. Thermal strain induced in structures due to diurnal solar radiation may appear as deformation in the foundations and be interpreted as settlement. Such phenomena have been noted on cooling towers with a ground beam foundation on piles.

In cases such as plate bearing tests and pile tests, the temperature changes may be reduced by careful screening and the use of suitable materials for datum and measuring systems. It has been noted, however, that a comparison between measured temperature and settlement records often indicates the cause of spurious readings. In Fig. 1 we show the temperature variation and response of measuring systems for a plate bearing test in a shaft and at the surface. Burland (1972) has observed the effect of temperature variations on a plate bearing test datum designed to minimise this effect and noted that a relationship could be obtained between temperature and change of reading, which was used to correct settlement results. However, a time lag of 200 min. was discovered between change of temperature and its maximum effect on the measured settlement, which was also taken into account in the correction.

Changes of temperature along an optical path is a well known factor reducing the accuracy of optical levelling. The accuracy of electro-optical distance measuring systems is also affected for the same reason. Similarly in systems using a

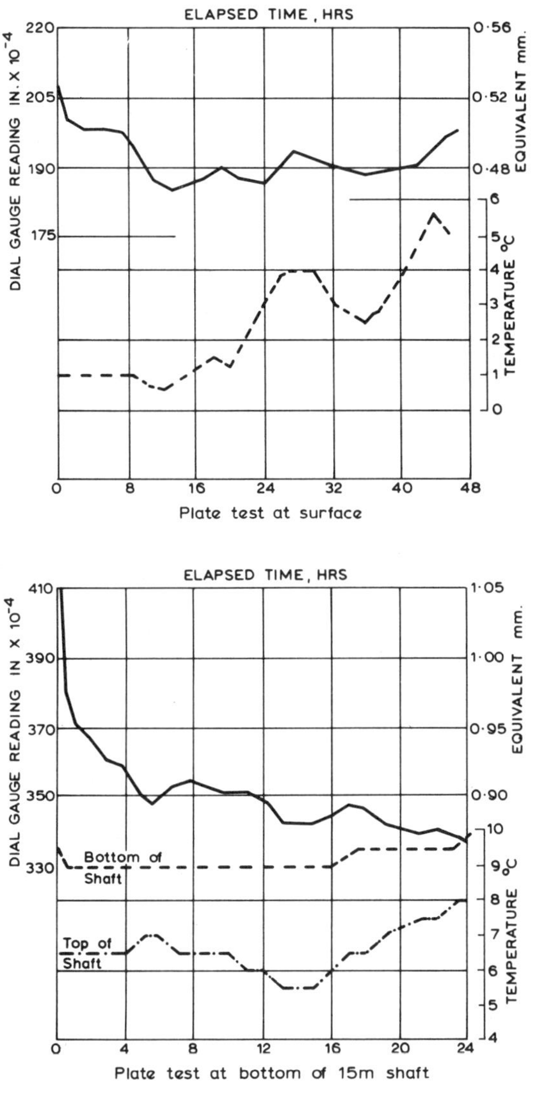

Fig.1. Temperature response in settlement measuring systems

laser beam as an optical lever, variations in refractive index is a factor limiting basic accuracies. For highly accurate schemes consideration should be given to maintaining the light path in a temperature and humidity controlled enclosure, but these measures would only be required when the accuracies called for are approaching strains of 10^{-8} (Harrison et al 1972). Laser based alignment systems are most accurate if used during the middle part of the day and in dry weather with a light wind, in order to obtain steady environmental conditions. It is also preferable in Great Britain to operate between mid-September and mid-April, as heavy cloud conditions reduce heat loss and thus minimise temperature changes. These comments are also true for photogrammetric studies for geological investigations or displacement monitoring. Harrison reports accuracies of 1 mm in vertical alignment following corrections over a 6 mm range in steady atmospheric conditions.

Variation of temperature may affect both ground and the instrumentation system in other ways. At low temperatures ice wedges may develop at undesirable positions. One such example occurred a few years ago when the measurement of the tilt response of a television tower foundation to strong winds was being investigated. An electrical spirit level was mounted on a granite block resting on a concrete floor. The basement was not weatherproofed, snow drifted inside, melted and was re-frozen, thereby raising the block to give spurious readings. Cementing down the block removed the problem.

High temperatures and dry weather may cause surface strain and deep cracks in clay soils. Surface strain measurements between points installed without consideration of the depth to which this phenomenon may occur at the particular location can be quite misleading. On the other hand, freezing conditions can cause vertical swelling of silts due to the formation of ice lenses.

Barometric fluctuations in so far as they affect local groundwater levels (e.g. tidal response) may be seen in some cases. Relative measurements from hydraulic settlement are not affected but open column 'u' tube micrometer water levelling instruments may give spurious results due to quite local pressure fluctuations near one column. This coupled with temperature gradients between sensing points limits the accuracy of this technique of relative settlement measurement. Wet and dry bulb temperatures, atmospheric pressure and temperature are observed in order to apply the appropriate refractive index correction when using electro-optical distance measuring devices, this correction being applied during calculation of the results or in some cases electrically during the measuring process (Hewlett Packard 3600 A/B Instruction Manual).

Humidity may have a direct effect upon the instrument itself due to the formation of water globules within a closed system. Not only is there the possibility of corrosion with time but,in the case of vibrating wire strain gauges, the measurements can be affected and to overcome this problem dry nitrogen gas may be circulated through the system, but care should be taken to prevent the formation of ice.

Other climatic influences such as rain and wind have obvious direct effects on measuring systems, calling where appropriate for a proper examination of the records. Electric storms may damage electrical installations and precautions should be taken accordingly.

Generally meteorological uncertainties give rise to second order effects but, when using a microwave device to measure across the large pit at Kimberley, Watt (1970) reports that such variations amounted to 1 to 1.5 cm. Comparison of this instrument with other measuring systems for determining structural movement in Britain has been reported by Penman and Charles (1971).

GROUNDWATER AND WATER CONTENT

Changes both in the water content above the phreatic level and variation of the phreatic level itself give rise to dimensional effects.

Soil movement due to changes in water content and the consequent damage to buildings, particularly in clay soils, is well known. Ward (1953) has reported changes of vertical movement at three grassland plots in the south of England due to the effect of weather, type of clay, and vegetation. Using a datum placed at 2.5 m below ground level, vertical movements were observed at the surface and at different depths. The maximum vertical change, naturally at the surface, was as much as 50 mm during some three years of measurements, and even at a depth of 1.5 m below the surface vertical changes of around 2.5 mm were observed. The effect of trees was also discussed. Recently Penman (1971) has described the location of monuments to avoid surface disturbance effects.

The effect of changes in the groundwater level is illustrated in Fig. 2, which indicates the consequence of the changes in effective stress in the ground between the phreatic level and the datum point. Such effects were noted during long term plate testing at Heysham Nuclear Power Station and during settlement observations on another site. In the former case it was possible to vary the groundwater level by pumping from an array of well points and so derive a correction factor, as shown in Fig. 3. The tidal response to a similar system is shown in Fig. 4. It is evident that under these circumstances, where settlements are being measured, the long term trend must be extracted from these quite large semi-diurnal movements. This can be done if pore pressure is monitored or by studying settlement data on a harmonic basis. In any case continous or near continuous data collection is essential if the time dependent trends are to be evaluated during a reasonable period of time.

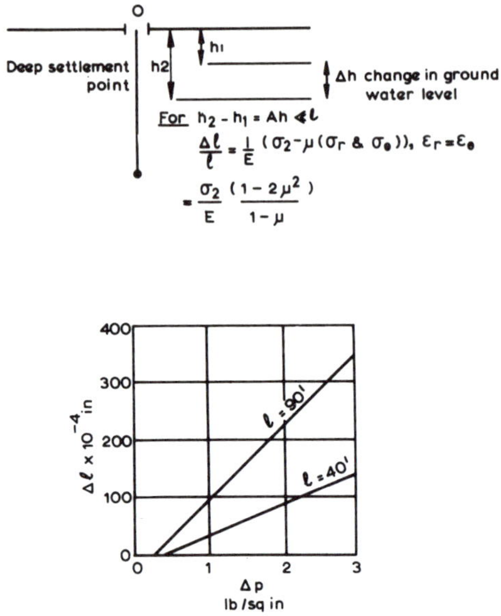

Figure 2 Simplified response of settlement point to change in ground water level

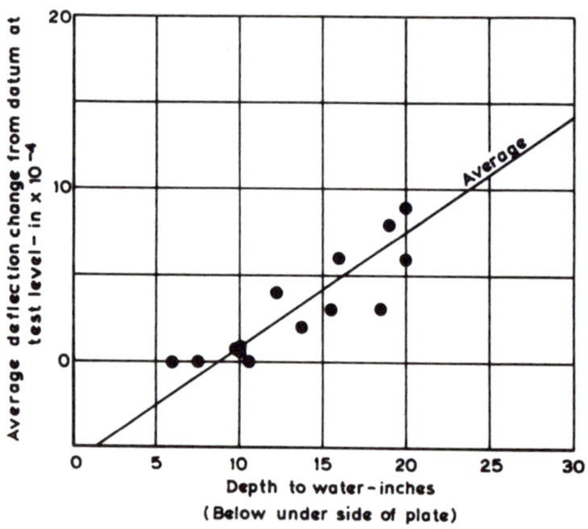

Figure 3 Water level corrections, deflection vs. water level over control period

GROUND VIBRATION (wind, traffic and other sources)

Consideration of ground vibration in this paper is limited to its effect on static measuring systems. Vibrations in general are characterised by their frequency and amplitude, levels of which may be to some extent predicted after consideration of the site environment. Background vibration is more marked at near surface installations than in underground situations. Meteorological, marine and man-made activities produce microseisms which are surface waves of low frequency (10-15 second period) and as such require special instrumentation to register them.

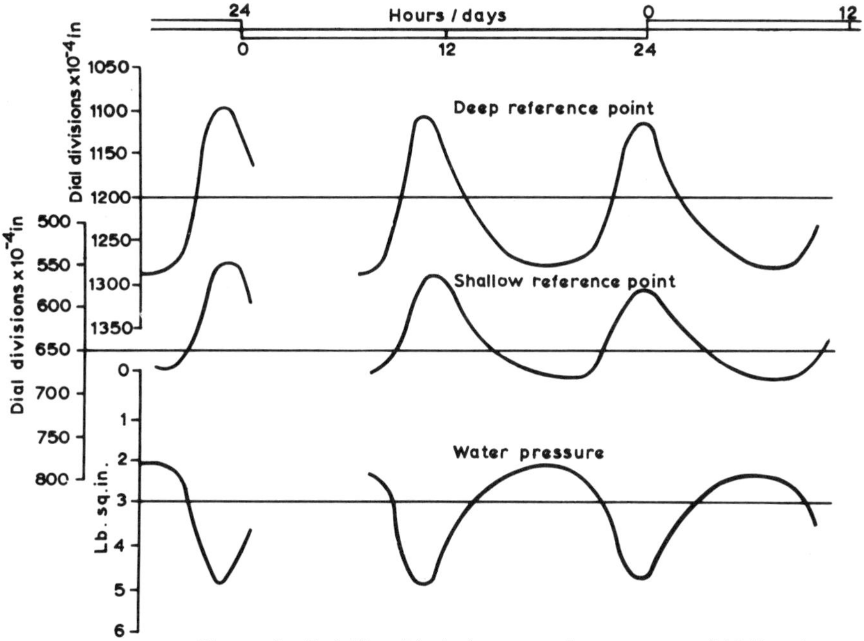

Figure 4. Relationship between water pressure (tidal) and displacement readings on settlement reference points

Possible origins of near surface vibration may be listed as:

Wind generated vibration which is generally controlled by wind speed, is easy to identify being non-harmonic and falling into the frequency range 100 Hz- 10 KHz and, even under quiet conditions, surface strains of one part in 10^6 may be attributable to wind in trees (King 1971).

Land traffic and aircraft generated vibration are easily identifiable and fall into the range of particle velocities of 3×10^{-2} mm/sec for traffic vibration and 1×10^{-1} mm/sec for low flying aircraft (Jarman 1960). Measurements along-side aircraft runways have indicated correlation between noise level (in decibels) and ground response with measured particle velocities of 7×10^{-2} mm/sec at about 100 m from the centre line of the runway.

Industrial/construction activity vibrations are in general non-continuous from sources such as forging or piling hammer, quarry blasting, etc. and particle velocities as high as 5 mm/sec may be experienced within 50 m of a machine.

Other non-continuous sources of vibrations include ground coupling from sonic boom and explosive generated air shock.

A source of vibration may be 'labelled' by its Fourier spectrum or more usually by its power spectrum showing the relative power in different frequencies. Examples of a spectrum from traffic vibration and from severe impact are given by Grover (1970). For permanent deformation of the ground, an input of high acceleration such as an earthquake is required or, under certain circumstances, sustained continuous vibration may suffice. In some cases ground movement as a result of dynamic loading (Skipp et al 1972) or settlement of a vibrating foundation (Fitzherbert & Barnett 1967) is the object of the measurements. In general the background vibration rarely exceeds 5.0 mm/sec, but when it does the vibration environment becomes an engineering problem in its own right and instruments could not be expected to function unless specially designed.

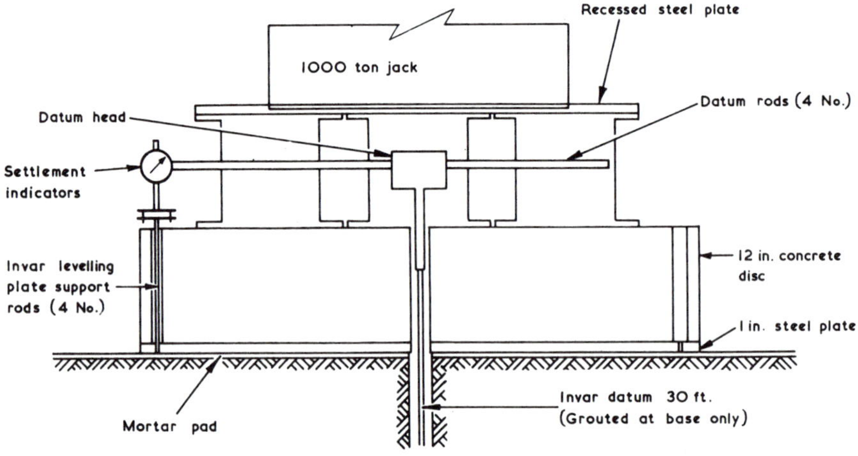

Fig. 5. Settlement measuring system - Heysham

Problems associated with vibrational effects on measuring systems include not only the displacement problem as discussed above but also the possibility of components of the measuring system resonating with some ground frequencies. An example of this was noted at Heysham Nuclear Power Station site as discussed earlier. Fig. 5 shows the settlement measuring system which consisted of inverted pendulums which had to be restrained from lateral excursion and which were found to have unwanted response to vibrations induced by nearby site plant. Strain meters, which may have natural frequencies in the region of 30 Hz, have been known under certain circumstances to suffer resonance problems while being used to monitor fault movements.

ELECTRICAL NOISE

Electrical currents, both AC and DC, may be ground borne and be natural or man-made in origin. Where man-made, they are often harmonics of the 50 Hz mains supply and as such can have complex wave forms. In general, apart from electrical resonance between ground borne currents and measuring apparatus, problems in this field are confined to effects on the electronics used in recording and displaying the data. The seriousness of the problem depends initially on the signal/noise ratio which compares magnitude of background noise with expected signal input from the measuring apparatus. To some extent the electronic design can remove background noise by filtering, screening of cables, and proper earthing to prevent earth loops, although where the signal/noise ratio is low, sufficient noise reduction cannot always be achieved. In this case at the initial commissioning stage of the system, analysis of input wave form should include identification of the unwanted background noise. In some cases the noisy electrical background favours the more sophisticated instrumentation systems in which data is transmitted by cable or radio links in frequency or pulse code form. Such a system was recently used in a settlement study on a cooling tower and is discussed elsewhere (Sherwood and Currey 1973).

In the case of DC noise, the problems are as noted above, although in general this is low level background noise and may be filtered out. Under certain circumstances it can cause induced corrosion within the electronic equipment and thus produce malfunction.

INADVERTENT DISTURBANCE

This category has been included because in it we can lump all those familiar hazards which abound on construction sites and of which, if one is not continually on guard for, may lead to serious misinterpretation of data. The example of 'ice wedging' has been referred to and might be included in this category. Instrumentation is often vulnerable, however one attempts to protect it. Physical displacement of settlement points or extensometer plugs is common. It is one of the advantages of a continous data collection system that spurious readings can more easily be pin pointed and discarded.

STABILITY AND ENVIRONMENT

All measuring systems are prone to deterioration with time. The damp and often corrosive circumstances are unfavourable towards electrical instruments unless properly protected. This calls for regular check procedures on each significant component in the system. Electrical systems have sources of error which may be classed as:

(a) Zero drift and sensitivity change in the basic sensor.

(b) Zero drift and sensitivity change in the signal conditioning unit if incorporated.

(c) Zero drift and sensitivity change in the data collection/display unit.

Checking of zero drift in a basic sensor is difficult, be it an LVDT or an electrical spirit level, a vibrating wire exensometer or even an electrical resistance strain gauge. It may call for the removal of the transducer from a mounting and calibration by location in a stable jig. For spirit levels, reversal on a plane surface or a suitable mounting is a relatively simple operation. Such provision has been made on the slip indicators currently being installed on a hillside. Checks on sensitivity require that known displacements be imposed on the measuring element from time to time analysing the resulting data. Checking the signal conditioning is best done by the provision of 'dummy' sensors, usually of high stability resistors which can be switched into the system replacing the operative sensors. The checks on the data collection/recorder elements call for the appropriate signal generators.

The fundamental stability of sensing elements is very high. Electrical spirit levels may have drift amounting to a few seconds of arc per month dropping to a few seconds a year with ageing. It is usually the dimensional stability of the mounting which is in question. Recently electrolevels in non-kinematic mountings were recalibrated after two years service by the same basic method (small angle generator beam) but by different bodies. There were two faulty units but the remainder were within 3% of their factory calibration. With more stable mounting this can be improved.

CONCLUSIONS

Ground movement measuring systems may be classified as in Table I and the significance of the environmental effects, as described in Table II, can be appreciated qualitatively with reference to the first table.

MEASUREMENT SYSTEM	ENVIRONMENTAL FACTOR			
	CLIMATIC EFFECTS	GROUNDWATER & WATER CONTENT	GROUND VIBRATION	ELECTRICAL NOISE
DIRECT METHODS 1. MECHANICAL	Surface strain-temperature change Temperature effect on materials Temperature gradient effect on deep datum Convection currents in borehole Heat flux in deep mines	Effective stress change Surface strain due to water content change above the phreatic surface	Resonance of measuring equipment Dynamic strains set up	Not applicable
2. OPTICAL	Corrections must be carried out for air temperature, humidity and pressure changes	Not directly applicable - affects datum stability	Resonance in the mountings Dynamic strain in the mountings	Not applicable
3. HYDRAULIC	Barometric pressure not the same at each end	Not directly applicable - affects datum stability	Resonance unlikely	Not applicable
4. ELECTRICAL	Temperature compensated over a given range	Not directly applicable - affects datum stability	Dynamic strains set up and some devices act as accelerometers Resonance of measuring equipment	Interference on records depending on signal/noise ratio. Resonance.
INDIRECT METHODS 5. ACCOUSTIC EMISSION *	No significant effect	No significant effect	Ground vibration recorded	Interference on recording apparatus

* may only be used to predict 'coarse' movement

Table II Effect of environmental factors on measuring systems

	RANGE OF SYSTEM		
	COARSE (C)	INTERMEDIATE (I)	PRECISE (P)
Displacement Strain	10–100 mm 10^{-1}–10^{-2}	0.1–10 mm 10^{-2}–10^{-4}	0.01–0.1 mm 10^{-4}–10^{-6}
Examples of use	Embankment settlement. Active slip displacement.	Steps in consolidation settlement. Convergence and deformation of tunnels and chambers. Slip warning.	Critical foundations Machine foundations. Fault movement. Precise bench marks.

Table I Ranges of movement measurement systems

ACKNOWLEDGEMENTS. The authors are indebted to the Directors of Soil Mechanics Limited for permission to publish the information given in this paper.

REFERENCES

BULLARD E.G. & NIBLETT E.R. Terrestrial Heat Flow in England. Mon. Not. Roy. Astr. Soc., Geophysics Suppl. V6, pp 222-238, 1951.

BURLAND J.B. Private communication, 1972

FITZHERBERT W.A. & BARNETT J.H. Causes of Movement in Reinforced Concrete Turbo-blocks and Developments in Turbo-block Design and Construction. Proc. I.C.E. 1967, 36 (February), 351-393.

GOULD J.C. & DUNNICLIFF C.J. Accuracy of Field Deformation Measurements. Proc. Fourth Pan-Am Conf. on SM & FE, San Juan, Puerto Rico, June 1971.

GROVER F.H. Vibrations on Structures, U.K.A.E.A. A.W.R.E. Report No. 032/70, H.M.S.O. 1970.

HARRISON P.W., TOLMON F.R. & NEW B.M. The Laser for Long Distance Alignment - a practical assessment. Proc. I.C.E. 1972, 52 (May) 1-24.

JARMAN B. Some Field Experiments on Seismic Wave Propagation. Ph.D. Thesis, University of London 1960.

KING G.C.P. The Siting of Strainmeters for Teleseismic and Tidal Studies. Recent Crustal Movements, Royal Society of New Zealand, Bulletin 9, pp 239-247 1971.

MEIGH A.C., SKIPP B.O. & HOBBS N.B. Field and Laboratory Creep Tests on Weak Rocks. In press. Int. Conf. SM & FE, Moscow 1973.

PENMAN A.D.M. & CHARLES J.A. Measuring Movements of Engineering Structures. Building Research Station Current Paper 32/71, Garston England 1971.

SEARLE A.B. & GRIMSHAW R.W. The Chemistry and Physics of Clays and Other Ceramic Materials. Ernest Benn Ltd., London 1959.

SHERWOOD D.E. & CURREY B. Experience in Using Electrical Tiltmeters. In press. B.G.S. Symposium on Field Instrumentation 1973.

SKIPP B.O. Discussion Session. Conference on Insitu Investigations in Soils and Rocks. I.C.E. May 1969, p. 43

SKIPP B.O., CURREY B. & WOODFORD C. Vibration Problems in the Demolition of a Cooling Tower. Conf. on Theoretical and Applied Structural Dynamics, Southampton I.S.U.R. 1972.

WARD W.H. Soil Movement and Weather 1953, Proc. 3rd Int. Conf. on SM & FE, Vol 1, pp 477-482.

WATT I.B. Control for Early Warning of Potential Danger in Open Pits, 1970. Proc. Symp. on Open Pit Mining pp 107-113.

CASTLANDHILL PIEZOMETERS

A C Paterson

Partner, Bullen and Partners, London

INTRODUCTION

A length of the northern approach road to the Forth Road Bridge had to be carried through a cutting some 90' deep.

At an early stage in the design, when the cutting was expected to have a depth of some 35'-40', it was proposed to adopt a side slope of $1\frac{1}{2}$:1. When the depth was increased to 90', it was initially decided to introduce two 15' wide berms at depths of 30' and 60' thus reducing the side slopes to an equivalent continuous slope of 1.83:1.

When the cutting had reached a depth of 60', cracks were noticed at the original ground level about 100' back from the top of the new slope. Although the cracks were not considered to be conclusive evidence of an incipient slip, a further investigation was carried out into the general stability of the cutting.

Ground Conditions

Bore holes indicated that the cutting was being formed in a stiff over-consolidated fissured boulder clay. The horizontal pressures which had been induced by the high over-burden pressure and which had remained after the over-burden pressure had been removed, were being released by the excavation for the cutting, resulting in an opening of the joints in the fissured clay. Open joints of this nature would greatly increase the rate of change of moisture content and, therefore, of the consequent increase in the pore pressure.

Slope Analysis

Undrained triaxial compression tests, without pore pressure measurement, only establish the total stress parameters for the clay. The results of such tests are applicable only to a soil mass in which the conditions are the same as those in the sample during testing. This might have been the case immediately after the cutting was formed but as soon as the pore pressure changed, the total stress parameters would no longer be applicable. Should the pore pressure increase, the factor of safety would be reduced. In stiff overconsolidated clays, the use of total stress analysis leads to an exaggerated factor of safety and the calculated strength corresponding to failure conditions has been considerably lower than the average measured undrained strengths.

In those clays where over-consolidation has produced a highly dilatant structure, much of the strength measured in the undrained compression tests is due to the large negative pore pressures set up during the shearing process. In the soil mass, however, any negative pore pressures are able to dissipate with time and thereafter, the pore pressure acting on the failure plane will be determined solely by the ground water conditions. Under these circumstances, a more reliable estimate of the factor of safety may be made in terms of effective stresses using shear parameters determined in a series of drained tests or undrained tests with pore pressure measurement. Slope stability coefficients have made effective stress analysis a comparatively simple matter.

Samples 4" in diameter and 8" long were subjected to triaxial compression tests with pore pressure measurement at Glasgow University. The results of these tests gave the effective stress parameters of the clay C'-ϕ' the apparent cohesion and angle of internal friction in terms of effective stresses. The results are extremely consistent as shown by the plot of Mohr circles on Figure 1. giving $C' = 1000$ lb/ft^2, $\phi' = 20^\circ - 22.5^\circ$ and $\gamma = 145$ lb/ft^3. The results of the effective stress analysis are illustrated in Figure 2, where curves are given relating the angle of slope to the factor of safety for various conditions. This figure also shows the relationship obtained by total stress analysis and by the assumption that $\phi = 0$.

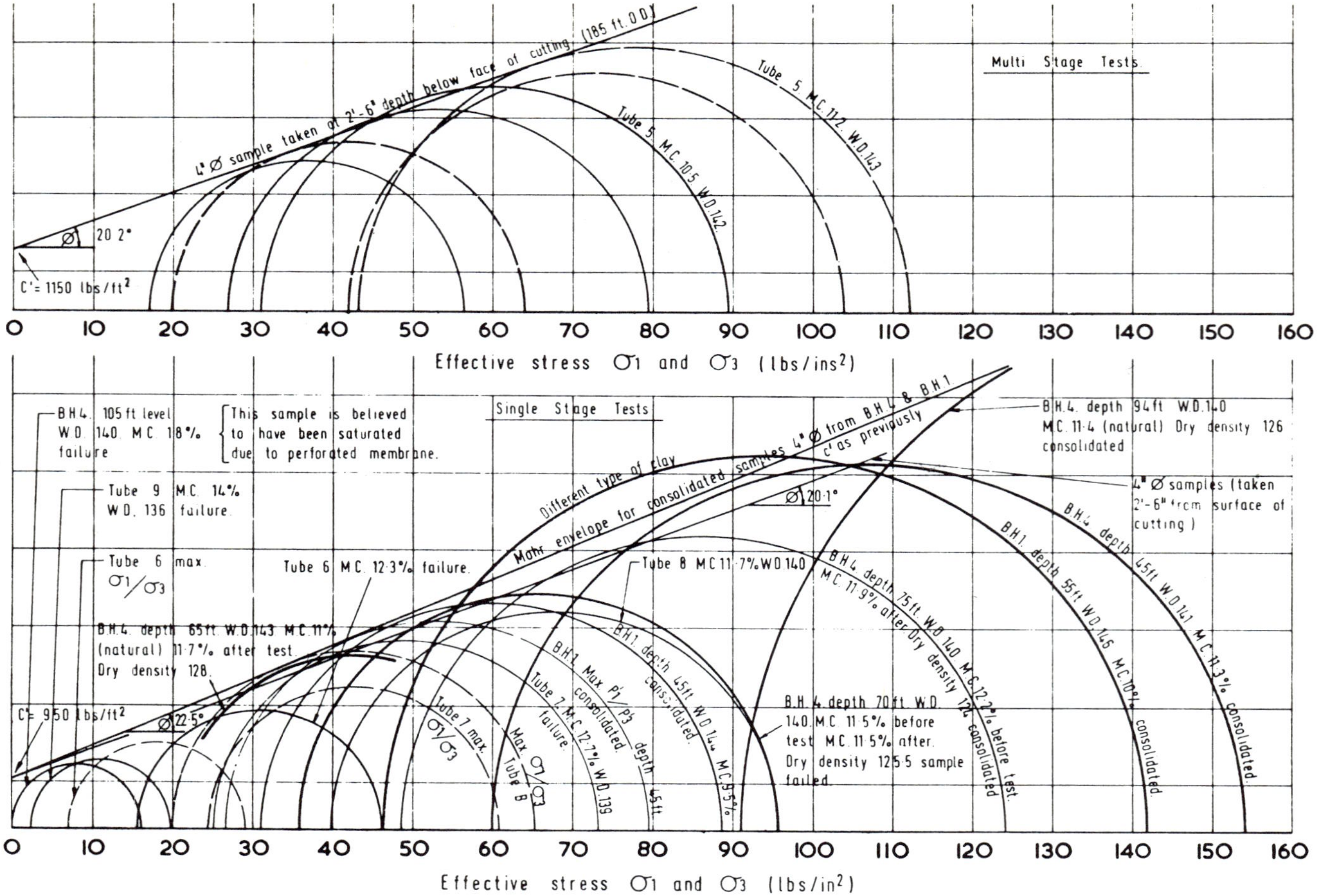

FIG. 1. MOHR CIRCLES FOR UNDRAINED TRIAXIAL COMPRESSION TESTS WITH PORE PRESSURE MEASUREMENT.

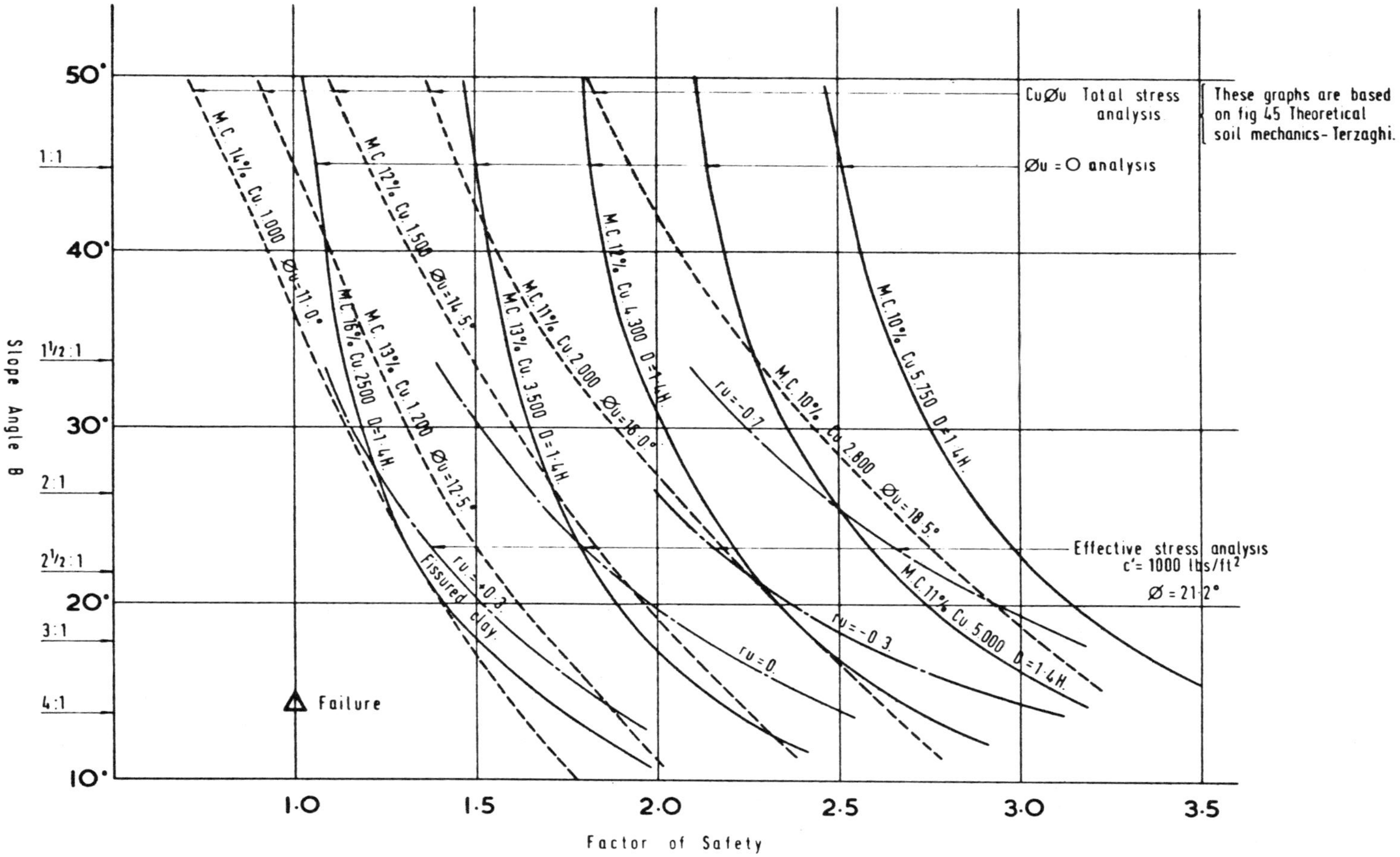

FIG. 2. BY EFFECTIVE AND TOTAL STRESS ANALYSIS.

Pore Pressure

The factor of safety of the slope produced by an effective stress analysis is very much influenced by the average pore pressure assumed to be present in the potential failure zone. The pore pressure is expressed in terms of the pore pressure ratio $r_u = \frac{u}{\gamma h}$, where u is the pore pressure, γ is the bulk density of the soil and h is the depth from ground level to the point being considered. The value of r_u thus varies with the depth and a mean value is normally taken when analysing the stability of a slope. In the present instance, it was thought that r_u would fall within the range of -0.3 to +0.3 for some considerable time ahead.

It was appreciated that it was not possible to forecast with any claim to accuracy, the rate at which the pore pressure would increase as this depended so much upon the seepage system through the system of hair cracks and joints. In view of this, it was recommended that a system of Piezometers should be installed which would enable the distribution of pore pressure to be measured on completion of the cutting and periodically thereafter. By extrapolation from these measurements the future pore pressure could be anticipated and any remedial measures taken in advance if found necessary.

Factor of Safety

British Standard Code of Practice CP.2003 for Earth Works recommends a factor of safety of between 1.2 and 2 for the slopes of cuttings, depending upon the consequences of a failure, the reliability of the data obtained and the degree to which such data is representative of the overall soil conditions. The accuracy of the factors of safety given in Figure 2 is limited by the lack of homogeneity of material, the simplifying assumptions necessary to enable an analysis to be made, and the extrapolation beyond the published stability coefficients which was necessary in the present case. In addition, Figure 2 does not take into account the effect of tension cracks forming at the top of incipient slip circles with consequent reduction in factor of safety but this effect is comparatively slight and the reduction in the present case lies between 3% and 5% depending upon the slope.

It was therefore, thought advisable to relate a high positive pore pressure to a minimum factor of safety provided that this did not imply an excessive factor of safety under more probable conditions. To achieve this, a factor of safety of about 1.3 would be appropriate when considering a pore pressure ratio of +0.3.

Piezometer Installation

In the light of the foregoing, it was agreed that a series of piezometers should be installed in the Eastern side of the cutting to allow the distribution and variation of pore pressure to be checked and anticipatory remedial measures to be taken, should these be seen to be necessary. By monitoring the pore pressure, it was considered that steeper slopes could be accepted with consequent savings in cost and loss of agricultural land.

Figure 3 shows details of the installation. A cross section was selected where the cutting was at its maximum depth and the positions of the piezometers chosen to straddle the most likely slip circle.

The piezometers were manufactured by H Maihak A G of Hamburg and supplied by Smail Sons and Company Ltd of Glasgow. They operate on the vibrating wire principle, the transmitter being located in the borehole and connected by cable through a position selector to a portable analogous tuning receiver.

The transmitter contains a measuring wire under tension, which oscillates in a damped natural vibration after excitation from the receiver. The square of the natural frequency of the measuring wire is proportional to the quantity to be measured. The natural frequency is transmitted to the receiver as an electric current.

The end of the transmitter contains a fine porous filter which withstands the soil pressure but allows the pore water pressure to be applied to a diaphragm connected directly to the measuring wire.

The position selector is installed permanently in a water tight chamber at ground level, as shown in Figure 3.

The portable analogue tuning receiver is brought to site when measurements are to be taken, which is at present annually. Power is supplied by two car batteries. The receiver contains a standard

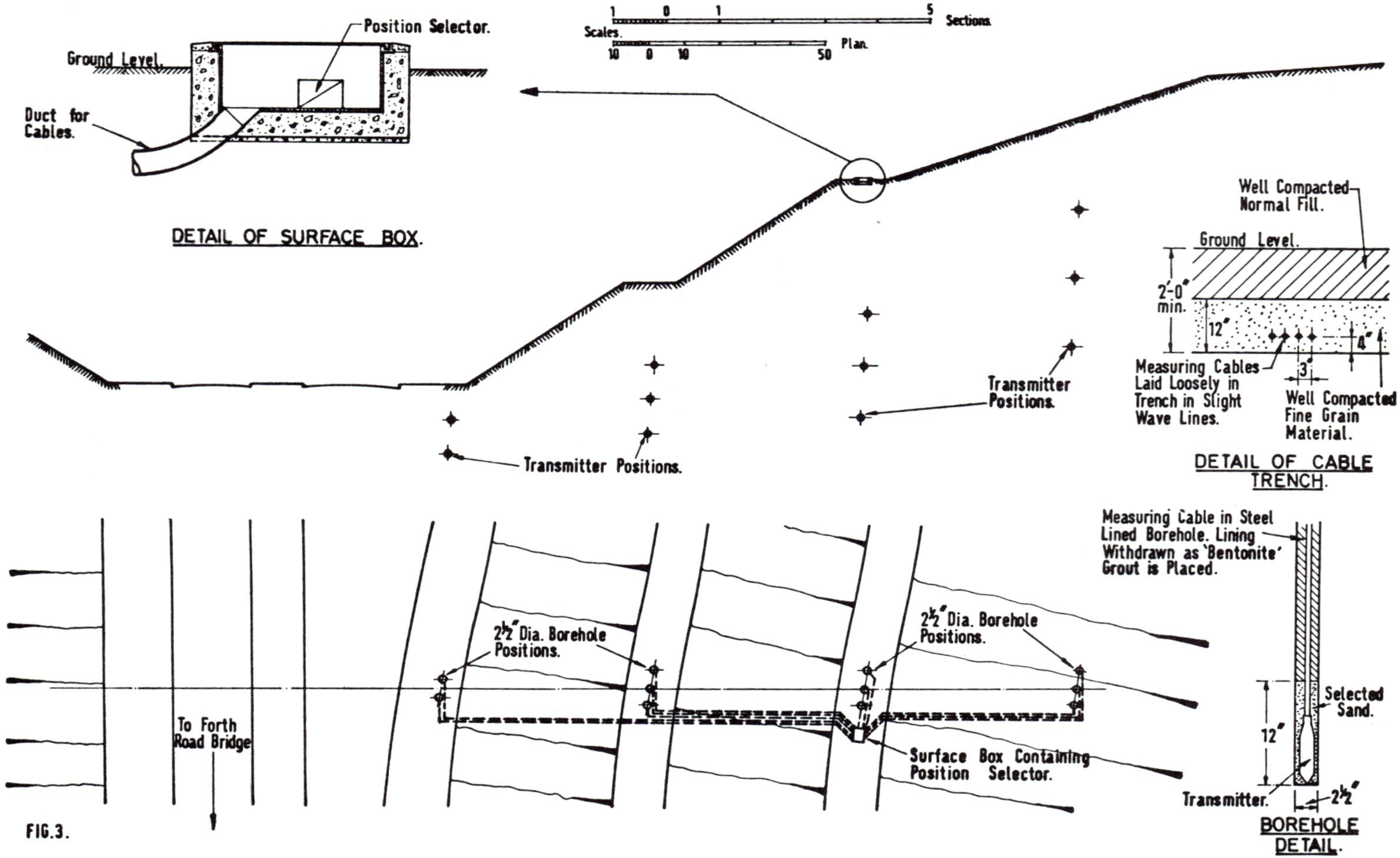

Position Selector.
Ground Level.
Duct for Cables.
DETAIL OF SURFACE BOX.
Scales.
Sections.
Plan.
Transmitter Positions.
Transmitter Positions.
Well Compacted Normal Fill.
Ground Level.
2'-0" min.
12"
4"
3"
Measuring Cables Laid Loosely in Trench in Slight Wave Lines.
Well Compacted Fine Grain Material.
DETAIL OF CABLE TRENCH.
Measuring Cable in Steel Lined Borehole. Lining Withdrawn as 'Bentonite' Grout is Placed.
Selected Sand.
12"
Transmitter.
2½"
BOREHOLE DETAIL.
2½" Dia. Borehole Positions.
2½" Dia. Borehole Positions.
Surface Box Containing Position Selector.
To Forth Road Bridge

FIG.3.

wire, the tension in which may be varied to tune its natural frequency to that of the incoming signal from the transmitter. The adjustment in tension is read on a dial and the scale reading multiplied by a calibration constant gives the required measurement.

The results of the piezometer readings over the period 1964 to 1972 are given in Figure 4. The expected trend is apparent but the range of change is reassuringly slow.

Acknowledgements
The author wishes to thank the Royal Borough of Dunfermline for permission to publish these results and Mott Hay and Anderson who originally asked Bullen and Partners to study the cutting.

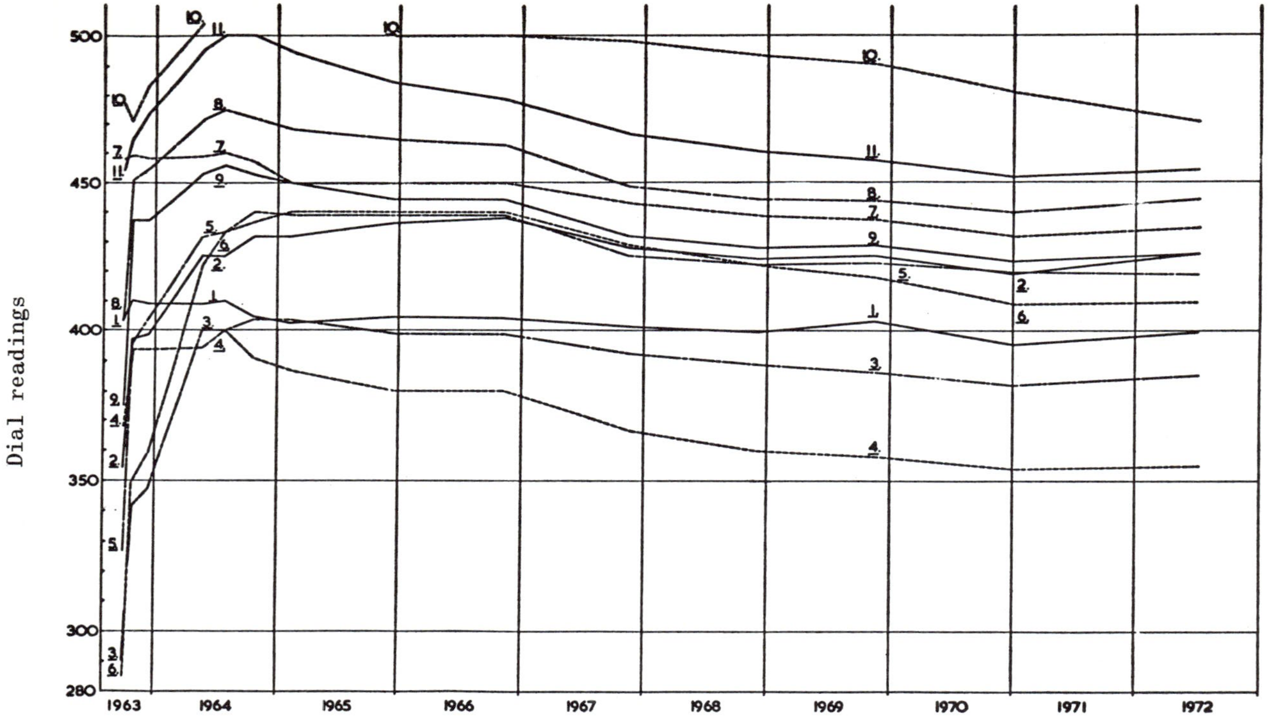

Note:

Average calibration
100 dial readings =
4.25 ft. water

PIEZ. NO.	NO LOAD READING.	PIEZ. READINGS. FOR F.O.S. OF 1·3
1	+471	+ 51
2	+494	−286
3	+480	−570
4	+480	−160
5	+476	−354
6	+491	−669

PIEZ. NO	NO LOAD READING.	PIEZ. READINGS. FOR F.O.S. OF 1·3
7	+493	+ 93
8	+505	− 65
9	+497	−203
10	+494	+324
11	+488	+148

PIEZOMETER READINGS 1963–1972.

FIG. 4.

MEASURING MOVEMENTS OF EMBANKMENT DAMS

by A D M Penman,
and J A Charles,
Geotechnics Division, Building Research Station

INTRODUCTION

The measurement of the movements of embankment dams can conveniently be divided into two parts: (1) measuring movements within the body of the fill, and (2) measuring the movements of points on the surface of the embankment. Both types of measurement have been made for many years and the purpose of this paper is to give a brief review of some of the earlier methods and to describe recent work by the Geotechnics Division of the Building Research Station in measuring the deformations of Scammonden and Llyn Brianne, currently Britain's two highest dams.

Embankment design has tended to concentrate on providing adequate factors of safety against collapse and has required information about pore pressures and total pressures in fill and foundation materials. There has been less emphasis on observing the movements of embankments, although settlement measurements have always been considered of interest. Various theoretical methods have been used to estimate settlements, but until recently there have been no methods for predicting the more general deformations of a dam. Development in analytical methods using finite element techniques now allows deformation predictions to be made during the design stage and this has increased interest in the measurement of movements.

INTERNAL MEASUREMENTS

Instruments completely buried are usually more successful in machine-placed fill than those with connections brought up to the surface which have to be extended as the fill is raised. Foundation settlements under small embankments have been measured by vertical rods brought up through the embankment, but in addition to being obstructions, the rods were liable to accidental damage. Spangler (1933) overcame these difficulties by using a simple remote reading water overflow gauge. A modification of this gauge was used by Skempton and Bishop (1955) to measure foundation settlements at Chew Stoke Dam. Further modification made at the Building Research Station led to a gauge first used to measure settlements within the fill of Selset Dam (Bishop and Vaughan, 1962) and since used extensively in earth and rockfill embankments.

The gauge consists of a buried container connected by three flexible plastic tubes, laid in a trench across the fill, to an instrument house. One of the tubes connects a transparent standpipe, fastened to a wall of the instrument house, to a small circular weir inside the container. Water added to the standpipe will cause an overflow from the weir into the buried container until the water level in the standpipe falls to about the level of the weir. The meniscus in the standpipe and at the weir develops sufficient pressure to prevent the water level in the standpipe falling completely to the level of the weir, but the few millimetres difference observed in laboratory tests on the apparatus remains fairly constant. This means that any reduction in level of the equilibrium water surface in the standpipe from its initial position indicates a true settlement of the buried container.

To obtain accurate measurements the water in the tube should be free from bubbles, be all at the same temperature, and have reached a static condition without backsurge. In practice, air free water is taken from a piezometer de-airing unit and twice the volume of the tube is passed into the buried container, which must be of adequate size to accept this water without immersing the weir. The tube is then connected to a full stand pipe, through a system of leak-proof, no-volume change cock-valves, and a measurement is made of the equilibrium water level. The time for equilibrium to be reached depends on the size of the tube, 6 mm internal diameter is about the maximum size which will still allow air bubbles to be swept out of the tube and this size reduces the equilibrium time to a few minutes for the lengths typically required for dam installations. Repeat readings are normally taken until they give agreement to within ± 1 mm.

Any difference of air pressure on the overflow and standpipe water surfaces will destroy accuracy. The second tube transmits the air pressure of the instrument house to the top of the buried container and it is important that it should be kept free from drops of water or other obstructions. A third tube is used to empty the container. This is done when the cock-valves on the water level tube have been turned off, by applying air pressure through the top tube.

This gauge transfers the vertical position of the inaccessible buried container to an accessible standpipe. Settlement of the standpipe itself, with the instrument house, is then measured by surveying methods, from a stable datum placed outside the zone of influence of the dam.

In situations where it is not possible to have the buried container and the standpipe at the same level, the air pressure in the standpipe can be increased or reduced by a column of water. The equilibrium level of the water surface in the stand pipe is in this way lowered or raised by exactly the height of the additional water column. To maintain accuracy this height has to be measured precisely. A line drawing showing a typical installation in a dam is given in Fig 1.

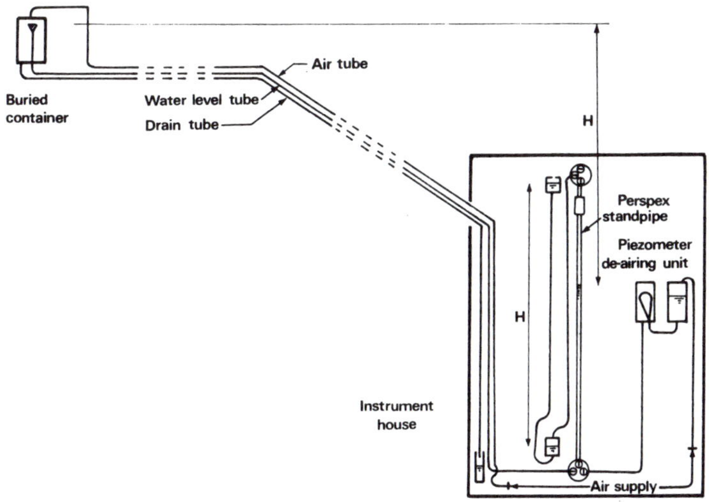

Fig 1 Water overflow settlement gauge

Another type of remote reading settlement gauge which can be buried in trenches during construction developed at the Transport and Road Research Laboratory has been described by Irwin (1967). It uses mercury in a small plastic tube fitted with electrical contacts and laid to a slight slope. Gas pressure pushes the mercury up the tube and the equilibrium pressure at each contact is used to measure its elevation.

Although the USBR cross-arm settlement gauge (USBR 1963) has a vertical column of telescoping steel tubes, the difficulties of obstruction and damage have in some cases been overcome by burying the top of the gauge. When the fill was raised enough for the next extension, connection was made by boring to uncover the buried end. The depths to the numerous cross arms were measured by a steel tape attached to a latching torpedo. The latches locked under the ends of all the tubes carrying cross arms and the action of resting on the bottom closed the latches so that the torpedo could be withdrawn. The BRS vertical plate gauge (Penman, 1962) is a similar apparatus which uses hard plastic tubes, steel plates instead of cross-arms, and an electric induction coil to locate the plates. Neither of these vertical gauges can conveniently be used to measure settlements in the upstream part of a dam after impounding.

An extra fitment for the USBR cross-arm gauge allows it to be used to measure horizontal strains over a 6 m gauge length. The fitment consists of two horizontal telescopic arms attached to the vertical gauge and connected to large plates on either side. Cables transmit any relative motion between the plates to counter-weights hanging inside the vertical gauge and their position is found by the latching torpedo.

A remote reading inclinometer, lowered into tubes placed vertically in the fill will measure horizontal movement in two directions (Wilson and Hancock, 1965). Settlements can also be measured with this type of installation by using a latching torpedo. These installations suffer the disadvantages of other vertical gauges in machine-placed fill.

Horizontal strains have been measured by remote reading electrical gauges using a variety of sensing units. One developed at BRS, described by Kennard, Penman and Vaughan (1968) used a vibrating-wire sensing unit and measured the relative movement between two cross-arms placed 1.8 m apart in the fill, to an accuracy of 0.5 mm. Chains of gauges connected end to end have been used to measure strains along dam axes.

Various types of rod or cable gauge have been used to measure the horizontal movements of buried plates. At Selset (Bishop and Vaughan, 1962), a steel wire in an alkathene tube was used to transfer the movement of buried plates to an instrument chamber on the downstream slope, where the wires passed over pulleys and were tensioned by weights. Unfortunately initial bends in the wire and tubes, which were both delivered to the site in coils, introduced too much friction into the system. A later apparatus built into Kainji Dam, Nigeria, successfully used straight steel rods, and this design of apparatus was used at Backwater Dam (Geddes, Rocke and Scrimgeour, 1972). Cables 3 mm diameter, tensioned by 57 kg weights were used at Oroville Dam, and a similar arrangement was used at Geehi Dam (Hosking 1972). USBR cross-arm gauges have been used in Mexico in horizontal attitudes, with the latching torpedo pulled into the gauge by a cable passing round a pulley at the inner extremity of the gauge.

A most satisfactory apparatus for measuring movements in fill is the horizontal plate gauge. It consists of a continuous access pipe passed through metal plates and placed in a trench during construction. Sensing units put into the pipe detect the plates and measure their horizontal and vertical positions in relation to a reference plate attached to an instrument chamber at the end of the pipe. The movements of the chamber have to be measured to obtain the total movements of the plates.

The horizontal plate gauges installed in Gepatsch Dam have been described by Lauffer and Schober (1964). They used a cable to pull the sensing units into the pipe. It was permanently installed in the gauge and passed round a pulley at the inner closed end of the pipe. A sensing unit located the plates by an electromagnetic system. The vertical position of each plate was measured with a water overflow gauge; the overflow was contained in the sensing unit.

The horizontal plate gauge developed at BRS and used at Scammonden and Llyn Brianne dams has been described by Penman, Burland and Charles (1971). It uses a pneumatic duct motor to pull the sensing units into the pipes and an induction coil to detect the metal plates. These cause an inductance bridge to be thrown off balance and movements of the indicating meter give the position of a plate with precision. Accuracy is lost due to stretch of the steel tape attached to the sensing unit. Tension measurements however have shown that the force required to move the sensing unit through a given plate does not vary greatly for repeated readings and indicates an accuracy of ± 2 mm in defining the plate position. Temperature variations could also affect tape length, but these were found to be small. At Llyn Brianne the range over the whole length of the gauges was only 12–14°C in the bottom gauge, 8–11°C in the middle one (reflecting winter fill conditions) and 13–15°C in the top gauge. The accuracy obtained with the overflow sensing unit is also ± 2 mm in defining the elevation of a plate. The positions of the horizontal plate gauges installed at Scammonden and Llyn Brianne are shown in Fig 2 and Fig 3 shows a gauge being installed at Llyn Brianne Dam.

EXTERNAL MEASUREMENTS

Finding positions for reference points on a dam site is not always easy. They have to be founded on ground which will not be caused to move (in terms of the accuracies required) by any natural phenomena such as creep or landslip, or any of the work for the dam or the stress changes due to the weight of the dam and impounded water. Ideally they should be at such positions in relation to the dam to suit the methods of measurement that are to be used and should be visible one from another. They can be useful in setting out work for the dam. Precise surveying by triangulation has been successful in measuring small movements of concrete dams (Willm and Schum,

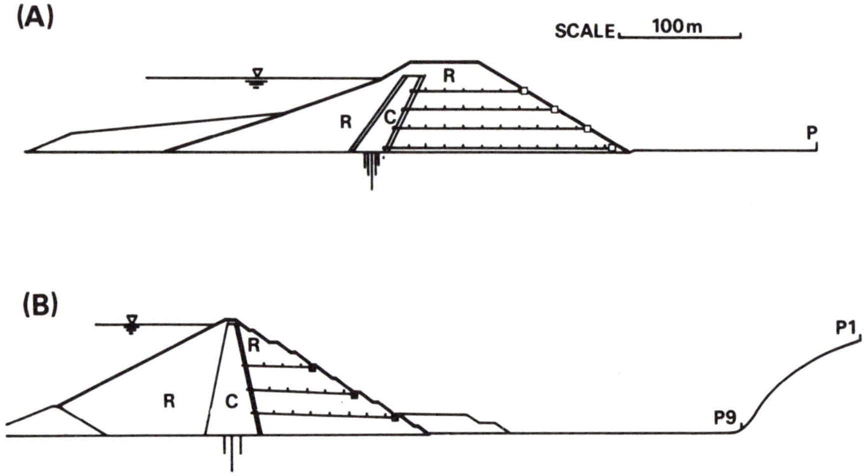

Fig 2 A Scammonden Dam C Clay core
B Llyn Brianne Dam R Rockfill
P Reference pillars

Fig 3

Installation of horizontal plate gauge

1972), often in relation to reference points several kilometres from the dam. The potential accuracy of theodolites can easily be wasted through inadequate methods for attaching the instruments or targets, or accidental damage to any of the fixing points between sets of readings.

The size of a triangulation network is usually calculated from the measured length of a base line. If a tensioned invar tape has to be used to measure the line, the length that can be found on a dam site is often relatively short. Fortunately inaccuracies in the size of the network have little effect on the accuracy of measured movement and are only noticeable if alternative methods of measurements are used to give the network size.

Electronic distance measuring instruments can be used to measure long distances over rough terrain. Recent developments have improved their accuracy so that not only are they suitable for measuring base-line lengths, but also for determining the coordinates of a point on the dam by trilateration.

The Mekometer, an EDM instrument developed at the National Physical Laboratory, was used in prototype form to measure movements of Scammonden Dam, by direct observation from one reference pillar built 150 m downstream from the dam (see Fig 2). A tripod was not used to support the instrument. Instead, in the interest of accuracy, the pillar, shown in Fig 4, was extended to a convenient height above ground level and fitted with a Kern centring plate type 196 which would accurately relocate an instrument on the pillar. Similar centring plates were attached to the roofs of the instrument houses to accept the targets. The Mekometer had a quoted accuracy of 3 ppm of measured distance $\pm$ 0.1 mm. The pillar was protected from shallow ground movements by an outer cylinder. The centring plate was protected by a heavy cover which could be locked shut.

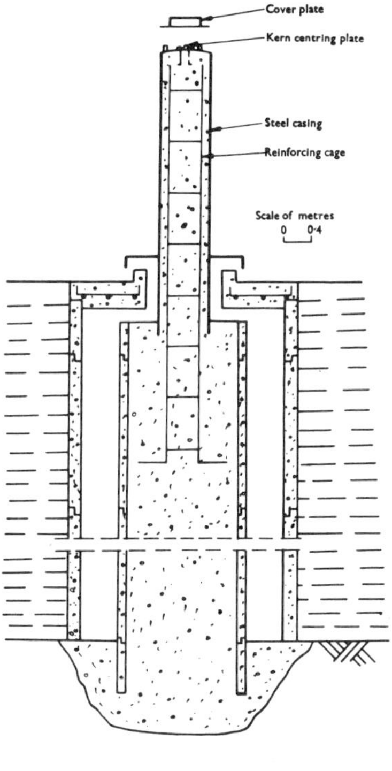

Fig 4
Reference pillar

An inverted pendulum shown in Fig 5 was installed in the instrument houses at the toe of the dam as an additional means of measuring horizontal movements. It was placed in a borehole 30 m deep, which was difficult to drill sufficiently

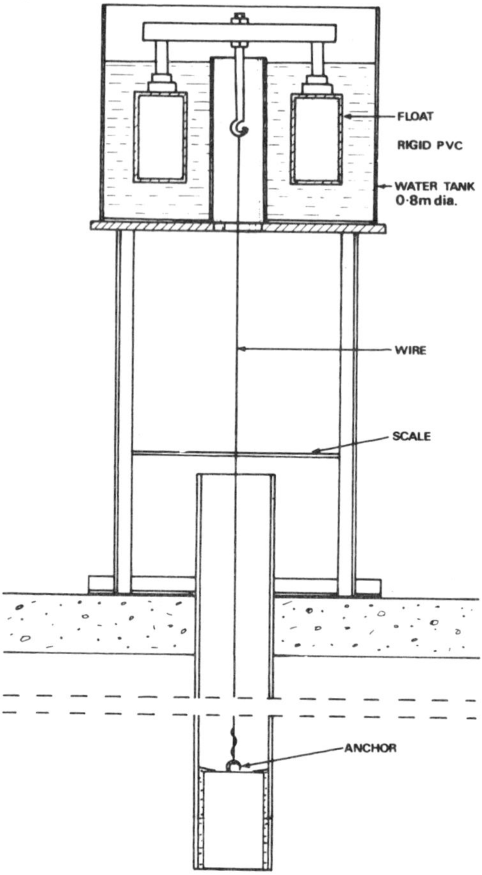

Fig 5

Inverted pendulum

vertical. This aspect of the installation is not easily satisfied.

The circular float was completely immersed and applied tension to the wire through four vertical support tubes connected to the arms of a horizontal cruciform. The dimensions of the tubes were chosen so that they had the average density of water. This kept the tension in the wire constant over a range of positions of the float and would have enabled the upper end of the wire to be used as a level datum, although this was not done. This instrument house, founded in bedrock, showed quite small horizontal movements, but there was reasonable agreement between the values given by the inverted pendulum and those obtained from the Mekometer measurements.

At Llyn Brianne Dam a number of reference pillars were founded on bedrock which was close to the surface on the valley sides. Figure 6 shows the positions of these pillars and it can be seen that a 90° bend in the course of the river just downstream of the dam has meant that pillars 1 and 9 could be

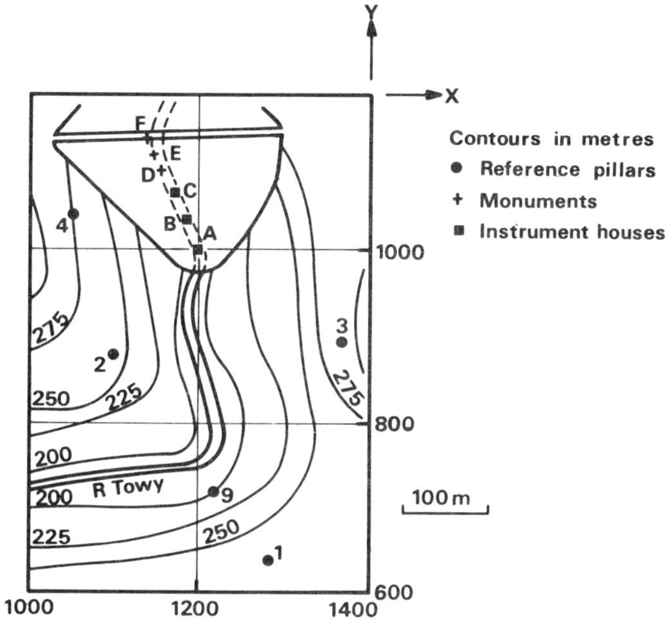

Fig 6 Plan of Llyn Brianne Dam showing positions of reference pillars

placed in line with the instrumented section, with pillar 9 in the valley bottom and pillar 1 at approximately crest level (see also Fig 2). Pillars 2 and 3 are on the right and left valley sides respectively at high levels. Not shown in the figure are bench marks which were placed on the right valley side at the level of each instrument house and monument on the slope of the dam. Reference pillars, instrument houses and monuments were all fitted with Kern centring plates, type 196, so that instruments as well as both theodolite targets and cube corner reflecting targets for the EDM instrument could be accurately re-located.

It should be mentioned that the height of the EDM instruments and their reflecting targets was measured from the centring plates to their optical centres each time they were set up so that allowance could be made for the height difference. The Kern targets used with the DKM2A theodolite have the same height so that no correction was necessary.

The arrangement of the reference pillars and bench marks has made it practicable to measure the movements of the instrument houses and monuments using three different surveying instruments, namely the Tellurometer MA100, an EDM instrument with a quoted accuracy of 2 ppm $\pm$ 1.5 mm, a Kern DKM2A one second theodolite and a Wild N3 precise level. Measurements with the Tellurometer were made from pillars 2, 3 and 9 and measurements with the theodolite from pillars 2, 3 and 4. Pillar 4 was very near the dam and it was necessary to check its position from the other pillars. Each house and monument was levelled in from the appropriate bench mark.

The accuracy of movement measurements can be assessed by comparing results obtained with these different instruments. Because the measurements with the Tellurometer took a complete day the survey with theodolite and level could not be done on the same day. This made a comparison of the different methods difficult as at times during construction the instrument houses and monuments were moving several millimetres per day. However, during the period December 1970 to March 1971 very little fill was placed and movements were small. Consequently measurements during that period together with measurements made in July 1972 some time after the end of construction (November 1971) can be expected to give a realistic comparison of the different methods.

The data derived from the survey of the movements of instrument houses and monuments has been analysed in three ways to give as clear a comparison as possible. Firstly, movements have been calculated from a three-dimensional trilateration. Distance measurements from the three reference pillars 2, 3 and 9 to an instrument house enable the coordinates of the instrument house in all three dimensions to be calculated with the help of a computer program. The second approach is a two-dimensional trilateration plus levelling. If the precise levelling is used to calculate the height coordinate of an instrument house then slope distances measured from a reference pillar to a house can be reduced to horizontal distances and with two such distances measured from pillars 2 and 3 a two-dimensional trilateration can be performed. The third method is a three-dimensional triangulation which uses the horizontal and vertical angles measured with the theodolite from pillars 2 and 3. The vertical heights quoted from this method are the average of the two values calculated from the vertical angles measured at pillars 2 and 3 respectively.

Table 1 shows the coordinates and movements of instrument house B as calculated by the three methods. The maximum difference in calculated movements is only 5 mm. A similar maximum difference applies to the actual coordinates except for the height coordinates calculated from precise levelling. These are on average 10 mm greater than those calculated from theodolite and EDM measurements. This is almost certainly due to a discrepancy between the levels of the pillars from which theodolite and EDM measurements were made and the bench mark from which the levelling was done. The settlements calculated from the different measurements are in good agreement. The

TABLE 1

CO-ORDINATES AND MOVEMENTS OF INSTRUMENT HOUSE B
AT LLYN BRIANNE DAM

Date	Method	X	Y	Z
7.12.70	3-dim trilat	1 189 118.0	1 031 940.3	227 614.4
	2-dim trilat and level	1 189 116.3	1 031 943.8	227 626.2
	3-dim triang	1 189 118.3	1 031 940.7	227 615.9
17. 3.71	3-dim trilat	1 189 120.9	1 031 937.5	227 592.4
	movement mm	+3	-3	-22
	2-dim trilat and level	1 189 119.8	1 031 940.0	227 599.8
	movement mm	+4	-4	-26
	3-dim triang	1 189 116.9	1 031 940.6	227 591.1
	movement mm	-1	0	-25
11. 7.72	3-dim trilat	1 189 191.2	1 031 472.5	227 392.6
	movement mm	+73	-468	-222
	2-dim trilat and level	1 189 190.2	1 031 474.8	227 399.7
	movement mm	+74	-469	-226
	3-dim triang	1 189 193.5	1 031 473.0	227 392.6
	movement mm	+75	-468	-223

CO-ORDINATES OF REFERENCE PILLARS

Pillar	X	Y	Z
2	1 103 607.0	881 727.0	263 373.0
3	1 369 451.8	893 932.1	296.717.2
9	1 222 799.6	717 786.2	199 608.0
BM at level of House B	---	---	225 717.0

success of the three dimensional trilateration approach in measuring settlements reflects the difference in elevation between the three reference pillars.

The initial survey of the reference pillars was made using a base length between reference pillars 1 and 2 measured with a Mk6 Geodimeter. Subsequently it was possible to check this distance with the NPL Mekometer. These measurements are listed in Table 2 together with the measurements of this base length that were made when the Tellurometer MA100 was on the site. The Tellurometer readings taken on seven different occasions over two years give a mean value of 299.3584 metres. This is 1.5 mm greater than the Mekometer measurement and 10 mm greater than the Geodimeter measurement. No Tellurometer measurement varied from the mean by more than 2 mm. When considering the quoted accuracy of these EDM instruments the consistency of all the measurements is surprisingly good.

Table 2

Measurement of distance between reference pillars 1 and 2 at Llyn Brianne Dam

Instrument	Quoted Accuracy	Measurement Date	Distance 1–2 metres
Geodimeter Mk6	2ppm ±10mm	Dec 1970	299.348
Mekometer Mk3	3ppm ±0.1mm	Nov 1970	299.3569
Tellurometer MA100	2ppm ±1.5mm	Aug 1970	299.3580
		Sept 1970	299.3599
		Dec 1970	299.3578
		March 1971	299.3581
		June 1971	299.3601
		Oct 1971	299.3569
		July 1972	299.3582

Surface movements of Llyn Brianne Dam were also measured during construction by photogrammetry from two 100 m long base lines, one upstream and the other downstream of the dam. Pillar No 1 (Fig 6) formed one end of the downstream base line. Photographs were taken with a Wild P30 phototheodolite and were analysed by Hunting Surveys using a Wild A7 Autograph. Three dimensional displacements were determined to an average accuracy of ± 0.05 m. The work, described by Moore 1973, gave a valuable general pattern of the movements of both sides of the dam.

MOVEMENTS AND STABILITY

Points on the slope of a dam move vertically downwards and horizontally outwards during construction. At Llyn Brianne Dam as the height of fill increased, the ratio of horizontal to vertical increments of movement increased (Penman and Charles, 1973) and in dams which have reached an unstable condition such as Muirhead Dam (BRS Report, 1942) increments of horizontal movement have been very much greater than those of vertical movement. The rate of horizontal movement could also be expected to increase if an unstable condition was approached. As a guide Penman (1971) listed four embankments with acceptable movements which ranged from 5 mm per month to 38 mm per month and two with unacceptable movements of 25 mm per day.

A more meaningful approach to the problem would seem to be to relate lateral movements of the embankment slopes to the amount of fill placed on the embankment which has caused the movement. Table 3 gives rates of lateral movement in mm per metre of added fill for the downstream slopes of a number of dams.

Scammonden Dam (Mitchell and Maguire, 1968; Winder, 1969) has a very wide crest and the clay core is in the upstream part of the embankment (Fig 2). Consequently the clay core has a negligible effect on the stability of the downstream rockfill shoulder. Using angles of shearing resistance obtained in large diameter (0.30 m) triaxial tests, circular arc stability analysis gives a factor of safety of 2.2 for the downstream rockfill of Scammonden at the end of construction (Charles, 1972). Llyn Brianne Dam (Carlyle, 1968 and 1969), Gepatsch Dam (Lauffer and Schober, 1964) and Blowering Dam (Hunter and Bacon, 1970) are rockfill dams with central clay cores. The stability of the downstream shoulders of these dams will be greatly influenced by the clay and the factor of safety will be less than that of the downstream rockfill at Scammonden. Maximum downstream movements expressed as a percentage of dam height are similar for the three central core dams and approximately twice as large as the Scammonden value. Rates of movements as the embankment neared completion were much greater for the central clay core dams.

Movements measured at a number of earth dams during construction are included in Table 3. At Chew Stoke (Skempton and Bishop, 1955), Derwent (Rowe, 1970) and Backwater (Geddes, Rocke and Scrimgeour, 1972) horizontal movements expressed as a percentage of dam height were smaller than those measured in the rockfill dams with central clay cores. Muirhead (Banks, 1948) and Galisteo (Catanach and McDaniel, 1972) are rare

TABLE 3

Dam	Type	Height m	Average slope	Max measured hor movement to end of construction m	Max horizontal movement as % of dam height	Max rate of hor movement measured mm/m height of fill
Gepatsch	rockfill with central clay core	153	1:1.5 d/s	1.0	0.65	13
Blowering	rockfill with central clay core	112	1:1.9 d/s	0.73	0.65	10
Llyn Brianne	rockfill with central clay core	90	1:1.75 d/s	0.61	0.68	20
Scammonden	rockfill with upstream sloping clay core	70	1:1.8 d/s	0.25	0.36	4
Galisteo	Earthfill	48	1:3.2 u/s	0.50	1.0	53
Backwater	Earthfill	43	1:3 d/s	0.07	0.16	3
Derwent	Earthfill with clay core	36	1:2 to 1:15 d/s	0.05	0.14	1
Muirhead	Earthfull with puddled clay core	22	1:2.9	1.5	6.8	330
Chew Stoke	Earthfill with puddled clay core	13	1:2.5 d/s	0.05	0.38	5

examples of dams where horizontal displacements of markers on the slopes were measured during virtually failure conditions. The total horizontal movements of Galisteo should be larger than the quoted values as measurements were only made during the placement of the last 9 metres of fill and excessive movements were prevented by raising the toe weighting berm as movements increased.

At Derwent (Buchanan, 1970) construction was stopped with the embankment three metres below crest level because high pore pressures indicated that the factor of safety had fallen to 1.33.

Where the stability of an embankment slope is in question the measurement of lateral deformation as construction proceeds would seem a practical method of assessing the nearness of an unsafe condition.

RECOMMENDATIONS

Generalisation is difficult because no two dam sites are alike, but the following points may prove of value when planning the measurement of movements of embankment dams.

1. <u>Internal measurements</u>
(a) The horizontal plate gauge provides a simple, reliable and accurate means of measuring vertical and horizontal movements in one plane of a large number of discrete points.
(b) Strains in the direction of the dam axis can be measured by chains of electrical strain gauges, each gauge measuring the movements over a gauge length in the range 2-5 m.
(c) Where it is not practicable to carry horizontal plate gauges into the upstream fill, settlements in this part of the dam can be measured by water level overflow gauges.

2. <u>External measurements</u>
(a) The most accurate EDM instrument commercially available at the present time is the Tellurometer MA100 (2ppm $\pm$ 1.5 mm).
(b) Given a suitable network comparable accuracy is given by trilateration with this instrument and triangulation using a one second theodolite, such as a Kern DKM2A.
(c) Reference pillars should be founded on stable ground with an unobstructed view at all times of other pillars and points to be measured on the dam. If built at an early stage they can be used when setting out the dam.
(d) On many dam sites the practicable locations of reference pillars will be very restricted and will make either an angular or distance method of measuring movements the more appropriate.

(e) Vertical movements can be measured by levelling using a precise level such as the Wild N3. This work is simplified if bench marks are established on the valley side away from the dam at the level of each point on the dam whose settlement is being measured. Where three reference pillars can be sited with considerable differences in elevation vertical movements can be obtained from three-dimensional trilateration. Settlements can also be monitored from appropriately sited reference pillars by vertical angles measured with a one second theodolite.

(f) Suitable locating plates should be built on to the pillars, instrument houses and monuments to position the instruments and targets exactly. These should be protected by vandal-proof covers.

ACKNOWLEDGEMENT

Research into the behaviour of embankment dams forms part of the work of the Geotechnics Division of the Building Research Station. All the work has been carried out in close cooperation with the consultants for the dams. Observations at Scammonden and Llyn Brianne were carried out in conjunction with Messrs Rofe, Kennard and Lapworth and Messrs Binnie and Partners, to whom the authors are particularly grateful. Mr D Burford of BRS gave valuable assistance with the field measurements.

REFERENCES

Banks, J A (1948): Construction of Muirhead Reservoir, Scotland. Proc 2nd Int Conf SM & FE, vol 2, pp 24-31.

Bishop, A W and Vaughan, P R (1962): Selset Reservoir: design and performance of the embankment. Proc ICE, vol 21, pp 305-346.

BRS Report (1942): Report of investigation on the stability of the earth dam at Muirhead Reservoir, Dalsy, Ayrshire. Building Research Station, DSIR.

Buchanan, N (1970): Derwent Dam - construction. Proc Inst Civ Engrs, vol 45, pp 401-422.

Carlyle, W J (1968): The River Towy scheme for the West Glamorgan Water Board. Water and Water Eng, vol 72, no 871, pp 371-382.

Carlyle, W J (1969): Llyn Brianne Dam, South Wales. Civ Eng and Pub Wks Review, vol 64, pp 1195-1200.

Charles, J A (1972): Discussion on Observed and predicted deformation in a large embankment dam during construction (Penman, Burland and Charles). Proc Inst Civ Engrs,vol 51,p730

Catanach, R and McDaniel, T N (1972): Lateral deformation of a dam embankment. Proc Conf on Performance of Earth and Earth

Supported Structures, vol 1, pp 867-883.

Geddes, W G N, Rocke, G and Scrimgeour, J (1972): The Backwater Dam. Proc Inst Civ Engrs, vol 51, pp 433-464.

Hunter, J R and Bacon, G A (1970): Behaviour of Blowering Dam embankment. Trans 10th Cong on Large Dams, vol 3, pp 217-242.

Irwin, M J (1967): A mercury filled gauge for measuring the settlement of foundations. TRRL Report LR 62.

Hosking, A D (1972): Monitoring of earth-rock dams. Symp on Monitoring of Dams. Australian Geomechanics Soc, Sydney, pp 4/1-38.

Kennard, M F, Penman, ADM and Vaughan, P R (1968): Stress and strain measurements in the clay core of Balderhead Dam. Proc 9th Congress on Large Dams, Istanbul.

Lauffer, H and Schober, W (1964): The Gepatsch rockfill dam in the Kauner valley. Proc 8th Congress on Large Dams, Edinburgh, vol 3, pp 655-660.

Mitchell, P B and Maguire, F J (1968): Scammonden Dam. Civ Eng and Pub Wks Review, vol 63, no 748, Nov, pp 1217-1221.

Moore, J F A (1973): The photogrammetric measurement of constructional displacements of a rockfill dam. The Photogrammetric Record, vol 7, no 41.

Penman, A D M (1962): Contribution to discussion on Selset Reservoir. Proc ICE, vol 23, p 737.

Penman, A D M (1971): Discussion on Derwent Dam. Proc Inst Civ Engrs, vol 48, pp 481-483.

Penman, A D M, Burland, J B and Charles, J A (1971): Observed and predicted deformation in a large embankment dam during construction. Proc Inst Civ Engrs, vol 49, pp 1-21.

Penman, A D M and Charles, J A (1973): Constructional deformations in a rockfill dam. Proc American Soc of Civ Engrs.

Rowe, P W (1970): Derwent Dam - embankment stability and displacements. Proc Inst Civ Engrs, vol 45, pp 423-452.

Skempton, A W and Bishop, A W (1955): The gain in stability due to pore pressure dissipation in a soft clay foundation. Trans 5th Congress on Large Dams, Paris, vol 1, pp 613-638.

Spangler, M G (1933): The supporting strength of rigid pipe culverts. Bulletin 112, Iowa State Col of Agriculture and Mechanical Arts, vol 32, no 37, pp 85-86.

USBR (1963): Earth Manual. United States Dept of the Interior, Bureau of Reclamation, 783 pps.

Willm, G and Schum, M (1972): Note on the application of geodetic methods to the determination of the movements of dams. International Commission on Large Dams, Bulletin No 23.

Wilson, S D and Hancock, C W (1965): Instrumentation for movements within rockfill dams. Instruments and Apparatus for Soil and Rock Mechanics, ASTM STP 392, pp 115-130.

Winder, A J H (1969): The Scammonden Valley scheme. Water and Water Eng, vol 73, no 877, pp 91-102.

AN INCLINOMETER FOR MEASURING THE DEFORMATION OF BURIED STRUCTURES WITH REFERENCE TO MULTI TIED DIAPHRAGM WALLS

S.H.E. Phillips and E.L. James

Geotechnical Research Engineers, Cementation Research Ltd., Rickmansworth, Herts.

INTRODUCTION

There has been an increasing interest in Britain during recent years in the use of inclinometer systems for the precise measurement of lateral movements. Inclinometers were originally developed some years ago, but their use has generally been restricted to the measurement of horizontal movement of earth masses. For this application, where relatively large movements and inclination changes are expected an instrument capable of measuring gradient to within 1×10^{-3} radians is adequate.

The measurement of the deformations of structures such as retaining walls requires a much more sensitive instrument. This paper describes an inclinometer which was originally developed specifically to measure the deformed profile of a multi-tied diaphragm wall. The object of that investigation was to monitor the behaviour of the wall at each phase of construction in order to calculate bending moments in the wall and thus check the design assumptions. Particular attention has been given here to the instrument design and operating techniques required to achieve the necessary accuracy. Naturally, having developed the instrument for this particular application, many other uses are envisaged and are being employed. These include the measurement of the deformations of piles under the influence of either applied lateral forces or adjacent excavations.

FACTORS WHICH INFLUENCE THE ACCURACY OF INCLINOMETER SYSTEMS

The use of inclinometer observations to determine the bending moments in vertical structures poses special problems over and above those encountered when using the instrument to measure horizontal movement through a potential failure zone in an earth mass.

The calculation of movements, whether of earth masses or of structures, involves the integration of gradient change data. This process of integration tends to cancel the influence of local errors.

Bending moments are obtained by differentiating the gradient change data. Small discontinuities of the data profile can lead to gross misrepresentation of bending moments.

In a structure, successive values of slope should give a smooth curve. However, where the curvature is small as in stiff walls and measurement accuracy is low the scatter of individual readings may easily mask the trend of the curve. Even when this trend is obvious, the 'best' curve must be drawn through the points. The range of scatter of the individual readings about the probable curve is a measure of the precision of observations.

This consideration, together with the stiffness of the wall, which will determine the curvature of the wall under the applied bending moment, determines the necessary precision which is required of the inclinometer system. For this application the retaining wall, on which the measurements were to be taken, had a stiffness, expressed as an EI value of the order of 28×10^9 lb ins^2/ft. width. The applied bending moments were considered to be around 50,000 ft lbs/ft width. A maximum gradient change of 4×10^{-3} radians was anticipated. This meant that a precision of about 2×10^{-4} radians on this gradient change was considered desirable from the inclinometer system in order to achieve the required accuracy in the bending moment calculations.

The overall precision of an inclinometer system is naturally dependent upon the precision of the slope measuring device itself but it is also determined by a large number of other factors related to the physical characteristics of the inclinometer system. These are outlined below.

(a) Gauge length and depth interval between readings

The basic principle of an inclinometer is to measure the gradient between two fixed points. A series of such measurements is taken so that a particular profile may be made up of a series of straight lines. Since the transfer of slope values to deformation requires a summation process, the errors can accumulate. Therefore, in instances where measurement of lateral deformation is required, the number of readings can affect the overall accuracy. On the other hand a curve is more accurately defined the greater the number of chords that make it up. Other considerations such as the time required for large numbers of observations, must also be taken into account and a compromise made when determining the gauge length of the instrument.

For the maximum accuracy the depth between successive readings should be equal to the gauge length of the inclinometer. Significant errors may arise if the spacing is greater than the gauge length because the deflection profile is not composed of straight lines or smooth curves between measuring points. For quick checks however readings may be taken at intervals of two or three times the gauge length.

(b) Repeatability of measuring position

An error in repositioning the instrument in the duct with each set of readings will considerably reduce accuracy. This problem is accentuated when the curvature is large and its significance also increased with the increase in the inherent precision of the inclinometer.

Not only is this repeatability of positioning of the instrument in the duct important as far as the depth is concerned but also as far as instrument azimuth is concerned. The inclinometer must be firmly held at all depths against a slot, face or ridge so that all possible rotation about the axis of the duct is eliminated.

(c) Casing twist

It is essential that the inclinometer is operating in the desired plane of measurement. If the installed ducting is twisted then the inclinometer, securely located against rotation relative to the duct, will follow the twist. This means that the readings taken will reflect the resultant of the respective components of the movement in both the desired plane and the plane normal to this.

(d) Cross sensitivity of the instrument

It is most probable that the total movement of the structure being measured will not take place in the direction in which measurements are being taken. It is therefore essential that any movement in the plane normal to this should not influence the measurements taken in the desired plane.

(e) Installation procedures

The installation of the duct for the measurement of the deformation of structures can be carried out in one of three ways. These include, cast in situ with the concrete, grouted into either a preformed or a drilled hole or firmly attached by clamps to the structure. In each, great care must be taken with both the clamping and duct jointing. The joints should not allow any twist of one duct relative to another and should be well protected to prevent ingress of grout or dirt into the duct. The slots, faces or ridges against which the instrument bears to maintain its desired orientation should be kept clean and free from dust and dirt. Any dust etc. picked up by the wheels or slides of the instrument will reduce the precision.

When using clamps or joints, which rely on friction, care should be taken that the clamping pressures do not deform the duct locally. Such deformations can lead to errors in calculations due to the slight differential rotation of opposite duct faces or slots.

(f) Inherent precision of the instrument

When an inclinometer is being used in foundation structures or below ground level, there is normally a variation between the atmosphere and duct temperatures and the latter also varies with depth. This phenomena is particularly marked in fresh concrete which is still curing. For this reason the inclinometer sensing system must be compensated for temperature changes.

It is essential to carry out a systematic calibration check of the instrument on a routine basis irrespective of the nature of the sensing device to ensure that the system is operating satisfactorily. Each instrument, especially those with high precision, should be supplied with a calibrating stand suitable for use throughout the whole range of the inclinometer. Where possible this should be combined with a complete servicing check on the operation of the instrument. This includes cleaning and oiling the wheels, bearings and springs etc.

(g) Measurement procedures

There are further factors besides the physical characteristics of the inclinometer system which influence the overall precision and which are related to the measurement procedures. Obviously there are certain general requirements for obtaining the most accurate observations. The use of the same observer, instrument, instrument cable and recorder for all the series of readings on a particular project is highly desirable.

The greater the number of sets of readings of any one profile the more accurately defined that profile should be. However the minimum number should be two, the instrument orientation for the second set should be 180° from that of the first. As well as providing greater precision than does one set of readings, this technique provides an easy check on the accuracy of the observations. Providing the opposite sides of the duct against which the instrument bears are parallel, over its total length, the sum of the two readings taken at the same depth in the two orientations of the instrument should be constant. Therefore by summing the two readings at the time of observation, any gross errors are immediately detectable.

One further advantage of taking the two readings at instrument orientations 180° apart is that the average of the two is the vertical reading of the instrument. Using this value for the vertical reading of the inclinometer eliminates the task of checking the vertical reading prior to each set of readings and the value is not dependent upon a spirit level and sophisticated levelling equipment on the calibration frame.

THE INCLINOMETER SYSTEM

Having determined that the precision of the instrument required to achieve the desired accuracy in the bending moment calculations should be 2×10^{-4} radians, it became essential to eliminate, as far as was practicable, all the sources of error outlined above.

Figure 1 shows two views of the instrument. It was designed to operate inside a 45 mm internal square extruded aluminium duct of 3 mm wall thickness. The twist of the 4 metre length of duct is kept to less than 1°. The sides of the ducting provide a smooth face on which the wheels of the inclinometer run. All the wheels on the instrument are ball races. The reference plane is defined by three fixed wheels arranged in a triangular configuration such that the gauge length between axes of rotation of the upper single and lower double wheels is 200 mm. These three wheels are held firmly against the inside face of the duct by further wheels which are mounted in spring loaded arms operating against the opposite side of the duct. The instrument is located in the plane normal to the measuring plane by two fixed and two sprung wheels. In this way the azimuth of the instrument is rigidly and accurately located.

The sensing device is a cantilever strip with a heavy pendulum mounted at one end and with the other rigidly fixed against movement normal to the reference plane. Any change in gradient between the fixed reference wheels is therefore registered as bending of the cantilever. This is monitored by four electrical resistance strain gauges mounted on it in a fully temperature compensated four-arm Wheatstone Bridge arrangement. Inside the watertight container the pendulum is surrounded by silicon oil to provide damping of the pendulum and insulation and protection for the strain gauges and connections.

The cantilever suspension itself is a 0.25 mm (0.010") thick phosphor bronze strip. This has a very low torsional rigidity and in order to reduce the cross sensitivity of the instrument, which could be brought about by movement normal to the preferred plane of measurement, the suspension upper clamp is mounted on ball races. This arrangement restrains movement of the pendulum in a plane normal to the base plate, but permits rotation in the plane at right angles to this.

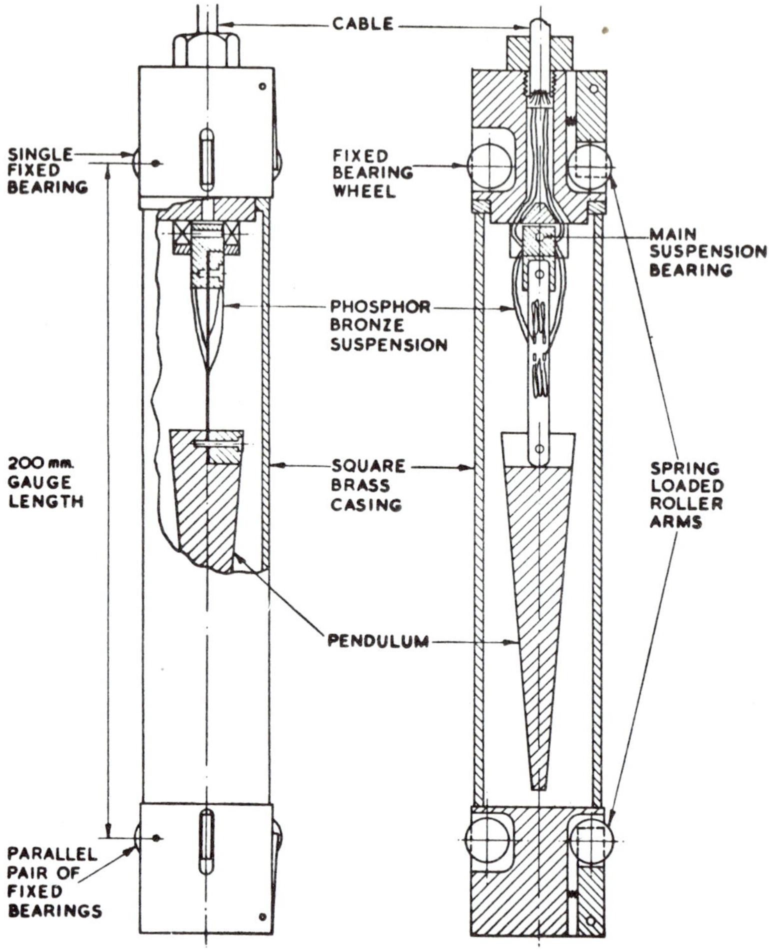

Figure 1 Sectioned drawing of inclinometer

The instrument is connected to the measuring unit by a P.V.C. sheathed cable containing the four conductors and a steel straining wire to support the weight of the inclinometer when it is being raised or lowered. The cable is graduated in 20 cm lengths and is wound on to a reel with a plug take off to a digital strain indicator giving a direct read out in strain units. Direct reading is to 2 microstrain, but interpolation to 1 microstrain can easily be achieved.

The calibration frame consists of two steel strips set normal to one another to form an inverted 'T'. A further plate mounted on a roller and to which the inclinometer is attached, is sprung loaded against a micrometer head which is secured into the upright of the frame. Therefore a change in the micrometer setting causes a change in gradient of the inclinometer base plate. In this way it is possible to measure a lateral displacement between the wheels on the base plate as a strain reading.

Over a range of the instrument, $\pm 5^{\circ}$, the calibration is linear and the repeatability is to within $\pm$ 1 microstrain. This corresponds to an error of 5 seconds of arc (3×10^{-5} radians).

ACCURACY DETERMINED BY FULL SCALE LABORATORY TRIALS

A 12 metre length of ducting (3 sections) was erected in a 13.7 metre deep steel cased shaft and held essentially vertical by eight specially prepared brackets. These were firmly clamped to flanges welded into the casing of the shaft and were spaced at intervals of approximately 1.5 metres. Each bracket held the duct firmly between two rollers so that some rotation of the duct about the brackets was allowed in the plane of measurement. No lateral movement of the duct was permitted. Two of the eight brackets were designed so that a predetermined displacement could be applied to the duct at that point by replacing the roller by one of a different diameter.

The preparation of the ducts was carried out exactly as it is done for a site installation. Prior to the erection of the ducts, each end of each length was milled square. This ensures that when two lengths of duct are butted together they are both in the same plane and there is no gap at the joint to allow dust and cement to enter. The whole duct was then coated with a protective paint. This is essential when casting into cement grout or concrete in order to prevent the chemical action between the aluminium and the cement. The ducts were then installed in the shaft. The ducts were jointed together with a friction type clamp which allows alignment of the external faces of the duct to be made. Further sealing protection is given to the joint by applying a very viscous mastic over a strip of thin adhesive tape.

The initial profile of the duct was recorded in the same manner as for a duct installed in a diaphragm wall. The inclinometer was passed down the duct taking measurements at 200 mm intervals in one plane of the duct. The readings were repeated when drawing the instrument up the duct. The whole procedure was again repeated with the instrument turned through 180°. This means that each profile was produced from four readings of the inclinometer taken at every instrument position in the duct.

Profiles were recorded for several configurations of the displaced duct. A computer program was used to convert the data to a series of deflections by numerical integration. The initial profile of the duct was stored on disc and subsequent profiles were plotted relative to this. Fig. 2 shows the profile of the duct for a displacement of the duct by 3.177 mm at the uppermost and the fifth bracket, towards the centre of the shaft. The displacements are shown relative to the lower end of the duct. This particular profile is shown here as this change in gradient is fairly representative of the maximum encountered in a diaphragm wall installation.

The profile of the duct as shown on the original plot from the computer shows that the calculated movement of brackets 1 and 5 are 3.135 mm and 3.150 mm respectively. It is also evident that the bottom of the duct has not moved relative to the fixed points (brackets 2, 3, 4, 6, 7 and 8).

As anticipated this high level of accuracy was not maintained when the duct was given excessively large deformations. It is believed that the source of error lay in the exact positioning of the instrument at a given elevation. Cable tension, which is a maximum when the inclinometer is being drawn up from the full depth gives rise to an extension of only 12 mm in a 12 metre length which will not give significant error. Deterioration and small movements of the cable markers had occurred since the previous calibration of the cable. Movements were gradual and accumulative along the cable. Checks carried out during the laboratory trials indicate that there is an error of 100 mm. in 15 metres. This accounts for the reduced accuracy of this particular profile where, as has previously been indicated, the error in the results due to non-repeatability of measuring position is accentuated when the curvature is large and the inherent precision of the instrument is high. However this is an uncommon occurrence in structural applications of the instrument.

Further tests on the effect of temperature variations were carried out on the instrument. Theoretically the electrical resistance strain gauges in the four arm Wheatstone Bridge circuit should be independent of temperature changes and indeed no apparent drift in the reading was observed when the instrument was subjected to a temperature rise of $10^{o}C$. However, it has been noticed on several occasions that the down and up readings on the same face are more consistent

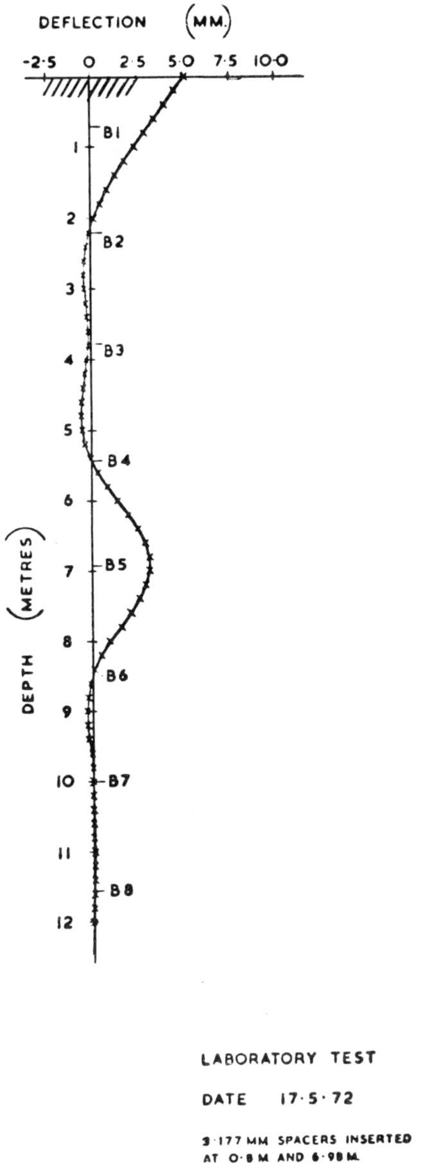

Figure 2 Results of laboratory test

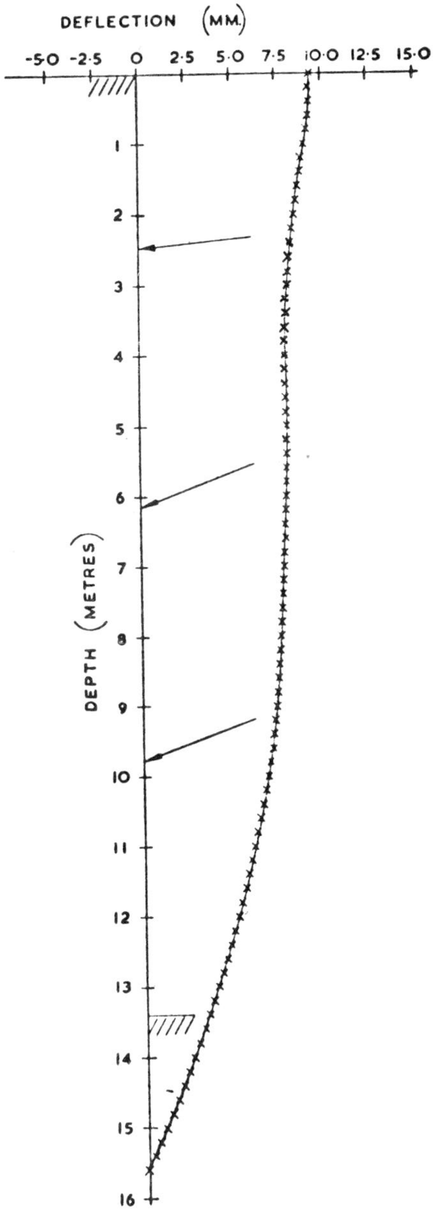

Figure 3 Typical profile obtained for a diaphragm wall

when the instrument has been left in the duct for half an hour before readings are taken in order to reach the duct temperature.

DIAPHRAGM WALL APPLICATION

A typical deformed profile of an anchored diaphragm wall at its final excavation stage is shown in Fig. 3. The duct was installed in the reinforcing cage along the neutral axis of the wall as the cage was being lowered into the trench. The techniques used to obtain this profile were exactly as have been described previously. Further instrumentation of the wall included the use of load cells to measure the anchor load variation at each stage of construction and geodetic surveying techniques to measure overall wall translation.

This particular profile is one that was measured twice as a check on the inclinometer reading repeatability. The maximum divergence between the two 15 metre profiles was 0.16 mm.

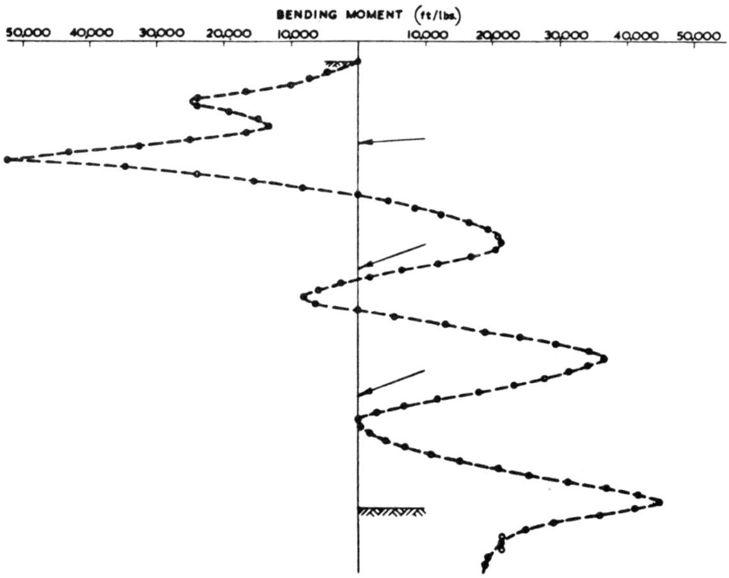

Figure 4 Bending moment diagram developed from the basic inclinometer data used to obtain Figure 3

The bending moment distribution in the wall corresponding to this profile has been calculated from the gradient change data (Fig. 4). This shows a logical variation of bending moment over the full depth of the wall. The influence of anchoring at three levels is clearly shown by the varying magnitudes of the discontinuities at the anchors. The anchor stressing superimposes high bending moments on the wall. For the lower anchor levels this bending moment is of opposite sign to that existing in the wall prior to stressing and the effect is to reduce the bending moment. An increase in bending moment however is experienced at the top anchor.

The bending moment discontinuities occur at slightly lower levels than the design anchor elevations. The difference is up to 1 metre and appears to be a direct function of anchor inclination. This will also be partly influenced by the depth of excavation below the anchor level that has occurred before anchor installation.

The top discontinuity in the bending moment distribution is probably due to the influence of the guide wall at the rear of the crest of the wall. The bending moment peak occurs at the bottom of the guide wall. The same phenomena has been seen in other diaphragm wall panels.

CONCLUSIONS

By careful attention to those areas where errors in inclinometer systems can occur an instrument and technique has been developed which is capable of measuring the gradient changes of a stiff concrete wall to an accuracy of 1×10^{-4} radians. From the data obtained it was possible to calculate satisfactorily the imposed bending moment distribution on the wall.

At relatively high gradient changes, which are not usually found in a stiff reinforced concrete wall the precision of the instrument is limited by the accuracy to which the instrument can be relocated with depth in the duct.

ROCK INVESTIGATIONS FOR CAMLOUGH UNDERGROUND POWER STATION

B. T. Seddon

Chief Engineer, Binnie & Partners, London

INTRODUCTION

The 230 MW capacity Camlough Pumped Storage scheme, which is under construction for the Electricity Board for Northern Ireland, is located 3.5 km West of Newry in County Armagh. Lower storage will be provided by the existing Camlough reservoir and upper storage by a new reservoir on the slopes of Slieve Gullion mountain 215 m above Camlough. The underground power station is located vertically below the upper reservoir as shown on Fig. 1 with a minimum rock cover of 185 m above the centre line of the cavern. A plan and sections of the power station cavern are shown on Fig. 2. It will be 100 m long, 35 m maximum height and 22 m span with a span to rise ratio of 4.0 in the arch.

Slieve Gullion mountain represents the deeply denuded relics of a large volcano. During the tertiary geological period the area around Slieve Gullion became a centre of intense igneous activity. This commenced with the emplacement of a ring dyke some 10 km in diameter. At later stages intrusion of granophyres and dolerites occurred and in the last stages of igneous activity some hydro-thermal alteration of dolerites occurred. The area is broken by faults including that which led to the formation of Camlough.

Site investigations comprising detailed geological mapping, drilling, boring and trenching were carried out during 1969 to establish the feasibility of the project. These investigations included an 80 m deep drillhole on the line of the power

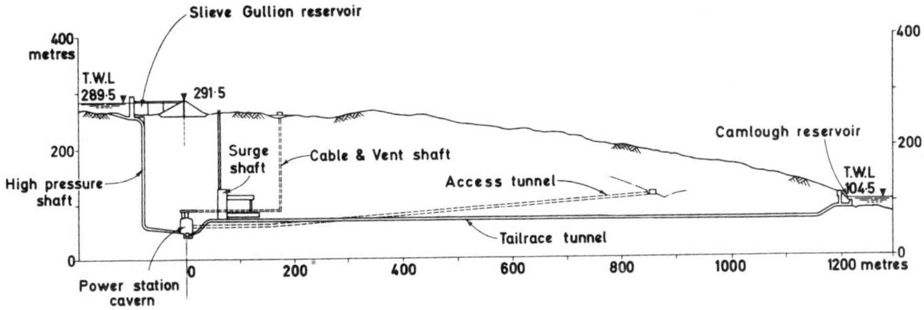

Figure 1 Camlough Pumped Storage Scheme

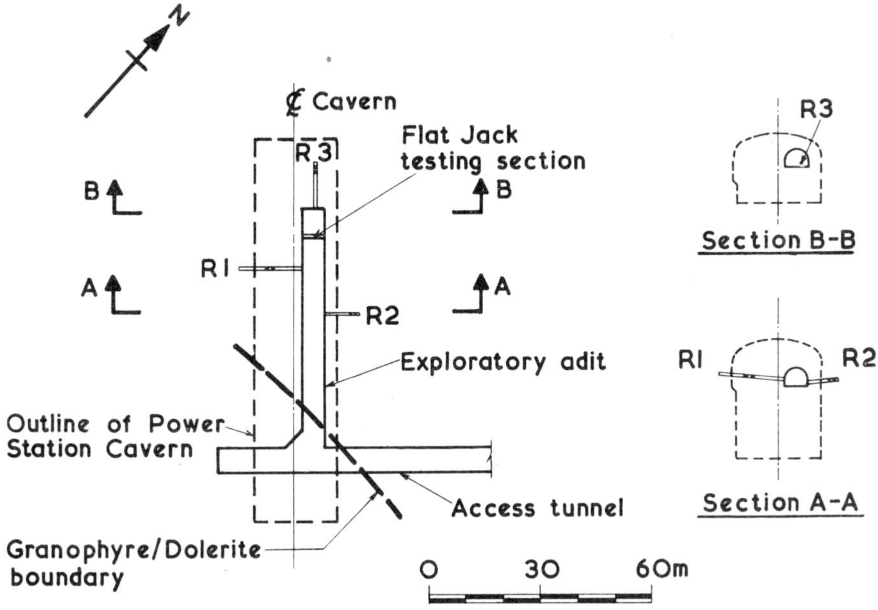

Figure 2 Plan and sections of underground cavern

station access tunnel and holes of similar depth on the upper dam site above the power station.

The site investigations indicated that generally good tunnel and shaft sinking conditions could be expected but local zones of altered rock might be encountered and it would be essential to locate the main underground cavern clear of any such zone. The cost of a series of 200 m deep drillholes or of a 1000 m long sloping exploratory adit from the ground surface into the cavern area was considered unjustified and it was decided to proceed with the excavation of the power station access tunnel to within 150 m of the proposed location of the main cavern under a preliminary contract. From the inner end of this tunnel an exploratory adit would be driven into the cavern area to enable geological mapping, drilling and rock investigations to be carried out. Some flexibility in the location and orientation of the power station was maintained at this stage.

Work on this contract commenced early in 1971 and as tunnelling conditions within the body of the mountain proved consistently good, the tunnel was excavated full size into the proposed power station area and then at right angles parallel to the axis of the power station as an exploratory adit.

The investigations in the power station area were carried out in November and December 1971 and comprised geological mapping, drillholes and stress measurements with C.S.I.R. triaxial stress plugs and with flat jacks. Drilling was carried out by a specialist sub-contractor to the access tunnel contractor, geological mapping and site stress measurements were carried out by the staff of the consulting civil engineers, and the Rock Mechanics Department of Imperial College, London provided measuring equipment and analysed the results.

While excavation of the access tunnel was in progress, further site investigations were carried out to provide data for the detailed design of structures. These investigations included a 215 m deep drillhole on the line of the pressure shaft 80 m upstream of the power station cavern.

ROCK DRILLING AND GEOLOGY

Twelve exploratory holes were drilled to the limits of the proposed cavern excavation, to the bottom of the pressure shaft, and into the surge shaft areas. The drillhole logs and geological mapping showed that most of the length of inner end of the access tunnel and of the exploratory tunnel were excavated in dolerite eruption breccia except the final section of the access tunnel which was in granophyre. Generally speaking these rocks were fresh, strong and medium grained. The approximate line of contact between granophyre and dolerite is shown on Fig. 2.

A fractured dyke zone 3 m wide was found in the exploratory adit about 20 m from the access tunnel and in the final length of the access tunnel. This caused a small amount of local overbreak in the exploratory adit. Several shear zones with small vertical displacements occur within and along the margin of the dyke. Most joints plotted dip at steep angles and have strikes which fall in the range of N80°E to N140°E.

The hole from the ground surface on the line of the pressure shaft and 80 m upstream of the power station centre line was drilled mainly through strong dolerite but a 10 m zone of heavily mineralised and jointed dolerite eruption breccia was encountered at about the level of the main cavern roof. The bottom 10 m of hole was drilled into strong granophyre.

ROCK STRESSES

Values for the directions and magnitudes of the principal stresses of the in situ stress field were obtained from the results of tests carried out with the C.S.I.R. triaxial rock stress measuring cell (Leeman 1969). In this technique three strain gauge pads are cemented to the surface of an EX drillhole extended 40 cm from an NXCU drillhole. Three electrical resistance strain gauges are mounted in each pad and a dummy gauge bonded to a short length of EX core is provided for temperature compensation. The strain gauge readings are taken on a Huggenberger strain balance after the cement has hardened. The end of the EX hole is then plugged to prevent the entry of water. The hole is overcored, the cylindrical core removed and the stress relieved strain gauge readings taken from each gauge.

Two tests were carried out in each of three holes drilled into the rock mass a sufficient depth to be unaffected by the excavation of the exploratory tunnel. The positions of these holes, which were 6.5 m minimum depth are shown on Fig. 2. Drilling was carried out with a Crealius Diamec 250 diamond core drill provided with aluminium drill rods. The EX drillhole was cleaned with a spray of water repellant fluid before the cell was inserted.

All three drillholes were in dolerite eruption breccia laced with brittle quartz veins and the cores were prone to mechanical fracture along these veins. Four tests were unsuccessful and had to be repeated, two because of fracturing during the overcoring operation, one through inadequate bond and the fourth through lack of seal in the plug fixed into the EX hole before overcoring. Cores for measurement of elastic modulus of the rock were obtained from holes R1 and R2.

Flat jack measurements

The flat jack measurements were designed to measure the stress around a cross section of the exploratory adit and to determine the elastic constants of the in situ rock. Two steel buttons were cemented to the rock a fixed distance apart using a dial gauge extensometer mounted on an invar bar. An invar bar with two index marks provided a reference length. A slot normal to the rock face was then cut with a 60 cm dia rock saw, the flat jack was inserted and the oil pressure raised in stages until the index marks were the original distance apart. The flat jacks were 5 mm thick, 30 cm radius and 17 cm deep. Other equipment included a hand operated oil pump and pressure gauges suitable for pressures up to 69 N/mm^2. A steel scaffold was provided to support the equipment during rock cutting and jacking.

The location of the flat jack measuring section, which was considered to be beyond the influence of the adit end on the stress field, is shown on Fig. 2. The shape of the adit at the measuring section was made as regular as possible by pre splitting with drillholes at 40 cm centres before firing the round to excavate the rock. The shape as excavated and the positions of the flat jacks are shown on Fig. 3. A more regular shape of section might have been obtained with greater experience of pre splitting.

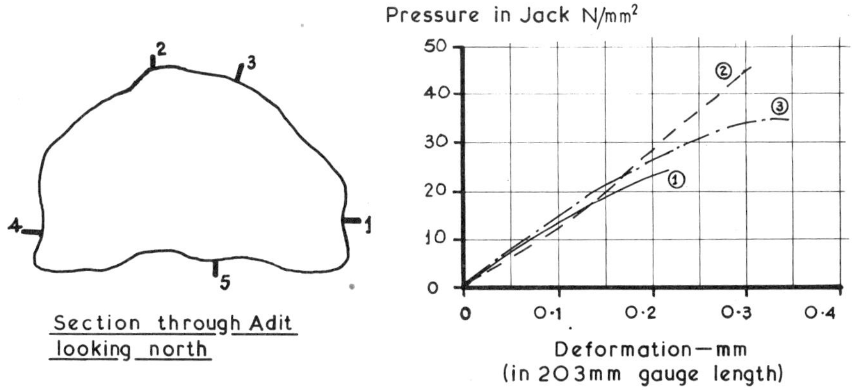

Figure 3 Flat jack positions and measured deformation

Slot numbers 2 and 3 in the crown of the section were in regions of high tangential stress. During jacking of slot number 2 a section of rock at the edge of the slot failed explosively and the jack burst. A similar failure occurred at slot number 3. Both failures occurred just before the index marks reached their original spacings, thus the tangential stress had to be gauged by extrapolation.

It was originally intended to cut nine slots spaced around the section but after the first three slots were cut and jacked, testing had to be suspended for 5 months. When a resumption of work became possible two slots were cut in the walls close to the floor but it was apparent that a general relaxation of the section had occurred. Thus while the slots were cut into solid rock the surrounding rock was loose. Relatively low rock stress values were obtained and in the case of slot number 5 failure occurred at $10.4\,N/mm^2$. It was considered that the benefit to be gained from further work did not justify the expense of maintaining services in the adit and testing was discontinued.

The strains measured at slots 1, 2 and 3 during jacking are shown on Fig. 3.

Strength and elastic constants of rock

Values of Young s modulus were obtained from uniaxial testing of cores from drillholes R1 and R2 and from the results of the in situ flat jack tests. The results of the uniaxial tests are shown on table 1. The mean value obtained from the flat jack tests was 122 x 10^3 N/mm^2 which was considered to be high for dolerite.

Table 1
Elastic Constants

Uniaxial Tests	E N/mm^2 $x10^3$	V (Poisson's ratio)
R1 cycle 1	66. 1	0. 263
R1 cycle 2	67. 6	0. 249
R2 cycle 1	82. 2	0. 272
R2 cycle 2	82. 2	0. 283

The uniaxial strength of the rock as determined by destructive tests on specimens obtained from drill cores, was obtained at 110 N/mm^2.

Calculation of principal stresses from C. S. I. R. tests

The values of the principal stresses produced from 5 of the 6 C. S. I. R. tests and using the highest values of the elastic constants obtained from the uniaxial tests are shown in table 2. The results of test R2a were neglected as creep in the strain gauge readings before overcoring was comparable to the strains produced by the overcoring process. Tests R3a and R3b both have readings substituted from a redundant gauge for those of a gauge which ceased to function during the overcoring process. The Eigen value technique was utilised to obtain the best fit values from the results of the tests.

Values for the horizontal and vertical stresses calculated from the test results are shown in table 3.

Table 2
Principal Stresses from C. S. I. R. Tests

Hole No.	Test	F1 N/mm^2	F3 N/mm^2	F2 N/mm^2
R1	a	53. 2	9. 2	17. 4
R1	b	31. 6	13. 6	17. 5
R2	b	18. 4	4. 1	7. 6
R3	a	70. 0.	11. 3	35. 5
R3	b	19. 5	-12. 1	11. 0
R1, R2, R3	a & b best fit	27. 0	11. 6	22. 9

F1, F3, F2 = Maximum, Minimum and intermediate principal stresses

Table 3
Horizontal and Vertical Stresses

Hole No.	Test	FH1 N/mm^2	FH2 N/mm^2	FH3 N/mm^2
R1	a	52. 6	14. 9	12. 2
R1	b	30. 6	18. 1	13. 9
R2	b	18. 1	5. 9	6. 0
R3	a	25. 9	55. 6	35. 0
R3	b	1. 3	12. 5	4. 5

FH1 = Horizontal stress in direction of hole R1
FH2 = Horizontal stress normal to direction of hole R1
FV = Vertical stress

The above values for the vertical stress compare with the value of 5. 5 N/mm^2 calculated from the depth of overburden and the unit weight of rock. The ellipsoid of principal stresses for the best fit values from the C. S. I. R. tests is shown on Fig. 4.

From a consideration of the results of the individual C. S. I. R. tests it seemed reasonable to make the following assumptions in calculating the rock stress field after excavation :-

(1) The maximum principal stress acts horizontally and transversely to the axis of the power station.

(2) The minimum principal stress acts vertically.

(3) The ratio of the greatest to the least principal stress is about 4. 0.

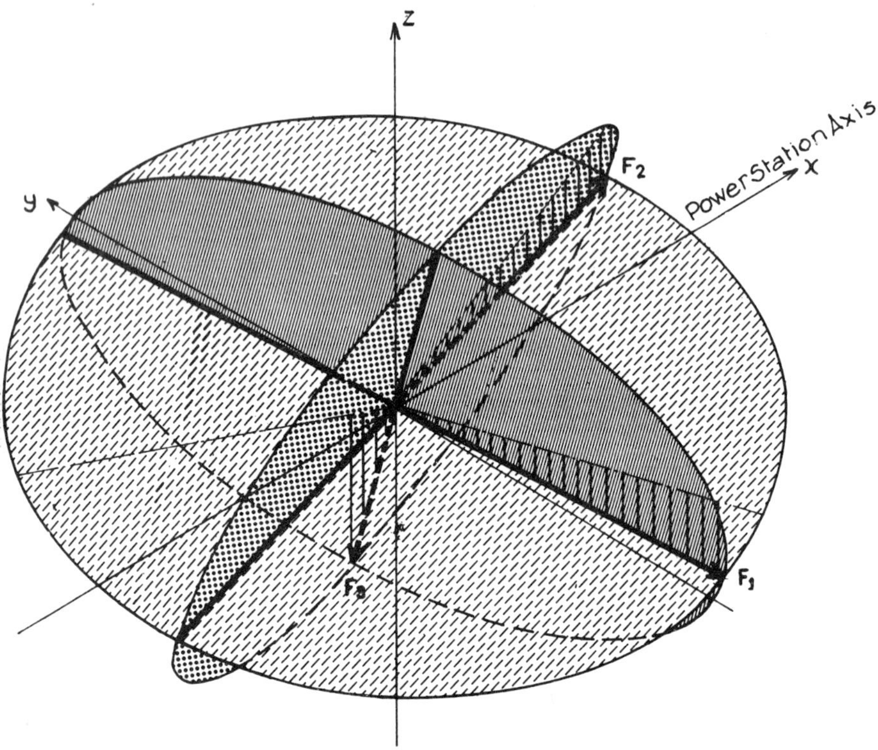

PRINCIPAL STRESS	MAGNITUDE N/mm²	BEARING * SHOWN-·-·-	DIP
F1	27·0	71°	13°
F2	22·9	339°	8°
F3	11·6	209°	74°

* Bearings measured from Power Station Axis

Figure 4 Principal stress ellipsoid

The results obtained from the Eigen value technique applied to all the holes together were not considered because it is inherent in the process that the principal stress value produced are less than those obtained from individual tests and are very conservative.

Calculation of the maximum principal stress from the flat jack tests

Values of the jack pressure required to restore the pre-slot gauge length and the corresponding values of rock stress normal to the jack are given in table 4. The rock stress normal to the jack was estimated from the jack pressure using equations given by Alexander (1960) and by Hoskins (1966). It was assumed that test 1 was carried out in the plane of least principal stress and tests 2 and 3 were carried out in the plane of the maximum principal stress.

From a consideration of results published by Obert, Duvall and Merril (1960) it was concluded that the ratio of the stresses measured at the crown of the exploratory adit and in the vicinity of test 1 was consistent with the conclusion drawn from the C. S. I. R. tests that the ratio of the greatest to the least principal stress is about 4. 0. The Obert and Duvall figures also indicated that for this ratio of principal stresses the stress concentration factor which is the ratio of tangential stress to maximum principal stress would be in the range of 2. 5 to 3. 0 in the crown of the adit and 1. 0 to 1. 5 in the vicinity of test 1.

Derived values of the maximum principal stress are shown in table 4. The minimum principal stress taken as one quarter of the average of these values is in reasonable agreement with the estimated vertical pressure.

Table 4
Maximum Principal Stress from Flat Jack Tests

Test No.	Rock stress Normal to Jack N/mm^2	Jack Pressure at cancellation N/mm^2	F1 N/mm^2
1	21. 3	22. 8	17. 5
2	41. 1	43. 5	15. 5
3	41. 1	43. 5	15. 5

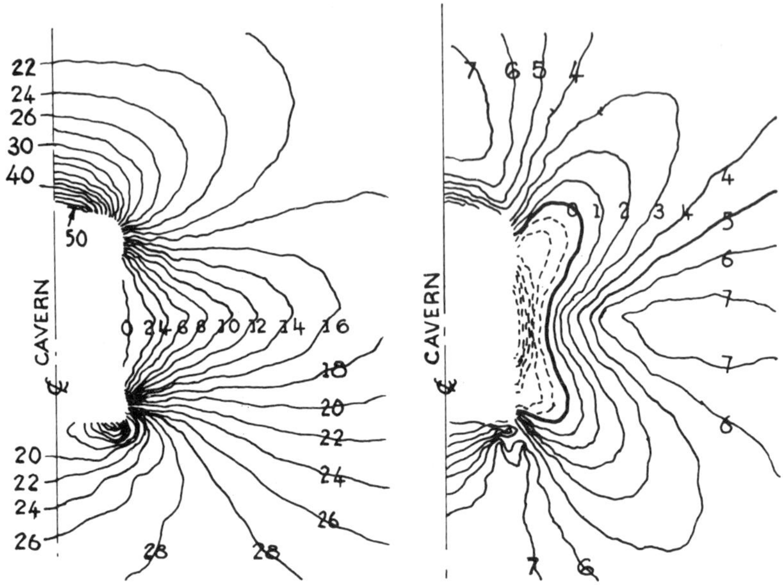

Figure 5. Maximum and minimum principal stresses

Stress distribution around excavation

The stress distribution and displacements in the rock surrounding the central cross section of the power station were calculated by finite element analysis for excavation, down to the arch springings and for complete excavation with the assumptions set out above. The minimum principal stress was assumed to be 5.5 N/mm^2 as determined from the depth of overburden.

Apart from small local areas of stress concentrations at the junction of the wall and floor the highest compressive stresses predicted from the analysis occur horizontally in the crown of the arch and the magnitudes for the two stages of excavation are as follows :-

Arch excavation	2.9 N/mm^2
Complete excavation	4.9 N/mm^2

These figures can be compared with the uniaxial strength of rock determined at 110 N/mm^2.

Mathematical analysis carried out prior to the completion of the C. S. I. R. tests indicated that tension would develop in the roof arch if the horizontal in situ stress was less than about half the vertical stress.

The full stress field around the opening is indicated on Fig.5 which shows contours of maximum and minimum principal stress. The analysis indicates a zone of tensile stress between 10 metres and 5 metres deep in the walls of the excavation at this section of the cavern and as a consequence the length of the wall rock bolts has been adjusted.

ACKNOWLEDGEMENTS

The author wishes to thank the Electricity Board for Northern Ireland for permission to publish this paper.

Consulting civil engineers for the Camlough Pumped Storage Scheme are Binnie & Partners who are associated with Kennedy & Donkin. Rock stress measurements were carried out by D. Finlayson, BSc. Analysis of the test results was carried out by D. Moy, BEng, MSc, of the Rock Mechanics Department of Imperial College London under the supervision of Professor E. Hoek, PhD, MSc.

REFERENCES

1. Alexander, L. G. (1960). Field and Laboratory Tests in Rock Mechanics. Third Australian-New Zealand Conference on Soil Mechanics and Foundation Engineering pp 161-168.

2. Hoskins, E. R. (1966). International Journal of Rock Mechanics and Mining Science. Vol. 3, pp 249-264.

3. Leeman, E. R. (1969). The 'Doorstopper' and Triaxial Rock Stress Measuring Instruments developed by the C. S. I. R. Journal of the South African Institute of Mining and Metallurgy. Vol. 69, Number 7, pp 305-339.

4. Obert, L, Duvall, W. I. and Merril, R. H. (1960). Design of Underground Openings in Competent Rock. United States Department of the Interior. Bureau of Mines. Bulletin 587.

EXPERIENCE IN USING ELECTRICAL TILTMETERS

D. E. Sherwood and B. Currey

Soil Mechanics Limited

INTRODUCTION

Electrical tiltmeters, which produce an electrical signal proportional to the angle of tilt of the device, have been developed both in the fields of metrology and aircraft instrumentation for a variety of applications. Consequently a wide range of basic instruments are available with various angular ranges and sensitivities.

In introducing these instruments to the geotechnical field, methods of relating angular measurement to, for instance shear displacement, settlement and foundation deformation have been devised. In addition, systems have been designed both for remote automatic data collection and computer controlled data acquisition. These systems have recently incorporated self checking circuits so that the problems of long term instrument performance are now reduced to the stability of instrument datum and the sensitivity of the electric spirit level itself. Manual methods of checking these aspects of performance have been developed and are currently in use.

THE ELECTRICAL TILTMETER

The electrical tiltmeter consists of two distinct parts, an electric spirit level,which is the measuring device, and a signal detector. The spirit level vial incorporates three electrodes as shown in Figure 1 and can be regarded as a variable resistance because the presence of the bubble affects the electrical resistance between the electrodes.

The vial is permanently enclosed in a mounting assembly, which provides protection and allows the spirit level to be mechanically coupled to the required measuring location. Thus, as the mounting assembly is tilted along its axis, the bubble within the vial moves and causes a change in the ratio of electrical resistance across AB and BC.

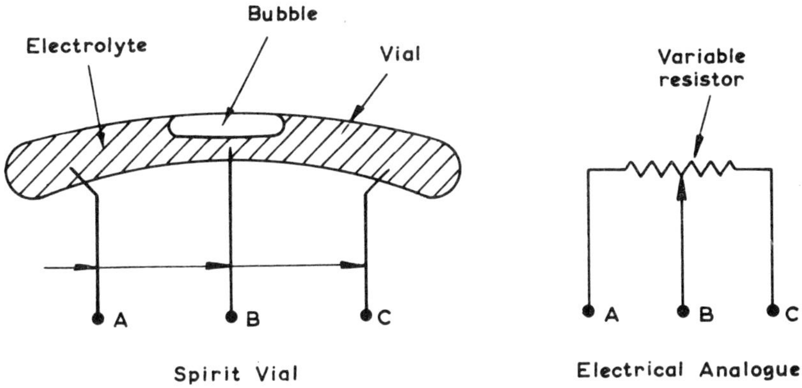

Figure 1 The electric spirit level vial and its electrical analogue

The signal detector may be regarded as two standard resistors,which together with the vial variable resistance, are connected in a Wheatstone Bridge configuration as shown in Figure 2. When a voltage is applied across the electrodes A and C, the centre electrode B takes up a potential dependent on the angle of tilt of the device. This is compared with the fixed potential at D between the standard resistors (R) by measuring the voltage across BD. This voltage difference (V) is then amplified to give a suitable output voltage proportional to the engineering units required.

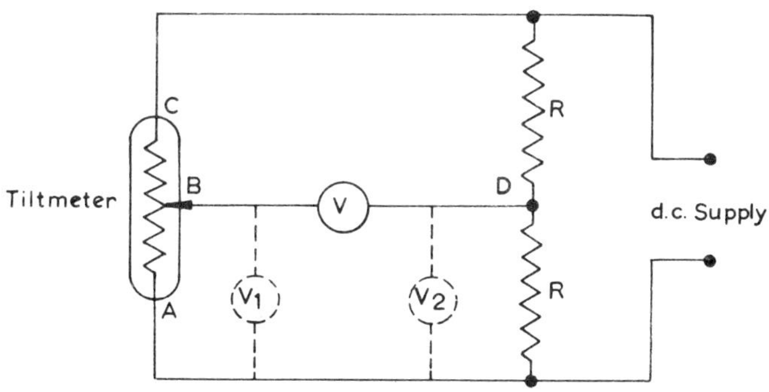

Figure 2 The Wheatstone Bridge D.C. detector circuit

If the electric spirit level is horizontal then the resistances across AB and BC are in the same ratio as the standard resistances (R), $V_1 = V_2$, and their difference, V=0. If the device is tilted, the ratio of the resistances across AB and BC is different to that of the resistances (R) and $V_1 \neq V_2$. Then V will be either positive or negative depending on the direction of tilt and its magnitude will be proportional to the angle of tilt.

The Direct Current Wheatstone Bridge circuit has been described here as its principle of operation is more readily understood. In practice, its use would cause polarization of the electrolyte within the vial resulting in instability of the electric spirit level. Consequently detectors are based on an Alternating Current bridge circuit as shown somewhat simplified in Figure 3, which operates in a similar manner.

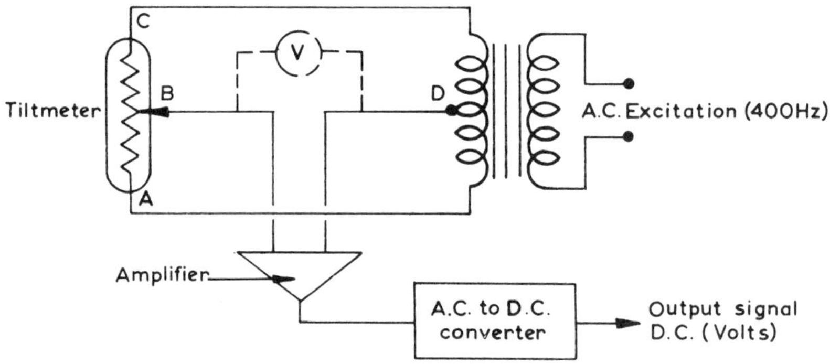

Figure 3 A simplified representation of an A.C. Bridge detector circuit

Range and Resolution

Currently available tiltmeters have a range of angular measurement from $\pm0.15^{\circ}$ in very precise instruments to $\pm30^{\circ}$ in less sensitive devices. An American instrument, initially designed for use in inertial navigational systems has biaxial sensitivity with a resolution of 0.1 arc seconds over a linear range of ±50 arc seconds. Instruments developed in the British aircraft industry have linear ranges of $\pm0.7^{\circ}$ and $\pm7^{\circ}$ with related resolutions of 2 arc seconds and 3 arc minutes. Another device has a linear range of $\pm30^{\circ}$ with a resolution of 15 arc minutes. The cost ranges from about £10 for the lowest resolution instruments to in excess of £1000 for the most precise device.

The experience reported has been largely gained in using the $\pm0.7^{\circ}$ range level, although the ±50 arc second instrument has been used in the U.S. for measurement of earth crustal tilts in earthquake research and subsidence observations in oil fields and during water abstraction.

Accuracy and Stability

Comments here are referred to the electric spirit level with a linear range of $\pm0.7^{o}$ and a resolution of 2 arc seconds.

Stability of the instrument datum is quoted by the manufacturer as 2 arc seconds per month but it has been found in practice,that as the devices age, the datum stability improves substantially. Indications are,that after 12-18 months of operation, a datum stability of about 5 arc seconds per year will be achieved. Sensitivity over a period of two years in a hostile environment has been found to vary from $\pm0.3\%$ to $\pm3.0\%$ with 60% of the instruments varying less than $\pm1.0\%$ from their initial sensitivities.

The temperature coefficient of a matched detector, electric spirit level unit is $\pm0.2\%$ of full scale deflection per ^{o}C in the temperature range $-10^{o}C$ to $+60^{o}C$. The time taken for the bubble to reach equilibrium (to within 3 arc seconds) after a 5° step change in tilt is 5 seconds, while the natural frequency of the device is approximately 1Hz. Rotation of the level about its longitudinal axis (referred to as 'cross roll') causes both a change of sensitivity and a shift of instrument datum. A $\pm5^{o}$ cross roll causes a maximum datum shift of ±50 arc seconds, while $\pm0.5^{o}$ roll produces a change of ±10 arc seconds. Although cross sensitivity of the instrument can be calibrated, a biaxial instrument is to be preferred where significant cross roll is likely to occur, particularly when the axis of rotation may not be coincident with the instrument longitudinal axis.

ERRORS IN RECORDED DATA

The data recorded from an electric tiltmeter monitoring system may be affected by four potential sources of error.

1. Changes in characteristics of the electric spirit level.
2. Changes in characteristics of the detector and recording electronics.
3. Dimensional instability of mountings and mechanical linkages.
4. Movement of physical reference datum.

The minimisation of these effects is largely a question of thoughtful design and the inclusion of checking facilities within the system. Methods of dealing with changes in characteristics of the detector and recording electronics are described later. Errors due to 3 and 4 above are largely outside the scope of this paper but aspects which should be considered include temperature, corrosion, long term stress relief, ground water level changes, mining subsidences, etc.

From our earlier consideration of a typical electric spirit

level, we can identify four individual sources of changes in characteristics on an instrument.

a) Long term datum drift.

b) Long term sensitivity changes.

c) Daily or seasonal temperature changes.

d) Datum shift and sensitivity changes due to cross roll.

It has been found that the best method of checking for datum drift is to mount the spirit level in such a way that it can be physically reversed without disturbance of the mounting assembly, as one would with a builder's level. Then the drift is calculated as half the difference between both readings, taking due account of mathematical sign changes. The sensitivity of the level can be checked by moving the instrument through a series of known angles and comparing the electrical outputs with the original calibration curve. How this facility has been built into typical installations is described later.

The problems of temperature changes and cross roll must be considered for each particular application. In some cases the effects of temperature may be such that they fall within tolerable 'noise levels', while in other applications it may be necessary to either control the temperature changes or to measure them and make appropriate corrections in analysis. Similarly the effects of cross roll may be insignificant and mechanical parts of systems are designed with this object in mind.

DATA COLLECTION SYSTEMS

The complexity of the data collection system depends on the level of monitoring required and the environment of the site. In its simplest form it can be a small portable detector, while at the other end of the scale, a computer any distance from the site can with appropriate programming direct system checks, collect data at various time intervals, and analyse the data as acquired. Figure 4 shows various systems with which we have recently been involved.

<u>Portable detector</u> This is simply a hand carried unit of about 15cm x 15cm x 10cm with an internal battery supply, into which is plugged the tiltmeter output lead. The data is recorded visually from a meter in the unit after a warming up period of up to 10 seconds. The detector is quite cheap and is suitable where observations are required at infrequent intervals and access is convenient. The arrangement is shown diagrammatically in Figure 4 (a). As a portable detector is designed and constructed to match a

particular type of electric spirit level, it is essential to ensure that a matching unit is used throughout monitoring. This is best achieved by using incompatible plugs and sockets on instruments of different types.

Automatic remote data collection The system described in Figure 4 (b) was designed primarily for operation within an electrically noisy environment such as a Power Station. Each tiltmeter is mounted adjacent to a signal detector and the output voltage of this unit is converted to a frequency signal by the voltage to frequency (V/F) converter. The data in this form is relatively unaffected by the noise of the environment and is suitable for transmission over long lines as attenuation of the signal is now unimportant. Automatic monitoring of the installation is effected by a pre-programmed data logger and data is recorded on punched paper tape in a form compatible with computer facilities for analysis. Some flexibility can be built into the data logger for variation of scan speeds, scan duration, and intervals between scans. As the instruments are continually energised, they can be scanned at a rate of about 1 per second, although this rate is largely dependent on the speed at which the data logger operates rather than the tiltmeters.

The major disadvantages of this system for more general use are its relatively high cost due to a signal detector and V/F converter being required for each channel, and automatic check systems cannot be built in economically. However, two pairs of standard resistors, which simulate zero and full scale output of the electric spirit level, can be incorporated in each detector unit. These can be manually switched into circuit to replace the level and thus the drift of all the electronics in each channel with the exception of the level can be checked when required.

Automatic Remote Data Collection and Control by Computer In less noisy environments several refinements of the system just described can be made and these are shown diagrammatically in Figure 4 (c). All the electronics, with the exception of the electric spirit level of course, can be conveniently placed at a central monitoring point. The system could be operated by a data logger but, in the case described here, a more sophisticated logic ability was required and a remote computer was connected by land line to control the installation.

The signal from the spirit level is transmitted to the central unit via cables of considerable length, where a relay scanner unit is located. Upon direction from the computer, each level is in turn switched to the signal detector, and its voltage output is fed to an analogue to

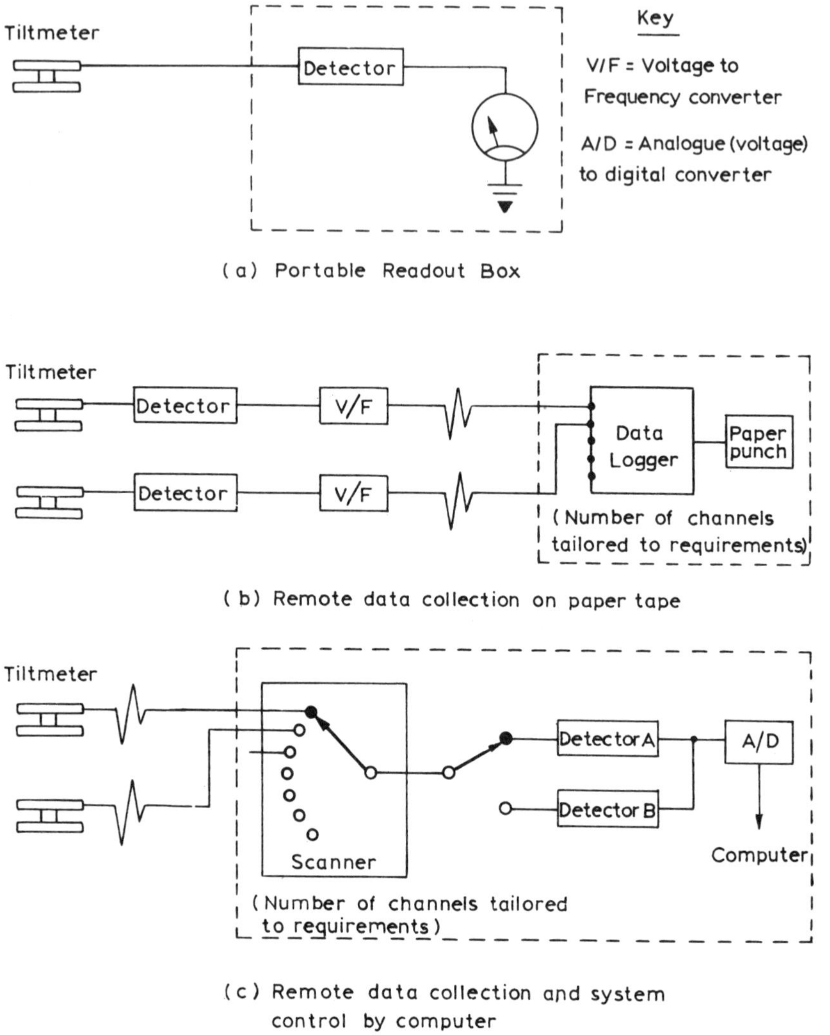

Figure 4 Various data collection systems for electric tiltmeter installations

digital converter to prepare the data for receipt by the computer. Two detectors are used so that, should a fault develop, half the system can continue to operate. The introduction of the scanner has the advantage that automatic system checks can be carried out with little additional hardware. These are achieved in this case by adding two pairs of standard resistors (which again simulate the full scale and zero output of the electric spirit level) at the central monitoring point. At the commencement of each scan for data, simulated outputs are switched to each detector in turn and consequently drift of all the electronics with the exception of the level is observed and corrected at the time of each data recording. The computer can check the trends of the received data and shorten the time between successive scan periods if changes of trends appear to be occurring. Tiltmeters can be scanned at the rate of up to 2 per second, the limitation being the time taken for stability to be achieved after relay switching.

CONVERSION FROM ANGULAR TO LINEAR MEASUREMENTS

For some applications, it is necessary to relate angular measurement to linear displacement by mechanical devices for say the measurement of settlement or shear displacement. These can take various forms but two typical patented installations are described below.

Settlement In measuring settlement to a high degree of precision, it is most important to have a stable datum to which measurements can be related. The system which has been developed utilised a fairly conventional deep benchmark and is shown in Figure 5. As the foundation settles, the reference beam AB on which an electric spirit level is mounted tilts and from the geometry of the simply supported beam AB an appropriate calibration for settlement of B relative to A can be computed. The 'T' shape of the reference beam is purely to achieve stability and the dimension BB is chosen with the effects of cross roll in mind. The resolution of the combined system is dictated by the choice of instrument and the length of beam AB.

A $\pm 0.7^{o}$ linear range instrument has been used in this work and on the assumption that a long term reproducibility of 5 arc seconds can be achieved, the range and resolution of the system has been computed in Table 1.

Beam Length (m)	Resolution (mm)	Range (mm)
1.0	0.025	25
1.5	0.037	37
2.0	0.050	50

Table 1 Settlement installation resolution and range

The precision which can be achieved together with the

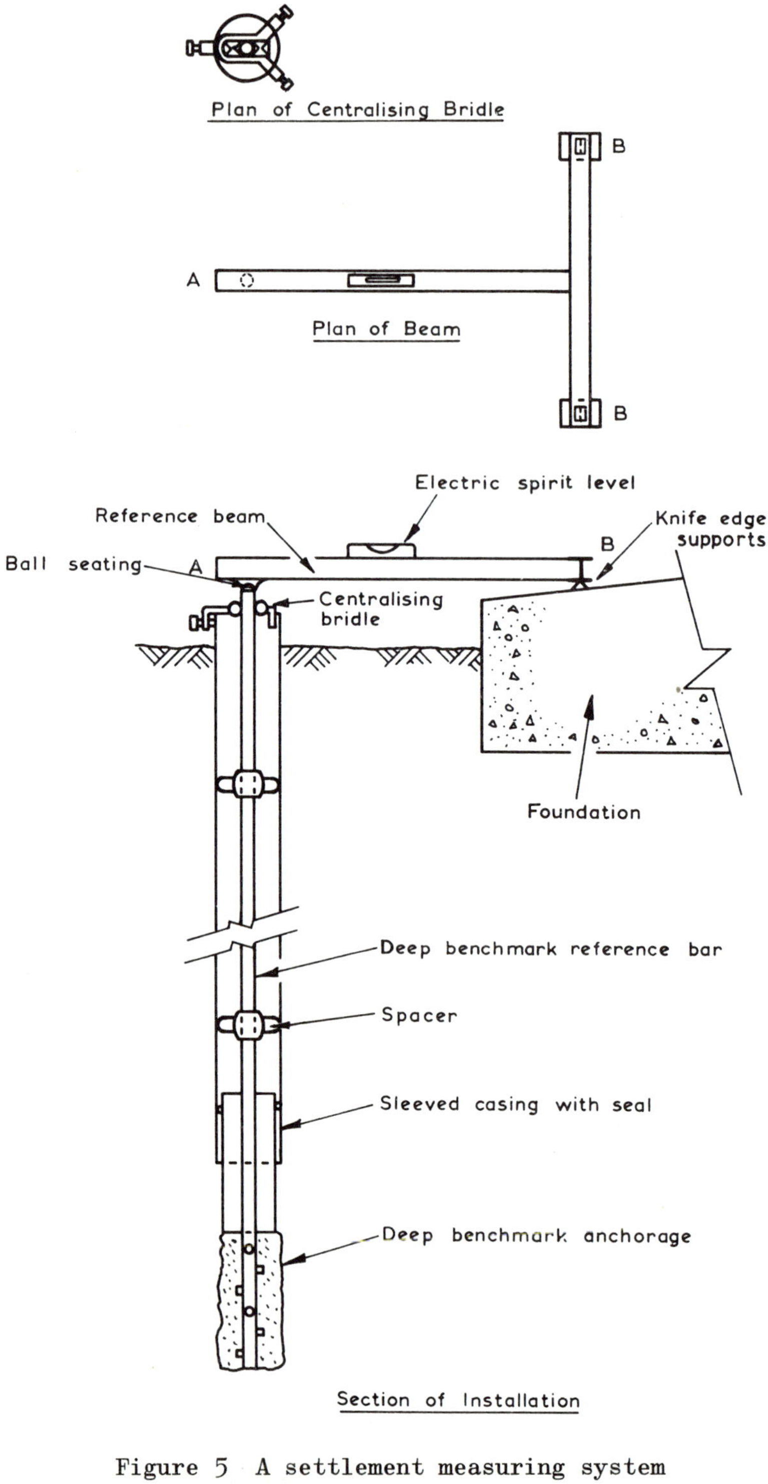

Figure 5 A settlement measuring system

system's ready adaption to automatic recording makes this arrangement excellent for monitoring foundation deformation under transient loading such as may result from traffic or wind. If a high precision system is required for relatively long term monitoring, it is necessary to control the immediate environmental temperature of the reference bar and beam. Datum drift checks can be achieved by mounting the level in such a way that it can be reversed. Sensitivity checks can be carried out if a micrometer device is incorporated in the ball seating at 'A' so that the reference beam can be given various artificial tilts.

Shear displacement A system designed to measure shear displacements within potentially unstable soil or rock is shown in Figure 6. Here a cantilever rod is anchored below the level of the slip surface and is constrained to bend with the movement of a surface reference casing. Thus the rod is subjected to elastic bending as a cantilever with a point load and from bending theory it can be shown that

$$\delta = \tfrac{2}{3}.\theta L \quad \ldots\ldots\ldots\ldots\ldots\ldots \quad (i)$$

when δ = horizontal displacement at the load point.
θ = angular deflection at the load point.
L = length of cantilever from point of fixity to the load point.

It can be seen that the resolution and range of this type of shear detector is dependent on the tiltmeter chosen and on the cantilver rod length 'L', although in some cases the range may be limited by borehole diameter. The effects of cross roll are minimised by aligning the longitudinal axis of the level with the expected direction of movement.

A similar electric spirit level to that described for settlement has been used for this work and again assuming a long term reproducibility of 5 arc seconds, the range and resolution of the system has been computed in Table 2.

Rod Length 'L' (m)	Resolution (mm)	Range (mm)	Minimum diameter of borehole (mm)
3.0	0.05	25	75
5.0	0.08	41	100
8.0	0.13	66	125
10.0	0.16	82	150
15.0	0.24	123	175
20.0	0.32	164	225

Table 2 Shear detector resolution and range

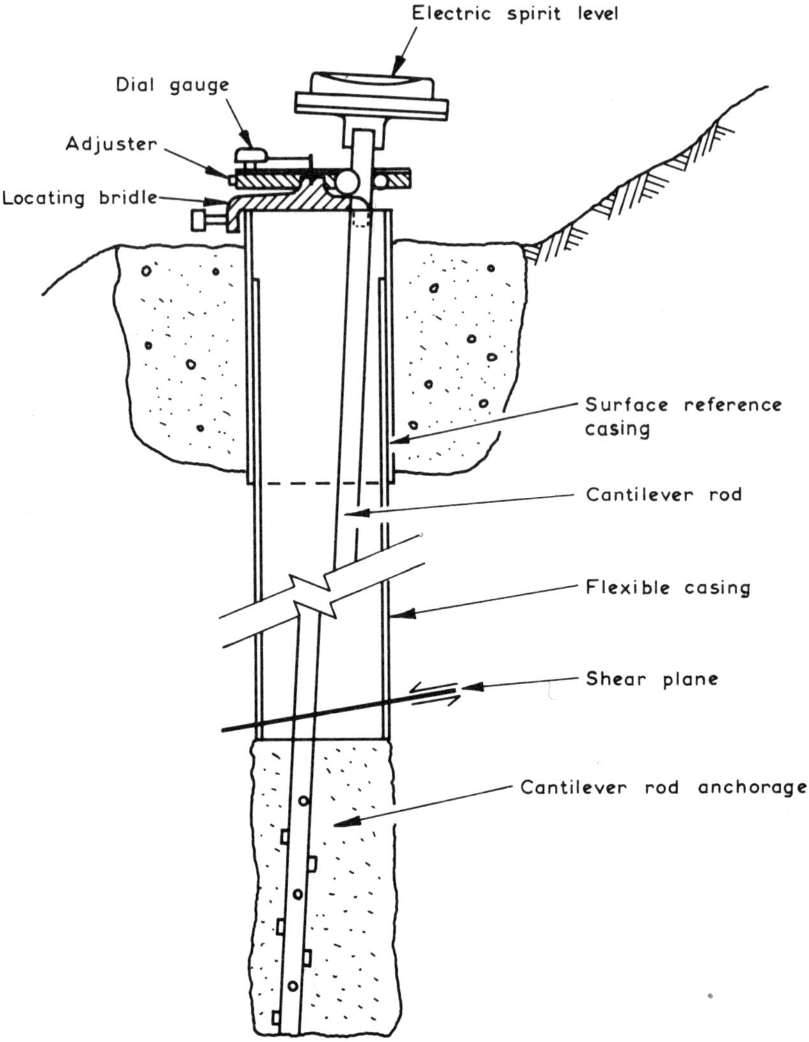

Figure 6 An electric tiltmeter shear detector

Where an automatic recording system (as described earlier) is to be used for several shear detectors, it is convenient to condition the electric spirit level signals so that an equal horizontal displacement at each borehole would produce an equal analogue signal to the data recording system. Consequently the precision of the complete installation is controlled by the least sensitive unit (i.e. that with the longest cantilever rod). The minimum rod length is chosen with the minimum acceptable range of movement in mind so that the linear range of the instrument is not exceeded. The precision which can be achieved, however, allows us to instrument slopes to detect very small movement or changes of trends in movement, and this information may be most useful in understanding the onset of movement prior to failure.

Datum drift of the electric spirit level can be checked by mounting the level in such a way that it can be reversed. Sensitivity checks can be carried out by moving the point of restraint at the surface reference casing against the dial gauge as illustrated in Figure 6.

INSTRUMENTATION INSTALLATIONS USING A TILTMETER

Near Norden Bridge, Great Harwood, three trunk gas mains were laid in adjacent trenches, across an area of slipped soil. Three shear detectors were installed to observe the occurrence of any movements which might fracture the mains. Data collection is by a portable readout unit. Readings commenced in June 1970 and are continuing successfully.

The foundation block of Penn Hill television mast was found to have tilted slightly in 1970, introducing a substantial moment in the mast. After remedial jacking, two tiltmeters (with portable readout unit) were installed at right angles to each other on the foundation block. As the movements being observed were very small, errors due to cross roll of the instruments were insignificant. Readings commenced in May 1971.

Sixteen tiltmeters and two settlement systems were installed in 1970 on the pond wall of a large cooling tower to collect data on the deformation and settlement of the foundations under wind loading. The locations of the instruments are shown in Figure 7. For analysis purposes, the pond wall was regarded as a series of hinged members, the hinges coinciding with joints in the wall. Consequently, it is possible to compute the settlement at any point of interest and the angular distortion between adjacent blocks around the pond wall. Temperature effects on the settlement gauges were minimised by using invar for the upper 3 metres of the reference bar and backfilling the borehole with soft grout. In addition the reference beam and electric spirit level were protected from wind and direct sunlight.

Data was recorded on a system similar to that shown in Figure 4 (b). Difficulties were experienced in commissioning and operating the voltage to frequency converters, although the operational problems were largely due to the very hostile environment in the vicinity of the tower. The installation was finally commissioned in February 1971 and removed after successful monitoring in May 1972.

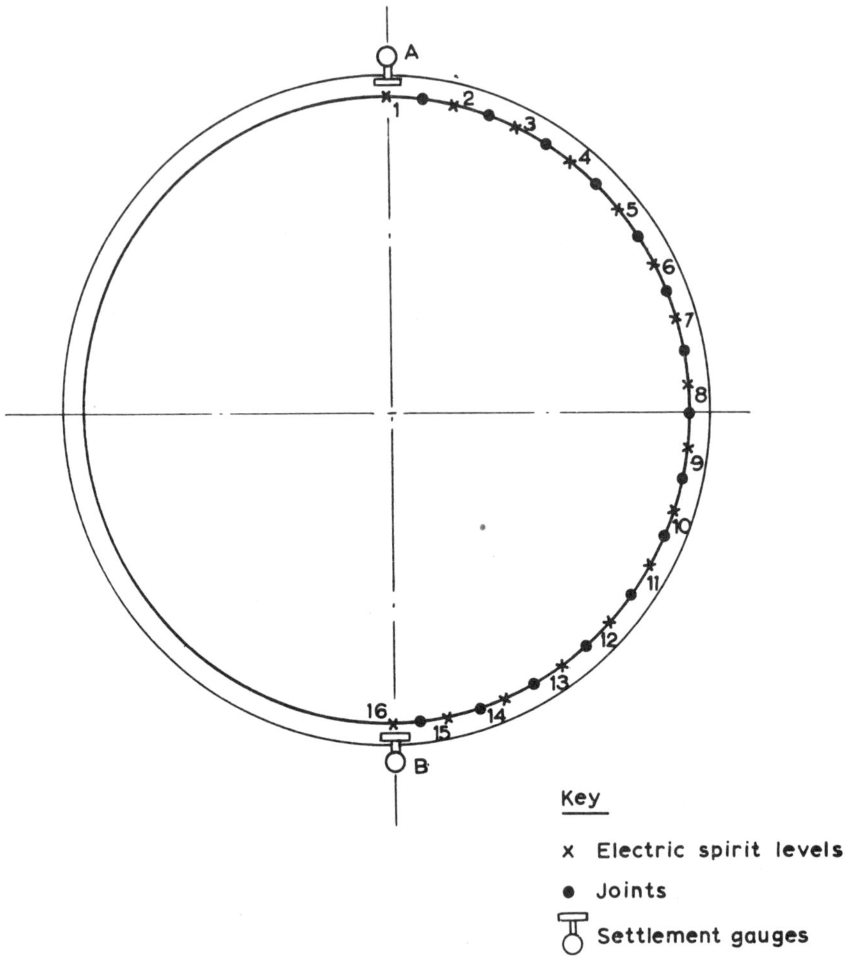

Figure 7 An installation on a Cooling Tower pond wall consisting of tiltmeters and settlement gauges

We have used electrical tiltmeters on several occasions as horizontal accelerometers. They are particularly attractive for some applications because of their low frequency response and have been used with a conventional ultra-violet galvanometer recorder to detect 'ground roll' due to the explosive demolition of large structures. As the natural frequency of tall buildings is normally well below 2Hz, the instrument has also been employed to investigate sways stimulated by wind loading.

Finally a system of eighteen shear detectors is currently being commissioned on a rockslope in South Wales. The area has undergone substantial movement, both in ancient and recent time, and the purpose of the installation is to provide early warning of any changes which might suggest the onset of further instability. Data collection and system control is carried out by a digital computer via a land line link as described in Figure 4 (c). The tiltmeters are interrogated by the computer at 90 minute intervals and the data is compared with preset alarm conditions to provide information on both proper functioning of the equipment and also warning of possible movement. The alarm levels for movement are continually updated in view of the previous history and are designed to detect small changes in trends. If serious changes take place, appropriate warning is given and the interval between interrogations is reduced to 5 minutes so that sufficient check data is quickly collected.

CONCLUSIONS

1. The electric tiltmeter is a high precision instrument which can be used to precisely measure settlement, shear movement, foundation rotation or distortion, low frequency ground movements or structural sway.

2. Being a completely electrical instrument, a suitable remote automatic data collection system can be developed to meet most requirements. This is particularly important when short duration incidents (such as wind loading) are being studied, or when frequent monitoring is required over long periods of time.

3. Data can be collected in a form compatible with available computer facilities so that large quantities of data can be recorded and analysed to very refined levels with little difficulty.

4. Tiltmeter monitoring systems can be readily adapted to computer control if appropriate computer programmes are written.

5. The problems of long term electronic drift of detection systems can largely be overcome by building in 'self-checking' logic circuits. The datum drift and sensitivity stability of the tiltmeter itself can be checked on site if appropriate mechanical arrangements are included in the design of the instrument mountings.

ACKNOWLEDGEMENTS

This paper is presented with the permission of the Directors of Soil Mechanics Limited.

THE MEASUREMENT AND INTERPRETATION OF GROUND VIBRATION

B. O. Skipp and M. Marriott

Soil Mechanics Limited

INTRODUCTION

The measurement and interpretation of ground vibration is becoming a more important aspect of environmental studies. In addition to vibration arising as a side effect from some human or natural activity, there has been an increase in the use of vibration for pile driving, casing extraction, compaction and breaking. There are of course other uses of vibration, particularly in geophysics, where the transmission time of vibrations through the ground is measured.

Geophysicists and seismologists have developed a range of instrumentation ranging from the simplest displacement gauges or accelerometers to most sophisticated arrays of detectors. The response of structures to vibrations has been extensively studied, particularly by aerospace engineers, and in recent years a number of approaches which had their origins in statistical physics have been applied in earthquake engineering. As yet the techniques familiar to the aerospace engineer, seismologist and earthquake engineer have made little headway with those vibration problems affecting the geotechnical engineer.

This paper will consider some of the tendencies in ground vibration measurement and interpretation which are inspired by developments outside the compass of conventional geotechnics. It will not be possible to deal in depth with those concepts and techniques for which there is already an extensive literature; the emphasis will be upon procedures available and some examples will be given of recent ground vibration measurements. The frequency content and magnitude aspects of vibration are considered rather than wave propogation, although wave propogation may on occasion have to be taken into account. A further important aspect is the relationship between ground motion and distance from source, and this will be dealt with by reference only to experimental data.

SOURCES OF GROUND VIBRATION

Some compilations of ground vibrations from various sources have been made quite recently but a comprehensive classification with proper descriptions has yet to be attempted. Ground vibration may be natural or man made, although some industrial/human sources are difficult to distinguish from natural events.

Natural periodic movements include such geophysical phenomena as earth tides (semi-diurnal) and microseisms. These are not of significance except to delicate installations such as particle accelerators. Earthquakes are common occurences in many parts of the globe and they may be classified as (a) single shock with short periods prevailing (0.2 secs), (b) long irregular motions with periods of 0.05 to 6 secs and equal distributions of power throughout the frequency range, (c) long ground motion with clear periodicity. Accelerations in excess of that due to gravity have been recorded. There are also low level natural sources of ground noise which one may usually regard as 'background' arising from such diverse origins as barometric pressure surges, wind or running water. Natural vibrations can be approximated to either periodic or random processes.

Man made vibrations can be classified as (a) periodic vibrations from fixed sources, machinery-compressors, engines, etc., (b) aperiodic vibrations from fixed sources such as piling hammers, forging hammers, explosions, etc., (c) combination of aperiodic and periodic vibrations from moving sources etc. such as road traffic, railways, (d) a ground coupling of air noise such as sonic boom.

In practice, because the transmission of vibrations throughout upper layers of the ground is complex, different wave types separate out with distance from a source, so that the ground motion may appear as a single period sinusoidal motion, even though near the source ground motion appears more random in character.

Whiffen (1972) has described traffic vibration and Grover (1971) has given power spectra from ground vibrations from industrial traffic. Air noise may be coupled to the ground giving a significant vibration near aircraft for example; up to a particle velocity of 0.01 in/sec in a vertical ground motion in the range 5-15 Hz.

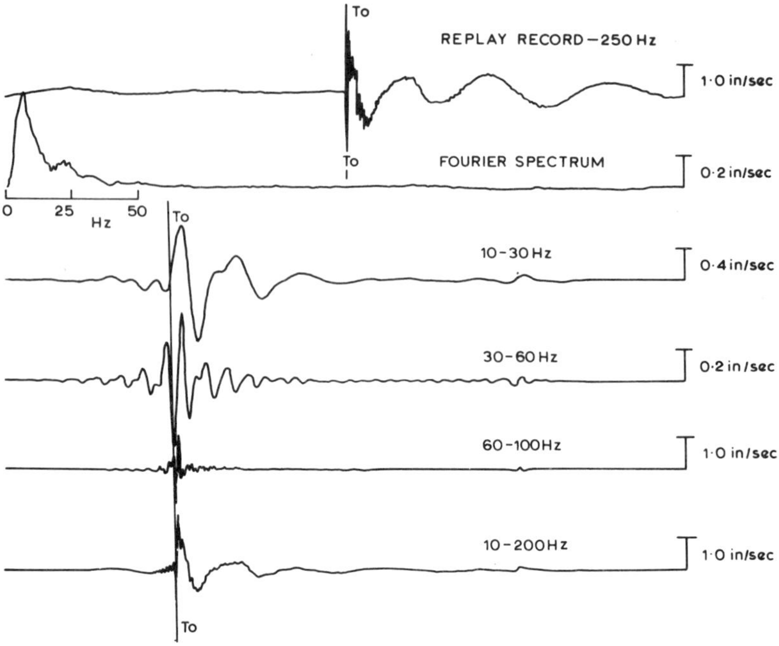

Figure 1 Example of ground vibration tape playback. Explosive source 7 m distant from charge of 10 lb gelignite.

DESCRIPTION OF GROUND VIBRATION

A vibration process requires certain minimum information to fully describe it, depending on whether it is (a) periodic harmonic, (b) periodic complex, (c) stationary random, (d) shock transient, or (e) non-stationary random.

Periodic harmonic Periodic vibrations with single period are fully described by a period (T) or frequency (f) and a magnitude which may be the amplitude of displacement (u), velocity ($\dot{u}$), or acceleration ($\ddot{u}$). A description based upon the peak values of these quantities relates only to instantaneous conditions. Time history for one cycle is included in the description if one uses the absolute average amplitude, $u_A = 1/T \int_0^T |u| dt$, or the Root Mean Square (RMS) amplitude, $u_{RM} = \sqrt{1/T \int_0^T u^2 . t dt}$

Periodic complex Periodic complex vibrations consist of a number of discrete sinusoidal vibrations. They may be separated and a spectrum drawn up in which the different frequencies appear as discrete lines. This kind of vibration is commonly encountered especially from reciprocating machinery.

Stationary random vibrations Many vibrations encountered can be regarded as having random characteristics, that is individual particle motions are never exactly repeated even if one measures over an infinite duration. To describe such vibrations we must use probabilistic terms. The most commonly used is the 'probability density':

$$Pu = \lim_{\Delta u \to 0} \frac{P(u) - P(u + \Delta u)}{\Delta u}$$

Here p(u) is the probability density, Pu the probability that an instantaneous amplitude exceeds u, and $P(u + \Delta u)$ is the probability of occurrence of instantaneous amplitude values exceeding $u + \Delta u$. All values of p(u) for all values of u when plotted give a 'probability density curve' which, when integrated between two values of u_2 and u_1, gives the probability of occurrence of instantaneous amplitude values between interval $u_2 - u_1$. Probability density describes an 'instantaneous' condition. In order to bring time history and frequency content into the description, the 'auto-correlation' function is used (see Robson 1957 for definitions). $\Psi_r = \lim_{T \to \infty} \frac{1}{T} \int_0^T f(t + r)\, dt$

From this function may be derived the 'mean spectral density' function, $W(f)$ when $\lim_{T \to \infty} \frac{1}{T} \int_0^T f^2(t)\, dt = \int_0^\infty W(f)\, df$

The left hand side measures power in the process in time terms and the right hand side in frequency terms. This equation may be presented in the form $W(f) = \lim_{B \to \infty} \lim_{T \to 0} \frac{1}{BT} \int_0^T u^2{}_B(f,t)\, dt$

describing that part of average power of the signal function u(t) contained in an infinitely narrow frequency band B with centre frequency f, averaged over an infinitely long time. In practice B and T are finite but their choice will not be dealt with here. It is this function which is usually obtained by direct spectral analysis from a vibration record.

Shock and transient vibration Shock is the transfer of kinetic energy to a system in a time short compared with the natural period of that system. Shock can be described by a Fourier form $F(f) = \int_{-\infty}^{+\infty} F(t) e^{-j\omega t}\, dt$ Although such Fourier integrals are necessary for a full description, the response of a single degree of freedome system is usually used in assessing the effects of a particular shock and in describing 'strong motion' earthquakes.

For transient phenomena, the duration of the input may be several times the natural frequency of the system excited. There may be a single period or an oscillation.

Non-stationary random vibration All stationary random vibrations in practice are non-stationary because their statistical properties change with time. The theoretical analysis of non-stationary random vibrations is difficult and requires vast amounts of data. Some kind of time averaging is usually used and this means that measuring equipment must have response and averaging times short compared with trends in the non-stationary random data.

MEASURING SYSTEMS

The different types of vibration impose specific requirements on measuring systems which, in general, consist of a sensor, recorder and analyser. The analyser may or may not be 'on site'. Vibrations are either regular or predictable, or unpredictable in time. In this paper we will be concerned primarily with events of the first category, although some instruments used could serve both categories. It is not intended to describe such instruments in detail as they are well covered in the literature (Halverson 1971).

Periodic harmonic ground vibration, such as might arise from some continuously running rotating but ill balanced machine, can be adequately measured by a simple transducer with a linear response to the vibration, either in terms of displacement, particle velocity or acceleration. Display may be as a meter reading. There are various sensors and their principles have been described. Such an instrument might be based upon a geophone or piezo-electric crystal accelerometer. It is sufficient for the sensor to have a natural frequency remote from that of the excitation and be calibrated to a harmonic source. The results from measurements made with a Dawe Vibration Meter near a Muller vibrator when extracting bored pile casing are shown in Fig. 2.

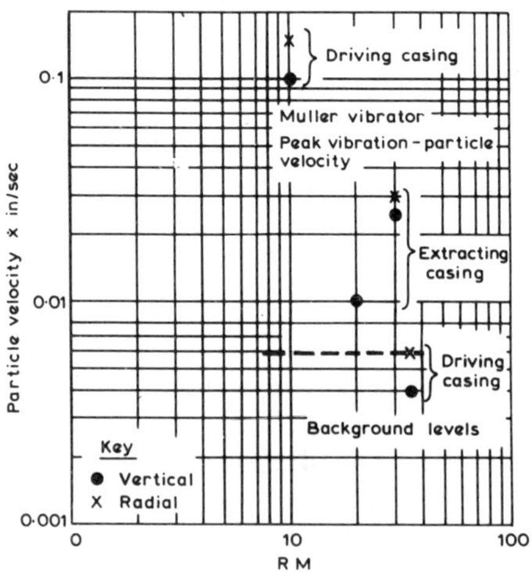

Figure 2 Peak particle velocities near Muller pile vibrator extracting 20 ft casing from London Clay. Dawe Meter Readings.

In order to cope with complex motion, it is necessary to have a trace. The example in Fig. 3 shows the near harmonic ground motion at varying distances from an 8 ton vibrating roller on gravel soil, recorded on an A.B.E.M. 'Trio' unit.

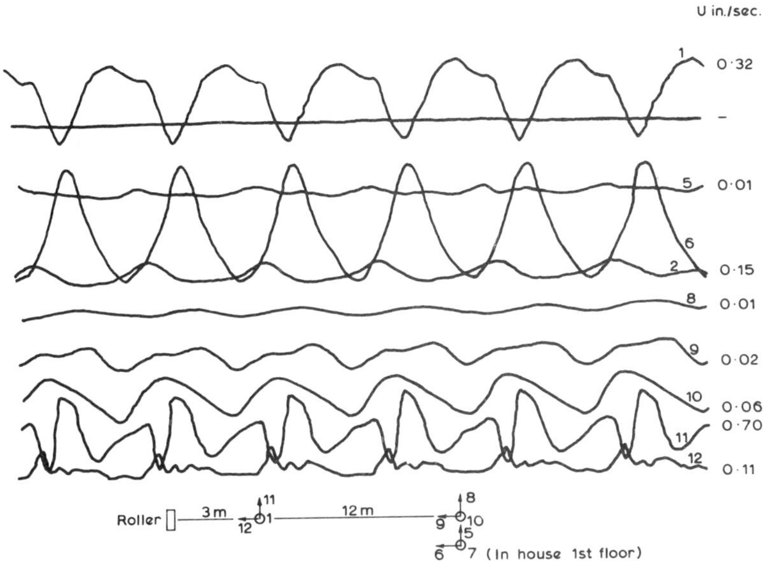

Figure 3 Particle velocities near a vibrating roller (A.B.E.M. 'Trio')

The proper measurement of random motion and shock requires a response to frequencies of from zero to infinity and the recording of time history of motion. This is not possible and therefore most measuring systems can be found wanting. Perhaps the nearest to ideal are the piezo accelerometers of very wide frequency range with their associated charge amplifiers. A useful device is an accelerometer with a built in amplifier, which gives a D.C. signal and is linear over a range useful in most geotechnical applications.

There is much in favour of the conventional geophone, especially when used with heavy overdamping and galvanometers (Willmore 1961). Geophones are cheap, robust, waterproof, readily available, give high output, do not require energising from separate leads and are available on 'shelf' in fully protected three component response assemblages. Specifications were written for a blast recorder some years ago which have now been fulfilled by systems using geophones. For measurement of explosive and impact generated ground vibration, a damped geophone and galvanometer system with F.M. tape recording has been used recently and is linear from 3–200 Hz.(Figure 4). Calibration is by measuring displacement optically, mechanically or against a sub-standard accelerometer. Usually they are calibrated with harmonic vibrations over a frequency range. Rarely is the response to transients allowed for, although it is known to have effects, particularly at lower frequencies in some systems. Absolute calibration at high particle velocities and low frequencies is difficult and requires large electro-dynamic or hydraulic vibrators.

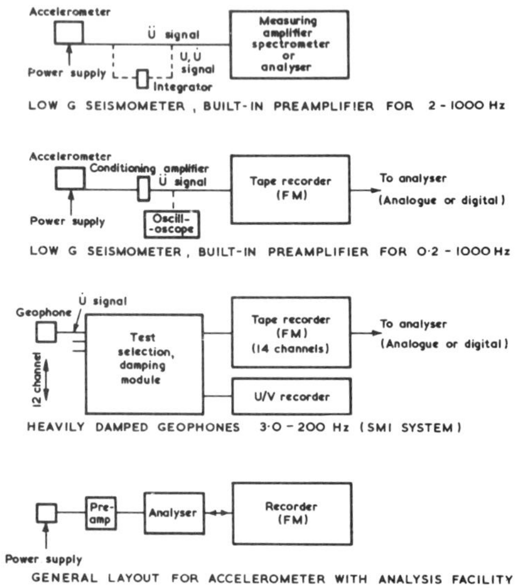

Figure 4 Block diagram showing vibration measuring systems

The Ultra Violet galvanometer recorder (UV), with up to 12 channels, has achieved widespread acceptance. A number of purpose built units with fewer channels are on the market (Table 1). A wide variety of tape recorders (F.M.) are available. Some are battery operated; others require a field generator. They are expensive, ranging from £1000 to £7000 for 4 to 14 channels, but their increasing use is enabling spectral analysis techniques to be more widely adopted. For longer term vibration monitoring, there are various self actuating strong motion seismographs (Halverson 1971), telemetry continuous recording systems and event recorders. Such monitoring is, as yet, not common in U.K. experience within the civil engineering field.

The object of all these transducers is to give a true record of the motion of a particle of the ground. Ideally a unit should have the same stiffness and density as the ground but this is not possible. Some instruments when mounted on soils have natural frequencies of their own. Typically a 25 lb weight unit with base radius 6.0 in. would have a vertical natural frequency on sand of about 150 Hz. For faithful response to intense shocks the 'stiff' inclusion (for that is what a geophone or accelerometer usually is) will not be adequate and soft pressure sensitive devices may be required. In Table 1 currently available systems are listed together with their basic capabilities. No single system excepting piezo-accelerometers covers the full ranged desired, although damped geophone/galvanometer systems approach it.

SPECTRUM ANALYSIS

There are now available a variety of spectrum analysers, the best known of which are analogue narrow band analysers with either fixed or constant percentage band width. If used with magnetic tape records on which the signals are proportional to acceleration, the outcome is an amplitude of acceleration spectrum. Presentation as power spectral density requires that the squared R.M.S. value measured with a narrow band filter be divided by the band width (Broch 1972). Signal analysis by digital procedures may be undertaken on special computer assemblages (Hewlett Packard Fourier Analyser) and also in real time (Spectral Dynamic Analyser). The input is usually in magnetic tape form. There are also a number of general computer programmes available for Fourier transforms and subsequent construction of the power spectra (Spect. Cambridge University). If working with response spectra, desk procedures starting from analogue records (UV paper or film) may be used. Usually the trace is read by a digitiser and then the data is handled by a computer.

INTERPRETATION OF DATA

Having obtained ground vibration data either as a meter reading, a UV or magnetic tape record, it is necessary first to manipulate it into a form which is relevant to the problem in hand.

The ultimate objectives of a ground vibration study could be:

1) Assess risk of structural damage to services or assemble evidence that vibration is a cause of structural damage.

2) To establish that ground vibration is likely or not likely to cause malfunction in some component of an industrial activity.

3) To establish ground vibration characteristics in order to predict the response of some structure yet to be built or establish parameters to enable isolation to be put into effect.

4) To relate some excessive but not necessarily damaging structural vibration to a specific source via ground transmission.

5) To assess the conditions for human well being.

Firstly let us consider the assessment of damage risk. At present empirical criteria based upon full scale tests with explosive generated vibrations require only a measure of the peak particle velocity. There are corresponding criteria using other parameters (Steffans 1966). For a continuous motion one may either use these criteria, which are far from being generally accepted for sustained ground motion, or make some attempt to determine structural response, usually as a maximum response. It has been suggested that stress levels in floors be used as a limiting criterion. This would require response analysis at design stage or experiments after construction. In principle, calculation of the linear elastic response of a simple structure to a simple periodic motion presents no difficulty, but often it has a great density of modes. Similarly the response to complex periodic ground motion requires that the components be identified and the response combined using superposition. Continuous records of ground vibration on magnetic tape is preferred, when such analyses are in view. It is possible to establish the response of linear elastic structures to a stationary random vibration. The methods are well described in the literature (Robson 1957). It requires a determination of the mean square spectral density of the excitation and the complex frequency response function. Even with a random element, there may for example have developed a 'pseudo harmonic' surface wave of several cycles duration. If the apparent period is long compared with the period of the structure under consideration, the response may be estimated as for sustained harmonic input (Fig. 1).

Even with the armoury of techniques from the theories of random vibrations, accounting properly for the structure/ground interaction may present difficulties. Estimating the response of structures to shock stimulation may be simplified by the concept of shock response spectra (Clough 1962). These are obtained by applying the shock pulse to a linear undamped single degree offreedom system and plotting the maximum response of the system as a function of the system's natural frequency. The application of spectra of this type is familiar to earthquake engineers both in the design of shock isolators and also as a description of ground motion obtained from an accelerograph by simplifying it to a set of arbitrary motions in a time domain (Newmark, Rosenbluth 1971) This approach is often preferred to deriving the Fourier spectrum, although the two may be similar, with the response spectrum being close to an upper bound.

The interpretation of response of a buried service (gas main, cable, sewer) presents special problems. For gas mains a simple elastic analysis requiring knowledge of transmission velocities has been suggested (McClure 1964). Criteria for sustained vibration would need to take fatigue into account and assume constant contiguity of a service and ground. We do know (Skipp et al 1972) that buried gas mains sustain particle velocities well above levels which would cause distress to surface structures and recent measurements with sensors on buried mains have shown marked reduction when compared with surface measurements. Theoretical studies assuming elasticity and soil interaction models have been undertaken quite recently.

Where the objective is to avoid malfunction of equipment in service, it is often not possible to establish critical vibration magnitudes and a 'comparative' approach may be adopted in which the maximum ambient level is established due to normal (or perhaps slightly abnormal) operation and this is taken as a caution limit.
Identification of frequencies can be of considerable qualitative value. Low frequency content is often under-estimated by simple inspection. Fourier analysis is preferred

Where thereis an intention to undertake passive isolation of a structure (or machine), it is desirable to describe the ground vibrations in such a way that rational isolation design can be carried out. There may also be cases where isolation of a unit is to be undertaken prior to the introduction of a vibration source (e.g. laboratory adjacent to a road yet to be built). In the former case, identification of discrete frequencies to be avoided in the natural period of the isolated element is a first step. Where the vibration is not periodic, a power spectrum will identify bands likely to cause trouble and provide, if needed, the starting point for a more sophisticated response analysis.

The responses of human beings to vibration is currently of great interest, although data is still sparse. Vibration exposure curves have been proposed (Broch 1972). They are given in terms of R.M.S. acceleration levels and frequency, producing equal fatigue-decreased proficiency. Elastic response analysis of the human skeleton has also been attempted.

SOME EXAMPLES OF VIBRATION MEASUREMENT

Four vibration studies are referred to in which diverse instrumentation was used to cover the frequency and magnitude range expected. In all the examples the relation between magnitude and distance was of interest. The cooling tower studies were interpreted in peak particle velocity terms with reference to empirical 'caution' levels. For the explosive trials a rough assessment of displacement was made within a given rather broad band of frequency.

Demolition of cooling towers Measurements of ground vibration set up on the impact of cooling towers demolished by explosively removing supporting legs have been described by Grover (1970), Skipp et al (1972) and Skipp (1973). In Fig. 5 the responsesof heavily overdamped geophones at a distance of 100 m and 150 m from impact point are given together with a relative spectral density curves for signals recorded at 100 m and another at 15 m from impact (Fig. 6). The foundations of that tower were on a ground beam on dense gravels overlying Coal Measure Sandstone. Records were taken on a Thermonic FM recorder. One notes the reduced high frequency at distance and the dominant low frequencies.

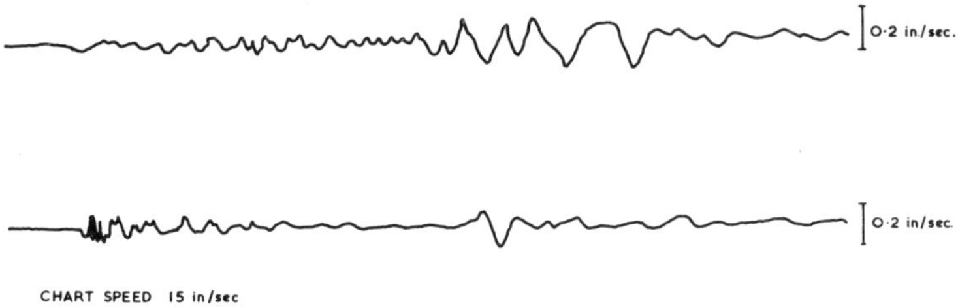

Figure 5 Geophone response, Thornhill

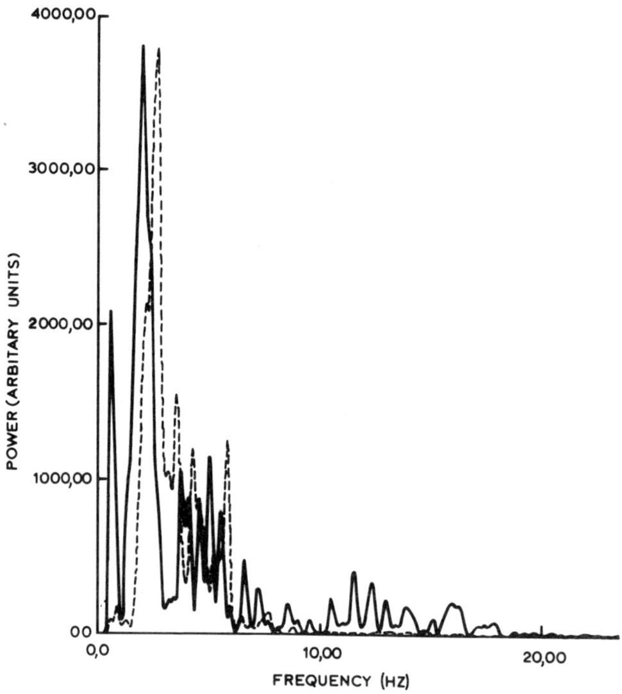

Figure 6 Relative power spectral densities, Thornhill

At two other sites, Cardiff (Glamorgan) and Ince (Cheshire), ground response was measured using heavily overdamped geophones and two Strong Motion Seismographs. Examples are given of a normalised response spectra from the Strong Motion Seismograph (Fig. 7). In Fig. 8 the peak particle velocities from both systems of measurement for a total of five towers is shown. The Ince cooling towers were founded on piles driven through 10 m of soft deposits into Triassic Sandstone. The Cardiff cooling towers were founded on piles driven 10 m through soft clays and loose to medium gravels to Keuper Marl.

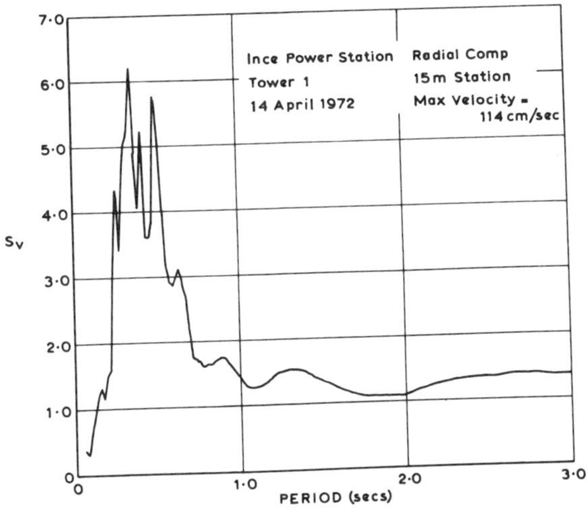

Figure 7 Normalised velocity response spectrum, Ince

The highest particle velocities were associated with frequencies of 2–10Hz from the seismograph. Geophones gave lower peak particle velocities, but it is evident that significant energy is at frequencies too low for geophone response. The evaluation of risk in all cases was based upon empirical levels and upon a prior study in the case of Cardiff of traffic generated levels (up to 0.1 in/sec; 2.5 mm/sec). Buried public services at Cardiff were within 30 m and at Ince concrete cooling water culverts were within 20 m of impact. Experience of the high particle velocities sustained by a nearby (10 m) buried gas main at Thornhill served as a reassuring experience.

<u>Explosions in wet sand</u> A series of experiments has recently been conducted in which heavily overdamped three component geophones were located up to 14 m below ground level on saturated sands. Explosive charges, ranging from 1 lb to 10 lb, were fired at the surface. The signals were recorded on UV paper and simultaneously with an Ampex 1300 FM tape recorder (14 channels). The response characteristics of the instrumentation have already been quoted. Data was examined to establish a 'site law' using both the UV records and

filtered tape (Fig. 1). The relation between peak particle velocities and distance from geophone determined from tape records as shown in Fig. 9. The tape recorded data was examined with the Hewlett Packard Fourier Analyser after actively filtering off signals above 250 Hz. All particle velocities are given as $\dot{u}^2 = \dot{x}^2 + \dot{y}^2 + \dot{z}^2$. Digital filters, the transforms of which were stored, were convoluted with the input signal, filtering off all but the bands of interest. With the assumption of an effective mid band frequency, a displacement was calculated. The power spectrum could have been obtained using this same analyser.

A Strong Motion Seismograph was also located at the surface. The peak particle velocities recorded were close to those given by the 5-15 Hz filter band from geophones at a comparable distance. At Ince and Cardiff the higher particle velocities recorded by the seismograph were consistent with the lower frequencies present, to which the geophones did not respond. Comparisons of the response in spectral terms between geophones and accelerographs are currently being made and will be published in due course.

CONCLUSIONS

Measurements of ground vibration between 0.1-1000 Hz and 0.001 -30 in/s are possible with available techniques, although no simple commercial system covers this range. For complete coverage it is advisable to combine systems using for example damped geophones and Strong Motion Seismographs or low 'g' piezo accelerometers. In order to take advantage of modern signal analysis, structural response computation methods, broad band sensors with magnetic tape recorders are needed. For many engineering purposes, given adequate frequency response, interpretation in peak velocity terms is adequate.

REFERENCES

BROCH J.T. (1972) Mechanical Vibration and Shock Movements. Bruel & Kjaer, Naerum, Denmark 1972.

CLOUGH R.W. (1962) Earthquake Analysis by Response Spectrum Superposition. Bull.Seism. Soc. Amer. 52(3) 647-660.

GROVER F.H. (1970) Vibration in Structures, A.W.R.E. Report 032/70, H.M.S.O.

HALVERSON H.T. (1971) Modern Trends in Strong Movement (Strong Motion) Instrumentation. From Dynamic Waves in Civil Engineering, John Wiley & Sons 1971

McCLURE G.M., ATTENBURY T.J. & FRAZIER N.A. (1964) Analysis of Blast Effects on Pipelines. Ground Pipeline Div., A.S.C.E., Nov. 1964, 25-47.

NEWMARK N.M. & ROSENBLUTH E. (1971) Fundamentals of Earthquake Engineering. Prentice Hall, New Jersey.

ROBSON J.D. (1963) An introduction to Random Vibration. Edinburgh University Press.

SKIPP B.O., CURREY B.& WOODFORD C. (1972) Symp. on Theoret. and Applied Structural Dynamics. I.S.V.R. Southampton.

STEFFANS R.J. (1966) Some Aspects of Structural Vibration in Vibration in Civil Engineering, Butterworth 1966.

WHIFFEN A.C. & LEONARD . (1971) A Survey of Traffic Induced Vibrations. R.R.L. Report LR 418.

WILLMORE P.L. (1961) Some Properties of Heavily Damped Electromagnetic Seismographs. Geop.J.Roy.Astr.Soc. V 4.

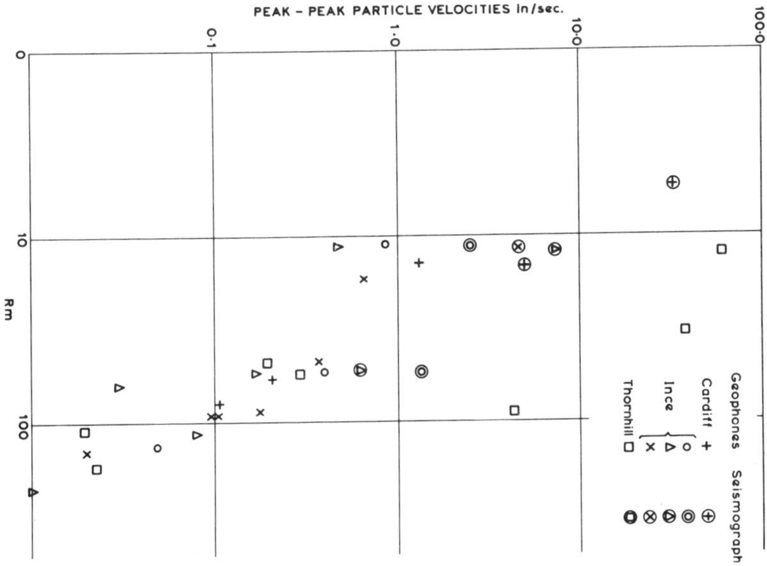

Fig. 8 Particle velocities, cooling towers

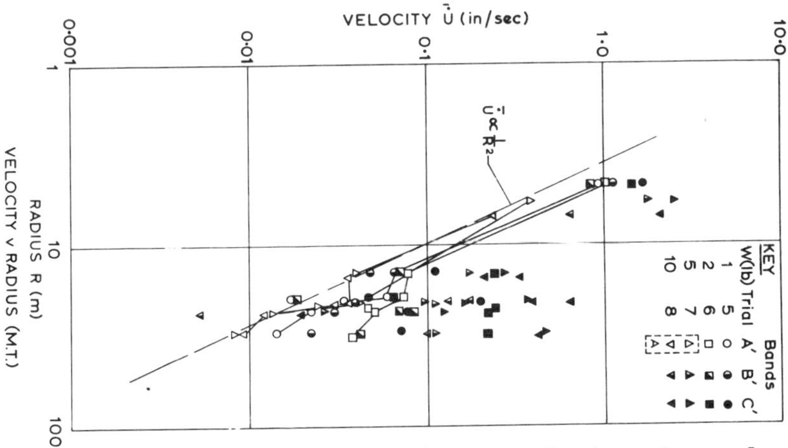

Fig. 9 Particle velocities, explosions in sand

INSTRUMENTATION CURRENT IN GEOTECHNICAL VIBRATION MEASUREMENT

CHARACTERISTICS	Cambridge vibrograph	Sprengnether blast seismograph	Sprengnether velocity vibration seismograph v5 1000	ABEM Trio (U/V)	ABEM Tellus	Dawe 1431A	Dawe 1433	Kinemetrics* VM1	B&K* 8306	SM1 system	Sinco S – 2	SMA1 (kinemetrics)
SENSING PRINCIPLE	Mechanical, spring/weight stylus	Inverted pendulums	Electro magnetic geophones	Electromagnetic (geophones)	Mechanical	Electro magnetic (geophones)	Piezo – accelerometer	Electro magnetic accelerometer, force balance accelerometer	Piezo – accelerometer	Electro magnetic (geophone)	Electro magnetic geophones	Electro magnetic accelerometers
PARAMETER	U	U	$\dot{U}$	$\dot{U}$	U	$\dot{U}$	$\ddot{U}$	$\ddot{U}$	$\ddot{U}$	$\dot{U}$	$\dot{U}$	$\ddot{U}$
CHANNEL	1	3	3 Components 1 timing	13	1	1	1	1	Single component	12	6	4(2V, 1H) 1
SIGNAL CONDITIONING	–	–		Attenuator cards amplifiers to galvanometer	–	Amplifier integrate & differentiate for U & $\ddot{U}$	Cathode follower integration, amplifier, attenuator	Filters(low pass)	Built inpreamplifier 10v/g ± 2% 10,000 pC/g ± 2%	Attenuator galvanometer matching	–	
PARAMETER RANGE	0.01 mm resolution magnification to 5:1 to 8 mm	x 50 x 500 magnification	0.2 – 20 n/s per inch record (2.75m)	Depends on geophone & galvanometer (0.001 - 1.0 in/sec)	10 – 500 $\ddot{U}$ > 7g.	0.003 in sec to 30 in/sec	3 in/sec^2 to 3 x 10^4 in/sec^2	60 aB or 80 aB	10^{-5}g – 1g	0.01 – 6.0 in/sec	0.5-1.0 in/sec per in. trace	0.01g – 1.0g 0.005 – 0.5g 0.025 – 0.25g
FREQUENCY RANGE	1 – 10 Hz 10 – 100 Hz	3 Hz ⟶	2 – 200 Hz	Usually 7 - 300 can be changed	15 – 170 Hz	6 – 100	1.6 – 10Kg or 1.6 – 1 Kg	0.1 – 25 Hz	0.2 – 1000 Hz	3 – 200 Hz	6 – 150	0 – 25 Hz
DATA PRESENTATION	Acetate film indentation	Photographic film	Standard photographic paper	Daylight develop paper	Waxed paper chart, carculum	Meter with jackplug to analyser	meter with jackplug to analyser	Stylus, wax strip	Various	U/V with parallel magnetic tape.	Direct write photographic strip	Photographic film
COMMENTS	Obsolescent but many in use not suitable for response analysis blasting, peak criteria.	Strong motion difficult to analyse fully blasting etc., for use with peak criteria	Wide range general use blasting and engineering, used with peak velocity criteria	Low frequency limits, A – D conversion tedious suitable for blast, multiple records can be analysed	Strong motion (monitor) structural monitoring not suitable for response analysis.	Frequency limits general use with FM tape for analysis rapid survey unit	Wider range general use with ancilliaries can be used in response analysis	Limit on higher frequencies general vibration analysis on peak criteria. Not suitable for full response use.	Various combinations with ancilliaries give very good ranges, & flexibility, but sophisticated, all purpose capability.	Wide range, flexible, suitable for further analysis. Suitable for most applications below 200Hz with response analysis.	Limited frequency range, self contained but not convenient for further analysis without A – D conversion general purpose.	Limited frequency range self contained remote on operation self trigger. Suitable with special reader for response analysis.

NOTES: This table is not complete, many accelerometers (piezo) are available which can be built into vibration measuring systems.
There is a wide range of analysis set ups (Bruel Kjair, Spectral Dynamics, Hewlett Packard)
Frequency range are approximate (flat to 10%).

THE MEASUREMENT OF PORE PRESSURES WITH PIEZOMETERS

P.R. Vaughan

Lecturer in Civil Engineering, Imperial College, London

INTRODUCTION

The field measurement of pore pressures with piezometers has developed over some 40 years. Such measurements are used to check assumptions made in effective stress analyses and to monitor the effectiveness of drainage and de-watering works.

The pore fluid in soil may be water or a mixture of gas and water. A piezometer is a hollow filter element, embedded in the soil, sometimes within a sand pocket. The cavity within the filter is usually filled with water, the pressure of which is measured and presumed to be equal to the pore fluid pressure. A correct measurement involves equalization of the pore fluid pressure with the cavity pressure and the accurate measurement of the cavity pressure. In partly saturated soils equalization may be to either the water or the gas pressure.

Piezometers cannot be used to measure pore pressures below absolute zero, which may occur in partly saturated or unloaded saturated soils, as the free water in the cavity will not sustain tension. Indirect methods of measuring pore water tensions in partly saturated soils have been developed (Richards:1969a, Escario:1969). The author knows of no measure ments of pore water tensions in saturated soils. Problems due to the formation of air bubbles in the piezometer cavity increase as the cavity pressure approaches absolute zero.

Pore pressure measurement with piezometers is considered in this paper in three sections : equalization between the cavity pressure and the pore pressure in saturated soils, equalization in partly saturated soils and measurement of the cavity pressure. Principles, sources of error and current uncertainties are reviewed. Equipment details are not discussed. Bishop <u>et al</u> (1969) reviewed recent practice and gave a more comprehensive bibliography than can be included here.

EQUALIZATION IN SATURATED SOILS

Richards (1969b) has pointed out that equalization may be inhibited by temporary osmotic effects in a fine grained soil. Bolt & Lagerwerffe (1965) examine these effects. They are usually ignored and the likely extent of errors in piezometer measurements due to them needs further study. Where they may be ignored, the pressure in the piezometer cavity will equalise to a steady uniform pore pressure by the processes of seepage and consolidation, provided that the installation of the piezometer does not influence this pore pressure. This requires that the connections to the piezometer do not form a preferential seepage path. Vaughan (1969) considered the influence on a steady pore pressure of a borehole backfilled with material of higher permeability than the surrounding soil and found that the backfill could be of the order of ten times more permeable than the soil before the influence on the pore pressure became significant.

Theory for the equalization of an out-of-balance cavity pressure with a steady pore pressure has been developed by Hvorslev (1951), for an incompressible soil, and by Gibson (1963) for a compressible elastic soil obeying the Terzaghi consolidation theory. Such theory may be used to estimate soil properties from the results of equalization tests or to estimate the time taken for sufficient equalization to enable a pore pressure measurement to be made. Gibson (1963) showed that the 'compressible' and the 'incompressible' theory give similar results at high degrees of equalization (which is illustrated in Fig.1) and at these high degrees of equalization a large error in estimated response time only implies a small error in pore pressure measurement. Thus the simple 'incompressible' theory is generally adequate for engineering purposes. It may be expressed as an equalization ratio, ϵ, where

$$\epsilon = \frac{u_g - u_m}{u_g - u_{mo}} = \exp\left\{-\frac{F.k.t.}{V.\gamma_w}\right\} \qquad .....(1)$$

where u_g is the pore pressure, u_m is the pressure in the cavity at time t, u_{mo} is the initial pressure in the cavity F is the intake factor which depends on the size and geometry of the piezometer, k is the coefft. of permeability of the soil, V is the volume factor (or compressibility) of the piezometer system defined as the flow into the piezometer for unit pressure change in the cavity and γ_w is the unit weight of water.

The influence of the volume factor on response was examined by Penman (1960). Table 1 shows response times calculated from equation 1 for some typical piezometer installations. The influence of trapped air is significant if rapid response is required. A particular problem lies in ensuring that sand

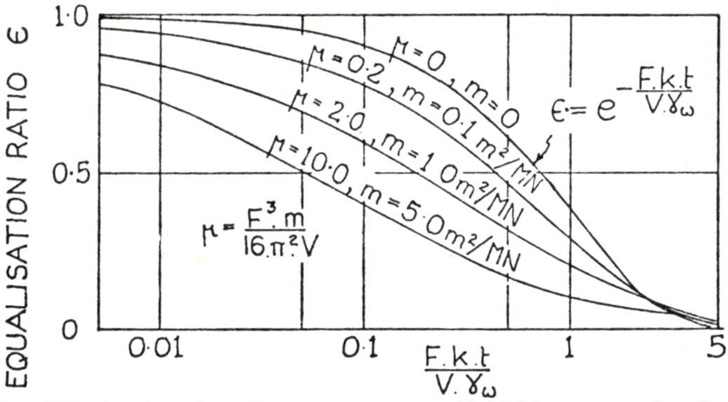

Coefficient of volume compressibility, m, calculated assuming F = 180cm , V = 1.76 cm^5/gm.

Figure 1 Influence of soil compressibility on theorectical piezometer response (after Gibson : 1963)

pockets used with borehole piezometers are saturated. Response times are generally rapid enough for engineering purposes unless open stand pipes are used in clay soils.

There are four field situations in which piezometer response must be considered. These are :-

Measurement of a pore pressure varying due to drainage In these circumstances the pore pressure must be equalized locally and, provided the piezometer is small in relation to the pore pressure gradient, the theory previously described

Piezometer system	V cm^5/gm $\times 10^{-4}$	F cm	t hours (ϵ=0.05)
Borehole piezometer, sand pocket 60cm x 15cm dia. 1.5cm dia. standpipe	17,650	180	820
Ditto with 150m 3mm bore nylon tube & 3mm Hg manometer	35	180	1.6
Ditto with electric transducer in the piezometer cavity	0.05	180	0.002
Ditto with transducer with 100 cm^3 air in the cavity	400	180	19
Fill piezometer 10cm x 4.5 cm dia, with 150m nylon tubing & 3mm bore Hg manometer	35	40	7.3
Ditto with transducer in place of manometer	7	40	1.5

Coefficient of permeability of soil, k = 10^{-10} m/s

Table 1 Typical piezometer response times

may be applied. The lag of the cavity pressure behind the pore-pressure can be estimated by representing the changing pore pressure as a sequence of instantaneous changes and applying the principle of superposition. It should be noted that rate of change of the pore pressure and the response time of the piezometers are both controlled by the permeability and compressibility of the soil. Thus the lag depends only on the intake and volume factors of the piezometer. It can be shown that adequate measurements can be made even with a standpipe piezometer of slow response irrespective of the soil permeability, provided it is not close to a drainage boundary where rapid changes in pore pressure can occur. With piezometers installed in boreholes, remoulding and smear adjacent to the borehole may reduce the permeability locally. The response of the piezometer will then become slower in relation to the rate of change of pore pressure.

Pore pressure changes due to undrained loading The rate of such pore pressure changes is not controlled by the soil prop erties and they may be rapid. The deformability of the piezometer will be different from that of the surrounding soil and thus the total pressure and pore pressure changes near the piezometer will be atypical. The equalization of the cavity pressure to the true pore pressure will depend on the rate at which this perturbed pore pressure zone equalizes, as well as on the response time of the piezometer as defined in equation 1. No theoretical study of this problem is known to the author but it is probable that the changes in average total stress will be small compared to the changes in principal stresses. Therefore the perturbations of the pore pressure distribution are likely to be smaller than the perturbations of total stress which, for instance, cause difficulties when earth pressure cells are used.

Soderberg (1962) examined the equalization of a zone of perturbed pore pressure around an expanded, impermeable cylinder, using the Terzaghi-Rendulic consolidation theory. This is approximately equivalent to equalization of a perturbed pore pressure zone around a long cylindrical piezometer of zero volume factor. Depending on the amount of plastic yield of the soil, Soderberg found that 80% equalization at the cylinder surface (ϵ=0.2) occured at a time equal to approximately $10.r^2/c$ and 95% equalization at $30.r^2/c$. r is the radius of the cylinder and c the coefficient of consolidation. These figures should be conservative for typical piezometers of finite length, as three dimensional effects will accelerate equalization. Comparison of the above figures with equation 1 indicates that equalization of a typical standpipe piezometer will be at a similar or slightly slower rate than equalization of the perturbed pore pressure zone. For all other types of piezometer equalization of the perturbed pore pressure zone will be much slower than response according to equation 1.

Equalisation of the perturbed pore pressure zone will be accelerated if the piezometer (or its sand pocket) is made as small as possible. Installation in a small diameter borehole may be facilitated if the tip is placed in a porous grout such as plaster of paris, which ensures an intimate contact with the soil and the exclusion of air. A rigid probe piezometer may be especially prone to error during rapid loading, as stress changes may be concentrated at the end of the probe where the pore pressure is measured.

Apparent undrained pore pressure changes observed after loading ceases may indicate the magnitude of lag during loading, although they may be due also to creep in the soil. A study of such changes, as recorded by piezometers subject to typical engineering rates of loading, suggests that errors due to non equalization during loading are not of great significance. However, if measurements are made where loading is more rapid, such as near driven piles, greater errors may result and more study of the problem would be desirable.

Initial equalization around a borehole piezometer This problem is similar to that described in the previous section. The perturbed pore pressure zone is due to the drilling of the borehole. The stress and pore pressure changes will be larger and the pore pressures will be reduced. If the piezometer responds more rapidly than the zone of perturbed pore pressure can equalize and if the cavity pressure is initially higher than the pore pressure around the borehole (as may occur due to grout pressures or due to initial filling of a standpipe), the piezometer may appear to equalize to a pressure lower than the true pore pressure. This is illustrated in figure 2, in which the initial equalization of a rapidly responding pneumatic piezometer is shown. The equalization of an adjacent standpipe piezometer is shown for comparison. The slower overall equalization suggests that the response time of the piezometer is more important than the equalization of the perturbed pore pressure zone. Pore pressures around a driven piezometer will be increased and the process described above will be reversed. The overall rate of equalization can be accelerated by reducing the diameter of the borehole.

Equalization after in-situ tests or 'de-airing' Piezometers with hydraulic connecting tubes may be used for falling and constant pressure seepage tests, for hydraulic fracture tests to estimate the minor principal stress (Vaughan : 1972) and they may be de-aired by circulating water under a differential pressure. All these operations create a zone of perturbed pore pressure around the piezometer, the size of which is dependant on the time the disturbing pressure has been applied, or, in the case of hydraulic fracture tests, on the volume of water forced into the soil. The equalization of the perturbed pore pressure zone may control the overall equalization of the piezometer, as described previously. The equalization of a

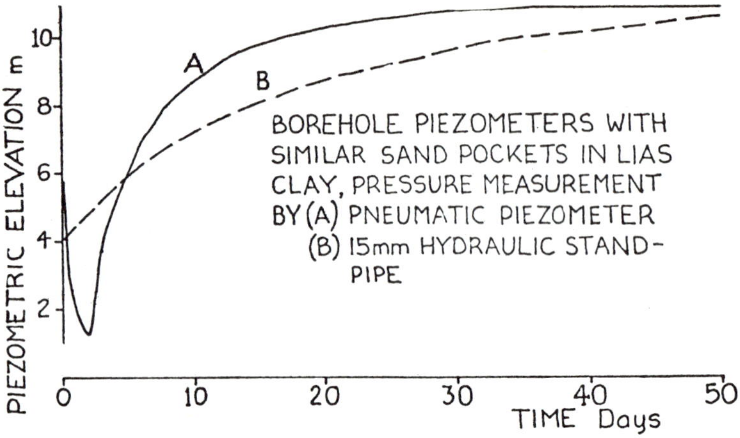

Figure 2 Initial equalization of borehole piezometers of varying response times (data from R.J Chandler)

piezometer to which an excess pore pressure has been applied for varying periods of time is shown in Figure 3. In the experience of the author, equalization after a hydraulic fracture test is comparable to that after a constant pressure seepage test (Vaughan : 1972).

EQUALIZATION IN PARTLY SATURATED SOILS

In partly saturated soils (fills, soils above the water table and soils in which gas is generated from organic matter) the pressure in the piezometer cavity may equalize to either the water or the gas pressure, the gas pressure being the higher due to capillarity. The gas or air may be continuous or it may exist in discrete bubbles, depending on the degree of

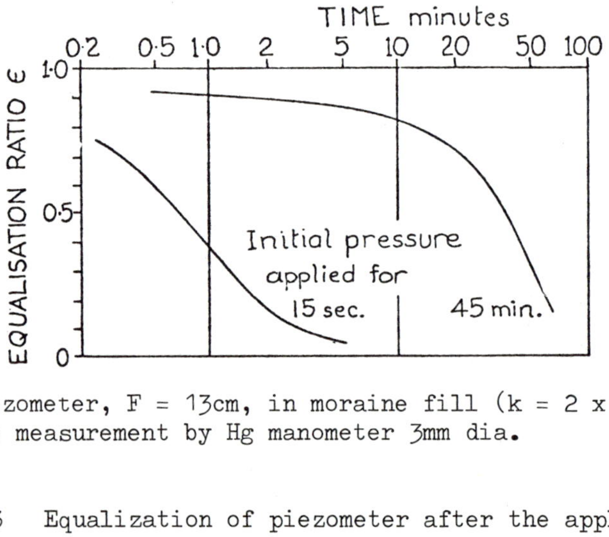

Disc piezometer, F = 13cm, in moraine fill (k = 2 x 10^{-7}m/s)
Pressure measurement by Hg manometer 3mm dia.

Figure 3 Equalization of piezometer after the application of a constant pressure for varying times.

saturation. If the air is continuous, the air permeability is high, the pore air pressure will tend to equalize to atmospheric pressure and the pore water pressure will be sub-atmospheric. If the porewater pressure is greater than atmospheric the air will generally be discontinuous.

In partly saturated soils it is essential to know whether pore air or pore water pressure is being measured. It is much more useful to measure pore water pressure, as changes in this pressure usually control changes in effective stress and it is also the pressure which directly responds to drainage of water. The principles of measuring pore water pressure are discussed in detail by Bishop et al (1964). If the cavity pressure equalizes to the pore water pressure the rate of equalization will be controlled by processes similar to those described for saturated soils.

If pore water pressure is to be measured, the piezometer filter must be of sufficiently fine pore size for it to be able to resist by capilarity at its surface an air-water pressure difference (the air entry value or blow-through pressure) greater than the pressure difference in the soil; otherwise the water in the filter and the piezometer cavity will be replaced by air and, unless this air is continuously removed, the pore air pressure will be measured. Spontaneous and continuous removal of air may occur in standpipe piezometers as air bubbles rise in the standpipe.

If the piezometer cavity is enclosed, then, even with a high air entry value filter, air will still enter the cavity by diffusion. This process occurs according to the general diffusion equation and Henry's law of solubility. Thus air flows from bubbles at high pressure to bubbles at low pressure. Diffusion of air through water is very slow. If an air bubble forms within the piezometer cavity its pressure will tend to that of the water in the cavity, it will be less than that of the air in the surrounding soil and diffusion of air into the cavity will continue. While the cavity still contains free water the porewater pressure will be measured. Once the cavity has emptied of water, the air pressure, which is being measured, will slowly rise as more air diffuses into the cavity until it equalizes with the pressure of the air in the surrounding soil. If the air is flushed from the cavity the cavity pressure will again equalize to the pore water pressure.

If the air is continuous (which implies sub-atmospheric pore water pressure) diffusion is only required across the thickness of the filter, whereas if the air is discontinuous diffusion must also occur through the surrounding soil and it will be much slower. The mass of air required to fill the cavity will depend on the cavity pressure and the lower the pressure the more rapidly the cavity will empty. Thus breakdown of pore water pressure measurement may be expected to be much more rapid when pore water pressures are sub-atmospheric than when they are above atmospheric. Practical experience (Bishop et al : 1969) confirms this prediction. Bishop et al (1964) quote the total flow of air into a high air entry value piezometer recording pore water pressures just above atmospheric in boulder clay fill. An inflow rate of $3cm^3$/year is given for the first four years of operation. Assuming that all this air was entering the piezometer cavity, breakdown of measurement should not have occured for at least 5 years. Fig. 4, after Bishop et al (1964), shows the pressure recorded by a similar piezometer in rather dry shale fill with continuous air. Breakdown is indicated as de-airing restored the measurement of sub-atmospheric pressure. The measured pressure rises smoothly as if responding to water inflow and breakdown is not apparent unless de-airing is performed.

Torblaa (1966) discusses the use of a thick high air entry value filter with a sealed cavity electric piezometer to inhibit breakdown by lengthening the diffusion path. Such units may operate without breakdown for many years if pore water pressures above atmospheric are measured. If breakdown occurs when sub-atmospheric pressures occur initially, correct reading may be restored if pore water pressures rise sufficiently rapidly for the air in the cavity to be compressed and for free water to enter it again. Breakdown should also be inhibited if the volume of water within the filter is initially as large as possible. However, provision for de-airing is always an advantage, as breakdown cannot be detected directly and correct reading may not be restored naturally. If sub-atmospheric pressures are to exist for very long, current experience suggests that provision for de-airing is essential. The provision of tubes for de-airing does not preclude measurement of pressure in the cavity by electric or pneumatic transducers. Careful thought should be given to such a system before it is used. Cavitation within the tubes, with the resulting flow of water into the cavity, may inhibit equalization, as well as preventing breakdown by supplying free water to the cavity.

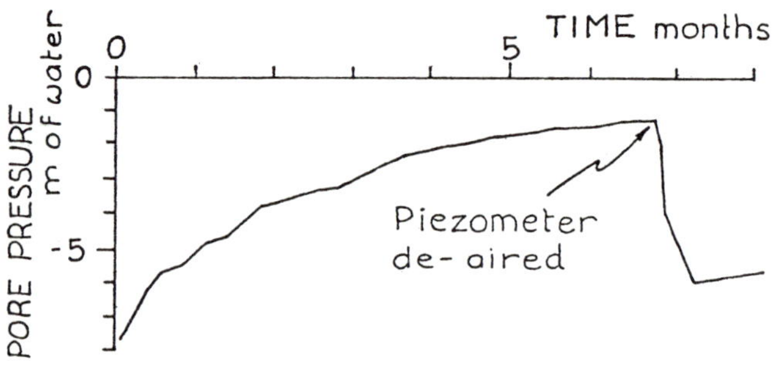

Figure 4 Breakdown of a sub-atmospheric pore water pressure measurement.

Two additional factors should be considered in providing piezometers for partly saturated soils. The seals at the piezometer tip must be capable of resisting the same 'blow-through' pressure as the filter. Where practicable, a blow through test should be performed on the assembled tip before it is installed. If the piezometer has de-airing tubes its internal shape should be 'clean' so that all air bubbles can be flushed out.

MEASUREMENT OF PRESSURE IN THE PIEZOMETER CAVITY

Pressure measurement may be via water-filled tubes connecting to an external measuring point, by an electric transducer in the piezometer cavity or by a variety of pneumatic pressure sensing systems in the cavity.

Hydraulic measurement   The simplest form of hydraulic measurement is the open standpipe. The pressure to be measured must be above atmospheric. Provided that the standpipe is of sufficient diameter to allow air bubbles to rise freely (in the author's experience a 'clean' diameter of about 12mm is required) the system is self de-airing. Access to the top of the standpipe is generally required for plumbing the water level, although this can be determined remotely using an air bubbling system (Vaughan : 1965) which enables the top of the standpipe to be buried.

In twin-tube hydraulic piezometers, the cavity is connected to a remote measuring point by the two water-filled tubes, which are available for de-airing. The cavity pressure differs from the pressure at the measuring point by the head difference between them;the tubes must be air free for a true correction to be applied. The entry of air into the tubing is insignificant if the correct tubing is used. Nylon 11 tubing, sleeved with polythene, is impermeable to both air and water and such tubing will operate at positive pressures for many years without de-airing being required. A difference in measured pressure between the two measuring tubes will often indicate the pressence of air in them. Air in the cavity will not cause such a difference. For trouble-free operation, pressures throughout the connecting tubes should be above atmospheric and the system will not operate at all if tension occurs in the tubes. This places a restriction on the relative levels of the piezometer, the tubes and the measuring point.

Other disadvantages of the system are : very fast response times are prevented by the compressibility of the tubing; exposed parts of the system may freeze and anti-freeze solutions can cause osmotic pressures; the tubing and the piezometer cavity must be filled initially with de-aired water, while this can be done off-site in part it complicates installation; joins in the tubing (which should be avoided if possible) and joins with the piezometer tip are potential causes of constrictions and blockages, they pull apart under tensile forces and small coils should be formed in the tubing on each side of couplings to reduce the tensile forces which can be applied to them ; tubes should be checked for continuity and leaks before installation ; dirty water can cause blockages and should not be used ; pressure measuring systems and de-airing units tend to be large and expensive and to require large and expensive gauge houses, although modern developments allow the use of small terminal valve units (which can be buried in small chambers to avoid freezing) together with portable measuring and de-airing systems.

Other advantages of hydraulic systems are : in-situ tests can be performed, both to determine soil properties and to check the piezometer ; automatic recording is possible ; pressure measuring systems are external and calibrations, etc., can be checked; piezometer tips and connecting tubes are cheap ; there are no buried moving parts and long term reliability is very good.

A variety of de-airing systems have been developed for hydraulic piezometers, some automatic. From the author's experience, de-airing systems should allow the following: the maximum pressure applied should not be sufficient to fracture the ground at the piezometer ; a small outflow from the piezometer into the ground should occur during de-airing ; pressure changes in the tubing should be minimised. A back pressure on the return tube is generally necessary to meet these requirements. It is often forgotten that occasional de-airing can be performed from two portable water containers at the appropriate levels.

Electric measurement   Electric piezometers usually have no provision for de-airing. The pressure in the cavity is measured by determining the deformation of a diaphragm between it and a second chamber. Usually, this chamber is sealed with a known pressure in it, although it can be vented to atmospheric pressure. Because the deflection of the diaphragm is small, very fast response times can be achieved (see Table 1). Where the piezometers are permanently buried and where the signal must be transmitted long distances, vibrating wire strain gauges are usually used to sense the diaphragm displacement. Such strain gauges are very stable and the signal is suitable for long distance transmission as it is uninfluenced by the resistance of the cables. For short term observations where transmission is over short distances, there is a growing tendancy towards the use of mass produced transducers with resistance strain gauges.

Advantages of electric measurement include : convenient installation, as only a single cable needs to be laid out ;

the equipment is not subject to freezing ; small and convenient terminal units can be used, with portable reading units ; there are no restrictions on the relative levels of the equipment. Disadvantages include : the high cost of the piezometers, connecting cables and the reading units ; the piezometer requires a calibration which cannot be checked after installation unless special provision is made ; the buried and inaccessible part of the equipment is sophisticated and long-term reliability must always be in some doubt as no check tests can be performed.

Pneumatic measurement Pneumatic systems involve two air-filled tubes connecting the measuring point to the piezometer, which contains a valve operated by the difference between the applied pneumatic pressure and the pressure in the piezometer cavity. Either the valve shuts when the input air pressure equals the cavity pressure, in which case the maximum pressure built up in the return tube gives the cavity pressure, or the valve opens, in which case the maximum pressure sustained in the input tube gives the cavity pressure. In most systems, the operation of the valve involves a small calibration factor. Operation of the valve also requires a small volume change within the piezometer cavity. Potential errors may arise due to this volume change, their magnitude depending on the volume factor of the system. Such errors may be overcome, at least in part, if the valve has a 'snap' action triggered by the pressure operating before the valve moves. If there are no hydraulic tubes connecting to the cavity (as can be provided for de-airing) the response time will be very short. The connecting tubes must be kept free of water. They must be of suitable impermeable material and dry gas must be used to operate the system. Other problems with plastic connecting tubes are as described for hydraulic connections.

Advantages of pneumatic systems include : they should not freeze ; portable reading units can be used, with small and cheap terminal valve boards ; installation is more convenient than with hydraulic systems, as the connecting tubes do not require filling with water ; tubing is cheap ; there are no relative level problems (unless de-airing tubes are used) ; while the buried valve may have an uncheckable calibration the effect of this is usually small and the main pressure measurement is external and checkable. Disadvantages include : the usual absence of de-airing facilities ; operation is slow and time consuming ; if the connecting tubes are long head losses in them during operation may become significant ; dirt entering the system may prevent valve operation ; piezometer tips are relatively expensive.

Pneumatic systems are a comparatively recent development and there is little experience of long term operation. Malfunction of valves due to corrosion, sticking, perishing of plastic parts and the like seems to be the most likely potential cause of failure. Until experience is obtained, the use of pneumatic systems for long term measurement requires caution. There is some possibility of checking the operation of pneumatic systems by in-situ tests on the circulating tubes.

REFERENCES

Bishop, A.W.,M.F.Kennard & P.R.Vaughan (1964) Developments in the Measurement and Interpretation of Pore Pressure in Earth Dams. Trans. 8th. Congress Large Dams, 1, 47.

Bishop,A.W.,P.R.Vaughan & G.E.Green (1969) Report on Speciality Session 'Pore Pressure Measurements in the Field and in the Laboratory'. Proc. 7th.Int.Conf. Soil Mech. & Found. Eng. 3, 427

Bolt,G.H., & J.V.Lagerwerffe (1965) Consequences of Electrolyte Redistribution during Pressure Membrane Equilibration of Clays. Soil Science, 99, 3:147.

Escario,V. (1969) Discussion. Proc. 7th. Int. Conf. Soil Mech. & Found. Eng. 3, 433.

Gibson, R.E.(1963) An Analysis of System Flexibility and its Effect on Time-lag in Pore-water Pressure Measurement. Geotechnique, 13, 1:1

Hvorslev,M.J. (1951) Time-lag and Soil Permeability in Ground Water Observations. Bull. No. 36, U.S. Waterways Experiment Station, Vicksburg.

Penman, A.D.M. (1960) A Study of the Response Times of Various Types of Piezometer. Proc. Conf. Pore Pressure & Suction in Soils, Butterworths, London, 53.

Richards, B.C. (1969a) Psychrometric Techniques for Measuring Soil Water Potential. C.S.I.R.O Div. Soil Mech. Australia, Tech. Report No. 9.

Richards, B.C. (1969b) Discussion. Proc. 7th. Int. Conf. Soil Mech. & Found. Eng. 3, 433.

Soderberg,L.O. (1962) Consolidation Theory applied to Foundation Pile Time Effects. Geotechnique, 12, 3;217.

Torblaa,I. (1966) Poretrykksmaller Utfort vet Dam Hyttejuvet med Fonskjellig Maleutstyr. Norwegian Geotech. Inst. Pub.68,25.

Vaughan, P.R. (1965) Field Measurements in Earth Dams. Ph.D. Thesis, University of London, 1, 76.

Vaughan, P.R (1969) A Note on Sealing Piezometers in Boreholes. Geotechnique, 19, 3:405.

Vaughan, P.R. (1972) Discussion. 5th. European Conf. on Soil Mech. & Found. Eng., Madrid, 2, 72.

SOME NEW TECHNIQUES IN FIELD INSTRUMENTATION APPLIED TO A BORED PILE TRIAL IN GAULT CLAY

A.G. Weeks

The Materials Engineer,
Kent County Council

R.B. Dixon

Assistant Bridge Engineer,
South-East Road Construction Unit, Kent County Council
Sub Unit

INTRODUCTION

The proposed route for the 18 Km M.25 which will run from Wrotham Heath, west of Maidstone, Kent, to the Kent - Surrey boundary, is located almost entirely on the Gault Clay outcrop. As the majority of the proposed structures will be carried on bored piles penetrating the Gault Clay, it was decided that a piling trial to assess the bearing capacity factors for bored piles in this material would be justified.

Two sites, one at Brasted and the other at Kemsing, were selected as representing the two major variations in the Gault Clay which had been disclosed by the motorway site investigation.

The paper outlines the tests that were undertaken and details are given of the new instrumentation techniques which were used during the project.

PROGRAMME OF SITE TESTING

The piling trials at both sites included four test piles and ten anchor piles consisting of straight shafted bored piles of nominally 500 mm diameter. The piles varied in length from 9.10 m to 19.7 m at Brasted and from 15.2 m to 25.9 m at Kemsing - full details of each pile are given in Table 1 and layout at each site is shown on Figure 1.

PILE REF.	PILE LENGTH (m.)		Instrumentation in Pile	REINFORCEMENT		
				% of pile area	Length (m.)	
	Brasted	Kemsing			Brasted	Kemsing
A	9.1	15.2	None	0.6	6	6
B	12.1	18.2	Base Load Cell	0.6	6	6
C	15.2	21.3	None	0.6	6	6
D	18.2	24.3	Base Load Cell	0.6	6	6
1	19.7	25.9	Inclinometer Tube	5.1	19.7	25.9
2	19.7	25.9	None	2	19.7	25.9
3	16.7	22.8	Base Load Cell	2	16.7	22.8
4	13.6	19.7	None	2	13.6	19.7
5	10.6	16.7	None	2	10.6	16.7
6	19.7	25.9	Inclinometer Tube	5.1	19.7	25.9
7	19.7	25.9	Inclinometer Tube	2	19.7	25.9
8	16.7	22.8	Inclinometer Tube	2	16.7	22.8
9	13.6	19.7	Inclinometer Tube	2	13.6	19.7
10	10.6	16.7	Inclinometer Tube	2	10.6	16.7

Table 1 Details of Piles

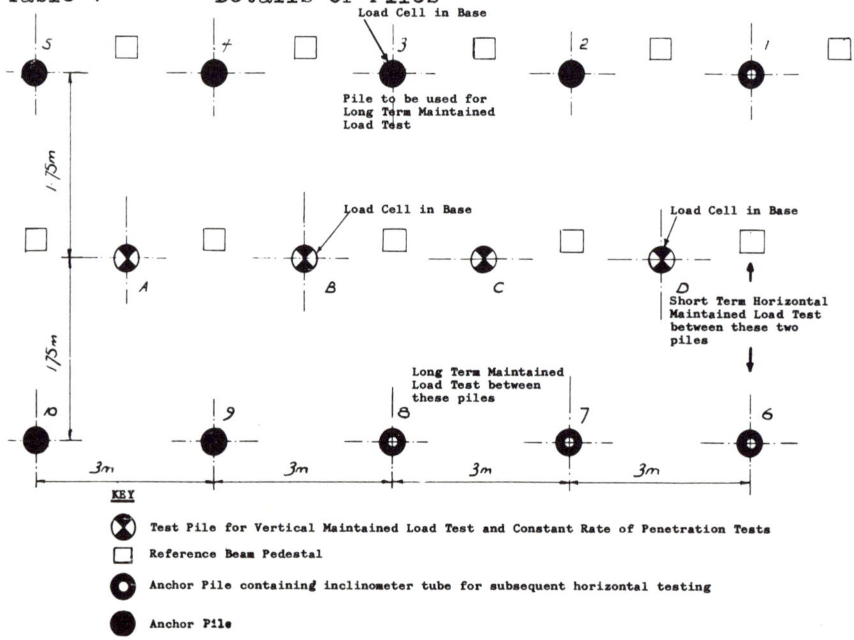

Figure 1 Plan of Piles & Location of Tests

The four test piles were subjected to short-term maintained load and constant rate of penetration tests for which the procedures outlined by Whitaker & Cooke (ref. 5) were adopted.

The Maintained Load (M.L.) test consisted of loading the test piles in six equal increments, each increment being one tenth of the calculated ultimate load for the pile. Each increment of loading was maintained until the pile head settlement was less than 0.10 mm per day. After removing entirely the load applied during the M.L. test a period of at least 48 hours was allowed for the pile to recover before the Constant Rate of Penetration (C.R.P.) test. This test was carried out by jacking the pile into the ground at a constant rate of 0.75 mm per min until the penetration was d/10, i.e. 50 mm. The load was applied by means of a hydraulic jack obtaining its reaction from a steel framework attached to four anchor piles (see Figure 2) whilst the magnitude of the applied load was measured using a load cell. The load cell, which was dumb-bell shaped, was of the electrical foil resistance type having two separate measuring circuits. To ensure reproducibility a spherical upper bearing face resting on a mild steel end cap was used. This end cap was renewed between each loading cycle (see ref. 1).

Figure 2 Test Rig for M.L. & C.R.P. Tests

In two of the four test piles at each site and one of the anchor piles 1000 KN load cells were incorporated at the base to enable separation of the shaft and base resistances. The load cells were modelled upon the design of Whitaker, Cooke & Clarke (refs. 6, 7 and 8) and the procedures for placing the cells into the pile bores came from the same source. One procedural change made was the inflation of the peripheral rubber sleeve by filling the cell with oil via the strain gauge cable tube and achieving a sufficient pressure by filling the cable tube when held at ground level. All the six load cells installed during the trial functioned without trouble. Upon completing the testing of the test piles a long-term maintained load test, using the likely working load for the pile, was started on the anchor pile which had been constructed with a load cell in its base.

Four of the remaining anchor piles, containing centrally placed inclinometer tubes down their axes, were subjected to horizontal load tests; the details of the equipment are described later. Two of the piles were given short-term tests during which the horizontal load was increased in steps of 20 KN until failure occurred. The other two piles were subjected to long-term tests, during which horizontal loads of the same magnitude as the likely working loads, were applied.

The two long-term tests at each site are still in progress.

PROBLEMS OF INSTRUMENTATION

At the design stage two problems were apparent:-

1) The maintenance of a series of constant loads for long periods of time on sites which were not continuously manned, and the ability to carry out tests with such varying output demands as the M.L. and C.R.P. tests.

2) The application of horizontal forces to a pile and the measurement of the resultant horizontal deflections.

During the trial a further problem arose from the movements of the reference beams used to monitor the vertical settlements of the piles.

The techniques used to cope with these situations are detailed in the following sections.

NEW TECHNIQUES

Load Application

Electricity was readily available at both sites and, as the forces to be applied were of unknown magnitude until some preliminary testing had been completed, an electro-hydraulic system was chosen. The testing sites were isolated and not continuously manned and as some tests were expected to last for nine months the design emphasis of the equipment was placed on its long-term durability and that components of the pressure maintaining equipment should be able to be replaced without loss of pressure at the jack and also on equipment safety. The pressure maintaining unit contained an integral stand-by pumping system to accommodate any possible electrical or mechanical breakdown.

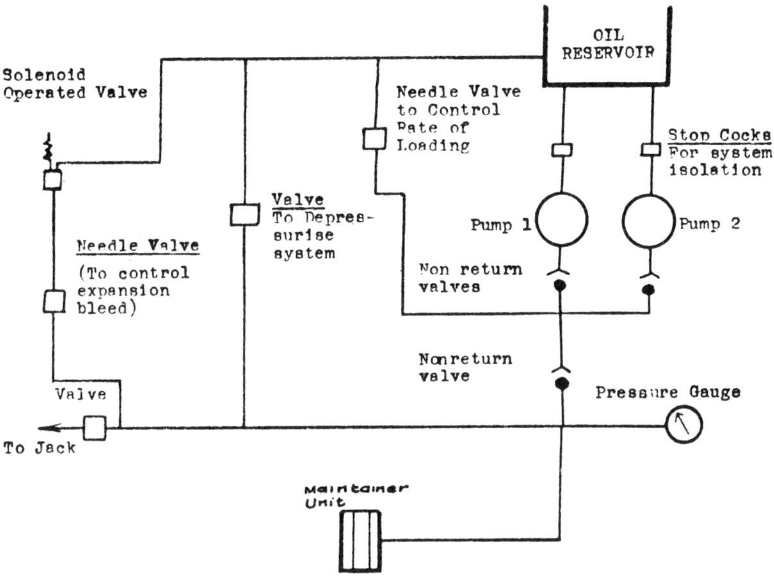

Figure 3 Circuitory of Pressure Maintaining Unit

Figure 3 shows the electrical and hydraulic circuitory. The pressure maintaining unit utilised a hydraulic servo. A master cylinder of small area was loaded by weights and was connected through hydraulic lines to a jack of larger area. In the "at rest" position the master piston was supported by hydraulic pressure, if the pressure in the system dropped, the master piston and its attendant weight pendant fell allowing a magnet on the weight pendant to actuate a reed switch, which through relays, activated Pump 1; if this pump could not replenish the hydraulic needs of the system the pendant fell further and the magnet triggered a second reed switch, activating Pump 2; also isolating completely Pump 1 which could not be used further without manual intervention. On the return of the piston to the "at rest" position the pump was switched off by a further switch. To minimise sticking in the master cylinder the piston had no seals and was manufactured to fine tolerances. It was also rotated continuously by an induction motor. If a temperature rise in the system caused an expansion of the hydraulic oil then the master piston would rise activating, through relays, a solenoid valve which allowed hydraulic fluid to bleed back, through a needle valve (at a pre-set rate) to the oil reservoir, until the piston returned to the "at rest" position. The output from the pumps bifurcated to the jack and to the oil reservoir, a needle valve on the reservoir return line establishing a fine control over the output to the jack. Applied load to the pile was measured by a force transducer, its magnitude being adjusted by varying the weights on the pendant; fine adjustment was achieved using lead shot.

The unit was capable of producing the required output for constant rate of penetration testing by running both motors simultaneously and isolating the constant pressure control unit. The coarse adjustment of the rate of movement of the jack was by means of the needle valve on the console whilst the fine adjustment was controlled by a second operator using the needle valve adjacent to the jack and referring to a dial gauge with driven pacer shroud bearing on the pile under test.

The following safety devices were incorporated in the equipment. An adjustable pressure relief valve was attached to the jack and a needle valve was fitted between the jack and the relief valve; prior to a load being applied the needle valve was closed and the pressure relief valve adjusted to vent at about five per cent in excess of the expected applied pressure. Pressure was removed and the needle valve opened; the valve keys were then taken off. A bell warned that the oil level in the reservoir was low and an electric cut-out actuated by a manometer capsule prevented the safe working pressure of the hydraulic lines being exceeded.

Lateral Testing

Investigation of the lateral bearing capacity of the piles utilised two anchor piles of similar length redundant from the short-term vertical test programme. The loads were applied 6 m below ground level after first removing the granular Head material overlying the intact Gault Clay. An inclinometer tube was incorporated centrally within each of these piles and the piles were reinforced with 4% of their total area of mild steel reinforcement. The large amount of reinforcement was required to ensure a ground failure and not a failure of the pile. It was necessary to record the profile of the deformed pile as accurately as possible and, at the time of the investigation, the most accurate inclinometer available was the "Digitilt Model 5031" manufactured by The Slope Indicator Company: This type of instrument was purchased for the investigation. The fact that the pile was so densely reinforced made the placing of the inclinometer tube centrally within 23 m long pile difficult. The reinforcement was made up in cages 4 m long, each cage was lifted separately into the pile bore, lapped and then welded to its neighbour. It was felt that if the tube was secured to the reinforcement the risk of it being damaged whilst the cage was being assembled and placed was too great. The inclinometer tube, which was a circular aluminium alloy extrusion 2.5 mm thick and 81 mm diameter, in sections 3.05 m long, had four equally-disposed grooves. When the tube was assembled it was found to be too flexible to be installed alone. Various methods of insertion of the tube were investigated; the method finally adopted, however, was to position at the base of the empty bore a steel spider which had three legs of slightly smaller radius than the pile bore and having a central vertical spigot. A hollow rock drill string (external diameter 76 mm) was lowered by crane onto the spigot. This drill string then acted as a mandrel for subsequent installation of the inclinometer tube. When the pile was to be concreted the weight of the tube and mandrel was held by a crane. The method was found to be effective when the tubes were subsequently monitored by the inclinometer and they were found to have a smooth profile. The piles were loaded by means of a strut incorporating a load cell and hydraulic jack as illustrated in Figure 4; the force was maintained by means of the constant pressure unit described elsewhere.

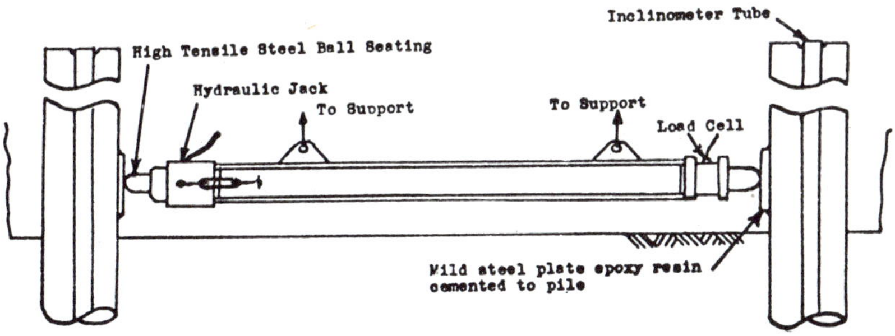

Figure 4 Arrangement of Horizontal Test Rig

A horizontal force was transferred to the piles by means of a high tensile steel semi-spherical ball seating, bearing on mild steel plates which was cemented to the piles with epoxy resin.

The piles were loaded in increments of 20 KN and each increment was maintained until lateral movement had ceased. The intact profile of the pile was first recorded on orthogonal axes, the probe was then rotated through 180° and the profile redetermined; arithmetic averaging removed the instrument error. This procedure was repeated for every observation of the subsequent deformed profile.

Calibration checking of this type of instrument required equipment which was not available outside the U.S.A. It was possible, however, to check the instruments reproducibility at intervals during the tests and its stability over a four day period was checked half-way through the testing programme.

Settlement Measurement

The test programme was developed using results of investigations into the properties of London Clay as guidance. Settlements in the short-term were expected to be of a similar magnitude and the settlement monitoring was of a similar layout to that of previous researchers. The vertical displacement was monitored by dial gauges attached to a reference beam 3 m long, mounted on stub piles. When the first pile was loaded it was apparent that settlement per increment of load was much smaller than that anticipated (in the order of 0.2 mm) and that the movement of the datum piles was significant relative to this movement. It was decided to use a longer beam, which was 14 m long, for the subsequent tests.

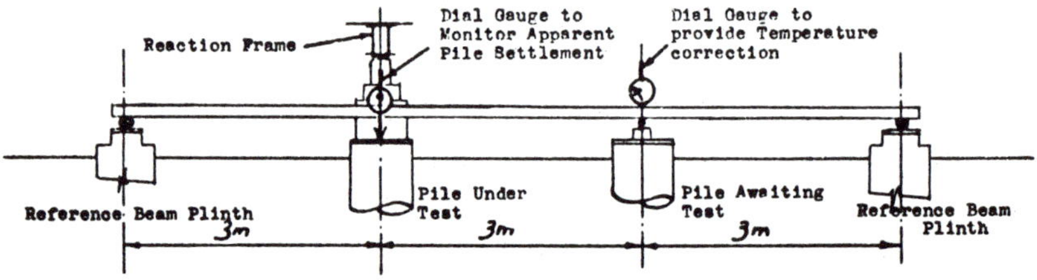

Figure 5 Diagram of Settlement Measurement using the Long Beam

This beam, which is shown in Figure 2, was supported by high tensile steel rollers resting on a high tensile steel plate, mounted on stub piles. The rollers were well greased and both the supporting plates were levelled accurately using a Wild N3 level. This long beam unfortunately introduced an error largely attributable to a temperature effect which was of such a magnitude as to swamp the settlement being measured at the pile. To overcome this temperature effect the 14 m beam was mounted on datums such that the test pile and a pile awaiting test were located at the third points of the beam span, as in Figure 5. Whilst the pile was under test apparent deflections were recorded simultaneously for the test pile movement relative to the beam and the beam movement relative to the pile awaiting test. The deflection data recorded at the pile, which was not loaded, was used to correct the apparent settlement data recorded at the pile under test. The two sets of deflection data and the corrected data are shown in Figure 6.

Prior to the commencement of each test the reference beam was deflected manually at its mid point and the point of contact of the reference gauge was moved along the line of the reference beam until its deflection agreed with the deflection which was recorded at the pile. The system was not particularly satisfactory as it was reliant on the symetrical behavior of the beam and, in addition, measurements were being made using Dial Gauges attached to a 250 mm cantilever arm projecting to one side of the beam.

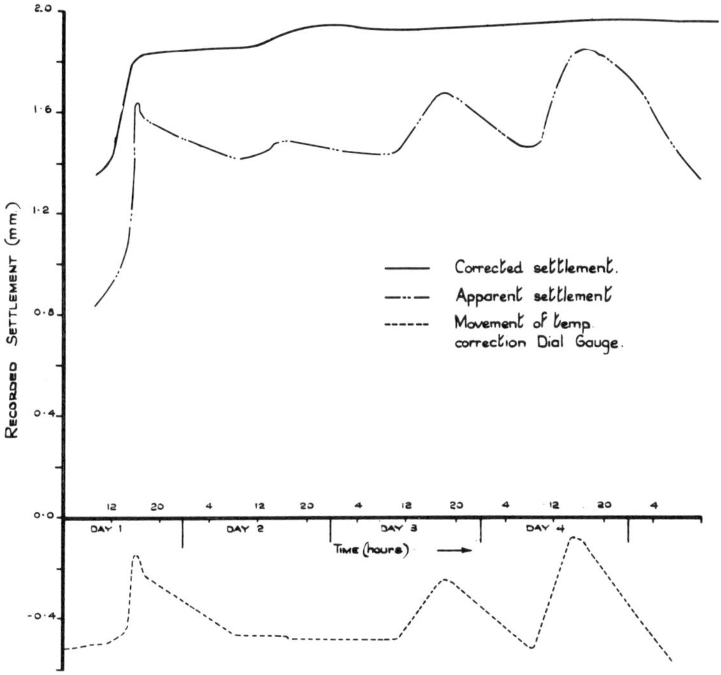

Figure 6 Graphs of Settlements relative to 14 m Reference Beam

The system was adequate to complete this series of tests but as a further series of tests was anticipated at a remote site it was decided to explore the possibility of an auto-recording system. An approach was made to the Optical Metrology Department of the National Physical Laboratory (N.P.L.) as it appeared that a light beam oriented system would obviate a lot of the difficulty experienced. It was fortunate that at the time of our approach to N.P.L. a general purpose auto-recording alignment device was under development which could be modified to suit our application.

The equipment, which was in its prototype form, (see refs. 2, 3 and 4) consisted of three units: A light source which took the form of a low powered Helium Neon Laser mounted on a datum; a lens for focussing the laser beam and an electro-optic position sensing detector. The beam from the laser was focussed by the lens to a small circular spot in the plane of the photo-electric detection unit as shown in Figure 7. Any movement of the lens would result in a proportionate movement of the image spot on the photo-electric detector. The location of this image spot was processed electronically resulting in its amplified movement being displayed on a chart recorder.

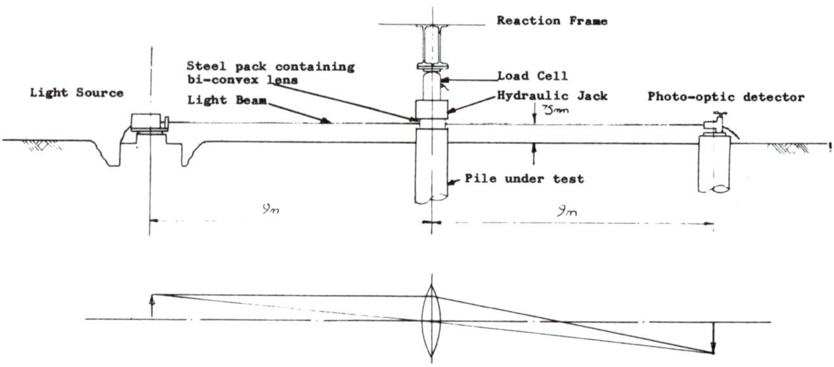

Figure 7 Arrangement & Principle of Optical Settlement Measurement

In order to remove interference from ambient light, a modulated system was employed and the signal/noise ratio was further enhanced by the use of optical filtering. The electronic processing of the signal from the position sensing photodiode included automatic gain control to compensate for any variation in laser output. Both in principle and design the optical system provided maximum immunity from sources of error resulting from tilt of the individual components. The equipment relied for reference on the position of the three units and did not utilise the light beam at all as a reference. The datums for the equipment were a redundant pile and a 5 cu.m concrete base, located 12 m apart so that the test pile was mid-way between them.

Figure 8 Laser on Base

Figure 9 Assembly at Test Pile Head showing Lens Mount

Figure 10 Detector Head Mounted on a Pile

The laser was mounted on one datum (see Figure 8) and the sensor on the other (see Figure 10). The lens was mounted in a brass mount which was able to slide in a vertical slot in the mild steel pack surrounding it (see Figure 9). The brass mount was spring loaded against the bearing plate on top of the pile, thus obviating any crushing of the pack being recorded as an apparent deflection of the pile. Thus the deflection measurement was monitored at the point where it was required, i.e. the centre of the pile. The settlement was displayed on a 250 mm bed chart plotter which was located in an instrument house remote from the test; a second pen on the recorder provided continuous monitoring of the signal strength arriving at the sensor. The intensity trace was particularly useful when the site was unattended, since the beam was broken on occasions by animals. These interruptions could easily be isolated from sudden pile movements since signal intensity dropped simultaneously.

The instrument was calibrated such that a pile deflection of 1 mm produced a full width deflection on the chart plotter. Increments of load were applied and maintained until settlement decayed to 0.1 mm per day, usually taking about forty-eight hours. The trace resulting on the chart recorder had a noise oscillation superimposed on it, of amplitude equivalent to a pile movement of 0.015 mm. In addition there was a diurnal variation which was difficult to allocate but it is thought that it could be partly attributed to:-

1) Small variations in the applied load.

2) The proximity of the laser beam to the ground, this could also be the reason for noise.

3) Temperature sensitivity of the prototype electronics since the instrument house was not maintained at a uniform temperature. The electronics have since been modified to overcome this.

The position of the laser beam was dictated by the design of the present rig. On future applications the beam would be repositioned by either elevating the three units or excavating a trench beneath the light path.

CONCLUSION

The equipment discussed proved to be generally satisfactory but for future work it is suggested that improvements should be looked for in the area of data logging of all the measurements taken.

It is felt that it is important to have some form of real time visualisation for site control, backed up with recorded data of a form which can be subsequently processed directly by a computer.

ACKNOWLEDGEMENTS

The authors wish to thank the Department of the Environment, South East Road Construction Unit, Director B.F. Edbrooke, B.Sc., M.I.C.E. and Kent County Council, Roads Department, County Surveyor A.D.W. Smith, B.Sc. (Eng)., F.I.C.E., for permission to present this paper. The photographs of the laser equipment are reproduced by permission of the Director of the National Physical Laboratory.

Special thanks are due to Messrs. Harrison, New, Tolman, Debnam and Jenkins of the National Physical Laboratory and Messrs. Cooke and Price of the Building Research Establishment for their continued help and encouragement throughout the work. The help of colleagues, Messrs. Ahmed, Hitchcock, Morgan, Morsley and South in connection with work on site is acknowledged.

REFERENCES

1. Debnam R., Jenkins R. (1971) The Influence of End Caps on End Loading Conditions. IMEKO Conference Paper No. 4.

2. Harrison, P.W., Tolman, F.R., New B.M. (1972) The Laser for Long Distance Alignment - a Practical Assessment. Proc. Instn. Civil Engrs. 52 (May) 1-24.

3. New B.M. (1972) Personal communication.

4. New B.M. (1973) A Versatile Electro-optical Alignment System for Field Application - to be published.

5. Whitaker T. and Cooke R.W. (1966) An Investigation of the Shaft and Base Resistances of Large Bored Pile in London Clay. Large Bored Piles. Instn. Civil Engrs.

6. Whitaker T., Cooke R.W. and Clarke G.W. (1962) 100 Ton Load Cell for Pile Loading Tests. Engineering, Lond., 194 (5040), 693-695.

7. Whitaker T. (1964) Load Cells for Bored Piles. Proc. Instn. Civil Engrs. 29 (Sept.), 151-154.

8. Whitaker T. (1964) Load Cells for measuring the Base Loads in Bored Piles and Cylinder Foundations. Build. Res. Statn. current paper, Engng. Series No. 11.

MEASUREMENT OF GROUND STRAIN AND TILT ARISING FROM MINING SUBSIDENCE

B.N. Whittaker and D.J. Forrester

Lecturer and research student, Department of Mining Engineering, University of Nottingham.

MINING SUBSIDENCE

Mining subsidence is the mass lowering of a body of strata overlying mine workings whose extracted area is sufficiently great as to permit collapse of immediately overlying strata and subsequent propagation of the movement to the surface. In the United Kingdom the ratio of extraction width/depth for mine workings needs to be greater than 1/10 before a discernible depression is formed at the surface.

Basic principles

Mining in the U.K. is mainly from fairly thin, flat, stratified deposits. This generally requires the use of longwall mining methods which results in a particular form of surface subsidence, namely the trough type, as illustrated in Figure 1. The maximum vertical displacement of the surface occurs perpendicular to the centre of the underground extracted area, and its magnitude is proportional to the extracted seam height and extraction width/depth ratio. The vertical displacement increases as this ratio increases up to a critical value of 1.2-1.4, and beyond which there is no further increase. The lateral and vertical displacements of the surface extend well beyond the lateral limits of the extracted area. Tensile strains exist outside these limits at the surface and compressional strains within them, the total sum of surface strains being equal to zero. Surface tilt is a maximum at the transition point.

Principal effects on surface structures

The effects of mining subsidence on surface structures are threefold, firstly the lowering of the land itself, secondly the relative lateral movements (or surface strains) and thirdly the tilting of the ground. The strain and tilt effects can be transitional or permanent in nature depending upon the position of the structure relative to the underground workings. Although subsidence may give rise to land drainage water flow and sewer drain gradient problems, the effects of strain and tilt are more severe on surface structures such as tall and long buildings, chimneys, factories containing precision equipment, railways, power transmission lines, and public service installations. An example of a relatively unsubstantial structure as affected by mining subsidence is shown in Figure 2.

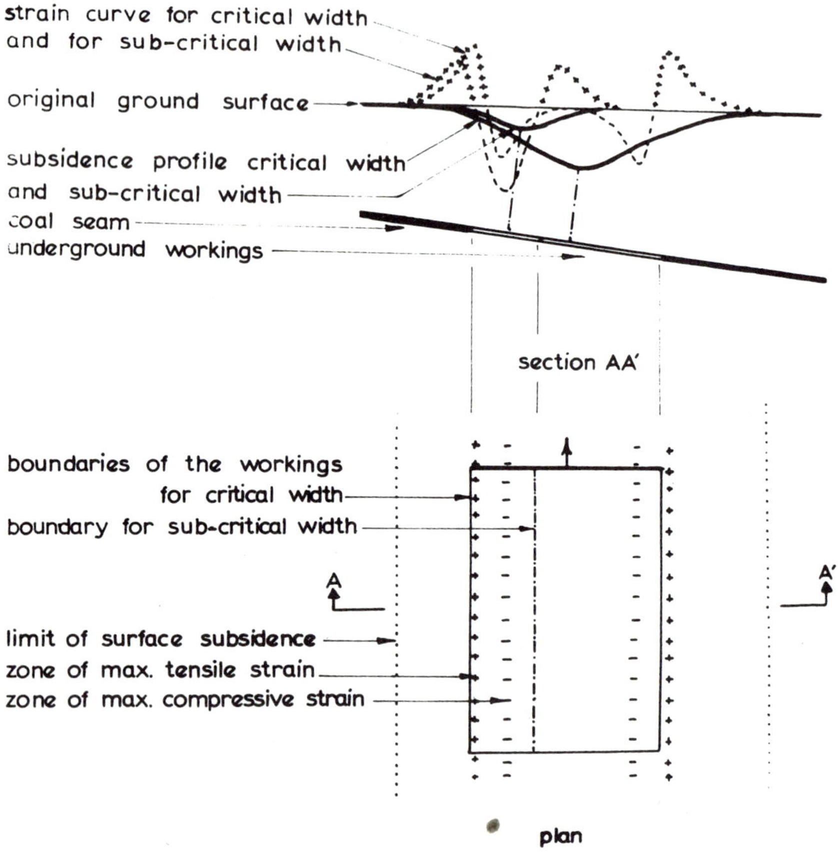

Figure 1 Mining subsidence and surface strains

REASONS FOR AND DEVELOPMENT OF NEW MEASUREMENT TECHNIQUES

The measurement of mining subsidence effects presents difficulties due firstly to the duration of movement, which can take from a few weeks to over a year before movement has completely ceased; and secondly to the often small magnitudes of the quantities to be measured, for example the maximum strain developed may only be 1½ mm/m. The traditional methods of subsidence measurement involve the use of accurate levelling for vertical displacements and steel tapes for linear displacements. Such measurements were sufficient to establish the main principles of trough subsidence and its prediction as presently recognised in the U.K. Wardell(1954), Orchard (1959). For more detailed analysis of case studies, however, it was felt that the above techniques needed to be supplemented by new methods. Such methods were considered to require accuracy together with consistency, a high degree of sensitivity to detect the onset of initial movement, a means whereby tilt can be measured directly rather than be derived from traditional subsidence measurements, and a combination of convenience and practicability. In order to fulfil these

Figure 2 Compressional strain effect due to mining subsidence

requirements, two techniques were envisaged, the first to measure tilt directly and the second to increase the accuracy and ease of ground strain measurements.

A TECHNIQUE FOR DIRECT TILT MEASUREMENT

In principle this technique involves the mounting of a sensing head, which indicates tilt directly, on special field stations which can be installed where and when required.

Standard equipment and its adaptation

The sensing head used was the B.A.C. Electrolevel, the operational principles of which are as follows: the head consists of a glass vial with a curved upper surface and three platinum electrodes, rigidly and symmetrically fixed along the lower surface, mounted in a casing. This provides both protection and a reference plane. The vial contains an electrically conducting liquid and a small air bubble. The electrodes are connected by a detachable and flexible lead to an alternating bridge circuit and moving coil galvanometer powered by standard voltage batteries. The components are mounted in a robust and compact case for convenient use in the field. A change in vertical angular position of the sensing head alters the electrical resistance of the electrodes and this displayed on the galvanometer scale which is calibrated in mm/m tilt. There are three scale ranges which have full scale deflections and respective reading accuracies of 0 to $\pm$10 $\pm$0.1 mm/m, 0 to $\pm$2

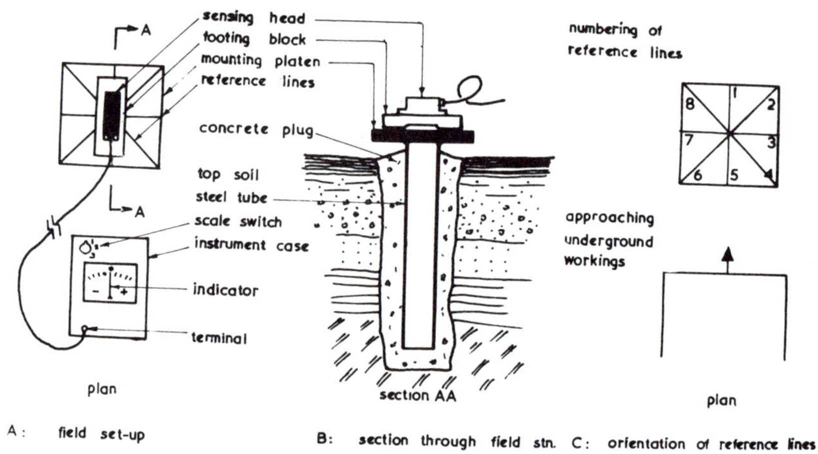

Figure 3 Method of measuring ground tilt

$\pm$0.02 mm/m and 0 to $\pm$0.5 $\pm$0.01 mm/m.

Field station design The sensing head required a simple permanent base in order to be used with the final design of field station. This base consists of a rectangular steel block with four accurately ground foot-pads and the sensing head rigidly attached to the upper surface. The field station provides a stable, easily accesible, cheaply produced and simple-to-install reference base for the sensing head. The station consists of an accurately ground cadmium plated platen which is rigidly attached to a 10cm diameter x 120cm long galvanised steel tube. There are four reference lines at 45° spacing engaraved on the platen surface, one line of which is set parallel to the line of direction of advance of the underground workings. The station is set in concrete and accurately levelled to a maximum tolerance of ± 0.5 mm/m of tilt with the electro-level at the previously prepared site. These features are illustrated in Figure 3. The platen surface is protected from the elements, accidental and wilful damage by a close fitting rubber-lined lid attached by key-operated screws.

<u>Field measurement of tilt</u>

In order to establish a consistent and reliable procedure for field measurement, it was necessary to consider the implications of using such a sensitive instrument as the electrolevel on permanent field stations. There appeared to be three principal sources of error arising from the use of this technique.

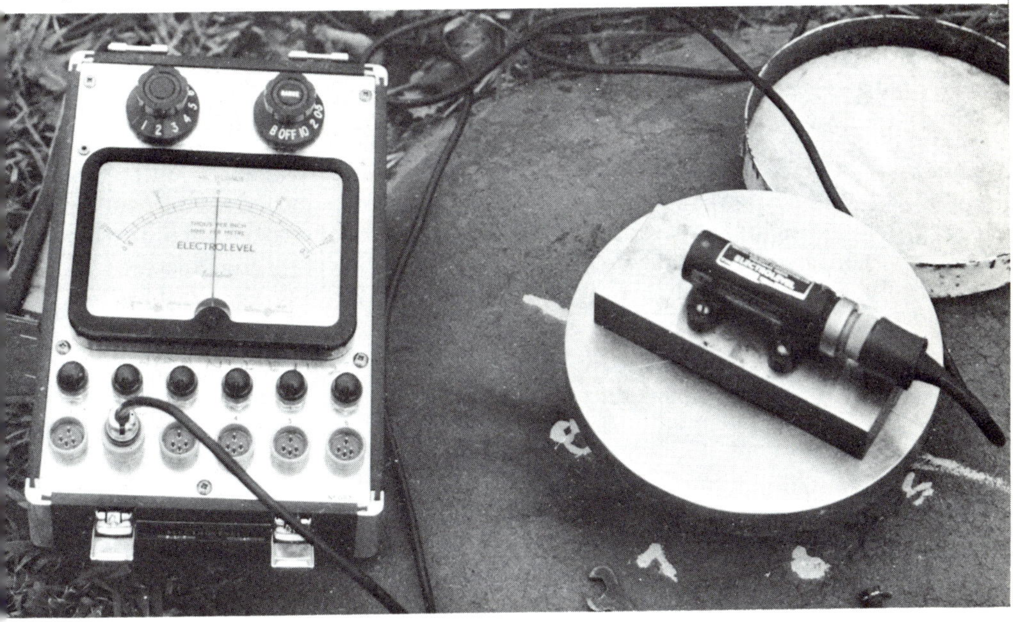

Figure 4 Electrolevel tilt measuring instrumentation

These are as follows: (a) instrument error - that possibly arising from the manufacture, assembly and time dependent drift of the instrumentation; (b) positioning error - that due to possible incorrect positioning of the sensing head over the station reference line; (c) surface contamination error - that due to the presence of dust particles between the contact surfaces. The following procedure was developed to take these points into account. In the case of (a) readings were taken on each reference line in both directions and the mean value determined. For each individual reading the contact surfaces were wiped clean and by taking the average of at least three acceptable values for each reference line, the effects of (b) and (c) above were minimised.

Measurement procedure For subsidence work, it is important to determine the change in tilt at a station caused by the advance of the underground workings in the direction of that advance. Thus one of the reference lines on each field station must be positioned parallel to that advance. A more complete picture, however, of the development of tilt is produced by measuring the tilt on four reference lines mutually orientated at 45° about the centre of the platen surface. To measure the tilt on each reference line, taking into account the previously mentioned errors, readings are made in both directions using the following procedure: the contact surfaces are wiped clean and the sensing head is firmly placed over the first reference line. It is then connected to the instrument case, the required scale range selected and a reading taken and booked. The field set-up is shown in

SITE: STAFF:
DATE: WEATHER:
STATION 11 REMARKS:

LINE	1st	2nd	3rd	4th	Av.	TILT
1-5	0·80	0·80	0·80		0·80	0·86
5-1	-0·93	-0·92	-0·94	-0·93	-0·93	
2-6	-0·33	-0·33	-0·33		-0·33	-0·33
6-2	0·33	0·33	0·33		0·33	
3-7	-1·33	-1·33	-1·33		-1·33	-1·26
7-3	1·20	1·20	1·20		1·20	
4-8	2·10	2·20	2·30	2·20	2·20	2·28
8-4	-2·35	-2·35	-2·35		-2·35	

SITE: STAFF:
DATE: WEATHER:
STATION 11 REMARKS:

LINE	value1	value2	initial	'zero'	TILT	chng
1-5	0·80	-0·93	-0·23	0·06	0·86	1·09
2-6	-0·33	0·33	-0·11	0·00	-0·33	-0·22
3-7	-1·33	1·20	0·18	0·06	-1·26	-1·44
4-8	2·20	2·35	0·02	0·08	2·28	2·26

B: SUMMARY SHEET (above)

A: FIELD BOOKING SHEET

Figure 5 Typical field booking of tilt measurements

Figure 4. This procedure is repeated until three readings agree to the required accuracy and their average then taken. This must be done for each reference line in both directions and the mean of the two average values is the tilt on that line, the sign of the tilt indicating a rise or dip. A sample booking and calculation of tilt for the four reference lines on one station is given in Figure 5. The change in tilt is the difference between the initial value, taken before underground workings began to influence the field station, and the later value.

Sign convention for tilt It is considered important to emphasise the sign convention used in the above procedure. This is possibly best explained by means of an example. The two ends of the sensing head were referred to as the 'head' and the 'tail' the latter being the end with the protruding cable. When the sensing head is placed on reference line 1-5, with the 'head' at 1, then on reading an anticlockwise swing of the needle indicates a rise along that line in direction 1; conversely, if the sensing head is placed on the same line with the 'head' at 5 then the needle would swing clockwise indicating a dip in direction 5.

Presentation of results

The results for a particular field investigation are generally presented in two ways. Firstly, as tilt-development curves by plotting change in tilt against time, the latter being expressed as the ratio of working advance/depth; secondly as tilt profile curves by plotting total change in tilt against lateral displacement of the field stations. For these latter curves several field stations are required to be positioned across the direction of underground advance. Some typical

Figure 6 Ground strain measuring instrumentation

results are shown in Figure 9(a).

A TECHNIQUE FOR GROUND STRAIN MEASUREMENT

This form of instrumentation was developed to give a rapid and accurate means of observing strain between special field stations on the surface. It was decided to use a standard steel tape of fixed arbitrary length and to design a suitable means of tensioning it quickly and precisely together with a practicable means of observing relative changes between two field stations.

Field station design

In view of the periodic nature of the measurements required for the subsidence observations it was decided to use steel tubes concreted in the ground for permanent field stations. Past experience with this design of station had shown it to be most suitable for this type of work. The tops of these

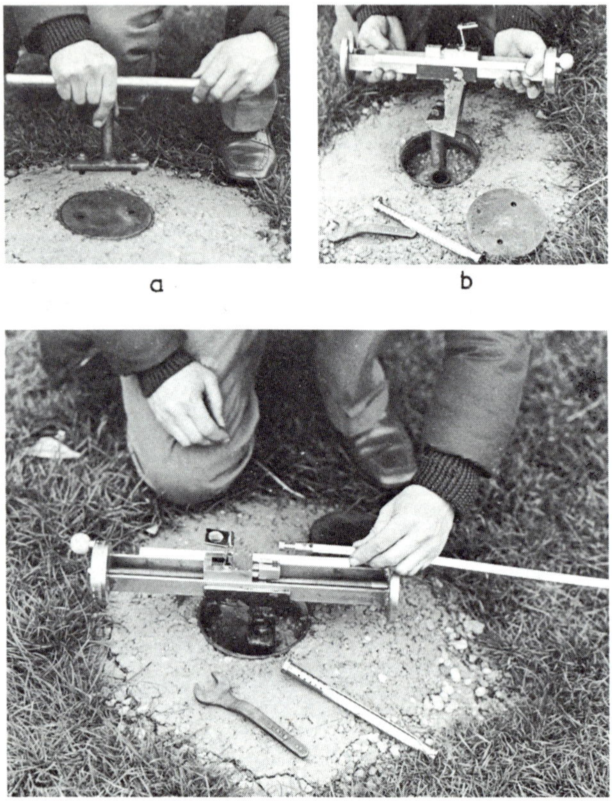

a b

c

Figure 7 Measurement of ground strain

tubes were accurately machined internally to accept special registration plugs which were designed to carry the tape suspension, tensioning and measuring components. Each field station was located truly vertical and below ground level and was protected by a short length of larger diameter tube placed around the top of the inner tube and secured in the concrete. A coverplate screwed into this outer tube in order to protect the inner one. The design ensured that the registration plugs could be rotated so that linear measurements could be made in any direction.

Extensiometer design features

The extensiometer is designed so that the tape can be tensioned and read in one operation. Traditional surveying practice for accurate linear measurements such as a base line involves tensioning and measuring as separate operations which leads to a potential source of error. The combination of these two operations was achieved by compressing a spring by a definite amount with the extensiometer so that when the correct tension was achieved the instrument was ready for reading without risk of change in tension. The amount of

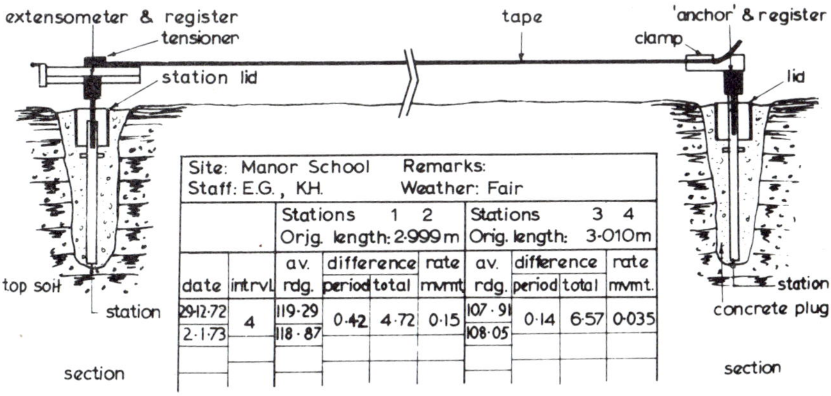

Site: Manor School		Remarks:							
Staff: E.G., KH.		Weather: Fair							
		Stations 1 2 Orig. length: 2·999 m				Stations 3 4 Orig. length: 3·010m			
		av.	difference		rate	av.	difference		rate
date	intrvl	rdg.	period	total	mvmt	rdg.	period	total	mvmt.
29·12·72 2·1·73	4	119·29 118·87	0·42	4·72	0·15	107·91 108·05	0·14	6·57	0·035

Figure 8 Method of measuring and booking ground strain

compression is observed by the coincidence of two finely scribed lines; a magnifying glass is incorporated to assist observing when the correct tension is achieved. The tape tension applied is 90N. The extensiometer has two scales; the first is located parallel to the main screw tape-tensioning shaft and has a 200 mm scale graduated every 2 mm. The second scale is circular and located on the end of the screw shaft; it is graduated 0 to 200 in 100 divisions. One rotation of the shaft is equivalent to 2 mm which means that one graduation on the circular scale is equivalent to 0.02 mm. The screw shaft carries a rider, which incorporates the tensioning spring, associated coincidence lines and a recessed seating for the tape-tensioning fitting to be located.

Measuring procedure

The coverplate protecting the registration plug sleeve was flush with the upper surface of the concrete in order to prevent obstructing land machinery. Figure 6(a) shows the coverplate just prior to removal. This design of coverplate tended to discourage wilful interference. Figure 6(b) illustrates the method by which the extensiometer was mounted at a field station; the split shaft and expander both en-

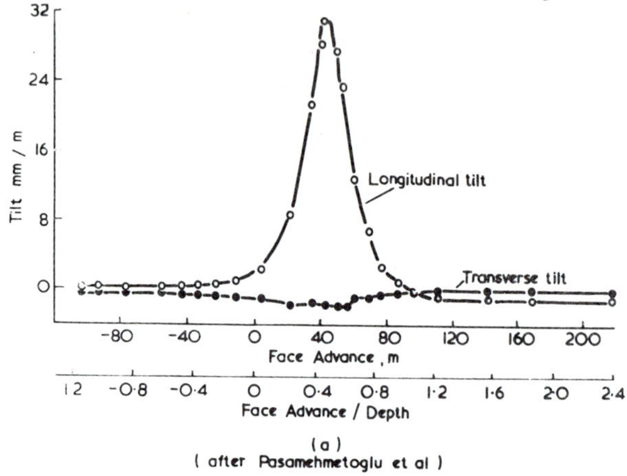

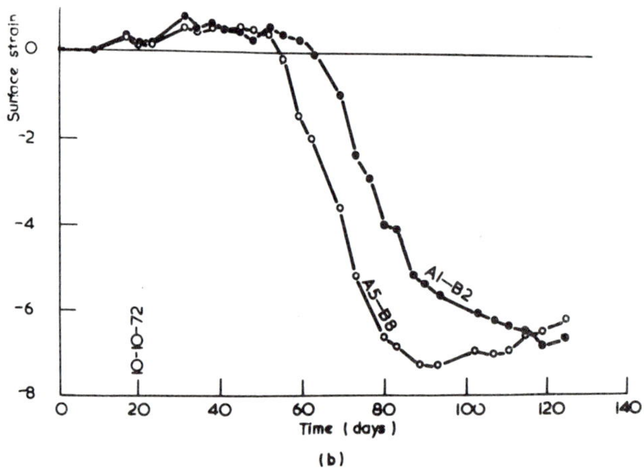

Figure 9 Typical ground tilt and strain results

abled the extensiometer to be firmly locked in any position. Figure 6(c) shows the tape about to be located on the extensiometer. When the tape is attached to the extensiometer, the latter is brought into final alignment by sighting down the tape. Figures 7 and 8 show the field set-up during the reading and booking of the extensiometer. In view of friction being present in an instrument of this type, it is necessary to apply the tape tension by rotating the tensioning handle in one direction only. Temperature is recorded simultaneously with each displacement reading. Figure 9(b) shows some typical results obtained with this form of instrumentation.

CONCLUSIONS AND RECOMMENDATIONS

The forms of instrumentation described in this paper have significantly broadened the scope of field investigation work in the case of mining subsidence and its effects. The onset of ground movement can be detected more precisely with such equipment which is both portable and easily operable by one person. Specific case studies can be investigated with a greater degree of precision. It is recommended that where such measurements are to be taken in the field, particular attention is given to station design and installation. The field station designs described have undoubtedly contributed greatly to the success of the experimental work.

ACKNOWLEDGEMENTS

The authors wish to express their gratitude to Professor H.J. King, Head of the Department of Mining Engineering, Nottingham University for his valuable comments and advice. Thanks is also due to Mr. C.H. Shadbolt, Regional Subsidence Engineer, N.C.B., and his staff for their collaboration and assistance with the field work. Gratitude is also expressed to the National Coal Board for their financial assistance given to this project.

REFERENCES

Wardell, K. (1954) Some observations on the relationship between time and mining subsidence. Trans.Inst.Min.Eng. Lond., 113.

Orchard, R.J. (1957) Surface effects of mining - the main factors. Trans.Inst.Min.Eng. Lond.116, 941.

Pasamehmetoglu, A.G. King,H.J. and Whittaker,B.N. Time dependent aspects of mining subsidence (to be published).

INSTRUMENTATION FOR THE KING'S LYNN SOUTHERN BYPASS

P. F. Wilkes

Norfolk County Council

INTRODUCTION

The instrumentation for the King's Lynn Southern Bypass has been carried out in three phases. First a trial embankment was constructed which was extensively instrumented with piezometers, inclinometer tubes, settlement gauges of several types, slip surface indicators, lateral movement and heave and thrust gauges. Two papers have been written on various aspects of the trial embankments, Wilkes (1972a) and (1972b). As can be seen from Fig. 1 the route of the bypass crosses the River Great Ouse some 200m downstream of a major slip in the river banks, described by Skempton (1945). In view of this history of instability, the Great Ouse River Authority expressed concern at the proposed piling methods for the new bridge foundations in the river banks. In order to demonstrate that the piling would not be injurious to the bank stability, a piling trial was carried out in the summer of 1971, for which piezometers and inclinometers were installed. The last phase was the installation of various instruments; piezometers, inclinometer tubes and two types of settlement gauge, for monitoring the construction of the bypass. This work was carried out as part of the Contract for the Bypass.

SITE CONDITIONS

These are described in detail in the three papers already mentioned, i.e. Skempton (1945), Wilkes (1972a) and (1972b). For the sake of completeness however, a brief description is as follows: the major portion of the bypass crosses an area of alluvial deposits overlying Kimmeridge Clay. The clay lies at a depth of approximately 13m in the vicinity of the River Great Ouse, rising to 6.5m at the Saddlebow Intersection and to 3m at Hardwick, the top of the Kimmeridge being weathered up to 3m in depth. At the eastern end of the bypass the Kimmeridge is overlain by Lower Greensand, the contact between the two being concealed by Glacial Deposits. The alluvium consists of three main groups; estuarine deposits of finely bedded clayey silts and fine sands, overlying fen deposits which consist of fen peat and soft silty and organic clays. Lenses of glacial drift occur within the latter layer.

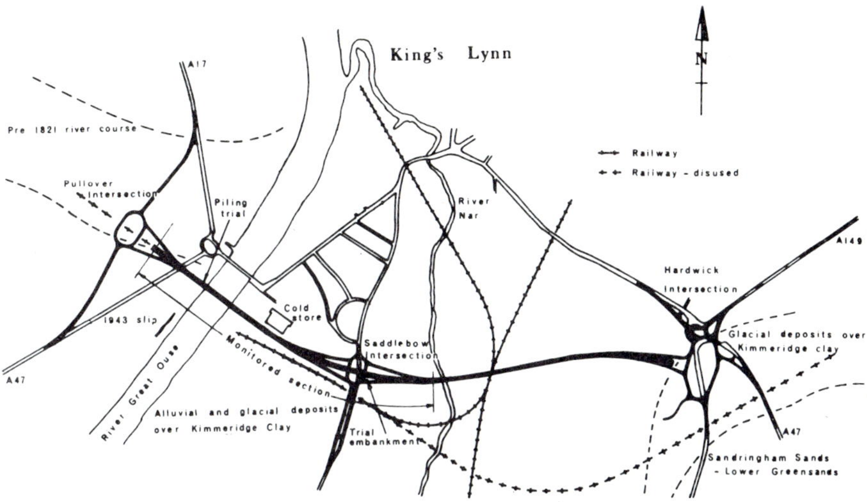

Figure 1 Plan of Bypass

INSTRUMENTATION

During the planning stages of the three phases of instrumentation, it became apparent that few published accounts existed on the problems associated with installation of the type of instrumentation required. It is therefore felt that perhaps an account of the difficulties that arose and also the methods of installation used at King's Lynn may be of some interest.

Piezometers

All piezometric installations were of the hydraulic type, using a push-in type of piezometer tip, Wilkes (1970), installed in a 75mm diameter borehole, drilled with a continuous flight auger. The borehole finishing 300mm above the final level of the piezometer tip, which was then pushed to the required level and the borehole was grouted with a 3:1 bentonite and cement grout. With this method it was found that the malfunction due to piezometer damage was negligible. Additional piezometers were later installed in hand auger holes, initially the piezometer tips were placed within a sand filter and at the bottom of the boreholes and sealed with a layer of punned bentonite and cement balls. However after problems with the leads parting during the punning operation, the original method was reverted to, i.e. the tip was pushed beyond the base of the hand auger hole which was then grouted. The success of this method for the trial embankment led to its adoption for the other phases of instrumentation, except that the Contractor was given the option of drilling the boreholes with either a continuous flight auger of between 75mm and 150mm in diameter, or a 150mm shell and auger. In both cases the Contractor was required to take precautions in drilling the last 500mm of the borehole to minimise the zone of disturbance in the vicinity of the ultimate position of the piezometer. In the case of the piling trial the Contractor elected to use a shell and auger rig as the majority of piezometers were positioned in the river bank below the high tide level. For the monitoring installation the Contractor commenced with a 75mm continuous flight auger, but due to problems in keeping the borehole open for sufficient time to allow the piezometer to be placed and grouted, the diameter was increased to 100mm. The placing tool was also modified to enable the grout to be pumped through the shaft, so that the boreholes remained open for the minimum amount of time, and to minimise possible disturbance of walls of the borehole. This method of installation has proved rapid, it being possible to install an average of 10 piezometers per day. With the number of piezometers being placed it was considered desirable to have a check on the depth of installation of the piezometers; this was achieved by attaching a standard 10 metre length of leads to each piezometer prior to installation, the depth being checked by measuring the length of tubing remaining above ground. Once the piezometer tip is in the ground the next operation is the connection to the gauge house installation. Experiences at King's Lynn have shown that this is the most time consuming part of the piezometric installation. Leads have to be laid out and jointed and each pair has to be coded to enable repairs to be effected, when some machine involved in the main works severs the most important batch of leads at the most significant time! Setting up the gauge house, filling manometers and de-airing the system all take up a considerable amount of time, which must be taken into account in planning any programme of works. Coupled with this an allowance for a "settling in" period for the completed installation of at least fourteen days, or twenty-eight when tidal conditions affect the site, should be made. This period is necessary not only to enable background data to be obtained but to check that the system is fully functional. The majority of faults at King's Lynn occurred during this "settling in" period, these were traced to faulty fittings, allowing air to enter the system. The more common faults encountered were careless stripping of the polythene coating to nylon tubing of the leads, leading to cutting the surface of the nylon which eventually fractures at this point; faulty tightening or fitting of couplings; faulty valves. The manufacturer has since modified the type of valve supplied. It is appreciated that air or gases can enter the system from other causes. Gases can occur in solution in the soil or be formed by decomposition of organic material as in some types of peat. This was not however the case at King's Lynn. A certain amount of air also enters the system from the walls of the nylon tube but this is a gradual process which ceases when the nylon becomes fully saturated and can be reduced by soaking the tubing prior to installation De-airing the system can cause a few problems if care is not taken. Tidal conditions can induce considerable suction in the installation, the fittings of which are not designed to withstand high suction and air can thereby be drawn into the system. At King's Lynn the installation for the piling trial was subject to a normal tidal range of the order of 5 metres. This was exceeded at spring tides, and considerable suctions were produced, drawing air into the system in large quantities necessitating daily de-airing. This at first appeared to render the installation virtually useless for obtaining reliable readings during the pile driving. It was found however that by de-airing at low tide, i.e. when the external pressures were at their minimum, the intervals between de-airing were considerably extended. During the pile driving operations when it was not possible to de-air whilst pore pressure readings were being taken, some air was noticed in several of the leads. The length of each air bubble was carefully measured and noted and on completion of the test, the affected piezometers were de-aired and air bubbles of varying sizes were deliberately induced into the system, the lengths were measured and the effect on the pore water pressure was noted. Thus it was possible to correct the test data for air in the system. Another danger inherent in the de-airing process, particularly prevalent when an operational system is involved, is the production of residual pressures (or suctions) at the piezometer tips in excess of the actual pressures which will, until full dissipation of the residual pressure or suction, give erroneous readings. In the extreme, hydraulic fracture can occur, this phenomena has been described by Bjerrum et al (1972). At King's Lynn when hydraulic fracturing was deliberately induced at the trial embankment site the period for full dissipation of excess pressures was of the order of 24 hours. It can therefore be seen that a certain amount of care must be taken during de-airing to balance the applied pressures and suctions to ensure that the residual pressures induced at the piezometer tips by de-airing are minimal. Measurements of the pore pressures were made by means of the usual pairs of mercury manometers and also by the use of an automatic unit which was first used on the piling trial where, for comparative purposes, it was necessary to measure the tidal effects on the system prior to pile driving. This involved the collection of a considerable amount of data, readings being taken at half hourly intervals over several tidal cycles each of 28 days. Fig. 2 shows a set of typical tidal curves obtained for one piezometer. The automatic system is based on the apparatus developed by the T.R.R.L: a solenoid valve is connected in one lead from each piezometer, these valves are opened in sequence and the pressures are measured by a transducer, connected to a common busbar, the pressure being indicated on a digital display and also printed onto a tally toll. The equipment has subsequently been modified to take a total of 114 channels and is being used for the monitoring installation in the Saddlebow Intersection area. In order to reduce the length of hydraulic leads half of the piezometers are connected to a "remote" transducer unit some 350 metres from the automatic scanning unit and connected to it by multicore cable. Facilities are built in to enable the "remote" solenoids to be opened independently of the main unit for de-airing purposes. As can be seen from Fig. 3, the second tube from each piezometer is connected to a manometer. This connection is considered necessary in case of breakdown of the automatic unit due to power failure or electronic faults. The manometers also enables a rapid check to be made on the accuracy of any reading given by the automatic unit or to decide whether de-airing is necessary. The manufacturer's recommended method of installation has the second lead connected to a second 'busbar' with manually operated valves to enable comparative readings to be made. It was felt that this could lead to human errors with the large number of tips connected to the unit, particularly with two separate gauge houses involved, the use of the manometers has simplified operations. In both installations initial teething problems occurred in the electronic equipment, but after these had been resolved the unit functioned reliably.

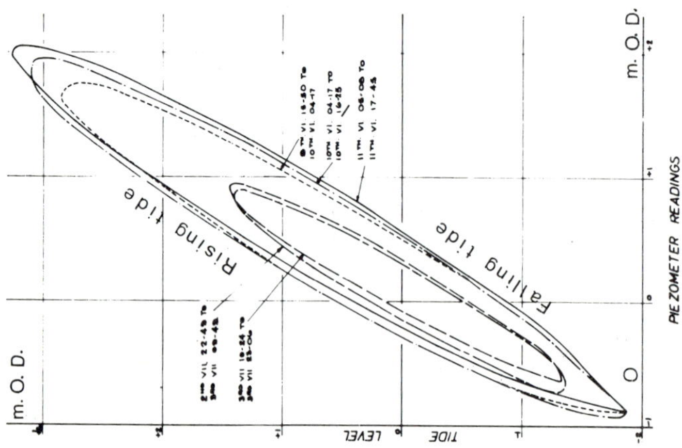

Figure 2 Typical tidal curves for Piezometer in river bank

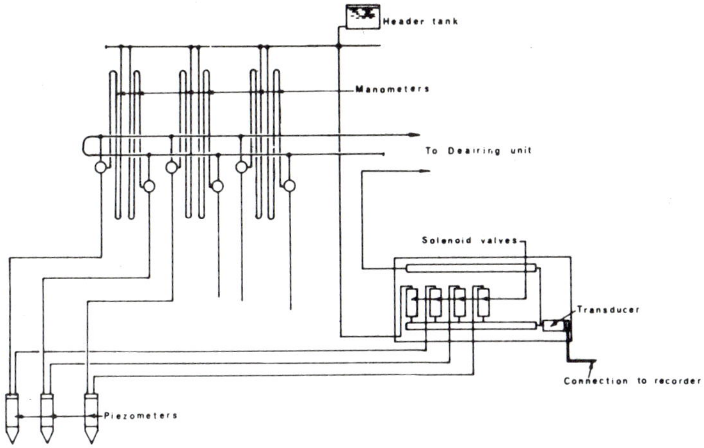

Figure 3 Diagrammatic arrangement for automatic recorder

Inclinometer Tubes

Two types of inclinometer tube have been installed, both manufactured by Messrs. Soil Instruments Ltd. The first was the now obsolete plastic type with internal keys and no serious problems were encountered with this type. The tubes were installed in 75mm diameter holes drilled with a continuous flight auger at the trial embankment site. The holes were filled with bentonite/cement grout and the tube pushed into the grout. At the piling trial 150mm boreholes were used and the tube was placed in the cased hole and grouted before removal of the casings. The other type is a new aluminium extruded tube, which was installed for the monitoring of the main works. Again 150mm diameter shell and auger holes were drilled, the boreholes were cased and the inclinometer tube inserted and grouted prior to withdrawal of the casing. However, after withdrawing the casing it was observed that some of the tubes had rotated so that the keyways were incorrectly aligned. Some tubes could be rotated to correct the alignment but with others this was not possible. This misalignment also occurred at another site. It is considered that the rotation is due to the borehole casing in the ground being allowed to rotate slightly whilst unscrewing a length of casing. In all cases the tubes were filled with water to overcome the buoyancy of empty tubes. Of the two types of tube it has been found that the plastic tube has the better behaviour characteristics. Considerable movement of the plastic tube was measured before fracture, whereas several of the aluminium tubes have broken with much less movement. A length of the aluminium tubing recovered from the ground was found to be considerably corroded after having been in position for three months; the anodised coating had almost completely disappeared and deep pitting of the metal had occurred. It is understood that the bentonite/cement grout could be the cause and the manufacturer is now recommending that the tubing should be coated with epoxy paint prior to installation. It is thus unfortunate that the plastic tubing is now out of production due to manufacturing problems. Where an inclinometer tube was positioned to pass through the fill it was considered that a transition should be provided between the more rigid fill material and the surface of the softer subsoil. This was achieved by the use of a box filled with bentonite slurry. As considerable settlements of the earthworks were anticipated it was also felt necessary to sleeve the inclinometer tubes through the fill and a length of 100mm diameter plastic tube was used with the void between the two tubes being filled with bentonite slurry. (These are shown diagrammatically in Fig. 5). The sleeving is perhaps not necessary when sufficient allowance for movement can be provided at the joints between the lengths of tubing. The inclinometers themselves have not proved to be troublefree, the original model suffers from considerable zero drift which must necessarily be adjusted daily, and the later model with digital output has developed a number of faults. A printer unit can be connected to the digital unit, which has also given some trouble. The manufacturer has given good service in correcting these faults as they have occurred.

Settlement Gauges

Three types of settlement gauge have been installed

Incremental settlement gauges These have measuring points at the top of the Weathered Kimmeridge, at the top and bottom of the peat layer and at ground surface. These gauges were installed in a 200mm diameter borehole. For the trial embankment the measuring points consisted of a metal ring cemented to the bottom length of the tube as the lower point, with expanding aluminium alloy 'arrow' plates keyed into the walls of the borehole at the top and bottom of the peat, and a metal plate placed at ground level. A telescopic access tube enabled a probe to be lowered to detect the position of the plates. It was found that the lower ring did not give a true indication of the movement of the weathered Kimmeridge and was probably influenced by movement of the tube to which it was cemented. In one or two cases it was possible that joints in the telescopic tube fouled the 'arrow' type plates. For the monitoring installation all points below ground surface were of the expanding 'arrow' type, and the access tube was made of lengths of single sized tubing with cemented joints. These latter were not allowed to be within 300mm above and 1.8 metres below any 'arrow' plate. The ground plate was as before, with the telescopic tubing arrangement being used through the fill, as can be seen diagrammatically in Fig. 5. This system has proved to be reliable and rapid to read.

Hydraulic settlement gauges Sixteen of these were installed beneath the trial embankment and were of the basic B.R.S. overflow weir pattern, with standpipes in the gauge house. Since the level of the weirs was below the level of the gauge house floor it was necessary to apply a known suction to the upper ends of the standpipes. This was applied by means of a water column in a moveable pot, Fig. 4. Eight standpipes were connected into the upper, fixed, container with brass compression fittings. Other fittings were included by the manufacturer to facilitate filling and setting up. Problems arose from leakage into the upper container producing unstable conditions within the partial vacuum system. Considerable maintenance was necessary to rectify these faults but even so, some of the readings obtained were subsequently found to be suspect. It was also found that the reading of a number of gauges of this type was time consuming, particularly as more care and patience is required with these units than with other instruments. It is therefore considered that this type of instrument is, in this form, not ideally suited to highway work. However with a re-designed suction system, and in an installation where more time is available for readings to be taken, there is no reason why they should not be more successful.

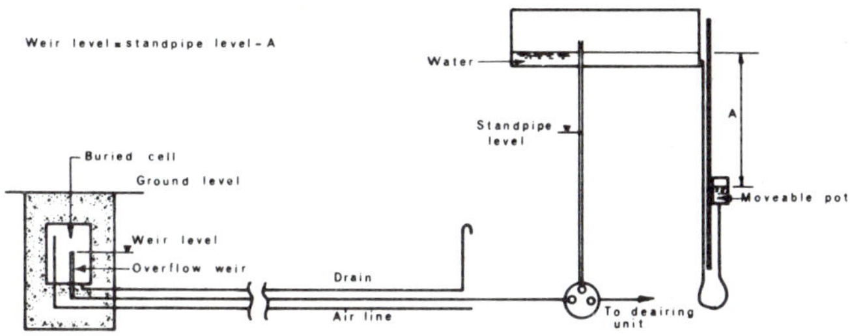

Figure 4 Diagrammatic arrangement of hydraulic settlement gauge

Horizontal settlement gauges This is a misnomer, describing the position of the access tube, which is laid horizontally in a trench beneath the embankment. A number of metal plates are positioned at strategic points across the width of the embankment, Fig. 5. A double probe which consists firstly of a detector for the metal plates which measures any lateral movement, and is followed by a second probe containing a small water-filled bladder which is connected to a pressure transducer in the cable drum; the level of the probe relative to the transducer datum being obtained by the changes in pressure in the bladder. At the time of writing there has been insufficient operating time to fully assess its potential. It is possible to obtain a full profile of the tube but this is time consuming, and at King's Lynn it has been decided only to measure the full profile at intervals. The main readings are taken at the metal plates, their lateral and vertical movements being plotted, which has considerably speeded up the process of reading. To enable more accurate measurement of lateral movements a steel tape was specified instead of the standard graduations on the cable as the cable has a certain amount of elasticity, unfortunately the consumption of tapes is rather high. Apart from minor 'teething' faults the instrument problems have mainly arisen in deairing the bladder system.

Self checking measured data

At King's Lynn the instrumentation is designed as far as possible to be self checking to ensure that all instruments are behaving in a compatible fashion. This is achieved in the case of settlement gauges, by having a mechanical fixing between the ground plate of an incremental settlement gauge and a lateral movement plate of a horizontal settlement gauge. Thus the level of that particular point is measured by two different instruments. The end lateral movement plates are similarly linked to the inclinometer tubes, allowing vertical movement between the two to occur. Lateral movement of a point near the top of the tube is thereby measured by two different methods. It is considered that this self checking is necessary in addition to individual standard instrument checks, to ensure that the equipment is fully functional under field conditions. Some precautions taken to check the readings obtained from the automatic pore pressure unit have already been mentioned, i.e. having a manometer connected to the second lead from each piezometer. In addition to this provision, each of the two gauge houses has a header tank connected to one channel. On each scan therefore two known readings are obtained, which, as the header tanks are positioned vertically above the transducers, can be checked with a steel tape. Variation from the measured distance indicates the probable presence of air in the transducer plumbing. It must be emphasised that with such sophisticated instrumentation it is considered necessary that every opportunity should be taken to ensure that such self checking methods are 'built-in' from the planning stage of any installation.

DESIGN PROBLEMS

In order to discuss the monitoring installation, a few words on the design problems of the bypass may assist in understanding the philosophy of the instrumentation. The site investigation report suggested a number of possible design methods for various areas and the trial embankment was constructed to test the efficiency of some of these and to enable full scale behaviour to be determined. However only methods directly concerned with the instrumentation will be considered in this paper. In the Saddlebow Road area and in the vicinity of the River Ouse, the height of the earthworks is considerably greater than the calculated stable height, construction being achieved by building the embankments with berms and in two stages allowing for a consolidation period of 275 days to follow each stage. Surcharge is also being applied with the second stage to assist consolidation. One section of the trial embankment was built to such a height that failure occurred and this also caused what is considered to be a sympathetic failure of the adjacent control section. From the analysis of these failures it was possible to deduce the safe height of first stage earthworks and the berms which were accordingly constructed to this measurement. The settlement measured beneath the trial embankment was greater than that deduced from the site investigation test results. This presents a problem in certain areas, as part of the route of the bypass follows the old Midland and Great Northern Railway line. The railway rose on embankment from the former South Lynn Station area to cross the River Great Ouse and to the west of the river was again on embankment, but the width of the old railway earthworks were considerably narrower than the width of the new highway embankment. A problem of differential settlement thereby occurs due to the preconsolidation of the substrata below the railway embankment. Another problem highlighted by the trial embankment was the considerable sub-surface lateral movements measured at the toes of the embankment by the inclinometer tubes. The bypass runs close to a cold store which has piled foundations, in order to prevent damage to these foundations;a sheet pile cut off wall was designed to contain these lateral movements. The problems of stability of the river banks have been mentioned previously

REQUIREMENTS OF THE MONITORING INSTALLATION

The basic philosophy of the monitoring instrumentation is the 'observational method' as discussed by Professor Peck in the ninth Rankine Lecture, Peck (1969). The areas instrumented are those where single stage filling approaches critical height and all areas filled in two stages. Fig. 1 shows the length of bypass being monitored and Fig. 5 shows the typical arrangement of instruments.

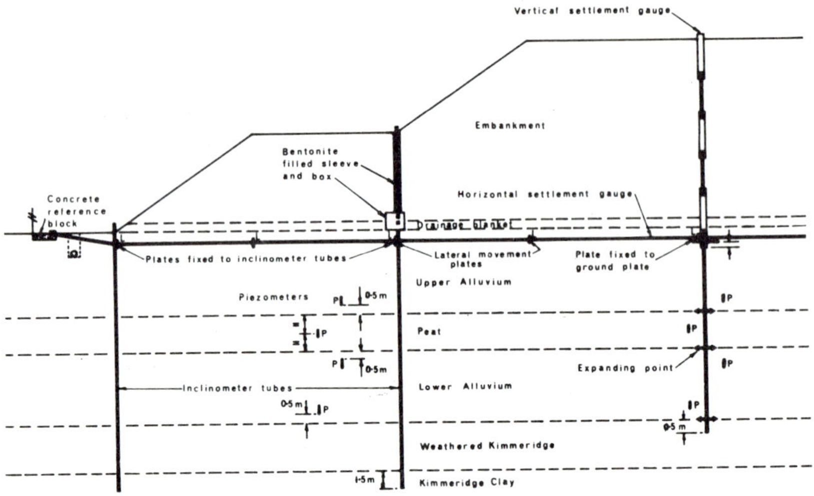

Figure 5 Typical arrangement of monitoring instruments

The piezometers are primarily intended to give information on dissipation of pore pressures, particularly during the first consolidation period. The Contractor will be allowed to place fill during this period provided that sufficient pore pressure dissipation occurs. This will assist his programme of Works, as although the contract period is 36 months, the construction of certain two stage areas is considered critical to the completion time. From the induced failure of the trial embankment it was possible to deduce the probable zone of slip surfaces; the lower piezometer in the outer groups was positioned in the vicinity of this zone in an attempt to obtain warnings of incipient failure by increases in measured pore pressures. Pore pressures are being recorded during filling operations and plotted against time. These pressure plots are combined with a plot of fill height against time, the fill height being converted to the equivalent height of water to give a direct comparison with the pore pressures. Standpipe piezometers have been installed clear of the affects of the works to enable changes in natural water table level to be monitored and a relation to be established with the monitoring installation. From the pore pressure records obtained from the trial embankment it was obvious that the detection of potential failure by this means was unlikely unless a piezometer happened to be on or near to the failure zone. Thus some other means had to be found to monitor the construction of the bypass, particularly in view of the low factor of safety, 1.25, to which earthworks were designed. It was felt that the inclinometer tubes could be the best means, as the behaviour of the complete profile, i.e. from the Kimmeridge to the surface, would be monitored. Hughes (1969) discusses culvert elongation as a possible means of monitoring earthworks and it was felt that the inclinometer data could be used in a similar fashion. It was noted that at King's Lynn, lateral movements were greater at some depth below ground level, than at the surface. By selecting a point at a level where maximum movement occurs and plotting the movement of this point, Δs, against time t a curved plot is obtained, Fig. 6. This method is not however an ideal form for monitoring. To allow for variations in the rate of fill Δs was divided by the fill height H, and it was found that by plotting $\Delta s/H$ against t a straight line was obtained, again as in Fig. 6, whilst stable conditions occur. As failure approaches, however the points deviate from this line, i.e. point A Fig. 6, as shown in the curve for tube 206 on the failed section. The curve shown for tube 401 in the sand drained section, which did not fail, does not show such a marked deviation. It is nonetheless considered that the plot shows that for the depth of fill placed over this section maximum stable conditions have been achieved since some deviation is shown beyond point A. The initial points on the $\Delta s/H$:t plot for this tube however do not lie on the straight line, and it is suggested that this is possibly due to the alluvial material being lightly overconsolidated.

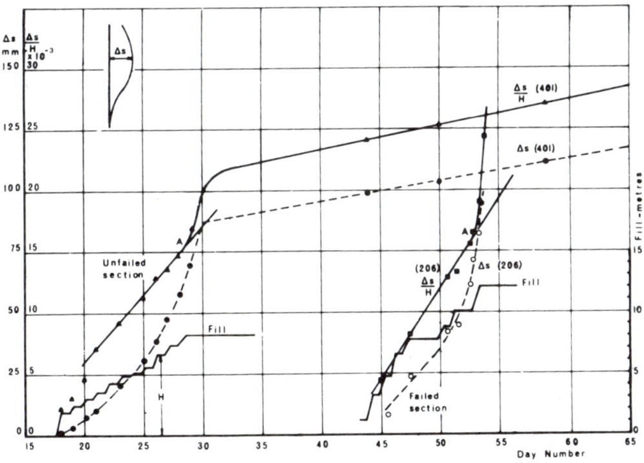

Figure 6 Plots derived from two inclinometer tubes, at a depth of 3 metres

On this basis the inclinometer tubes are being utilised as the primary instrument for detection of instability. Should an inclinometer tube break or be damaged, similar information can be obtained from the movement of the lateral movement plates on the horizontal settlement unit. As mentioned earlier the movement at the surface is not so great as a point below ground and a certain amount of sensitivity is lost. Inclinometer tubes are being used to detect any subsoil movements behind the sheet pile cut-off wall previously mentioned to check that the lateral movements are being contained and that there is little or no danger to the cold store foundations. They have also been installed in the River Great Ouse banks, in order to detect bank instability during pile driving operations. The measurement of the settlement profile with the horizontal settlement gauge is considered necessary to detect areas of differential settlement, particularly that due to the partial preconsolidation mentioned earlier. Additional surcharge can then be placed, subject to the observance of stability criteria, to speed up the consolidation process in such areas in an attempt to minimise the long term effect on the finished Works. In view of the relatively large settlements measured during the filling operations for the trial embankment, of the order of 0.25m, a considerable volume of fill material could be 'lost' beneath the embankment. Consequently the horizontal settlement gauges are being used to measure this volume to allow payment to the Contractor for the additional filling required. Incremental settlement gauges have been installed to enable the consolidation of the various layers to be determined and their behaviour checked against the results of the trial embankment. These gauges also act as a check on the horizontal settlement gauges.

ACKNOWLEDGEMENTS

The author would like to thank P. Deavin, M.C., C.Eng., F.I.C.E., F.Inst.H.E., County Surveyor, Norfolk County Council and F. J. S. Best, C.Eng., F.I.C.E., M.I.Mun.E., F.Inst.H.E., Director, Eastern Road Construction Unit for permission to publish this Paper.

REFERENCES

Bjerrum L, Nash J.K.T.L. Kennard R.M. and Gibson R.E. (1972) Hydraulic fracturing in field permeability testing. Geotechnique XXXII, 2, 319-332.

Hughes, J.M.O. (1969) Culvert Elongations in fills founded on soft clays. Canadian Geotechnical Journal VI, 2, 111-117

Peck, R.B. (1969) Advantages and limitations of the observational method in applied soil mechanics. Geotechnique 19, 2, 171-187.

Skempton A.W. (1945) A slip in the west bank of the Eau Brink Cut. Proc. Inst. Civil Engineers May 1945, 267-287.

Wilkes, P.F. (1970) A note on the installation of piezometers in small diameter boreholes. Geotechnique 20, 3, 330-333.

Wilkes, P.F. (1972a) An induced failure at a trial embankment at King's Lynn, Norfolk, England. Proc. A.S.C.E., Conf. on Perf. of Earth and Earth Supptd. Structures - Vol. I, part I.

Wilkes, P.F. (1972b) King's Lynn trial embankment. Inst. Highway Engineers XIX, 8, 9-16.

THE BEHAVIOUR OF STRESS MEASURING DEVICES IN ROCKS OF VARIOUS PROPERTIES

F. T. Williams

Senior Lecturer, Sheffield University

INTRODUCTION

The determination of stress levels in field situations in rock masses has been the subject of much discussion and experimentation over the past twenty years or so. Much is understood on the problems involved but it is clear that many instruments may be providing results of limited accuracy especially if they are used without a full appreciation of their capabilities or of the exact circumstances in which they operate in a predictable manner.

It is important to realise the requirements demanded of a stress measuring system in relation to the reasons for which the stress is required and particularly the period of time over which measurements from one installation are to be collected. This is most particularly the case where the rock material involved has appreciable time dependant strain properties.

In situ stress determinations, in which a "one-off" measurement of the existing stress levels is obtained by a stress relief technique normally involves time measured in minutes only, when the rock behaviour on change in stress becomes approximately elastic. In this circumstance, provided the appropriate values of the elastic parameters are known and due account is made of such matters as grain size etc., stresses may be determined with sufficient accuracy from the measurement of rock strain.

Longer term stress change measurements (days) become necessary if these are required in conjunction with in situ modulus of deformation measurements. Normally these operations are undertaken using various methods of loading the rock mass and measuring deformations of the rock at its exposed surface and at various points within the mass.

In fractured and/or layered rock formations numerical analysis allows the results to be interpreted to provide a variety of deformation moduli appropriate to each recognised layer. Assumptions are inevitably made regarding the behaviour of each rock component, the validity of which might be usefully verified by independent stress measurements at strategic points in the rock mass while the loading is in progress.

Stress measurements extending over periods of years rather than minutes or days become necessary when monitoring long term effects at installations involving stressed rock. The variations on this theme including dam foundations, tunnels, shafts and other excavations are legion. The results may be required for a number of reasons varying from the provision of data from which future designs can be developed, to giving warning of dangerous conditions. In these cases it is most essential that the time dependent properties of all the materials involved in the measurement system are understood and taken into account, otherwise the value of the results may be nullified. However, even if the numerical values of such measurements are not fully understood they may still provide warning of possibly dangerous changes. The establishment of a 'normal' range of values for a given installation may be sufficient, when any unexpected departure beyond this range can be employed to indicate the necessity for vigilance and investigation.

The chief problems which have to be overcome in the measurement of rock stress changes arise from the material itself. Stress determinations are theoretically straightforward in linear elastic solids. In some cases the situation in which this may be approximated exists in rock engineering situations. In general though, rock behaves in a non-linear manner, has time dependent properties and is discontinuous. In addition to this, problems may arise due to weathering, temperature changes, and rapid alteration on contact with air and/or water when drilling for the installation of instruments is being performed. Further difficulties associated with the texture of the rock, its grain size and anisotropic properties also require consideration.

STRESS MEASUREMENT PRINCIPLES

Stress, being a concept rather than a physical entity, is not directly measurable in the manner of load or volume and it is generally the manifestations of the existence of stress which are the basis of its determination. In nearly all cases this involves the existence of strain although it is frequently the secondary effects of strain (such as magnetostriction, photoelasticity etc) which are used as the basis of its measurement. An exception to this is a"constant dimension" device in which internal pressure can be varied to restore the dimensions of a cell which is exposed to stress, when the pressure required to achieve this is assumed to be equal to the required stress.

In the greater proportion of engineering experimental stress analysis, measurements are confined to the exposed surfaces of the parts under investigation. Use is made of such devices as electrical resistance strain gauges of an extremely wide variety of types and what may be described as "wide area" methods involving techniques such as brittle laquer and photoelastic coatings.

All such methods depend for their accuracy upon knowledge of the stress/strain/time properties of the material of the part concerned in addition to calibration factors, correction factors etc which are concerned with the strain determination technique.

Rock engineering problems are invariably concerned with large masses in which the distribution of stress within the mass is one of the chief questions. This fact, together with the permanent damage suffered by rock during excavation, subsequent weathering processes and the essentially discontinuous nature of rock masses generally renders exposed surface measurements of extremely limited value.

An essential feature of rock stress measurement is therefore the necessity of performing measurements within a drilled hole, - the simplest and cheapest method of reaching a chosen point within a rock mass. Due regard has to be paid to the existence of the borehole and the effect of its presence on the stresses that it is desired to measure.

What is required therefore is a device installed in a hole drilled in rock which gives an output which will allow the determination of the stress that would have existed at that point had the hole never been drilled.

THE RELATIONSHIP BETWEEN ROCK STRESS AND METER STRESS

Let us assume that the meter is a device having dependable linear elastic properties, the output of which is proportional to meter strain (and in consequence, meter stress). If meters having a variety of elastic moduli are available what value should we choose?

First let us consider the elastic properties of the rock. In general, rock stress/strain behaviour is not linear so that the modulus is dependent upon stress level and on whether the stress is increasing or decreasing. The extent to which rock materials exhibit this property is very variable, rocks having higher moduli, say exceeding 6×10^5 bar, frequently possessing reasonably linear elastic properties, while the softer materials, in particular those having moduli less than 2 x 10^5 bar, are distinctly unpredictable in their behaviour, when a single "modulus" value has little meaning.

The most easily understood method of obtaining a stress dependent output from any position within a borehole lies in some form of borehole deformation gauge system capable of measuring changes in borehole diameter and displaying the results on a meter external to the hole. Many such devices have been designed and used. Ideally these meters offer no resistance to the movement of the rock at the borehole sides, i.e. they have no stiffness, E' = O. In this case the accuracy of any determination in terms of rock stress depends totally upon knowledge and the dependability of the elastic properties of the rock. In practice devices of this character can be used effectively in high modulus rock materials only.

The stresses developed in inclusions at various modular ratios E'/E (dashed parameters refer to inclusions, others to host materials) have been the subject of many investigations over past years, some of these being specifically directed to the assessment of the behaviour of rock stress measurement devices (Goodier 1933, Muskelishvili 1953, Wilson 1961, Sezawa and Nishimura 1963, Barron 1965). Of particular interest is the two dimensional solution for a circular inclusion in a host body, the interface between the two being continuously connected. The Sezawa-Nishimura solution for this gives principal stresses in the inclusion:

$$\sigma_1' = \frac{p}{2}(D + H) + \frac{q}{2}(D - H) \qquad (1)$$

$$\sigma_2' = \frac{p}{2}(D - H) + \frac{q}{2}(D + H) \qquad (2)$$

in which σ_1', σ_2' are principal stresses in the inclusion,

p, q are principal stresses to be measured in the host body, i.e. the stresses that would have existed at the site of the inclusion in its absence.

D,H are constants (Coutinho 1949)

D and H are functions of the elastic moduli and Poisson's ratios of the inclusion and host materials. They do not include position coordinates, thus a homogenous state of stress exists in the inclusion provided it is solid.

Equations (1) and (2) can be conveniently re-written

$$p = K\sigma_1' + k\sigma_2' \qquad (3)$$

$$q = K\sigma_2' + k\sigma_1' \qquad (4)$$

in which $K = \frac{1}{2}\left(\frac{1}{D} + \frac{1}{H}\right)$

and $k = \frac{1}{2}\left(\frac{1}{D} - \frac{1}{H}\right)$

Wilson re-defined K and k as

$$K = \frac{1}{M} \cdot \frac{(1 + \nu')(3 - 4\nu')}{8(1 - \nu)(1 + \nu)} + \frac{5 - 4\nu}{8(1 - \nu)} \qquad (5)$$

$$k = \frac{1}{M} \cdot \frac{(1 + \nu')(1 - 4\nu)}{8(1 - \nu)(1 + \nu)} + \frac{4\nu - 1}{8(1 - \nu)} \qquad (6)$$

in which $M = E'/E$

It is the variations of these values which finally determine the performance of a gauge.

DEPENDENCE OF THE STRESS FACTORS ON THE ROCK POISSON'S RATIO

That rocks show a variability of Poisson's ratio dependent upon stress level, direction of stress change and duration of loading has been shown by numerous investigators and even negative values at low stress levels have been noted, in special circumstances (Ingles, Lee and Neil 1972). In general, reported values of Poisson's ratio for rocks at low stresses vary from zero to 0.25 and at near fracture this may increase to the range 0.6 to 0.7 in the case of the lower modulus rocks.

As an example of this, in the case of Darley Dale sandstone Poisson's ratio varies from 0.03 to 0.25 as uniaxially applied stress increases from zero to 400 bars, rising to 0.6 at 600 bars. The value of the tangent modulus of this rock varies from 0.57×10^5 to 1.94×10^5 bars over the same stress range.

An example of a harder rock, Cornish granite has an elastic modulus the tangent value of which varies from 4.96×10^5 to 5.46×10^5 bar over a stress range of zero to 700 bars while the Poisson's ratio varies between 0.21 and 0.28.

A solid inclusion of an elastic material will probably possess a Poisson's ratio of approximately 0.25, hence the following Table will show the values of the stress factors K and k for the rock types quoted in terms of modular ratio:-

TABLE 1 EVALUATION OF STRESS FACTORS

($\nu' = 0.25$ throughout)

Stress (bars)	E (bar x 10^5)	ν	K=	k=
			Darley Dale Sandstone	
0	0.57	0.03	$\frac{0.313}{M} + 0.628$	$\frac{0.157}{M} - 0.113$
300	1.86	0.25	$\frac{0.334}{M} + 0.670$	0
600	1.94	0.60	$\frac{0.488}{M} + 0.830$	$-\frac{0.342}{M} + 0.437$
			Cornish Granite	
			K=	k=
0	4.96	0.21	$\frac{0.328}{M} + 0.658$	$\frac{0.026}{M} - 0.025$
300	5.44	0.27	$\frac{0.337}{M} + 0.670$	$-\frac{0.013}{M} + 0.014$
700	5.46	0.28	$\frac{0.340}{M} + 0.673$	$-\frac{0.020}{M} + 0.021$

It will be observed that the stress factor K is insensitive to variation in ν particularly at lower values and in fact varies by less than 10% for all values of ν less than 0.3. This covers the greater proportion of situations in which rock stress measurement is likely to be practised, but it must be noted that errors could arise in softer rocks at high stress levels.

The stress factor k has generally low values, particularly where $\nu \simeq 0.25$, ($k = 0$ when $\nu = 0.25$ and only has values comparable with K at values of ν approaching zero and $2\nu'$ in the cases quoted. Over a large range of rock Poisson's ratios therefore, k is small and although more sensitive to ν is unlikely to cause serious error if ν is assumed constant but again the warning must be sounded that low and high values of rock Poisson's ratio must be considered more carefully.

DEPENDENCE OF THE STRESS FACTORS ON THE MODULAR RATIO

The chief difficulty in rock stress measurement as has been noted earlier is the variability of its modulus of elasticity amounting to an indeterminate value in some cases. This has resulted in the development of borehole gauges having high moduli, it being clear that the relationship between both K and k on the one hand and M on the other is hyperbolic and at a sufficiently large value of M the variability of E becomes unimportant.

The values of the stress factors for Darley Dale Sandstone in which a gauge having a Poisson's ratio of 0.25 has been inserted is shown in Fig. 1. If variations in elastic properties other than modulus are ignored then it is clear that the adoption of a modular ratio having a minimum value or 4 will ensure that any variation of rock modulus during the period of a test will not result in a variation of K greater than 10% - even if E approaches zero. The meter is then virtually a stress meter and the principle involved in this choice of meter modulus is normally referred to as the "hard inclusion theory".

However, if such a ratio is chosen on the basis of rock modulus at 50% of its failure strength ($E = 1.86 \times 10^5$ bars) this would lead to the use of a meter having a modulus of at least 7.44×10^5 bars, and K at stresses in the region of 300 bars would be 0.754. At very low stresses, however, it is necessary to consider variations in both the modulus and Poisson's ratio of the rock and K then becomes 0.647, a variation of more than 14%. It is necessary in this case to increase the modular ratio to 6 to reduce the errors to under 10%, K then becomes 0.643 at near zero stress and 0.727 at 300 bars. Even then if the stress level achieved in the rock rises to near failure K rises to 0.815 and errors of 12% or so appear.

Although the corresponding values of k are small they are not zero as might first be supposed on the basis of $\nu \simeq 0.25$. With a modular ratio of 6, k at low stress becomes -0.10 and at near failure rises rapidly to 0.37.

The position in the case of hard rock materials such as Cornish Granite is simpler, in that the lesser variations of both E and ν with stress allow the adoption of values of M less than 1 without giving error of K in excess of 10%. At this value of M, K has a value of 1.007 at 300 bars, 0.958 at near zero stress and 1.013 at 600 bars. The corresponding values of k are +0.001, -0.011 and +0.011.

The modulus of a meter most suitable for use in Darley Dale sandstone is thus 11.16×10^5 bars which instrument would also be entirely suitable for Cornish Granite situations, although the gauge calibration factor would differ between these rocks, as this depends upon the stress factors.

The normal rock stress measurement problem involves a rock mass already under stress in which either future stress changes or the existing stress level is required. On drilling a hole into such a mass the existing stress is changed. In rock generally it is clear that a gauge

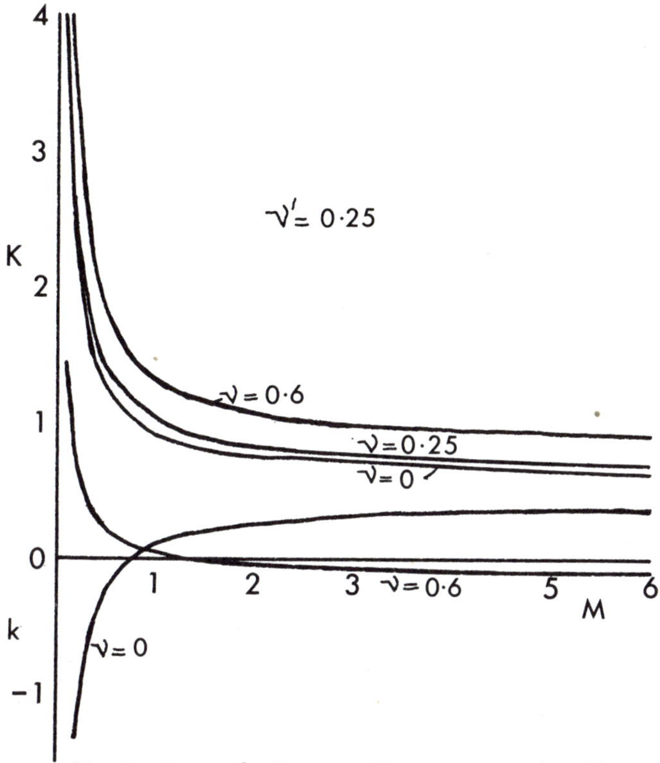

FIG 1. Variation of Stress Factors with M and Poisson's Ratio

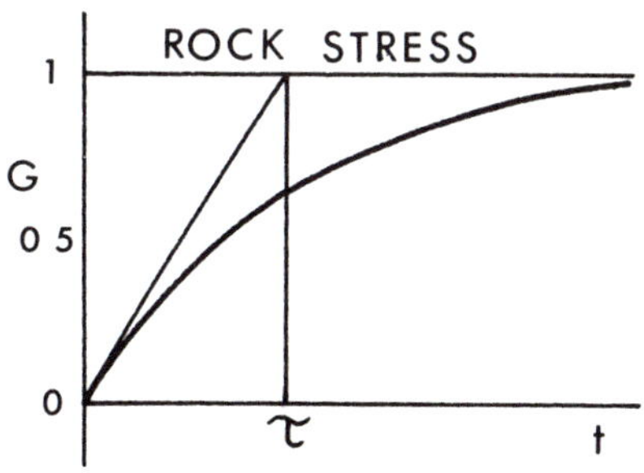

FIG 2. Variation of Gauge Output with Time

τ = Time Constant

installed in such a hole is surrounded by a zone in which variable values of modulus and Poisson's ratio are operative and it is essential that any gauge used must be as nearly independent of these variations as possible if useful results are to be obtained.

INCLUSION STRESSES IN TIME DEPENDENT HOST MATERIALS

Upon installation of a meter in a hole drilled in a stressed mass, the inclusion will instantaneously experience no stress unless a device having pre-stressing capabilities is used. If the rock mass is purely elastic and is not suffering changes of temperature or weathering effects involving volume change the inclusion will continue in this state until the stress level within the host mass varies. It has been shown that with suitable choice of modular ratio taking into account variations of rock Poisson's ratio such stress changes will be measured by the gauge.

If the rock mass has time dependent strain properties, then assuming that the time lapse between hole drilling and meter installation is small, displacements will occur in the rock mass following gauge installation which are not related to stress changes and stresses will appear in the inclusion. It has been shown (Hult 1963, Hult, Kvapil and Sundqvist 1966 Berry and Fairhurst 1966) that the stress level within the inclusion will eventually rise to a steady state equal to that expected, had the material behaved elastically and the stress in the system increased from zero to the existing value. The increase is of an exponential character (Fig.2) and the "time constant" of the system is given for linear visco-elastic host materials (Kelvin-Maxwell) by

$$\tau = \frac{1}{C}\left(\frac{c_1}{E} + \frac{c_2}{E'}\right) \tag{8}$$

where c_1 and c_2 are constants dependent upon dimensions of the meter and drilled hole and

$$C = \frac{1}{\eta_1} + \frac{1}{\eta_2} \tag{9}$$

in which η_1 and η_2 are the "dashpot viscosities" of the Maxwell and Kelvin elements of the material. Few rocks can be compared precisely with this linear reheological model but a sufficient approximation may be made to determine the order of magnitude of τ.

If this value is sufficiently small it becomes possible to determine in situ stresses without the necessity of a destressing process. Stress changes in a rock mass having time dependent properties will only be possible (without developing methods of correction) if the changes measured occur in a time which is either small or very large compared with τ. In the latter case the meter will follow the value of the in situ stress rather than directly measure the change.

The value of the time constant can be kept at a minimum by the use of a high modulus inclusion. It is difficult to analyse generally the behaviour of stress measuring devices in host materials subject to time dependent strain. The chief reason for this is the very large variability of these properties within rock materials, the fact that they are frequently of a non-linear character and that there is some lack of knowledge of these properties.

In the case of a rock mass in which the load is raised in a short time from zero to some value the immediate response of a gauge can be judged from elastic considerations. The effect of maintaining the stress at the raised value upon the gauge can be estimated by assuming that the Poisson's ratio is increased to 0.5. The use of a sufficiently high modulus inclusion will result in a constant reponse while the stress remains constant even when time dependent movements are occurring.

PRACTICAL RESULTS

A comparison of some results obtained by practical tests on an inclusion meter consisting essentially of a glass annulus are illustrated in Fig. 3. It will be seen that the form of the theoretical curve follows generally the experimental results and that most of these fall within a 10% range. It must be remembered that some doubt must exist in respect of rock moduli and Poisson's ratios, which in the case of the curve is assumed constant.

Each experimental point represents several experiments and the figure serves to illustrate that in practical rock mechanics problems, experimental calibrations are required to check any theoretically determined response factor.

Several investigators (Dhir 1966, Hawkes 1968, Skilton 1971) have published results obtained from similar meters in laboratory conditions where time dependent strain was occurring, the material involved including concrete, sandstone frozen sand and rock salt. In no case were any results obtained which were not consistent with the theoretical conclusions described previously. Some recent work by the author in the field (CIRIA, 1970) showed that in an impounded concrete dam the time required for a glass insert meter to achieve the stable stress level consistent with the concrete stress levels and the values of M and Poisson's ratio operating was approximately one year, Fig. 4.

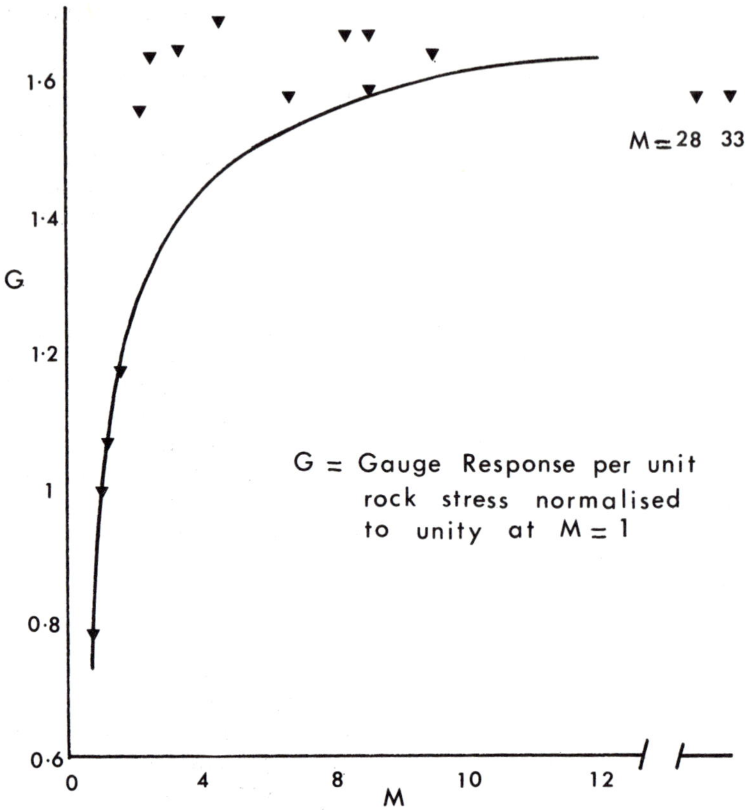

FIG 3 Effect of Rock Modulus on Meter Output

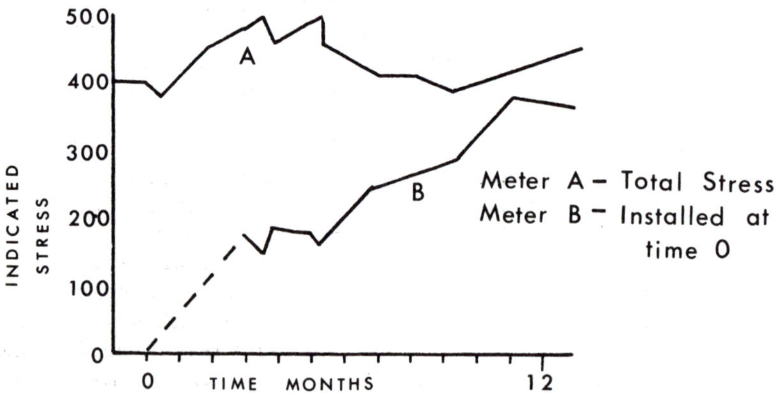

FIG 4 The Effect of Time Dependent Strain on Stressmeter Readings

CONCLUSIONS

1. In ideal elastic circumstances stress measuring devices in the form of inclusions in a mass are understood and their response is predictable.

2. In rock materials the results are only predictable if the modulus and Poisson's ratio can be determined in all stress conditions.

3. In practical circumstances stresses in the region of a meter are disturbed and are not readily determinable, also the elastic parameters will be variable and the rock may be anisotropic.

4. Dependence upon constancy of these parameters is reduced by ensuring a high value of modular ratio.

5. In addition, the experimental verification of theoretically determined gauge performance is advisable, particularly in low and mid range modulus rocks.

6. The behaviour of such a meter in rocks subject to time dependent strain is characterised by the time constant of the rock/gauge system.

7. This value determines whether the results from a proposed stress measuring installation will be affected by time dependent strains, and the manner or feasability of correction.

8. More research is required in order to elucidate fully the behaviour of inclusion meters in host materials subject to time dependent strains, having both linear and non-linear characteristics.

REFERENCES

Barron K. (Dec. 1965) "Glass Insert Stress meters" Trans A.I.M.E., Vol 232, pp 287-299

Berry, D.C. and Fairhurst, C. (1966) "Influence of Rock Anisotropy and Time Dependent Deformation on the Stress Relief and High Modulus Inclusion techniques of in situ Stress Determination" ASTM STP 402 pp 190-206

Coutinho, A. (1949) "Theory for an Experimental Method for Determining Stress not Requiring an Accurate Knowledge of Elastic Modulus". Int. Assoc. of Bridge & Struct. Eng. Congress 1949, Vol. 9, pp 83-103.

Dhir, R.K. (1966) "A Study of Photoelastic Stressmeters" Ph.D. Thesis, University of Sheffield.

Goodier, J. (1933) "Concentrations of Stress Around Spherical and Circular Inclusions and Flaws" Journ. App. Mechs., (Trans A.S.M.E.) Vol. 55 p.39.

Hawkes, I. (1968) "Stress Evaluation in low Modulus and Viscoelastic Materials using Photoelastic Glass Inclusions" SESA, Spring meeting 1968, Paper 1353.

Halcrow, Sir W. & Partners (1970) "Instrumentation Investigations at Clywedog Dam" CIRIA (London) Tech. Note 9.

Hult, J. (1963) "On the measurement of Stresses in Solids". Chalmers Tekniska, Hogskolas Handlingar, No.280, pp 4-21

Hult, J., Kvapil,R., & Sundkvist,H. (1966) "Function and Scope of Stressmeters in Rock Mechanics" Inst. J. Rock Mechs. Min. Sci., Vol. 3, pp 1-10.

Ingles, O.G., Lee, I.K., and Neil, R.C. (1972) "New Manifestations of Stress History" Report Uniciv.No. R90 School of Civil Engineering, University of New South Wales.

Muskhelishvili, N.I., (1953) "Some Basic Problems in the Mathematical Theory of Elasticity", Noordhoff, Groningen.

Sezawa, K and Nishimura, G. (1963) "The Stress under Tension in a Plate with a Heterogenous Insertion". Rep. Aero Res. Inst. Tokyo, Vol.6, No.68.

Skilton, D. (1971) "Behaviour of Rigid Inclusion Stressmeters in Viscoelastic Rock". Int. J. Rock Mech. Min. Sci. Vol. 8 pp 283-298.

Wilson, A.H. (1961) "A Laboratory Investigation of a High Modulus Plug Gauge for the Measurement of Rock Stress". 4th Symp. Rock Mechs. Penn State Univ., pp 185-195

FIELD INSTRUMENTATION OF VIBROFLOTATION FOUNDATION

H.Y. Wong

Research and Development Engineer, Cementation Research Ltd.
Herts. England.

INTRODUCTION

Vibroflotation is basically a mechanical method of soil densification. The equipment comprises an eccentrically vibrated steel mandrel which penetrates into the ground forming closely spaced holes which are backfilled with compacted sand or gravel when the mandrel is withdrawn. The behaviour of the treated foundation as a whole will, however, vary from soil to soil (e.g. Watt, de Boer and Greenwood, 1967; Thorburn and MacVicar, 1968; Webb and Hall, 1969; Greenwood, 1970). Sandy and gravelly materials will be considerably densified, and, together with the backfill materials, form an integral soil mass. Design under such conditions will be fairly simple, and can be based on the results of the standard penetration tests on the treated foundation. In weak cohesive soils, the soil itself will be little affected, and the backfill materials, in the form of stone columns, will take most of the foundation load. For silty materials, a certain degree of densification will occur, and a considerable portion of the foundation load will be taken by the soil. It is for the latter two types of soils that special research efforts are necessary.

For a silty material, as the portion of foundation load taken by the surrounding soil is no longer negligible, it is necessary to take this into account for an optimum foundation design. In view of the above, full scale foundation load tests were devised in which a fully instrumented strip footing on a coarse silt treated by vibroflotation was loaded to about twice its design load. Pressure cells of the diaphragm/electrical transducer type were installed in the footing slab with their active face flush with the footing base. They were arranged to measure the load intensity on the treated soil as well as that on the stone columns. Data of this nature can serve not only as guidance to foundation design in similar materials but also shed some light on some of the assumptions employed for design in weak cohesive soils. At the same time, the effect of the lateral restraint of the stone columns was also studied by comparing the load-settlement characteristics of columns forming part of a group, and isolated columns outside a treated area.

GENERAL DESCRIPTION OF LOAD TESTS

The testing site was a filled area in St. Helens. The fill material is known as burgy, and is a silt size material comprising mainly of fine glass residue, with the median grain size (d_{50}) equal to about 0.023 mm and d_{10} equal to about 0.007 mm. The depth of the fill material around the testing site was about 9 m. (30 ft.). Two load tests were carried out, each on two stone columns, 2.1 m. (7 ft.) centre to centre. Test No. 1 was on two stone columns inside the treated area whilst Test No. 2 was on two specially prepared isolated stone columns at about 2.4 m. (8 ft.) from the edge of the treated area.

The load was applied through a rectangular heavily reinforced concrete slab of dimensions 2.74 x 1.22 m. (9 x 4 ft.) and 229 mm (9 in.) thick, by means of a hydraulic jack, using kentledge as reaction. The exact load on the foundation was registered by a calibrated proving ring. The plan of the testing sites is shown in Fig. 1.

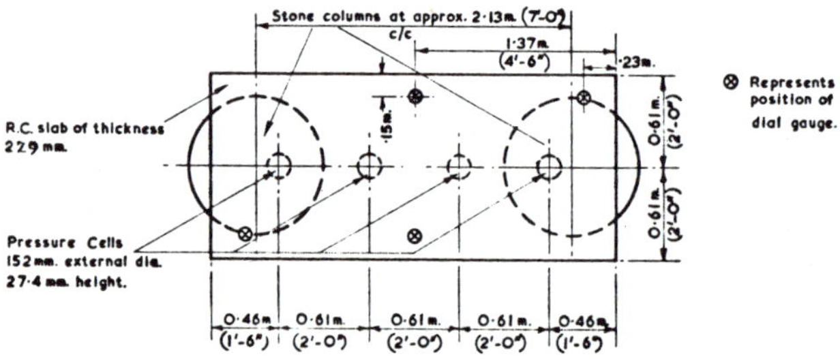

Figure 1 Location of reinforced concrete slab

In each test, load increments (or decrements) were in the range 21.5 to 43.0 kN/m^2 (0.2 - 0.4 ton/ft^2). For test no. 1, the load was first increased to 150 kN/m^2 (1.4 ton/ft^2) and then reduced to zero. After that, 3 cycles of loading and unloading were performed with a maximum load of 225 kN/m^2 (2.1 ton/ft^2). A creep test lasting for 64 hours was carried out at an average load of 172 kN/m^2 (1.65 ton/ft^2). For Test No. 2, 3 cycles of loading and unloading with a maximum load of 193 kN/m^2 (1.8 ton/ft^2), and a creep test lasting for 14 hours at an average load of 102 kN/m^2 (0.95 ton/ft^2) were performed. During each load increment (or decrement), readings on the four dial gauges (see Fig. 1) and from the four pressure cells in the slab were taken every 5 to 10 minutes. The load on the slab was increased (or decreased) when the average change in dial gauge readings over 5 minutes was less than 0.025 mm (0.001 in.).

DESIGN OF PRESSURE CELL

In order to measure the relative pressure distribution between the stone columns and the surrounding soil, special pressure cells were developed. These were set into, but with one face flush with the base of the reinforced concrete slab.

In developing a suitable type of pressure cell for the purpose, the following criteria were specified:

(a) the cell must be very sensitive with accuracy at least up to 1.4 kN/m^2 (0.2 p.s.i.) since pressures transmitted to the soil are usually small;
(b) the sensitive area in contact with the stone column must be fairly large as the stone column is made up of gravels with sizes ranging from 25-50 mm. (1 - 2 in.); and
(c) the cell must be very robust in view of the adverse site conditions.

Optimum cell dimensions were based on the following rules suggested by previous workers:

(1) the average hypothetical grain diameter should not exceed 2% of the cell diameter (Kallstenius and Bergau, 1956);
(2) the diaphragm flexibility (ratio of central deflection to applied pressure) should be less than 6.3 x 10^{-5} $mm/kN/m^2$ (1.7 x 10^{-5} $in./lb/in^2$) according to Kallstenius and Bergau (1956), and 1.1 x 10^5 $mm/kN/m^2$ (0.3 x 10^{-5} $in/lb/in^2$) according to Trollope and Lee (1961);
(3) the diameter to central deflection ratio should be at least 1000 as recommended by U.S. Waterways Experiment Station (1944);

With these considerations in mind, a diaphragm type pressure cell using electrical strain gauges was designed. The basic principles in using electrical strain gauges in diaphragm type pressure cells have been discussed in detail by Peattie and Sparrow (1954) and Reese, Brown and Dalrymple (1968). The basic features of the present pressure cell are as shown in Fig. 2.

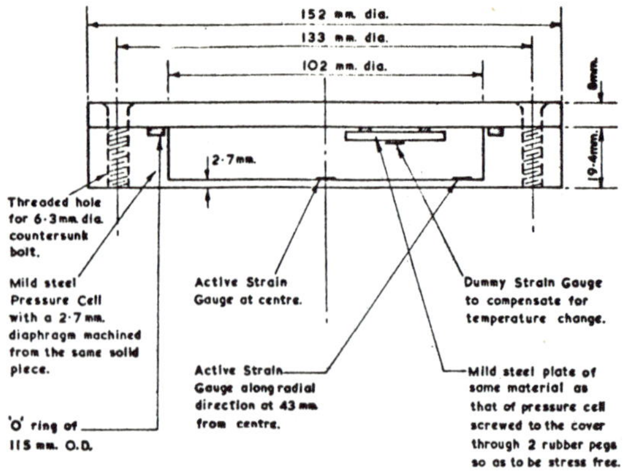

Figure 2 Main features of pressure cell

Various difficulties have been encountered in previous work. The first one is to prevent any water from entering the cell, and to bring the strain gauge leads out of the cell, while maintaining a waterproof system in order to avoid reduction of the resistance of the strain gauge to earth. This was overcome by coating the strain gauges with silicon rubber (Gagekote No. 5) and then filling the whole cell with an oil of very high electrical resistance. As well as acting as an insulator, the oil can also act as a medium through which an internal pressure can be applied to the cell. The second difficulty is that when the cell is used on sand and gravel, the increase in stiffness of the soil with respect to the cell tends to give an under-registration of the actual pressure with increasing load. This can be improved by increasing the oil pressure inside the cell during loading so as to increase the stiffness of the cell.

The stability of the strain gauges was checked from time to time before and after the field tests for a period of about three months. The electrical resistance relative to the ground remained high, but there was some fluctuation in the zero reading. This was probably due to the rapid temperature changes to which the active strain gauges on the diaphragm, and not the temperature compensating dummy gauges of the pressure cell, were subject when exposed in open space. Drifting of the zero reading was negligible during the period of field test when the diaphragm was in contact with the soil and was not subject to rapid temperature change. Typical calibration curves under mercury pressure, oil pressure, as well as other loading conditions are shown in Figs. 3a-c; the wiring, pressurising and read-out systems of the pressure cells are shown in Fig. 4.

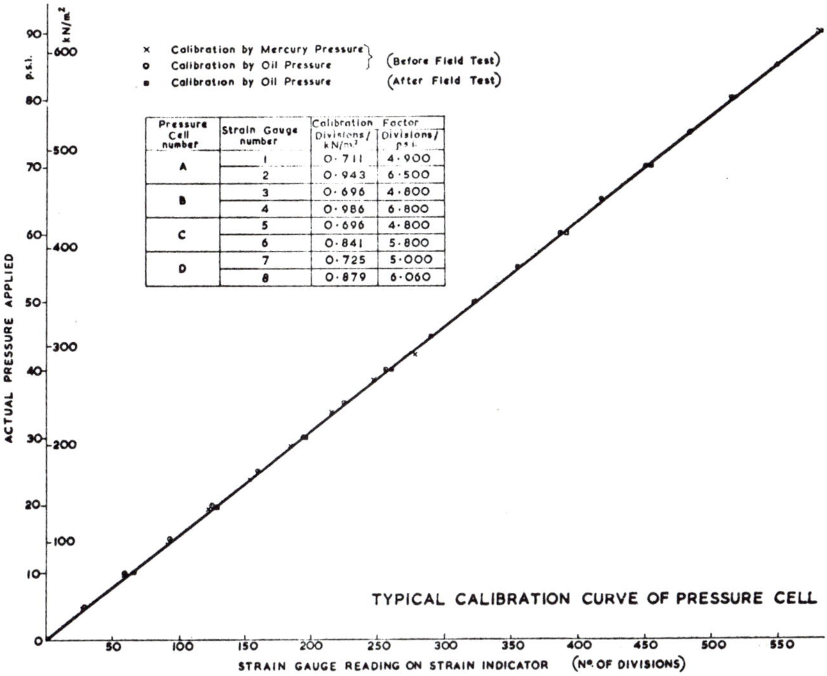

Pressure Cell number	Strain Gauge number	Calibration Factor Divisions/kN/m²	Divisions/p.s.i.
A	1	0·711	4·900
	2	0·943	6·500
B	3	0·696	4·800
	4	0·986	6·800
C	5	0·696	4·800
	6	0·841	5·800
D	7	0·725	5·000
	8	0·879	6·060

Figure 3a Typical calibration curve of pressure cell

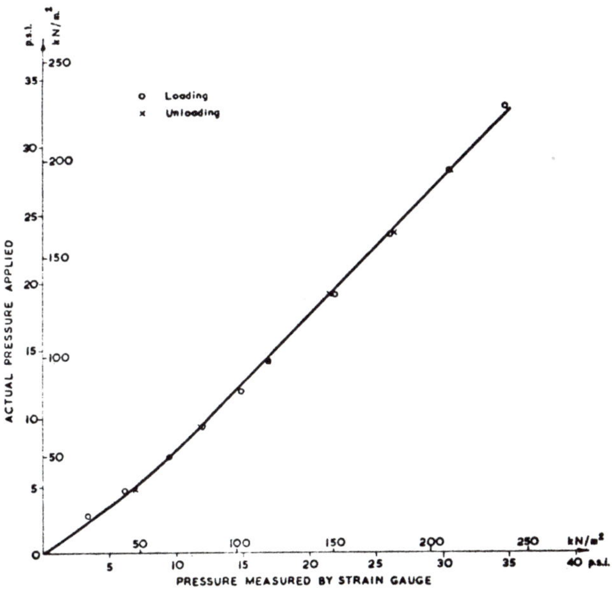

Figure 3b Calibration of pressure cell on clay

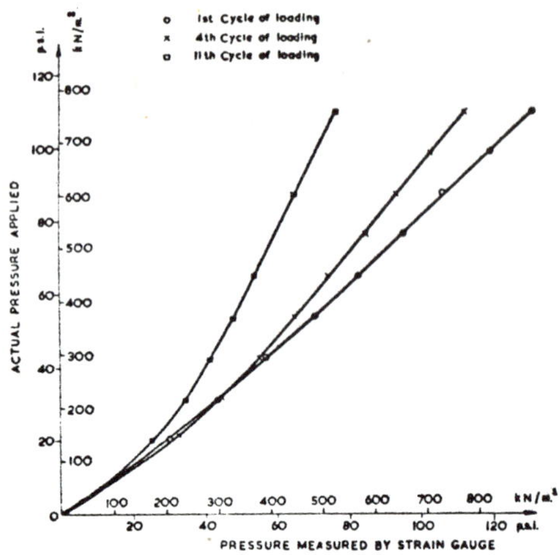

Figure 3c Calibration of pressure cell on sand

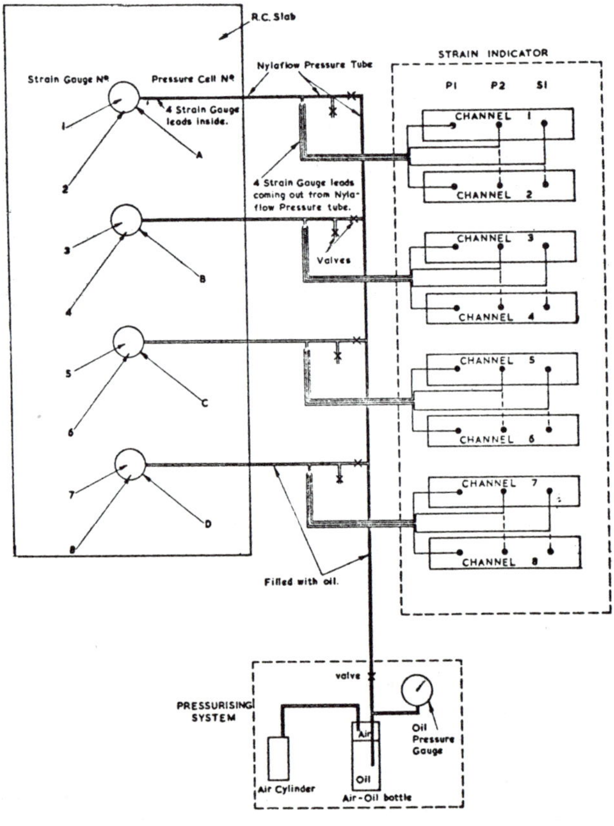

Figure 4 Wiring, pressurising and read-out systems

The results indicate that the pressure cells perform satisfactorily in fine-grained soils. However for soils such as coarse sand and gravel, the calibration curve changes along the under-registration direction with the increase in load magnitude and the number of loading cycles. This is probably due to the increase in arching effect with the increase in soil stiffness relative to the cell. Although improvements were obtained by increasing the internal cell pressure progressively so as to keep the cell stiffness relative to soil more or less constant, better results were obtained by placing a thin layer of fine-grained soil over the gravel, so that the pressure was distributed more uniformly.

LOAD-SETTLEMENT CHARACTERISTICS

The load-settlement characteristics of the two tests are similar. Test No. 1 inside the treated area (Fig. 5) shows less settlement than Test No. 2 on isolated columns outside the treated area (Figure 5).

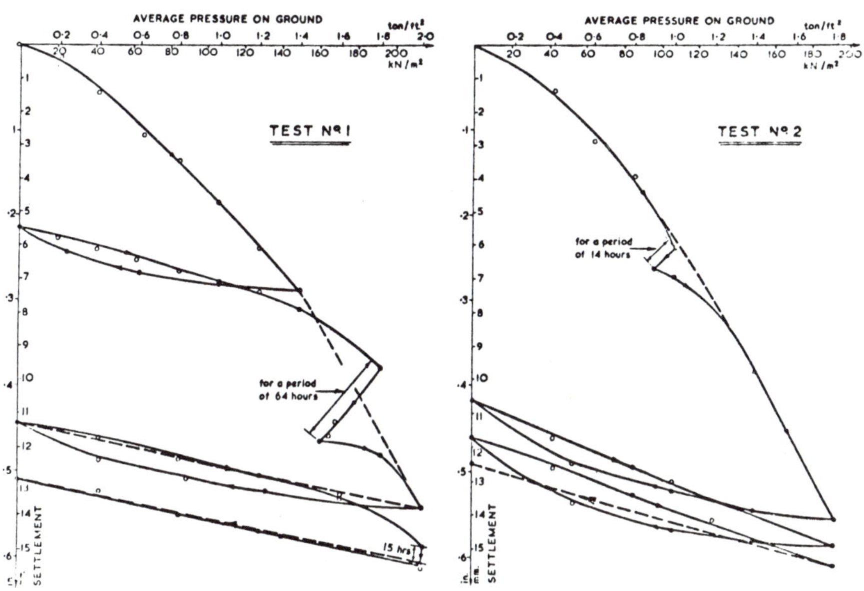

Figure 5 Load settlement characteristics

On the whole, the settlement is small: 15 mm (0.6 in.) at a unit load of 225 kN/m^2 (2.1 ton/ft^2) for the treated area, and 15 mm (0.6 in.) at a unit load of 193 kN/m^2 (1.8 ton/ft^2) for the isolated columns. These results indicate that the vibroflotation process has been successful in stabilising the ground.

Further comparison of the two foundations can be made by studying the rate of settlement under constant load. By plotting the amount of settlement versus logarithm of time at various constant loads (Figure 6), it can be seen that the rate of settlement in the treated area, under a unit load of 177 kN/m^2 (1.65 ton/ft^2) is approximately equal to that on the isolated columns at a unit load of 150 kN/m^2 (1.4 ton/ft^2).

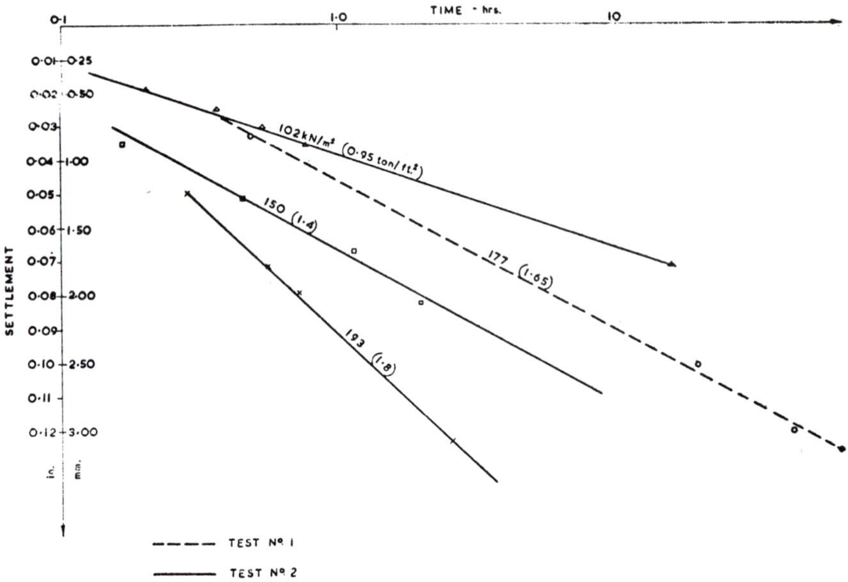

Figure 6 Creep tests under various constant loads

It may be concluded that the columns inside the treated area are stiffer than the isolated columns, because of the higher degree of constraint imposed on the treated columns by the surrounding soil and the adjacent columns. The difference is, however, not large. This might be expected since the centre line of the two isolated columns is only about 2.4 m (8 ft.) from the edge of the treated area.

PRESSURE DISTRIBUTION

The average pressures on the stone columns and on the surrounding soil, as measured by the earth pressure cells for the various loading cycles in Test No. 2, are shown in Figs. 7a-b. The general tendency is an increase of the amount of load taken by the soil, with the increase of total load on the whole foundation. This pattern of load transfer is consistently indicated by all eight strain gauges in the pressure cells in each of the three loading cycles.

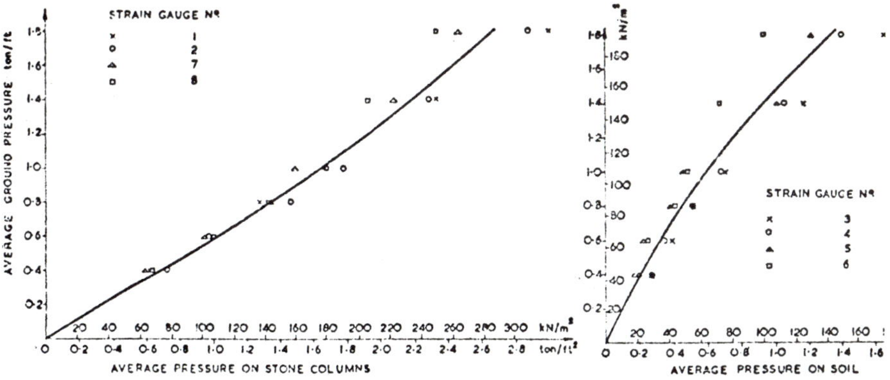

Figure 7a Pressure distribution between soil and column (test no. 2 - 1st cycle loading)

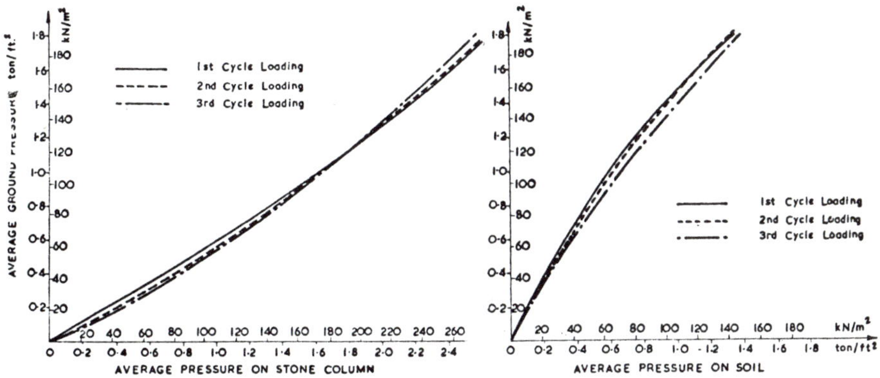

Figure 7b Pressure distribution between soil and columns (test no. 2 - comparison of 3 cycles)

This general pattern of load transfer is supported by the test results of Test No. 1. Because not all the pressure cells were working properly in this test, an absolute check on their reliability is not possible.

The improper functioning of some of the pressure cells in Test No. 1 was probably due to the imperfect contact between the diaphragm of the pressure cells and the ground surface which was only covered by a 10 mm layer of fine silt. In Test No. 2 contact was improved by covering the ground with a 30 mm layer of saturated silt in a slurry form. An initial load of 21.5 kN/m^2 (0.2 ton/ft^2) was applied overnight to squeeze out the excess water. The (initially) very soft nature of the saturated silt enabled a good contact between the pressure cells and the ground to be maintained at all times during the test. Any excessive settlement due to this layer should be negligible, as most of the settlement should have occurred during the application of the initial load. The thickness of the layer should also be negligible.

The available pressure distribution data of Test No. 1 are shown in Fig. 8, it can be seen that the same general trend as in Test No. 2 is followed and the validity of the results may be accepted.

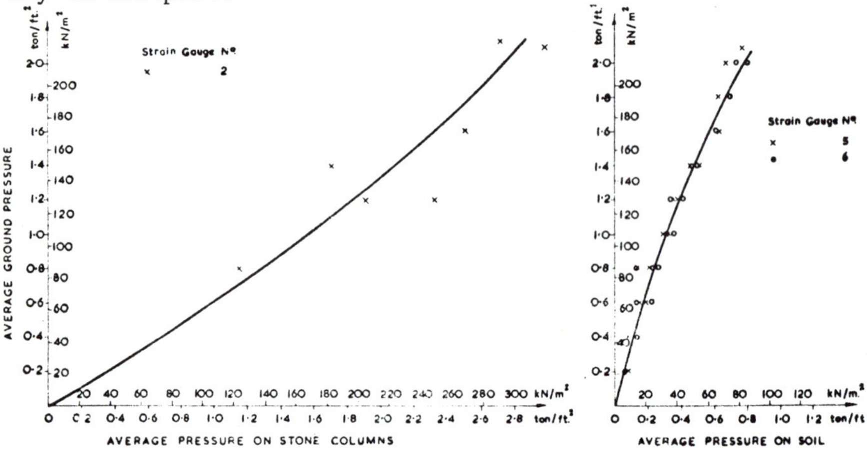

Figure 8 Pressure distribution between soil and column (Test No. 1)

Further justification of the pressure cell readings can be obtained by plotting the actual total load applied against the total load calculated from the various pressure values measured by the cells. Fig. 9 shows that nearly all the points fall on a 45° line passing through the origin, indicating a high degree of reliability of the pressure cells in that test.

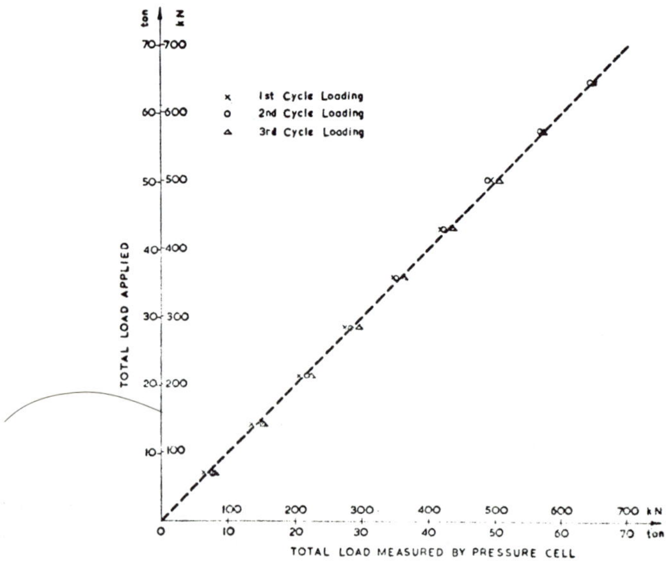

Figure 9 Actual load versus calculated load (Test No. 2)

The gradual transfer of load to the surrounding soil is summarised in Fig. 10. This may be explained as follows:

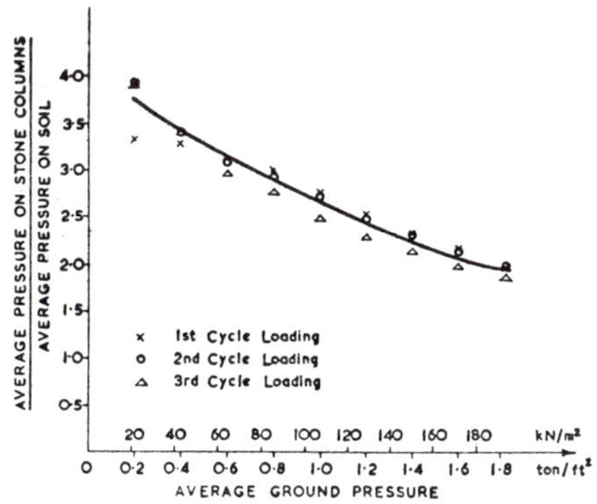

Figure 10 Ratio of pressure on column to pressure on soil at various loadings

During the first load application, the stone column, being a stronger material, will take most of the load. In doing so it will tend to dilate, especially near to the ground surface, compressing the surrounding soil. The soil, compressed horizontally by the columns and vertically by the slab, will increase in strength. At the same time, the stone columns will become weaker on dilation, with the result that more load will be transferred to the soil with increase in slab load. The relative amount of load that can be taken by the soil will be controlled by the initial stiffness of the soil relative to the stone columns and the degree of settlement allowed.

CONCLUSIONS

For the successful performance of any boundary type earth pressure cell, the effect of the cells on the original soil pressure distribution must be kept to a minimum, and the measuring system (which is the electrical resistance strain gauge type transducer in the present case) must be very stable during the testing period. As far as the above two factors are concerned, the present two series of tests indicate that the pressure cells thus developed can work satisfactorily when under well controlled conditions. Nevertheless, the poor response of some of the pressure cells in Test No. 1 does illustrate the importance of maintaining a good contact between the pressure cells and the surrounding soil materials.

The results of the tests show that for a vibroflotation foundation founded in silty materials in which considerable densification of the soil can occur, the stress on the soil (q_o) is no longer negligible in comparison with the stress on the stone column (q), and this ratio q/q_o tends to decrease with the increasing foundation pressure. For example at a unit foundation load of 83 kN/m^2 (0.8 ton/ft^2) q/q_o is approximately equal to 3. It follows that for foundation design in similar materials, the total foundation load (P) can be given approximately by

$$P = q(A_c + A_s/X)$$

where A_c is the area of the stone columns, A_s is the area of the foundation soil such that ($A_c + A_s$) is the total foundation area, and X is a positive number, depending on the type of soil.

Finally, it can be seen from a comparison of the load-settlement characteristics between Test No. 1 and Test No. 2 (Figs. 5 and 6) that the stone columns inside the treated area are stronger than the isolated columns formed in the untreated area. From this it may be inferred that for any treated area, the part near the centre will be stronger than that near the edge, provided other factors such as the spacing and diameter of the stone columns are the same.

REFERENCES

Greenwood, D.A. (1970). Mechanical Improvement of Soils below Ground Surface. Ground Engineering Symposium, London.

Kallstenius, T. and Bergau, G. (1956). Investigation of Soil Pressure Measuring by Means of Cells. Royal Swedish Geotechnical Institute Proc. No. 12.

Peattie, K.R. and Sparrow, R.W. (1954). The Fundamental Action of Earth Pressure Cells. Journal of the Mechanics and Physics of Solids, 2: 141-155.

Reese, L.C., Brown, C. and Dalrymple, H.H. (1968). Instrumentation for Measurements of Lateral Earth Pressure in Drilled Shafts. Centre for Highway Research University of Texas, Research Report 89-2.

Thorburn, S. and MacVicar, R.S.L. (1968). Soil Stabilisation Employing Surface and Depth Vibrators. The Structural Engineer, 46: 309-316.

Trollope, D.H. and Lee, I.K. (1961). The Measurement of Soil Pressures. Proc. 5th Int. Conf. Soil Mech., 2: 498-499.

United States Army Waterways Experiment Station (1944). Soil Pressure Cell Investigation. Interim Report, Technical Memorandum No. 210-1.

Watt, A.J., de Boer, R.B. and Greenwood, D.A. (1967). Loading Tests on Structures Founded on Soft Cohesive Soil Strengthened by Compacted Granular Columns. Proc. 3rd Asia Regional Conf. on Soil Mech. and Found. Engrg., 1: 248-251.

Webb, D.L. and Hall, J.W. (1969). Effect of Vibroflotation on Clayey Sands. Proc. Amer. Soc. Civ. Engrs., 95: 1365-1378.

SOME ORGANIZATION AND BENEFIT ASPECTS OF INSTRUMENTATION

Jack Wrightman

Engineer, Engineering Intelligence Division,
Highways Directorate, Department of the Environment

INTRODUCTION

All engineering requires a degree of organizing and the instrumentation of works is no exception, the amount of effort required varying with the size and nature of the project. The principles will be similar in each case. Some of the material in this paper could be considered as too obvious to need stating, and will be well known to engineers experienced in this type of work. Nevertheless the author has found from experience that many schemes for instrumentation yield disappointing results because of a lack of understanding of what is required in organizing and managing the work, and the problems such as protection which may arise. Effective instrumentation of works will depend as much upon attention to details of organization as upon attention to design, and the paper attempts to set out the needs for each stage of the work and how these can be fulfilled. The inherent problems in instrumentation often arise from a combination of the variety of types of activity and these can be divided as follows:

<u>Design work</u>

1. The choice of the various types of instrument required, and the establishment of the instrument patterns and reading stations, derived from a knowledge of the design philosophy of the main works.
2. The actual design of the instruments and the recording apparatus where new types of equipment are needed.
3. Programming of the instrumentation to fit the programme of the main works.
4. Specification and letting for instrumentation contracts and matters affecting the main contract.

Installation

5. The construction of reading stations.
6. The actual placing of instruments, their connections and recording apparatus.
7. The forming of borings, and backfilling.
8. The forming of other earthworks and backfilling.
9. Protection and malfunctioning of the installation.
10. Testing and commissioning.

Usage

11. a. Taking readings and analysis
12. b. Reporting and application of the data obtained
13. c. Dealing with any malfunctioning of instruments.
14. d. Maintenance.

DESIGN

It must first be established that instrumentation is justified, either on the grounds of necessity for controlling the rate of construction or for modifying the works; or for providing data giving future benefits. Schemes will often be a combination of these. This will determine the types and layout of instruments which are desirable and the design must then be checked to ensure it fits in with any constraints which may be imposed by the contract programme or site conditions. Also at this stage the method of financing should be considered. Where new instruments have to be developed, then financial arrangements will need to be flexible enough to permit this work in advance and also adequate time allowed for their design construction and testing. All the work needs to be indicated to the main contractor either on drawings or in a specification even though he may not be involved in their installation, as they may effect his temporary as well as the permanent works. In particular, observation points should be indicated when clear of the works as they can be affected by haul roads, stockpiles and the like.

Programme

Once the instrumentation pattern has been established then a programme for installation should be drawn up. This should be carried out with the probable programme for the main contract in mind which will enable the following to be established:-

1. The order of procuring equipment.
2. The order of installation including requirements for advance installation of reading stations and connections.
3. Staffing needs.
4. Co-ordination with any ground treatment and other construction which may affect the instruments.

5. Supply of contractor's services.

Contract matters

It is unlikely that the main contractor will be in a position to supply all the equipment and services required. In any case much of the work is the responsibility of the engineer. The type of procurement will be governed by the complexity and programme of the work, but probably the best form for a wide variety of work is by a nominated sub-contract placed,with specialists in this field. Their work could comprise items 2, 6, 7, 10 and 14 in the introduction, the main contractor normally being responsible for items 8 and 9. The responsibility for the construction of reading stations will vary with the type of work. A nominated sub-contract is especially useful where a foundation is to be instrumented since it allows for everything to be on hand early on in the works thus avoiding delays.

Where say an earth dam is to be instrumented and many of the instruments are to be placed in the fill, and work is also intermittent, then the nominated sub-contractor might best be restricted to supply and initial installation only.

Whilst item 3 is regarded as the engineer's province, the necessary expertise may not be available. This is only likely to be so where the area requiring instrumentation forms just a section of the main works, or where there are complex geotechnical problems. Such expertise is best sought on a consultancy basis rather than incorporation in a contract.

The placing of instruments will affect the main contractor's work and the documents must state this, and tell the contractor to allow for delays and special working arrangements occasioned by the instruments. The earthworks items such as trenching and backfilling, should also be specified and billed in the earthworks section. The provision of services and protection should also be allowed for in the rates, unless they are particularly extensive and can be specified in detail.

INSTALLATION

Ideally the placing of instruments and their connections to reading stations should cause no disturbance within the body of soil to be instrumented but of course this ideal is unattainable. The acceptable level of disturbance needs to be considered as this will determine the degree of stringency governing installation and backfill procedures. Well organized installation and especially backfilling operations will do much to keep disturbance and alteration of the condition of the soil surrounding the instruments to a minimum. Civil engineering operations with heavy plant are incompatible with the placing of comparatively delicate instruments and the installers, in arranging backfilling to proceed on and around the instruments will be constantly seeking to reconcile this.

In particular, consider the placing of a number of twin tube piezometers located at one elevation in the clay fill of an earth dam and connected to a reading station on the downstream face. When clay fill is placed in layers as a continuous operation it will generally be subject to only slight moisture content changes at the surface of each layer. Where instruments are placed over a period of several days, considerable changes in moisture content can occur both on the fill surface and in the trench faces, and in the arisings. In a hot dry climate it is doubtful if the arisings could be satisfactorily compacted back into the trench. Because of the chances of damage, removal of either dry or wet soil will be impracticable. Thus the trench for the connections could act to a certain extent as a drain and under such circumstances the instruments would give a misleading picture at least temporarily for both the development of pore water pressures during construction, and possibly seepage pressures later on impounding. Thus there is a need for speed consistent with good workmanship.

Installing instruments often requires intermittent periods of activity, and this will give rise to both staffing and labour problems. In Great Britain today it is doubtful if even on a large site the contractor would employ sufficient labour for the necessary excavation and backfill by hand. The best use must be made of construction plant and the work designed and organized to permit it's maximum use. This may require special adaptation of plant or specifying certain types of plant for different operations.

The staffing needs will be dictated by the number of instruments, the time available for the installation of any group of instruments, and the degree of complexity of the instruments. It is difficult to generalize on numerical strength, but for many jobs one engineer supported by a technician and tradesmen should suffice.

Consider the case where the instrumentation engineer is on the R.E.'s staff and the design requires many instruments placed intermittently. There will be a steady increase in work load due to readings and analysis, possibly until the point is reached where work load exceeds the time available. The necessary progressive increase in staff requirements should be ascertained from the programme. Consideration could be given to data logging equipment but this gives rise to a further source of malfunctioning and it is doubtful if the field staff could carry out any but the simplest maintenance. On site facilities will vary with the nature of the installation and also the remoteness of the location. They could include workshop for maintenance and fabrication, communication system for use during installation, transport and protection from weather.

Protection and malfunctioning

Instruments can be damaged or may function incorrectly due to the following causes:-

1. The effect of climate.
2. Damage from contractor's operations either from his negligence or insufficient cover.
3. Deformation of the works.
4. Deformation or deterioration of equipment.
5. Chemically aggressive groundwater or soil conditions.
6. Vandalism.
7. Damage during installation.
8. Various types of malfunctioning of electrical instruments.
9. Foreign matter causing blockages in tubing or causing failure of electrical circuits.

At the design stage the probability of damage and malfunctioning should be assessed and allowed for or the design modified. From experience and as a guide 10% of instruments are likely to be written off for various reasons and the pattern can be increased at critical points to allow for this.

Where frost can occur the connections and reading stations for hydraulic instruments must be adequately protected. In some instances protection can be given by using antifreeze, but any adverse effects of the antifreeze must be checked and the density of the mixture allowed for in calculations. Reading stations are often located in zones of the work liable to flooding and arrangements for protection should be made.

Many plastics when exposed to sunlight can deteriorate with time and the original properties may alter. Protection from exposure to direct sunlight should be provided for all such materials. Temperature changes may cause fluctuations in materials in reading devices giving rise to erratic readings.

Most damage to instruments seems to occur from the contractors' operations and there are two main reasons for this. The first is that insufficient protection may be given to buried instruments and damage from vehicles results, and the second is the contractor's negligence. The first can be overcome by specifying adequate backfilling to instruments and cables and ensuring this is performed as specified. The second cause of damage by plant can be minimized by ensuring at planning stage that the contractor will have adequate space for his operations. Also there must be adequate marking of the position of instruments as well as ensuring the contractor's staff are fully aware of these positions and the consequences of damage.

As earthworks are progressively constructed they will deform and connections to instruments will be placed in tension and may rupture. An estimate of deformation should be made during design and if thought necessary allowance made in the specification for cables and tubes to be laid with adequate slackening. The ability of the equipment to remain stable under the field conditions to be imposed should be checked by testing before installation. This can include resistance to attack by chemically aggressive soil or ground water.

Vandalism as a cause of damage mainly occurs after working hours and chiefly to reading stations but sometimes to exposed vertical tubing. In some urban areas exceptional measures such as reinforced concrete reading stations with steel doors and no windows may be required. Siting reading stations within a protected compound will do much to reduce vandalism.

Damage during installation is likely and where possible spare items of equipment can be included for in the contract. Financial provision for major repairs as required can also be allowed for in the instrumentation contract to minimize delays.

The stability of all electrical instruments should be checked and calibrated before installation and for recording instruments allowance made in the contract for checking their continued reliability at suitable intervals.

Testing and commissioning

The formality of testing and commissioning will depend upon the nature of any contract. There should be a specific requirement on whoever is installing the instruments to test that the instrument is functioning properly and at the same time can be readily identified and that all the installation details such as elevation are recorded. The zero or first reading should then be taken and recorded and the instrument regarded as commissioned.

USAGE

Item 3 in the introduction sets out the operations involved in making use of the instrumentation and these chiefly involve the engineer. It is the needs of such work which often tend to be dealt with less than adequately, possibly because the work is not subject to contract or for a variety of other reasons. For proper use to be made of the installation the following should be considered:

1. Staffing.
2. Reporting procedures.
3. Stationery.
4. Computer time.
5. Contract item for maintenance.

As a rough guide the following staff are required for simple instruments.

Staff	Number of Instruments
Can be read by R.E.s staff when trained about ½ day per week.	up to 20
One of R.E.s staff for 1 day per week.	20 - 50
Additional staff needed as 3 days work per week.	50 - 100
Full time specialist staff required	100 or more

Where the work is seasonal then the staff could for a time be under-employed, but there are frequently maintenance, measurement and other tasks. Some under-employment of site staff in winter in Britain is not unusual. The tasks of the full time specialist could also include attention to minor malfunctioning, measurement and records of the relevant increments of construction and routine maintenance.

The reporting procedures need to be set out for the Resident Engineer, and it is good practice to include the records, brought to a specified state of completion, in the monthly site report.

Stationery and drawings required should be designed, and computer time and programmes arranged, all in advance of site work and should be on hand at the start of the contract. The benefit of detailing methods of recording prior to the contract are, that important details are less likely to be overlooked where there is a specific requirement to record such data.

Reading devices such as data logging equipment will require maintenance and calibration from time to time and as the work will require specialists, the contract for supply should allow for such a service where this cannot be done by site staff.

BENEFITS

At the design stage of many works there is often doubt as to the reaction of soils to the proposed loading conditions, however thorough the site investigation. There is also the variable of weather during construction and it's effect on the moisture content of fill to be considered. Instrumentation provides the means of indicating the development of unexpected conditions or confirming the design assumptions. Provided contracts have been suitably drawn up, design can be modified to avoid failure or to permit savings in both materials and time. Instrumentation of trial or test sections of work prior to the main works can permit a more rational design.

In particular consider a length of road embankment traversing a layer of soft alluvial clay and forming part of a much larger project. For economy the bank construction will be programmed in stages, the period between stages permitting the dissipation of pore water pressure leading to an increase in the resistance to shear failure in the foundation soil. This will permit a higher bank to be built than is indicated by the initial soil conditions, and this may reduce bridge works or save excavation and replacement of the soft soil thus reducing costs. Such financial benefits may be wholly or partly offset by the longer period required for construction which delays the benefits from the investment. The possible use of various types of foundation drainage is also a factor in such appraisals.

At the design stage a programme can be drawn up for incremental construction but this needs to be done with some degree of conservatism as it will be based on an estimation of foundation behaviour derived from the site investigation data. Factors of safety for this kind of operation are necessarily low in comparison with other civil engineering design and are at their lowest during construction. Instrumentation for measuring the pore water pressure and its changes with time and loading, will enable the stability to be estimated during construction with much more confidence. The selection of the thickness of the first stage of filling and the timing of its placement is usually the critical feature in the programming. Most alluvial clays are slightly overconsolidated and also their initial macro-permeability is often much higher than measurements indicate due to the presence of a multiplicity of drainage paths, especially near ground level. Under quite low loads these drainage paths close up and become almost ineffective. Thus whereas the site investigation information will indicate one thickness, for the first stage, instrumentation will usually permit a useful increase in thickness especially about the centre line and may allow a speeding up of the rate of filling for this stage. This will give the benefit of reducing post construction settlement and possibly bringing forward subsequent stages of the filling. Thus without instrumentation the construction would have to be drawn conservatively possibly to such a degree that construction in stages was no longer financially worthwhile.

DISCUSSION

The human factor in the success of this kind of work is considerable and those involved will need enthusiasm for the task, resilience to setbacks, and will need to exercise firmness in dealing with site matters. Any instrument design should certainly be set out to minimize interference with the main works, but having conceded this the adoption on site of a pose that "the soils are there principally for the benefit of the instruments" should not be discouraged. The chief advantage to be obtained from good organization is in allowing those installing and using the instruments to give this their full attention under the best possible conditions which will minimize malfunctioning and allow the best use of the results.

ACKNOWLEDGEMENT

This paper has been offered by kind permission of Mr. A.D. Holland, Chief Highways Engineer, Highways Directorate, Department of the Environment.

Part 2: Discussion

Contents

Part 2: Discussion

Foreword

There were four sessions for discussion each with a General Reporter who opened the session with a brief introduction to his general report. Delegates who then spoke were asked to submit their remarks in writing with any necessary illustrations for publication. Some have failed to do so and some others have sent in written contributions although they did not speak. No differentiation between these types of contribution has been made in these records of proceedings and the order of the contributions is not necessarily that of the speakers during the discussion. The General Reporters have amended their reports in the light of the discussion and although they are given here at the beginning of the discussion, the reader will find that they sometimes refer to the subsequent discussion.

A. D. M. Penman
Technical Editor

Introductory address

Kevin Nash
Secretary General, ISSMFE

Introduction

It is now nearly ten years since the name of the Society was changed from the British National Society of the International Society for Soil Mechanics and Foundation Engineering (surely a mouthful if ever there was one) to the British Geotechnical Society. This change in name was accompanied by a slight change in emphasis: the Society was specifically to embrace Rock Mechanics as well as Soil Mechanics and the Committee membership was adjusted forthwith to contain people with rock interests.

The connection with ISSMFE remains as before – the Society is still the British National Society of ISSMFE and all the members of BGS are automatically members of ISSMFE – but in addition you all belong to the International Society for Rock Mechanics (ISRM) as well, and a tiny proportion of your subscription is sent to those two bodies.

The International Society for Soil Mechanics and Foundation Engineering

The embryo of ISSMFE was formed at the 1936 Harvard Conference under the genius of Karl Terzaghi. He was the first President and Arthur Casagrande agreed to act as Secretary until such time as another could be found. It was resolved that further conferences should take place every few years. The Society was to be run by an Executive Committee elected by National Committees. But it was not until twelve years later, in 1948, in a Rotterdam flattened by the war, that the Second Conference took place. By then there were National groups in 39 countries (including the United Kingdom) but the Society still remained embryonic. Professor Don W. Taylor was appointed Secretary and his draft Statutes for an International Society were passed by the Executive Committee at the next Conference in Zürich in 1953. These Statutes are similar to those which we now operate: the Society was named the ISSMFE and its object was to promote international co-operation in the field of soil mechanics. This was to be achieved largely by the periodic holding of Conferences. The official languages were to be English and French.

The name of the International Society

Even in that extract from the first Statutes we can find the seeds for future discussion. Was that name really the best we could find? This has been debated by the Executive Committee (on which BGS may have two representatives) at immense length, first in 1965 and again in 1971. The latest thinking on this matter is that International Geotechnical Society or International Geomechanics Society (I.G.S.) might be a great improvement – but this implies not only Soil Mechanics but also Rock Mechanics and Engineering Geology, and that such a name should be reserved for a united society of all three (if such should ever come about). In this connection it is interesting to note that a secretariat of ISSMFE, ISRM and IAEG (the International Association of Engineering Geology) has recently been established in Brussels. So far the Secretaries of the three Societies have met twice; their expenses have been paid by a grant from the Belgian Government but there is a long way to go before I.G.S. becomes a reality.

Official Languages of ISSMFE

The languages of ISSMFE have stirred the emotions for some time. These have been English and French since the early days but the Proceedings of the first conference was in English only. After that time papers might be in English or French with summaries in both languages, and simultaneous translation has been provided between the two. The cost of running the conference is borne by the host country and if they wish to provide additional translation services they are at liberty to do so. They may also provide the paper summaries in additional languages if they wish to do so, and it is probable that this will happen at Moscow.

Many National Societies have expressed the view that we should operate in English only. The point was last debated by the Executive Committee at considerable length at Mexico City in 1969 when it was decided by a large majority that English and French should remain the official languages of the Society. It may be noted en passant that the next President is to be the distinguished Parisian, Professor Jean Kerisel. During his period of office there is likely to be a French dimension not hitherto felt and this will be welcomed in many parts of the world.

The Society is now run on very democratic lines – one vote per National Society – and you will find our Statutes in English and French at the front of the 1972 List of Members, a copy of which was distributed to each person who was a member on 1st January 1972. The officers of the Society are shown in Table I.

International Conferences

The main achievement of the Society has been in providing a link between soil mechanics workers in different countries. Many people in this room must know personally their opposite numbers in the U.S.A. and also in Norway or Belgium or Thailand. I suggest that the International Society has played a great part in bringing this about – not just for the few, the élite, but for very large numbers of our members.

It is interesting to see the growth of the attendance at these conferences (Table II and *Figure 1*). You will all be familiar with the Proceedings of the seven. They were listed in the ISSMFE News for March 1972 indicating how and

Table I OFFICERS OF ISSMFE 1936–1973

Period	*President*	*Vice-Presidents*	*Secretary*
1936–48	K. Terzaghi (USA)	A. Casagrande	A. Casagrande
1948–53	K. Terzaghi (USA)	A. Casagrande	A. Casagrande
1953–57	K. Terzaghi (USA)	W. S. Hanna (Africa) A. E. Cummings (N. America) M. Vargas (S. America) K. Hoshino (Asia) A. W. Skempton (Europe)	D. W. Taylor
1957–61	A. W. Skempton (UK)	J. E. B. Jennings (Africa) R. F. Legget (N. America) A. J. L. Bolognesi (S. America) K. L. Rao (Asia) G. D. Aitchison (Australasia) A. Mayer (Europe)	A. Banister
1961–65	A. Casagrande (USA)	R. L. Mitchell (Africa) L. Zeevaert (N. America) A. J. Costa Nunes (S. America) K. L. Rao (Asia) K. S. Birrell (Australasia) L. Bjerrum (Europe)	A. McDonald
1965–69	L. Bjerrum (Norway)	B. A. Kantey (Africa) W. J. Turnbull (N. America) O. Moretto (S. America) J. G. Zeitlen (Asia) D. H. Trollope (Australasia) J. Brinch Hansen (Europe)	J. K. T. L. Nash
1969–73	R. B. Peck (USA)	M. P. dos Santos (Africa) D. H. MacDonald (N. America) G. Perez Guerra (S. America) T. Mogami (Asia) E. H. Davis (Australasia) E. E. de Beer (Europe)	J. K. T. L. Nash
1973–77	J. Kerisel (France)	J. W. de Graft Johnson (Africa) R. Marsal (N. America) V. F. B. de Mello (S. America) Z. C. Moh (Asia) P. W. Taylor (Australasia) A. Kezdi (Europe)	J. K. T. L. Nash

where they could be obtained. Those of you who have attended the conferences will know how very exciting it is to be able to see and hear the great fish in discussion with one another. Maybe here in London we have become blasé about this but at the Mexico conference I personally felt very privileged to have the opportunity to listen to Ralph Peck on the subject of deep foundations and to Alec Skempton and Laurits Bjerrum on slopes. It is simply marvellous to be able to strike up a friendship with someone from some remote part of the world whose interests and problems overlap with your own.

The disadvantage of the large conference is that intimate debate between the not-so-great becomes impossible. In theory it is splendid that any member should be able to jump up from the body to debate some topic. In practice (alas) his

Table II DISTRIBUTION OF ATTENDANCES AT ISSMFE CONFERENCES

	1st Conference (N. America)	*2nd Conference (Europe)*	*3rd Conference (Europe)*	*4th Conference (Europe)*	*5th Conference (Europe)*	*6th Conference (N. America)*	*7th Conference (N. America)*
Region:							
Africa	3	9	18	15	16	17	31
N. America	177	48	60	89	157	678	834
S. America	1	16	19	23	33	80	125
Asia	3	16	16	25	34	50	67
Australasia	—	3	4	7	12	11	19
Europe	23	357	427	615	773	350	321
Total	207	449	544	772	1025	1186	1397
Accompanying persons	—	147	178	256	326	279	533
Overall Total	207	596	772	1028	1351	1465	1930

pearls of wisdom may not be so regarded by the rest of the body and some control is required. The Specialty Sessions, initiated at the Mexico City conference, enabled smaller groups of enthusiasts to get together and this is being repeated at Moscow.

Regional Conferences

It is here that the Regional Conference comes into its own. The International Society is divided into six Regions and over each of these there is a Vice-President. One of his important duties is to assess the invitations received from the various member countries within his Region to act as host for the one conference which takes place between the international conferences. He also vets the proposed topic.

Thus every four years or so there is a conference in each Region — five in all, as North and South America combine to hold a Pan-American Conference. The first of these in Europe was held in Stockholm in 1954 and the series has been as shown in Table III. The next of this series has been arranged to be held in Vienna in March 1976 on the subject of Deep Foundations and Deep Excavations.

Table III EUROPEAN REGIONAL CONFERENCES

No.	*Year*	*Location*	*Number attending*	*Accompanying persons*	*Topic*
1	1954	Stockholm	65	11	Stability of slopes
2	1958	Brussels	84	25	Earth pressure problems
3	1963	Wiesbaden	172	—	Settlements and compressibility of soils
4	1967	Oslo	223	84	Shear strength of soils and rocks
5	1972	Madrid	503	105	Structures subjected to lateral forces

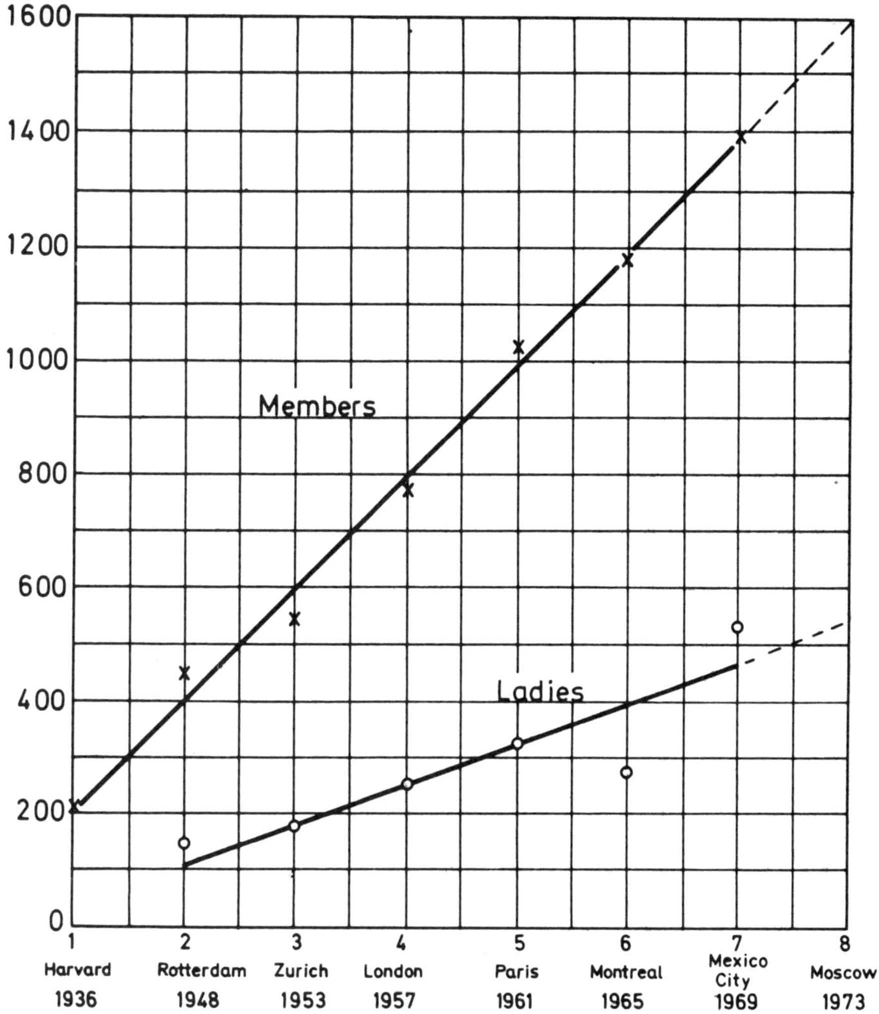

Figure 1 Numbers attending International Society for Soil Mechanics and Foundation Engineering Conferences, 1936–1969

Since 1961 the International Society has permitted a third language at these conferences – that of the host country – and this had undoubtedly been a great help to the home side.

It may be that in due course the U.K. may offer to act as host to one of these conferences and this must be made (well in advance) to the Vice-President of the Region. If, for example, you had in mind to do this for the slot after Vienna, I would suggest April 1979 and maybe you should be thinking in terms of 1400 participants with 300 accompanying persons.

The spin-off from such a conference may be increased if the Executive Committee is invited to meet at the same time as the conference itself. Perhaps we do not need this in Britain but undoubtedly Australia welcomed this opportunity of bringing in a number of distinguished engineers to their last Regional Conference in 1971. The current series of Regional Conferences throughout the world is now over until 1975: the previous ones have been as shown in Table IV.

Table IV REGIONAL CONFERENCES

Period	*Year*	*Region*	*Place*
1948–53	1952	Australasian (1)	Melbourne
1953–57	1954	European (1)	Stockholm
	1955	African (1)	Pretoria
	1956	Australasian (2)	Christchurch
1957–61	1958	European (2)	Brussels
	1959	African (2)	Lourenco Marques
	1959	Panamerican (1)	Mexico City
	1960	Asian (1)	New Delhi
	1960	Australasian (3)	Sydney
1961–65	1963	African (3)	Salisbury
	1963	Asian (2)	Tokyo
	1963	Australasian (4)	Adelaide
	1963	European (3)	Wiesbaden
	1963	Panamerican (2)	São Paulo
1965–69	1967	African (4)	Cape Town
	1967	Asian (3)	Haifa
	1967	Australasian (5)	Auckland
	1967	European (4)	Oslo
	1967	Panamerican (3)	Caracas
1969–73	1971	African (5)	Luanda
	1971	Asian (4)	Bangkok
	1971	Australasian (6)	Melbourne
	1071	Panamerican (4)	San Juan (Puerto Rico)
	1972	European (5)	Madrid

The Proceedings of the conferences provide a useful reference to the soils of the world and how to order them was also indicated in the ISSMFE News in Geotechnique of March 1972. Since 1970 the papers have all been abstracted in Geotechnical Abstracts produced by the German National Society and you ought all to be familiar with these and the Geodex Retrieval System, both of which are officially backed by ISSMFE.

National Society Conferences

The British soil mechanics group has initiated a number of conferences. Following the 1939 James Forrest Lecture, which was given by Terzaghi, and the four rather important lectures on soil mechanics given by Cooling, Skempton, Glossop and Markwick in 1945, the conferences shown in Table V were arranged. Only those marked with an asterisk were the direct responsibility of BGS but many of the same people were involved with the organisation of the other conferences. In 1948 great vision was shown in the launching of Geotechnique, and following that from 1961 there has been the world-famous series of Rankine Lectures. As a Society you appear to be flourishing and as a former Chairman (1959–61) I am delighted to see it so. May I just suggest that when you plan your future conferences your thinking as to timing and subject should be notified to ISSMFE.

Table V CONFERENCES INVOLVING SOIL MECHANICS IN THE U.K.

Year	*Subject*	*Published by*
*1950	Measurement of shear strength	Geotechnique
1951	Design and construction of foundations	Building Research Congress
1955	Correlation between calculated and observed stresses and displacements in structures	I.C.E.
*1960	Pore-pressure and suction in soils	Butterworths
*1963	Grouts and drilling muds in engineering practice	Butterworths
1966	Large bored piles	I.C.E.
*1969	In situ investigations in soils and rocks	B.G.S.
1970	Ground engineering	I.C.E.
1970	Behaviour of piles	I.C.E.
*1973	Field instrumentation in geotechnical engineering	Butterworths

Soil mechanics is prospering in many countries of the world. Our family is large – there are nearly 9500 members in 44 different National Societies ranging in size from 6 members in Ireland to 900 in West Germany. Despite the large numbers a close fellowship exists between them all. This I believe to be a very worthwhile pursuit and I sincerely hope that Britain will continue to play a leading rôle in the ISSMFE deliberations and councils.

On behalf of ISSMFE may I wish you every success in your present symposium.

Session 1: Principles of measurement

GENERAL REPORTER: J. R. F. Arthur

(University College, London)

The principles of measurement with which the opening session has been concerned are not the heart of the matter for the geotechnical engineer. Knight has maintained that we should have considered all aspects of a geotechnical problem, but our terms of reference for the first session were clearly much more specific than this and indeed excluded the key questions of why the measurements were needed (Peck 1970), and what relative value should be attached to them. A discussion of the principles of measurement naturally centred on the essentials of calibration and errors and seemed to make the right starting point for this symposium.

The fundamental statement that measurement is a comparison of unknown with known, suggests something simple and straightforward but, of course, in the case of geotechnical field measurements the comparison is often remarkably indirect. Success depends on the elimination of as many unwanted variables as possible together with the correct assessment of the influence of the remaining variables which are inextricably involved in the measurement. The elimination of unwanted variables is achieved by choosing the right physical phenomenon as the means of measurement and by good detailed design of the measuring instrument. The random variation in any type of measurement can be estimated satisfactorily by normal statistical methods, but insight, analysis and experimental skill are the means for estimating the consistent errors such as those caused by the very presence of the measuring instrument.

In order to establish a frame of reference for this report the list of fundamental attributes for a measuring system provided in the preliminary report has been extended. Some of the attributes are painfully obvious and I hope readers will tolerate their presence in the cause of completeness. Even these obvious attributes are essential to a basic description of a measuring system. The enlarged list is as follows:

(1) *The variable(s)* to be measured.
(2) *The physical phenomenon* employed in the measuring instrument.
(3) *Robustness.* The intrinsic ability to survive in its chosen environment standing up to impact, prolonged submergence, corrosive substances, temperature variations, etc.
(4) *Sensitivity.* The smallest change in the variable being measured which the instrument will consistently detect.

(5) *Response time.* The time the measuring system takes to complete a single reading of the variable.
(6) *Range.* The difference between the maximum and minimum quantities that can be measured by the same instrument without undergoing any alteration.
(7) *Stability.* The variation with time of the tolerance within which any measurement of the variable by the given instrument will lie in relation to the actual value of the variable.
(8) *Environmental calibration* (cf. cell action). The relationship between the actual measurement and the ideal measurement. This ideal measurement is the value the measured variable would have had if the presence of the instrument had been miraculously avoided. The instrument is taken to include any connections to it and any borehole and disturbed soil associated with these connections.
(9) *Accuracy.* The tolerance within which any measurement of the variable by the given instrument will lie in relation to the ideal measurement (as defined above), or preferably, twice the standard error of the measurements from the ideal measurements. Qualifications of Statements of Accuracy in terms of Attributes (5), (6), (7) and (8) will frequently be essential. Methods of using instruments will sometimes have to be included.
(10) *Data reliability.* This comprises the accuracy as defined in (9) under normal working conditions together with the ability to check for erroneous readings by comparison with a separate instrument installed in a similar position or the ability to recalibrate insitu and check a reference or zero reading *in situ.*
(11) *Data Presentation Chain.* The component chain through which a reading of the variable passes before it is read by a responsible person capable of interpreting the meaning of the data. The time for data to complete the chain, the probability of a break in the chain and the time likely to elapse before a break in the chain is discovered and repaired are important features of this chain.

Variables

Three types of measurement are common in geotechnics, namely deformations, pore pressures and total stresses. These three are listed in ascending order of difficulty and form a secondary structuring for this report which recurs in most sections dealing with individual attributes.

Deformations are relatively widely measured in present practice and a wealth of valuable information has been provided in this symposium. The key to successful deformation measurements, the establishment of a safe stable datum in the right position and at the right time (as early as possible) has been stressed by a number of participants, notably Penman and Knight. The location and measurement of localised shear strains within a soil mass which would allow the description of the formation of failure planes within the ground remains an unsolved problem (Bishop). For the moment, the best guide to the formation of failure planes comes from models (Roscoe 1970) and the stress range that these models can cover has been greatly increased by centrifugal model testing (Avgherinos & Schofield 1969).

Pore Pressures are again widely and successfully measured in the field. This is not surprising in view of the dominating importance of the Principle of Effective Stress in soil mechanics practice. The only uncertainty which might be claimed to be concerned with principle mentioned during the symposium concerned the significance of osmotic pressure in the field (Bishop, Vaughan).

Total Stresses are much the most infrequent measurements made in present practice. This is particularly true for cells buried within the soil mass and reflects the real uncertainties concerning the accuracy of measurements of this type. There is progress to be discussed on this problem under the attribute Environmental Calibration.

Physical phenomena

The range of choice of physical phenomena for making deformation measurements is large and well represented in the symposium proceedings. There is a real choice for the other variables pore pressure and total stress, but it is much more restricted. In choosing a particular phenomenon its influence on the remaining listed attributes will be important, but cost and other practical factors may be more important, and the descriptions of the use of various phenomena in the symposium will be valuable when the choice is being made.

Robustness

This quality is essential, and well appreciated as such, so that little need be written in explanation. There are many examples of the care that goes into achieving survival on site but perhaps the outstanding ones are the reference datum pillars and points presented by the Building Research Station team (Burland & Moore, Cheney, Marsland, Penman & Charles).

Sensitivity

This attribute is chiefly the concern of the instrument designer; the user is really only interested in accuracy which is a much more complex and comprehensive assessment.

Response time

Response time is clearly critical in dynamic measurements associated with moving traffic or earthquakes. Less obviously, it may be associated with the time required for a true reading to be attained in the soil after the disturbance of inserting an instrument. This is particularly important in pore pressure measurement (Vaughan). Marsland and Eason have provided an example of the reverse response time relationship in their plate loading tests which must be carried out quickly to provide the best results.

Range

Although the choice of the correct range may present the designer of an instrumentation programme with a difficult problem it does not really touch a principle. The link between range and accuracy is well illustrated by Hutchinson's session 2 discussion contribution describing an inclinometer capable of measuring large deformations.

Stability

The stability of an instrument depends to a great extent on the physical phenomenon used in the instrument. A vibrating wire instrument can be made stable much more cheaply than a gauge employing electric resistance strain gauges. Detailed design and standards of component quality also play an important role. There is clearly a case for grading instruments on the basis of their stability as Bishop has suggested in his contribution to this Session. This would save money and lessen the prospects of failure. Bishop also raised the problem of performance monitoring in earth dams for periods up to one hundred years. Here it seems that provision for replacing instruments may be needed as well as in-place calibration and reference point checking (see Attribute (10)).

Environmental calibration

The assessment of the influence of the presence of an instrument on the measured variable remains a most difficult problem when the instrument is within the earth mass or even on the boundary of the mass. Ashkenazi points out that in surveying, external means are used to check consistent errors (as opposed to consistency as he calls his random error). The difficulty is that in the case of buried instruments or installations the consistent error can be large but it is not always large and there are often no easily available external means to determine it. Unfortunately, this remark applies to two out of the three variables we are concerned with, namely deformations and total stresses. In the case of pore pressures most of the instrument's influence dissipates with time and then the results can be relied on (Vaughan).

In deformation measurements when the gauge length is long and strain gradients are low one may argue intuitively that the problems are likely to be less severe. However, Marsland has raised serious questions concerning the compatibility of borehole linings for settlement points installed in boreholes. It is clear that in soft soils there may be a danger of measuring the deformation of the borehole lining rather than the soil. His discussion is a good example of this particular type of problem and the uncertainty at present involved. Comparative measurement of settlement with an entirely independent means in a variety of soil types would reduce the uncertainty. It would have to be carried out over a prolonged period to estimate the influence of the change in strength of the grout linings with time.

Deformation measurements with short gauge lengths are probably best attempted by pairs of induction coils separated by the intended gauge length. This approach was mentioned by Dunnicliff and has been investigated earlier by Morgan (1972) using X-rays at Cambridge. The results were not encouraging for the larger strains which occur when the stress ratio exceeds three. These tests need to be extended to cover different soil types.

The problems of total stress measurement can be separated into those of boundary and buried cells. In theory, boundary earth pressure cells offer no problems, but in practice great care has to be taken to mount the cells to avoid the possibility of inducing stresses through the mounting. There is also the problem of ensuring that the stiffness of the cell corresponds precisely with the surrounding boundary. Lack of boundary stiffness correspondence may explain the under-registration of the Cambridge cell reported by Symons. The over-registration of the other two cells which use a balancing back pressure is predictable but the variation in calibration with soil type and condition which is reported by Symons shows how much care should go into environmental calibrations. Wroth has commented on the surprising lack of shear stress measurements; in spite of the danger of being partisan (Arthur & Roscoe 1961) I share this surprise and would suggest that the Cambridge cell has a further advantage in the measurement of the eccentricity of the normal load imposed on it. This ability is important along any boundary sustaining a stress gradient. Diaphragm cells may have their calibration altered in the presence of a stress gradient and it will be extremely difficult to allow for this. The information they provide will in any event be seriously incomplete.

Vaughan questions Wroth's assertion that a buried cell measuring both shear and normal components of stress would be an advantage. There is little evidence for either view and certainly it is desirable to develop both types and make a careful comparison, but the presence of stress gradients would probably again favour the Cambridge type of cell.

The work of Brown and Elsenaar strongly suggests that at stress ratios of three and below, total earth pressure cells do give relatively reliable readings provided they are carefully designed and calibrated. Further work is needed on the influence of placement techniques, varying types of soil and soil condition, but there does appear to be real progress in this low strain region which contrasts sharply with the uncertainties as the stress ratio and strains grow.

In this more general case, there appears to have been little progress since the relatively old work referred to by Chaplin. It was surprising and disappointing to learn from Thomas in Session 2 that it had not been considered justifiable to invest in a compression chamber to run an environmental calibration on the B.R.S. cell. In view of the permanent nature of this facility, it would be wrong to ascribe the cost to the calibration of a single cell. There would be years of work for a chamber of this kind in evaluating the influence of soil type, soil packing and stress ratio on buried instrument readings.

The push Glötzl type cell described by Massarsch is another example of a cell that requires careful evaluation, preferably by an X-ray technique, in the manner reported by J. M. O. Hughes in describing the promising new Cambridge device for measuring a stress-strain curve *in situ* (albeit without estimating anisotropy). F. H. Hughes in discussion has suggested that there should be more field research into hole making and certainly the Cambridge success shows that great improvements are possible. Whether it should be done in the field or the laboratory should surely depend on the possibilities of measuring the disturbance produced. In the case of sand drains the rate of water flow may provide an indirect measurement, but it will perhaps be difficult to compete with the laboratory X-ray method, especially as many inhomogeneities in soil structure can be provided by simulation or more expensively from block samples.

Brown in his discussion over-estimates the reliance I placed on the comparison reported by Irwin between earth pressure cell readings and the calculated overburden pressure. Such a comparison can only be an approximate check but when data is so short this type of check is valuable. The same view was taken by

Vaughan in his discussion of the use of hydraulic fracture as an independent method of assessing stress. However, full environmental calibration will often be required with controlled simulation in the laboratory of a wide range of anticipated field conditions.

Accuracy

The position of accuracy in this list is intended to emphasise that it is the quantitative summary of most of the attributes that precede it in the list. This was the reason for putting forward accuracy as the theme for discussion in the first Session. Dunnicliff asserts in his discussion that 'the role of accuracy *per se* is a limited one and one to which excessive attention is often paid'. This statement seems to imply that for the practical purposes of geotechnical instrumentation, existing instruments are more than sufficiently accurate and there was no engineering purpose in staging this Session of the Symposium. As participants will realise, such a generalisation is premature. Many statements of accuracy for buried instruments which do not include environmental calibrations within them are useless and environmental calibrations are rare. The practical alternative in some cases to such a calibration is a second estimate or measurement, which is made by another means and is independent of the environmental influence, to check on the first measurement. This is normal surveying practice but reports of this type of check are as rare as environmental calibrations for some classes of measurement.

In his discussion, Brown has provided valuable data showing the scatter found in environmental calibrations of total stress cells. These data illustrate the need to define accuracy in terms of a multiple of the Standard Error as advocated by Ashkenazi. Accuracy defined in this way will give the designer of field installations an excellent guide to performance provided it is backed up by a detailed description of how the accuracy given was arrived at.

In the absence of relatively large local shear strains, it is unlikely that the laboratory calibration of an inclinometer differs greatly from the environmental calibration. In these circumstances, the papers by Cornforth, Green and Phillips and James deal with inclinometer performance in a useful quantitative manner. Cornforth provides very useful statistics from a reasonably large series of repeated measurements by several observers in stable ground. This technique probably gives the best possible estimate of field accuracy. Green compliments Cornforth's work by carrying out a thorough investigation of two field inclinometers in the laboratory where variables can be isolated or accurate ancillary measurements taken. Excellent results are reported by Burland and Moore for the improved inclinometer developed from Green's work. The work on inclinometers seems to me to illustrate a fundamental approach to the estimation of accuracy in field instruments. The estimation is split into three stages; the first might be called the artificial calibration in which the instrument is put through its paces in the controlled situation of a laboratory and cross-sensitivities can be investigated selectively in the classical manner by isolating variables.

The second stage is the estimation of the standard error in the field by repetitive measurement in situations where the ideal measurement remains constant. The population of measurements or even populations of measurements required must be judged according to the variety of conditions in which the instrument will be used. The first stage laboratory work helps in this judgement because it indicates some of the factors likely to affect field performance.

If the presence of the instrument is likely to alter the quantity being measured there must be a third stage, namely environmental calibration.

Data reliability

It was indicated in defining accuracy that a qualification in respect of stability would frequently be needed. Such a qualification would usually limit the time for which the accuracy given would be maintained. In longer term measurements some form of repair such as de-airing is frequently needed (Vaughan, Walbancke). Even if no repair can be effected a check on reliability is needed so that ability to recalibrate in situ and check a reference or zero point on the calibration then becomes extremely desirable, and Dibiagio's discussion in Session 2 is especially welcome.

Data presentation chain

This chain has always been part of a measurement system but the paper by Sherwood & Currey and the discussion in Session 2 by Wilson show that a new phase of vastly increased complexity is upon us. The essential features of every chain were summarised in the list at the beginning of this report, but it must be recognised that automated systems encourage the building in of predetermined significance criteria.

The influence of these criteria on the primary aim of the presentation chain needs careful consideration; the chain's primary aim is, of course, to bring the data to a responsible person capable of interpreting it.

Conclusion

The symposium has shown that most of the fundamental attributes of measurement systems are receiving attention from individual researchers and engineers, but there are still surprising gaps in our knowledge of environmental calibrations.

Perhaps descriptions of measurement systems need more system themselves, at all events they are frequently incomplete. For instance, accuracy should preferably be defined in terms of the Standard Error and should always be accompanied by a full description of the calibrations made to define it.

REFERENCES

Arthur, J. R. F. and Roscoe, K. H. (1961). 'An earth pressure cell for the measurement of normal and shear stresses', *Civil Engineering and Public Works Review*, **56**, 659, 765–770.

Avgherinos, P. J. and Schofield, A. N. (1969). 'Drawdown failures of centrifuged models', *Proc. 7th Int. Conf. of S.M. & F.E. Mexico City*, 2, 497–505.

Morgan, J. R. (1971). 'Stress-strain behaviour of soils', *Proc. Roscoe Memorial Symposium*, G. T. Foulis, Henley-on-Thames, 496–497.

Peck, R. B. (1970). 'Observations and Instrumentation: some elementary considerations', *Seminar on Field Observations in Foundation Design, New York Met. Section, A.S.C.E.*

Roscoe, K. H. (1970). 'The influence of strains in soil mechanics', *Geotechnique*, **20**, 2, 129–170.

Session 2: Equipment design, installation, and performance

GENERAL REPORTER: G. E. Green
(Imperial College, London)

Introduction

Elaborate instrumentation programmes will fail to produce useful results if too many unsuitable sophisticated instruments are installed and operated by unskilled persons. The overall performance of an instrument is controlled by the three R's of instrumentation – resolution, reliability and ruggedness (Cornforth) and generally the most successful instruments are simple and robust (Little). Lovenbury and McKenna & Roy used simple direct reading piezometer systems whereas Wilks employed a logging system to monitor pore pressures and Sherwood & Currey described a computer-controlled slope movement early warning system. Peck's (1970) views on observations and instrumentation merit serious consideration. Field instruments described in the papers are classified in *Figure 1*.

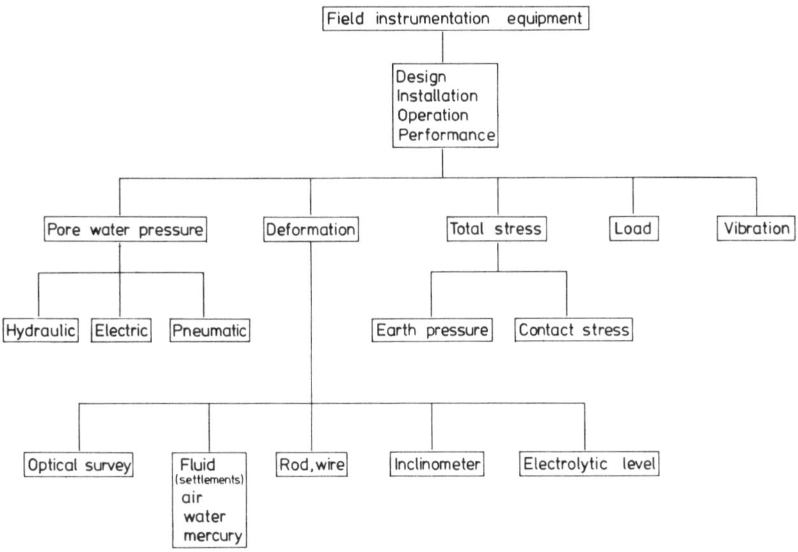

Figure 1. A classification of field instruments

Design

Good instruments must incorporate suitable quality components and materials. They will fail if care and attention are not paid to design details such as joints, seals, corrosion and assembly. Ideally, they should be designed to be stable and to continue to work for long periods under the most unfavourable conditions of temperature, water, humidity and vibration. Overseas, if no indigenous technical expertise exists, simple instruments are preferable since damage in transit is more likely and servicing more difficult.

Development takes time and it may be hazardous to try to introduce too many new ideas too fast. If one small screw corrodes or a seal fails the instrument may become worthless. Drift may occur due to poor design; for example Sherwood's discussion on the electrolytic level. Failure of two pneumatic piezometers due to silty flood water entering the gas lines (Cole's discussion) could probably have been prevented by filters in each lead – a modest design change.

I agree up to a point with Mr Marsland concerning the desirability of designing instruments compatible with particular problems and soil conditions; certainly this is possible within the BRS organisation. However in more general engineering practice, since the market is relatively limited, a wide variety of available equipment may imply too many 'one-off' instruments which are very expensive and unproven.

Since instrument cost is only a small part of the total instrumentation cost (Wilson's discussion) it would be very beneficial if they were better designed, tested and calibrated (Skipp and Thomas's discussion). A price increase coupled with more research and development by the manufacturers would be well justified.

Installation

A well-designed instrument must be properly installed and a strong personal commitment in the man doing this is desirable (Wrightman). It seems preferable to deal directly with the manufacturer rather than an agent since the latter may have only a limited knowledge of the instrument. Organisations in the U.K. such as BRS can maintain considerable installation expertise, transferable from job to job and equipment suppliers usually provide installation services; for others it may be more difficult. There is a need for details of grouting and backfilling techniques to be more available (McKenna & Roy, Marsland) and these were discussed in Session 1.

Operation

It must be possible to make simple on-the-spot field checks that data is reliable. The person taking the readings should know what to expect and be prepared and able to take action if the readings are unexpected. There are two separate problems:

(a) is the data reliable and

(b) what is the significance of the measured values.

These are of course related but our concern here is with (a). An automated system should be self-checking and contain references standards. If it goes wrong (as it usually will) it should be possible to fall back on manual operation which may mean some duplication and extra costs. Operation should be by the R.E.'s staff (Kennard) although in some cases a reading service provided by the supplier may be best.

Very elaborate automated systems were described by Mr Wilson which one can gain access to via the public telephone system, primarily for monitoring displacements but also for pore pressures and rock noise. These are ideal if reliable, but there does seem to be a danger that the man receiving the data may be very remote from the instruments and will be unable to check directly the validity of the readings. As an early warning system they are of great value. A very much simpler non-automatic system for monitoring piezometers was described by Dr Vaughan. This system seems to combine 'electronics' with simplicity and checkability in the best manner and appears very flexible and reliable for long periods.

Performance

Reliable data will be obtained from a well-designed instrument if it is properly installed and regularly read. Unreliable data is very costly and may be dangerous. Precision may often be of second order importance. Data must be processed, interpreted and presented in a concise intelligible form. It is unfortunately the case that too much data is often collected which nobody has time to analyse.

Pore water pressures

In a definitive paper, Dr Vaughan discusses many important factors affecting piezometers and current UK interest in tidal pore pressures is exemplified by Mr Marsland. Reliable results should always be obtained from demonstrably de-aired hydraulic systems whereas electric or pneumatic piezometers may read pore air pressure. Viggiani describes one instance where vibrating wire and hydraulic piezometers showed similar readings after two years in spite of the possible occurrence of gas. Various installation techniques are available: in a sand pocket, if gas is not present (Marsland); pushed in (Wilkes); cast in (Vaughan). Smear effects can be reduced by jetting (Marsland) or by sleeving the piezometer (Wilkes) so that in situ permeability tests are more reliable. Minimising the effects of smear and disturbance on piezometer response requires further field research; as also in relation to sandwick installation (Hughes' discussion).

Some of the factors in favour of electric or pneumatic piezometers were also discussed by Cole.

Hydraulic measurement. A recent development in hydraulic measurement consists of replacing many lengthy mercury manometer tubes by a single precision electrical pressure transducer (Irwin, Vaughan's discussion) resulting in a smaller gauge house, optional automation, easier de-airing and more rapid response. An automatic system was used by Wilkes reading only one limb of each piezometer in addition to the usual pair of mercury manometers. Irwin has also automated the constant head test which is very useful. Porous plastics are being used in hydraulic piezometer tips and an attempt has been made to 'clean up' the cavity of the high air entry Bishop tip following Vaughan.

Push-in piezometers, used earlier by Little, were used by Wilkes grouted up in flight-augered holes and also by George & Parry, avoiding the use of sand filters. Wilkes comments that considerable time is required to lay tubes, fill manometers and to de-air while George & Parry refer to their being simple to operate but time consuming. Lovenbury used a simple, reliable multiple point piezometer; this should perhaps be used more often.

Important principles involved in de-airing hydraulic piezometers were discussed by Miss Walbancke (Session 1) and by Dr Vaughan. Vaughan and Marsland both recommend careful de-airing before installation since subsequent flushing out is more difficult. A small excess pressure should be maintained at the tip (Penman, Vaughan) and low pressure difference, low feed-rate, automatic de-airing systems (Little, Vaughan) are preferable. However, a simple gravity feed system is sometimes all that is needed for shallow piezometers although it is not often used. The application of a large suction to the return tube may cause air bubbles to come out of solution which may subsequently be very difficult to remove. Each piezometer should be de-aired separately at a carefully chosen pressure and several initial de-airings may be needed in some cases. Considerable time may be required to de-air a piezometer properly. Portable de-airing units are often useful, e.g. on road-works. Where suctions are to be measured, or occur anywhere in the leads, frequent de-airing is essential (Vaughan, Marsland, George & Parry) but inconvenient.

Prevention of freezing appears difficult. Antifreeze solutions may cause osmotic effects (Vaughan) or floccing may occur which blocks the leads (W.E.S., 1971). Leads could usefully be identified more readily by better colour coding (or whatever) than at present (McKenna & Roy). Tubing quality (Little) must be carefully checked. MacKeller *et al* describe serious installation problems in hot climates including fungus growth in leads. Leads have also been attacked by rodents or insects (Little) or have been of unsuitable plastic (Kennard).

Electrical measurement. Pressure transducers, vibrating wire or resistance strain gauge, can be installed in the piezometer tip to measure the pore pressure. If permanent they cannot normally be recalibrated or de-aired but are useful where short-term, rapid response is needed (Vaughan). Dibiagio describes an elegant technique for *in situ* scale span and zero calibration checks suitable for diaphram type electrical piezometers; Eason describes tests to study temperature effects on strain gauge type pressure transducers. Removable electrical devices have been used in North America and self de-airing of the tip upon removal of the pressure transducer is possible. A push-in resistance strain gauge electrical piezometer was described by Hutchinson and Bhandari (1971). Experience in the UK with vibrating wire piezometers appears limited and suggests that despite their stability and signal transmissibility advantages they are not very popular. In contrast they are apparently used successfully in some countries, e.g. Italy (Viggiani's discussion).

Pneumatic measurement. Pneumatic piezometers are new to the U.K. but have been used for a few years in North America (e.g. Cornforth). Limited experience to date suggests that installation and flow rate problems may give significant errors (Little, Marsland). Head losses and flow rate effects in long leads (Vaughan) require consideration. It appears that flow rates, need to be very low to avoid errors and a 'bleed-off' reading technique is preferable. The tip usually incorporates a low volume change spool valve with or without a spring, or a flap valve. It should be possible to blow out water from the system by applying a high gas pressure without damage. Dry nitrogen seems best and advisable. At least one pneumatic piezometer has a third lead venting the water filled cavity, the usefulness of which is not clear. A special purpose de-aira but rather complicated, pneumatic piezometer was described by Marsland and where large tidal ranges occur an ingeneous solution to prevent air ingress at low tide is being tried by Strouts and Hallowes.

Pneumatic piezometers appear to possess a number of attractions especially over electrical piezometers and careful development and testing may prove fruitful. Readout may consist of manual valves plus a pressure gauge or an electrical pressure transducer and digital display. Automatic recording is possible.

Deformations

Ground movements are measured by optical survey techniques, adapted where necessary, or by a wide variety of fluid, rod, wire, pendula or level actuated devices, of varying accuracies and degrees of refinement (Dunnicliff, 1971).

Optical survey. Standard high quality survey equipment and techniques, adapted where required, are used for levelling, triangulation, trilateration and alignment. Cheney's paper on precise levelling is a mine of information. The design and construction of reference points with forced centering plates requires great care so that stable stations are established outside the movement area and points within the construction area remain undamaged (Burland & Moore, Penman & Charles, Cheney) and unaffected by spurious seasonal effects or 'noise' (Nixon *et al*, Cooke's discussion).

Skilled operation and experience may in some cases be necessary to obtain good results; measurements may be treated as 'specialised services'. Burland & Moore suggest that a measurement accuracy of 1 mm is needed for buildings and Ashkanazi, Cheney and Penman & Charles give triangulation and trilateration results – are their figures typical of what might generally be achieved? Are the best quality stations currently used sufficiently stable? How significant are temperature and refraction errors (Penman & Charles, Nixon *et al*) on the accuracy of EDM or laser based systems (Weeks & Dixon and Kitching's discussion)?

Haws *et al* describe an 'optical inclinometer' which appears simple and fairly accurate; is this a viable alternative to the normal electrical inclinometers?

Fluid settlement gauges. An air, water or mercury column is used to transmit a pressure over a distance. Difficulties occur if the difference in level between the measuring tip and the recorder is large, particularly if suctions are involved. A permanent gauge house can incorporate elaborate circulation and de-airing equipment but a simple portable unit is also needed which requires little de-airing. Difficulties arise in practice in de-airing, temperature variations, flow rate effects, mercury column break-down, electrical pressure transducer vulnerability.

The water overflow gauge (Penman & Charles) is apparently well developed. Yet both George & Parry and Wilkes experienced difficulties when the gauge was below the gauge house and suctions were involved. A positive gas back pressure (such as was used by Dunnicliff, 1968) would be better.

The mercury settlement gauge, single or multiple point, has an accuracy of

±2 mm and operates with up to ±3 m difference in elevation between recorder and cell (Irwin). Gauges of this type are currently being used over very long distances at Tarbela. What experience do others have of their use and how reliable are they? More recently a constant volume system water gauge has been designed to operate over greater head differences with similar accuracy (Irwin).

A 'mole' which can be drawn or pushed along a flexible horizontal tube was used by Penman & Charles, George & Parry, McKenna & Roy and Wilkes. Penman & Charles, employed a water overflow gauge drawn by an air-operated duct motor operated from a substantial gauge house, obtaining accuracies of ±2 mm. The others used a simple probe relying on the cable stiffness to 'rod' it along the buried tube. The probe incorporated a water-filled balloon connected by a water column to an electric pressure transducer at the external measuring point, at a higher level in the present instances. Unsatisfactory results were frequently obtained due to temperature fluctuations, cavitation (Coles' discussion), no on-site de-airing facilities. It might work better if: the balloon were back pressured by gas; the pressure transducer was installed in the probe; the buried tube could be installed so that the probe is always above the measuring point. A reliable 'mole' of this or similar type would be very valuable.

Rods and wires. Surface access points require protection during and after construction, especially in fill, and vertical devices are prone to damage (McKenna & Roy, Penman & Charles). Environmental factors affect accuracy and stability and their significance depends on the precision required (Nixon *et al*). Problems arise due to non-creep-free anchors for borehole installations, in accuracy of marked electrical cables and the inconvenience of using a steel tape.

Marsland describes a rod device for use in soft strata where large settlements occur and Marsland & Eason used a purpose-made system for settlement beneath a plate test in a borehole (see also Eason's discussion). A variety of hand-held micrometer rods have been used to determine horizontal movement, to about ±1 mm (Burland & Moore, Haws *et al,* Marsland). Simple devices of this sort may often be overlooked.

Extensometers for measuring internal strains in the laboratory are described by Brown and Penman & Charles mention the BRS vibrating wire type used in fill. Linear potentiometer type extensometers have recently been installed in Singapore having previously been used in Mexico and the U.S.A. Where access is not possible these seem useful.

Thin tensioned wires were used at Selset (Penman & Charles) but bends and friction were serious. Thicker wires, or rods, were used at Kainji, Backwater and Empingham dams (Vaughan's discussion). Whittaker and Forrester used a tensioned tape for surface strain measurements, this is affected by temperature and wind and the accuracy is discussed by Forrester. An inverted pendulum is described by Penman & Charles, Marsland and Nixon *et al* for measuring horizontal movements. These appear reliable and fairly accurate but require a vertical borehole and installation can be difficult. Can they be used to measure vertical movements or is creep excessive?

Little use now appears to be made of the U.S.B.R. settlement device (Penman & Charles) although it is very reliable (Little). A latching device on a steel tape can be used in inclinometer casing.

An inductance or reed switch probe is used to detect steel plates or magnet rings respectively. Their accuracy is largely limited by the tape or rods to which the probe is attached. The inductance probe was used successfully by MacKeller *et al* and Penman & Charles whereas Wilkes had some difficulties with the telescopic tubes fouling the steel plates. A number of 'shoot out' type anchors (Marsland, *Figure 6c*) have failed and tubing damaged beyond repair (McKenna & Roy). The reed switch and magnet system is an important recent development which is inherently very simple. For settlement an accuracy of 1 mm can be obtained with a removable steel tape and 0.05 mm with a permanent rod or tape (Burland & Moore). Burland discusses their stability with time and the influence of rotation or lateral positioning of the reed within the magnet ring. Installation in an inclinometer (Vaughan's discussion) eliminates a number of difficulties. Problems arise in selecting the best method of installation, particularly in soft soils and ensuring that anchor slip does not occur (Marsland, George & Parry and Hallowes' discussion). A number of techniques have been tried: spring anchors, shoot-out plates. Marsland describes an improved stiff-spring anchor with a novel explosive release. Care must still be exercised in selecting a suitable grout.

Inclinometers

Two inclinometers are generally being used in the U.K. incorporating either a rotary potentiometer (Cornforth) or bonded resistance strain gauges (Burland & Moore, Wilkes, Marsland). Both types are described, together with performance tests, by Green. In addition a more sensitive, limited range instrument is described by Phillips and James. Measurements under controlled conditions (Green) compare well with statistical analysis of field results (Cornforth) and data from a diaphram wall (Burland & Moore).

A new, and even more precise general purpose inclinometer is now available (Cornforth); whether it is very robust is uncertain. Calibration tests by Phillips and James suggest that small deformations can be determined to an accuracy of the order of 0.1 mm which is excellent if it can be maintained in practice. For best results it is suggested to (a) use a steel tape (Vaughan's discussion) (b) take repeat readings at close intervals and (c) use statistical analysis. However for many purposes this will be impractical. Cornforth's data, BRS, McKenna & Roy and Bundred's experience suggests that reliable results can be obtained but difficulties can be experienced (Kennard, Little, Wilkes) and errors can occur due to water ingress, temperature effects and maltreatment.

Automatic integration and/or recording of data on tape cassettes (Wilson's discussion) appears attractive but is probably only worthwhile if reliable. Experience so far appears limited, although computer data processing has been used extensively. Wilson describes an in-place inclinometer (similar to the Terrametrics Deflectometer) which is used as an early warning system.

Installation is fairly straightforward, vertical movement may need to be allowed for (Cornforth, Green) but difficulties can occur (Weeks & Dixon). Adequate protection is required (Burland & Moore). Vertical movements can be readily measured with the magnet/reed switch device (Vaughan's discussion).

Twist occurs in plastic casing (Green, Skermer's discussion) and extruded aluminium alloy has many attractions except that it readily corrodes in highly alkaline grouts (McKenna & Roy, Wilkes, Coles' discussion) and requires a protective coating (Green, Burland & Moore). It is also less flexible (Marsland) and a soft grout used locally is useful (Wilkes). For shallow holes in very soft ground the less accurate system described by Hutchinson may suffice.

Research is needed on protective coatings, and backfilling techniques – grouts, sand or plastic foams.

Electrolytic levels

The recently developed electrolytic level consists of a spirit level vial containing three electrodes linked in an AC energised Wheatstone bridge type circuit (Sherwood & Currey). Accuracy and stability are discussed by Cook & Price and Sherwood & Currey; typically a resolution of 2 arc seconds over ±0.7° is quoted. Small temperature effects occur, but can be reduced by using an opposed pair of vials. Cross-sensitivity is high but can be accounted for either in the design of the layout or in the reading procedure. Drift can be accounted for by reversing the level and performance is largely controlled by the accuracy with which the level can be replaced on its mounting (Whittaker & Forrester). Errors and precautions required to eliminate them are further discussed by Forrester and by Sherwood who also points out that a torque balance accelerometer might be better.

Precise measurements of vertical deformations around a pile were made by three techniques (Cooke & Price), the limitation in the system being the inability to replace the level accurately in the tube. Is this the only way to make this type of measurement? Whittaker and Forrester used a portable system for mining subsidence studies and achieved an accuracy of ±0.2 mm/m (Forrester's discussion). Sherwood & Currey describe a sensitive settlement device and a tiltmeter shear detector. The latter is used as an early warning device and will measure total relative movement between the top and bottom of the casing. The minimum borehole diameter required increases with depth and must presumably be straight. In contrast an inclinometer is not directly suitable as an early warning device but gives more detailed displacement measurements and may be easier to instal. What are the relative merits of these two systems? How reliable is the computer controlled shear detector system for the S. Wales rockslope?

Total stresses

It is difficult to make reliable total stress measurements either at a rigid boundary or within the soil mass due to the many design constraints on the cell and the near impossibility of fulfilling all of them. It is difficult to perform relevant calibrations (Rymsza's discussion), particularly on an embedded cell, field installation technique is probably critical and is hard to duplicate in calibrations. Thus the number of well documented reliable field measurements are few.

The cell action factor may be affected by shear stresses if the cell faces are not principle stress planes (Brown). The magnitude of this effect will depend on interface friction and the measurement principle, e.g. vibrating wire (BRS) or oil-filled capsule (Glötzl). Exact matching of a buried cell to the soil is probably impossible (Thomas & Ward, 1969) although attempts were made for Oroville dam (Sparrow, 1967). However simplicity has many attractions (Little) and an approximate but reliable measurement, easily obtained may be more satisfactory. What accuracy is really required for a total stress measurement?

The seriousness of temperature effects on total stress cells is difficult to assess. A fully embedded cell in the field is usually subject to small temperature fluctuations during and just after installation. A contact stress cell may be subjected to large temperature variations due to the heat of hydration of the cement during early curing. Laboratory calibrations under a fluid pressure will be irrelevant and even the tests described by Mr Symons (Session 1) which are, of course, much better, do not exactly duplicate the field constraints.

Bundred used the BRS cell for contact stresses and obtained higher stresses on the outer (B) than on the inner (A) faces in agreement with Uff (1970) and Vaughan & Kennard (1972). Similar tests were performed by Jones but no details are given. The differences seem probably due to the 'soft' rubber between the cell and the cast iron casing (Brown, 1969 and present discussion) and not to the presence of point loads. If the latter were the cause it is remarkable how consistently they occur. Thus, in contrast to Brown's view, the outer (B) face may more correctly represent the contact stress. Would a single active face gauge such as the strain gauged device used by Wong or the Carlson cell used in the U.S.A. not be better?

Wong's paper demonstrates the uncertainties associated with contact with sand (or burial) due to large hysteresis effects. Presumably rockfill would give similar problems. The influence of the variation in placement density was discussed by Symons in Session 1. Hawes *et al* used hydraulic cells under a heavy raft. Details of the effect (if any) of the initial temperature rise on the cell output together with their subsequent performance, would be of interest.

Cell action factors were obtained for fully embedded small strain-gauged diaphragm cells in controlled laboratory tests by Brown, who questions in the discussion the merits of large cells (e.g. the BRS cell). Large scale environmental calibrations are not yet forthcoming (Thomas's discussion) yet the idea of burying a 2.5 in diameter cell in fill is not attractive. If anything embedded cells should be larger than at present (Thomas).

BRS vibrating-wire earth pressure cells can work very well but are expensive and complicated and have fallen out of favour by at least one user (Little) who may have been unfortunate (Thomas's discussion). A fluid-filled metal flat-jack connected to a small low volume change pressure transducer (e.g. Irwin, Glötzl cell) is simple and therefore attractive. Temperature effects and cell action factors appears relatively unknown for this type of cell although the temperature data given by Symons in Session I are encouraging. Perhaps these may be the best type for general use.

Loads

Load measurement is relatively straightforward and may employ a vibrating wire (Bundred, Weeks & Dixon) bonded resistance strain gauges or more recently, photoelastic devices (Haws *et al*). The vibrating wire load cell has been extensively used (Chaplin's discussion) and is stable with time. The photoelastic device (Hooper's discussion) is mechanically simple but experience is as yet more limited. It requires visual access which is a restriction.

Vibrations

This specialised topic is discussed by Skipp & Mariott who outline the sensing heads and rapid response recording equipment required.

Summary

Well designed reliable field instruments are required that can withstand the hazards of field use. They must be properly installed, operated and the data analysed and interpreted. A number of reliable field instruments exist (Bishop & Green) but there is scope for improvement in ruggedness and reliability – resolution is often quite adequate.

REFERENCES

Dunnicliffe, C. J. (1968). 'The instrumentation of the Plover Cove main dam', *Geotechnique,* **18**, 3, 283–300.

Dunnicliffe, C. J. (1970). 'Equipment for field deformation measurements', *Proc. 4th Pan. Amer. Conf. Soil Mech.*, Puerto Rico, 2, 319–332.

Hutchinson, J. N. and Bhandari, R. K. (1971). 'Undrained loading, a fundamental mechanism of mudflows and other mass movements', *Geotechnique,* **21**, 4, 353–358.

Peck, R. B. (1970), 'Observations and instrumentation: some elementary considerations', *Seminar on field observations in foundation design and construction, New York Met. Section, ASCE.* (Reprinted in Highway Focus, **4**, 2 (June 1972), U.S. Dept. Transportation, Fed. Highway Admin.)

Sparrow, E. C. (1967) 'Stress cell systems for monitoring static and dynamic stresses in earth and rock-fill dams', *World Dams Today, Japan Dam Assoc.*, Tokyo, 255–259.

Thomas, H. S. H. and Ward, W. H. (1969). 'The design, construction, and performance of a vibrating-wire earth pressure cell', *Geotechnique* **19**, 1, 39–51.

Uff, J. F. (1970), '*In situ* measurements of earth pressure for a quay wall at Seaforth, Liverpool', *Proc.Conf. In situ Investigations in Soil and Rocks,* Brit. Geot. Soc., 229–239.

Vaughan, P. R. and Kennard, M. F. (1972), 'Earth pressures at a junction between an embankment dam and a concrete dam', *Proc. 5th European Conf. Soil Mech,*, Madrid, **1**, 215–221.

Waterways Experiment Station, (1971), 'Instrumentation of earth and rockfill dams: Earth-movement and pore pressure measuring devices', *Engineer Manual* EM 1110–2–1908.

Discussion. Principles of measurement

D. J. Knight (Sir Alexander Gibb & Partners). The question of accuracy of measurement has been raised, but it should be viewed within the context of the complete problem. For example, the accuracy to which porewater pressures in an embankment dam can be measured by means of closed-circuit hydraulic piezometers is only one of several factors affecting the value of the factor of safety for the condition analysed, and its contribution will vary according to the accuracy with which those other factors are determined. Those factors include (i) the densities of the various fill and foundation materials, (ii) the shear strengths of the various materials, and (iii) the seismic criteria, where relevant, e.g., the bedrock acceleration, the shear wave velocities of the various materials, and their damping factors. By reference to the upstream critical sliding surface on a typical embankment dam profile shown in *Figure 1* it may be seen that inaccuracies in measuring the above factors can easily outweigh inaccuracies in the measurement of the pore-water pressures. Relative accuracy is thus the principle at issue.

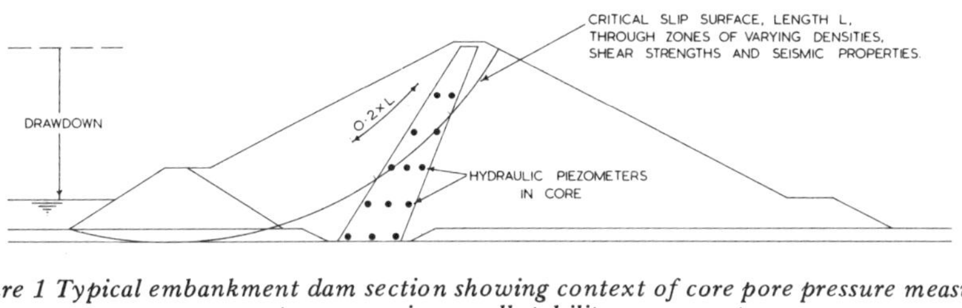

Figure 1 Typical embankment dam section showing context of core pore pressure measurement accuracy in overall stability assessment

Another principle of measurement which has been mentioned but which deserves repetition is that of providing, where possible, more than one type of instrument for the measurement of critical elements of behaviour. For example, vertical deformation may be measured with internal hydraulic settlement cells, installed during construction, with deformation tubes installed either during or after completion of construction, and with surface survey stations. By such means a check may be kept on the adequacy of embankment dam crest level design provisions for post-construction settlement.

A third principle, stressed by Dr Penman, is the importance of obtaining initial readings. This may be illustrated by reference to *Figure 2* showing a portion of the longitudinal section along the axis of a proposed high embankment dam where the foundations change in a short distance from solid rock to deep compressible clay to a possibly rigid but 'hinged' lava flow overlaying the clay. An impervious core across the central clay zone would be subject to concentrated stressing at the place arrowed. Not only is instrumentation necessary

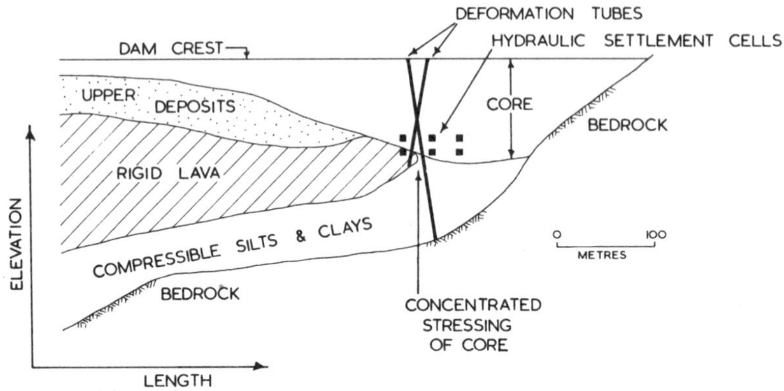

Figure 2 Example of embankment dam core foundation transition requiring vertical strain observations from start of construction

to measure deformations in that area, but its installation and operation *ab initio* is essential in order that early behaviour is not missed. If this is not done then a completely misleading view could be obtained of both total and relative strains. This particular case is also an example of the need for applying the previous principle of providing independent means of measuring the same behaviour.

A. D. M. Penman (Building Research Station). In connection with accuracy of field measurements, I want to draw attention to the inaccuracies which can be caused through missing zero readings. We often find ourselves installing instruments because the fill has reached the right level or some other part of construction has reached the right stage etc, without having instrument houses, or stable reference pillars etc that may be necessary to allow zero readings to be taken. This applies to all types of instrumentation but to give a specific example, if you want to measure movements of an embankment dam as we have been doing recently with a horizontal plate gauge, you need to know the movements of each plate from time zero, when they are first installed.

Since the gauge only measures the plate movements relative to the terminal instrument house, the movements of the house itself have to be measured, by the methods described by Dr Ashkenazi, from surveying pillars built on the valley sides at a safe distance from the dam. To obtain zero readings, it is therefore imperative that both the pillars are there and their position established, and that the instrument house is complete, with a surveying target on its roof *before* the plate gauge is installed in the fill. If the house has to be built *in situ*, the contractor has to make special arrangements to ensure that it can be completed before the next metre of fill has been placed and the level is right for the plate gauge.

To avoid difficulty we have had instrument houses pre-cast, and towed into position when the fill had reached the level for their foundation. The gauge enters the house about a metre above floor level, and while this is being placed we could take measurements to establish the position of the house and get its exact position when the gauge was installed to establish the initial positions of the plates. The principle applies not only to plate gauges, but also for all types of field instrumentation, pore pressures, total pressures, etc. I do emphasise that it is essential to read the zero condition as the instrument is put in and I think that consultants responsible for instrumentation of any sort of civil engineering jobs have to keep the contractor under very strict control to ensure that the facilities for taking the zero readings are available at the time when the instruments are installed.

A. W. Bishop (Imperial College, London). The General Reporter has referred rather briefly to pore pressure measurements. I would like to raise the question of the part played by osmotic pressures. Are they significant? I have heard of no evidence that they are significant, except in one case when piezometer leads were filled with anti-freeze and subsequently gave pore pressures greatly in excess of the total principal stress incurred on that particular project. Is there any evidence from piezometers, possibly where there is a difference of salinity of water, of osmotic pressures or is it a particular area in soil science of interest to the agriculturist but of no relevance to engineering?

The General Reporter also referred to the difficulty of tracing the initiation of a shear zone or propagation of a slip surface which will show up eventually from inclinometers. Is there any better way of continuously monitoring this, possibly by a modification of the slope indicator type of technique with a different type of casing? In relation to studies in progressive failure it is of great interest to know whether the movement is confined to a single surface or a very thin zone and at what stage it begins.

As a final point I would like to make some remarks about the durability of instrumentation. There seem to be two quite different classes of problem that we have been considering. One is the long-term performance of structures, e.g. the observation of settlements for 100 years as obtained from the late London Bridge, or, under the Reservoir Safety Act, the monitoring of the performance of a structure over the whole of its useful life, which may be of the order of 100 years, with reliable readings required by the Act every 10 years. How many of our instruments are suitable for this type of work, and now that this type of legislation may be increasing in its scope, should we be looking for the development of instruments which will remain reliable over these long periods?

The second area is the monitoring of deformation and stresses, pore pressures and so on during construction or for research work where we are concerned maybe for only weeks or periods up to 3 or 4 years. It seems that in the developments of instruments there is some confusion about the life we are aiming at. We may need throw away instruments that will last for a few months, and can then be dispensed with, or those in which we know the electronics will cease to be reliable after a certain period of time, but we also need those to which we can return over a very long period at intervals, possibly getting readings where the response time, or if it is strain, where the method of observation is a bit tedious but where we can be 100 per cent sure that it is comparable with a reading taken either months or even 10 years before.

In this connection, it is useful when developing a new type of instrument, to put it in alongside some other simpler or proven type of equipment to see how the performance of the two compares. Too often it appears that someone is so enthusiastic about their new piece of equipment that they put it in to the exclusion of all other types of observation, and then you are left either comparing it with theory or justifying it from intuition. I think more comparisons would be worth while, even if they have to be paid for as a research contribution to the project.

I. Hawkes (Creare Incorporated, USA). I would like to draw attention to US Government development work to measure stress change in rock cheaply. We need a broad picture of behaviour in underground work rather than high accuracy at a small number of discrete points and have developed a low price gauge. It follows the principles outlined by Dr F. T. Williams in his paper to this Symposium. It is a cylindrical type gauge with a vibrating wire sensing unit. The $\frac{3}{4}$ in long wire is anchored by an extrusion clamping process which with a jig fixture is a very quick and simple to do. We have built about 200 of these gauges: they meet the exacting fire regulations of the mines. They are installed in boreholes in the rock with pairs of folding wedges at either end so that they can be pre-stressed against the sides of the holes. Changes in readings then indicate stress changes in the rock. At present these gauges cost £20. The gauge has been calibrated embedded in slabs of rock subject to plain strain. By placing a lot of these gauges in a mine, the whole pattern of stress change can be observed giving the broad picture which is what we want to check the satisfactory behaviour of the mine.

A. Marsland (Building Research Station). It is necessary to ensure compatibility of instruments and the ground in which they are installed. I will illustrate some of the problems which arise by taking the case of settlement points installed in boreholes. The majority of settlement devices consist of several anchorages

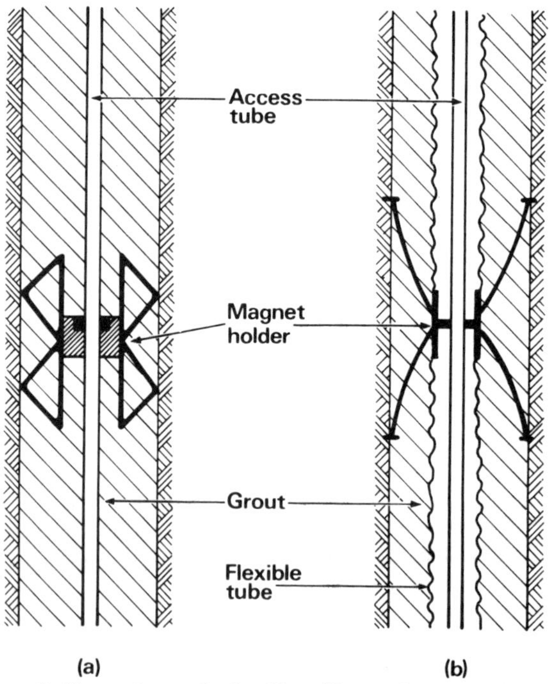

Figure 3 Alternative methods of installing settlement devices

(measuring points) connected by rods and tubes to the surface to allow either direct measurement or the insertion of measuring devices to measure changes in elevation of the measuring points. Typical anchorages currently being used are illustrated in *Figure 10* in the paper by Burland and Moore, and *Figure 6* of my own paper to this Symposium. It is the general practice to install the measuring points at appropriate levels in the borehole and then backfill the borehole with a 'suitable' material. Take the case of a magnetic settlement device of the type illustrated in *Figure 3a* where a grout has been used to backfill the borehole. Even a relatively weak grout will lock the magnet holder to the access tube and unless the ground is stronger than the access tube this tube will either fracture or act as a local reinforcement which may cause the magnet holder to move less than the surrounding ground. Access tubes with sliding joints which have been used with some instruments are also suspect for the same reasons, particularly in cases where the ground is subjected to large displacements.

More recently some instruments of the type shown in *Figure 3a* have been installed in boreholes backfilled with a thick suspension of bentonite. This provides the necessary flexibility around the instruments and the access tube provided the grout does not become badly contaminated with silt and sand particles during injection. Until more data is available on the long-term stability of boreholes containing bentonite, I consider that this method should only be regarded as a short-term solution. A more positive solution is to surround the access tube with a wire-reinforced flexible tube as illustrated in *Figures 3b and 23.* Such a tube prevents the hole closing on to the rigid access tube, but does not restrict the movement of the surrounding ground. The space between the walls of the borehole and the flexible tube must still be filled with a grout having properties as near as possible to those of the surrounding ground. Unfortunately it is not as easy to meet this requirement as it may seem since there is often a serious lack of reliable data on both the properties of the ground and possible grouts. In many cases grouts are being used which are several times stronger than the surrounding ground. To provide long-term stability a grout must contain a component which will give it some permanent rigidity, otherwise there is a danger of settlement of the grout which could drag down devices installed in the borehole. At present portland cement is the usual material used to produce this set.

Limited work done at the Building Research Station suggests that the strength and compressibilities of grouts containing cement are mainly controlled by the water/cement ratio. Seven days strengths of some typical grouts are given in *Figure 4a.* Ranges for the intact and large-scale strength of London Clay are given as an indication of the high water/cement ratios required to ensure that the grout has similar properties to the surrounding ground. I would also like to draw attention to the substantial increases in strength with time of mixtures containing Portland cement. The increases in strength with time of two different concrete mixes over times from 3 days to 5 years are given in the top part of *Figure 4b.* These show an almost linear increase of strength with the log of time. The present writer has only obtained results for grouts for times up to 28 days, but the typical curves given in the lower part of *Figure 4b* show a similar trend. Since there is a tendency for grouts to develop their strength more slowly it is possible that the ratio of the 1 year to 7 days strength may be greater than for concrete and cement paste, particularly when the grouts contain substantial quantities of pozzolana such as pulverised fuel ash. The results at present available suggest that it is unwise to make on the spot predictions of the possible long-term compressibility and strength of grouts from inspection and 'feel' of grout exposed at the top of boreholes a few days after

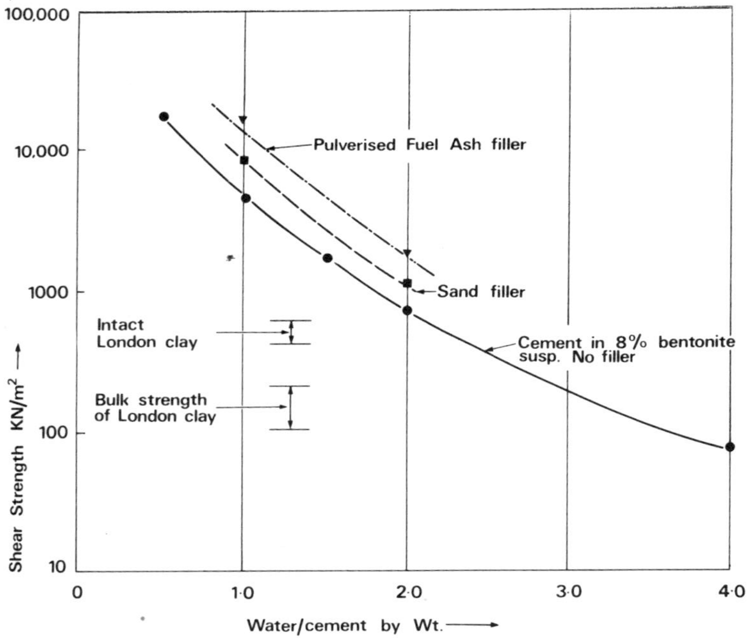

Figure 4a Seven day shear strengths of typical bentonite cement grouts (unconfined compression tests under water 38 mm by 76 mm long specimens)

Figure 4b Increase of shear strength (triaxial tests) of concretes and grouts with time

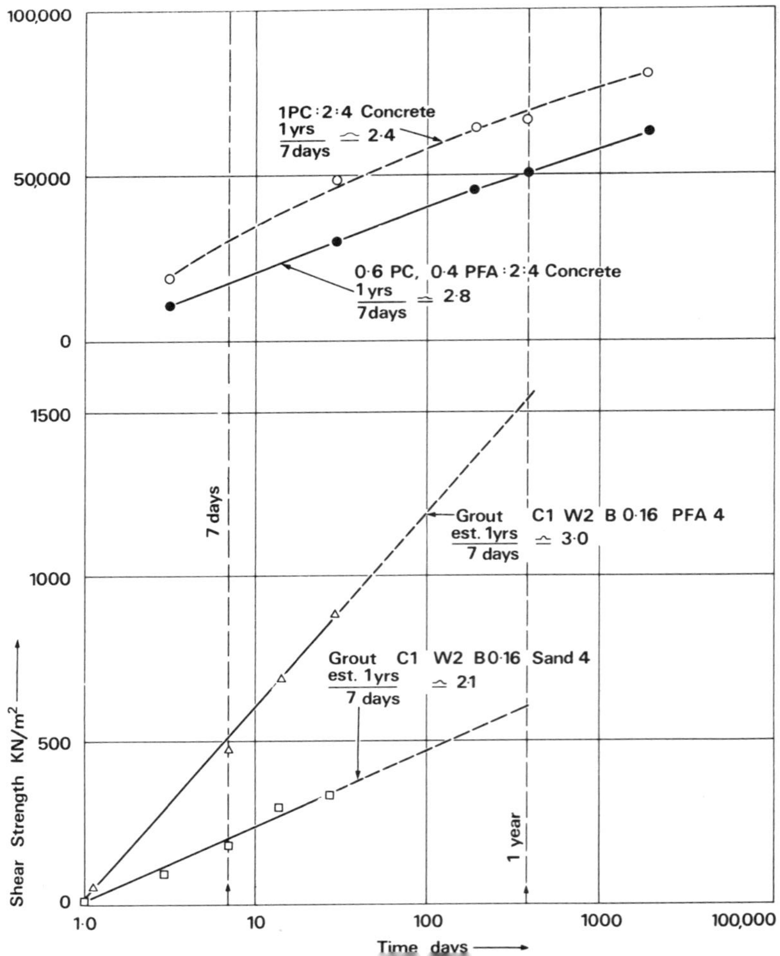

installation. This grout may also be much weaker than the grout lower down the borehole as a result of dilution with water already present in the borehole.

In spite of the difficulties in choosing a suitable grout I consider that the only reliable way of backfilling a borehole containing instruments is to pump a suitable grout through a pipe to the bottom of the borehole. It is preferable for the grout to be softer than the surrounding ground. On the basis of present knowledge I would suggest that the grout should have a 7-day strength of about one quarter the strength of the ground given by the lower limit of the results from laboratory tests.

C. J. Dunnicliff (Soil and Rock Instrumentation, Inc., USA). A question has been asked about making dynamic measurements of deformations. There is an apparatus which is semi-proven in the USA, developed by Prof Ernest Selig of the University of New York at Buffalo, called a soil strain gauge and which is commercially manufactured by Bison Instruments, Inc., Minneapolis (Selig and Grangaard, 1970; Selig, 1971). It uses induction coils placed one above the other with coil separation in the range 1 to 16 in to measure deformation between the coils, e.g. during compaction. There is no mechanical connection between the coils, and therefore the system is likely to be capable of measuring short-term movements without falsification due to instrument inertia.

REFERENCES

Selig, E. T. and Grangaard, Jr. O. H. (1970). 'A New Technique for Soil Strain Measurement', Materials Research and Standards, *MTRSA*, **10**, 11, pp. 19–21.

Selig, E. T. (1971). 'Soil Compaction Evaluation with Strain Measurements', Report on Research Project 761 for New York State Dept. of Transportation and U.S. D.O.T. Federal Highway Administration.

I. F. Symons (Transport and Road Research Laboratory). As part of the Laboratory's current research programme, a pilot scale retaining wall is being constructed to enable measurements to be made of the earth pressures produced by compaction plant in a range of fill materials. The work will also include an investigation of the effect of wall movements on the measured pressures. In this programme a study is being made of the performance of a number of different types of earth pressure cell in soil.

To investigate the measurement of boundary stresses, the cells are mounted flush with the surface of a rigid base (*Figure 5*) on which a cylindrical soil

Figure 5 Rigid base containing a recessed pot to house the pressure cells

specimen is prepared. The specimen is then enclosed in a rubber membrane and placed inside a pressure vessel (*Figure 6*). Connections in the side of the vessel enables the cell to be vented to atmospheric pressure and provide an exit for the leads. The pressure vessel is filled with water and the cell calibrated under a hydraulic pressure.

Figure 6 Soil specimen mounted in the pressure vessel

Tests are being carried out using three types of earth pressure cell. Firstly the hydraulic cell described by Irwin in his paper to this Session. The second type is a pneumatic cell developed and patented by the Ponts et Chaussees Laboratory at Angers in France described by Levillain (1972). This cell also consists of a thin diaphragm attached to a backing plate containing two holes for connections, one located near the edge of the diaphragm and the other at the centre of the cell (*Figure 7*). The method of operation is to apply a

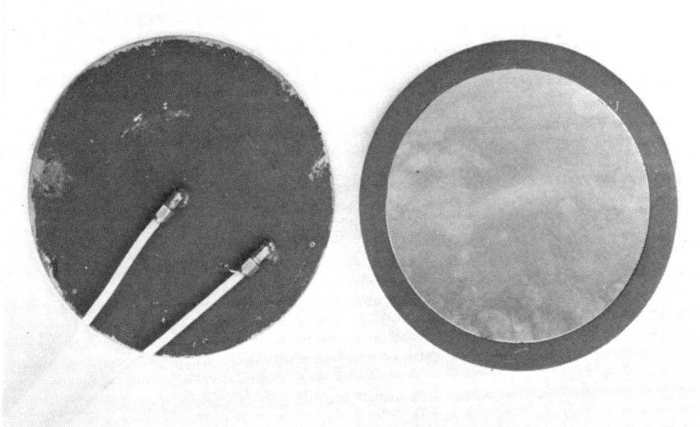

Figure 7 Pneumatic earth pressure cell

pressure to the outer connection until a small flow of gas occurs from the central connection. The applied pressure is then adusted until a small and constant rate of flow is obtained. In the tests carried out at the Laboratory the applied pressure is recorded when a constant flow rate of 5 litres per hour has been attained. The third type of cell being studied consists of a Cambridge contact stress transducer fitted with an enlarged top plate to reduce the range of the cell. To enable this cell to withstand large dynamic stresses induced during the compaction of the soil a locking system has been fitted.

The cells are being calibrated using a sandy clay and a washed sand at different density and moisture content conditions. The results of tests on all three cells using lightly compacted washed sand are shown in *Figure 8*. At low applied pressures the cells give a slight over-read in relation to the air calibration. At higher applied pressures all three cells and particularly the contact stress transducer, give some under-read.

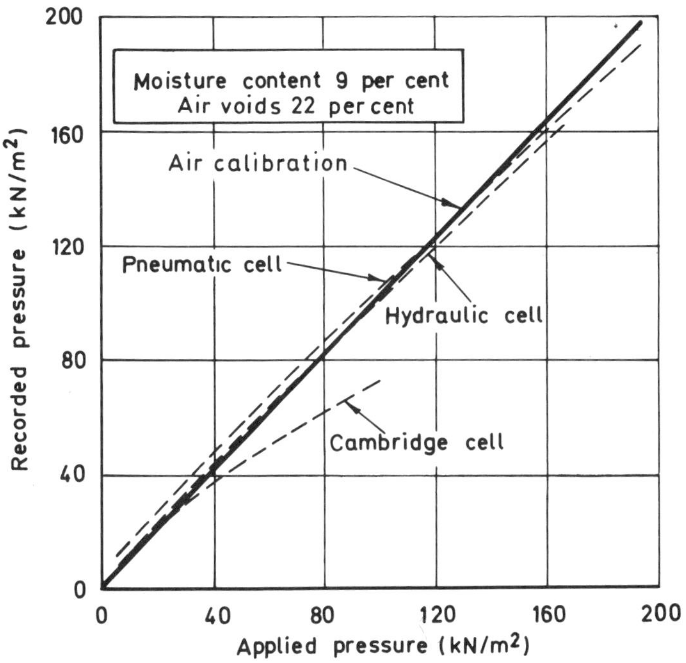

Figure 8 Tests of different pressure cells using lightly compacted washed sand

The results of tests to examine the effect of different degrees of compaction on the performance of the pneumatic cell using washed sand are shown in *Figure 9*. At the higher density a greater over-read is recorded over the entire pressure range. The soil calibrations are nearly linear and approximately parallel to the air calibration. Significant over-read is apparent at low applied pressures for both density conditions.

Tests have also been carried out to study the effect on the soil calibrations of different density distributions within the specimen. This was achieved by varying the distribution of the loose material to be compacted so as to produce specimens with a difference in density between the central core and the surrounding material. The calibration (*Figure 10*) carried out on the specimen having a relatively low density above the cell shows appreciable under-read particularly at the higher pressures and this suggests that arching is taking place within the specimen as a result of the lateral density gradient.

Figure 11 gives the calibrations obtained with the contact stress transducer using lightly compacted sandy clay at different moisture contents. The calibration on the specimen prepared at the higher moisture content, corresponding to about 1.25 times the Plastic Limit, lies close to the air calibration whilst the result with the specimen at the lower moisture content, at about the Plastic Limit, shows appreciable under-read at higher applied pressures.

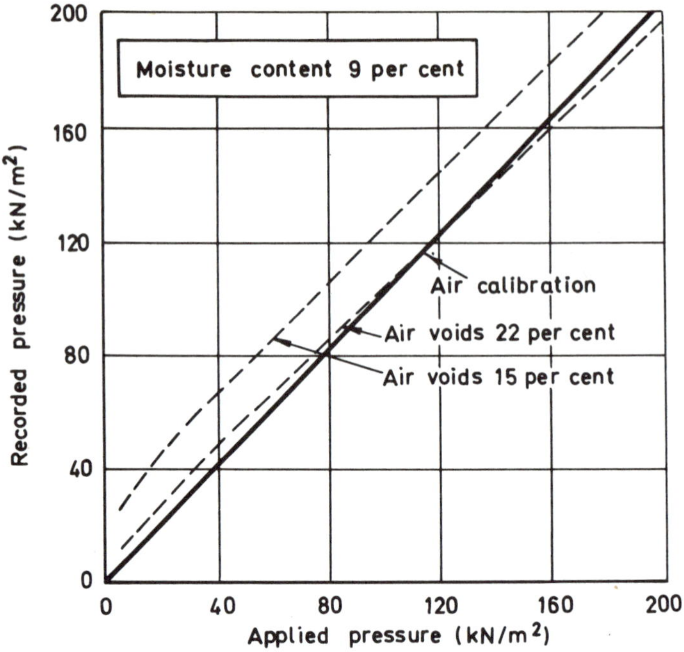

Figure 9 Pneumatic cell – effect of different degrees of compaction for washed sand

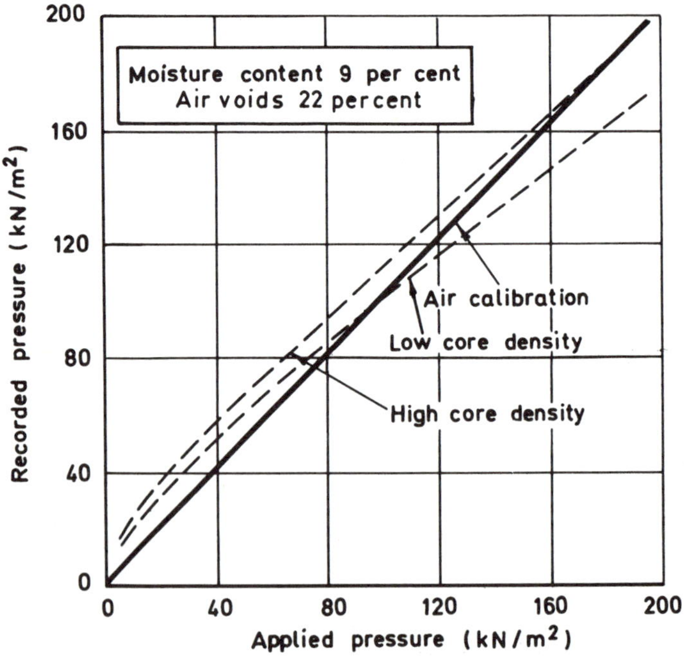

Figure 10 Pneumatic cell – effect of different density distributions for washed sand

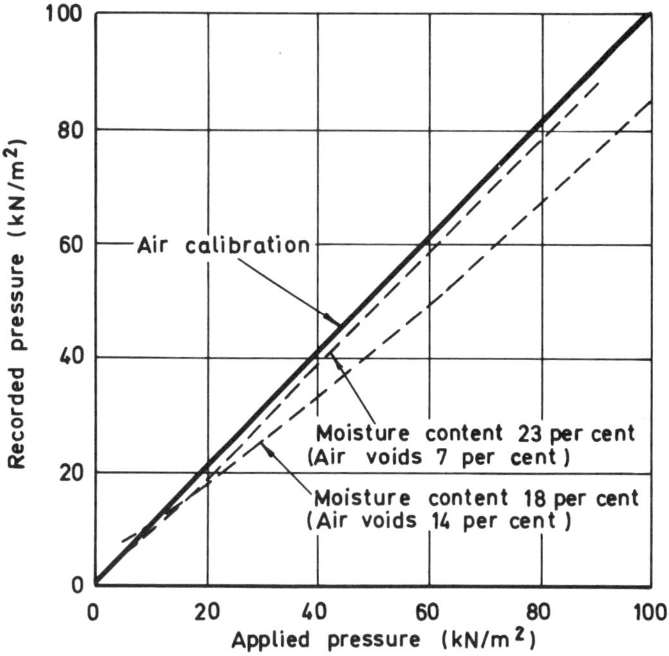

Figure 11 Cambridge cell – effect of different moisture contents for lightly compacted sandy clay

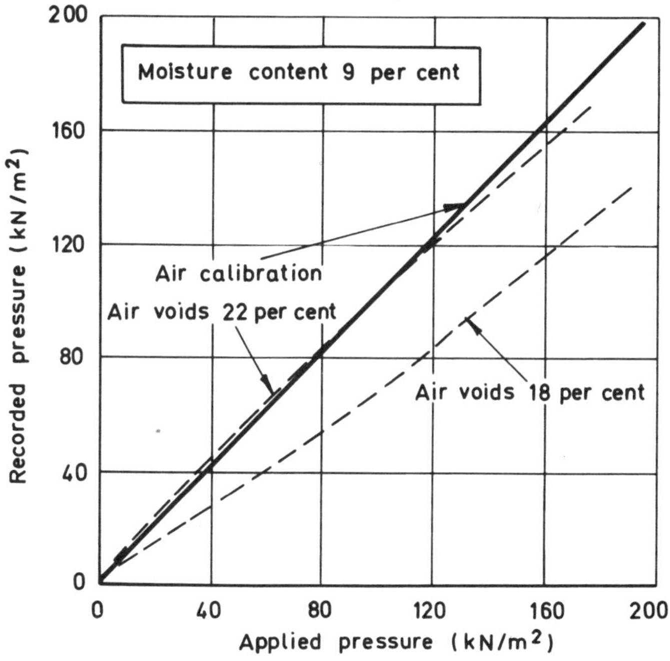

Figure 12 Hydraulic cell – effect of different degrees of compaction for washed sand

Figure 12 gives the results of tests carried out with the hydraulic cell using different degrees of compaction for washed sand. The higher density specimen shows an appreciable under-read whilst the results for the lower density specimen lie close to the air calibration.

Figure 13 gives the results of tests to examine the effect of different degrees of compaction on the hydraulic cell using sandy clay. The marked curvature of the calibration for the higher density specimen together with the over-read at the higher applied pressures is difficult to account for.

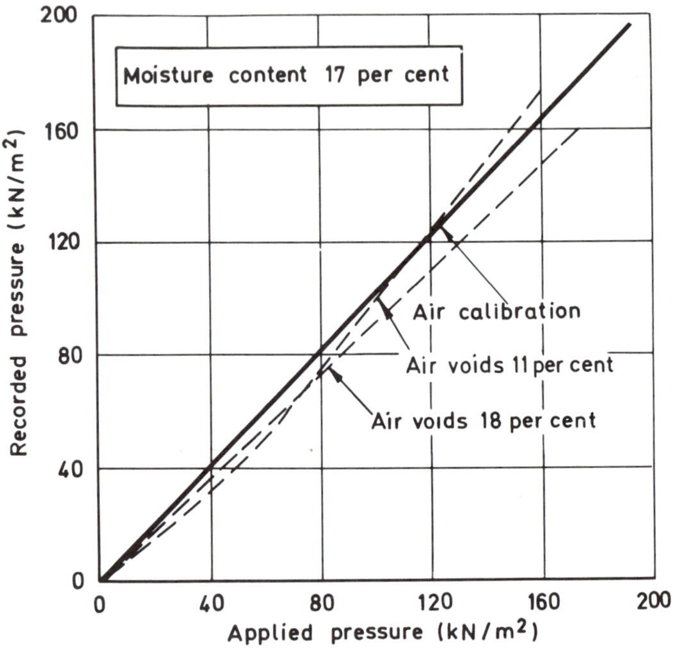

Figure 13 Hydraulic cell – effect of different degrees of compaction for sandy clay

These preliminary results indicate some of the practical difficulties involved in carrying out meaningful calibrations of cells in soil because of the large number of variables involved. However unless the effect of method of compaction, density, density distribution and moisture content on the performance of pressure cells can be determined and allowed for, the accuracy of field measurements appears somewhat questionable, particularly at low recorded pressures.

In his introductory notes to Session 2, Dr Green has enquired about the temperature characteristics of the hydraulic type cell. A limited number of tests have been carried out at the Laboratory on the hydraulic cell to study the effect of temperature using a specimen of natural sand compacted to 11 per cent air voids at a moisture content of 15 per cent. These tests gave a coefficient of 0.4 kN/m^2 per degree Centigrade at an applied pressure of 30 kN/m^2 and 0.6 kN/m^2 per degree Centigrade at an applied pressure of 60 kN/m^2. These tests were carried out with the cell mounted flush with the surface of the rigid base using ranges of temperature from 10°–28°C and 13°–26°C respectively.

REFERENCES

Levillain, J. P. (1972). 'Capteur de pression totale à contre-pression pneumatique et débit contrôle'. Bulletin de Liaison des Laboratories Ponts et Chaussées No. 58.

C. P. Wroth (University of Cambridge). My comments relate to the measurement of shear stress in the field. There appear to be few records in the literature of attempts to measure shear stress, and I was disappointed to find none in the papers presented to this Symposium (other than a reference by Bishop and Green to the work of Bozozuk which is reported below). It is important that for a proper understanding of the behaviour of many complex foundations and structures in the field that at selected points within the soil or at the boundary of the soil, the *complete* stress regime is measured. An obvious example is in the design and performance of friction piles and piled foundations, where major assumptions are made about the skin friction generated on the cylindrical surface of the pile, and how this frictional shear stress changes with time during service of the pile. Another example of the importance of shear stress is in the behaviour of retaining walls.

I suggest that considering the success and sophistication of other forms of measurement in the field (e.g. pore-pressure, normal total stress, and displacement) it is a major deficiency in the development of soil mechanics not to have put similar effort into the measurement of shear stress.

A version of the earth pressure cells developed at Cambridge during the last decade, has been used successfully for measurement of shear stress in the field by Bozozuk (1970). The cell was designed by Hambly and is described by Bransby (1973). Seven cells were installed in the ground under a test fill near Ottawa, and the results obtained are reported by Bozozuk and Leonards (1972).

Small circular telemetric cells working on the same principles have been developed by Prange (1971) for the measurement of free field normal and shear stress within soil masses.

REFERENCES

Bozozuk, M. (1970). 'Field Instrumentation of Soil'. *Proc. of Conf. on Design and Installation of Pile Foundations and Cellular Structures.* Lehigh University. pp. 145–157.

Bozozuk, M. & Leonards, G. A. (1972). 'The Gloucester Test Fill'. *Proc. of A.S.C.E. Conf. on Performance of Earth and Earth-Supported Structures,* Purdue University, 1. pp 299–317.

Bransby, P. L. (1973). 'Cambridge Contact Stress Transducers'. *Lecture notes for course on Research Techniques and Equipment in Soil Mechanics,* Cambridge University.

Prange, B. (1972). 'The State of Telemetry in Soil Mechanics'. *Proc. Roscoe Memorial Symposium,* pp 476–488, Foulis.

P. R. Vaughan (Imperial College). Dr Wroth has referred to the combined measurement of shear stress and normal stress by earth pressure cells. Such observations seem most desirable where boundary cells are used, but when cells are embedded, knowledge of cell action factors due to interference by the cells with the stresses being measured is sparse. If plane strain can be assumed, stresses at a point can be defined if four observations are made, three of them parallel to the known plane of the intermediate principal stress. It is usual to place an additional cell with these three so that the redundant observation can be used as a check. If normal and shear stresses are observed by the same cell, four cells can be replaced by two, with attendant advantages in compactness of installation. However, until more is known about cell action factors, both for normal and shear stresses, it is impossible to judge which arrangement is likely to give best results.

Despite the severe problems involved, total stress observations can be very useful. They may help to demonstrate basic mechanisms of behaviour, the results may sometimes be used directly in check stability analysis (as when observations of horizontal thrust are made in embankments), initial stresses in the ground may be observed, results may provide a valuable check on analyses of the finite element type, which should predict stresses as well as displacements correctly, and they are of particular value in monitoring cracking of embankment dams.

Recently, hydraulic fracture tests on embedded piezometers have been suggested as a supplementary method of total stress measurement. Experience suggests that the minor principle stress (σ_3) can be determined in this way. Such tests have been used to observe lateral stresses in soft clays (Bjerrum and Anderson, 1972) and to monitor stress changes in embankment dam cores (Vaughan, 1972; Penman, 1972). Similar techniques have been developed in rocks (Haimson, 1968).

The principle involved is that if a pressure is applied sufficient to cause and propogate a crack in the soil, this crack will form in the plane of σ_3. Formation of the crack is indicated by a sudden increase in flow. If the hydraulic pressure is then reduced a sudden reduction in flow occurs as the crack closes and the 'close up' pressure at which this occurs is presumed to be σ_3. It is most convenient to determine this pressure by a falling head test performed immediately after fracture has been induced. *Figure 14* shows a comparison of 'close up'

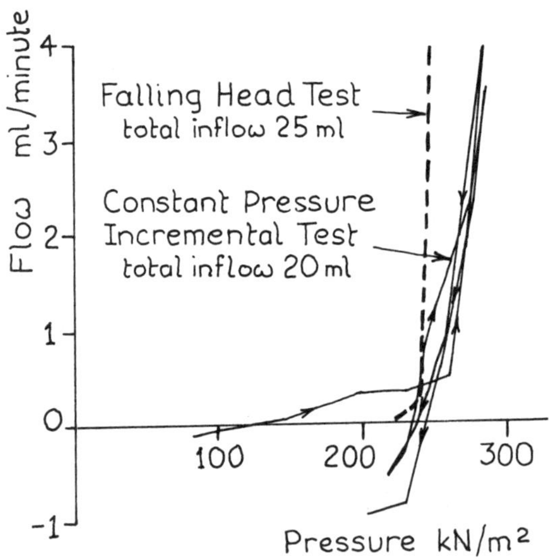

Figure 14 Close-up pressure from constant and falling head hydraulic fracture tests

pressures determined by incremental constant pressure tests and by a falling head test using a mercury manometer, in a compacted boulder clay fill. In these tests an unambiguous 'close up' pressure is demonstrated. A potential advantage of the test is that the crack is formed well away from the perturbed stress zone caused by the rigid piezometer. It has been demonstrated (Vaughan, 1972) that the test can be performed with small total inflows of water and without permanent damage to the soil.

The exact relationship between the 'close up' pressure and σ_3 is difficult to examine, as the mechanism is complicated. During the test, components of flow occur by seepage from the piezometer, by seepage from the walls of the crack and by change of volume of the crack. Definitive examination in the laboratory is very difficult, as the whole of the cracking mechanism must be contained within a volume of soil which remains at constant overall stress throughout the test. Thus very large specimens are required. An alternative method of checking the test may be to perform it within an overconsolidated soil with Ko greater than unity. Under these circumstances σ_3 is the overburden pressure, which can be calculated accurately.

The general utility of the hydraulic fracture test is currently uncertain and more experience with it is required. Recently, Wernek, of Imperial College, has performed tests in a dry compacted plastic clay fill in which suctions exist. Current results show unrepeatability, with the 'close up' pressure increasing for each successive test. It is thought that this is due to irreversible swelling causing a pressure bulb to build up in the soil. Subject to difficulties of this type, the test may become a valuable alternative method of examining total stresses in the field. It may also be used in conjunction with earth pressure cells and to check cell action factors.

An alternative type of hydraulic test is worthy of examination. Bjerrum *et al* (1972) examine the 'blow off' pressure at which the soil separates from the face of a piezometer when the hydraulic pressure applied within it is increased. This 'blow off' pressure, which is indicated by an increase in flow, is equal to the total stress on the face of the piezometer. Tests based on this principle might form a valuable and cheap method of checking boundary stresses. A restriction on tests of this type clearly must occur if the normal stress on the face of the piezometer is not uniform, as 'blow off' may occur at the minimum contact pressure rather than the average.

It is unlikely that stresses within, say, a compacted fill will be uniform locally. There is some evidence that non-uniformity remains constant in magnitude or decreases with increasing average stress. For instance, the difference in pressure observed on two faced pressure cells tends to decrease with increasing pressure. Thus non-uniformity of stress probably causes most problems when small pressures are measured, as in the observations of earth pressure described by Bundred and Jones in papers to this Symposium. This suggests that much larger pressure sensitive pads should be used for such observations. Such pads should be large in relation to local pressure variation and should measure total load independently of this variation. If shear loads were also measured, as suggested by Dr Wroth, final results could be of enhanced value.

The general reporter for session 1 has pointed out that the presence of instrumentation may interfere with the measurement being made. The principles governing such interference need consideration. Interference may occur when the instrument or the connections to it modify the total stresses or the pore pressure at the point of measurement. Different types of interference may arise when different measurements are attempted.

When measuring a pore pressure, errors can arise due to a change in total stress caused by the presence of the piezometer, which causes in turn a change in pore pressure. Since pore pressures equalise such errors are temporary. A permanent error can arise if the connections to the piezometer cause a preferred seepage path. When piezometers are embedded in fill, the connections can generally be buried in the same soil as surrounds the piezometer. When a piezometer is installed in a borehole, the backfill of the borehole may form a seepage path. An approximate solution for this problem has been derived (Vaughan, 1969). The results for a typical borehole piezometer are shown in *Figure 15*. It can be seen that the error in the measured pressure only becomes serious when the permeability of the borehole backfill is more than two orders

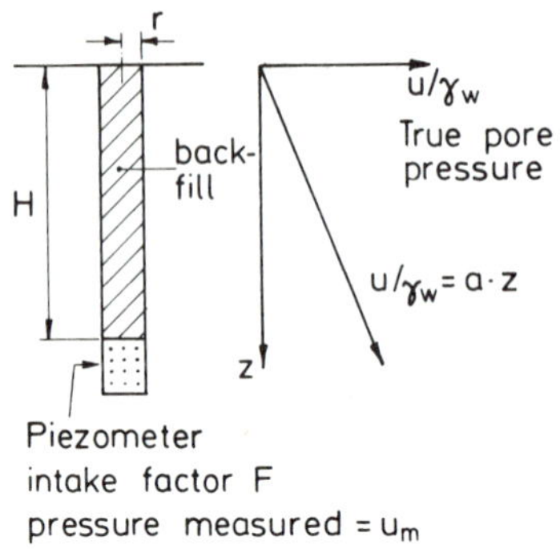

Permeability of soil = k

Permeability of backfill = kg

$$u_m/\gamma_w - a\cdot H \approxeq e_o\,(1-a)$$

$$e_o = \frac{1}{\sqrt{A+B}}$$

$$A = 2k/kg\cdot r^2 \ln(H/2r),\quad B = \frac{F}{\pi\cdot r^2}\cdot\frac{k}{kg}$$

For H = 20 m, r = 0·1 m

kg/k =	2	10	100	1000
F = 0, e_o =	0·2	0·5	1·5	4·8 m
F = 1·0 m, e_o =	0·05	0·2	1·0	4·2 m

Figure 15 Error in a borehole piezometer observation due to a borehole backfill of higher permeability than the surrounding soil

of magnitude greater than that of the surrounding soil. The minimum permeability which is likely to be encountered in natural soils is about 10^{-11} m/s. Thus a backfill of a permeability of the order of 10^{-9} m/s is needed.

In British practice the borehole is commonly grouted with cement bentonite grout, with a grout pipe placed down the hole to ensure that it is completely filled. With a cement/bentonite ratio of 1:1 a permeability of about 5 x 10^{-10} m/s can be achieve with a water content of about 200 per cent. With typical British bentonite, a grout of this water content can be used provided that it is mixed and pumped before the bentonite is fully hydrated.

The interference of an embedded pressure cell with the total stress being measured is a well known problem about which knowledge is limited. Preferential drainage along the connections to such a cell may also modify effective and total stresses at the cell. Where boundary cells are used total stress interference can be avoided, but effects due to preferred drainage must also be prevented.

When measurements of displacement and strain are made errors may arise due to interference with total stresses or pore pressure. Both may cause a typical soil deformation and an error in measurement. Interference due to a change in total stress occurs when the instrument is stiff in relation to the soil in which it is embedded. Axial stiffness in rigid equipment can be avoided by suitable sliding joints, but stiffness in bending is more difficult to overcome. A particular problem arises from the use of aluminium alloy tubes for inclinometers. When bending of such tubes occurs primarily at the more flexible joints, some interference is indicated. Such equipment is unlikely to give accurate results in soft soils when shear strains vary significantly between joints. More study of this problem is desirable, particularly in relation to the identification and observation of narrow shear zones.

Preferred drainage may also interfere with strain and displacement observations, although some local drainage adjacent to the instrument can often be tolerated without significant error resulting. Installation in boreholes requires particular attention and it may be necessary to grout in the equipment in a manner similar to the installation of a piezometer. When this is done the stiffness of the grout may also cause interference. *Figure 16* shows coefficients of volume compressibility for 1:1 cement grout of varying water content. These results show that interference is unlikely in moderately stiff soils but that it may occur in soft clays. Reduction in the cement content results in very long setting times.

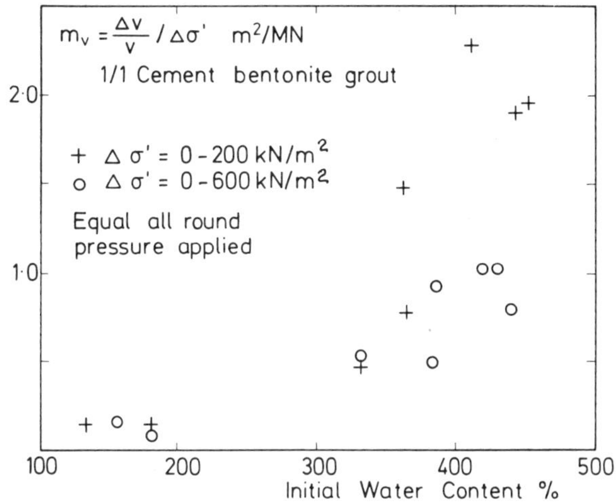

Figure 16 Compressibility of cement bentonite grout

REFERENCES

Bjerrum, L. & Anderson, K. H. (1972). '*In situ* measurement of lateral pressures in clay. *Proc. 5th European Conf. S.M. & F.E.* Madrid, **1**, 11–20.

Bjerrum, L., Nash, J. K. T. L., Kennard, R. M. & Gibson, R. E. (1972). 'Hydraulic fracturing in field permeability testing'. *Geotechnique*, **22**, 2, 319–332.

Haimson, B. (1968). 'Hydraulic fracturing in porous and nonporous rock and its potential for determining *in situ* stresses at great depth'. *Tech. Report No. 4–68*, July 1968, U.S. Corps of Engineers, Omaha, Nebraska.

Penman, A. D. M. (1972). Discussion. *Proc. 5th European Conf. S.M. & F.E.*, Madrid, **2**, 87.

Vaughan, P.R. (1969). 'A note on sealing piezometers in boreholes', Geotechniques, **19**, 3, 405–413.

Vaughan, P.R. (1972). Discussion. *Proc. 5th European Conf. S.M. & F.E.*, Madrid, **2**, 72–75.

S. F. Brown (University of Nottingham). An essential requirement for the *in situ* or 'environmental' calibration of earth pressure cells is that the free field stress should be accurately known. Arthur (1973) in his general report refers to techniques such as that used by Irwin (1973) wherein the cell readings, based on a 'bench' calibration, are compared with theoretical values which depend on the accuracy of the predictive theory being used for the stresses in the earth mass. Since one of the objects of developing pressure cells is to check the validity of such theories this procedure is, generally, unacceptable.

Determination of the 'free-field' stress must always rely on a measure of theoretical prediction. It is hence essential that the regime used for the calibration tests is simple so that this prediction can be carried out with confidence. The usual arrangement has been to test the cell in a cylindrical soil sample to which known stresses can be applied and in which, when attention is paid to compaction and end effects, the field stress can be determined with reasonable accuracy. It is clearly desirable that the cell be small in relation to the soil sample and while this presents no real problem for small instruments, it is clearly difficult for larger ones.

Experiments in which properly calibrated pressure cells have been used to determine the vertical stress distribution at various depths in a uniform soil mass (Brown & Pell 1967) resulting from a circular surface load, have shown that linear elastic theory predicts the stresses quite accurately for cohesive soils. This 'Boussinesq' theory could, therefore, be used with some justification as a basis for larger scale *in situ* pressure cell calibration tests. The Building Research Station cell, because of its large size, is the type of instrument that could possibly be calibrated by this means. Care should be taken, however, to ensure that it is not installed at a location where the stress gradient across the cell is excessive.

Arthur (1973) has emphasised the importance of errors and the need to quantify them; this being the main object of environmental calibration tests.

Brown (1973) has indicated that errors in registration due to 'cell action' can be predicted theoretically though no quantitative information was given about the scatter in experimental results.

Figure 17 shows results from a number of calibration tests carried out on two similar instruments following the procedure described by Brown. These tests were performed by a person with some experience of installation and the large amount of scatter, due to placement effects, must be regarded as realistic.

It is significant that the mean cell registrations for the TRRL/LVDT cell are accurately predicted by the theory outlined by Brown, but that the other two situations shown, wherein successively fewer tests were involved resulted in poorer prediction.

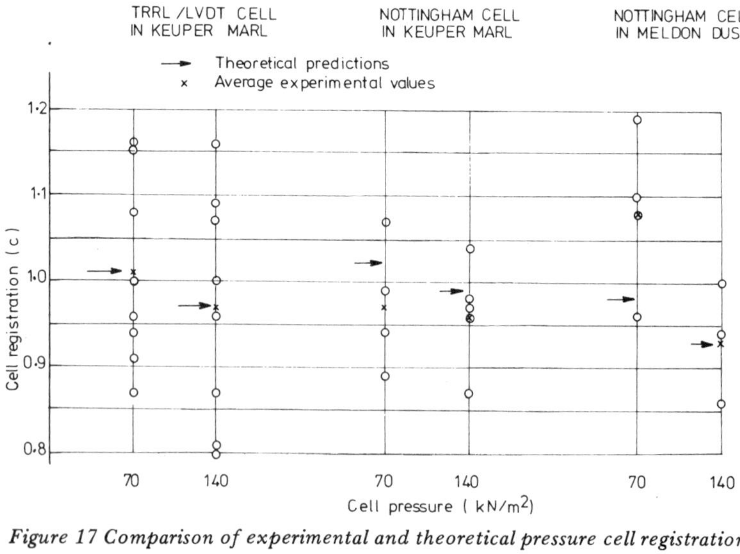

Figure 17 Comparison of experimental and theoretical pressure cell registrations

In view of the errors involved in placement technique, it, therefore, seems necessary to carry out plenty of tests in the laboratory (about 7) and install enough instruments in the field to obtain as many duplicate measurements as possible. Reference has been made in discussion to Professor Peck's remarks about one, two or three field measurements giving increasing confusion. It is suggested that while this may need be true for three results, six or seven are likely to present a more optimistic picture. The results in *Figure 17* also lend further confidence to the theoretical method for predicting cell registration outlined in Brown's paper.

REFERENCES

Arthur, J. R. F. (1973). General report on Session 1 of this symposium.

Irwin, M. J. (1973). 'Instruments developed by the TRRL for studying the behaviour of earthwords', *Symposium proceedings,* Part 1, pp 194–206.

Brown, S. F. and Pell, P. S. (1967). 'Subgrade stress and deformation under dynamic load', *Journ. S. Mech. and Found. Div., ASCE,* 93, SM1, pp 17–46.

Brown, S. F. (1973). 'The measurement of *in situ* stress and strain in soils', *Symposium proceedings,* Part 1, pp 38–51.

P. M. W. Elsenaar, (Rijkswegenbouwlaboratorium, Delft). In the Netherlands a working committee was formed to gather information on earth pressure cells for use under road constructions. After comparative tests, together with the Bundesanstalt in Cologne, a large calibrating device, known as 'Plantema cell' was used (*Figure 18*). This cell is a triaxial cell, diameter 0.90 m, height 1.00 m.

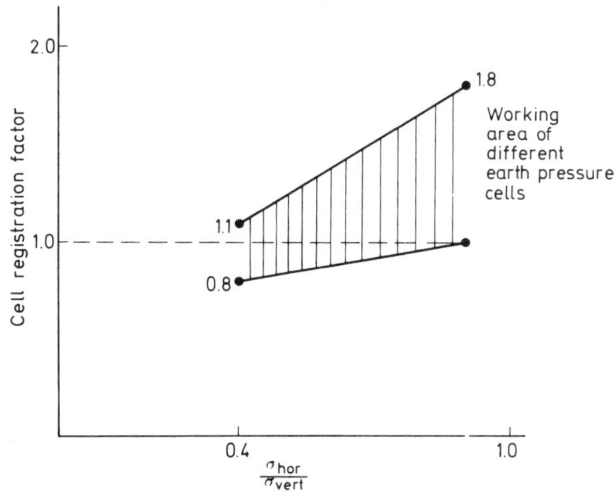

Figure 18 Data for the Plantema Cell

Tests were started with several types and sizes of earth pressure cells. The aim of this research was at first to compare the outputs of the different pressure cells under various conditions in fine graded sand which is used as subbase in our country. From these results we learned that the problem was not to be found in the pressure cell but in the interaction between cell and surrounding sand. I would like to underline one of the results of research in Nottingham as reported.

The cell reading depends on vertical and horizontal stresses. When horizontal stresses increased the cell tended to give lower readings (*Figure 18*). In the Laboratory of Soil Mechanics in Delft computer calculations have been carried out by means of finite elements methods. This was done to compute the stress distribution around the pressure cell. However the problems have not been solved at this moment. The difficulties in measuring earth pressures are signalised and theoretical investigations should be carried out. A report on the first results of our tests will be published this year.

Rainer Massarsch (Royal Institute of Technology, Sweden). The increasing importance of field investigations for both research and construction work requires new equipment which to a greater extent is adapted to the variety of practical problems. Therefore at the Royal Institute of Technology in Stockholm, Sweden new methods for field observations have been developed in cooperation with manufacturers of geotechnical equipment.
Total stress cell. The measurement of total horizontal stresses *in situ* is still an unsolved problem in site investigation. A spade-like total stress cell has been developed for soft clays using the well known Glötzl measuring system. The cell can be pushed into the ground and it causes only minor disturbances because of its low volume-area ratio (*Figure 19*). An oil-filled, flexible steel

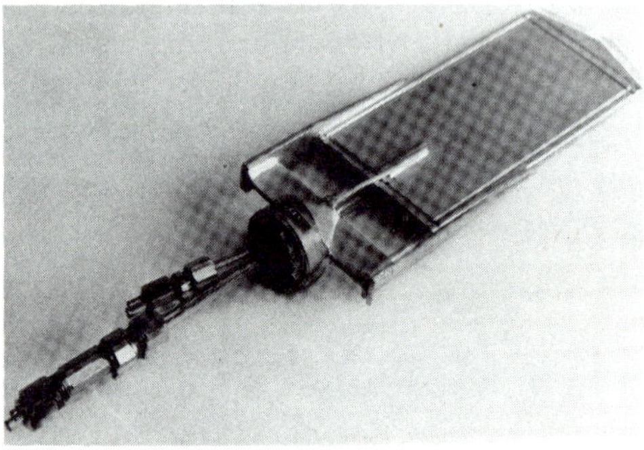

Figure 19 Total stress cell type Glötzl

cell is used which is connected to an air or oil operated valve (*Figure 20*). The instrument is connected to a pump. By increasing the pressure at a slow and constant rate, the sensitive steel membrane in the valve will open at a pressure slightly exceeding the pressure on the cell and can be read on a manometer. Only small deformations ($\sim 5\mu$) of the membrane and of the earth pressure cell are necessary to balance the pressure in the valve. Temperature effects can be neglected for practical purposes. The air or hydraulically operated system is simple, it is not affected by electrical interference and is independent of any power supply.
Multiple piezometer. For measuring the pore water pressure in clay a new piezometer has been developed by Geotech AB, Askim, Sweden using an improved vibrating wire gauge. The signal from the gauge is amplified at the pick-up in the ground and then transmitted through one thin steel wire to a frequency counter at the ground surface. By amplifying the frequency signal at the piezometer, it can be transmitted over longer distances and is less sensitive to interference. The thin steel wire can be attached to the outside piezometer rod, which makes it possible to connect together a number of piezometers on one unit as shown in *Figure 21*.

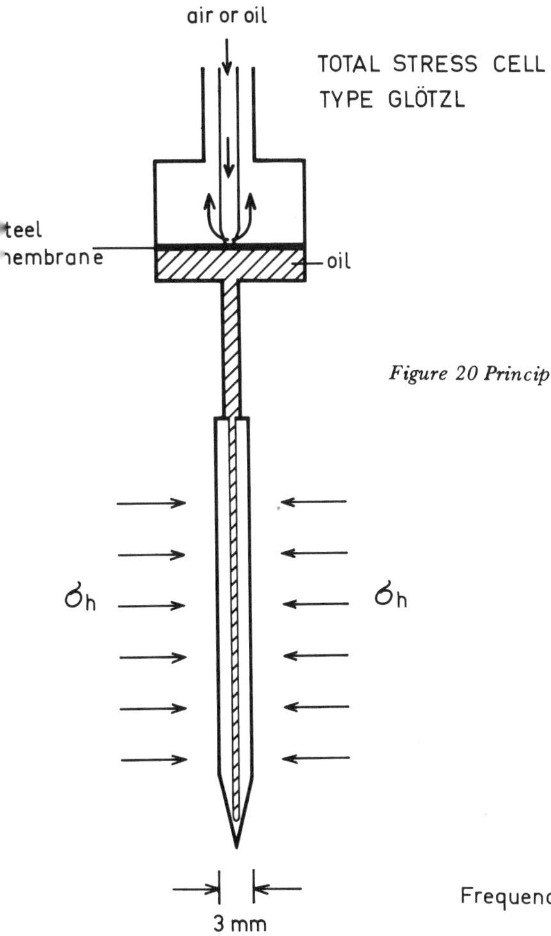

Figure 20 Principle of the oil or air operated Glötzl gauge

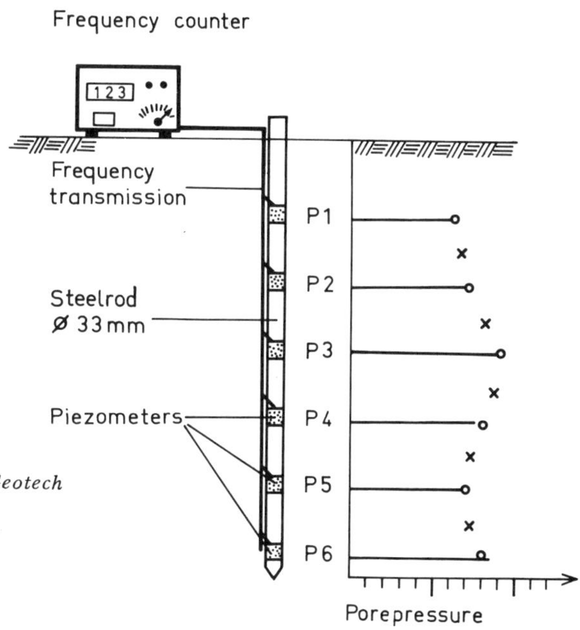

Figure 21 Multiple piezometer type Geotech

The new multiple piezometer can be pushed or bored into the ground similar to conventional piezometers thereby reducing the number of required piezometers and thus the installation time and the costs. Pore pressure readings can thus be made at several levels in one hole at the same time. By pushing the multiple piezometer in steps to several levels, a large number of pore pressure readings can be obtained faster and at a lower costs than with conventional methods.

Stereo cameras. The need for measuring equipment in foundation engineering which gives surface deformations in vertical and horizontal directions initiated a test programme using the stereo photography method. After promising results in laboratory model tests, stereo cameras have been used at a construction site near Uppsala at the river Fyris, where deformations of an old stone wall were measured during excavation and pile driving out in the vicinity. In order to evaluate available equipment and to find the best measuring arrangement, several camera types were used, including the recent Hasselblad Lunar Electric camera.

The test programme has not yet been concluded but the preliminary results are promising. By using the stereo method, a great number of photographs can be obtained in a short time and at low costs. The evaluation of the stereo photographs can be made anytime. Measurements can be concentrated on the part of the model which is of special interest.

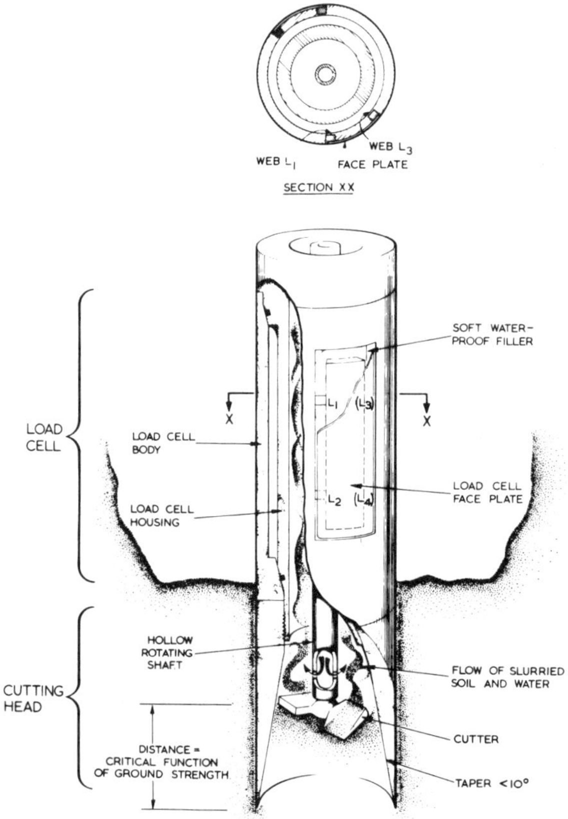

Figure 22 Instrument for measuring the in situ lateral stress

J. M. O. Hughes (University of Cambridge). I would like to describe a method for measuring *in situ* horizontal stresses developed at Cambridge University. The instrument shown in *Figure 22* is essentially a miniature tunnelling machine which acts vertically, drilling its way into the ground whilst removing the material displaced by it. When the instrument is in position the soil adjacent to it is almost undisturbed. Radiographic observations in the laboratory have shown that the disturbance can be reduced to less than 0.5 per cent radial strain.

Mounted flush with the outside wall of this instrument are sensitive load cells which record the total radial stress exerted by the soil on the wall. Initially the stress recorded by the cells is slightly above the *in situ* stress, probably due to the small disturbance but with time the stress creeps back to an equilibrium state which is believed to be somewhere near the *in situ* stress. The results obtained using this instrument compare favourably with those obtained by the hydraulic fracture technique. In *Figure 23* the results obtained by both methods are compared at a site near King's Lynn which has been described by Wilkes at this conference.

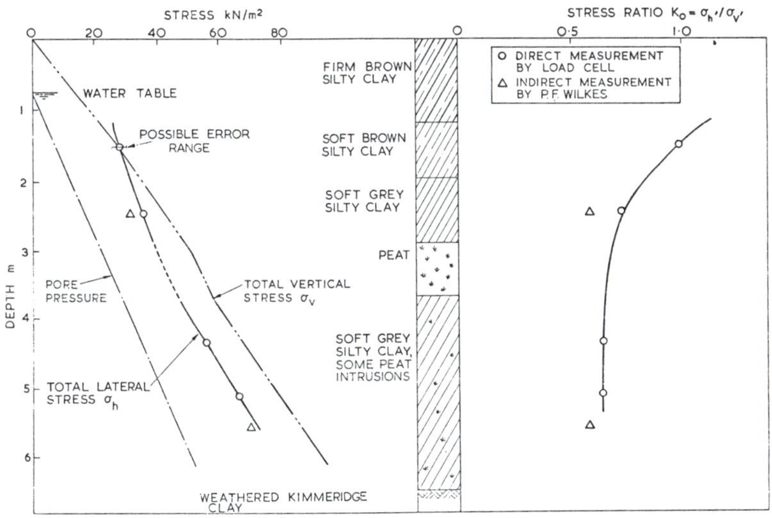

Figure 23 Measurement of the lateral stress at different depths using the instrument shown in Figure 22 and the hydraulic fracture techniques

The same drilling method can be used to insert a cylindrical rubber bag into the soil such that the soil surrounding the bag is essentially undisturbed. By expanding the membrane against the essentially undisturbed soil many of the soil properties can be obtained *in situ*. (Wroth & Hughes 1973).

REFERENCE

Wroth, C. P. and Hughes, J. M. (1973). 'An Instrument for the *in situ* Measurement of the Properties of Soft Clays'. *Proc. 8th Int. Conf. S.M.F.E. Jon. Eng.* Moscow. **1.2.** 487–494.

C. J. Dunnicliff (Soil and Rock Instrumentation, Inc., USA). In his preliminary report the Session Chairman asks: 'How should accuracy be defined? What accuracies can be expected? What accuracies do we need, where do we need them and why?' A few years ago the writer attempted to answer some of these questions, as applied to deformation measurements, and co-authored a state-of-the-art paper on this subject (Gould and Dunnicliff, 1971). The writer believes a brief review of the paper would be relevant to the objectives of this Symposium.

In surveying the literature to summarise information on accuracy, it was necessary to adopt criteria for a numerical statement of accuracy. Accuracy is synonymous with 'degree of correctness'. When making field measurements in geotechnical engineering, it is in general neither practicable nor necessary to apply a precise statistical definition of the term. In many cases a geotechnical engineer is satisfied to know that accuracy of his measurement is 'about' such and such a value. Accordingly, in reporting the experience of others, the paper does not attempt to apply confidence limits in a strict sense since data are seldom available for statistical analysis. A plus-minus range stated without qualification is used to mean the measurement 'probably' or 'generally' differs no more than this amount from the true value, and the writer considers this simple minded approach to be adequate for most of our purposes.

The paper discusses the factors which characterise and influence accuracy, and the accuracy required of measurements on buildings and in embankment dams. Specific instruments are considered in seven categories: conventional survey, geodetic methods, vertical-tube settlement gauges, remote settlement gauges, horizontal movement gauges, inclinometers, extensometers. In each category, the various instruments are described and accuracies reported.

Major conclusions of the paper concerning the role of accuracy in field observation of displacement are:

1. Accuracy is only one factor to be considered in selection of the appropriate device or technique. Ruggedness and durability can be dominant requirements and these may be at odds with sensitivity of a device. In this regard, the rule is: 'the simpler device is likely to have the best chance for success . . . the more complex the equipment, the greater the likelihood of malfunction' (Peck, 1970). Probably the factor of overwhelming importance to accuracy is the care and skill of the personnel responsible for installation and observation.

2. Devices used to measure movement on or within soil masses typically have a precision in the range of ±0.005 to ±0.04 ft ($\frac{1}{16}$ to $\frac{1}{2}$ in). These include multi-point vertical-tube gauges, many remote settlement gauges, horizontal movement devices and earlier types of inclinometers. Absolute accuracy of the movement measurement may be less than precision of the device if ordinary surveying methods are employed to convert observations made with respect to a local reference to absolute movement relative to a distant fixed point.

3. Absolute accuracy of observations, even if less than the range of ±0.005 to ±0.04 ft, is usually much greater than the accuracy intrinsic in the methods for analysis or prediction of displacement of earth masses or the estimate of stress state derived from field observations. Even the application of the flexible and reliable finite element procedure at this state-of-the-art cannot provide an accuracy equal to that afforded by current techniques for field observations.

4. Determination of internal strain between measuring points frequently is more precise than the movement measurement because systematic errors are diminished. Also in this case, methods of analysis are not able to take advantage of the potential accuracy of the observations. An exception is the straightforward case of strain determined by buried extensometers to detect a trend to internal cracking in embankment dams.

5. Much higher precision is required of devices such as hose levels, inclinometers or extensometers placed on or within structures or rock masses. Measure-

ments of displacement in this instance often must be within ±0.005 to ±0.1 in for a meaningful analysis of stresses or stability, roughly ten times as accurate as measurements within soil masses. In certain situations, precision of a device may be more important than great durability since the critical stage of performance can occupy only a short span of time. It should be recognised that in many cases displacement measurements of a structure are not sufficiently precise to permit reliable determination of stresses, and strain gauges attached to structural elements must be employed.

To sum up, when planning and executing an instrumentation programme the role of accuracy *per se* is a limited one, and one to which excessive attention is often paid. The factor which overrides all others is 'tender loving care' on the part of all concerned.

REFERENCES

Gould, J. P. and Dunnicliff, C. J. (1971). 'Accuracy of Field Deformation Measurements', *Proc. 4th Panamerican Conf. on Soil Mechanics and Foundation Engineering,* San Juan, 1, 313–366.

Peck, R. B. (1970). 'Observation and Instrumentation, some elementary considerations', *Lecture Notes, Met Section ASCE, Seminar on Field Observations in Foundation Design and Construction,* New York, April 13.

V. Ashkenazi (Nottingham University). In surveying and in other fields of technology dealing with measurements, accuracy is defined in terms of standard or r.m.s. (Root-mean-square) error and usually denoted by σ. The standard error of a directly measured quantity is calculated from

$$\sigma = \pm\sqrt{\left[\left(\sum_i v_i^2\right)/(n-1)\right]},$$

where v_i is an individual residual error, i.e. the difference between a single measurement and the average (most probable) value of that quantity.

However, the above formula (which has often led to widespread misunderstanding and confusion) gives only a measure of the consistency of the readings or the repeatability of the observations, and not the overall accuracy of the measured quantity. For example, a distance measuring instrument may have a consistency or resolution of 0.1 mm, but the overall σ of a distance measured with that instrument would certainly be larger to account for the uncertainties remaining after applying calibration, centring and atmospheric corrections.

In theory, the overall σ of a measurement could be obtained by the square-root of the sum of the squares of the σ's of the contributions of all the individual (and uncorrelated) factors affecting the instrument and the measurement. In practice, the overall σ is either guess-timated or obtained from external means (as opposed to the internal consistency of the readings). For instance, the departure from 180° of the sum of the angles of a triangle or the circuit misclosure in a levelling layout would give an indication of the overall σ of an angle or of the difference in level between two height points respectively.

In the context of measurement of spatial deformations, it is best to use the concept of standard error as follows. Statistically, 95 per cent of all possible errors in a series of measurements are likely to be less than 2σ, i.e. twice the estimated standard error of the observations. Similarly, there are 99 per cent chances that an individual error is less than 2.5σ. In statistics 95 per cent and 99 per cent chances are considered as being 'probably significant' and 'significant' respectively.

Thus, if one adopts 95 per cent chances as being sufficiently significant in the case of engineering measurements then 2σ becomes the criterion which one uses to decide whether the differences between two or more periodical values of a quantity can be attributed to random observational errors or are indicative of a significant change (or displacement or deformation) in the value of that quantity. For instance, if the σ of a series of linear deformation measurements is estimated to be, say, 10 mm then a difference of 20 mm or more between two consecutive measured values would be considered as probable displacement, whereas a smaller difference could be attributed to observational errors.

It is up to the geotechnical engineer to estimate the order of magnitude of the deformations which are likely to occur between two successive series of observations, to deduce from it the maximum amount of uncertainty which he is prepared to tolerate in the measured values of these deformations and to ask for an observational procedure leading to a σ which approximately equals half the value of that maximum uncertainty. It is up to the surveyor to consider the existing geometrical station-target layout and to adopt the most practical combination of angle and distance measuring instruments which is likely to achieve the required accuracy. A method leading to a bigger σ (lower standard of accuracy) would be useless whereas one resulting in a substantially smaller σ is bound to be unnecessarily wasteful and expensive.

H. J. Walbancke (Imperial College, London). McKellar *et al* and Vaughan in their respective papers to this Symposium have pointed out the importance of de-airing hydraulic piezometers measuring negative pore pressures.

Study of the rate of post-construction pore pressure equilibration in compacted clay embankments has formed part of a recent research project for the Water Resources Board. This has included the continuation of the pore pressure observations in a dam at Peterborough. The embankment, which was designed by Rofe, Kennard & Lapworth, is built in a worked-out brick pit for the C.E.G.B. Land Reclamation Scheme.

The 16 m high embankment, completed in 1963, is composed of Oxford Clay and a little Kellaway Sand and Clay. The 'redeposited' clay used for the core and the downstream toe is the spoil from the brickpit and hence at a rather higher moisture content. The 30 piezometers installed have high air entry value ceramic filters (Bishop *et al,* 1960) with polythene coated nylon 11 leads and a mercury manometer read-out system. Since construction, pore pressure observations have been made at intervals. The complete early records made by Mr Wrightman, the site engineer, have since proved invaluable.

The sections in *Figure 24* show the pore pressures at the end of construction and after eight years. Impounding was complete about eighteen months after the end of construction. The equilibration of pore pressure over this period is of interest. The core has consolidated and the high pore pressures have dissipated sideways into the dryer shoulders and the positive pressure zone in the downstream toe has spread. A zero pressure line from the upstream face has progressed about 5 m in 6 to 7 years. These results show equilibration of pore pressures in a compacted clay embankment can be a slow process and that to observe pore pressures for monitoring or forward prediction, measurement of negative pressure over long periods is necessary.

Errors in recording negative pore pressures can occur due to evacuation of the cavity or air in the leads. Only air in the leads is immediately obvious because of the difference in measured pore pressure on the two manometers.

For piezometers measuring negative pore pressures, Vaughan states that

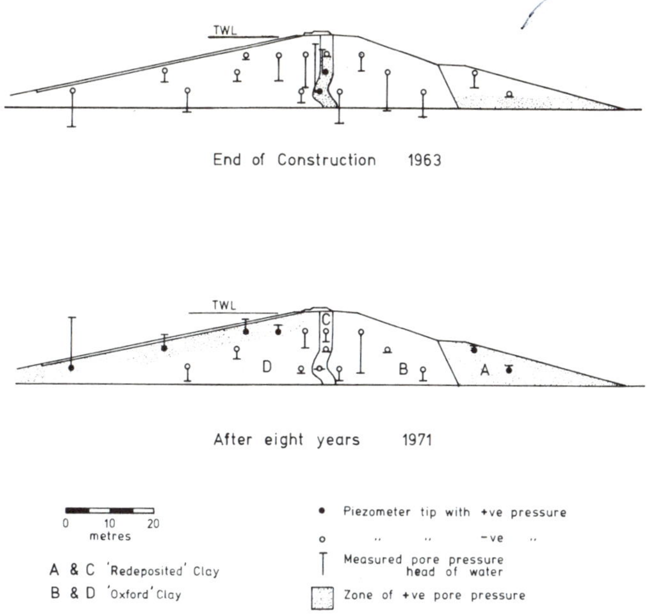

Figure 24 Peterborough Dam – embankment pore pressures

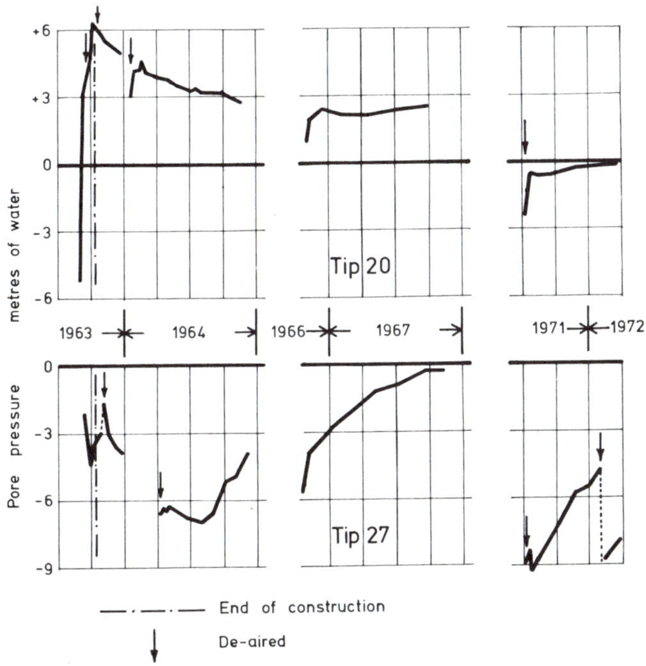

Figure 25 Pore pressures – Tips 20 and 27

de-airing is essential. An example of this can be shown in the records of two of the piezometer tips from Peterborough, one measuring positive and the other negative pressures. *Figure 25* Tip 20, with positive pore pressure, is uneffected by de-airing except for a short term perterbation. On the other hand, Tip 27, with large negative pressures, shows the necessity of de-airing to maintain the correct cavity pressure. The breakdown is however, not apparent until the de-airing has been carried out.

Once any air remaining after installation is removed, the error in transmission of pressure due to air in the leads is insignificant. Tip 27 is a good example of this behaviour, as despite an error of some 9 metres in cavity pressure, the difference in the pressures recorded by the two manometers never exceeded 0.1 m. This gives a false impression that the piezometer is recording correctly.

If after installation, the system is not fully de-aired, flushing out becomes difficult. This is shown very clearly for Tip 16 (*Figure 26*) which required de-airing four times before the air was finally cleared from the leads and the error became insignificant. It is of interest to note that no significant errors have occured again in 9 years of use.

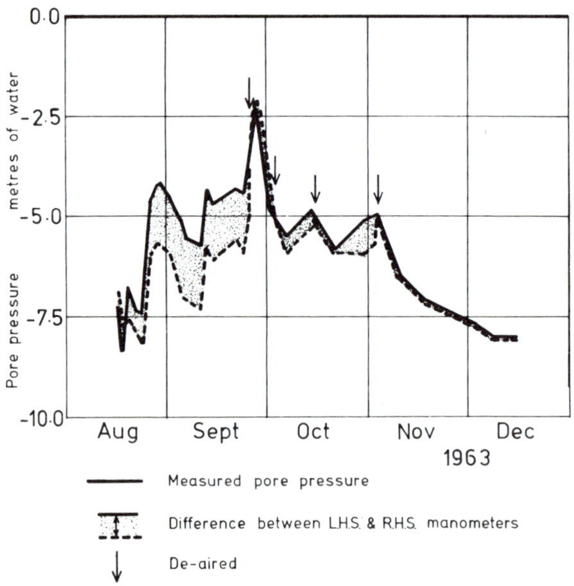

Figure 26 Pore pressure – Tip 16

Marsland has also pointed out the need for piezometer tips to be saturated with de-aired water on installation. Under field conditions this is difficult to ensure and failure results in considerable flushing being required to remove the air as described above. To help to eliminate this problem, recent practice has been to de-air the tips under vacuum in the laboratory as Marsland suggests. When possible the previously de-aired leads are attached before leaving the laboratory, alternatively the couplings are sealed. The tips then travel to site wrapped in a wet cloth and sealed in polythene. To prevent breakage they are given a further protective jacket. In this way they remain de-aired as the covering can be kept in place until the moment of installation.

In his paper, Little mentions the negative pressures measured in Foxcote Dam which was constructed in 1956. High air entry piezometer tips have recently been installed in the downstream slope and negative pore pressures are still being recorded 17 years after construction.

REFERENCE

Bishop, A. W., Kennard, M. F. and Penman, A. D. M. (1960). 'Pore pressure observations at Selset Dam', *Proc. Conf. Pore Pressure and Suction in Soils*, Butterworths, London, 91–102.

Discussion. Equipment

S. D. Wilson (Shannon and Wilson, Inc, USA). Simple instruments are usually the best, but often also the most time-consuming to read. The cost of instrumenting a particular project usually involves three elements of cost:

(a) The cost of the instruments;

(b) The cost of installing instruments;

(c) The cost of taking readings and evaluating the data.

Often (a) is the least and (c) by far the greatest, therefore anything that can be done to increase the reliability of the instruments is well worthwhile since they cannot be repaired or replaced once installed. Also, any simplification in reading and interpretation is significantly worthwhile. In the USA therefore, there is an increasing trend towards automation. For example, the data from inclinometers is often recorded directly on to punched tape in the field, but since the equipment is heavy and cumbersome, the move now is to record on to cassette-type tape which in turn can be fed directly into computers which will give you the results immediately. But there is still the time-consuming work of lowering and raising the inclinometer in the holes. To avoid this work we have developed a new instrument called an in-place inclinometer: several of these have been installed in recent months in the United States.

This device consists in effect of a series of inclinometers in tandem, which are lowered down a cased borehole. They are spaced at 5–20 ft intervals and connected by rods. It is the inclination of the universally joined rods which is measured, by a force-balance servo mechanism of the type used in inclinometers for many years. Readings are taken with a rather involved control box which automatically samples each one of the sensors at a rate of one complete set of readings every two seconds for 24 hours a day. It takes each of the readings and compares that reading with the initial reading and if the deviation exceeds any pre-set amount (which can be varied) it stops the circuiting, samples that particular device again three times to make sure there is no mistake, and then activates warning sirens and flashes warning lights so that people can get out of the excavation before it caves in.

In addition to this, the apparatus displays in inches the deflection change of all the sensors so that movements may be plotted. All the sensors can be removed from the hole if repair is necessary, or at the end of the project, they can be recovered to be used elsewhere. An installation has recently been made in a set of circular cofferdams in West Virginia by the Corps of Engineers on a major lock installation where the excavation is being made below the bottom of the river in laminated limestone, and there was concern about possible failure along a clay interbed. The control unit is connected to the telephone network so that results can be obtained, through a special punched tape device which can be connected to the telephone anywhere.

As another example, there is an installation to monitor the performance of the reservoir slopes on the left abutment at Libby Dam in Montana. Here, due to rock conditions, it was necessary to monitor continuously a large number of points within 3000 ft upstream of the left abutment. The installation does not record all the time, but once a day it records on punched tape the output of 50 electrical piezometers, 165 extensometers, 10 rock noise detectors, and several other devices. Rock noise detectors are rather new in soil mechanics. If the rock becomes highly stressed to the point when it may begin to fail the level of noise in the bedding planes will increase: it is picked up by very sensitive microphones in boreholes, amplified, put into square-wave signals, chopped off, converted into numbers and every 15 minutes this device automatically scans 10 of these counters, picks off the numbers and sends them by teletype to an office of the Corps of Engineers, 300 miles away, where it is watched 24 hours a day to ensure the safety of this abutment.

J. N. Hutchinson, (Imperial College, London). As earlier speakers had questioned whether the usual plastic or aluminium inclinometer tubes actually followed the deformation of soft ground, Dr Hutchinson demonstrated a very flexible inclinometer tube used at the BRS in the early sixties. It is of 50 mm nominal bore and consists of thin, multilayered plastic with helical wire reinforcement. It can withstand considerable radial pressures but it is very flexible laterally and can easily compress or extend.

An inclinometer, mounted between spring-loaded skids, was designed to be able to negotiate bends in this tubing of about 1 m radius. This flexibility was achieved only at the cost of accuracy, which was about ±20′ in angular measure. In a test set-up this gave a lateral accuracy of about ±30 mm at 3 m depth. The successful application of this system to the measurement of the velocity profile with depth through a shallow mudslide is described by Hutchinson, (1970). 'A coastal mudflow on the London Clay cliffs at Beltinge, North Kent'. (*Geotechnique,* **20**, 412–438.)

N. A. Skermer (G. E. Crippen & Associates Ltd, Canada). This discussion concerns the effect of casing twist on the measurement of horizontal deflections in inclined movement gauges using inclinometers tracking in grooves. At Mica Dam in Canada movement gauges up to 244 m long are inclined up to 0.25 on 1 from the vertical. The twist in a 1.5 m length of plastic casing at Mica is not more than about 1° and the top of each section was aligned parallel to the reference axis at installation. The gauges are surveyed by theodolite, and if the inclination is then plotted with respect to the top of the casing, the base in bedrock would appear to have deflected horizontally by amounts up to 3 m. Such movements are fictitious and in reality twist in the casing grooves in which the inclinometer rides, causes a cumulative component of the inclination to be registered as deflection in the plane at right angles. The geometry of the twist is illustrated in *Figure 1.* Attempts to measure satisfactorily the casing twist at Mica have been unsatisfactory so far, and reliable observations have been obtained only in the vertical gauges.

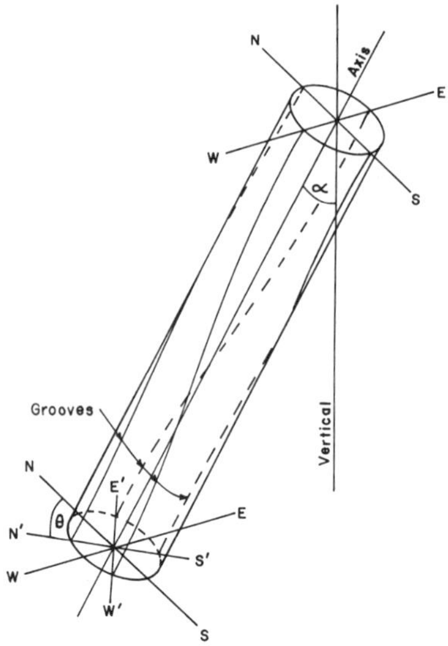

Figure 1 Geometry of the twist at Mica dam

Incidentally the statement by Gould and Dunnicliff (1971) that successive spirals can be made to cancel by installing lengths of casing end to end is incorrect.

REFERENCE

Gould, J. P. and Dunnicliff, C. J. (1971). 'Accuracy of field deformation measurements', *Proc. 4th PAN Conf. SMFE,* Puerto Rico, 1 p. 346.

P. R. Vaughan, (Imperial College). The importance of reliability, simplicity and ruggedness of field equipment has been stressed quite rightly at this symposium. Where sophisticated equipment is necessary, considerable development is required. However, more sophisticated equipment can reduce the engineering effort required in taking and processing readings. Some risk of malfunction of such equipment can be tolerated when it is accessible and when back up facilities are provided. Special equipment for particular jobs can be used with greater confidence when it is formed from well developed and proven components. Four examples of equipment illustrate some of these principles.

Pore pressure reading equipment for hydraulic piezometers. Such equipment should combine reliability, accuracy, convenience and compactness. Modern pressure transducers and digital reading equipment offer advantages. Transducers with small volume change/pressure change characteristics allow pressures in many piezometer tubes to be read quickly by connecting them in turn to the transducer. Such a system, for automatic recording is reported to the symposium by Irwin. However, automatic recording can usually be justified only for short term high frequency observations. In a typical earth dam, for instance, perhaps two hundred readings will be required, spread over, perhaps, five years. Electric equipment may not prove reliable for periods as long as this and checking and maintenance may take longer than manual observation.

Figure 2 shows equipment recently developed by Soil Instruments Ltd., for the Empingham dam. It is based on the assumptions that automatic equipment could not be justified, that observations should be presented as piezomettic elevations, hand plotted to ensure regular inspection, and that the observations should be obtained in this form with minimum time spent on readings and subsequent calculation. The equipment has the following features. A valve system connecting to 200 piezometer tubes (A); a pressure transducer (B); rigid stainless steel tubes connecting the valves with the transducer (C); a digital read out box (D), with zero and scale adjustment facilities and a variable offset so that pressures are presented as piezometric elevations; a small screw allowing the stiffness of the system to be checked, (F); and a mercury manometer and a datum pot (G), allowing the scale and zero of the read out to be checked and adjusted at any time. A strain meter can be substituted for the digital read out, if necessary. Readings can be taken rapidly by operating each valve in turn.

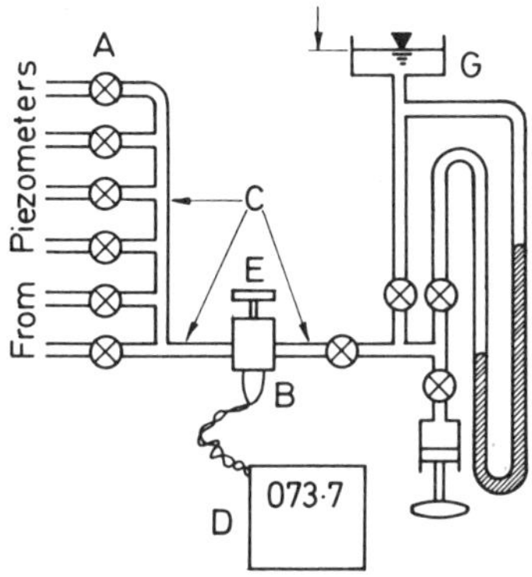

Figure 2 Hydraulic piezometer read-out system

De-airing equipment for hydraulic piezometers. A major advantage of hydraulic piezometers is that they can be de-aired if necessary. A major disadvantage lies in the need to fill the piezometer and tubes with de-aired water at the time of installation. If this is done successfully, the hydraulic tubes remain effective as an accurate means of transmitting pressure, even when subatmospheric pressures are measured and de-airing is necessary to flush air from the piezometer cavity. Miss Walbancke has demonstrated this in another contribution.

The principles involved in circulating water around a twin tube hydraulic piezometer are illustrated in *Figure 3*. It is desirable to maintain a small excess pressure (Δh) at the tip, to prevent air-saturated water being drawn from the soil; to avoid applying pressures which may rupture the soil; and to avoid a major drop in pressure in the return tube, which may cause air bubbles to come out of solution and, in some installations, cavitation. To fulfill these requirements it is often necessary to use a small differential pressure with a back pressure on the return tube. In these circumstances time for complete circulation may be long and automatic operation is helpful. *Figure 4* shows conventional three cylinder de-airing equipment adapted to allow the use of a back pressure and automatic operation. This equipment, manufactured by Soil Instruments, Ltd, requires a compressed air supply pressurising both the inflow and outflow cylinders through pressure reducing valves (R). Both the inflow

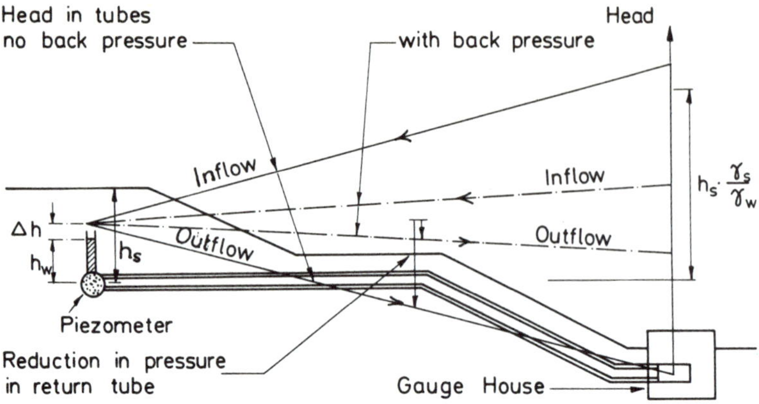

Figure 3 Head in circulating tubes of hydraulic piezometer during de-airing

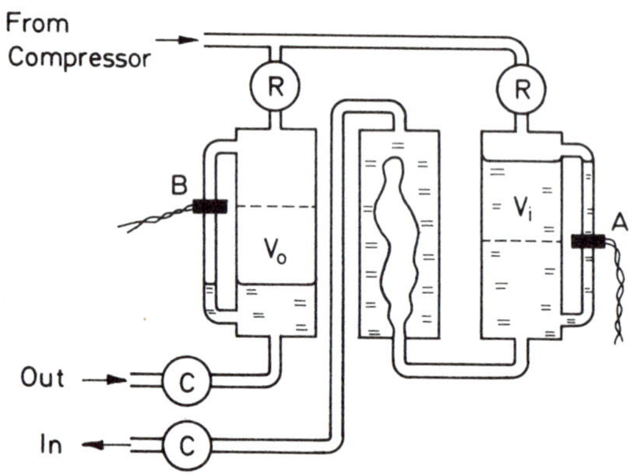

Figure 4 Automatic circulating system with back pressure

and outflow cylinders have water level sensors (A) and (B) which shut two solenoid valves (C) when either a preset volume V_i has flowed into the system or a volume V_o has flowed out. The system can be operated manually if required.

Combined inclinometer and settlement gauge. When deep inclinometer installations are used in embankments it is convenient to use them also for the measurement of settlement. Accurate knowledge of the position of joints in the tubing as it compresses is also useful. During construction it may be necessary to operate with the top of the casing at varying heights above a fill surface. It is convenient to measure settlements and inclinations with the same instrument but it is difficult and expensive to provide and handle a cable of sufficient longitudinal stability and with adequate markings for convenient and accurate settlement observations. *Figure 5* shows a modification of the inclinometer described by Green in a paper to this symposium, made in an attempt to solve some of these problems. A reed switch settlement device (of the type described by Burland & Moore in their paper to this symposium) is incorporated in the inclinometer, which is lowered on a steel tape, the electric cable being held below the inclinometer by a weight and pulley. The cable and tape are lowered into the hole together. Subsequently, the inclinometer can be operated independently of the cable until readings are complete. The tape is kept on a drum on an arm temporarily attached to the top of the casing. A datum mark for the tape is on this arm, which can be operated at any angle θ with the casing, depending on the height of the top of the casing above ground level. The arm telescopes automatically so that the distance L between the datum and the

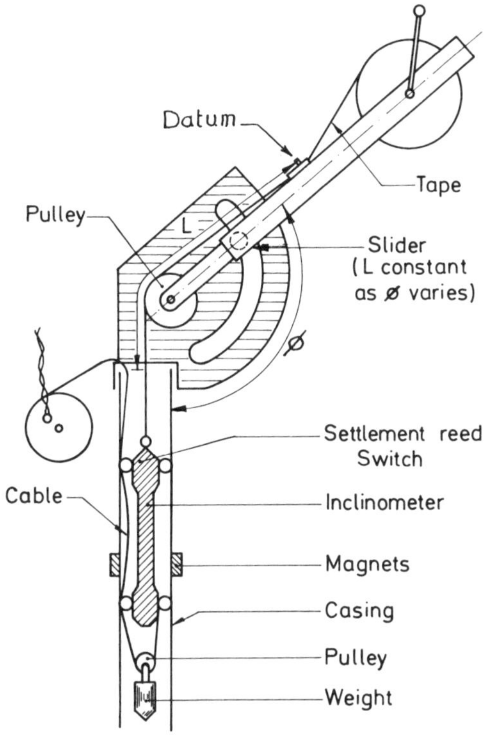

Figure 5 Inclinometer and settlement gauge

top of the casing remains constant. All the equipment is kept on a small trolley and neither the tape or the cable need be laid on potentially muddy ground during operation.

Wire extensometers for lateral deformation observations. Where the failure mechanism of a structure such as an embankment involves lateral displacements, measurement of these displacements plays an important part in the monitoring of performance. If such measurements are used as part of construction control, rugged and reliable equipment is required, with a rapid availability of results. *Figure 6* shows a very simple wire extensometer for measuring horizontal strains and displacements, which meets these requirements, The relative displacement of a series of plates can be measured with a pocket rule, or with more sophisticated equipment, which remains accessible. The observation point can be remote from the structure, if required. The relative movements of the wires can be used to activate a warning system when a predetermined displacement is reached. When the system is extending the wires are self-tensioning. When the system is compressing the wires may be tensioned by hand but, since the wires are quite stiff, the errors involved are small. Errors due to general settlement are small, but kinking of the tubes at shear discontinuities

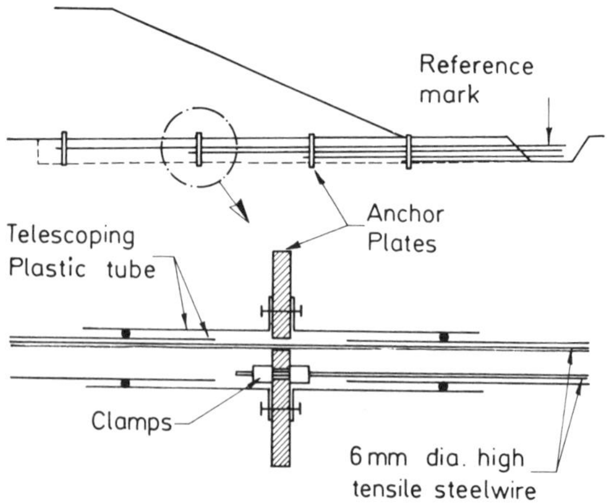

Figure 6 Wire extensometer placed on embankment foundation

can cause major errors. Experience of the use of these gauges, with gauge lengths up to 100 m, in two embankments, suggests that the accuracy of relative movement observations is better than ±5 mm.

A. D. M. Penman (Building Research Station). When de-airing hydraulic piezometers the pressure at the tip should be kept below the overburden pressure to avoid hydraulic fracture. This is a recommendation we have always made – it perhaps needs modifying, replacing 'overburden pressure' with 'minimum total stress' – however, it is interesting to see that the process of increasing the pressure to cause hydraulic fracture has become a method for measuring total stresses. The approach has been described by Vaughan (1970) and Bjerrum *et al* (1972). If the pressure at the tip is increased above the existing pore pressure, water will flow into the soil according to its permeability, but at a critical pressure, a crack will form in the soil adjacent to the tip and the flow will increase rapidly. Vaughan found that the critical pressure can be more accurately defined during pressure reduction, when he found that the flow suddenly decreased as the crack closed. He has used a mercury column

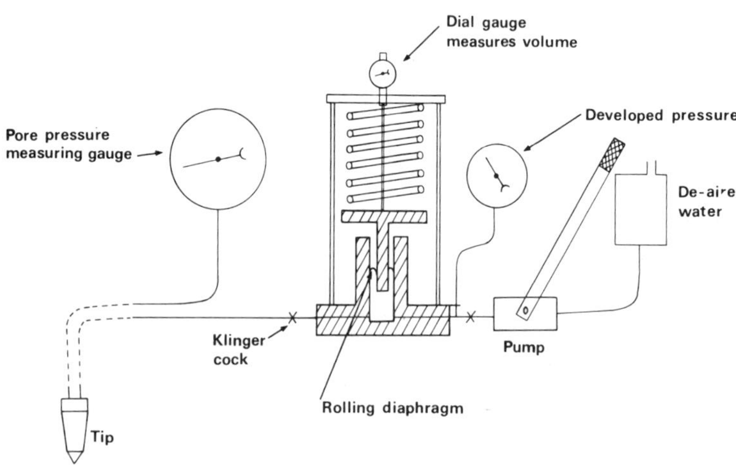

Figure 7 Critical pressure test apparatus

which fell as water escaped from the tip, giving a reduction of pressure and we have used this system with a double burette volume gauge to measure flow.

With the aim of a more compact, portable apparatus to make these tests, we have designed the spring-loaded device shown by *Figures 7 and 8*. The piston, loaded by a strong spring, is sealed with a rolling diaphragm and its movement measured by a dial gauge reading to 0.025 mm. De-aired water from a plastic container is pumped with a hand pump to move the piston against the spring and the developed pressure shown by a bourdon gauge. The apparatus is connected through a Klinger valve to one of the leads to a tip while the other is left connected to the pore pressure measuring guage to measure the pressures developed at the tip. This gauge must have a big enough range

Figure 8 Taking gauge readings

to measure the cracking pressure. At time zero, the Klinger cock is opened and both the flow to the tip and pressure developed are recorded. If the pressure developed in the apparatus initially was sufficient, the pressure at the tip rises to a maximum and then falls as a crack forms and the flow of water remains high. As the piston moves, allowing the spring to extend, the pressure falls until the crack closes, when the rate of pressure fall and the flow of water both suddenly decrease. A plot of pressure *v* flow gives the value of the critical pressure by the abrupt change of slope of the curve.

The apparatus has been used to assess the total pressures in the core of Llyn Brianne Dam and has been referred to by Carlyle (1973).

REFERENCES

Vaughan, P. R. (1970), 'Cracking of clay cores of dams.' *Report by Penman, Proc. I.C.E.* **46**, pp 115–117.

Bjerrum, L., Nash J. K. T. L., Kennard R. M., and Gibson, R. E. (1972), 'Hydraulic fracturing in field permeability testing,' *Geotechnique,* 22, No. 2, pp. 319–332.

Carlyle, W. J. (1973), 'The design and performance of the core of Brianne Dam,' *Trans 11th Int. Con. on large Dams,* 3, pp. 431–455.

C. Viggiani (University of Naples, Italy). In connection with a large test embankment on soft clays, 60 vibrating wire piezometric cells and 45 twin tube hydraulic piezometers have been installed at depths between 8 and 30 m below ground level. Based on the experience gained during the installation and two subsequent years of operation, the following comments may be added to the discussions held at the Symposium.

Sealing piezometers in boreholes. In each borehole two instruments have been installed at different depths. After various trails, the most effective seal was found to be a cement-bentonite grout (1 cement; 3 bentonite) pumped in a fluid state from the bottom of the hole. The effectiveness of the seals of the hydraulic piezometers was tested some days after the installation by performing a falling head permeability test from one of the two piezometers installed in a borehole and verifying that the water level in the other one was unaffected; only in two cases was sealing revealed to be unsatisfactory, (*Figure 9*).

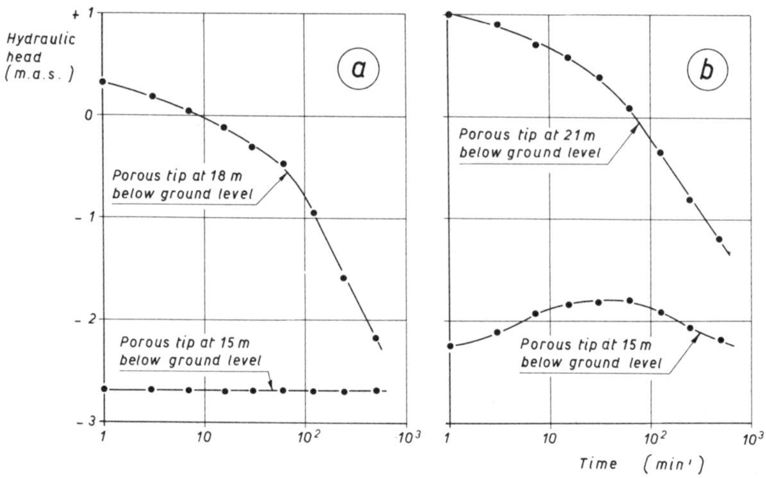

Figure 9 Falling head tests to verify the sealing of hydraulic piezometers in boreholes (a) effective seal; (b) ineffective seal

Influence of gas. The occurrence of gas within the clay layer under investigation is claimed by Croce *et al* (1973) to explain the measured excess pore pressures; it is also proved by the need for repeated de-airing of hydraulic piezometers, following different readings on the two tubes. In spite of this the vibrating wire piezometric cells behaved satisfactorily. *Figure 10* shows that at the end of embankment construction and two years later the excess pore pressures measured by vibrating wire and hydraulic piezometers are practically equal. It seems thus that the oil saturated porous tip of the Maihak piezometric cells has been effective in this case in preventing gas intrusion.

Initial zero reading. A time of the order of 15 to 20 days has been found to be needed to get a reliable zero reading after installation, i.e. to allow a complete dissipation of the excess pore pressure caused by boring operations and hydraulic tests for seal effectiveness. In this connection it is noted that the permeability of the clay, measured by means of constant head *in situ* tests, is of the order of 2×10^{-7} cm/sec; the consolidation coefficient from laboratory tests on small samples averages 3×10^{-3} cm^2/sec.

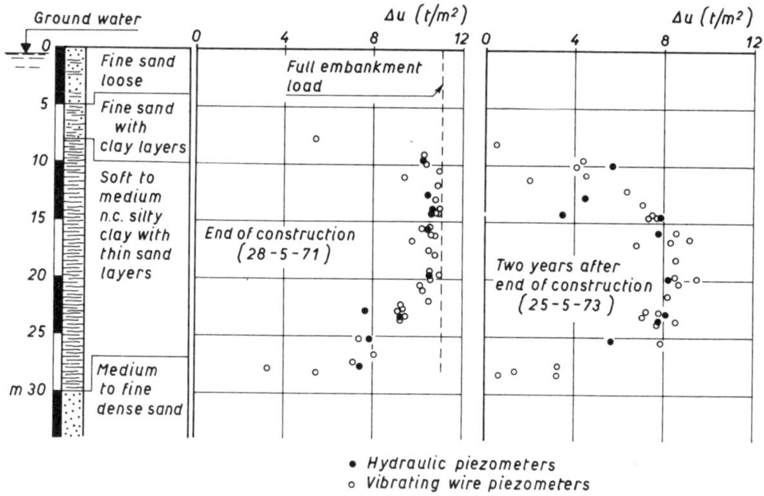

Figure 10 Excess pore pressure in clay below the central part of the test embankment

Damage from electrical storms. Damage to the vibrating wire piezometers from lightining strokes, putting them out of use, have been reported at this Symposium and elsewhere (Hosking and Hilton, 1963). In the present case an electrical storm actually occurred one year after the end of construction, and it seriously damaged the read out instrumentation (automatic scanning and digital printing unit). Nevertheless the piezometric cells have not been affected and are still in use by means of a manual portable measuring unit. Of course, a calibration with both manual and automatic units had been carried out at the installation time.

Relative merits and shorthcomings. Hydraulic piezometers have been found to be relatively inexpensive but somewhat lengthy and tedious in operation, mainly because of the need for frequent de-airing. They have the advantage of allowing *in situ* permeability testing and are fairly reliable as control instruments. On the contrary vibrating wire piezometers are rather expensive, but the possibility of automatic recording has proved invaluable during the relatively rapid construction stage. Both types of instrument have performed quite satisfactorily with regard to time lag, accuracy and stability. After 2½ years from the installation, 51 out of 60 vibrating wire and 29 out of 45 hydraulic piezometers are still in use. The relatively high percentage of hydraulic piezometers failure is believed to be caused by the large settlement (about 1 m) of the test embankment.

REFERENCES

Croce, A., Calabresi, G., Viggiani, C. (1973), 'In situ investigations on pore pressures in soft clays, *8th Int. Conf. Soil Mech. Found. Eng.*, Moscow.

Hosking A. D., Hilton, J. I. (1963), 'Instrumentation of earth dams on the Snowy Mountains Scheme,' *4th Australia-New Zealand Conf. Soil Mech. Found. Eng.*, Adelaide.

E. DiBiagio (Norwegian Geotechnical Institute). People working in instrumentation worry about many things. Two of the things that we worry about at NGI are: (a) protection of electrical cables, as John Hughes has mentioned at this symposium, and (b) the stability of instruments, particularly the zero point.

During the past three years we have been working with a vibrating-wire membrane type piezometer that has the capability of checking the zero point while it is in the ground. It has the traditional layout of a stiff metal encased diaphragm with the vibrating wire normal to it and attached at the centre. The body of the gauge is ventilated through a tube in the connecting cable to keep the back of the diaphragm at atmospheric pressure. This arrangement allows the slope of the calibration curve to be checked by appling known gas pressures to the back of the diaphragm and observing the resulting change in frequency of the vibrating wire. This concept is well-known; I think I first read of its use at the Building Research Station in Dr Cooling's Rankine Lecture in 1962 (*Geotechnique*, **12**, *no* 2, pp 75–104), but it does not give the zero reading of the wire since it is impossible to ascertain when the diaphragm has equal pressures on both sides. Our instrument is able to give this value through a special design of the diaphragm.

The metal diaphragm is perforated and has a separate flexible neoprene membrane covering the side subject to the pore water pressure. Under normal working, the pore pressure presses the neoprene against the metal diaphragm and its deflection, measured by the vibrating wire, indicates the value of the pore pressure. To check the calibration and zero reading, gas pressure is applied to the back of the diaphragm through the ventilating tube. Readings of the gauge-wire frequency are taken at various pressure increments until eventually, when the gas pressure is slightly greater than the pore pressure, the neoprene will move away from the metal diaphragm, leaving it with equal gas pressure on both sides and the frequency of vibration of the guage wire reached an asymptotic value. Any further increase in the applied gas pressure will not cause any further change in the frequency of the gauge wire and this limiting frequency value is the zero frequency for the instrument.

On the basis of the assumption that we can learn from the mistakes of others, I would like to mention another problem we have had with electrical piezometers. We have had two installations where electrical piezometers gave data of such a nature that one could only conclude that the instruments were inconsistent and erratic or drifting badly. We now believe that this type of difficulty can arise with metal piezometers due to galvanic action and the formation of fluctuating gas pressures in the soil because of dissimilar metals in the piezometer and/or in the connecting pipes. In an attempt to eliminate this problem we insert an insulating ring or bushing between the piezometer and the connecting pipes and we avoid dissimilar metals in the piezometer itself.

A further difficulty with electrical gauges, particularly where there are long lengths of connecting cables, is the catastrophic effects of electrical storms. We have had damage to systems in Norway where instruments have been permanently put out of action and I know that the same thing has happened in Australia and in the United States. In a region of thunderstorm activity you must provide some form of overvoltage protection, you must know what excess voltage your instrument can tolerate and you must have a suitable grounding system.

C. B. Strouts and G. R. Hallowes (Binnie and Partners). Difficulties arise when trying to obtain long-term reliable readings from pneumatic piezometers subject to varying degrees of saturation. In partially saturated silty sands (as in the Wash) very large capilliary tensions can develop. The resulting suction on the ceramic of the piezometer tip tends to draw water out of the tip, and air or gas in solution inside the tip comes out of solution and is trapped inside the tip especially if the ceramic has a high air-entry valve. If the pore pressure then increases, water flows back into the ceramic, compressing the free gas inside which does not readily go back into solution. Thus an increasing quantity of free gas is trapped inside the tip of any piezometer in fine sand where the pore pressure varies from positive to negative (for example with the rise and fall of the tide.) Evidence of this effect has been obtained in laboratory tests, simulating a cyclic variation in pore pressure, on a type of pneumatic piezometer which cannot be de-aired.

So as to prevent this accumulation of trapped air, some piezometer tips for the Wash Barrage Study were installed in 'buckets' containing fully saturated coarse sand which cannot transmit capilliary suction because of the relatively large pore spaces. In addition, each 'bucket' (0.5 m length of 75 mm P.V.C pipe capped at the bottom) was connected to a nylon tube, leading to the surface, which can be used to keep the bucket topped up with water. This arrangement of a piezometer tip within a bucket appears to have prevented air from accumulating. It has of course, prevented the measurement of pore pressures less than about 0.3 m positive head; however, negative pore pressures measured in a partially saturated sand are unlikely to have a significant meaning, and this loss is therefore not important.

F. H. Hughes (Cementation Ground Engineering Ltd.). As a member of the firm which undertook much of the site investigation, some of the instrumentation and, of course, the Sandwicks at the Canvey Island site, I wish to refer to the paper by George and Parry, which discusses a cost comparison of drainage wicks.

Speakers have already referred to the problem of installing instruments and a device was illustrated which used wash boring cutters to place an earth pressure cell. It is apparent that there is a common problem between installing instruments and sandwicks. I mean, of course, the problem of disturbance. It is true to say that whatever method we might use to make the hole, disturbance is caused. We must accept this. The problem lies in which method to adopt in a given soil to cause the least disturbance.

I would say that is is not always obvious which method to use and I feel that too many engineers reject say, a driven method for another system in certain soil conditions without real field evidence. I was interested to note that a number of contributors have used driven methods to place piezometers. I suggest that there is room for field research on hole making, and placing systems to complement papers based mainly on laboratory work, and a historical review of completed jobs.

An interesting cost comparison of drainage wicks is reported in the paper. This did bear out the monetary savings due to the use of wicks. However, the comparison was based on the use of two quite different systems.

The Sandwicks were installed by wash boring and casing, whilst the paper drain would be placed by means of a driven displacement mandrel oversize to the placed drain. In fact, only the wash bored sandwick type was selected for the field trial and this, the authors will agree, proved successful despite, as it turned out, a too rapid loading – the loading rate was in excess of that used in the non-treated area and of greater height. The wash boring system was chosen and used by mutual agreement, a driven or vibrated system could have been employed.

The Sandwick has the advantage, in common with a number of instruments, of being usable in a hole made by many methods. For a deep hole in soft alluvium, a small diameter driven or vibrated tube is usually cheaper to place than a wash bored one. Thus, if the wash-bored system is replaced by a method using a driven casing, then the cost falls, and saving from the use of Sandwicks increases. The system becomes more competitive.

In the comparison of systems set out in the paper the same wick spacing of 1 m was used for both methods. In the Sandwicks field trial the actual spacings used were significantly wider, being 1.5 and 2.0 m on a triangular pattern.

I do not discount driven methods for small diameter drains as we have known these to be successful, notably on large sites in India and Yugoslavia, but as I have said, I would like to see field research into the problem. Indeed, my firm would like to cooperate in instrumented field trials using different placement methods. To do this we need the use of a soft alluvial site where earth moving is under way.

G. R. Hallowes (Binnie & Partners). I wish to refer to experience with a borehole extensometer placed in the boulder clay foundation of a clay trial bank which was approximately 10 m high. Five magnets were installed in the borehole, the lowest magnet (Magnet 1) being approximately 18.5 m below stripped ground level. The diameter of the hole was 140 mm at the level of Magnets 1 and 2, and 169 mm at the level of Magnets 3, 4 and 5. The magnet holders were type C, shown in *Figure 10* of Burland and Moore's paper to this Symposium, pushed down over flush-jointed access tube through which a torpedo containing a reed switch sensor was subsequently lowered. The bottom magnet was not in rock but was assumed to be below the level of significant settlement and used as the datum for settlement measurements.

Measurement was by reading a tape, graduated in millimetres, from which the torpedo was suspended. Relative settlement (i.e. settlement measured relative to the bottom datum magnet) has been plotted in *Figure 11* shown using the same format as that in *Figure 12* of Burland and Moore's paper. Unfortunately, the resulting plot does not show such smooth curves as those obtained by Burland and Moore.

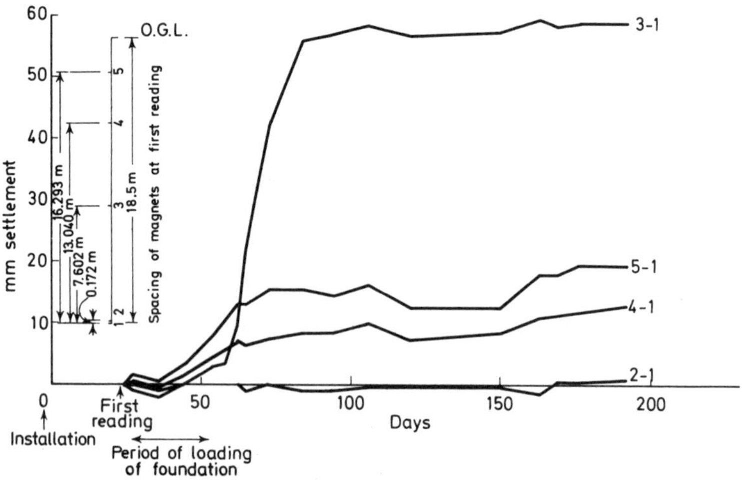

Figure 11 Resulting plot of readings

Due to a blockage in the access tube, which was only cleared after manufacture of a special tool for the job, only one zero reading could be obtained before loading of the foundation commenced. However a good idea of the accuracy of the system can be obtained by examining the relative behaviour of Magnets 1 and 2. Magnet 2 was installed only 172 mm above Magnet 1 which was approximately 18.5 m deep. It can therefore be assumed that there was no relative movement between these two levels and that the relative movements recorded by the extensometer were a measure of the inaccuracy of the system. As can be seen from *Figure 11*, the records show movements varying from a relative heave of 2 mm to a relative settlement of 1.5 mm. Between day 106 and day 163, Magnets 3, 4 and 5 also show unexpected 'heave' which is likely to be the result of inaccuracies in the system or of reading. This 'heave' or inaccuracy is as much as 4 mm in the case of Magnet 5.

It therefore appears in the light of this experience that the accuracy claimed by Burland and Moore (about 1 mm) when simply lowering a reed switch sensor at the end of a graduated steel tape is rather optimistic. However, it should be emphasised that for our purposes the order of accuracy obtained was acceptable.

It can also be seen that Magnet 3 moved exceptionally between day 58 and day 84. Since neither Magnet 4 nor Magnet 5 moved similarly, this large movement of Magnet 3 can only be explained by its coming loose in the borehole and settling downwards through the grout. It is interesting that this should have occured so long after installation and grouting.

Burland and Moore mention that the grout used to backfill a borehole containing instruments should have a stiffness similar to the surrounding soil. However the design of such a grout is glossed over as is so often the case. If anybody here has the formula to provide grout of the required stiffness for different soil conditions I would be interested to hear it (see Marsland p. 530 Ed.). We did not feel that we could reproduce the original stiffness and therefore adopted a fairly weak grout, to avoid any possibility of forming a concrete pile which would restrain the magnets from settleing with the surrounding soil. We felt that it was reasonable to rely on the springs of the magnet holders to fix the magnets in the surrounding soil, and that the purpose of the grout was to maintain the stability of the sides of the borehole so that there would be no collapse of material which might knock the magnets out of position. The grout mix was 1:2:6 (cement: bentonite: water) and it was tremied down the hole. The fact that Magnet 3 was able to move as it did shows that this grout did not restrain the movement of the magnets but suggests that the design of spring anchorages of the magnet holders could be improved.

J. B. Burland (Building Research Station). I would like to make a contribution arising out of Mr Hallowes remarks relating to possible sources of error in using these magnetic devices. A cross-section through a ring magnet shows that the position of the reed switch as it enters the magnetic field is important. If it comes in along the edge it will close at a different position from that if it came in along the centre line. This forms a source of error if the reed switch is not accurately guided. Many of these ring magnets are not symmetrically magnetised, so that the shape of the force field may vary and lead to errors unless the reed switch is accurately guided. The magnets should be specified to be symmetrically magnetised.

Another important source of error is tilt of the reed switch. It should retain the same attitude as it passes through the magnet each time. It should also face the magnet in the same way each time and it should not be rotated in the tube in relation to the magnet between successive readings. It is most important that the reed switch should not be left in a magnet, because it will take on a certain amount of magnetism itself, and this will cause it to close at a different position. It should therefore be stored away from a magnetic field. The reed switch is in a very delicate glass capsule and if the stress in the glass changes due, for example, to putting the switch in a medium which may creep, this may cause a significant change in the point of closure of the switch. It is important, therefore, to mount the switches in such a way that the stress on them does not change.

In connection with long-term drift, BRS carried out an experiment which has now been running for 18 months and found no detectable drift on the particular units we were using. We are still closing to within 0.002 in of the zero readings taken 18 months ago. The reed switch does not close at the same point going in as coming out of the magnet so it is very important, where more than one person is taking readings, to ensure that you know whether you are reading as it goes in or as it comes out. We normally read as the reed switch is going into the magnet.

B. W. Kitching (Allen, Gordon & Co.). Accurate measurements of the movements of the crest of the Upper Glendevon Dam in Perthshire are being made using a technique developed by the National Physical Laboratory and described by Harrison, Tolman and New in the Proc. I.C.E., 1972, 52 (May) 1–24. An application of the technique is described by Weeks & Dixon on page 431 of Part 1 of this symposium.

The equipment at the Dam consists of four units: a laser powered from a car battery; a small hole mounted immediately in front of the laser and through which its beam is directed; a zone plate holder into which zone plates of the correct focal length for each of the seven movements positions are placed; and a screen on to which the beam from the laser is focussed by the zone plates to form a small bright spot, the horizontal and vertical position of which is recorded for each zone plate position. Ten sockets are embedded in the dam for accurate location of the units. Movements are measured relative to and at right angles to the line between the small hole and the screen which are 362 m apart. It is important to appreciate that precise re-location of the laser is not necessary.

Spot movements on the screen magnify zone plate movements. For example, a zone plate half-way between the small hole and screen would have half the movement of the spot on the screen.

A sight line only 300 mm above the roadway surface has been adopted in order to simplify procedures although it was appreciated that there would be some loss of accuracy because of greater atmospheric distortions at low levels.

Figures 12 and *13* show the horizontal and vertical movements of the highest part of the dam (46 m from stilling basin floor to crest) recorded between October, 1972 and May, 1973. *Figure 12* also shows the reservoir water level which rose 20 m to overflow level during the period.

The following are comments following the first eight months use of the system:

(1) There is little difficulty in obtaining satisfactory observations of the horizontal positions even when atmospheric conditions are not good. On two out of twenty-one visits to the dam, conditions were too bad to obtain observations for some zone plate positions furthest from the screen.
(2) Satisfactory observations of the vertical positions are only obtained during

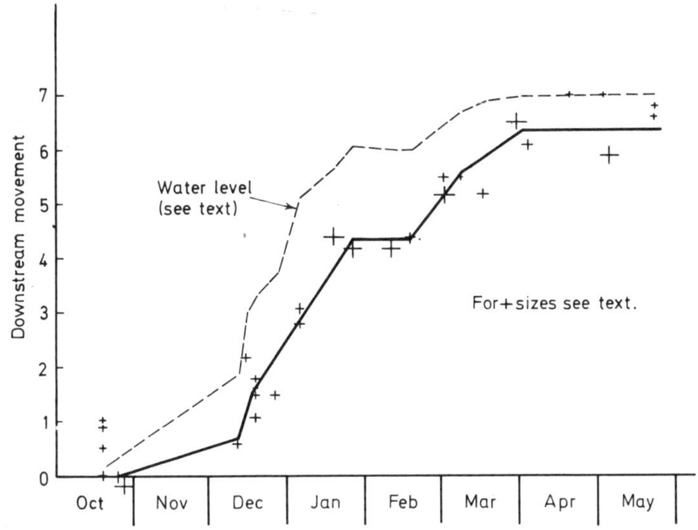

Figure 12 Horizontal movements in mm

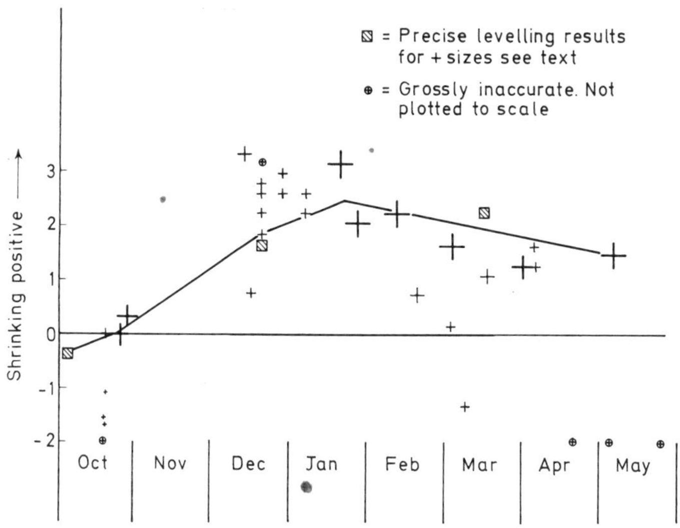

Figure 13 Vertical movements in mm

suitable weather conditions. It would be of great interest to know to what extent raising the sight line would have improved the incidence of suitable conditions; it is thought probably not much.

(3) The ideal weather conditions are overcast with a strong wind. In winter good vertical results are often obtained when the weather is not 'ideal'. In summer it is much harder to get good vertical results and present indications are that they may only be obtained before sunrise in 'ideal' weather

(4) There is no difficulty observing during 'ideal' weather. One demonstration of the equipment in use is enough to make a competent observer.

(5) The amplitude of the oscillations of the spot on the screen provides an excellent guide to the reliability of the observations. The greatest amplitude occurs with the zone plate position furthest from the screen (242 m). This is never better than 0.7 mm and at worst is about 5 mm. Quite good vertical results are obtained when the amplitude of this spot is less than about 1.8 mm. In *Figures 12* and *13* the largest crosses are for the smallest amplitudes.
(6) Although greater amplification of the movement is obtained the further the zone plate is from the screen, greater accuracy is obtained the nearer it is to the screen because atmospheric distortions are less.
(7) Two observers will usually agree on the mean spot position within 0.3 mm when it is oscillating 3 mm.
(8) The spot size varies between about 6 mm for the zone plate 242 m from the screen and 1.5 mm for the zone plate 41 m from the screen.
(9) A set of observations take $1\frac{1}{4}$ hours at the dam.
(10) It seems reasonable to claim that the errors in the movements shown in *Figures 12* and *13* are less than 0.5 mm.

R. W. Cooke (Building Research Station). One of the topics for discussion in this Session is the stability and reliability of equipment available for field investigations. I should like to comment on the stability of the laser alignment device developed by the National Physical Laboratory and used for observations of pile settlement in the work described by Weeks and Dixon in the paper 'Some new techniques in field instrumentation applied to a bored pile trial in Gault Clay'. Dr Weeks and Mr Dixon are to be congratulated on being the first investigators, as far as I know, to combine a laser with an electro-optic detector in such a way that tilt of the transmitter has no effect on the measurements. The system appears to be extremely sensitive and not difficult to install and operate.

In the first part of their investigation the authors used a conventional reference beam supported on shallow plinths relatively close to each test pile. *Figure 6* of the paper showed that daily movements of the reference beam could be as large as 0.4 mm and the authors stated that the variation was comparable with the pile settlement under each increment of load. For this reason they investigated the potential of the laser system since this could be operated with much larger distances between the transmitter and detector and, in addition, was less likely to be affected by environmental changes.

In the pile loading tests currently being carried out by the Building Research Station it has been found very difficult to establish a reference datum stable to the order of 0.1 mm over day and night periods, irrespective of changes in the soil moisture conditions and ambient temperature. It would, therefore, be very interesting if Dr Weeks and Mr Dixon would present figures demonstrating the stability of the laser which could be compared with data obtained in the BRS study using simple cantilevered trusses. Earlier experiments by BRS to measure ground movements during the installation and loading of a pile had shown that, in order to be unaffected by the test pile, the reference supports needed to be placed at distances greater than fifteen test pile diameters away from it. In the present investigation, because of the existance of an observation trench and an instrument hut at one side of the test pile, it was necessary to site the whole of the reference system on the other side. For this reason the cantilever form shown in *Figures 14, 15* and *16* was chosen. The two triangular trusses formed of 50 mm o.d. dural tube were supported by roller bearings from a cross-rail attached to steel scaffold poles cast into three 300 mm dia

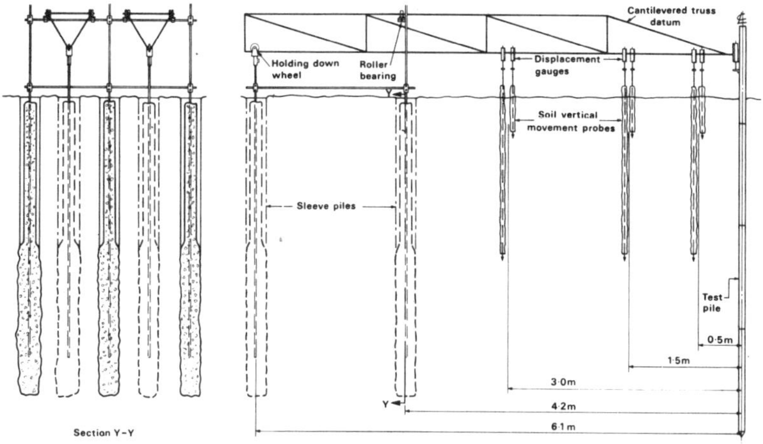

Figure 14 Side elevation and section through the cantilever truss datum

Figure 15 Free end of the truss showing the vertical movement probes and displacement gauges

bored piles. The ends of both trusses remote from the test pile were held down by V-pulleys carried by two more piles. The piles which were all approximately 4 m long, were sleeved from the surrounding clay and topsoil for the uppermost 2 m to eliminate any forces due to shrinkage or swelling of the clay. The whole test area was protected from the weather by a permanent polythene enclosure as shown in *Figure 16*.

Observations of pile settlement and of vertical movement in the soil were made relative to the trusses at twenty positions and some of the observations over a 40 hour period when no loads were applied to the pile are plotted in *Figure 17*. They showed that over 12 hour periods at night variation along the trusses was minimal and, assuming that no soil movements actually occurred, stability to better than 0.1 mm was obtained. The stability at the end of the trusses given by relative movement between them and the pile was better than 0.15 mm during the same periods. By day, because of the larger temperature variations, somewhat larger relative movements were recorded and for this reason, when very accurate observations of soil movement corresponding to pile loading were required, much of the work was carried out at night, In short term tests with loads up to the design load it would be reasonable to assume that the readings were accurate to within 0.05 mm. This level of accuracy was necessary in order to relate pile settlements to the corresponding soil displacements measured by the horizontal inclinometers described in a paper to this Symposium (Cooke and Price). Dr Weeks and Mr Dixon may care to comment on whether any form of reference datum including its supports, bearing in mind the requirement that these must be set up outside the soil zones affected by both the test and the reaction anchorages, can be expected

Figure 16 The sleeved pile supports for the truss datum and the polythene enclosure covering the entire test area

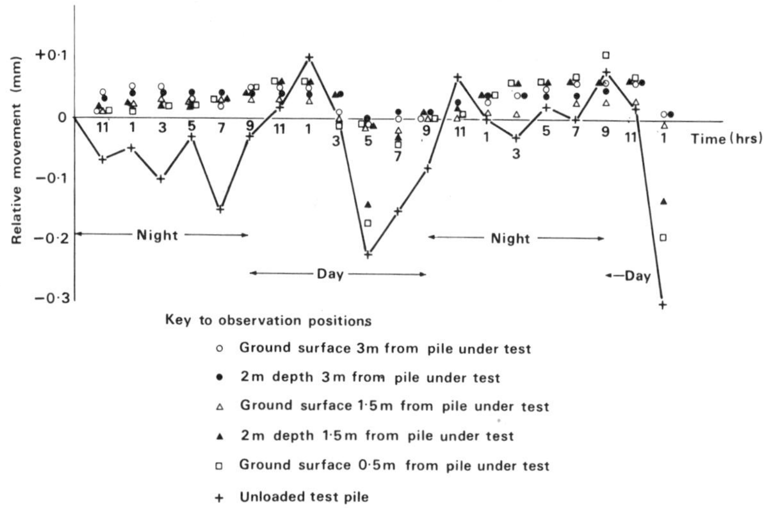

Figure 17 Relative movements between the reference system, soil probes and the unloaded pile during a 40 hour period

to be stable to better than 0.2 mm over long periods. If the laser system is intrinsically more reliable than any form of beam then its supports must be of comparable stability and the levels of these and the movements recorded at the foundation under test may need to be corrected for the effects of temperature fluctuations. Practising engineers will not be interested in this standard of accuracy but, for certain purposes, research workers are and they need to be satisfied that all related observations are made to approximately the same accuracy if the data is to be interpreted meaningfully.

D. E. Sherwood (Soil Mechanics Ltd.). I can best answer some of the points raised by Dr Green in his general report by describing our recent experience in using electrolytic levels in a little more detail. We were dissatisfied with the performance of our original levels as our aim had been to maintain long term monitoring installations within ±1% error due to sensitivity charges and datum drift. We consequently carried out very careful examination of the instruments used (as shown in *Figure 18*) in the cooling tower monitoring described in our paper to this Symposium, and identified several problems with the method used for mounting the level.

(1) The rigidity of the vial holder tube was unsatisfactory. Variations in torque on the pillar screws gave variations in sensitivity.
(2) Long term stress relief and stress corrosion is likely to take place around the trunion mounting and in the base plate causing datum drift errors.
(3) Torque applied to the connecting plug set up stresses along the vial holder tube which caused both sensitivity charges and datum drift.
(4) No absolute datum check was possible without involving a fresh batch of mounting stresses in relocating the instruments.

We have overcome these problems by using a simple V-block mounting as shown in *Figure 19*. This has the additional advantage that reversal of the instrument can readily be carried out as a check on datum drift. If drift is detected the instrument can be properly calibrated on a granite block in the laboratory without involving errors in remounting. We have also had to pay

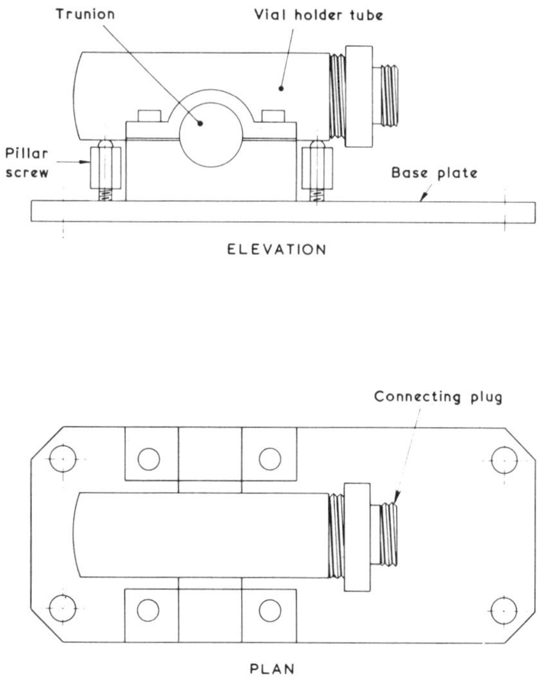

Figure 18 Trunnion mounted electrolevel

much closer attention to protecting individual components of the mounting assembly to avoid galvanic action between dissimilar metals, and to reducing possible stress concentrations to an absolute minimum. Test to date on the new instruments lead us to believe that long term operation will be well within the ±1% tolerance we wish to achieve.

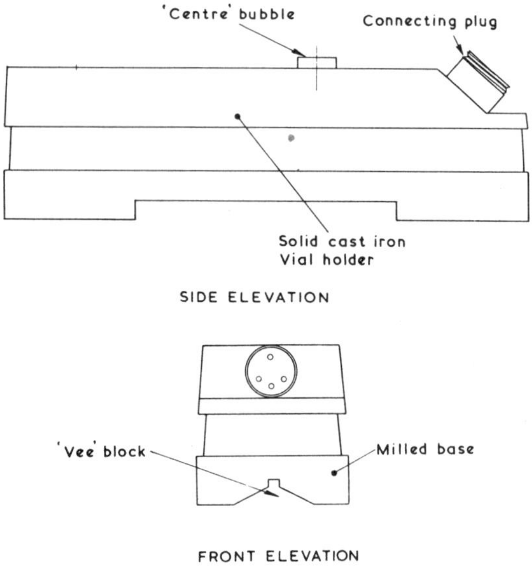

Figure 19 'Vee' block mounted electrolevel

In concluding, I would like to emphasise that what we require is an instrument which will measure angles conveniently to the required accuracy, and produce an electrical signal proportional to angle. Automatic monitoring can then be readily arranged both of angular and linear displacement as described in our paper. A generation of torque-balance accelerometers have recently been introduced which are competitive in price with the electric spirit level, and fulfill the same function to an accuracy about an order better than the level. Reliability and long term stability of this type of instrument is potentially much better and no doubt they will now be given serious consideration for the type of functions described.

D. J. Forrester (Nottingham University). My contribution concerns the points raised by Dr Green in his general report, regarding the accuracies of the two measuring techniques described in the paper by Dr Whittaker and myself. However, I would first like to point out that these techniques are used to examine ground behaviour on undermining and that vertical subsidence data must be obtained in addition to tilt and strain data to gain a clear picture of the total ground movement.

Regarding the tilt measurement using the electrolevel, tilt is measured by placing the instrument on purpose designed field stations (*Figure 4*, p. 441). The authors have found that the errors of this technique fall into three categories: Positional, Surface Contamination and Instrument Errors.

Positional errors. As tilt is measured in several distinct directions, reference lines on the field stations are used as a guide for positioning the instrument.

If great care is not taken in this operation, error occurs in the reading such that the larger the value of tilt, generally the larger the error. This error can only be minimised by very careful positioning for *all* readings or by modifying the design of station so that the instrument is keyed in position for each reading. This and more sophisticated modifications increase the cost of each field station.

Surface Contamination Errors. To ensure the success of this technique, good surface contact between instrument and field station is essential. For this reason the surfaces are thoroughly cleaned before *every* measurement. The presence of small particles such as sand grains, dust or drops of rain cause poor surface contact, and alter the tilt of the instrument relative to the station causing a false reading. Such errors usually appear clearly in the set of readings and are ignored in the measuring process.

Instrument Errors. Those errors which occur due to the nature of the electrolevel instrument and associated electronic circuitry. Experience with this type of instrument has shown these errors to be very small. However, datum drift can be appreciable and it has been found essential to take readings in opposite directions on each line and to mean the two to obtain the value of tilt. This practice has been found to eliminate any error due to this cause.

To summarise the affect of these errors, the authors have found that an accuracy of ±0.2 mm/m for the value of tilt on any field station reference line using this technique can easily be attained, provided that great care is taken during measurement. For small values of tilt, using the fine scale on the meter, an accuracy of ±0.1 mm/m can also be attained.

Regarding errors using the strain measurement technique, these are of five types. Those due to:

(1) poor fit of the registers in the field stations;
(2) friction in the sliding parts;
(3) the angle of the tape;
(4) the effect of temperature; and
(5) the effect of wind.

Each will be dealt with in turn.

(1) Error due to poor fit of registers in station. This is avoided by using the expanding mechanism built into each register whereby the register can be clamped centrally and firmly in each station.

(2) Error due to friction between sliding parts of the instrument. This is the largest source of error and apart from severe modification to the instrument it can be absolutely minimised by using the following measurement procedure. Firstly the instrument must be kept clean of dirt etc., secondly certain points must be kept carefully lubricated and thirdly all readings must be made by tensioning the tape from one direction only. Providing the same procedure is used throughout any set of periodic measurements, the values of strain deduced will contain minimal error. Series of readings for any one line agreeing to better than ± 0.05 mm can be obtained in this manner.

(3) *Angle of the tape.* The angle of the instrument can be altered to align the tape using a reference line marked on the end plate of the instrument. Providing this alignment is made no error is induced in the readings due to this cause. The system is designed so that true slope distance is measured on non-horizontal bay measurements.

(4) *Effect of temperature.* The technique described in the paper makes use of standardised stainless steel tape with a manufacturer's temperature coefficient, standardised at 20°C (68°F). Temperature corrections of significance only occur on measurements of greater than 2 m. Invar tape or wire could be used to eliminate this correction but would be expensive.

(5) *Effect of wind.* This is a factor which is common to all distance measurements using tapes. It is found to take two forms, firstly, vibrations can be set up in the tape making the tensioning process difficult and secondly, the tape can be forced out of a straight line giving rise to false readings. It has been found that if measurements must be made in windy conditions additional personnel must be used to shelter and steady the tape immediately beford a reading is made.

To summarise, field experience has shown that experienced personnel using the techniques outlined above can obtain an accuracy of ±0.1 mm for bay lengths of up to 9 m.

E. M. Gosschalk (Sir William Halcrow & Partners) said that it would be of much interest to hear of development work on instruments to measure two parameters which presented difficulties: firstly, tensile stresses in rock and secondly, *in situ* stresses in rock in a direction parallel to the axis of an access borehole.

It is recorded with thanks that a number of participants subsequently suggested to Mr Gosschalk helpful lines of investigation and inquiry.

S. F. Brown (University of Nottingham). The earth pressure cell developed by the Building Research Station has been used extensively both for free field and, in its special casing, boundary measurements.

Both Arthur (1973) and Green (1973) in their general reports emphasise the need for what they term 'environmental calibration' of this and other pressure cells, particularly those intended for free field measurements but such tests do not seem to have been done for this instrument. Thomas and Ward (1969) in their paper describing the BRS cell mentioned that such tests were to be conducted and it would be interesting to know the outcome of these.

The important characteristics of the BRS cell are very similar to those of the Nottingham cell described by Brown (1973). It has an aspect ratio of 0.14 compared with 0.17 and the relationship between flexibility factor (F) and soil modulus (E_s) for the 700 kN/m^2 model is $F = 0.67 \times 10^{-4} E_s$ compared with $0.59 \times 10^{-4} E_s$ for the Nottingham cell, E_s being in kN/m^2.

The good and reasonably predictable performance of the small Nottingham cell in 'environmental' calibration tests may, therefore, lend some confidence to measurements made with the BRS cell, though the possibility of some 'scale effect' as mentioned by Green (1973) obviously exists. Installation procedure in laboratory tests can, however, reproduce field conditions and adds to the importance of such tests.

It is perhaps relevant to query whether such a large instrument as the BRS cell is worthwhile in view of the difficulty of assessing its *in situ* performance. The TRRL's LVDT pressure cell (described by Wilson & Pimley 1972) which has a diameter of only 70 mm and the potential for short or long term measurements may be a useful development. Calibration tests reported by Dr S. F. Brown in Session 1 have been carried out on the TRRL instrument and indicate good performance for the short term tests involved.

In using the boundary version of the BRS cell, Bundred (1973) reports similar differences between the registrations on the inner and outer faces that were noted by Uff (1969). At that time a possible explanation for this phenomenon was presented by Brown (1969) which may be related to the different stiffnesses of the soil and the rubber lining of the cast iron casing. Because of the errors caused by point loads which may be applied to the outer face as a result of the presence of large particles, it was concluded that the inner face readings are likely to be the more reliable provided they can be adequately calibrated. A possible calibration procedure similar to that described in Session 1 by Symons was outlined at that time (Brown 1969). Wong (1973) indicates that he has carried out this type of test on his pressure cell, though no details are given in his paper.

The question of cross-sensitivity of the BRS cell was also raised by Brown (1969) when it was stated that this was of particular concern for measuring stresses on or near the minor principal plane. Jones (1973) has used the BRS cell for both vertical and horizontal measurements behind a retaining wall so the question of cross-sensitivity would seem to be relevant to his tests. It would be interesting to know if this effect has been investigated.

REFERENCES

Arthur, J. R. F. (1973), *General report on Session 1 of this symposium.*

Green, G. E. (1973), *General report on Session 2 of this symposium.*

Thomas, H. S. H. and Ward, W. H (1969), 'The design, construction and performance of a vibrating-wire earth pressure cell', *Geotechnique*, 19, 1, pp. 39–51.

Brown, S. F. (1973), 'The measurement of *in situ* stress and strain in soils', *Symposium proceedings, Part 1*, pp. 38–51.

Wilson, D. S. and Pimley, B. (1972), 'Earth pressure measurements in pulverised fuel ash behind a rigid retaining wall', *TRRL report, LR 893.*

Bundred, J. (1973), '*In situ* measurements of earth pressure and anchor forces for a diaphragm retaining wall', *Symposium proceedings, Part 1*, pp. 52–69.

Uff, J. F. (1969), '*In situ* measurements of earth pressure for a quay wall at Seaforth, Liverpool', *Proc. Conf. on In Situ Investigations in Soil and Rocks*, pp. 229–239.

Brown, S. F. (1969) *Proc. Conf. on In Situ Investigations in Soil and Rocks,* Discussion in session D, pp. 243–245.

Wong, H. Y. (1973) 'Field instrumentation of vibroflotation foundation', *Symposium proceddings, Part 1*, pp. 475–487.

Jones, C. J. F. (1973), 'Field measurements of earth pressures against motorway retaining walls and bridge abutments', *Symposium Proceedings* pp. 207–219.

H. S. H. Thomas (Building Research Station). In answer to Mr Brown's question, the evaluation of the cell action factor for the BRS total earth pressure cell at various attitudes and for a range $\sigma_1 : \sigma_3$ ratios has not been done. The feasibility of building a triaxial compression cell for samples of 1 m diameter and 2 m length for this purpose has been studied and some design work done. The cost would be in the region of £50 000 to £100 000 and the expenditure of this amount of public money was not considered justifiable. The rockfill testing that would have used the apparatus is being done with a large oedometer. Edge effects in this type of apparatus severely limit its usefulness for the cell action problem.

The question of scale effect was raised. If there is no effect then experiments such as Buck's with model cells in a small apparatus would serve. If, however, a scale effect does exist the size of desirable testing apparatus will increase alarmingly because already we should be using larger earth pressure cells. Our own cell was designed for, and worked satisfactorily in clay. If measurements are to be made in coarse material the point loads between the large particles and the sensitive surface must be averaged over a large area. In other words, a large cell is required for measurements within a soil mass or, in the case of a wall, the load on a large section should be measured.

I would like now to comment on Mr Little's unfortunate experience with a former manufacturer of the BRS earth pressure cell. Many users, as well as ourselves have had consistent results from these instruments. Binnies were the first to have the cells made for them rather than make their own, and like the buyer of a new model of car, were vulnerable to other people's mistakes.

May we examine why mistakes occur in manufacturing some instruments rather than others and how they can be avoided?

Two examples of the use of scientific apparatus are:

(1) The research worker operating apparatus which he has designed himself using his own training and skill to solve a particular problem;

(2) The technician using one of the mass produced instruments supplied by a manufacturer with long experience of, and a reputation for, satisfying the requirements of that technology.

I suggest there is a range of cases in between these two in which the responsibility for producing a good instrument, and subsequently using it appropriately, is shared between the maker and his client. Instruments for measurements in civil engineering are made in small batches often to a design unfamiliar to the workman. Their operation may be delegated to someone unfamiliar with the task. This is a situation in which more care than usual must be taken to see that all staff on both sides know what is expected of them. Good communications are required within each organisation as well as between the organisations. Also as in civil engineering practice, it behoves the client, or his agent, to examine the product to ensure that his requirements are correctly met. To this latter end an inspection and test schedule has been prepared for the BRS cell indicating what should be checked at the various stages of manufacture, Particularly we recommend that the client's representative be present at the final sealing together of the two halves of the cell. Mr Little asked about calibration curves. Since the calibration of the cell is one of the items specified, any cells with non-linear calibrations are rejected. We would not expect all cells to have exactly the same slope of calibration line. The cost of the inspection is about £10 per cell.

Summarising, the needs are for quality control of the instruments and for education of the users. This Symposium has already provided much of the latter, I would like to stress the former.

REFERENCE

Buck, G. F. (1961), 'An interim report on cell action studies connected with research on pressure measurements in sands.' *Proc. Midland Soil Mech. Soc.*, 4, pp. 96–105, 1961.

T. K. Chaplin (University of Birmingham). My first group of comments concerns Dr Hawkes' remarks about vibrating wire gauges. The 'non-linearity' which he referred to (on the basis of his graphs) is due to plotting the results directly against frequency. In 1952 when working at the BRS I had to correlate readings taken with two very different control sets (with Muirhead reference gauges of which one was very heavily modified), and from then on plotted all results in terms of units of $(\text{frequency})^2/1000$. For details see Chaplin (1956).

The high stress in his wires to which Dr Hawkes referred (100 000 lb/in^2) would normally be expected to cause large losses by relaxation. Though he has assured me since that his own procedure for forming end grips by extrusion and the choice of very high strength piano type wire have given him negligible creep relaxation losses, one would normally expect creep losses to occur in other circumstances. In particular the stainless steel wires (chosen to resist corrosion) might well show more creep than music-type high-tensile steel wires. It is essential to carry out many tests to check creep losses.

Dr Hawkes' very ingenious automated resonant plucking method could give rise to troubles when applied in other circumstances. About 1954—55 when Mr Wakeling and myself were calibrating the extensometers for the Ashford Common tunnel (see p. 32 and *Figure 5* of Ward, 1955) we had to use my own method of resonant plucking with continuously adjustable variation of frequency. The very bad response of many extensometers when using d.c. plucking (they sounded like a badly cracked bell in the headphones) was due to the presence of *two or three* natural frequencies at any one tension of the wire. By resonant plucking those three frequencies were precisely separated and measured. There are likely to be only a minority of vibrating-wire gauges (except perhaps for long wires at fairly high tensions) which have a fairly pure vibration. Only that minority would be suitable for application of any method of maintained vibration (using a pick-up and amplification) or of any method of automatic 'resonant' plucking where the excitation frequency is based on the frequency measured just before. Even when a wire starts with a pure response and a steady decay in amplitude after plucking ends, after straining it can develop a markedly unsteady response. An unsteady decay in amplitude almost always means that the wire has developed more than one natural frequency.

My second point concerns earth pressure cells in relation to arching and stress reduction of cells embedded in sand. When there is inwards movement of the sand across the cell face towards its centre, as during isotropic consolidation, it is virtually impossible for arching to develop. If the incremental strain conditions are changed to zero lateral strain increment in the sand parallel to the cell face, the response can be greatly reduced by arching. When analysing test results on Redshaw cells by Hu (1960), the response (cell action factor) for each unit of normal pressure added under zero lateral strain conditions was found to vary inversely as the square root of the cell pressure reached (under isotropic changes) when these conditions were begun, see pp. 143 to 145 and *Figure 3.7* of Chaplin (1961). The very simple explanation was that the stiffness of the sand arch varied as the square root of the pressure, and then it controlled the bending of the diaphragm on further loading. Clearly this phenomenon depends on the stiffness of the active part of the diaphragm and the cell as a whole.

So many of the curves which have been shown in this Symposium have only given *loading* responses. It would have been most enlightening to see how those cells behaved on unloading, as this is a real test of the system. One can also get anomalous results when a cell is placed in a sand mass with the minor principal stress acting on the cell face. A virtually linear curve at 45° on isotropic loading is no guarantee of good behaviour under field conditions.

REFERENCES

Chaplin, T. K. (1956), 'Conference on the correlation between calculated and observed stresses and displacements in structures.' Final volume, 381–384, *I.C.E.*

Chaplin, T. K. (1961), *An experimental study of the settlement of footings in sand.* Ph.D. Thesis (unpublished), 2 vols., University of Birmingham.

Hu, G. C–Y. (1960), *An experimental investigation of the distribution of pressures in a homogeneous sand section and two-layer systems.* Ph.D. Thesis (unpublished), University of Birmingham.

Ward, W. H. (1955), 'Techniques for field measurement of deformation and earth pressure.' Conference on the correlation between calculated and observed stresses and displacements in structures. Preliminary volume, 28–40, *I.C.E.*

B. Rymsza (Warsaw University of Technology, Poland). The author wishes to make several comments in relation to the application of load cells in measurement of the distribution of earth pressure on retaining structures with particular reference to the variation of this pressure with the type of deformation of the structure, (*Figure 20*).

Proper calibration of the load cells is of prime importance in the study of variation of the earth pressure during a decrease from the 'at rest' condition (E_o) to the active thrust condition (E_a) or during an increase to the passive resistance condition (E_p).

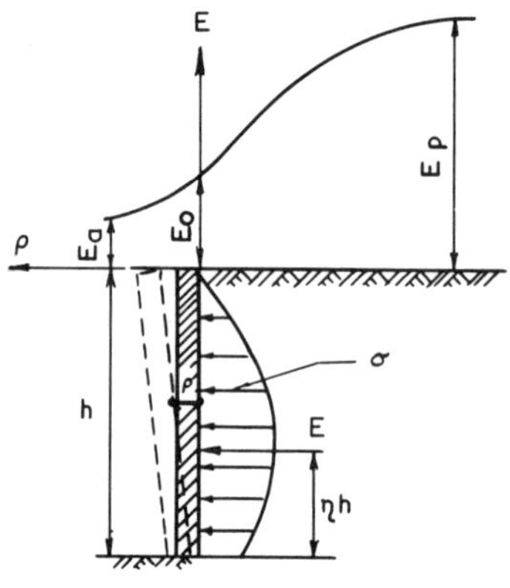

Figure 20

For simple load cells, in which deformations of the diaphragm are significant, different calibration conditions result in different relationship (calibration curves) between the diaphragm deflection f_m and applied pressure σ (*Figure 21(a)*). It must also be borne in mind that even when the hysterisis effects are taken into consideration in calibration of the load cells in soils, i.e. different calibration curves are used for loading (continuous line a in *Figure 21 (a)*) and unloading (lines *b* in *Figure 21 (a)*) – the results will still be subject to certain errors introduced by the 'disturbance' of the stress field in the soil due to the deformation of the diaphragm; the type of disturbance and the size of its zone of influence depending on the degree of divergence between the mechanical proper-

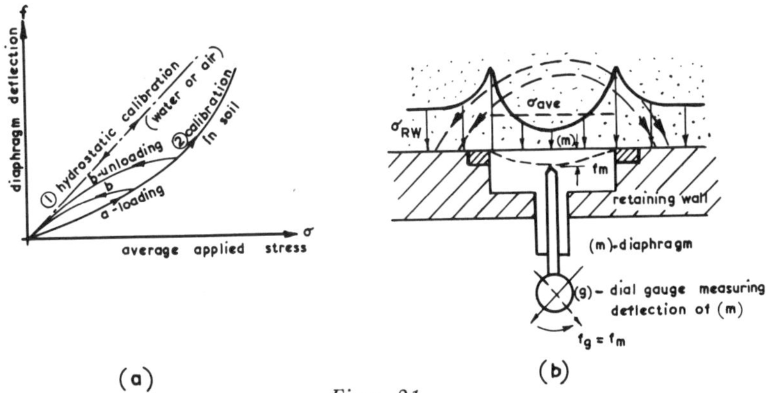

Figure 21

ties of the soil and the load cell, on the mode of deformation of the diaphragm, on the stiffness of the retaining structure and other similar factors. Under such circumstances, schematically illustrated in *Figure 21(b)*, it cannot even be said that the average stress $\sigma_{ave} = F(f_m)$ determined on the basis of the deformation of the diaphragm $f_g = f_m$ according to the calibration curve, corresponds to the pressure σ_{RW} exerted by the soil on the actual face of the retaining structure.

A considerable reduction in the disturbance of the stress field in soil in the immediate vicinity of a load cell can be achieved by filling the cell with oil of an appropriate type and thus decreasing its deformability. With the same accuracy of measurement as in the air filled load cells ($f_g' = f_g$), considerably smaller deformations of the diaphragm are obatined ($f_m' << f_m$). In view of the practical incompressibility of fluids the ratio of the area of the diaphragm subjected to the earth pressure, A_m, to the area of the dial gauge piston, A_g (measuring deformation f_g') becomes now the governing factor (*Figure 22 (a)*). Such a cell satisfies requirements for field instrumentation outlined by the chairman of this session i.e. it is simple in construction and is relatively cheap.

Research is currently being carried out in Poland using oil filled load cells coupled to an automatic balancing system which continuously equalises the internal oil pressure with the external diaphragm loading. To obtain zero deformation condition ($f_m = 0$) under a continuously changing earth pressure (external diaphragm loading) it is necessary to achieve instantaneous balancing of this loading by the internal pressure ($\Delta\sigma = \Delta e$). The problem which has not yet been satisfactorily solved and presents considerable difficulties in the

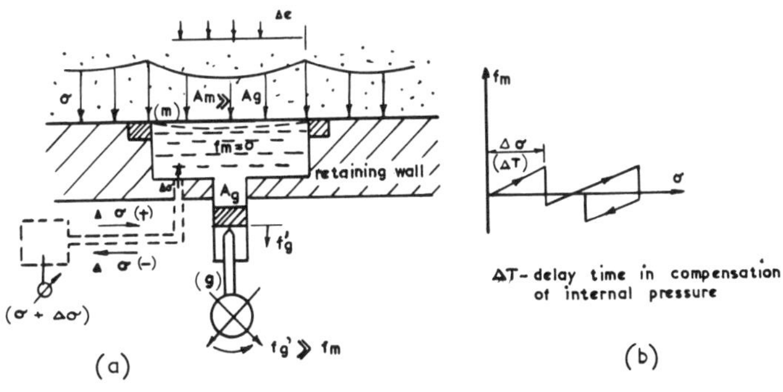

Figure 22

instantaneous balancing of the diaphragm loading: the author has not yet come across a sufficiently sensitive automatic balancing system which completely eliminates diaphragm deformation. As can be seen in *Figure 22(b)* insensitivity of the balancing system introduces oscillation of the diaphragm deflection and hence of the soil pressure in contact with it.

The author would be grateful to hear if anybody from the audience has perhaps solved this problem and would be pleased to discuss the problem and exchange ideas with people interested in this field.

J. Bundred (Greater London Council). I was interested to hear of the need for environmental calibration and for the duplication of instrument readings in order to establish clearly the reliability of performance and the degree of accuracy of readings obtained. May I comment on these and other aspects of field instrumentation as related to GLC experience, bearing in mind the *total context* of the Wandsworth field study – which attitude, as Mr Knight so rightly suggested, should be central to the philosophy of each instrumentation project.

One would prefer to have seen a duplication of the diaphragm wall monitoring, which is still continuing and for which the reliability of instrumentation over a five year period (only the first two years of which have been reported in my paper to this Symposium due to space restrictions) is, I believe, not in doubt. The methods of installation and zero calibration were environmentally opposite, with the environment for the earth pressure cells being firstly that of a bentonite filled trench and secondly that of a stiff clay/concrete interface. Remote placement was designed and carried out successfully for this dual environment, the major source of unassessible local error being believed, if anything, that due to the a typical presence of embedded placement material in the concrete wall and not so much that due to the presence of the cells at the interface, good contact having been assured. The initial differences '*in situ*' between the A and B face responses for each cell (a) in bentonite, (b) during concreting, (c) one month or more later, when compared with the A–B trial jacking (i.e. 'prestress') calibration results, can give, one might suggest, a measure of the accuracy of cell response environmentally.

The anchor load cells were likewise installed under controlled conditions with duplicated independent initial readings being provided during the anchor stressing and locking off. These were, of course, additional to the installers' and GLC instrument readings already mentioned.

Observing Dr S. F. Brown's remarks on cell action factor, scale effect, etc., I consider the BRS boundary cell to be well suited in general to a stiff fissured London Clay and concrete interface regime; I hope that Mr A. Marsland and Mr H. S. H. Thomas would support this view. (In passing, Mr Marsland's assistance in the determination of the fissured clay's parameters must be acknowledged.)

To operate a vibrating wire earth pressure cell reliably long-term, a dry nitrogen atmosphere and continuous feed safeguard against internal corrosion or water ingress is essential, long-term maintenance being a first priority in a hostile environment. The causes apparently of Mr A. L. Little's dissatisfaction with the BRS cells seem to relate to a most unfortunate series of accidents, as the GLC's limited experience suggests that, given guaranteed cells and observer dedication, temporary difficulties are surmountable and the cell's performance can be completely satisfactory; this despite such possible distractions as 'zero error', long term gauge drift, or random variations in cell response.

To Dr G. E. Green's comment, I must state that I always believed that the BRS modified earth cell's A face gives the more uniform and reliable (if marginally lower) pressure reading with less possibility of random scatter than is apparent in B face readings. The environment of either an 'active' or 'passive' stress regime also plays a part in determining the nature of a cell's responses and the relationship between A and B face readings. Referring to Dr J. B. Burland's remark, I feel also that the possible 'prestress' effect should not be over-rated since one needs to distinguish, in any case, between random variations in the actual discrete earth pressures from variations in the response of each cell.

The BRS cell is partly self-checking by virtue of its dual face readings, which also tend to provide a useful 'early warning system' for an a typical local stress condition. On *Figure 3*, of my paper the pressure cell No 3 record between November 1969 and August 1970 provides a clear, if somewhat dramatic, response to particular local conditions of a very deep hole in the adjacent excavated underpass and its performance accords as expected with a known and measurable external environment during this period and subsequently.

In the same way that no two prestressed anchors give the same initial load readings, why should one expect a faulted, stiff fissured clay (also highly over consolidated, slightly silty with extensive claystone bands and some slickensides) to produce comparable measured local earth pressures on opposite sides of a bentonite filled trench at nearly the same depth? A vital factor in demonstrating consistency and reliability of cell performance is that individual cells show a similar pattern of stress changes even though the range of readings may be large.

One should be realistic and consider always whether the results obtainable are consistent with the objectives and limitations of an investigation, taking a global view; e.g. the analysis however sophisticated can only be as good as the data obtained. In the Wandsworth case, despite reservations, the results obtained were mainly adequate for the limited objectives in the total context considered. They provide valuable data and offer useful guidelines for the future.

There are particular external indicators of the efficacy of an instrumentation project; indeed unplanned, involuntary checks on reliability, ruggedness, performance, accuracy, are often forced on one by human and other agencies. The following suggestions are given by way of illustration:

When piezometers and inclinometers are well established, they can conveniently be torn out, dug up, or just fractured. If earth cells function reasonably, they can either be drowned in a flood or contrariwise exposed to sunlight, which at least has the advantage of making them accessible for inspection after some use. Anchor load cells may be grouted up solid; failing this, an oxy-acetylene torch can act quicker than normal long-term corrosion to damage instrument leads. Such occurrences, provided as easily assessible localised effects, may assist one to check the validity of prior results obtained at well defined discrete points under known changes of external conditions, as indicated previously for earth pressure cell No 3. Reinstallation stimulates recalibration, recorrelation, etc., and even a useful duplication of some instrumentation.

Again, taking a global view, there are overall checks which can be applied to assess the validity of instrumentation results. At Wandsworth the resolved horizontal resultant of the measured forces on wall panel 1, ϵH, measured $+ve$ (positive) acting westwards into the underpass, was utilised for this purpose and also proved the value of construction monitoring. For example, ϵH changed from a $-ve$ to a $+ve$ quantity during the final underpass excavation in 1969 until maximum excavation depth was achieved, when a near zero 'equilibrium' balance of +14 kN/m was recorded.

For the Wandsworth wall, the design method rather than strict theory was qualitatively and part quantitatively checked, with adequate success. Thus, for retaining walls, relatively simple field measurements may allow the designer to gain an indication and assurance of adequate structural performance and may also allow him to deduce the mechanism of wall behaviour, thus crystallising his intuitive design experience.

As costs are of interest, estimates of the magnitude of significant items are given in the Table.

In conclusion, the author's experience is that the arch enemies of a field instrumentation project are dampness leading to corrosion, accidental damage or vandalism and lack of adequately skilled operational and truly committed staff. There is little doubt that the most successful instruments are simple and robust.

G. L. C. WANDSWORTH BRIDGE SOUTHERN APPROACH UNDERPASS – INSTRUMENTATION COSTS FOR DIAPHRAGM WALL PANEL 1, ETC.

The following instruments were installed/used at the costs indicated

(a)	7 earth pressure cells	£2 450
(b)	6 anchor load cells	940
(c)	4 slope indicators	325
(d)	2 Piezometers	232
(e)	Control Manhole including terminals, nitrogen cylinder and feeds through valve systems	328
(f)	RSJ for earth cells	357
(g)	Borehole in front of wall including continuous sampling	505
(h)	Kier's costs in connection with *in situ* shear box tests	106
	Total to December 1968	£5 243
	Add running costs at £200 per annum for four years to December 1972	800
	PT/C Anchored Diaphragm Expt Field Study *Total*	£6 050 say

B. O. Skipp (Soil Mechanics Ltd.). Some of the quality control problems associated with the manufacture in small numbers of rather specialised instrumentation have been mentioned particularly by Mr Thomas. It has been our experience that caution must be experienced in accepting instruments even from large Companies. There is always the need to carry out proper proving tests in the laboratory and the field prior to installation and commissioning of an instrumentation system. Such calibration/proving tests are often not undertaken in a suitable form by either manufacturers or the providers of the devices. It is especially important that the pre-site verification be carried out not only upon a single device as a 'bench test' calibration but on at least a representative portion of the complete system. This may mean calibrating load cells with the appropriate end plates suited to their field application. Tilt meters for example, often require a proper evaluation of datum drift and temperature response, when linked up to their associated electronics. It has been our experience that to undertake this type of pre-installation study proper time and money must be allowed and the appropriate laboratory facilities provided. These may well be available within Universities and Government Research Establishments but with the increasing involvement with specialised sub-contractors in this field, one should expect them also to have some such facilities.

A. Marsland (Building Research Station). I would like to comment on some of the questions raised by Dr Green in his general report. My suggestion for the need to pay more attention to the detailed design and installation of instruments to suit particular problems and soil conditions does not imply that I consider it necessary to design special instruments for every project. It is, however, essential that the design of instruments and their method of installation should fulfil the requirements for each case including compatibility with the ground and the appropriate degree of accuracy. If the available instruments do not fulfil these necessary requirements then, as I see it, we have no alternative but to design new instruments. I will illustrate this need by reference to two recent examples from my own experience.

Twin-tube hydraulic piezometers are unreliable in situations where large suctions develop in the connecting tubes and in this situation pneumatic and electrical piezometers are usually more suitable. Unfortunately standard pneumatic and electrical piezometers which cannot be de-aired often give readings appreciably higher than the pore water pressure when gas is present in the ground. We have recently had a case in organic soil where a pneumatic piezometer gave an excess pore pressure equivalent to 7 m head of water while adjacent hydraulic piezometers only recorded an excess pore water pressure of 1.25 m of water. In our present project on flood defences we need to measure pore water pressures in organic soils situated on the seaward side of the flood banks (*Figure 1* in my paper). To permit adequate de-airing it is necessary to fit additional twin pipes into the porous cavity of the electrical or pneumatic piezometers which are brought to ground level and terminated with valves so that de-airing can be carried out at low tides. It is necessary to keep the volume change of these tubes to an absolute minimum and heavy gauge metal tubes are required with this arrangement. The design for a 'combined' piezometer described in my paper in which air-operated valves are fitted adjacent to the piezometer permits the use of flexible tubing between the piezometer and the gauge house. I would, however, like to emphasise that more complicated instruments of this type should only be used where simpler instruments are unsuitable or as a check for research purposes.

Problems with multi-point settlement devices in soft alluvial soils also necessitated the design of a more appropriate system. The main requirements were for a more positive fixing of the measuring points in the ground and provision of a means of isolating the rigid parts of the instrument from the surrounding ground. The system was based on the multi-point system using magnets described by Burland and Moore which was originally designed for use in stiff clays. The rigid access tube was isolated from the ground by surrounding it by a thin flexible p.v.c. tube reinforced with helically wound wire. The addition of this flexible outer tube necessitated the development of a complete new system of installation. The final arrangement is shown diagrammatically in *Figure 23*. An essential feature is a small explosive device to cut the monofilament nylon cord or wire which holds back the stiff springs during installation. The use of this simple arrangement has permitted the use of much stronger springs which provide a much more positive fixing of the magnet holder into the borehole. The complete unit of several measuring points is assembled as the anchor weight is gradually lowered into the borehole. When the anchor weight is firmly bedded at the bottom of the borehole the lowest magnet holder is finally positioned by applying a slight tension to the flexible tube and the springs of this lower unit are then released by the explosive cutter. This procedure is repeated for each magnet holder in turn working upwards from the bottom of the borehole. Trial installations have been made in a 10 m deep borehole in soft alluvial soil and in a 50 m deep borehole in back-filled rock waste. Some further development is required but these initial trials are very encouraging. Installation is reasonably quick as exemplified by the fact that the 10 magnets were installed in the 50 m deep borehole in under 3 hours. A note giving a full description of the system together with suggestions for methods of installation is being prepared for publication in 'Geotechnique'.

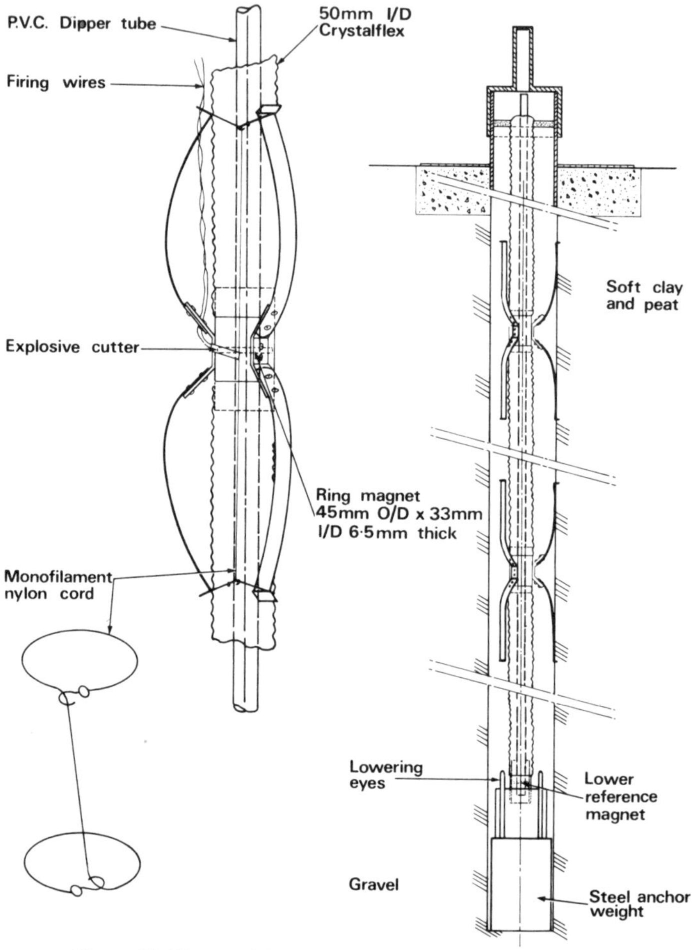

Figure 23 New multipoint magnetic settlement system

The inverted pendulum described in my paper has been installed in a 200 mm diameter borehole having a depth of 18 m. When the pendulum was pushed off centre by 20 mm it returned to its original position to within ±0.1 mm in 6 minutes. This accuracy is appreciably better than was anticipated and our experience to date shows that inverted pendulums should be considered as a viable means of providing stable position datums in areas some distance from stable ground.

K. W. Cole (Ove Arup & Partners). In its 5½ miles (9 km) length, the Gateshead Western By-Pass in Co. Durham traverses a variety of subsurface conditions, ranging from soft peaty clays and deep deposits of post-glacial laminated clays, to shallow, unrecorded mineworkings. The total length of embankment is 14 000 ft (4300 m) and of this 35 % is raised 20 ft (6 m) or more above original ground level. No single, isolated length of embankment greater than 20 ft (6 m) in height and traversing a consistent soil succession exceeds 1500 ft (450 m) in length. In all, there are nine such isolated lengths of embankment and the underlying soil conditions range between the extremes shown in *Figure 24*.

A comprehensive site investigation was conducted under the Engineer's direct supervision. The strength and settlement (compression and drainage) parameters applicable to the soils beneath each of the embankment sections were measured, generally by standard laboratory tests, but 10 in (250 mm) diameter piston samples of the laminated clays at Location Type 1 were taken and laboratory tests made to measure horizontal permeability.

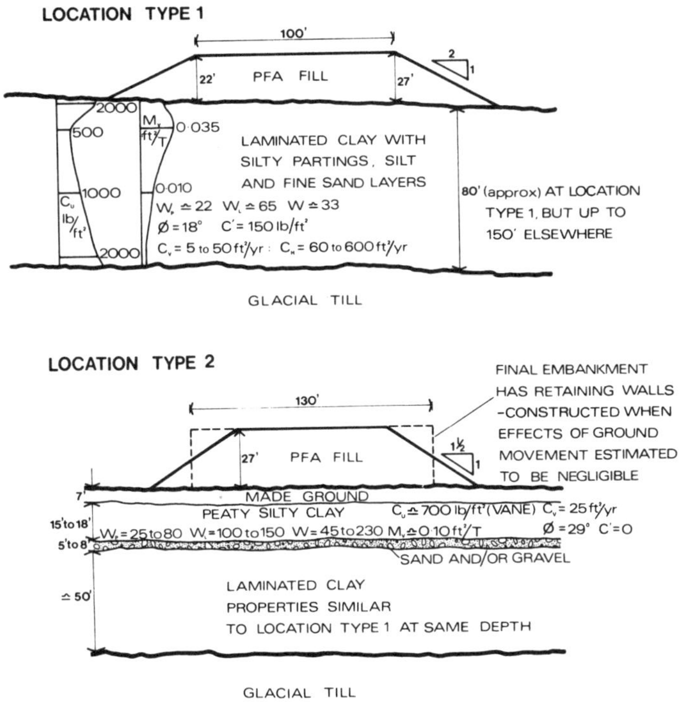

Figure 24 Typical instrumented sections

For various reasons, including cost, trial embankments (there would have had to be at least four to cover the range of soil successions) were not practical and a 'pre-contract' to construct the critical embankments was not feasible. Thus, the engineer had to make the forecasts described in the contributor's discussion for Session 3 (p. 623) on the basis of his analysis of the measured soil strength and drainage characteristics. In order to confirm as far as practical that the Engineer's forecasts were not widely in error, and to give opportunity to amend the rate of raising the fill, if this seemed desirable, each isolated length of embankment was instrumented, generally at one section and on one side only, the longest length being instrumented at two sections some 500 ft (150 m) apart.

The desirable qualities of the instrumentation were listed as follows. (i) Reliability (stability, repeatability); (ii) Ease of handling in adverse conditions; (iii) Ease of reading; (iv) Rapidity in response; (v) Security from damage (both by the Contractor and by vandals). In addition, the data obtained should be readily interpretable by site staff.

The third paragraph of Dr Vaughan's Introductory Notes to Session 3 adequately summarises other considerations uppermost in our minds at that time. Our guiding principles were thus reliability, simplicity and security.

The basic layout for all the instrumented sections was as shown in *Figure 25*. It was hoped that, by providing precast concrete manholes with standard manhole covers as the instrument 'houses', the security aspect could be achieved. The instruments shown in *Figure 25* were supplemented by:

(1) Shallow settlement stations at and near the toes of the batters – to give back-up data should the inclinometers fail;
(2) Rod settlement gauges in the central part of the embankments at two sections to give back-up recording of the absolute settlement, as this was considered to be vital at those sections. As was foreseen, these would suffer from the vagaries of site conditions (*Figure 26*).

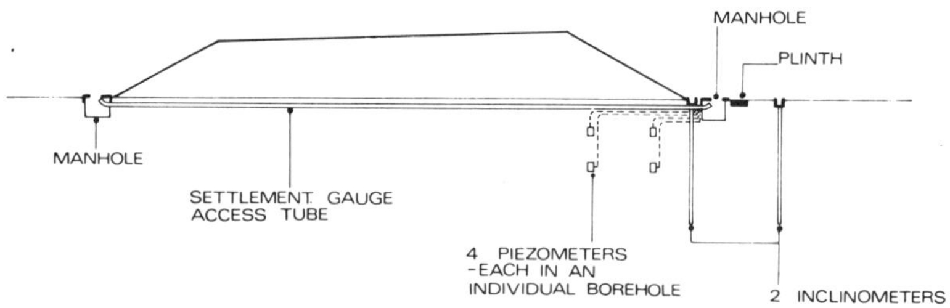

Figure 25 Typical instrumented layout. Example. Location Type 1

At an early stage, discussions were held with leading instrument manufacturers to see if their ranges included, or could be extended to introduce, the desired instruments. As a result of these discussions, the following specifications were evolved.

(a) Piezometer. Electrical or pneumatic (dry inert gas) type, capable of recording pore pressure to within ±0.20 ft (±60 mm) of the true head of water within a small zone of saturated soil. Each piezometer to be installed in a separate borehole and to have a rapid response time and low volume factor. Leads from piezometers to terminate in manholes with easy to make and break connections for a portable read-out unit (see *Figure 27*).

Above ground gauge houses and mercury pressure systems were considered to be far too vulnerable to be acceptable.

Figure 26

(b) Inclinometer. An instrument capable of measuring and recording its own inclination with respect to the local gravitational field to an accuracy of ±2 minutes up to 5° and ±10 minutes up to 20° while being lowered and raised inside a casing having guides to fix its azimuth. Automatic computation of displacement was available in one model on the market and this added facility was judged to be well worthwhile in obviating a tedious step in calculating

Figure 27

Figure 28

displacements from inclinations (the inclinations are nevertheless recorded as a precaution against error in the computer). The casing must be capable of absorbing quite large vertical ground strains without affecting its ability to follow horizontal strains and allow the passage of the inclinometer. Because of the delicacy of the read-out instrument, a portable shelter was included in the equipment specified (see *Figure 28*).

(*c*) *Settlement gauge.* An instrument to measure vertical movements and capable of producing a continuous profile to an accuracy of ±0.10 ft (±30 mm) across the width of the base of an embankment was required. This definition required a tube type gauge installed in a trench prior to filling, the tube terminating in the same manholes as the piezometer leads (see *Figure 29*). A probe at the end of a protective sleeve enclosing the fluid and air venting leads is

Figure 29

pushed or dragged by a cord (see *Figures 30* and *31*) through the tube to various distances or 'stations' from the manhole and the difference in head between the probe and the read-out unit standing on a level base alongside the manhole is recorded. The elevation of the base is then checked periodically by optical levelling to enable the absolute levels at the stations be be calculated.

Figure 30

Soil Instruments Ltd. were successful in competitive tender for the Nominated Sub-Contract with their offer to supply and install pneumatic type piezometers (using nitrogen gas instead of air), inclinometer equipment (with digital read-out) and settlement gauges of the access tube and probe type.

Figure 31

The piezometers were not permitted to be installed until they had first been demonstrated to record the pressure differential of a known head of water (by being first saturated and then immersed in a special section of inclinometer tube put down outside the R.E.'s office) and by giving a sensibly stable recording of that pressure over a period of 24 hours. They were then kept immersed and installed in water-filled borings by, if practical, pushing the tip 18 in (0.45 m) into the ground below the bottom of the boring, or by surrounding the tip with a sand zone sealed above by cement/bentonite grout to ground level. Each piezometer was installed in a separate borehole and the heads laid at the base of a trench at least 18 in (0.45 m) deep leading to a terminal unit sealed into the wall of the manhole (see *Figure 27*). Before final acceptance, the piezometers were checked to see that they gave readings close to those of nearby standpipe piezometers installed during the site investigation.

Of the twenty-eight piezometers installed, all but two are apparently still functioning correctly. Several have given records of excess head reaching up to 30 ft (9 m) of water during filling and returned to the initial readings consistent with the completion of consolidation settlement, as also indicated by other measurements. Others still record decaying excess heads consistent with continuing settlement. The two which have failed did so after recording for some 200 days and failure was almost certainly caused by repeated flooding of the manhole permitting silty water to enter the gas lines.

The inclinometer tubes are of the anodised aluminium type. One installation suffered almost immediate failure, due to corrosion of the tubing at the coupling, and subsequently two tubes have become obstructed. The failures are believed to be due to alkaline attack from the cement/bentonite grout used. Clearly there is a call for development of a very flexible continuous plastic tube of the early BRS type, but the problem of twist must be overcome.

The tube type horizontal settlement gauge is generally satisfactory in operation. Two defects are that:

(1) The flexible access tube is too small in diameter and slight compression under the load of the fill makes it hard to pull the probe through;

(2) The instrument works on negative head (suction), which is fine when the instrument is freshly de-aired, but leakages inevitably occur and the instrument rapidly loses accuracy. The suggestions made by Dr Green should be seriously researched; the contributor feels the most likely answer lies with either a pressure transducer or a pneumatic piezometer at the probe.

The by-pass was instrumented with 28 piezometers, 9 settlement gauges and 15 inclinometer tubes for a sum of about £12 000, or 0.2 % of the main contract sum, including approximately £1500 for read-out instruments and operation manuals. The amount cannot be precisely stated, as the main contractor did not price the individual items for attendance on the instrumentation subcontractor, the provision and installation of some twelve manholes and various rod settlement gauges and shallow settlement stations. The cost of providing the full-time services of an instrumentation engineer with his own transport is also not included.

Examples of the data obtained from the site measurements and the manner in which these were used to monitor the stability of the earthworks are given in the contributor's discussion in Session 3.

The contributor acknowledges the assistance of his colleagues, past and present, who assisted in developing the specifications and supervising the work. Drs L. Barden and W. B. Wilkinson helped to guide the work on the measurement of permeability.

B. J. Eason (Building Research Station). I would like to discuss the effect of temperature change on load, pressure and displacement systems employed in field investigations.

The first of these is the multi-point settlement measurement system described in the paper by Marsland and Eason. The apparatus, normally located within the hollow loading plate, is shown in *Figure 32* mounted on a calibration rig. This reproduces the settlement movement of all four points at once. *Figure 33* shows the effect of temperature on the calibration of this arrangement when allowed to come to equilibrium in a constant temperature room set to −6°C, +5°C and +20°C. The electrical read-out unit was outside the controlled room and maintained at 18°C. The temperature sensitivity of the DC-DC LVDT

Figure 32 Multipoint settlement measuring system mounted on the calibration rig

transducers is less than 0.025 mm per °C. To reduce error, temperatures will be measured during subsequent experiments, although the temperature of the ground in a deep borehole is around 10°C and equipment probably attains equilibrium quickly, before commencement of a loading cycle.

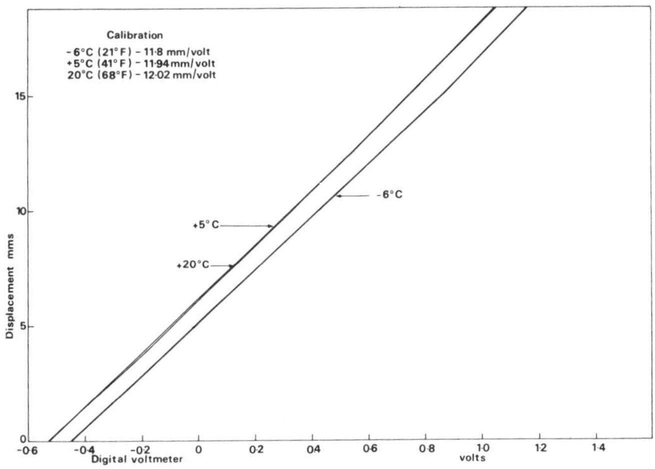

Figure 33 Effect of temperature on calibration of multipoint measurement system

The second measurement device is an oil-filled load cell of 250 kN capacity, fitted with a fully temperature compensated double strain gauge bridge. Thus data logger readings and manual readings can be taken together during a test.

The load cell and associated cable 20 m long were placed in a cubicle, temperature controlled from +30°C down to −15°C, and readings of the zero taken on a micro strain indicator. If it is assumed that normal calibration would have been carried out at 20°C in a laboratory, then *Figure 34* shows how the change in micro-strain reading with temperature at zero load, would have appeared as an apparent increase or decrease in load during a test.

The hysteresis may be due to the difficulty of maintaining the correct temperature of all the components at the same time in a small cubicle. The error may seem to be small, but on occasions high capacity load cells may be used to accurately measure small load changes where the final load is expected to be high. The presence of oil in the cell may have produced some end loading as the oil contracted giving an apparent increase in measured load. The manufacturers have no means of checking their load cell behaviour at low temperatures. How well then do our load cells behave in practice in varying temperature conditions?

The last device is the pressure transducer consisting of a diaphragm connected to a force rod and spring cross over and fitted with unbonded strain guages. The transducers show a temperature effect, not in this case due to outside temperature changes, as the transducers should be effectively lagged,

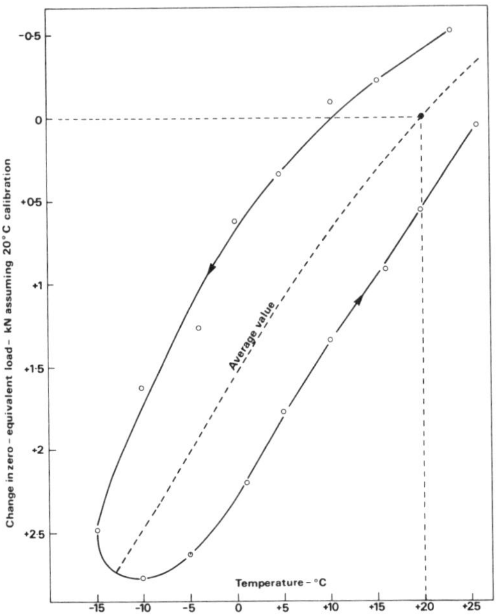

Figure 34 Variation of output with temperature at zero load for an oil-filled 250 kN capacity load cell fitted with bonded strain gauges complete with 20 m of cable and fully compensated

but due to the excitation current warming the gauges themselves. *Figure 35* shows the effect of duration of excitation on the proportion of final output under continual excitation. It indicates that at switch on, the transducer is giving a reading 2.5 % less than the true value and that almost 3 minutes are required to obtain the full response. For the bonded gauge pressure transducer this figure may be of the order of 5 % and rather a longer time is required to reach equilibrium.

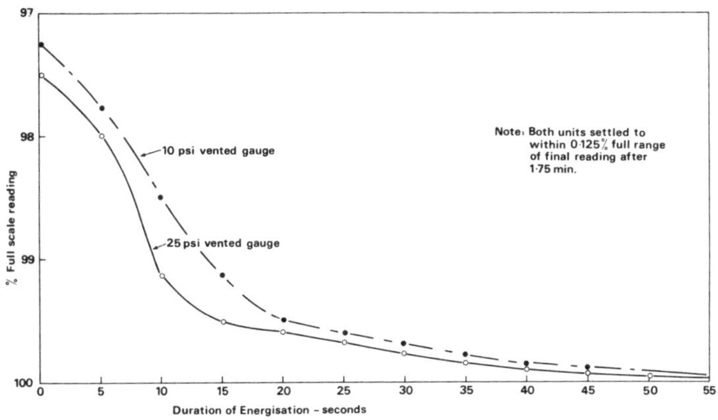

Figure 35 Change of residual output with duration of energisation of pressure transducer fitted with unbonded strain gauges

We have found, however, that readings are consistent if the unbonded pressure transducer is excited for an exact time, e.g. 3 sec, this being the excitation period for a battery operated magnetic tape data logger. Calibration should, of course, be done under the same conditions. This performance probably reflects the high standard of manufacture of these items, and the fact that all the electrical components contain a high proportion of platinum which has a linear temperature resistance relationship.

The question therefore arises, how does our expensive instrumentation perform during extreme conditions of cold or heat, and should we measure temperature more often? Alternatively should we design both instruments and read-out units to be better compensated for temperature changes?

J. A. Hooper (Ove Arup & Partners). For some applications, photoelastic load cells are a suitable form of field instrumentation. Load cells of this type were installed in a piled-raft foundation in 1967, and have functioned successfully during the subsequent years. Details of this instrumentation have been recently published by me: (Hooper, J. A. (1973) 'Field instrumentation for long-term measurement of pile load and raft contact pressure'. *Civ. Engng. Pub. Wks Rev.,* **68**, pp. 438–446, May).

T. Ramamurthy (Indian Institute of Technology, Delhi). There is a general trend to use more and more sophisticated field instruments for better accuracy. Accuracy is certainly costly. Experience indicates that embedded electrical field instruments do not function effectively in the long run. Mechanically operated instruments are much more reliable. When the instruments are embedded, some of them fail to function in spite of many precautions taken and some give very erratic results. On the basis of the results of a few instruments we make our interpretations and predictions, sometimes very valuable information is lost from a region in which we are most interested. There seems to be only a few people who know the art of placing instruments properly so that they function as desired and make proper interpretations. Therefore, it becomes necessary to develop simple and rugged instruments which could function under adverse conditions.

The cost of instrumentation in a particular job varies from 1 % to 4 % of the cost of the project. Even after some failures of instruments, to get enough data for realistic interpretation, the instruments must not only possess resolution, reliability, ruggedness and be simple and robust but also cheap and easy to handle and enable quick reading, even if it means sacrifice of some accuracy.

The idea of adopting two or three types of instruments for the same purpose for the sake of verification of measurements is certainly desirable, but this is to be done within the funds allocated for field instrumentation.

Sessions 3 and 4. The applications of instrumentation in geotechnical engineering and the interpretation of their results

Sessions 3 and 4 of the Symposium were devoted to discussion of these topics. They cannot be separated exactly and the two general reports which follow involve considerable common ground. They incorporate most of the material in the preliminary notes circulated to delegates as an aid to discussion, revised after consideration of the discussions. The two reports are followed by some joint concluding remarks by the two general reporters.

Session 3: The applications of instrumentation

GENERAL REPORTER: Dr. P. R. Vaughan
(*Imperial College of Science and Technology*)

Introduction

In geotechnical engineering, the determination of soil and rock properties by laboratory and *in situ* tests, the prediction of the behaviour of masses of soil and rock under engineering loading by analysis and the observation of this behaviour by instrumentation are essentially complementary techniques. While the need for field observations as a means of checking and improving engineering predictions has always been appreciated, the last decade has seen a major increase in the use of instrumentation, as described by Dr Cooling in his closing address. The principles controlling successful measurements of stress and displacement have been studied, and much equipment has been designed and developed. Many of the contributions to this Symposium have discussed these aspects of the subject. The broader aspects of the use of instrumentation have also been considered, in particular in Session 3 and Session 4. The purpose of this report is to review the use of instrumentation in Geotechnical Engineering.

Organisation of Successful Instrumentation

The successful use of field instrumentation involves many factors in addition to an understanding of the principles of measurement involved and the use of properly designed equipment. Some of these are:

(1) The purpose of the observations must be defined and the methods by which they will be interpreted considered.
(2) The methods of observation and the location of the equipment must be decided.
(3) Equipment must be properly constructed, tested, installed, operated and maintained.
(4) The results must be processed, plotted, interpreted and made use of.

The purpose of observations, the methods of observation, the location of equipment and the interpretation of results are interrelated. The crucial uncertainty is the basic mechanism of behaviour. Where this can be presumed the other factors follow. Where the mechanism is uncertain, the planning of instrumentation to determine it is difficult and requires careful judgement. Even when the mechanism is assumed from previous experience, it is wise to try and provide sufficient instrumentation to check that this mechanism develops, and to consciously and specifically check that this happens.

The provision, testing, installation, operation and maintenance of equipment and the processing, plotting and interpretation of results involve major organisation requirements. These organisational and contractual aspects of instrumentation are the subject of papers by Kennard and by Wrightman and are discussed by Cole, Gray, Kennard, Knight, McKenna, Patchet, Sherwood, Snelgrove and Vaughan *et al.* A range of views are presented. A number of points can be made in summary.

The overall effort required to achieve success should not be underestimated and it should be planned for. The staff and engineering facilities required must be provided. Staff must have the necessary experience or training. Their enthusiasm and commitment to success are important and they should be informed about the purpose of the work they are required to do and the intended function of the equipment which they are to install and use. George and Parry, McKenna and Roy, and Wrightman, in their papers, and Vaughan *et al*, in discussion, give information on staffing requirements. The speed with which information obtained must be processed, interpreted and acted upon is important, as discussed in the paper by Kennard.

Great caution should be exercised if competitive tenders and fixed rates are used for instrumentation work. Any tendency to reduce quality of workmanship in order to work within a low cost may result in failure of the installation and a complete waste of money. It is dangerous, and usually uneconomic, to assume that duplication of equipment will allow lower standards to suffice. As stressed by Dr Cooling in his closing remarks, unnecessary sophistication reduces the likelihood of success. Sophistication may sometimes be necessary to achieve the required result but extensive development work is desirable. Sophisticated equipment may also be of value in reducing operating time but only if there are adequate guards against equipment failure. A few, well-chosen, simple and reliable observations are more valuable than the spasmodic and dubious results of over-ambitious and ill-considered instrumentation schemes.

Where instrumentation is incorporated in engineering construction, the compatibility of the instrumentation proposals with the construction should be considered carefully. These problems are considered subsequently. The minimisation of interference with construction should be a major factor considered when instrumentation is planned and equipment designed.

There has been some discussion at the Symposium of the best contractual arrangements for instrumentation. Whatever system is used the necessary professional expertise must be provided and, therefore, paid for. It may only be effective if it is integrated into a continuous process of design, planning, installation, operation and interpretation, with responsibility for each stage clearly assigned and with those responsible for each stage fully informed about the complete process and with a strong interest in the success of the whole operation.

Typical Uses of Instrumentation

The circumstances in which instrumentation may be applied vary widely. The problems which arise and the benefits which may be derived tend to vary according to the stage in the construction process in which the results are used, and it is convenient to summarise these stages.

Observations during Site Investigation

Observations of ground water pressures by piezometers form an essential part of site investigation, and instrumentation techniques may be involved in *in situ* tests to determine soil and rock properties. Numerous examples have been given to this Symposium. With the growing use of deformation analyses, the determination of *in situ* total stress is of increasing importance. Such observations in rock are discussed in the paper by Sheddon and in soils in the discussion by Hughes and Massarsch.

Preconstruction Trials and Research Construction

When circumstances and the programme for final construction allow its use, a preconstruction trial is a powerful tool, provided that the scale of the trial is sufficient to reproduce the stresses operating in the prototype. Such trials may be intended to investigate foundation and fill properties, structure-soil interaction, the influence of construction techniques and construction problems. The type of trial may vary from simple loading tests to full-scale prototype construction and the function may vary from basic research to direct testing of specific structures. Preconstruction trials offer a major advantage in that construction can be subordinated to the needs of instrumentation and, often, specialist staff can be provided who are not subject to distraction by other construction work. Parry & George point out that trials provide an opportunity for testing instrumentation intended for use in the eventual structure.

The use of preconstruction trial embankments is growing. Embankments may be used for loading tests for foundation problems, as described in the papers by George & Parry, and Wong. More usually they are used to test the performance of permanent embankments on weak foundations. The paper by Bishop & Green gives a comprehensive review of the use of trial embankments and the benefits which may be obtained. The papers by Marsland, McKenna and Wilkes also refer to trial embankments and they are discussed by Johnston, Marsland, Massarsch, Pawsey, Sanders & Fookes, and Vaughan *et al.* Some general points may be made. Such trials are in no way a substitute for site investigation, as the area of foundation which may be tested must be small in relation to the final structure. The cost of such trials can be reduced where they can be used as part of temporary or permanent works. While the behaviour under load of large volumes of *in situ*, undisturbed material can be observed, the extent to which soil parameters can be derived depends on the availability and suitability of methods of analysis. This places a constraint on the use of trial data in different circumstances and with different methods of analysis. Trial construction is particularly valuable if construction is taken to or near to failure. True safety margins and failure mechanisms can then be examined.

The use of trial construction in foundation engineering is extensive. Surface loading tests may be performed, as described in the papers by George & Parry, and Wong, and sub-surface loading tests, as described by Marsland & Easton. Individual foundation elements, such as piles, may be tested at full or reduced scale, at working loads or failure loads, with varying amounts of instrumentation. In the simplest tests only load-settlement observations are made. In more sophisticated tests boundary stresses may be observed or deduced, structural strains observed and information obtained about the behaviour of the soil near the foundation element. Examples of tests of this type are given in papers by Cooke & Price and Weekes & Dixon, and in discussion by Jamiolkowski, Morris & Roy and Price. The amount and type of information obtained controls the extent to which mechanisms of behaviour can be checked and methods of analysis checked and improved.

Trial retaining structures (see, for instance Uff, 1969) are rare and no examples have been reported to the Symposium. No examples of trial excavations are reported, either, although there have been recent examples (Chinsman, 1972; Kwan, 1971). Opportunities for observing excavated slopes are quite numerous, as advantage may be taken of quarries and clay pits (as mentioned by Burland & Moore), and temporary slopes may be formed within large excavations, within borrow pits and the like.

Basic soil behaviour may be deduced from the behaviour of such slopes more readily than from the behaviour of supported excavations, where the interaction of soil and structure is more complex.

An interesting example of a loading test on an experimental pressure cavern in rock is given by Thompson. While soil and rock problems are similar in principle, the difference in approach in rock, dictated by the much higher modulus and the extent to which it is self-supporting, is well illustrated by this example. The examples in the Symposium drawn from rock mechanics have greatly increased its value.

Observations of Main Works during Construction and Operation

Most instrumentation is used for these purposes and is generally subject to the constraints of installation and operation during normal construction site conditions discussed elsewhere. Such instrumentation may have three general functions. These are:

(1) Specifically to allow the design to be modified during construction, or to allow the rate of construction or loading to be controlled to avoid failure or to restrict post-construction settlements. The use of instrumentation for this function has been discussed extensively at this Symposium in the papers by Cooke & Ingold, George & Parry, McKenna, McKenna & Roy, Marsland and Wilkes and, in discussion, by Cole, Hutchinson, Johnston, Jones, Marsland, McKenna, Parry, Snelgrove, Symonds, Vaughan *et al* and Wilkes. Embankment construction and

the controlled filling of oil storage tanks are the subjects of these presentations. Whitaker and Patchet discuss comparable observations in rock excavations.

Problems are: the engineering effort required in successfully obtaining and processing the data; the short time in which data must be acted upon; the integration of the instrumentation and construction programmes; the uncertainties introduced into the construction contract; and the interpretation of the results.

(2) Instrumentation may be used to confirm design assumptions, to give general confirmation that the structure is performing satisfactorily, with the implicit option that modifications may be made if this is found not to be so.

(3) Instrumentation may be used to obtain data of a research nature for use in future construction of a similar type. The research function of such observations is restricted by the fact that they can only be made at working stresses. In trial construction the behaviour near and at failure can be observed, much enhancing the value of the information.

Observations fulfulling these last two purposes have been made extensively in embankment dams. Such observations are described and discussed in papers by Little, MacKellar *et al* and Penman & Charles. Observations of loads and deflections of excavations of temporary and permanent support systems for excavations, and of fill retaining structures are also relatively common. Such observations are reported in papers by Bundred, Burland & Moore, Jones and Phillips & James, and by Bundred and Jones in discussion. In the discussions, Jones comments on the value of such observations in relatively small scale structures and their possible use in improving design methods.

In contrast to most other types of structure, the instrumentation of working building foundations has been very restricted, apart from simple settlement observations. Cooke and Burland comment on this in the disscussion. The Symposium has been notable in that two examples of foundation instrumentation are reported; in the paper by Haws *et al* and by Hight in discussion. As a result of this paucity of instrumentation, there has been negligible verification of design predictions or improvement of design methods by the direct observation of full-scale behaviour. A number of factors may have contributed to this.

Individual foundation contracts are small and for many clients instrumentation is expensive and cannot readily be supported within a small contract. Also, delays to foundation construction from installing instruments could be very expensive. Since design is generally to limit settlements, the worst that is likely to happen due to design errors is excessive settlement, which may cause superficial damage only. Foundation engineering has developed successfully on simple, often empirical, bases which may not have produced an attitude sympathetic to instrumentation. For instance, piling is often provided by specialist contractors on the basis of simple performance tests on single test piles. The designer has little incentive to think about overall foundation behaviour. Structural engineering, with which foundation construction is associated, has no history of full-scale observation, and relies heavily on codification and design rules, usually based on theory, experience and model tests.

It is clear that even small improvements in design methods, which might well result from the instrumentation of full-scale foundations, could lead to better foundations and a substantial reduction in overall expenditure. The contributions to this Symposium suggest that this need is recognised and it is noteworthy that the two examples quoted are supported by a research organisation, thus meeting some of the difficulties outlined previously. It is to be hoped that the problems which may well arise in this new and difficult area do not deter these praiseworthy efforts. It may be that simple, axi-symmetric foundations should be instrumented first, as both the cost of instrumentation and the difficulties of interpretation might be less.

Special Problems of Long-Term Observations

Post-construction observations may be intended to check design assumptions or make research observations during initial operation, after which they may be discontinued. They may be intended also for permanent monitoring, to give advance warning of malfunction. Special problems associated with post-construction and long-term observations are the durability of equipment and the provision of staff to observe and maintain it and to interpret the results. These problems were discussed by Knight. Careful thought should be given to the selection and interpretation of observations suited to long-term monitoring. Some associated problems are discussed in the papers by Nixon *et al* and Sherwood & Curry, and in discussion by Skipp. A distinction should be made between observations intended to show that some form of malfunction is occurring and observations to determine the type of malfunction. Any form of instability is likely to be preceded by increasing strains and displacements and simple superficial deformation measurements using survey techniques of the type discussed by Ashkenazi and Cheney may be most effective for monitoring of the first type.

Instrumentation for Problem Diagnosis and to monitor External Effects

Instrumentation may be installed after a problem has occurred, to examine its nature and to aid the design of remedial works. Cornforth *et al* describe the instrumentation of a slide in sand, induced after similar slides had become a construction problem, the aim being to determine the mechanism of failure.

Lovenbury describes the use of piezometers to determine the position of a leak in a dam core, partly as part of a monitoring system and partly used for diagnostic purposes.

Instrumentation may be of particular value in monitoring the effects of external agencies such as adjacent construction, de-watering, mining and sources of ground vibration. Burland & Moore discuss observations of deformations adjacent to excavations, which may affect nearby structures; Whitaker and Forrester describe the observation of ground deformation due to mining, and Hutchinson discusses an interesting consequence of such movements; Skipp and Marriott describe methods of measuring ground vibrations.

Advantages and Disadvantages of Instrumentation

The general advantages which have been derived from instrumentation in geotechnical engineering can scarcely be questioned. Engineering progress is made mainly by learning from experience. Design theory provides a framework within which this experience can be evaluated. Field observations of the performance of structures constitute the experience and instrumentation allows the factors controlling performance to be evaluated separately, thus much increasing the value of the observations.

Two examples may be quoted. The development of excess pore pressures during construction of earth works was first observed more than thirty years ago and the process of consolidation has been understood for a similar period of time. Controlled consolidation during construction, bringing about an increase in strength and a reduction of post-construction settlements, has been developed as an engineering technique over the last quarter of a century, with monitoring of pore pressure decay and settlement by instrumentation. Understanding of rates of consolidation of *in situ* soils and of secondary consolidation effects has improved. Numerous papers and discussions at this Symposium describe applications of this process, which can provide major economies in construction and allow successful construction on difficult sites. A second illustration is the use of observed strut loads in strutted excavations to establish simple empirical design rules, which allow quantitative design in this difficult problem of soil-structure interaction. Information is also available from which new analytical methods can be checked.

The advantages and disadvantages of instrumentation to any one particular project require careful consideration. Certain disadvantages of instrumentation become apparent, as discussed by Gray. There is, firstly, the cost of providing, installing and operating the equipment, which can form a significant part of construction costs. Perhaps more important when full-scale construction is instrumented, instrumentation requirements and construction requirements are often incompatible. The programme for construction will seldom be ideal for instrumentation purposes and delays may result. The presence of instrumentation may interfere with construction. Provisional elements in designs, intended so that advantage can be taken from field observations during construction, introduce uncertainties into construction programming and costs. Rates of construction controlled and varied from field observation may cause further uncertainties and influence the economics of earth-moving operations.

Wrightman, in his paper, and McKenna and Wilkes, in discussion, comment on advantages which can arise from control of construction by instrumentation. Perhaps most important is the added security given by checking design assumptions, with modification of constructon if necessary. McKenna points out that repairs of local failures, maintenance, and delays to completion, can be very costly. Reduction of the risk of such problems occurring is financially attractive, but difficult to cost precisely. With structures such as dams, where failure can be disastrous, no justification on financial grounds is necessary. Secondly, as pointed out by Wilkes, instrumentation may allow the safe construction of earthworks in place of much more expensive structures. A similar point is made in the paper of George & Parry. On the other hand, the modification of design during construction from the results of instrumentation offers only limited savings, as illustrated by Vaughan *et al* in discussion. However, where instrumentation is provided for security and to minimise the risk of failures and delays, design changes may allow the cost of the instrumentation to be recovered with, perhaps, small additional savings.

It is difficult to draw conclusions which are generally valid. Some general points are made in the final concluding remarks. The factors contributing to success are summarised at the beginning of this report. Unsuccessful instrumentation is a waste of money and effort.

The purpose of instrumentation should always be thought out and the advantages and disadvantages weighed. Disadvantages due to interference with construction should always be considered and minimised. More attention to this problem seems desirable. The overall effect of instrumentation on construction costs, including delays and interference, needs to be considered, not merely the direct cost of the instruments and their installation. More cooperation between construction and design is desirable. Wrightman points out the need to integrate instrumentation into overall planning and programming. If the most efficient overall system meeting the desired ends is evolved, the total cost should be minimised. Finally, it seems desirable that contract arrangements should reflect the real overall cost of instrumentation, and, as commented on by Cole, allow for the effect of programme modifications resulting from observations.

REFERENCES

Chinsman, B. W. E. (1972). '*Field and laboratory studies of 'short-term' earthworks failures involving the Gault Clay in west Kent.*' Ph.D. Thesis. University of Surrey.

Kwan, D. (1971). 'Observations of the failure of a vertical cut in clay at Welland, Ontario.' *Canad. Geotech. Jour.*, **9**, 283–298.

Uff, J. F. (1970). '*In situ* measurements of earth pressure for a quay wall at Seaforth, Liverpool Proc. Conf. *In situ* Investigations in Soils and Rocks.' *Brit. Geotech Soc.*, p. 229.

Session 4. The interpretation of field observations

GENERAL REPORTER: Dr. J. B. Burland *(Building Research Station)*

Introduction

This conference is concerned primarily with the techniques of field instrumentation and the majority of papers are rightly devoted to design, installation and performance of instruments for a variety of problems. However instrumentation is seldom an end in itself and if the results are to be of value they need to be carefully interpreted. The final session is devoted to disscussing this topic.

Field measurements may be used for essentially three purposes:

(1) To check that an engineering structure is behaving adequately in terms of its design requirements.

(2) To determine the mechanism of behaviour in order to gain a physical understanding of the problem.

(3) To determine, by back analysis, the *in situ* properties of the ground.

Most practical jobs will entail an element of all three but usually one will dominate. Sometimes a simple monitoring system is established and the behaviour turns out to be so different from that envisaged that the correct mechanism has to be discovered and only then can the adequacy of the structure be assessed.

Discussion in this session is most conveniently dealt with by considering specific types of problem such as retaining walls, foundations etc. Interpretation of observations for each type of problem depends on how well the mechanism of behaviour is understood and to what extent the problem is amenable to theoretical analysis.

One of the strong impressions formed from reading the papers is the emphasis that is placed on deformation measurements. This is perhaps not surprising since deformation is a physical reality whereas stress is to some extent a philosophical concept particularly in soil. Nevertheless, the emphasis on deformation measurements places a responsibility on the engineer to ascertain what is acceptable under a given set of circumstances. It was therefore suggested that a major theme for discussion in this session should be the interpretation of deformation measurements for stability and performance.

Analysis

In order to interpret the results for cases 1 and 3 above it is necessary to have some analytical or empirical approach to the problem however simple. For example it is unthinkable that design is carried out without either making use of past experience of attempting to idealise and analyse the problem in some way in an effort to ensure that the design requirements will be met. Moreover, if the observations for problems of the second type are to be of future use some form of analysis will usually be required if only for the purposes of comparison. Hence analysis forms an essential part of interpretation.

Modern methods of numerical analysis, such as the finite element method, offer great scope in the analysis and interpretation of complex field problems. However, there is a danger of loosing a physical 'feel' for the problem if the analytical method incorporates too many variables. The degree of sophistication requires careful consideration and should be kept to a minimum for the problem in hand.

The debate about the use of deformation to monitor stability is at an early stage but it is clear that if deformation is to be used, some prior knowledge of the form and magnitude of the movements is required in relation to the soil properties and the class of problem.

Loading Trials and Embankments

Bishop and Green list the situations which may dictate the use of loading or embankment trials. These include establishing the *in situ* properties of a wide range of natural soil types and fills which are difficult to test in the laboratory or by *in situ* methods, problems of analysis, and the effectiveness of various geotechnical processes.

A proper interpretation of the compressibility and consolidation characteristics of the ground requires measurements of vertical displacement at various depths, preferably at the boundaries of well defined strata, together with pore pressure and permeability measurements.

The problem of interpreting the onset of local rupture is much more difficult since it can develop with relatively small movements particularly in a stiff brittle material. The work of Hoeg *et al* (1969) shows that in soft clays the onset of local failure may be accompanied by a sharp increase in the rate of pore pressure change with load and this feature may well be a more sensitive indicator than displacement measurements.

Results of settlement against pressure for various embankments given by McKenna in his *Figure 11* illustrate the difficulties of using displacement observations to indicate the onset of failure. There is little indication from the results for the three failures to indicate at what pressure collapse will occur. It is noteworthy that McKenna and Roy base the decision to suspend construction of two embankments on observations of pore water pressure rather than excessive movement.

George and Parry have opted for deformation measurements as a means of monitoring the performance of storage tanks during filling. Wilkes argues for the use of inclinometer measurements as a means of monitoring performance. He has found that the relationship between $\Delta s/H$ (Δs is the maximum horizontal deflection and H is the fill height) and time is approximately linear for stable construction and that a sharp increase of slope is indicative of the onset of failure. This method may well be more sensitive than settlement observations but may need careful interpretation when the rate of construction is not uniform.

Much of the discussion on Session 4 is devoted to the important problems of monitoring and interpreting the behaviour of embankments on soft ground so as to detect the onset of failure. It seems that there is no simple or unique approach to this problem.

The traditional method is to measure pore water pressures and carry out a stability analysis. The use of stability charts described by Symons (p. 698) and Cole (p. 623) greatly facilitates this approach. The method suffers from the fact that the effective strength parameters for the soil must be known. These parameters, in particular the effective cohesion c', are difficult to measure accurately in lightly overconsolidated clays. Moreover Bishop has drawn attention to the fact that in normally consolidated sensitive clays the apparent factor of safety in terms of effective stress may be much higher than the actual factor of safety.

More recently deformation measurements have been used as an aid in monitoring stability. A number of contributors including Bishop, Wroth, Marsland and Hutchinson have pointed out the difficulties of employing deflection criteria and the force of these arguments is very apparent when figures of lateral deformation at failure quoted in this Symposium range from 10–1000 mm.

The main difficulties arise with sensitive and brittle materials for which neither limit analysis nor numerical methods of stress analysis can be relied upon to give accurate predictions of failure. When monitoring the behaviour of structures on such soils it therefore seems that the emphasis should be on the detection of incipient failure rather than attempting to define absolute criteria of stability or performance. Under these circumstances one is looking for a change in behaviour indicating the development of local failure. The most promising methods appear to be to look for changes in the relationships between increase in pore pressure or lateral deformation with change in total stress (or embankment height H).

For pore pressures, the method of presenting the observations is to plot u against H or $\Delta u/\Delta H$ against time. As mentioned previously for lateral deformations Wilkes has been successful by plotting $\Delta s/H$ against time. The results for King's Lynn presented by Snelgrove show that the values of $\Delta u/\Delta H$ increase by a factor of about 3 at the time that the values of $\Delta S/H$ increase sharply so that in this case both methods appear to give satisfactory results although the deformations appear to be very much more sensitive. It should be emphasised that these methods indicate the onset of local yield but give no information as to when failure will take place.

It is evident that there is a need for many more field measurements of pore pressure and deformation before definitive criteria for monitoring and interpretation can be laid down. The use of well instrumented trial embankments for studying the behaviour of the ground up to failure is undoubtedly the best method of establishing criteria for the monitoring of the full scale structure. Centrifuge tests of the type described by Dr Bassett look very promising and it will be interesting to compare the prototype behaviour with the predictions presented in his contribution. It is important not to overlook the fact that the centrifuge test is carried out on a sample of soil that has undergone a fairly severe sampling operation before undergoing reconsolidation in the centrifuge. It is known that the stress-strain properties of sensitive clays can be significantly affected by such a procedure.

Foundations

As mentioned by Dr Vaughan in his report for Session 3, instrumentation of building foundations has been particularly restricted. Yet the long-term benefits for subsequent construction in a given locality can be great. Traditionally instrumentation of building foundations has involved measuring settlement. Interpretation of surface settlements in terms of soil properties is not easy because the underlying ground is seldom homogeneous. The measurement of settlement and pore pressure at various depths greatly increases the value of the observations since this permits an accurate back analysis of undrained stiffness, compressibility and the coefficient of consolidation at various depths.

Hawes *et al* describe the instrumentation for measuring the compression of the ground beneath an exceptionally large building in London and the results should give information about the stiffness of the London Clay and the Woolwich and Reading beds. Marsland and Eason have measured the settlement at various depths beneath large diameter plate loading tests and the results show that the stiffness of the clay immediately beneath the plate is very much reduced. The results may be applied to a variety of problems. For example, the average stiffness obtained from the settlements of the plate would be applicable to the base of a pile. The greater stiffness at depth would be more applicable to settlement calculations for a large near surface foundation where the bulb of pressure extends to some depth. It is notable from Marsland and Eason's results that the load-settlement curves are only mildly non-linear. It seems doubtful that the incorportion of non-linearity in the back analysis is required, although non-homogeneity is clearly important (Burland, Sills and Gibson, 1973).

Pender has drawn attention to two possible sources of error in the interpretation of the plate loading test results given by Marsland and Eason; the first is the assumption that the ground can carry tension and the second is the presence of the extensometer borehole vertically beneath the centre of the plate. Both of these are known to be of minor importance. The depth correction factor used by Marsland and Eason is based on an analysis by Burland (1970) in which the presence of the shaft is accounted for. The depth correction is very small and the inclusion of no tension would therefore have a negligible affect particularly if self weight is taken into account. With regard to the influence of a central hole this factor was investigated analytically in connection with the interpretation of the Mundford tank test results (Burland *et al* 1973) and its influence was found to be slight.

In his discussion Atkinson has compared the stiffness of London Clay obtained from the back analysis of full scale behaviour with values measured in the laboratory. The comparison shows that the laboratory values under-estimate the stiffness by a factor of five even when high quality samples and careful testing techniques are employed. Experience at the Building Research Station is that when routine commerical sampling and testing methods are employed for stiff fissured materials the *in situ* stiffness can be under-estimated by as much as 10 to 20 times. It is hardly surprising that experienced structural engineers frequently complain about the accuracy of settlement predictions. Dr Atkinson's contribution underlines the value of careful field measurements of foundation behaviour for determining the *in situ* properties of the ground in specific localities and soil strata.

Piles

Four papers describing widely differing types of instrumentation for piles are presented. At the fundamental end Cooke and Price are concerned with the detailed study of the deformations of the ground around driven piles. In another paper (Cooke and Price, 1973) the authors describe the interpretation of the observations for working loads using a finite element method of analysis. The analysis is self consistent in that it accounts, for the observed axial load distribution down the pile as well as patterns of ground displacement. Such independent checks are important when back analysing results. In his discussion Mr Price gives some results of measured load distribution in adjacent piles which suggest that the principle of super-position applies under working loads. This finding is of great importance in extending results from tests on single piles to group behaviour.

The measurements described by Weeks and Dixon are aimed at providing information for the desigh of bored piles on Gault Clay. Field observations on the lateral displacement of piles and their interpretation are scarce and the discussion by Jamiolkowski is therefore most welcome. As is so often the case in the interpretation of field observations the importance of non-homogeneity in the form of increasing stiffness with depth is apparent.

Hawes *et al* describe instrumentation for measuring the load distribution beneath a large pile cap. The results are probably highly dependent on such factors as the stiffness of the raft and the distribution of load imposed on it, the stiffness of the ground at various depths and the influence of drainage. Interpretation and analysis of observation will therefore not be easy.

Excavations and Retaining Walls

The papers by Bundred and Jones deal with the measurement of stress at discrete points on retaining walls. As well as the difficulty of interpreting the actual instrument readings it is often difficult to interpret the results of stress measurements due to inherent scatter and the limited number of measurements that economy dictates. In the case of Bundred's measurements one wonders what influence the magnitude of the initial 'prestress' has on the subsequent readings. For example cells 1 and 2 are nearly at the same depth but have widely differing initial stresses. During the four months subsequent to installation and prior to excavation both cells show similar stress changes but there is no tendency for the absolute values of stress to converge.

The measurements presented by Jones indicate that classical earth pressure theories are inapplicable for materials compacted behind rigid abutments and retaining walls. Comparison with elastic predictions using finite element methods appear promising.

Phillips and Jones use the observed deflected shape of a tied back diaphragm wall to obtain the bending moment distribution. The approach involves obtaining the differential of the gradient of the deformed profile at various points down the wall. The smoothness of the bending moment diagram in *Figure 4* of this paper is remarkable and it would be interesting to know if this entailed drawing a smooth curve through the experimental points for the deformed profile.

Measurements of movement around excavations are important for establishing mechanisms of behaviour. The results presented by Hawes *et al* in their *Figure 5* of the movements of a tied back wall are interesting in that they reveal a completely different mechanism of behaviour from that of a beam on spring supports (as assumed in the design). In the simple analysis the deflection at the bottom of the wall is small whereas in practice it is nearly a maximum. This results from the movement of the ground at depth into the excavation and is due as much to the release of vertical stress as to the release of horizontal stress. Huder (1969) has measured similar behaviour and the mechanism can be reasonably accurately reproduced by means of a stage by stage finite element analysis (Ward and Burland, 1973).

The back analysis of observed deformations around excavations not only involves the use of numerical methods but also a knowledge of the initial *in situ* horizontal stresses prior to excavation. Cole and Burland (1972) analysed an excavation in London Clay using laboratory determined values of K_o. The deduced E values were in reasonable agreement with other *in situ* values. Such back analyses are important if results in a given locality are going to be of maximum value. The assumptions used in the back analysis should be maintained in future predictions as the theoretical method is primarily a means of 'scaling' results from one project for use on another.

Slopes and Dams

The monitoring of pore pressures in slopes and dams in order to ascertain the factor of safety against failure is standard practice. Attention is now being paid to deformation and Mackellar *et al* and Penman and Charles describe instruments for measuring the deformation within and at the surface of embankment dams. Finite element analysis employing quite simple stress-strain formulations show good agreement with observed movements during the construction of rockfill dams and it appears that the mechanisms of behaviour are fairly well understood. The figures quoted by Penman and Charles of observed rates of lateral movement with height of fill also provide valuable quantitative information. Under these circumstances the use of deformation measurements to monitor stability and control design show considerable promise particularly as rockfill is unlikely to exhibit brittle behaviour under the stress conditions imposed within embankment dams.

The discussion by Vaughan, Wernek and Hamza on the trial bank for the Empingham dam is of particular interest and raises a number of important points. The measurements are clearly serving all three purposes mentioned at the beginning of this report. The measurements have revealed the underlying mechanisms of behaviour, some of them unexpected, and have confirmed the basic design parameters. A detailed analysis is being carried out to enable reasonable predictions to be made of the deformations of the main embankment. As mentioned previously such predictions are essential if the observed movements of the main embankment are to be correctly interpreted.

Vaughan *et al* draw attention to the difficulties of deriving stress-strain characteristics by back analysing observed movements. The process can be exceedingly tedious and, at times, baffling. In the General Reporters experience it is necessary to keep the stress-strain model as simple as possible initially (a point emphasised by Dr Wroth) and to carry out the trial and error procedure in an orderly and systematic manner. The best procedure appears to be to work inwards from the observation points most remote from the region of stress application. This is because the detailed stress-strain properties in the vicinity of highly stressed regions do not appear to influence greatly behaviour remote from these regions – a finding which seems to accord with St Venant's principle. The suggestion that the observations should include the measurement of the complete strains at a few key locations as well as the overall displacements is worth considering and accords with the views expressed by Dr Wroth. The development of special instrumentation may not be necessary since the displacements of three neighbouring points, preferably approximating to the vertices of an equilateral triangle, are sufficient to define completely the strains in the plane of the three points.

Mr Knight's discussion is concerned with the long term monitoring of slopes and dams and the difficulties of such work. A major difficulty is the organisational one of taking readings regularly for many years, keeping the results updated and presenting them in a form which is easily digested. Interpretation is particularly difficult and it is essential to have independent checks in order to detect faulty instruments and errors in measurement. As mentioned by Dr Skipp it is also necessary to separate general trends from inherent variability. As for embankments on soft clay absolute criteria of stability or performance are difficult to establish so that one is normally looking for changes in performance which indicate the need for more detailed measurements and analysis.

Mining Subsidence

Whittaker and Forrester describe a simple extensometer and tiltmeter. It is presumed that tilt is defined as the change of slope of the ground surface. The tiltmeter will only measure this quantity when there is no vertical shear strain and this seems a risky assumption. For example it is not difficult to envisage a situation in which vertical lines remain approximately vertical during subsidence in which case the instrument would register approximately zero tilt whereas the slope of the ground surface may change significantly. In his discussion Dr Whittaker has suggested that this behaviour is only likely to occur if the ground actually fractures. However vertical shear strain can occur without rupture, particularly in the softer materials near the surface. The complete interpretation of the near surface behaviour of the ground requires the relative horizontal and vertical displacements of neighbouring points at ground surface together with relative horizontal and vertical displacements at depth beneath each point.

REFERENCES

Burland, J. B., Sills, G. C. and Gibson, R. E. (1973). 'A field and theoretical study of the influence of non-homogeneity on settlement.' *Proc 8th Int Conf Soil Mech and Found Engg*, Moscow. (In the press.)

Cole, K. W. and Burland, J. B. (1972. 'Observation of retaining wall movements associated with a large excavation.' *Proc 5th Europ Conf Soil Mech and Found Engg*, April 1972, vol 1, pp 445–453.

Cooke, R. W. and Price, G. (1973). 'Strains and displacements around friction piles', *Proc 8th Int Conf Soil Mechand Found Engg*, Moscow. (In the press.)

Hoeg, K., Andersland, O. B., and Rolfsen, E. N. (1969) 'Undrained behaviour of quick clay under load tests at Asrum', *Geotechnique*, 19, 101–115.

Huder, J. (1969). 'Deep braced excavation with high ground water level', *Proc 7th Int Conf Soil Mech and Found Engg*, Mexico 1969, vol 2, 443–448

Ward, W. H. and Burland, J. B. (1973). 'The use of ground strain measurements in Civil Engineering', Building Research Station, Current Paper 13/73.

Discussion. Application

R. H. G. Parry (Cambridge University). I have been concerned with sealing the beds and sides of drainage channels in the fens. We have used impervious sheeting to line the channels, and to observe the effect on the ground water flow we wanted to put in a number of simple and inexpensive piezometers. This requirement led to the design of a standpipe piezometer with a capped porous stone in the end. It is driven into position through the soft ground then the porous stone is exposed by putting a rod down the middle and pushing off the cap. The stone remains clean, there is no problem of smear, and the piezometer is sealed into the soil because the tube is flush.

In using instrumentation to control construction we need to know the limits of behaviour to allow before we stop the work. Driven piles were proposed for the King's Lynn by-pass bridge over the Eau Brink Cut, which is a famous cut because of the rotational slip analysed by Skempton (1945). The great Ouse River Authority were concerned about the effect pile driving would have on the stability of the bank, particularly in view of its past history, and a trial was made using seven piles and measuring amongst other things horizontal movements with inclinometers. After some consideration I decided that the movement should be limited to 100 mm before stopping the work, but during the trial the movements were only 1 or 2 mm – almost undetectable – and as a result we changed the restriction from 100 mm to 10 mm. When the actual work started, however, the 10 mm was soon exceeded and we compromised by bringing the limit back to 50 mm. The point I want to make is that if instrumentation is to be used in this way, it is necessary to know what tolerance to allow, and this is an area that I think needs a lot of investigation at the present time.

REFERENCE

Skempton, A. W. (1945), 'A slip in the West Bank of the Eau Brink Cut', *Jnl. Inst. C. E.*, 24, 7, pp. 267–287

D. B. H. Pawsey (Metropolitan Water Board). Bishop and Green (p. 28) refer to the possible benefit by way of eliminating design changes during construction of trial banks for bunded reservoirs with embankments of high length/height ratios, in particular the reservoirs at Chingford, Walton and Wraysbury. These embankments have core walls of compacted or puddled London Clay with gravel shoulders and rest on gravel overlying a substantial thickness of London Clay. The last of these, Wraysbury (1965), was commenced some 30 years after Chingford (1935) and Walton (1939), during which time there had of course been significant advances in soil mechanics. That is not to say that is is yet more of a science than an art.

The purposes and justification of trial banks will change as our knowledge and techniques develop. Bishop and Green make the same point. It is surely too simplistic to make even a first assessment on length/height ratio alone. There will be many factors to consider. Prior to the war, without the benefit of soil instrumentation, the trial would have provided at best a practical check on stability and settlements. It is interesting that there were trial embankments at both Chingford and Walton in the shape of existing reservoir banks in an adjoining field in each case; full height at Chingford, $\frac{2}{3}$ height at Walton, yet a slip during construction occurred at Chingford with the bank at $\frac{3}{4}$ height and a major redesign ensued.

Cooling and Golder (1942) described the slip which was in a stratum of soft yellow clay under the embankment at a place where construction had been too rapid, the embankment being raised 13 ft from about $\frac{2}{5}$ to $\frac{3}{4}$ full height in about a month, which in those days represented a considerable feat by the Contractor.

Tests at the time indicated that after only 8 months consolidation under the weight of the $\frac{3}{4}$ height embankment, the clay had increased in strength to the extent that it could have carried the full height embankment with an adequate margin. It seems that redesign was precipitous; it being necessary only to spread the raising of the embankment more uniformly over the construction period. If one thinks of the embankment which slipped as the second trial bank, it was singularly unsuccessful in promoting a best design and contradictory to the first which has stood for 60 years. Interpretation of results will also change as knowledge and techniques develop.

I am unaware of the design changes at Walton referred to by Bishop and Green as due to reassessment of foundation strengths. During the original prewar contract, following consultation with Terzaghi, a stratum of sandy clayey loam under part of the embankment was reduced in thickness but extra excavation had been anticipated and provided for in the contract. Later when a second contract was let, the first having been determined following the outbreak of war, compacted clay had been substituted for puddle of the core wall, I believe for the first time in this country, and the loam was stripped altogether from under the embankment to minimise settlement adjacent to the core. The design of the embankment was essentially unchanged.

Trial embankments are being used for Datchet Reservoir at present under construction essentially to confirm parameters for the rate of consolidation of the London Clay, with and without sand drains. From experience at Wraysbury the rate was expected to be faster than predicted from laboratory tests. The trials form part of the finished embankment and were built at the commencement of construction to approximately $\frac{1}{4}$ full height.

Trials for a similar purpose may have given some economy in the initial design at Wraysbury if carried out prior to the contract as part of the site investigation, but would not have obviated the changes in design found necessary during construction or provided information not available otherwise at that time. The design changes resulted from the discoveries first of a region where the surface of London Clay was higher than expected and subsequently of a tectonic shear plane. Both were localised features which unfortunately coincided.

However good the initial site investigation it will be supplemented during construction. The features were investigated using conventional technique and at that time results were available from instrumentation of the main dam. Unless a trial embankment had been most fortuitously sited it could have been of no advantage. A tectonic shear plane and a second bedding plane of weakness were also discovered at Datchet during initial investigations. Again these were fully investigated conventionally and trial embankment considered to be of no advantage in predicting the behaviour of these specific features.

Trial embankments can be a useful tool in the site investigator's kit but like the remainder of his techniques need to be used with discretion.

REFERENCES

Cooling L. F. and Golder H. Q. (1942), 'The analysis of the failure of an earth dam during construction.' *Jour. I.C.E.* **19**, pp. 38–55.

Figure 1 Showing the motorway embankment

Figure 2 Close-up view of the slip

C. J. F. Jones (N.E. Road Construction Unit). The M18 and M62 motorways to the east of the A1 in Yorkshire are being built on difficult subsoils, particularly peat and silts overlying the Bunter sandstones. The preliminary soil survey isolated the worst sites where excessive settlements and difficult construction conditions could be expected. It was decided to let a pre-earthworks contract to construct three embankments. Piezometer and settlement equipment, essentially the same as that described by K. W. Cole in his discussion on Gateshead, in Session 2, was installed in two of the embankments (see p. 589).

At 7 o'clock on the morning of 10 March 1973 a large slip occurred. It was witnessed by the contractor's foreman who saw the manhole which is the settlement house, swaying up and down. The whole slip took about 10 minutes and left the settlement house 9 ft above its original level. At the time the majority of the gauges were working satisfactorily, no filling had been placed for over a month and all the readings were quite steady. (Subsequent investigation has shown this statement to be in error on one point; a small quantity of fill was placed in the week preceding the failure, although its presence was not recorded on the piezometer equipment which gave steady readings in the month prior to the failure.) Although the slip was of considerable magnitude, the majority of the instruments are still recording. *Figures 1* and *2* indicate the size of the embankment and the extent of the slip. The causes of the failure are still being investigated.

P. R. Vaughan, M. L. G. Wernek & M. M. A. F. Hamza (Imperial College, London). Bishop and Green, in their paper to the symposium, refer to the trial bank for the Empingham dam, currently under construction for the Welland & Nene River Authority, for whom T. & C. Hawksley are consulting engineers. This bank has been constructed, as part of the main contract and incorporated in the main works, in an attempt to clarify and confirm design assumptions. Provision was made in the contract for modification of the final dam cross-section after completion of the trial bank. Construction of the dam continues and it is premature to assess final success but it is intended to describe here the approach adopted, some of the information obtained and current work on the analysis of this information. Advantages and disadvantages of this approach will be reviewed.

Figure 3 shows a diagrammatic cross-section of the embankment. The reasons for adopting this section will not be discussed here. Its stability is controlled

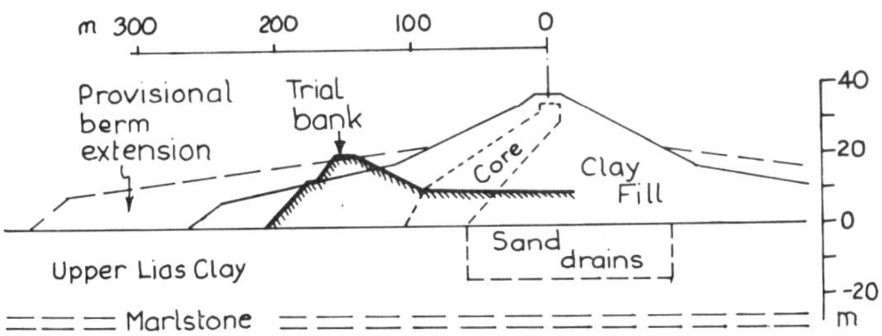

Figure 3 Diagrammatic cross-section of the main embankment

by the properties of its foundation, which consists of the over consolidated Upper Lias Clay, severely sheared and disturbed by valley bulging and peri-glacial frost action. When examined by limit analysis methods, the stability depends on the reliable undrained and drained strengths of the clay, on the rate of consolidation between the sand drains, on the properties of the fill and on the reliability of the analytical methods, bearing in mind the risk of progressive failure, anisotropy of shear strength and the like. The general site investigation utilised conventional tests on standard sized soil samples and the results were all subject to sample size effects. 260 mm dia plate tests were performed to examine the undrained strength and these gave results similar to the 100 mm dia triaxial tests. The application of this result to the design without further confirmation was not considered prudent. Further *in situ* tests were considered, but time was short and only the bulk undrained strength could have been examined. A conservative design could have been prepared but it would have relied excessively on judgment unconfirmed by quantitative information.

The problem was considered by the late Mr Hawksley and Mr Dyer, the engineers responsible for the dam, and by Professor Bishop, Mr McKenna and Dr Vaughan, acting as geotechnical advisers. As a result it was decided to rearrange the construction sequence and form a trial bank within the upstream berms during the initial construction. The contract was prepared on this basis, including provisional berms the extent of which was to be decided four months after completion of the trial. The arrangement of the trial bank is shown on *Figure 3*. With a steep slope, the shear stress on the trial bank foundation would be greater than that imposed by the final higher but much flatter embankment. This was confirmed by finite element analyses carried out at the University of Swansea. Thus the undrained strength and, with measurement of pore pressure, the drained strength could be checked. The properties of the fill, as placed could be established and the delay would enable the rate of consolidation of the foundation, examined initially by laboratory and *in situ* tests, to be checked by pore pressure measurements during the initial construction. The trial bank would also allow the deformation characteristics of

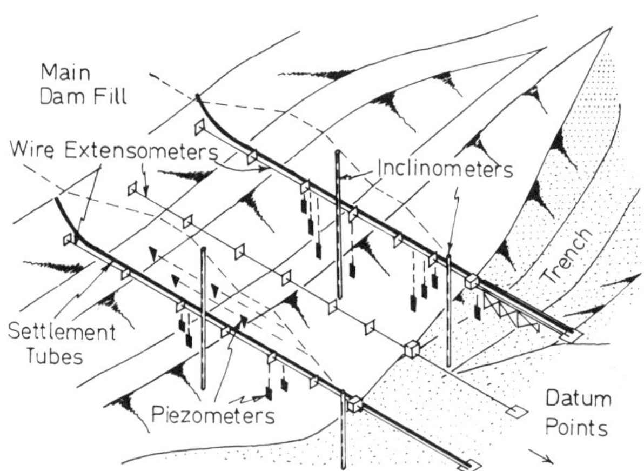

Figure 4 Arrangement of the trial bank

the fill and foundation to be checked and provide a check on methods of analysis. While it was hoped that complete failure of the bank could be avoided, only an acceptably small part of the foundation would be weakened by failure.

The final arrangement of the trial bank is shown on *Figure 4*. The bank was placed to one side of the valley centre line, to avoid the main valley bulge, which contained atypical material. A trench was cut into the side of the valley to give a sufficient length at the same height. Since the section was not uniform, instrumentation was placed on three cross-sections. Subsequent deformation observations showed that plane strain behaviour was effectively achieved. Before the bank was completed, several slips occurred in temporary borrow pit slopes. Back analysis of these slips, showing an average undrained strength 50 kN/m² (half the average of the 100 mm dia triaxial tests), gave partial conformation of the undrained strength assumed in the design and lessened the need to reach failure of the trial bank. In the event this did not fail, construction being terminated when large deformations had occurred. This enabled four months of creep measurement to be made.

The instrumentation at one of the sections and the pore pressures at completion are shown in *Figure 5*. *Figure 6* shows vectors of displacement of the foundation, together with horizontal strains at foundation level. *Figure 7* shows the strains developed at the two points where inclinometers and horizontal strain and settlement gauges cross. These strains are subject to some uncertainty, primarily due to the stiffness of the inclinometer casing.

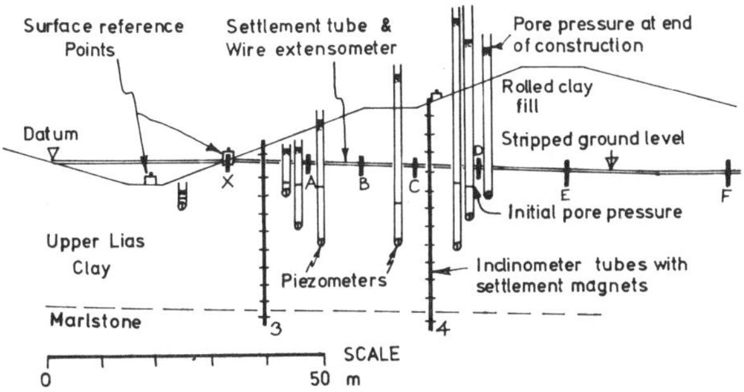

Figure 5 Cross-section of the trial bank showing instrumentation and pore pressures at the end of construction

The horizontal strains show an interesting trend which was initially unappreciated. As placed, clay fill compacted by heavy plant has high horizontal stresses. If the embankment slope is steep these stresses try to impose a high

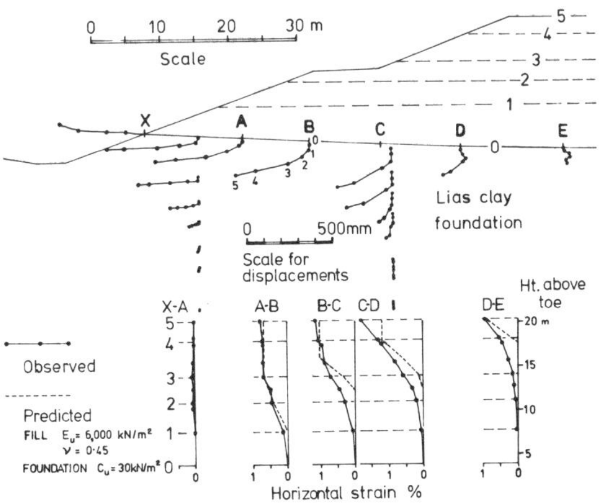

Figure 6 Displacements and horizontal strains of the foundation

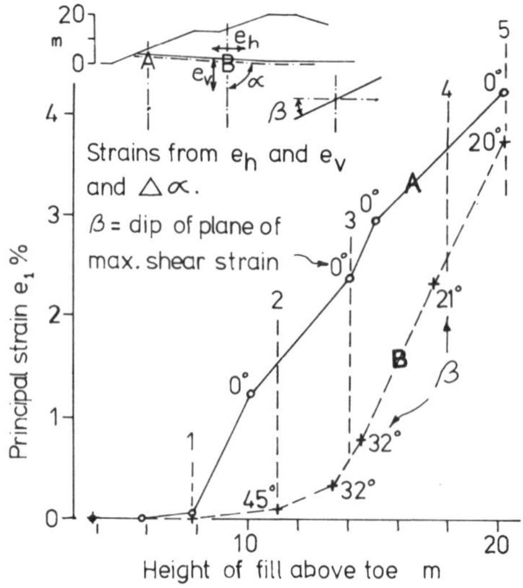

Figure 7 Estimated strains at the intersections of inclinometers and horizontal displacement/ settlement gauges

shear stress on the foundation. If its shear strength is exceeded yield and spreading of the fill will occur. A mechanism of this type seems to have operated in the trial bank, leading to large horizontal displacements while the bank remained stable. A simple analysis may be set up which reproduces this behaviour. If a strength is assumed for the foundation, the maximum horizontal force on any vertical section can be computed. If the vertical stress is assumed equal to the overburden pressure, the horizontal strains in the fill can be deduced from the appropriate laboratory compression test on the compacted fill. The horizontal strains computed by this method are shown on *Figure 6*, where they may be compared with the observed strains at foundation level. A similar trend is shown. A berm was introduced in the trial bank during construction as a superficial failure of the outside part was feared.

A major difficulty in interpreting the trial bank performance arises due to the undrained strength of the fill being some three times greater than the bulk undrained strength of the foundation. Some valuable qualitative information has been obtained. While distortion of the inclinometer tubes during the later stages of construction prevented the instrument from passing, the casings could always be plumbed. Thus, while large strains had occurred, there was no evidence of significant displacement along shear discontinuities. Two general conclusions follow. Firstly, progressive failure seems unimportant in this material, despite brittle behaviour indicated by large post failure displacements of the borrow pit slips. Secondly, as failure is preceded by large displacements, the monitoring of the displacements of the main embankment as it is completed should enable distress to be determined easily.

Analysis has been directed towards two ends. Firstly, preliminary analysis, using trial slip surfaces, has been performed to examine the undrained and drained strengths mobilised. The values used in the design have been confirmed, using the same methods of analysis as for the main embankment. Secondly, more sophisticated finite element methods are being used to analyse deformations. The aim is to develop the necessary techniques by which the main embankment may be analysed, and its deformation pattern determined prior to its completion. Such an analysis is most desirable to enable the observed deformations of the main embankment during the final stages of construction to be interpreted on a quantitative basis.

Non linear analyses are required to simulate the behaviour of the trial bank. However, in view of the absence of shear discontinuities in the foundation, it is not thought that strain softening need be considered. Non-linear analysis involves several problems. Firstly, the techniques used must ensure that the solution follows the nominated stress-strain curve with reasonable accuracy, either by using small increments or some iterative technique which converges economically. Secondly, some stress-strain function must be adopted. It is important to establish the sensitivity of the results to the function assumed, and therefore there must be flexibility in the way this function is specified in the programme. There is no obvious soil model, applicable to material such as the Empingham foundation, on which a stress-strain function may be based. Therefore a simple non-linear elastic function has been used.

Some results of preliminary linear and non-linear analyses are given in *Figure 8*. For the type of mechanism occurring in the trial bank foundation, horizontal displacements are controlled primarily by shear strain and the stress-strain relationship of the soil. Vertical displacements are controlled primarily by volume change, simulated in the finite element analysis by the Poisson's ratio adopted. In *Figure 8* the computed horizontal displacements are shown

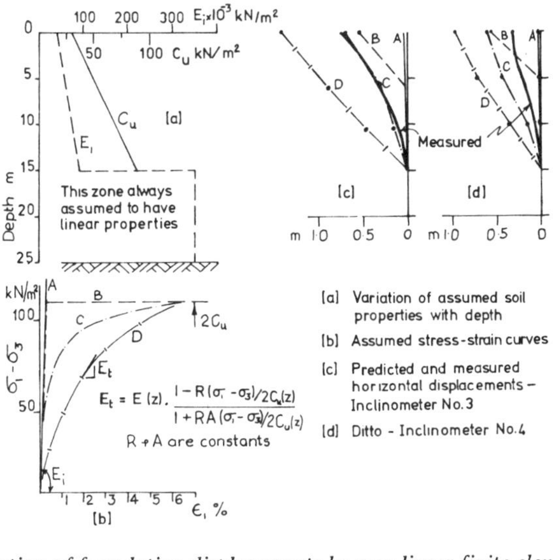

Figure 8 Prediction of foundation displacements by non-linear finite element analyses

for the two inclinometers and compared with the measured values. Work is proceeding on improving the agreement obtained. When attempting to back analyse the observed displacements to obtain stress-strain characteristics which may be used to analyse other problems, a number of difficulties arise. Both the magnitude and the distribution of the stress-strain assumptions may be varied in the attempt to improve the fit of the predictions with the observations. This process is illconditioned and time consuming, it depends heavily on judgment and there is no guarantee that a unique solution will be found. However, if it can be shown that, during certain phases of the construction, the strains in part of the problem are controlled primarily by one of the variables, it may be possible to quantify the process of back analysis. Current research is directed towards seeing if this can be done. Initially, an assumption of the distribution of properties with depth must be based on the site investigation. Preliminary analyses have shown that the foundation behaviour is not sensitive to the properties assumed for the fill. During the initial stage of construction, foundation displacements are controlled by the initial modulus assumed, enabling this to be checked by direct comparison. During later stages of the construction the displacements are controlled partly by the shear strength assumed and partly by the curvature assumed for the stress-strain characteristic.

Currently, methods of determining the shear strength by direct comparison of the results are being examined. It may be noted that there seems to be considerable advantage in measuring the complete strains at selected points in addition to displacements. These enable a specific check to be made on the results of analyses of the finite element type. The development of equipment specifically intended for the purpose seems desirable.

The extent to which the stress-strain information required for a realistic finite element analysis can be obtained from laboratory and *in situ* tests is also of considerable interest. The principal difficulty seems to be in reproducing the yield stress of the bulk material. This aspect of the problem is the subject of current research supported by the Science Research Council.

Finally, a preliminary and personal attempt to assess the advantages and disadvantages of the overall approach adopted at Empingham may be made. The advantages will first be considered. The principal assumptions involved in the stability analysis have been checked at near full scale prior to completion, partly by observations made during the initial construction of the main embankment. A much better understanding of the foundation behaviour has been obtained: in particular, the large prefailure strains developed and the absence of evidence of progressive failure has given increased confidence in the foundation behaviour; and insight has been obtained into the mechanisms controlling failure. This information has been used in the final design and the opportunity has arisen for the behaviour of the embankment to be analysed in more detail and for methods of monitoring the final stages of construction to be improved as a result. The information obtained should have considerable research value.

The approach adopted has had a number of disadvantages. The trial bank was constructed during the initial stage of the main contract when there were many other calls on the site staff. The programme for its construction had to be controlled by the needs of the job as a whole and not by the need for obtaining information. These problems were appreciated at the outset and the instrumentation was kept to a minimum. In the event, insufficient time was available for installation of equipment, checking it, taking initial readings and gaining experience in its operation. The information obtained during the early part of the construction was therefore of doubtful quality. Fortunately, these problems were overcome before the more interesting later stages of construction were reached. With hindsight, more attention should have been paid to ensuring that the equipment was mutually self checking. Interpretation of the results has been handicapped by the major difference in strength between the fill and foundation. While there are major advantages in a flexible approach to design, contractual and organisational problems ensue which require additional engineering effort and may result in extra costs.

Installation of equipment and its operation was performed by the resident engineer's staff, with assistance from Imperial College. Different people become responsible for the operation of different parts of the equipment. Observation and calculation and plotting of results probably required the equivalent of between two and three people engaged full time. Information was considered on site as it was obtained.

Costs involved cannot be apportioned accurately but the following is a reasonable guide. The contractor put in a lump sum of £25 000 to cover the inconvenience of forming the trial bank. Forming the foundation trench and tidying up afterwards cost £10 000. Instrumentation equipment and installation cost £15 000. Site engineering costs are estimated at £5 000, making an on site total of £55 000. Preliminary analysis and the final design of the main embankment section cost about £3 000. The remaining analytical work is being conducted mainly on a research basis. On a commercial basis the cost of the work currently envisaged might be of the order of £20 000. The cost of the main embankment is £3.7 m and the provisional element in the contract is approximately £200 000. About half of this will be spent, although not all the saving can be attributed directly to the results from the trial bank. It is uncertain if the preliminary design, including the provisional berms, would have been accepted as a final design without the proving of assumptions allowed by the trial bank. If it would have been, then the extra cost of the trial bank is roughly equal to the savings made. Had the initial design assumptions been shown to be unneccessarily conservative, greater savings might have resulted. However, the justification of the trial bank and the flexible design approach adopted must rest primarily on the security and the elimination of risk arising from the proof of the principal design assumptions at near full scale. One of the risks which arises in an embankment such as Empingham is that uncertainties about performance towards the end of construction may lead to expensive delays.

Gleeson (Civil Engineering) Ltd. are contractors for the dam. The site investigation was carried out by Soil Mechanics Ltd. Instrumentation equipment was provided by Soil Instruments Ltd. and Nuttall Geotechnical Services Ltd. The site work on the trial bank was performed under direction of the Resident Engineer, Mr A. J. H. Winder, and Messrs R. C. Bridle, P. S. Horswill, C. J. Sammons, G. A. Miller, H. J. Walbancke and J. J. M. Powell assisted with the observations. The final results are due primarily to their efforts, made under difficult circumstances.

K. W. Cole (Ove Arup & Partners). In the contributor's discussion in Session 2, (p. 589), the type of instruments installed to monitor the construction of nine sections of embankment at the Gateshead western by-pass were described. Stability analyses using undrained shear strength showed that, if the embankments with 1 in 2 slopes were built 'instantaneously', all would fail when raised above a height of about 15 ft (4.5 m). However, when using the drained shear strength parameters, and assuming no excess head of water above natural water table levels (virtually ground level), all fills had factors of safety in the range of 2 to 4 at full height. Thus, successful raising to the full height (with an additional amount of fill to compensate for settlement) would depend upon the rate of dissipation of excess pore water pressures and controlled rates of filling would be imperative.

With the possibility that the contractor could be simultaneously filling at all nine instrumented sections spread over a distance of 5½ miles (9 km), it was judged that all monitoring must be done on site by site staff without the need to undertake further analyses. They must therefore have in advance the monitoring charts applicable to each instrumented location.

Additionally, not long before the contract was put out to tender, the new Ministry of Transport Specification for Road and Bridgeworks was issued and was adopted for the by-pass. It is inherent in the philosophy of this specification that contractors tendering will be informed of the limitations to progress in constructing earthworks. The drawing showing each instrumented fill therefore had a note similar to the following:

'The rate of embankment construction shall not exceed 3′ 0″ in any week above a height of 15′ 0″.'

Predicting the permissible rate of construction of earthworks on the basis of soil investigation data only was acknowledged to be somewhat hazardous, and the specification preamble contained the following clause:

'Should the engineer instruct that earthworks on any length shall proceed at a slower rate than than shown on the drawings solely due to consideration of stability the said instruction will be regarded for the purpose of sub-clause 12(2) (of the I.C.E. Conditions of Contract) as arising from physical conditions that could not reasonably have been foreseen by an experienced contractor.'

The engineer therefore had both to forecast the permissible rate of construction and to indicate the consequences of his possible error in terms of slower permissible rate of construction. A bill item to cover this reduced permissible rate of construction for a proportion of the fill volume was then priced by the tenderers.

Possible methods of control of stability during construction were therefore examined, see *Figure 9*. The conclusion was reached that monitoring stress by way of pore water pressures would afford the most straightforward control, as the actual magnitude of measured pore water pressures could be related to

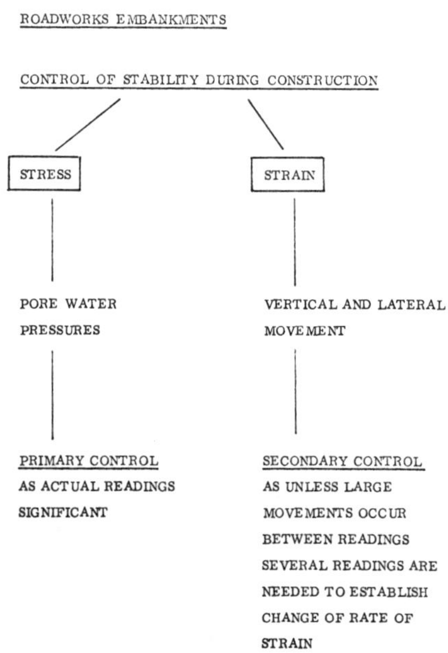

Figure 9 Roadworks embankments

a 'Factor of Safety' (F) number, whereas control by measurements of strain might require readings to be taken over several days before a potentially dangerous situation could be recognised with sufficient certainty to warrant stopping the contractor filling. Further, there appeared to be no consistent data on the amount of strain an earthwork could undergo before it reached the state of limiting equilibrium. Nevertheless, since the pore water pressure measurements might be erroneous, or the instruments fail, strain measuring instruments were installed and monitored as secondary control.

In order to be able to produce the monitoring charts for use of site staff, some simplification of the effective stress analysis was necessary. Circular arc stability analyses solving the equation:

$$F = \frac{R}{\Sigma Wx} \sum \left[\frac{[c'b + \tan\phi' \, (W(1-\bar{B}) + x_n - x_{n+1})] \ \sec\alpha}{1 + \dfrac{\tan\phi' \tan\alpha}{F}} \right]$$

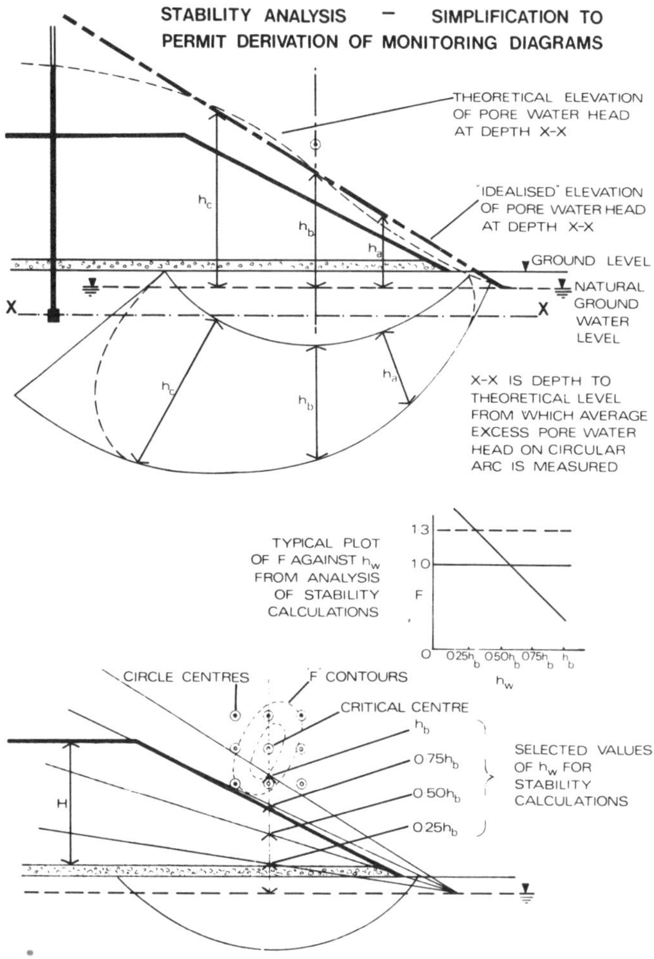

Figure 10 Stability analysis – simplification to permit derivation of monitoring diagrams

due to Bishop were made for ranges of circle diameters, each range being centred on the node of a preselected grid.

Piezometers give readings of pressure which are readily converted into head of water, the excess head h_w being that above the head of original ground water level. Thus, as may be see in *Figure 10*, the stresses in the ground due to the surcharge of a given height (H) of fill of known density can be idealised into a linear form, which gives for the same excess head h_b beneath the circle centre a virtually identical value of F. The lowest F value for a given excess head line is then computed, and this is repeated for a range of hypothetical excess head lines. A plot of h_w (the excess head beneath the critical centre) against F can then be made, as shown in *Figure 10*.

Since the height of fill, H, increases and the position of the critical circle centre and depth to the mean depth from which the excess head is computed also increases (in a generally uniform soil), the stability analysis is carried out over a range of H values (H_1, H_2, H_3) as shown in *Figuee 11*. The plot of H

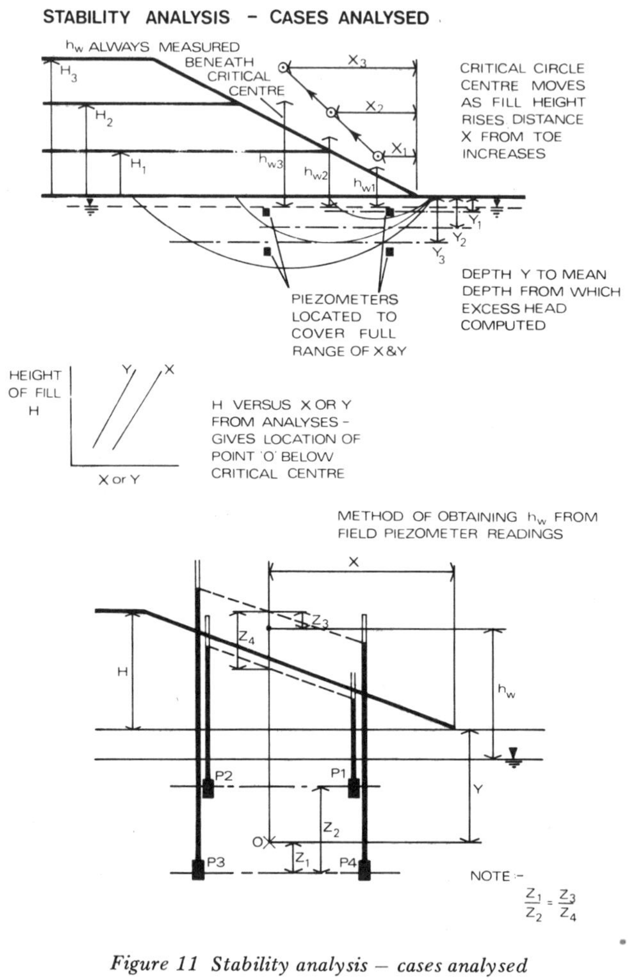

Figure 11 Stability analysis – cases analysed

versus X or Y then identifies the position of the mean depth beneath the critical centre for any value of H, the value of h_w being obtained from the field piezometer readings in the manner shown in *Figure 11*.

'A monitoring diagram of the type shown in *Figure 16* (which was devised by Mr R. T. Murray and given in Road Research Laboratory Technical Note TN 212, September 1967) is then referred to, giving values of F directly from H and h_w.' For the by-pass the value of $F = 1.3$ was selected as being the lowest value permissible without the need for continuous supervision and monitoring.

Should the soil succession not be amenable to circular-arc type stability analysis (because, at Location 2, for example, it can be predicted that the failure surface will lie close to the centre of the peaty silty clay layer), wedge type stability analyses covering a range of wedge shapes were made. As can be seen from *Figure 12*, the same linear excess pore water head h_w was adopted,

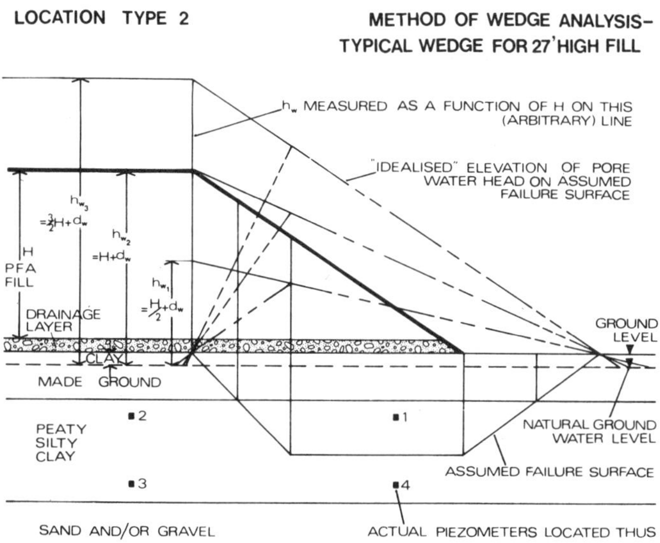

Figure 12 Location type 2. Method of wedge analysis – typical wedge for 27′ high fill

as this gives a close approximation to the theoretical excess pore water pressure on the failure surface, and also enables a range of h_w values to be tried in the calculations to give an F versus h_w relationship for a given H, similar to that shown in *Figure 10*.

Using the values of horizontal and vertical coefficients of consolidation obtained from the site investigation, calculations were made to determine the rate at which dissipation of excess pore water pressures would take place. For Location Type 1, this analysis was carried out in a two-dimensional field, taking account of the greater horizontal permeability, whereas for Location Type 2 only vertical drainage was considered. A step-wise calculation of increasing fill height, allowing time for sufficient dissipation to take place between increments to maintain F higher than 1.3, gave the maximum permissible fill rate.

An illustration of the contributor's discussion in Session 2 gives the soil parameters at two typical locations on the by-pass. In this discussion a typical layout of the instruments was also illustrated; with hindsight, the second inclinometer away from the toe could have been omitted for the purposes of control, although valuable data have been obatined from those installed.

Figure 13 gives the data from Location Type 2, a notable feature being the sudden change in the response of the piezometers to increasing height of fill after Day 79, as shown by the H versus h_w diagram. At this time, the settlement and outward movement also increased markedly. *Figure 17* shows the fill at Location Type 2, placed between two already constructed box bridge structures founded on driven steel H piles.

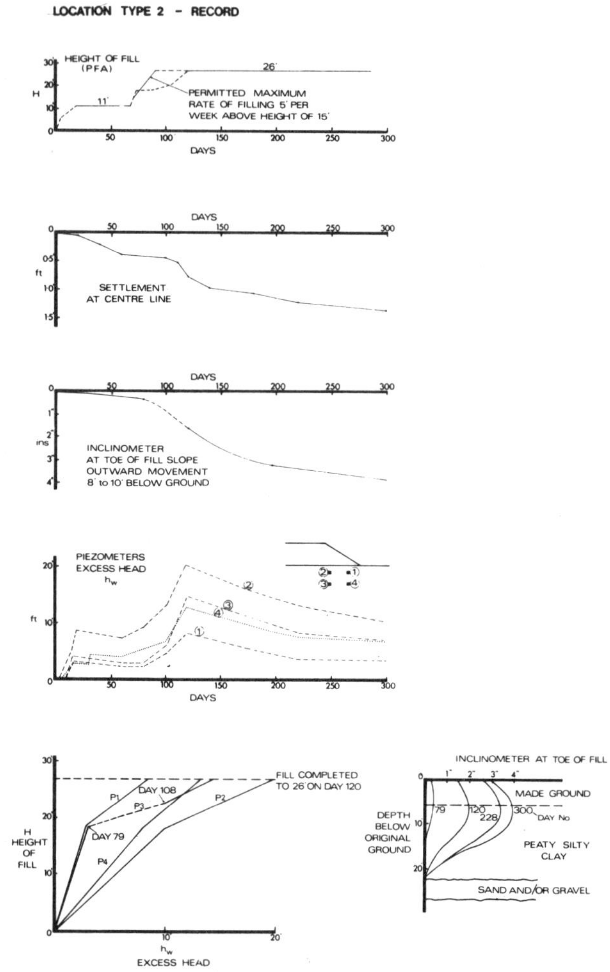

Figure 13 Location type 2. Record

At the first instrumented section of Location Type 1, pulverised fuel ash (PFA) was used as filling, as specified, with the instrumentation results shown in *Figure 14.* As can be seen from the stability monitoring diagram, *Figure 16,* F did not fall below 1.50.

The decision was therefore made to complete the remainder of the filling at Location Type 1 using common fill with a considerable cost benefit. The instrumentation results are shown in *Figure 15,* giving a low value of $F = 1.10$ on the stability monitoring diagram. The inclinometer registered a lateral movement as shown during the three months before filling commenced; no explanation of

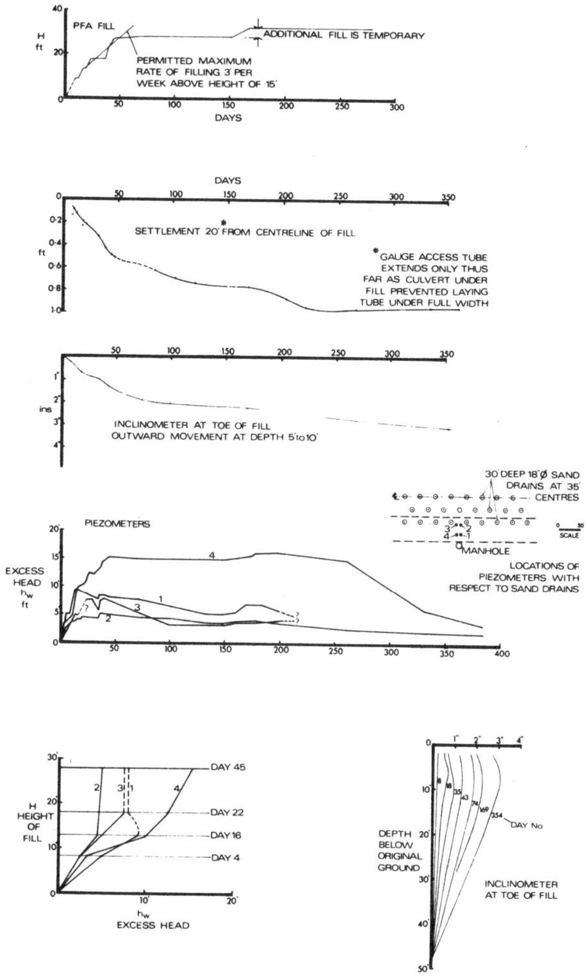

Figure 14 Location type 1. PFA fill

this has been found. *Figure 18* shows the 1500 ft (450 m) length of embankment concerned, the PFA fill being nearer the camera and the lighter coloured common fill beyond.

There is sufficient instrumentation on this site to justify the full-time attention of an engineer who has had one assistant and his own transport throughout. The installation, recording, maintenance arrangements and data production have therefore always been the direct responsibility of one man and this has undoubtedly paid off in the quality of the data available.

With the situation in the instrumentation world as it was at the time of tender (mid 1971), it was clear that only the leading manufacturers could develop and install the required instrumentation and take the proper responsibility for it. Thus only manufacturers were invited to tender for the supply and install contract, and stringent reliability trial procedures were laid down in the specification; instruments not passing these trials were rejected. As the

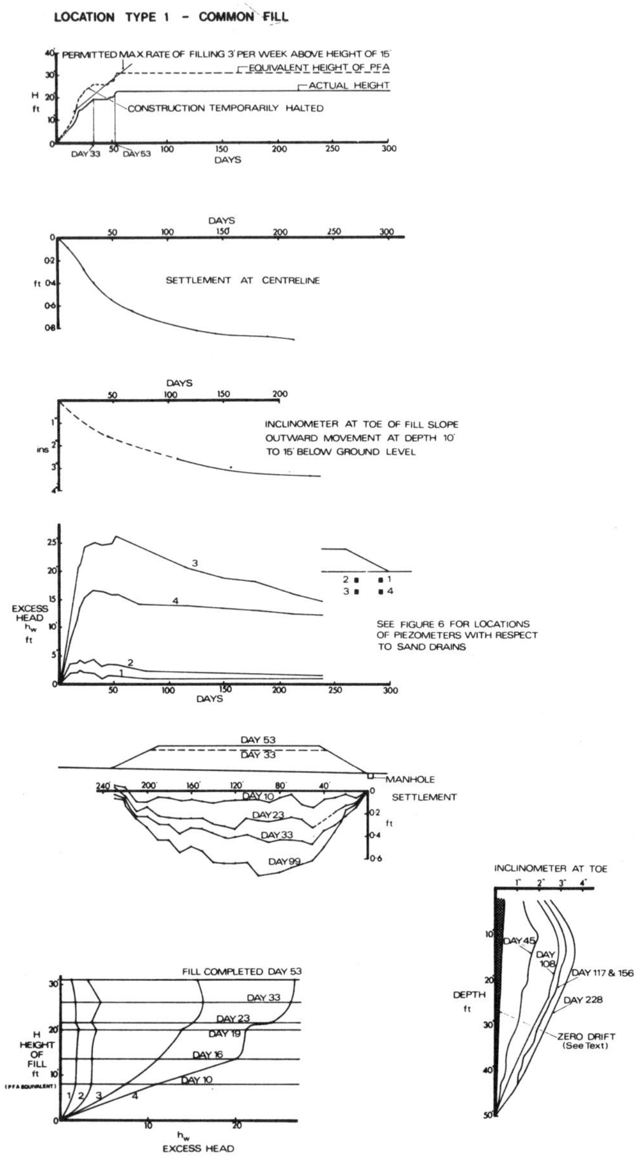

Figure 15 Location type 1. Common fill

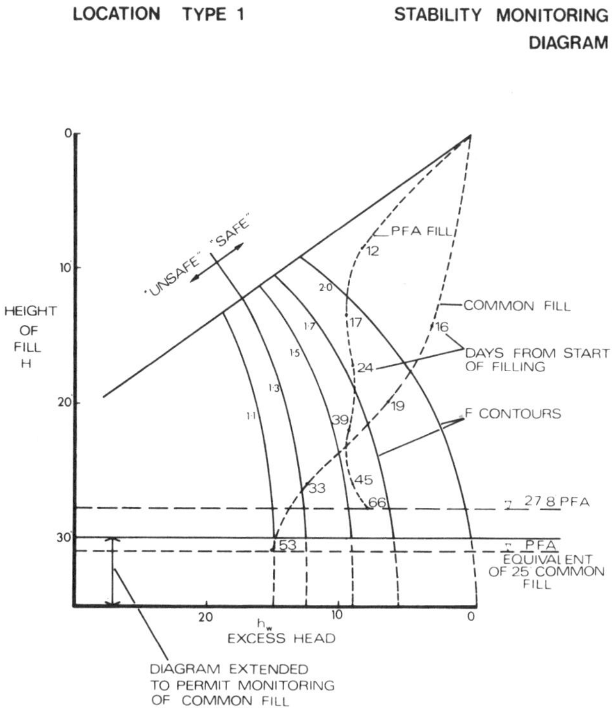

Figure 16 Location type 1. Stability monitoring diagram

Figure 17 Showing the fill at location type 2

Figure 18 The embankment showing the PFA fill and lighter coloured common fill beyond

inclinometer, settlement gauge and piezometer read-out units are all vulnerable and require servicing or repair, arrangements were made with the successful tenderer to have a 48-hour return to site maintenance arrangement, using British Rail as the carrier; this has generally worked well.

The assistance of the author's colleagues, past and present, and the willing help of manufacturers is acknowledged.

J. C. Gray (W. & C. French Ltd.) From the contractor's point of view, instrumentation in an earth embankment is a complication which he would prefer to be without because it tends to interrupt the flow of the work and it is particularly vulnerable to damage from the machines. If we go back 20 years or so when instrumentation started to be installed in earthworks, the earth moving plant generally in use at that time consisted of tractors hauling 8 cu yd scraper boxes at walking speeds. The total loaded weight of these machines was about 25 tons. It was not too difficult to work around the small amount of instrumentation which was being installed at that time. Today, typical plant for earthmoving consists of very large dump trucks capable of carrying 30 cu yd of material at 20 miles per hour weighing 60 tons or alternatively large dual engine rubber tyred scraper units, again carrying 30 cu yd, but at speeds of up to 30 miles per hour and weighing 100 tons. The vast increase in the weight and speed of machines tends to stress and deform the ground to far greater widths and depths than the older type of equipment. It is not surprising, therefore, that although we have progressed in our ideas and methods for making more robust gauges and taking more stringent methods to protect them that damage still occurs.

Rod settlement gauges are a particular nuisance in earthmoving: they are vulnerable to damage and can themselves damage the tyres of the machines. A tyre burst on a large machine costs £1200.

On the Edithmead Dunball Section of the M5 Motorway, my company built on the alluvial plain several miles of surcharged flyash embankments to induce settlement, the instrumentation cost included in the contract was £60 000 and as near as we can judge the probable cost of reading and interpreting the data from this instrumentation was of the order of £10 000. The cost to us in maintaining the instruments in working order was of the order of £15 000, making a total instrumentation cost of £85 000. The value of the earthworks associated with this instrumentation was approximately £2 000 000. Because the engineer had information about the stresses in the soil below the embankments, he had occasion to stop the work when slips appeared imminent and he also decided that it was prudent to redesign some of the bridge approaches. The instrumentation cost was only some 4¼ % of the earthworks cost and was probably adequately justified in this instance.

My company are also currently constructing the King's Lynn Southern by-pass which is the subject of Mr Wilkes' paper. The cost of providing and installing the instrumentation on the monitored section was £55 000 and our estimate of the engineer's costs in reading and interpreting the data over the period of contract is of the order of £20 000. Because of the complications of the bridge over the Great Ouse and the instrumentation that has had to be installed there, it is possible that these figures should be reduced for comparison purposes. The value of the earthworks within this monitored section is only £300 000 but in spite of this low figure our hidden costs associated with the instrumentation and the sensitive nature of the embankment have been £20 000 to date. We therefore consider that the instrumentation has cost £95 000 compared to an earthworks cost of £300 000 and are of the opinion that this is not particularly good value for money. Could not a more stable embankment with less instrumentation have been designed and built for less total cost?

M. K. Sanders (Engeol Geology Consultants Ltd.) and P. G. Fookes (Consultant). As part of the site investigation for a 175 km highway in Fiji, two trial embankments were constructed, one on bedrock, the other across a section of swamp land. The overall site investigation is discussed more fully in a paper by Lovegrove and Fookes (1971).

A large proportion of the road is being constructed over areas of soft alluvial material and peaty desposits associated with inland swamps and coastal mangrove swamps, and it was therefore considered necessary to evaluate the performance of embankments constructed on different foundation materials.

An area at Wainadoi to the eastern end of the road was selected as a suitable site for the construction of the trial embankment on swamp. There was abundant local supplies of suitable fill material and river gravel and as it was situated in an area of high annual rainfall (460 cm per year) the effects of climate could therefore be studied.

The soil used for the trial embankment construction was a Miocene siltstone, having a typical weathering profile formed by hot humid conditions (Sanders and Fookes, 1970) which extended to some 5 or 6 m below ground level.

A typical cross-section of the trial embankment indicating the ground conditions is shown in *Figure 19*. A layer of very soft, peaty clay with silt, about

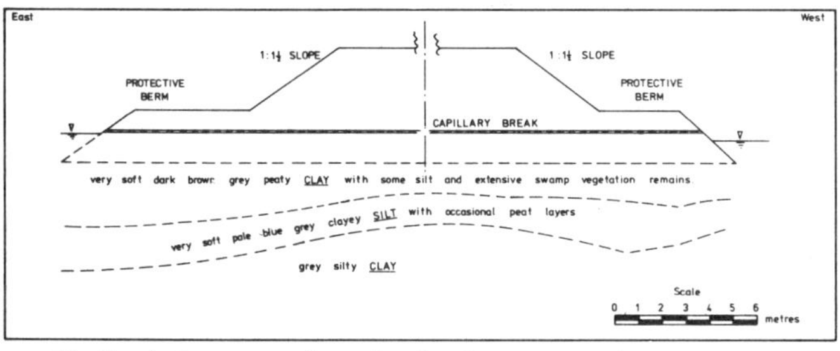

Figure 19 Typical cross-section of embankment indicating ground conditions

2 to 3 m thick, generally overlies a very soft clayey silt with occasional peat layers. Beneath this and over-lying the weathered siltstone bedrock is a soft grey silty clay. The standing water level of the site lies at or above ground level.

Laboratory tests on these alluvial sediments classify the upper material as OH on the Casagrande System, whilst the lower soil falls in the CI group. Natural moisture contents were generally greatly in excess of the liquid limit. Oedometer tests indicated both the materials to have high values of compressibility and of consolidation. A summary of these properties is given in Table 1.

Table 1 SUMMARY OF ENGINEERING PROPERTIES OF FOUNDATION MATERIALS

Geological description	*very soft dark brown grey peaty clay with some silt and extensive swamp remains. (alluvium)*	*very soft pale blue grey clayey silt with occasional peat layers. (alluvium)*
Natural moisture content	99.0 %	112.3 %
Liquid limit	101.6 %	44.0 %
Plastic limit	42.6 %	32.0 %
Plasticity index	59.0 %	12.0 %
Specific gravity	–	2.62
Bulk density	1.27 Mg/m^3	1.36 Mg/m^3
Coefficient of compressibility	1.88 m^2/MN	1.45 m^2/MN
Coefficient of consolidation	0.0777 cm^2/min	0.0527 cm^2/min

Five different types of base materials were proposed for use in the road embankment design and each tried in the trial embankment over a section of some 23 m in length and settlement and pore pressure characteristics were separately monitored. Two sections were constructed with mangrove fascines, one forming a rigid base and the other allowing flexibility. The base of the third section was formed with a 25.4 mm square mesh P.P.F. membrane, and the fourth section was constructed over a close woven polypropylene cloth. The fifth section had no underlying mattress but 0.6 m of free-draining gravel was placed directly on the area of the swamp, which was generally levelled.

The instruments used to monitor the settlement and rate of pore pressure dissipation were all locally made with the exception of the Casagrande piezometer tips, and were therefore simple in their design and construction.

The settlement gauges consisted of a 0.92 m diameter steel plate 6 mm thick welded to the end of a 20 mm inside diameter galvanised water pipe. A larger diameter (51 mm inside diameter) water pipe fitted loosely over the smaller pipe. The gauges were extended in 1.54 m sections by threaded sockets as the height of the trial embankment increased. At the completion of the embankment, both pipes of the gauges were closed with end sockets.

The instruments were quick and cheap to make, readily transportable, and easily installed. The rates of settlement of the embankment were determined from the continual survey of the exposed pipes from the settlement gauges. This had the disadvantage in that they were extremely vulnerable to damage by earth compacting equipment. Boreholes drilled after the completion of the trial embankment to the capillary break, a layer of free-draining gravel used as a marker horizon, indicated that the gauges had settled under their own weight. The use of a lighter alloy in their construction may, therefore, have been advantageous in view of the extremely soft nature of the material underlying the embankment.

The measurement of the rate of dissipation of excess pore pressures was made by the installation of nine Casagrande-type piezometers, attached to 12.7 mm diameter polythene tubing. To ensure an air-tight seal between the porous tip and tube, all the joints were covered with a thick coat of wax several days before installing the instruments. To prevent buckling of the plastic tube during the large settlements expected, it was protected by placing inside a 20 mm diameter water pipe.

In order to determine the ultimate bearing capacity of the alluvial soils, a section of the trial embankment was increased in height to attempt to induce a failure. To locate the position of any shear plane should failure of the trial embankment or of the foundation material have occurred, an inclinometer in the form of a length of 3.2 mm diameter solid steel rod was placed inside the porous tube of one of the piezometers and soldered to the end of a length of plastic-covered wire.

The height of the embankment was increased to about 7 m above swamp level which, spread over a foundation width of 27.7 m applied a pressure of approximately 70 kN/m^2. There was, however, no sign of failure under this pressure, and the usefulness of the inclinometer could therefore not be estimated.

Figure 20 indicates the effects of the foundation material on the settlement

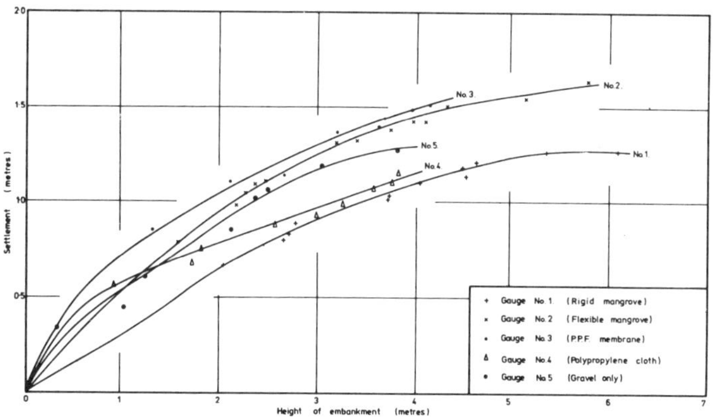

Figure 20 Relationship between settlement and height of embankment for various base materials

of the embankment and values of between 1.3 m and 1.7 m were recorded. The approximate relationship between these two parameters is given by the equation:

$$s = 0.2h + c$$

where s = settlement (metres)
h = height of the embankment (m)
c = imaginary intercept whose value depends on the type of mattress and generally lies between 0.3 and 0.7.

The true relationship is approximately parabolic but for the range of embankment heights between 2 m and 5 m may be considered linear, with the introduction of an imaginary intercept c on the settlement axis by backward extrapolation.

The rigid mangrove mattress appeared to reduce the settlement by up to 50 % for low embankment, but the effect was reduced for higher structures and the settlement of the rigid mangrove mattress was only about 20 % less than the other four sections of the 7 m high trial embankment. The other flexible mattresses did not significantly change the settlement characteristics of the embankment.

The trial embankment was constructed using local labour at a cost of about £7000, and the results enabled a more economical design for the road embankments to be used. It was considered that the extra expense in laying a rigid mangrove mattress to limit settlement was not necessary, and the cheaper technique of incorporating a close weave polypropylene base has subsequently been used.

The authors wish to thank the Secretary for Works, Government of Fiji, and the Consulting Engineers, Rendel, Palmer and Tritton for permission to publish this data; and also to thank, in particular, Mr J. A. Chandler and Mr G. W. Lovegrove for many helpful discussions.

REFERENCES

Lovegrove, G. W. and Fookes, P. G. (1971). 'The planning and implementation of a site investigation for a highway in tropical conditions in Fiji.' *Q. Jnl. Engng. Geol.* **5** 43–68.

Sanders, M. K. and Fookes, P. G. (1970). 'A review of the relationship of rock weathering and climate and its significance to foundation engineering.' *Eng. Geol.*, **4** 289–325.

John M. McKenna (Consulting Soils Engineer). I wish to discuss two points. First, the cost benefit ratio of instrumentation and, second, the best method of controlling the stability of an embankment during construction.

Cost Benefit Ratio of Instrumentation. Mr Gray of W & C French (Constrution) Ltd discussed the cost of the instrumentation on two contracts for which French are the contractors (p. 632), and he calculated that the total cost of the instrumentation was about 4½% of the cost of the earthworks for Contract 12 of the M5 Motorway in Somerset, and about 75% for the King's Lynn by-pass in Norfolk, a figure which Mr Wilkes disputed.

The figures Mr Gray quoted for the M5 Contract, on which I advised the SWRCU, are reasonably accurate for the purpose of this discussion. I would agree that the cost as a percentage of the earthworks can be used as a sort of rough yardstick, but I do not believe that it provides an accurate guide to the cost benefit of the instrumentation. On the contract in question instrumentation was installed to provide information on the two basic design problems of embankments on soft alluvium:

(1) The stability of the embankments during construction;
(2) The rate and amount of settlement under the surcharged side roads and the motorway embankments.

Although the stability problem is the major one during construction, it is only a temporary one, as once embanking has been completed, stability generally improves. The settlement problem is perhaps the more important, as it is virtually a permanent one, for the success or otherwise of the motorway will be judged by the amount of differential settlement which shows up in the pavement.

The settlement readings were used to calculate the volume of fill in the embankments. Accurate settlement records were required as the cost of an inch of fill over the complete length of motorway was £28 000.

The benefit of successful stability control is obvious. As the actual cost of repairing an embankment failure elsewhere on a motorway was about £60 000, the prevention of a single slip would equal the cost of the whole of the instrumentation on this contract. If the embankments had not been instrumented, it is probable that some six or seven failures would have occurred which would mean a total saving in this case of about £$\frac{1}{3}$ m to £$\frac{1}{4}$ m. I think this is the sort of yardstick which should be used when considering the cost of instrumentation. Dr Vaughan did this when discussing the trial bank at Empingham (p. 617).

The second point of my discussion and the one relevant to this session is what is the best method of controlling the stability of an embankment during construction. This problem arises under two conditions: The construction of a trial embankment; and the construction of the earthworks for a major motorway contract. In the case of a trial embankment, there is a large amount of engineering supervision available in proportion to the volume of fill being placed, and is obvious to all concerned, and in particular to the contractor, that the instrumentation is important. Because of the small quantities involved, duplication of the instrumentation is possible, and some damage can be accommodated without losing any control of the monitoring. As some risk of failure can generally be accepted, I consider that embanking should be controlled by strain measurements. With accurate surveying and undamaged equipment, the onset of failure can generally be identified by accelerating ground movements.

The picture changes for major earthwork contracts. As we have heard, contractors consider instrumentation to be a nuisance if it interferes in any way with their filling operations. As a result, there is a higher proportion of instruments unserviceable at the same time as the amount of engineering supervision is reduced. Construction of side road embankments can be erratic as plant spreads are moved around the site. Under these conditions I consider that the most reliable readings are obtained from piezometers, and that the stability of the embankments can be more easily controlled by pore pressure readings. Ground movements readings should be continued, however, as the piezometers may not be positioned in the weakest part of the foundations.

P. K. Wilkes (Norfolk County Council). I wish to comment on the points raised during the discussion, by Dr R. G. Parry of Cambridge University and Mr Gray of W. & C. French Ltd., and also discuss contract organisation as affected by instrumentation (see pp. 614 and 632).

Firstly to amplify the contribution by Dr Parry, regarding the Ouse Bridge problems. As it is intended to fully report the piling trial and also the problems arising during the construction of the piled foundations, only a brief résumé of the significant points that arose during the trial and construction period will be given.

The trial consisted of driving a group of five 560 mm nominal bore BSP cased piles, which were prebored to an average depth of 9 m, instrumentation consisting of piezometers and inclinometer tubes. The diameter of the pre-bored hole was a nominal 700 mm with the casing remaining in until after pile driving, the void left was filled with bentonite slurry immediately prior to extracting the casing. The excess pore water distribution at a level of −6 m i.e. 1.25 m below the preboring levels is shown in *Figure 21*. The main concern was the instability of the river bank due to the pile driving, but the inclinometer profiles did not reveal any detectable movement of the river banks apart from tidal movements. Pile driving was completed in two days and an average bank movement of about 10 mm was measured between consecutive high and low tides.

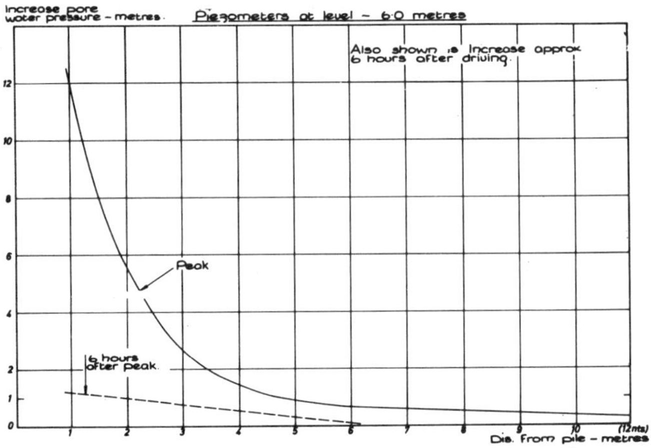

Figure 21 Increase in pore water pressure due to pile driving

During the driving of piles for the permanent works the time period concerned was considerably greater and a problem arose in determining what movements of the bank were due to piling and what movements were due to tidal effects, as a considerable variation in bank movements were recorded at different parts of the tidal cycle. *Figure 22* shows typical maximum variations determined from data from one inclinometer tube.

Considering next, Mr Grey's contribution, two esoteric questions were raised:

(1) What is the true cost of instrumentation?

(2) What are the cost-benefits?

It would appear that for both of these it is very difficult to give a full quantitative answer.

As far as the engineer is concerned the cost of instrumentation is the cost of equipment, installation, reading and analysis, plus any contractural costs. It is the latter that are perhaps the more difficult to assess. If the contractor, as stated by Mr Grey, considers instrumentation a nuisance, then do the increases applied to the various rates for instrumentation and its affect truly reflect the contractor's costs? Perhaps as a contractor gains the necessary expertise this point will not apply. Is the solution to this, as suggested by Mr Cole, to add items to the Bill of Quantities for delays incurred by the contractor due to instrumentation, with perhaps a credit item for acceleration? It is felt that this method could be very difficult to apply.

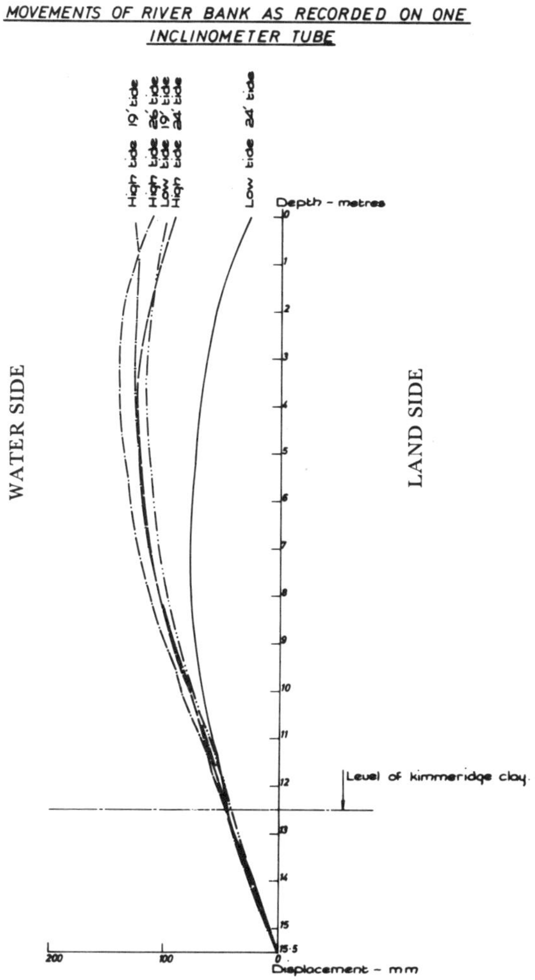

Figure 22 Movements of river bank as recorded on one inclinometer tube

The true cost benefit of the instrumentation is also very difficult to assess. Should this be judged on the overall savings, the overall contract price or merely to the length of earthworks being monitored, as suggested by Mr Grey? What are the true benefits that accrue due to instrumentation? This question can be somewhat imponderable, and King's Lynn Southern by-pass is perhaps a good example of this. Originally the scheme was envisaged as having a very large structural element with embankments of a maximum height of 3 m, due to the low strength and high compressibility of the alluvium. However, a very comprehensive soil survey followed by a trial embankment showed that the structural element could be reduced to a minimum, and embankments could be built to the necessary heights if precautions were taken. A conservative estimate shows that the cost of earthworks is about 25–30% of the cost of a suitable structure. Thus on this basis alone the cost of instrumentation is justified as it increases the designers confidence in the earthworks solution. This is particularly so in view of the consequences of failure – if the river banks are breached then there is a real danger that King's Lynn area could be partially flooded. A large piled cold store is adjacent to the route, so the cost of a failure in this area could be considerable. The proximity of a railway line and a housing estate also make the site of a grade separated intersection, with embankment heights up to 7 m, a high risk area as far as consequences of failure are concerned.

Furthermore the instrumentation allows quantative decisions to be made on placing of additional fill during consolidation periods, and also permits a rapid rate of filling, a metre per day for the first three metres followed by half a metre per day for the next metre, thus allowing the first stage filling to be completed in a minimum of five days. The piling trial and subsequent instrumentation have allowed the use of a better piling system for the Ouse Bridge. The contract instrumentation is also being used to monitor the behaviour of the river banks during the construction of the temporary works, as a requirement of the Great Ouse River Authority.

In the long term, settlement data could possibly effect a reduction in maintenance costs, as it will enable an assessment to be made of the best time to carry out any necessary remedial works. As has been demonstrated, some of the benefits are intangible and thus it is very difficult to obtain the full financial implications of instrumentation.

The cost benefit of the trial embankment is also not easily assessed, as what is the true cost value of the design data obtained? As one of the aims of the trial was to test the effectiveness of sand drains, which were found not to be economically viable, the cost of the trial itself has been saved several times over.

The final point I wish to make in this session is the importance of good organisation in the control of earthworks by instrumentation. Instruments installed for this purpose are only of value if the data obtained allows the Resident Engineer to make a decision the instant that it is required. Thus the instrumentation must be carefully chosen to enable rapid reading and ease of analysis; the site organisation must be such that the data can be assimulated rapidly and reproduced in a form suitable for the Resident Engineer to make his decision. As mentioned in the paper (p. 448), the inclinometer is being used as the primary control with the piezometric installation as the secondary and tertiary control elements respectively. The inclinometer used is a Soil Instruments Mk. II model with digital readout to which has been added a printed output, thus obviating the need for logging between readings and saving a considerable amount of time. The Contractor was required to supply a small 'desk top' computer with graph plotter, and the use of these has meant that the average time taken from measurement to handing over the $\Delta s/H$ versus time plot, as given in the paper, is of the order of one and a half hours. This speed is a necessity when road embankments are being constructed with a rapid rate of filling on these types of sub-soils.

In the critical area, where two stage filling is being undertaken the piezometric system is also automated which enables the production of the back up data for the Resident Engineer with almost the same rapidity.

Summarising then, the type of instrument must be carefully selected not only for the type of measurement but also for ease and speed of reading, modified for the latter if necessary, so that the requisite control can be achieved. The organisation of the site staff must be conducive to the production of the necessary control data, supplemented with equipment such as desk-top computers and plotting facilities wherever possible. The use of large computers can be kept for processing long and short term monitoring rather than the immediate control during construction.

D. J. Knight (Sir Alexander Gibb & Partners). Dr Vaughan has highlighted particular questions in his preliminary report and I should like to comment briefly on three of them under the general heading of organisational and benefit aspects, namely (1) contractual matters, (2) benefits and (3) organisation for long-term observations, each of them with particular reference to remote, overseas embankment dam sites.

Contractual matters. Both Mr Kennard and Mr Wrightman dealt in their papers with contractual arrangements for instrumentation, and two points are here selected for comment, namely, contractual responsibility for the installation of instrumentation, and the method of billing. My firm now prefers to make the main contractor responsible for installation but to specify that he obtains specialist assistance. The engineer, of course, retains overall supervision. The reason for this preference is that it obviates claims due to delays, and gives the contractor as much scope as possible for fitting the instrumentation installation into his overall programme. This procedure appears to have worked satisfactorily.

We have used both main methods of billing instrumentation. The first is billing a provisional sum under which in effect a supply sub-contract is let to a nominated supplier based on a separate specialist tender, after award of the main contract. The second is by requiring the main contractor to price individual instrumentation items detailed in the main bill. We usually adopt the second as far as possible with 'standard' equipment, but, according to the job, provide provisional sums for expensive items which are continually being developed, such as distance measuring equipment and seismometers, the purchase of which is then dealt with on a nominated supplier basis. Provisional sums are also included for additional instrumentation. Where instrument design modifications which are desirable occur after award then revised rates are negotiated as necessary for the items affected.

The main and real disadvantage of individual billing in the main contract is that contractors tend to go to one supplier for all items, whereby its is possible that some items may not be acceptable. This is a situation best clarified before award, although a main tenderer may not wish to decide on his specialist supplier before award. This objection is dealt with in a moment. An alleged disadvantage is that main contract tenderers will naturally accept the lowest offer for instruments from potential suppliers, but to this and the previous objection the reply is two-fold: firstly, each instrumentation item has to comply with a technical and functional specification; if it does then the cheapest is acceptable and if it does not the contractor is obliged to obtain an item which does; secondly, it is most unlikely that the placings of the two lowest main contract tenderers will be affected by differences in instrumentation costs. If it was then the matter would be examined in detail before award.

The advantages of individual billing for overseas work with international tendering are those of enhanced competition, a potentially greater choice of instruments, and satisfaction to the client, the owner, by manifest impartiality with respect to supply.

Benefits. It should be acknowledged more than it often appears to be that it is not, in general, practicable or even desirable that all the possible 'benefits' should be obtained from a given system of instrumentation, if by 'all' are included benefits to engineering knowledge generally. It is important that recognition be given to what can and what can not be done. There is a difference between UK conditions and overseas conditions in this respect. In the

UK, where promoters are often Government Departments or public bodies, it is possible for some instrumentation to be included within a given structure which is for a research or 'future general benefit' purpose only, and not strictly required for the particular structure. Provided that the promoter's consent is obtained to such work it will potentially lead to a general engineering benefit. Overseas, however, the purpose is rightly much more likely to be restricted to the benefit of a given job only, and thus to a particular promoter. The benefits are specifically limited to confirming the design and to monitoring behaviour. Only future publication, where agreed, will extend the benefits. Overseas clients are, however, more likely to accept research type investigations where the results could lead to economy in a second stage of a job. The trial embankment in Roseires reservoir described by Fitt *et al* (1967) is an example of this.

Organisation for long-term observations. The following remarks are stimulated by and are intended to amplify some of the points listed in the section of Dr Vaughan's preliminary report under the same heading.

Firstly, an instrumentation system should be designed as simply as possible compatible with its purpose. For example, in one of our overseas dams for which it was desirable to monitor internal seepage, standpipe piezometers only were used, being installed in the downstream shell and foundations, instead of hydraulic piezometers additionally being installed in the core. The standpipes were considered sufficient for the seepage monitoring, and these obviously are also much simpler to read. Survey stations were also provided for recording vertical and horizontal external movements.

The problem of staffing after the dispersal of the construction supervisory staff will arise. Local, trained nationals are clearly the most desirable, but this is not always possible to arrange. Where expatriates serving Government agencies and departments are available, for example survey, their skills can be enlisted. In some cases it may be necessary to apply to outside agencies for assistance, such as those representing the UN.

Reliability and maintenance of equipment is of obvious importance. On remote sites it is prudent to double up in the provision of key items such as slope indicator torpedoes, which can take a long time to repair if sent away.

Just as important as the purely technical aspects of instrumentation is the organisation of the actual business of specifying, taking, recording, submission and processing of the readings from a remote site. Our practice is to write an operations manual for the works, in which the instrumentation system is described within the context of their overall design, operation and maintenance, and to produce a schedule of instrumentation readings required, *pro forma* for each type of instrument for the recording of the readings, and plotting charts for the observations from each instrumented section. These charts are, as far as possible, designed with the plot for each instrument within a particular section located in correct relation to the others, with separate charts for piezometric levels and deformations; for example, upstream foundation piezometers' records would be found at the bottom left hand corner of the chart, core piezometers' records in the top centre and downstream shell piezometers top right. Plotting charts are provided as negatives, so that at agreed intervals prints of the plotted readings may be sent back to the design office for perusal and comment. Such a system has the advantage that the organisation responsible for the readings can plot the end product on to the charts and spot immediately any odd reading, which may then be repeated if necessary without delay. A 'feel' for the behaviour is thus hopefully engendered. After a period of say 5 years a new set of charts is issued from the design office, with perhaps a reduced future reading requirement, depending on the pattern of behaviour observed during the initial period.

The problems of establishing safety criteria, guarding against 'false alarms' and acting on warning signs in long-term monitoring are discussed in Session 4 under 'Interpretation'.

REFERENCE

Fitt, R. L., Marwick R. and Whitaker, F. (1967) 'The Roseires Dam, Sudan: planning and design'. *Proc. I.C.E.* 38 21–51.

D. E. Sherwood (Soil Mechanics Ltd.). Mr Kennard suggests that there are three methods of arranging instrumentation on large contracts; the *nominated sub-contractor*, the *nominated supplier* and the *provisional sum.* Jack Wrightman describes various approaches but seems to favour both the *nominated supplier* and *nominated sub-contractor* or a combination of both. These writers have presented their viewpoints as an engineer and major employer but I would like now to present the specialist sub-contractors problems as I see them.

First, let me add a fourth method to Kennard's list; *no special arrangement.* This is the procedure followed when the supply and installation of instruments is billed in the same fashion as the supply and placement of concrete, but the specification for the latter is much more definitive. I am sure we would all agree that this is the best method of getting an inoperative and inadequate installation. Certainly the 'strong personal commitment' which Wrightman has found to be desirable is bound to be absent.

Let us look then at the *nominated sub-contract.* It would be naive to believe that all the best instruments come from the same stable, but can an engineer be sure that a manufacturer/contractor will price for the use of another manufacturer's instrument when that is the most appropriate instrument to use? Can any contractor afford to price for anything but the cheapest instrument which meets the specification but may have uncertain long term performance characteristics? The latter is rarely specified in the competitive instrumentation market today, he will not stand a chance if he does otherwise. Instrumentation like many other things is very specialised and it is a question of getting what you pay for.

What then of the *nominated supplier*, who installs the instruments so carefully chosen usually from one stable but occasionally several stables? Not the main contractor but the lowest tenderer from a long and ill assorted list of boring sub-contractors assembled by the main contractors buying office. Should by chance a good sub-contractor emerge he may well have to install instruments with which he is not familiar leading to indifferent results. If the installation is to be carried out by the Resident Engineer's staff, will this staff have the experience particularly early in a contract when foundation instruments are being placed?

Finally, the *provisional sum.* On the face of it this looks the best method so far. The engineer can choose his supplier and sub-contractor and decide as the job proceeds exactly what he is going to do. There is just one problem though. The main contractor does not know anything about the instrumentation at the time of tender and does not want to know too much about it at the construction stage, unless of course the employer pays up on a host of claims for 'obstruction', 'charged working', 'work in small quantities' and a dozen other quantity surveyor terms. What then is the answer to the engineer's dilemma. Regrettably nothing is ever ideal but the following modus operandi should meet much of the criticism.

(1) Treat the main contract and instrumentation sub-contract as entirely separate at the tender stage.

(2) At the time of tendering, tell the main contractor what instrumentation is likely, where it will probably be placed, what assistance he will be expected to give with his on site plant, what protective measures he will be expected to take and what programme constraints apply for installation. But he must not be asked to price these commitments within his rates say for bulk filling as their extent cannot reasonably be assessed. Is it not reasonable, however, to include daily rate items for genuine obstruction or complication of major work items such as earthmoving and bulk filling based on instrumentation rig days or team days as appropriate? I believe that such practice would do much to curb the attitude that 'instrumentation only hinders the work'.

(3) Chose a competent specialist sub-contractor. In many cases he will require sufficient expertise to be able to advise the engineer on choice of instruments and methods of installation. It is not reasonable, however, to ask for this advice under the pressures of competitive tendering, and the sub-contractor should be chosen on the basis of tenders for which a selected list can compete on an equal basis; that is metre rates for drilling, boring, grouting, trenching, line laying, etc. and unit rates for rig mobilisation, moves and the like. It will also be possible in many cases to include the main contractor's tender programme with the sub-contract tender documents. The supply of instruments and in some cases the design and manufacture of special items, should be included as a Provisional Sum.

(4) Finally, when both contractors have been chosen on a basis of work for which they can reasonably be expected to compete, the purchase of the right instruments for the job can be dealt with. Points which would be considered are the engineer's requirements and preferences, the sub-contractor's expertise and preferences and the main contractor's programme for the works. It does not seem necessary to have the cost of individual instruments bound in contract as the vast majority will be be off-the-shelf items for which price lists freely circulate.

Whether the instrumentation contract is then organised as a nominated sub-contract or as a separate contract will depend largely on the nature of the project and the extent of responsibility which the engineer wishes to take for coordinating the works as a whole.

M. F. Kennard (Rofe, Kennard & Lapworth, Consulting Engineers). The contractural matters involved in supplying, installing, maintaining and observing instruments is as important as the design of instrumentation. Unfortunately little discussion took place on these matters, but Mr D. E. Sherwood referred to my paper in which several methods were set out (see p. 220).

My paper was from the point of view of the experienced consulting engineer, whilst Mr Sherwood's views were those of an experienced specialist contractor. Providing that the leading party, who may be the client's engineer, the consulting engineer, or the specialist contractor or sub-contractor, has continuity of experience in this particular field, satisfactory results may be achieved with any method. Unfortunately, the leading party does not always have sufficient experience and it is then that mistakes occur and wrong procedures are practised.

No method is perfect and the system should be chosen having regard to each particular job and its geotechnical problems. Mr Sherwood emphasises that the main contract and the instrumentation sub-contract should be entirely separate, and I agree that this is most important. I consider, however, that Mr Sherwood over-emphasises the obstruction problems, and daily rates by the main contractor would not appear to be necessary. Obstruction in one area can easily be avoided by working in another area for that day.

I have not encountered any claims by contractors on account of any instrumentation problems. Good experienced Resident Engineer supervision is required in any case, and the Resident Engineer must know the purpose of the instrumentation programme, and he and his staff should be involved in reading and plotting the results.

D. A. Cook (Bath University). Out of thirty-seven Papers presented to this Conference, it is perhaps a little surprising to find that only one deals specifically with the instrumentation of a building structure. Three papers indirectly relate to buildings in terms of measurements of diaphragm walls, survey observations and piling. By comparison, figures published by Small (1972), for Construction Output during 1970, indicate that by far the greater proportion of the national budget is directed to building construction, as opposed to road works, earth dams and major civil engineering works. Yet the emphasis in instrumentation seems to be directed towards the latter category. In this type of work relatively large deformations have to be accepted almost as commonplace, and this Conference has shown that their measurement can make a major contribution to the final finished design.

Nevertheless, engineers approach the design of building structures with no less care than larger works; although margins on Foundation Factors of Safety are inevitably higher because of the much more severe limitations on settlement. Seldom however are instruments ever used to 'back-check' on a design. This itself is a great pity; because an enormous wealth of information and experience is lost thereby. Not only is it a pity, it shows something of an anomaly. Few engineers who design a concrete mix would question the need for back-checking their design by cube tests, continued throughout the duration of a contract. Yet the variables here are no greater than the assumptions made from a soil survey.

In the context of available field information from full scale structures Mr W. J. Larnach, as Chairman of the South Western Association of the Institution of Structural Engineers, convened a meeting to discuss what contribution could be made to the Institution's working party on Soil-Super-structure interaction. The discussion revealed the most acute scarcity of reliable data on actual building performance in the locality.

What was equally revealing were some of the probable reasons why such information is so limited. For instance, it was felt that any attempt to go round on a survey of building settlements would meet with quite overwhelming resistance from both the owners and engineers responsible for the structure. The fear on part of both parties was the publicity which might be attached to findings of a disparaging nature. One must hope that sooner or later this mental 'bloc' to routine back-checking will diminish. At the present it restricts much of the scope for research, and for practising engineers to better their day to day experience, in that section of construction where the greatest volume of work is done.

REFERENCE

Small I. (1972) 'Broader Training of Civil Engineers', *Correspondence in New Civil Engineer* 2 Nov. 1972 pp. 20–21.

J. B. Burland (Building Research Station). Following Mr Cook's remarks I should point out that although 70% of construction cost might be going into buildings, they are all in relatively small units of cost and the cost of instrumentation is high. This is probably the reason why very little work on soil-structure interaction has been done. It is perhaps one of the most expensive types of research one can envisage and it's not surprising that owners are not eager to embark on a £50–100 000 research project for a building costing perhaps £¼ to 1 million. However, there are proposals to do work on soil-structure interaction.

Mr Cook has mentioned the group under the auspicies of the Institution of Structional Engineers. Close liaison has been established between this group and the Building Research Station who are now carrying out a detailed programme of work in this area. It is particularly difficult to find suitable buildings and if anyone has a building which they think would be worth instrumenting they may care to contact the Building Research Station.

D. W. Hight (Scott Wilson Kirkpatrick and Partners). The purpose of this note is to describe one of the foundation instrumentation schemes referred to by Dr Burland in his summing up of Session 4 and to comment on some details of the design and behaviour of the instruments. CIRIA have given Scott Wilson Kirkpatrick and Partners a research contract to instrument the foundations of Dashwood House in London. The foundations of this seventeen storey structure comprise a 1.5 m deep raft capping 0.485 m diameter bored piles; the piles are on a square grid at 1.5 m centres and they penetrate, on average, 1.0 m of Thames Ballast and 14.0 m of London Clay.

The object of the instrumentation scheme is to determine how the structural loading is shared between the raft, which sits on the Thames Ballast, and the pile group, and to determine how that portion of the load carried by the pile group is distributed amongst its members. Settlements are to be monitored at three floor levels to see how the settlement profile develops as stiffness of the structure increases and a borehole extensometer has been incorporated to determine the development of strains within the London Clay.

The symmetry of the structure and its loading and the uniform soil profile across the site has enabled us to instrument one quadrant only of the foundation. Fourteen pressure cells have been installed at the raft/soil interface to determine the distribution and magnitude of contact pressure. Four thermistors have been placed within the ballast adjacent to the pressure cells in order that temperature corrections of the pressure cell readings may be attempted. Nine load cell units have been placed at the head of selected piles to measure the total load carried by the piles. Instrument leads and cables have been led to access chambers set in the raft above each load cell position.

Each load cell unit consists of three 250 kN capacity vibrating wire cells placed concentrically between circular, 25 mm thick, steel plates of 457 mm diameter. The use of three cells in each unit was dictated by cost and speed of construction; it has the advantage that the unit will continue to provide information even if failure of one or two of the vibrating wire cells occurs. Nonlinearity and hysteresis in the load/(frequency)^2 calibration plots were largely eliminated by using a vibrating wire cell with a height to gauge length ratio of 3 and by fixing each cell at its top and bottom to the steel plates.

The pressure cells are of the oil-filled capsule type. To satisfy the criterion noted by Wong in his paper to this conference (p. 475) of having grain diameter not greater than two per cent of diaphragm diameter a 75 mm bed of sand was provided. The type of settlement points and borehole extensometer have been described by Cheney and Burland and Moore, respectively, in this conference.

The instruments have been in place between 3 and 6 months and construction of the first floor is now in progress. The load cell units showed zero drift as a result of handling and placing of the raft concrete. Further destressing and isolation of the unit from the vibration and temperature effects due to concreting might have improved their early performance.

The heating/cooling cycle resulting from hydration of the raft concrete has also affected the zero reading on the pressure cells. Volume expansion/contraction of the oil produced a cycle of loading on the pressure cell and its bedding. The behaviour of both was non-elastic and a lower pressure existed in the cells after cooling. The use of a pressure cell with a smaller oil volume or one which had been effectively 'scragged' to achieve recoverable behaviour might have reduced this problem.

To determine the cell action factor (cell registration) for the pressure cells, laboratory tests are to be performed which simulate the field installation procedure, including the heating/cooling cycle, since it is appreciated that the factor will change under cyclic loading of a granular bedding. In order to arrive at a temperature calibration for the cells, the *in situ* restraints to cell expansion/contraction are to be simulated.

The instruments are now responding to structural loading and their behaviour appears to be satisfactory; the problems described above seem only to have affected instrument zeros.

Rainer Massarch (Royal Institute of Technology, Sweden)
In situ measurement of horizontal total stresses in clay under an embankment.
The knowledge of the horizontal stresses, as expressed by the K_0-value, as well as the changes of horizontal total stresses is of importance in the analysis of many geotechnical problems. In the literature only few examples are reported from *in situ* measurements of total stresses in clay. A method suggested by Kenney (1967) gave reasonable results in soft clay but required complicated and expensive equipment. The hydraulic fracturing test as suggested by Bjerrum & Andersen, (1972) is simple but limited to normally consolidated clays. The boundary conditions are difficult to evaluate for a heterogeneous clay around the filter stone.

The inexpensive and simple Glötzl valve was used in the construction of a total stress cell which can be pushed into soft clays (see contribution by Massarsch, p. 546, Ed.). This cell was tested at the Skå-Edeby test field near Stockholm where the soil conditions are well known from earlier investigations. The measurements as described were carried out in connection with a full scale loading test where a new paper-plastic drain developed at the Swedish Geotechnical Institute has been used.

The soil profile at Skå-Edeby consists of 10 m thick varved glacial and post glacial, slightly organic clay with sulphide bands overlaying rock or moraine (*Figure 23*). The cell was pushed 5 m into the clay and then pulled up to check that no damage occurred due to installation. Then the cell was installed 5 m below the ground surface at the edge of a planned trial embankment. Nearby the total stress cell, a Geonor piezometer and an inclinometer pipe were installed (*Figure 24*).

The results of the measurements which were carried out under a period of 8 months are shown in *Figure 25*. The excess earth pressure caused by the disturbance due to installation of the cell dissipated within four days. During

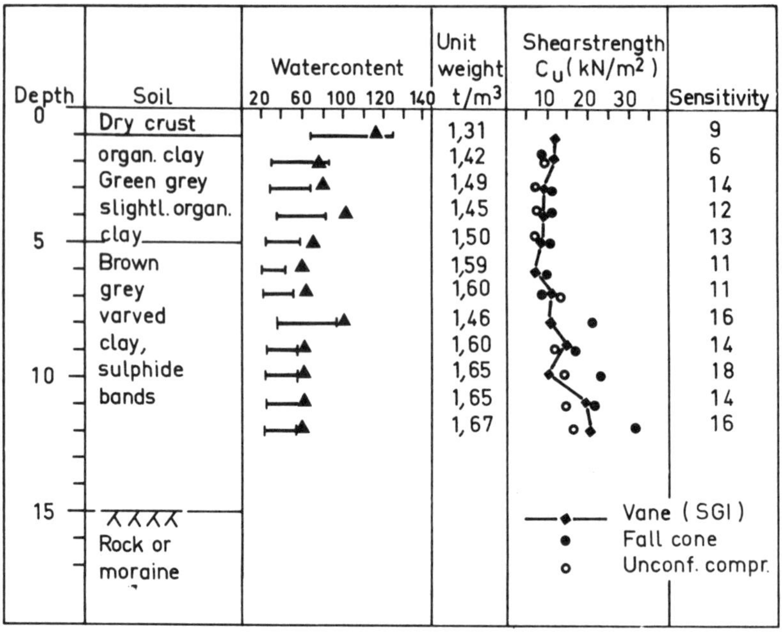

Figure 23 Soil profile at Skå Edeby

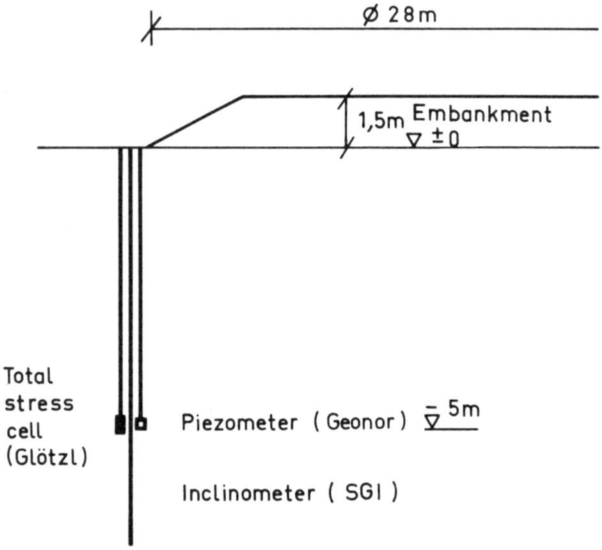

Figure 24 Measuring equipment at the edge of the trial embankment

the summer the paper drains were installed close to the cell which can explain the slight increase of the horizontal total stresses.

From the horizontal total stress- and porepressure measurements the K_o-ratio could be calculated. It varied between 0.43 and 0.48. After installation of the drains a 1.5 m high embankment was placed in layers causing an overburden load of 27 kN/m^2. Continuous readings were made during the filling procedure (*Figure 25*) and they show an immediate increase of the total stresses and the pore pressures. The inclinometer measurements indicated a sudden horizontal deformation in the clay due to loading.

The pore pressure increased during the filling procedure somewhat faster than the total horizontal stresses but both reached about the same peak value. While the total stresses remained constant during 7 months, the pore pressure decreased after the fill was completed. Also horizontal deformations due to the consolidation have been observed at the level 5 m below the surface by inclinometer measurements.

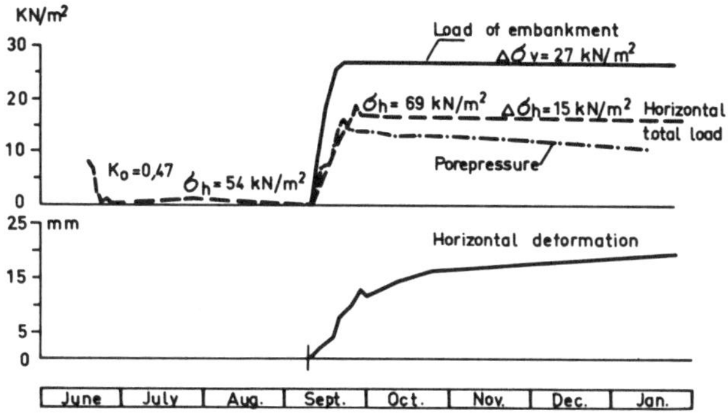

Figure 25 Results from earth pressure, pore pressure and inclinometer measurements 5 m below the ground surface

At the test field at Skå-Edeby, hydraulic fracturing tests were carried out which gave an effective stress ratio (K_o) somewhat higher than measured with the total stress cell.

Total stress cells type Glötzl have been installed at two additional constructions sites at a maximum depth of 10 m and they seem to give reasonable results.

REFERENCES

Bjerrum, L. and Andersen, K. H. (1972), '*In Situ* Measurement of Lateral Pressure in Clay'. *Proc. 5th European Conf. Soil Mech. Found. Engng*, Madrid, April 1972, 1, 1 pp. 11–20.

Kenney, T. C., (1967), 'Field Measurements of *In Situ* Stresses in Quick Clays'. *Proc of the Geotechnical Conference* Oslo 1967, Vol. 1, pp. 49–55.

B. O. Skipp (Soil Mechanics Ltd.). I would endorse Mr Cooke's plea for more measurements of buildings. It may interest the participants of this symposium to know that there are currently discussions with the International Standards Organisation on acceptable limits for vibration in structures. Some of these criteria are in terms of dynamic stress levels. It is however important to know what deformations are associated with various qualities of foundation performance, because without such information dynamic stress criteria may be of dubious value.

V. Ashkenazi (Nottingham University). Several speakers, notably Professor Bishop and the session reporter, Dr Vaughan, stressed the importance of methods for long term monitoring of displacements. A modern surveying network, based on a series of strategically placed pillars and involving the measurement of theodolite angles and high accuracy lightwave distances, is eminently suitable for monitoring displacements which are likely to occur in an engineering structure, such as an embankment dam and its immediate surroundings, both during construction and the long term post-construction period.

The network would include not only reference stations from which a large number of individual targets on the embankment walls could be observed, but would also incorporate points in close proximity to the other instrument and gauge housings used for monitoring surface and sub-surface movement. The periodical angle, distance and levelling measurements would lead to a series of 3-d coordinates based on an absolute cartesian system. With current (1973) equipment, it is not too difficult to achieve relative overall positional accuracies between stations of 3 mm or better.

The great advantage of such a surface reference system would be its absolute flexibility. Old stations might disappear under the water when the reservoir is full to re-appear again on those occasions when the reservoir is emptied, new stations might be added at will; the space between some stations which were intervisible (for observational purposes) may become obstructed by growth of dense vegetation. And yet it would always be possible to obtain the 3-d coordinates of the remaining and newly established stations which together form the current network.

Computer based statistical network adjustment and strength analysis techniques would provide not only the current values of the coordinates, but also estimates of their absolute positional accuracies (with respect to a chosen fixed origin) as well as relative accuracies (with respect to any other station in the network). These statistical estimates of accuracy can be forecast *before* the observations are carried out, by considering the approximate shape of the proposed layout of the network and the purported accuracies of the available observational equipment. Conversely, given the proposed layout of a network and the required positional accuracies, one can advise on the most practical combinations of surveying equipment which are likely to produce the required standards of accuracy.

The survey pillars and the other types of semi-permanent reference marks, which constitute the network stations, can be built anywhere which, in the opinion of the geotechnical or structural engineer, deserves to be incorporated directly in the primary monitoring system. A network of about ten to fifteen stations would be considered as suitable for monitoring displacements on and in the immediate vicinity of an embankment dam.

While all the primary network stations are fully interconnected through the geometry of the network, the observations of large numbers of secondary targets would only be made from and referred to one pair of stations. The extent of the area covered by the primary network would again be left to the discretion of the consultant engineer and would be based on considerations of stability of pillars built beyond the area immediately affected by the structure.

Surveying networks have and are currently being used in a variety of scientific and technological applications, from continental satellite networks connecting national triangulations, to the setting-out of complex motorway intersections and the very high precision layout of complex scientific apparatus such as radio-telescopes or synchrotrons. Chalk marks would probably be sufficient as targets for close range theodolite observations made to monitor the structural displacement of some bridges during construction, to accuracies of 10 mm or better.

S. J. Patchet (Newcastle upon Tyne University) described deformation measurements made in a new shaft more than 3500 ft deep, sunk for Cleveland Potash Ltd. The shaft was excavated 26 ft diameter and lined with a 2 ft thick lining, only 18 ft internal diameter. The 2 ft peripheral gap was filled with vermiculite to allow for rock creep which would tend to close the gap. Simple diameter deformation stations were installed as the shaft was sunk and the shaft diameter measured with a very simple gauge consisting essentially of a ¾ in diameter steel bar with grooves at 1 in spacings.

Rod extensometers were also installed in 15 ft deep holes drilled radially into the rock from the shaft. Several anchors were placed at intermediate positions in the hole, each with its own rod coming to the end reading point. The gap between the rock and the lining presented something of a problem and a special telescopic unit was used to bring the gauge through the lining when it was built, after the shaft had reached full depth. These gauges had an accuracy of ±0.0025 in and a range of 6 in which was achieved by using 1 in spacers. The system has to be extremely simple to get accuracy in the very difficult conditions during shaft construction when, in addition to all the construction plant etc., there was a continuous shower of salt water.

The deformations measured by these extensometers could be converted to absolute movements by relating the ends of the rods to the mine surveyors plumb lines. To facilitate this, the gauges were placed just behind the plumb lines.

The instrumentation for the whole shaft was put in in 7¼ hours and the shaft cost was reckoned at £200 per hour.

T. W. Thompson (Newcastle upon Tyne University) gave details of an underground experiment to measure the *in situ* properties of rocksalt carried out to obtain data for the design of underground storage cavities. A cylindrical cavity had been excavated in the rocksalt and provision was made to alter the temperature in the cavity and increase the pressure in it from 0 to 200 psi. Strain bars were put across the cavity to measure diametrical deformations and there were nine radial boreholes containing rod extensometers with anchors at 10 ft 6 ft, 2 ft and 9 in from the wall of the cavity. No one could be inside the cavity when the temperature and pressure was changed, so all the gauges had to be remote reading, and LVDT's were used. Pressure in the radial holes would have upset the stress distribution, so the rod extensometers were sealed with O-rings and vented through pipes to the atmosphere. Strain gauges 4 ft long were attached to the walls of the cavity, also using LVDT sensing units.

Temperatures were measured by thermistors. For near surface measurements these were embedded in the laboratory in a core taken from the rocksalt walls, which was then returned and cemented into the corehole. In total there were 54 transducers and 138 temperature probes in the cavity. A sophisticated logging system was used which gave the readings on punched tape which could be fed to a computer for analysis.

Troubles included dry joints and a peculiar fault with the LVDT's. Four of them were connected to a common supply lead. They were of the d.c.-d.c. type, with an internal oscillator supplying a.c. and an A to D converter for the output. Considerable trouble was caused by inductance fields which prevented the proper functioning of the oscillator and caused the readings to fluctuate excessively. This was overcome by putting capacitance across the supply leads.

With this instrumentation, it has proved possible to measure deformations to better than 10^{-4} in.

B. N. Whittaker (Nottingham University). I wish to add a few words of support to the contributions by Dr Patchett and Mr Thompson who spoke on rock instrumentation work currently being carried out by Newcastle University. The form of instrumentation described is based on measuring displacement. This is not surprising in view of its meaning being better appreciated than other parameters in addition to its measurement being more reliable.

Rock mechanics instrumentation schemes often use the borehole strain wire (or tape) and extensometer method of measuring displacement around underground excavations or in rock masses generally. It requires a hole of suitable diameter drilled deep into the rock mass. Anchor plugs, each with a wire attached, are then located at predetermined points along the hole. Up to ten plugs have been located in one hole. The plugs have a hollow central shaft in order to permit the wires from leading plugs to pass freely. At the mouth of the hole, the wires each have a tensioning adaptor. This is for location on an extensometer which is mounted on a special standpipe grouted to the strata at the hole mouth. Each wire is tensioned accurately in turn for a reading to be taken of the displacement of the wire relative to the borehole mouth. This form of instrumentation is a useful means of determining the stability of the roof and sides of underground excavations, the slopes of rock masses and support pillars in mines. The rate of movement of the rock at the borehole mouth is observed relative to a plug placed at some depth within the surrounding rock mass. This technique is now widely accepted as a valuable rock mechanics control method of ascertaining the relative stability of rock structures.

Mr Thomson has made a unique contribution to this Conference. He has described a scaled down field investigation of a gas storage chamber in salt. I should like to ask if this sort of approach to collecting basic design data can be applied to other forms of rock structures.

J. N. Hutchinson (Imperial College London). Dr Hutchinson expressed interest in the paper to the Symposium by Whittaker and Forrester on the measurement of ground strain and tilt arising from mining subsidence. He wished to confine himself to discussion of a topic not touched upon in the above paper, the effect of tilting of the ground on slope stability.

At the sides and ends of subsidence troughs, it appears that tilts of the ground surface of between 1 and 2° over slope lengths of 100 m or more are not uncommon. A maximum measured tilt of 1.8° is given, for example, in *Figure 9a* of the above paper. Where tilting increases the valleyward slope of the ground, the influence of a given tilt on the slope stability will tend to be the greater, the smaller the shear strength is on a potential slip surface. Renewals of movement in old landslides, on their pre-existing, residual strength slip surfaces, are thus likely to be particularly prone to occur in such circumstances.

The potential seriousness of such tilting may be illustrated by considering an old translational slide with a residual strength of $c_r' = 0$, $\phi_r' = 15°$ on its slip surface and a groundwater table coincident with the ground surface. If the slope at which this is just stable is tilted valleywards by 1°, a reduction in the safety factor against sliding of around 10% will ensue.

In view of the potential significance of subsidence-produced tilting Dr Hutchinson valued the work being done by Whittaker and Forrester. However, as the amount of tilt tends to increase with depth, from the ground surface down, and it is the tilt at the depth of the potential slip surface which is most crucial, he urged the Authors to extend their investigations in this direction.

A. Silvério (Laboratário Nacional de Engenharia Civil, Portugal). Dr Vaughan, in his Preliminary Report to Session 3, states '. . . the uses of instrumentation in rock engineering are relatively thinly covered in papers to the Symposium and further discussion of this field would be valuable'. The following notes on some of the methods developed by the Laboratório Nacional de Engenharia Civil (LNEC), Lisbon, Portugal, are a contribution towards filling up this gap.

Characterisation of deformability by dilatometer tests. As is well known, in most cases it is impossible to characterise the deformability of rock masses by laboratory tests. This is so because, to account for the heterogeneity and jointing usually occurring in rock masses, the specimens would have to be excessively large. Frequently, even the *in situ* jackloading tests yield unsatisfactory results, either because the rock mass in the tested zone is disturbed by excavation or because the tested volumes are not yet representative of the rock mass.

To avoid such difficulties a new flatjack technique was developed at the LNEC, which allows the quick testing of large volumes of undisturbed rock masses. On the other hand, the determination of the deformability of rock masses along boreholes was recognised to be of great interest from the early stages of geomechanical studies. Thus, a dilatometer was also designed to operate in rather narrow boreholes, as are usual in site investigation.

The dilatometer developed by the LNEC (see references) meets the following requirements: (i) it is suitable for test diameters from 76–82 mm and for vertical depths down to 200 m, and is still operative in boreholes holding water; (ii) applies radial pressures up to 200 kg/cm^2; and (iii) measures the radial displacement of the borehole wall to within 1 μm, along four diameters 45° apart. The instrument can be used in holes drilled in any direction, the maximum allowable depth decreasing as the direction of the hole approaches the horizontal.

Figure 26 is a diagram of the dilatometer longitudinal section. The external diameter of the apparatus is 70 mm in the zone that applies the pressure. This corresponds to a 6 mm clearance in relation to the most frequent diameter of an NX borehole, viz. 76 mm. By pumping water between the rubber jacket and the steel body, the wall of the hole is made to withstand a uniform radial pressure, over a length of more than seven times its diameter. Upon conclusion of each test, a relief valve, operated by the water pump, drains the water previously fed to the jacket, thus allowing the dilatometer to be moved into another test position.

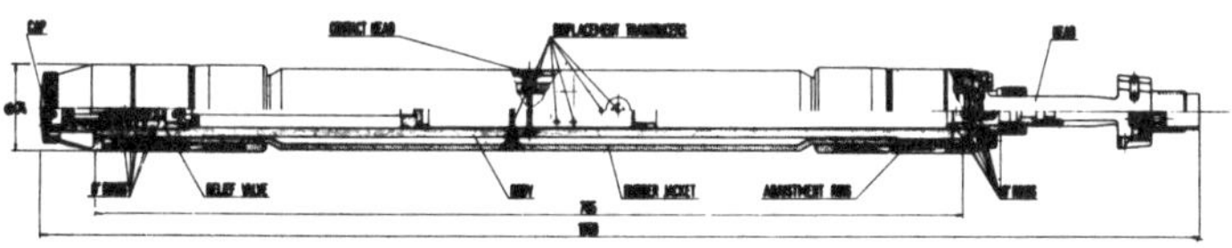

Figure 26

Eight feelers to follow the radial displacement of the borehole wall are bound to the jacket on generatrices at 45° intervals. The two opposite feelers moving along each direction are centred on cross sections 15 mm apart; the corresponding distance is 40 mm, for two feelers 45° apart. Each feeler is attached to the magnetic core of a displacement transducer with a 5 mm stroke and the accuracy 1 μm. Once connected to its reading unit the instrument has the overall sensitivity 2 dial divisions per micron.

Figure 27 shows the testing equipment, complete except for the steel positioning rods which install the apparatus inside the borehole, and define both its depth and azimuth at the test point.

Typical loading diagrams are presented in *Figures 28, 29, 30,* where the ordinates are pressures applied by the dilatometer jacket, and the abscissae the

Figure 27

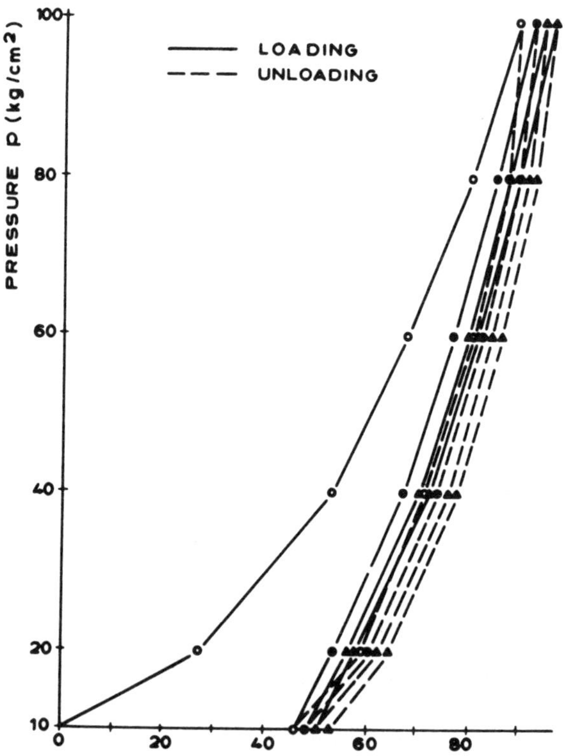

Figure 28

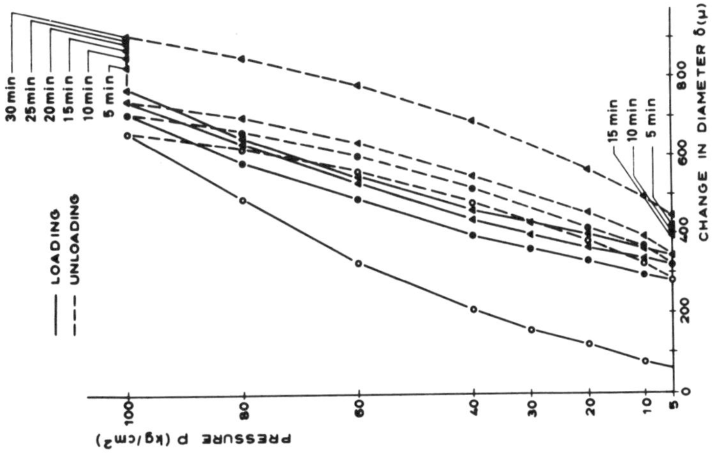

Figure 29

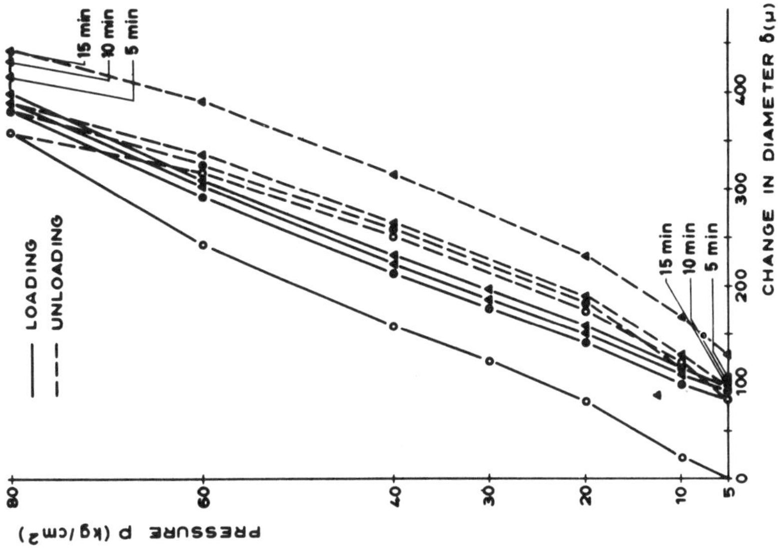

Figure 30

corresponding diametral deformations along one of the directions of measurement. Assuming the tested rock mass to be homogeneous, isotropic and elastic, the plotted diagrams yield moduli of deformability

$$E = (1 + v)\, d\, e$$

where v is the Poisson's ratio of the rock mass, d is the diameter of the test borehole, and e stands for the deformation under the applied pressure, to be read on every diagram, for the pressure interval selected.

In parallel with the classical deformability tests which use hydraulic jacks of a conventional type, it is common for the dilatometer tests to display large initial deformations mainly due to structural discontinuities in the material involved. The same cause holds for the large permanent set in the first load cycle, as well as for the decrease of the permanent sets in the successive cycles.

The $p - \delta$ diagrams are, in general, curves with the concavity directed towards the increasing pressure. However, this may not be so for the ascending branches of the hysteretic loops, especially the first one. S-shaped curves, reflecting the familiar plastic-elastic-plastic behaviour of rock materials, are also quite frequently found. *Figures 28, 29* and *30* are examples of the three diagram types. The first one refers to the approximate depth 57 m in a metasiltstone and slate formation; the remaining two belong to shallow tests, run at less than 1 m from the surface of the ground in a shale with a very small modulus of deformability.

The shape of the loading diagrams obviously depends on the details of the loading programme, as suggested by the nature of the rock mass to be tested. When testing a rock mass which deforms easily, small values must be given to the cyclic pressure peaks, and short intervals devoted to following the creep behaviour of the loaded rock, i.e., its deformation under constant applied pressure. On the contrary, some rock masses require high pressures, say peak values of the order of one or two hundred kilograms per square cm, to yield deformations which can be measured with proper accuracy. The latter call for long test periods if their creep behaviour is to be investigated. *Figure 31* shows

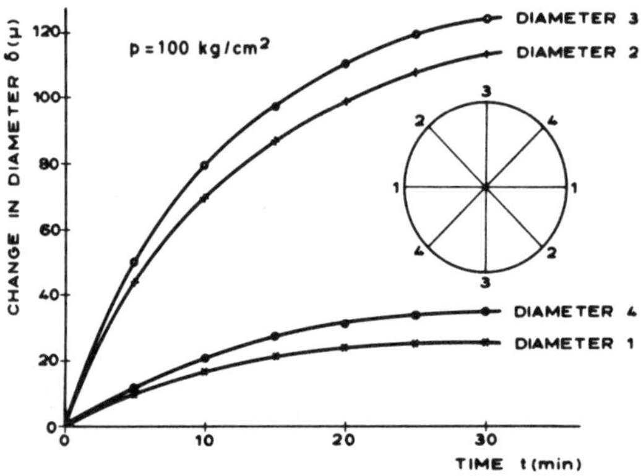

Figure 31

a set of dilatometer results for creep under the applied pressure 100 kg/cm^2, plotted as diametral deformations versus time. The tested rock, a shale drilled perpendicularly to the bedding planes, was seen to creep considerably in a 30 min time. Moreover, the diagrams suggested that two deformability domains prevailed under creep conditions, viz., a large-deformability domain whose directions included diameters 2 and 3, and a small-deformability domain which contained diameters 1 and 4.

The fact that the dilatometer applies a uniform pressure to the test borehole wall and measures deformations along several diameters suggests its use when studying the deformability anisotropy of rock masses. To this effect, the equation of a closed quartic may be set to describe the locus of the heads of the vectors which represent the moduli of deformability of the tested rock mass along the equatorial directions of measurement which meet at the test point in the borehole axis.

BHD TEST RESULTS ALONG A BOREHOLE

DEPTH m	E-ANISOTROPY ELLIPSE 10^3 kg/cm^2	ELLIPSE SEMI-AXES 10^3 kg/cm^2	COEFFICIENT OF ANISOTROPY	DEPTH m	E-ANISOTROPY ELLIPSE 10^3 kg/cm^2	ELLIPSE SEMI-AXES 10^3 kg/cm^2	COEFFICIENT OF ANISOTROPY
38.00		211 150	1.41	67.20		306 190	1.61
41.50		276 184	1.50	70.70		371 287	1.29
45.10		290 173	1.68	78.50		593 126	4.71
48.00		289 164	1.76	80.60		253 188	1.35
55.00		243 135	1.80	85.20		269 212	1.27
62.60		403 182	2.21	89.00		300 245	1.22

Figure 32

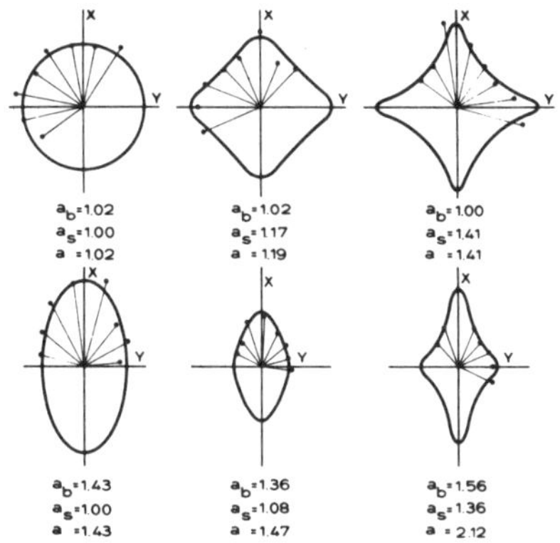

Figure 33

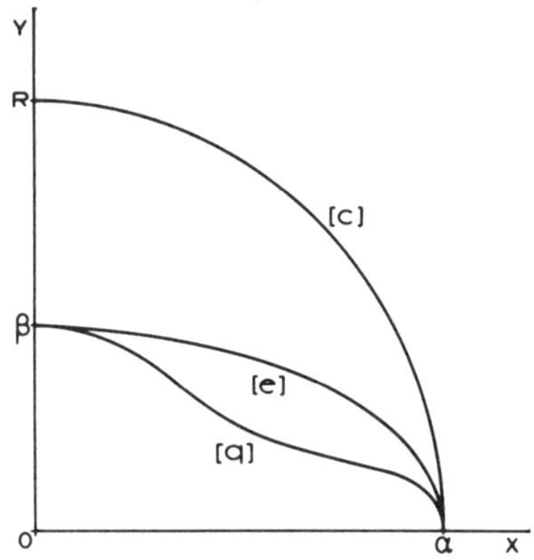

Figure 34

When no anisotropy-inducing surfaces are apparent, the anisotropy figures have turned out to be conics (ellipses or circles), into which quartics may in fact degenerate. As an example, *Figure 32* shows the most probable E-ellipses, determined by the least squares method for a deep borehole in a compact rock mass consisting mainly of siltstone. The listed coefficients of anisotropy are simply the ratios between the axes of each ellipse.

On the contrary, quartic curves have been produced by the computer for input data originated in rock masses which exhibited schistosity or foliation planes, or joint systems. They are illustrated in *Figure 33*, together with the conics referred to above. As regards the figure, it should be pointed out that, when the envisaged approach calls for measurements along more than four diametral directions, it is quite feasible to double-test each point in a borehole, i.e., to test each point twice while installing the dilatometer at azimuthal angles 22.5° apart.

The index quantities in *Figure 33* are more elaborate anisotropy ratios, defined in terms of the anisotropy figure ([q], *Figure 34*), as well as the circumscribed ellipse and circle ([e] and [c], *Figure 34*): the bulk-anisotropy ratio, $a_b = R/R_e$; the surface-anisotropy ratio, $a_s = R_e/R_q$; and the total-anisotropy ratio, $a = R/R_q = a_s\, a_b$. R, R_e, and R_q stand for the radius of the circle, the radius of the equal-area or equivalent circle to the ellipse, and the radius of the equivalent circle to the quartic, respectively.

When the immediate goal of a dilatometer test programme is to ascribe one mean value to every zone of the rock mass under consideration, testing is continued to the stabilisation of the average modulus of deformability over each zone. Once this has been achieved, a picture of the dependence of deformability on depth becomes readily available. Therefore, allowances being made for the geology, it is sometimes possible to extrapolate dilatometer results to depths beyond those reached by the instrument.

Figure 35 presents the results of a test programme carried out in seven boreholes with a view to gathering information on the deformability of the rock mass where an underground power house and tail race tunnel were to be excavated. The deeper sections tested in the boreholes, which extended to the maximum depth 94.50 m yielded quite satisfactory results. Now, a former survey had shown that the rock mass was a metasiltstone and slate interbedded with limestone, improving in quality with depth. This made it possible to conclude that, at the depth of the planned structures, no difficulties were likely to stem from the deformability of the rock mass.

Two instances of another kind of 'deformability *vs* depth' graph are shown in *Figure 36*, where dilatometer moduli and drill core moduli were plotted against the same ordinate axis. In the figure, depth refers to the position at which lay both the centre of each rock core previous to recovery and the

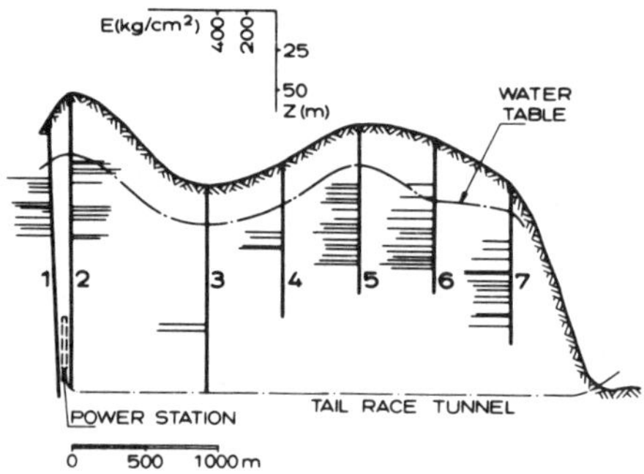

Figure 35

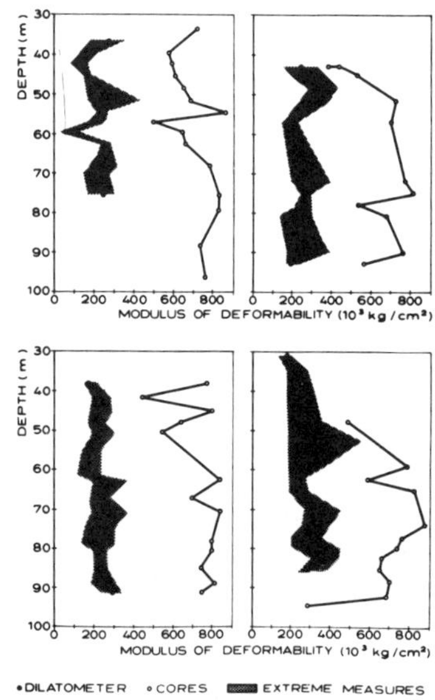

Figure 36

centre of the dilatometer while testing. The *in situ* measures represented were average moduli, plotted as full circles, and extreme moduli, bordering a hatched band; moduli of deformability from laboratory tests on cores were plotted as open circles. It is seen that dependence on depth is rather alike for dilatometer and core moduli, although no linear correlation of results appears to hold

A more meaningful comparison of results was achieved when the parallel *in situ* and laboratory measurements were performed along equivalent directions in regard to the anisotropy of the studied rocks. A reduction factor was then considered, when afforded the estimation of the oriented modulus of deformability at other points in the rock mass, on the basis of laboratory measurements.

So far, the LNEC has carried out more than 500 dilatometer tests, contributed by 13 programmes spanning a wide range of engineering interests. It appears that two consecutive tests along the same borehole is a fair estimate of the output of a 4-hour (half-day) working period. This figure may drop to about one half when the change from one borehole to another is comparatively important, so that an overall estimation implying 1 to 2 tested points per day seems quite a realistic one. The equipment is operated by two trained technicians, assisted by at least an unskilled worker. Rental rate of the equipment plus services is in the range U.K. £100 per day (as of May 1973).

On the strength of the experience gathered it can be stated that the dilatometer is a major tool in preliminary testing for deformability. The quality index it provides allows the division into zones of the rock masses whose deformability will have a bearing on the behaviour of the structure under consideration, be it the foundation of a dam or an underground work. This may well be a decisive step in working out a programme of larger sweep: it helps saving a number of longer and more expensive tests such as conventional jack and flatjack tests, and provides a clue as to the location of the ones to be actually performed.

Determination of deformability by means of thin flatjacks. Although the dilatometer appraises rock masses very little disturbed by the mere drilling of the test boreholes, it should be realised that it develops a complex state of stress in the neighbourhood of the borehole under test, the tensile components of which may cause fracturing, thus influencing deformations in a way difficult to assess. On the other hand, the meaning of the results may often be questioned on grounds of the small volume of rock mass actually involved in each test.

To extend the measurements to a more representative volume of the rock mass flatjacks have been in common use, which are accommodated in slots opened in the rock. Normally, flatjacks are inflatable vessels about 1 in thick, dished in appearance, composed of two circular steel sheets both of which are troughed at their peripheries and joined together by welding their external butting edges. Since the slots, obtained by drilling parallel holes and shearing the webs of rock left between them, are usually much wider than the jacks are thick, the hydraulic pressure fed to the jacks is transmitted to the rock mass by means of a concrete infilling. Thus, flatjacks have been expendable in use.

Both cutting the slot and installing the flatjack are much easier in the thin-flatjack techniques developed by the LNEC. As far as the determination of deformability is concerned, this technique (*Figure 37*) consists essentially of: (i) cutting by means of a disc saw one or more coplanar contiguous slots (depth h, thickness t), perpendicular to the test surface; (ii) inserting in each slot a large flatjack (effective height $h - a$); (iii) applying a pressure p in all the jacks; (iv) using deformeters at different points inside the flatjacks to measure the opening δ of the slots; and (v) determining the modulus of deformability of the tested rock mass, from 'p versus δ' plots.

Figure 38 shows two thin flatjacks which are the same in thickness (6 mm) and area of the large loading surfaces, but different in unloaded height, a (cf *Figure 37*). The magnitude of parameter a is selected in accordance with

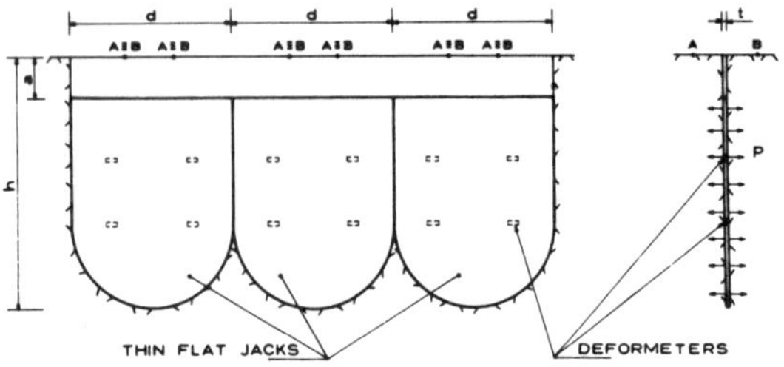

Figure 37

Figure 38

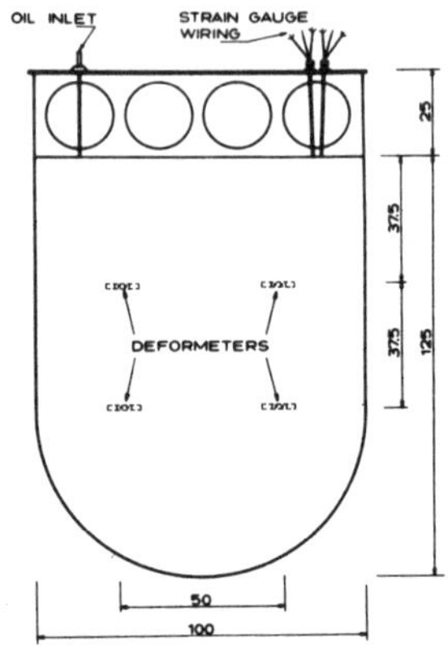

Figure 39

the fracturing of the rock mass, so as to avoid failure near the surface when pressure is applied by the jacks. As can be seen in *Figure 39*, the flatjack houses four deformeters, strain-gauge instrumented, which measure the actual thickness of the installed jack, thus allowing the opening of the slot to be followed throughout the test. Deformeters are available in several measuring ranges, in order to cope with the opening of the slots in rock masses spanning a wide deformability domain.

The calibration curve 'readings on a deformeter versus thickness of the jack (or opening of the slot)' is plotted with the help of the setup presented in *Figure 40*. This is simply a heavy frame provided with four deflectometers, where the thin flatjack is installed horizontally. The jack is then inflated with oil fed by a hand pump, while simultaneous readings are taken on both a strain indicator connected to the deformeter and the dial gauge whose pin seats on the upper loading surface, just opposite the inner deformeter. Since straight-line graphs are always obtained over the load range, calibration factors may be ascribed to the deformeters, which are supplied with each jack.

To cut the test slots a rather ingenious machine was developed (*Figure 41*). Its main feature is a large diamond-edged disc, 1 m in diameter and 7 mm thick, held by a tube which houses the system that rotates the disc. Due to its chain drive, indicated in the sketch in *Figure 42*, the device allows slots to be cut open far beyond the disc radius: it suffices to drill a previous hole that will receive the disc-holding tube. This is done by the cutting machine itself, which can be set either to saw or to drill (*Figure 43*).

Figure 41

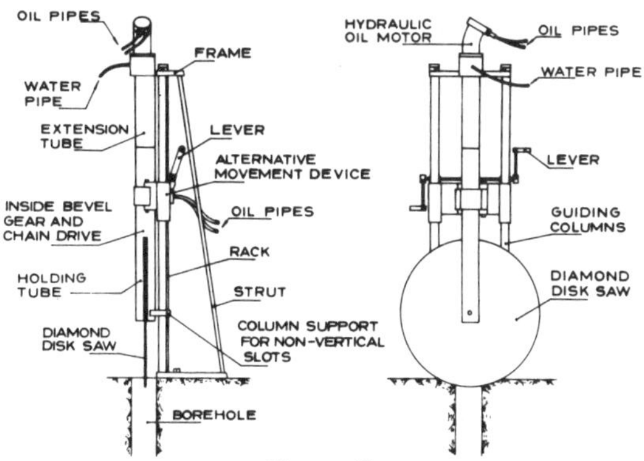

Figure 42

The machine is operated by an oil engine fed by a pump which in turn is driven by a petrol or electric motor (*Figure 44*). Long high-pressure pipes convey the fluid from the oil pump to the oil motor, and back to the pump. Thus, it is possible to leave the petrol engine in the open while cutting slots inside galleries. Besides the screw feed which moves the machine crossbar along its guiding columns there is a mechanism to impart transverse movement to the saw. In a later model both mechanisms can be actuated by hand or automatically. By adding extension tubes to the basic unit is is possible to cut slots with depths up to h = 2.5 m (*Figure 37*) in some 10 hrs or less, depending on the rock quality. It should be mentioned that the machine frame can be bolted on to a test surface in any direction.

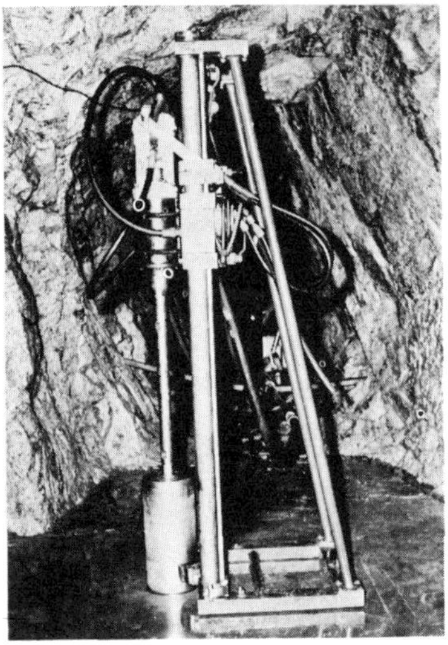

Figure 43

Figure 44

Figure 45 shows a thin flatjack being introduced into a slot 1.9 m deep where two other jacks are already installed. With the support provided by wooden half cylinders, or by mortar cast in the holes inside plastic bags, the

Figure 45

flatjacks will withstand pressures exceeding 100 kg/cm². In either case it will be possible to recover the jacks, once the test is complete.

Typical loading diagrams are presented in *Figure 46*, where the ordinates are test pressures applied by the jacks, from the seating pressure 5 kg/cm² to the maximum pressure 30 kg/cm², and the abscissae the corresponding values of the opening of the slot, measured at the point where a given deformeter is located. By making some assumptions as to the mechanical behaviour at the tested rock mass, each of the plotted diagrams yields a modulus of deformability

$$E = c\,e$$

where e stands for the increase in slot opening under the pressure interval selected, and c is a constant which has the dimensions of a length and whose value depends on the position, shape, and dimensions of the set of test jacks, as well as on the position of the measuring deformeter.

One 1-LFJ test takes little less than 1 week, while 8 working days is a rough time estimate for one 3-LFJ test. The equpment is operated by two trained technicians assisted by two unskilled workers. Rental rate of the equipment plus services is in the range U.K. £100 per day (as of May 1973).

Although amazingly economical for the volumes tested – cutting the slot is straightforward, installing the jack is quick, measuring the displacements is

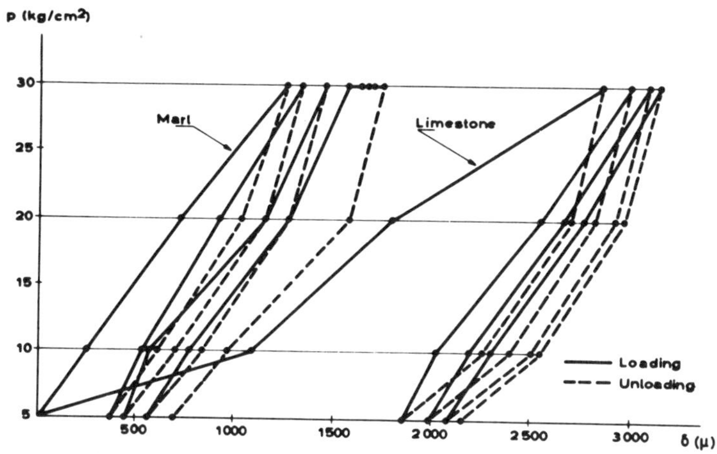

Figure 46

simple, recovering the jack is easy – thin flatjack testing is far longer and more expensive than testing with the dilatometer. Thus, a philosophy is needed to share the available means, when assessing the deformability of a rock mass. The adopted procedure frequently consists of the following steps: (i) run a large number of dilatometer tests along boreholes drilled through typical zones of the rock mass, down to the depths with an interest for the envisaged work; (ii) use the results to work out a divison of the rock mass into zones which may be regarded as uniform from the standpoints of both rock nature and deformability; (iii) perform a reasonable number of flatjack tests to characterise the large-volume deformability of some zones, and to establish a correlation between dilatometer and flatjack results; and (iv) use the correlation to estimate the large-volume deformability of the zones which were tested with the dilatometer only.

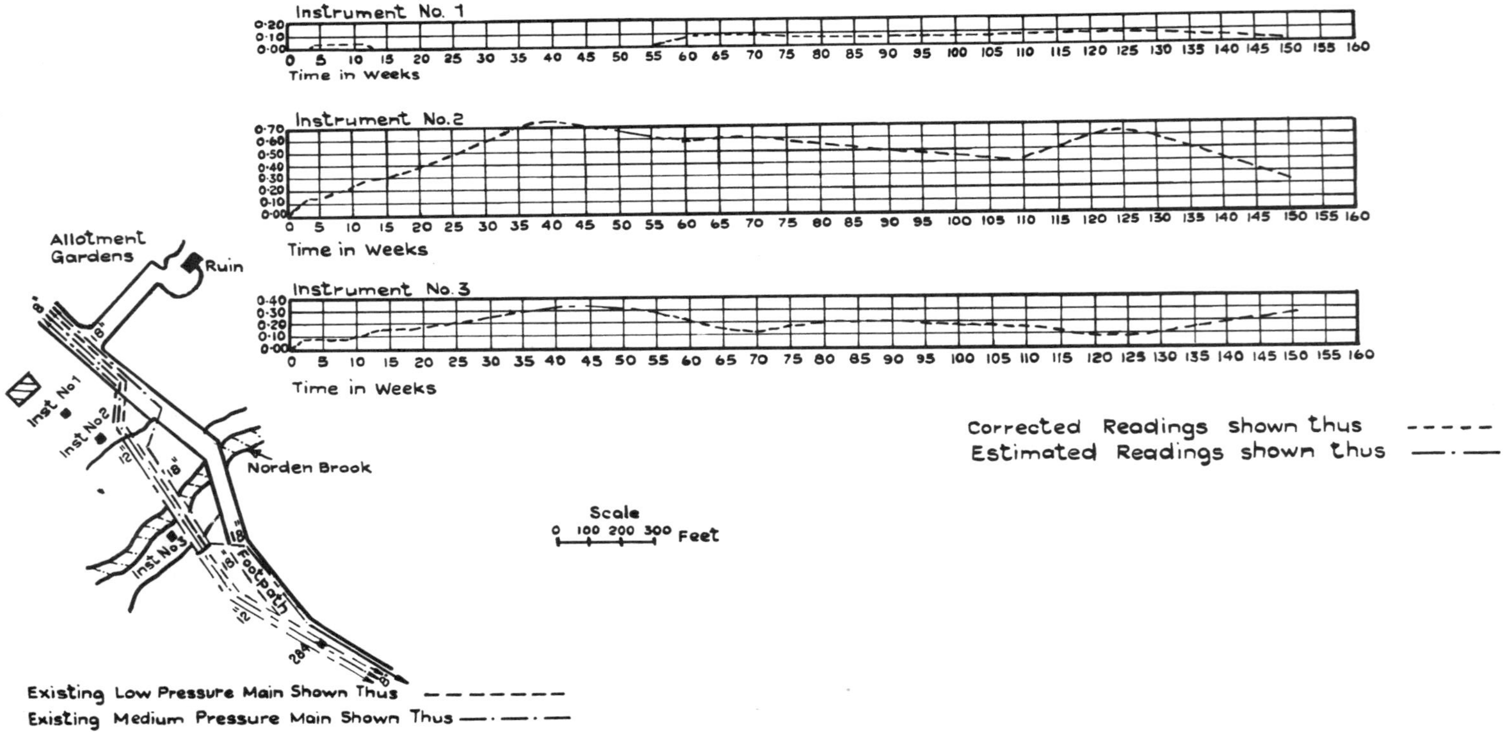

Figure 47 Graphs of ground movements from Tiltmeter readings. Reading started June 19 1970

REFERENCES

Rocha, M., Silveira, A., Grossmann, N., Oliveira, E. (1966), 'Determination of the deformability of rock masses along boreholes', *Proc. 1st Cong. Int. Soc. Rock Mech*, **1**, p. 697 Lisbon.
Rocha, M. (1968), 'New techniques for the determination of the deformability and state of stress in rock masses', *Proc. Int. Symp Rock Mech.* p. 289 Madrid.
Rocha, M. (1970), 'New techniques in the deformability testing of *in situ* rock masses', *ASTM Spec. Techn. Publ.* 477, p. 39 Philadelphia.
Rocha, M. and Neves da Silva, J. (1970), 'A new method of the determination of deformability in rock masses', *Proc. 2nd Cong. Int. Soc. Rock Mech.* **1**, p. 2–21 Belgrade.
Rocha, M., Silveira, A., Peres Rodrigves, F., Silvério, A. and Ferreira, A. (1970), 'Characterisation of the deformability of rock masses by dilatometer tests, *Proc. 2nd Cong. Int. Soc. Rock mech*, **1**, p. 2–32 Belgrade.

B. M. Evans (British Gas Corporation, North Western Region). The Tiltmeters described by Sherwood and Curry have been used in a glacial valley at Norden Brook, Great Harwood, Lancashire, to assess the ground strain taking place on the slopes of the valley.

In the valley four gas mains and a water main exist in close proximity to each other and the object of the measurements was to assess the amount of compressive or tensile strains in the ground and therefore the gas mains etc.

The measurements have been taken on the three Tiltmeters installed since June 1970 and the relevant engineering decisions made, the main problem was not the degree of accuracy of the measurements but the provision of sufficient data in order to define the extent of the problem etc. The details of the location of the Tiltmeter and the measurements are shown on *Figure 47*.

Discussion. Interpretation

Professor A. W. Bishop (Imperial College, London) (Chairman). The advanced notices from potential contributors to this Session suggest that there may be a lack of discussion on the degree of confidence in the stability of an earth structure which can be obtained either from pore pressure observations or from displacement observations. I propose, therefore, to make a few introductory remarks on this aspect of the subject.

When I first became involved in soil mechanics it was customary in the UK to analyse both foundations and embankment fills only in terms of total stresses (for example, Cooling and Golder (1942) and Banks (1948)). The problems involved in obtaining a reliable measurement of the undrained strength of boulder clay or till became apparent to me when I was invited in 1950 to assist Binnie & Partners in an examination of the stability of the Daer Dam, then under construction. The boulder clay fill, much of it very wet, contained boulders up to about 1 ft in diameter. It was clear that an effective stress approach was likely to be very much more reliable than one based on undrained strengths, and my calculations showed an almost linear relationship between factor of safety and pore pressure ratio $u/\gamma h$ (*Figure 1*). Piezometers were accordingly installed in the downstream fill and the dam was successfully completed.

Now the definition of factor of safety F used in the limit analyses in such cases infers that a state of limiting equilibrium exists when the mobilised shear stress τ_m is given by the expression:

$$\tau_m = \frac{c'}{F} + \frac{\tan \phi'}{F} \quad (\sigma - u)$$

where c' denotes the apparent cohesion in terms of effective stress;
ϕ' denotes the angle of shearing resistance in terms of effective stress;
σ denotes the total normal stress on the relevant element of the slip surface;
u denotes the pore pressure.

As a general rule the values of c' and ϕ' can be measured or estimated without great difficulty, but the analysis also requires an estimated or measured value of the pore pressure u. It is at this point that an important philosophical

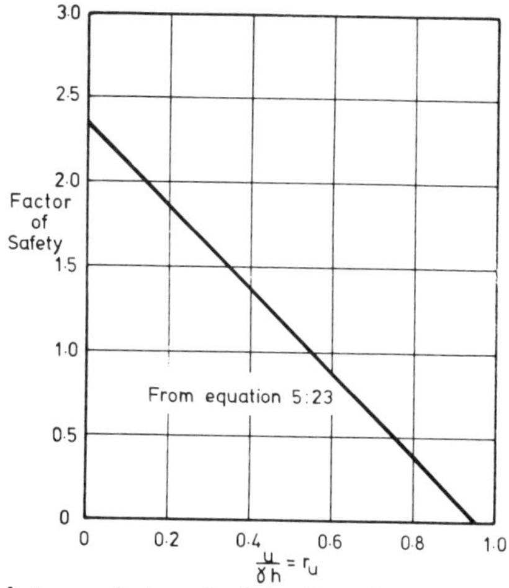

Figure 1 Relationship between factor of safety, F, and pore pressure ratio, r_n : Daer Dam, slip circle No 4 (after Bishop 1952)

question arises about the meaning of factor of safety, to which I drew attention in an earlier discussion of this problem (Bishop, 1952). Is the pore pressure an independent variable or is it influenced by the magnitude of τ?

Three cases may be distinguished, in which:

(1) *The pore pressure is an independent variable.* In relatively pervious soils subject to steady seepage or rapid drawdown the pore pressure is determined by the geometry of the boundaries, the relative permeabilities of the component soils and the rate of fluctuation of level of external water systems. Stress change, except at very high rates, has no influence on the magnitude of the pore pressure.

(2) *The pore pressure is controlled by the magnitude of the shear stress τ.* Where failure may result from an undrained stress change (i.e., one in which the time factor is small due to a low coefficient of consolidation or a long drainage path), the pore pressure is controlled by the changes in stress required to bring about failure. In principle the pore pressure can be predicted from undrained tests, but in practice an accurate prediction is much more difficult than is often realised, due to the necessity of following the correct stress path and applying the correct stress system (which will be plane strain in many problems) and due to stress changes arising in sampling. The same limitations, including the influence of test duration, apply to the estimate of undrained strength used in the alternative total stress analysis, which might otherwise be considered the ideal analysis for this class of problem.

(3) *The pore pressure is controlled both by the magnitude of the shear stress τ and by the rate of construction.* If the time factor is large enough for a significant change in pore pressure to occur, then the factor of safety given by an undrained analysis will give only one bound to the range of possible values. While this will in general be the lower bound for cases involving an increase in load, it will be the upper bound in such cases as the excavation of a cut, even in a normally consolidated clay (see *Figures 14a* and *17* of Bishop and Bjerrum, 1960). An excep-

tion may be fills on foundations containing thin layers of more permeable soil (Ward, Penman and Gibson, 1954; Bishop and Bjerrum, 1960). The effective stress analysis is the only way of assessing the intermediate and long-term stability.

However, the definition of factor of safety used above has certain limitations of which it is as well to be aware. These are apparent from an examination of the idealised stress paths illustrated in *Figure 2*.

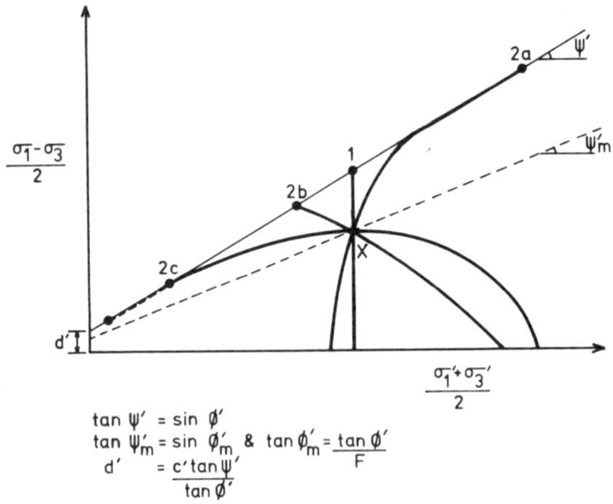

Figure 2 Range idealised stress paths for soils of low plasticity tested under undrained conditions

Let us assume that an effective stress analysis based on observed pore pressure indicates a value of factor of safety F as defined by equation (1), and that F has a value of the order of 1.5. The stresses in a typical element in this analysis may be represented by the point X. If we then consider an additional shear stress to be applied without any change in the mean normal stress $\frac{1}{2}(\sigma_1 + \sigma_3)$, along which stress path will the element proceed and what will be the strength at failure?

If the pore pressure is not an independent variable, then for the case of plane strain and a fully saturated soil ($B = 1$) we have:

$$\Delta u = \Delta\sigma_3 + A_p\,(\Delta\sigma_1 - \Delta\sigma_3) \qquad (2)$$

or rearranging:

$$\Delta u = \tfrac{1}{2}\,(\Delta\sigma_1 + \Delta\sigma_3) + (A_p - \tfrac{1}{2})\,(\Delta\sigma_1 - \Delta\sigma_3) \qquad (3)$$

The stress path will thus only proceed to point 1 in the special case of $A_p = \frac{1}{2}$, a value which could be realised in a soil with a certain degree of overconsolidation or compaction.

If it is assumed that the value of ϕ' does not drop at any stage of shear, which is a reasonable approximation for clays of very low plasticity, then the range of values which may be encountered in practice is illustrated by the three stress paths of *Figure 2*. The 'safest' stress path is that given by a heavily overconsolidated boulder clay or till ($X2a$) which tends to strain harden (Vaughan, 1971) and to show a negative incremental A value before failure. In this respect it is similar to a dense sand in which the suppression of dilatancy under undrained conditions leads to similar negative A values (Bishop, 1952,

Figure 8:2). The value of $\frac{1}{2}(\sigma_1 - \sigma_3)$ at point $2a$, which represents undrained failure, may exceed the constant pore pressure value 1 by a factor of as much as 1.5 or 2 (Vaughan, 1971). It is therefore difficult to resist the view that this higher strength should form the basis of design, provided:

(a) that the presence of the stone content of the boulder clay does not prevent the determination of this strength with reasonable accuracy, and;

(b) that the strains are permissible and will be monitored, since an effective stress analysis will show a factor of safety of 1.0 over all that part of the stress path which coincides with the effective stress failure envelope.

If, on the other hand, the clay is normally consolidated but relatively insensitive, the stress path will approximate to $X2b$, the value of A being positive and around 1.0. The reserve shear strength will be less than indicated by the calculated value of F, but there will at least be some reserve. If one is even less fortunate and happens to be dealing with a quick clay or other very sensitive soil, the stress path may be as $X2c$ and there may be no reserve of shear strength at all, although the nominal factor of safety in terms of effective stress is apparently ample. Examples of these typical stress paths are given by a number of authors including Bjerrum (1966), Bishop (1971), Bishop, Webb and Skinner (1965), Schofield and Wroth (1968), Skempton and Sowa (1963) and Vaughan (1971). The test data given by most of these authors relates to tests with axial symmetry rather than plane strain.

In more plastic clays the value of ϕ' drops significantly after a peak value has been reached. This effect combines with the pore pressure changes illustrated in *Figure 2* to control the shape of the undrained stress-strain curve. This shape is of crucial importance, particularly if one wants to take advantage of an apparent reserve strength of an overconsolidated clay. There are two basic shapes (*Figure 3*). Curve 1 is typical of boulder clay or till and even in the ring shear apparatus shows no significant drop in strength at very large strains.

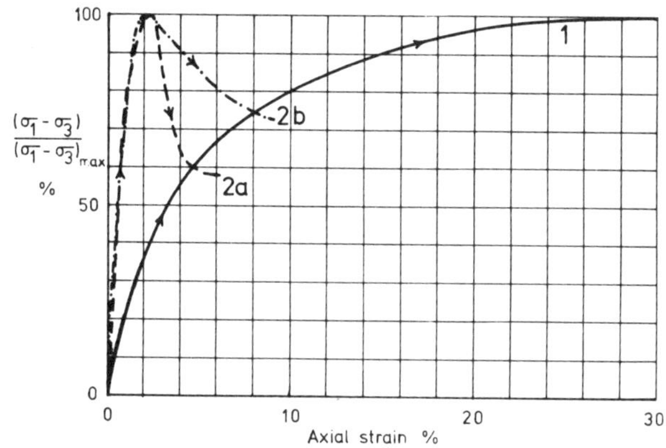

Figure 3 *Stress-strain curves for undrained test on:*
1. Boulder clay
2a. Heavily overconsolidated clay of high plasticity
2b Normally consolidated sensitive clay

Deformations are usually large and generalised before failure and as the brittleness index is zero a runaway slide is improbable. Curve 2a, on the other hand, is typical of relatively unweathered heavily overconsolidated clay of high plasticity and curve 2b of normally consolidated clay of high sensitivity.

Failure will tend to lie within a thin zone, deformations before failure will be small, and due to the high brittleness index a runaway slide is probable.

Examples of case 1 behaviour are the performance of Muirhead Dam (Banks, 1948), Otter Brook Dam (Linell and Shea, 1960) and Cow Green Dam (Vaughan, 1971). An interesting example of case 2 behaviour is given by Ladd (1972), in which a 21 ft high test bank on a sensitive clay foundation failed overnight with a crest settlement of 10 ft preceded by no apparent warning from either the displacement or pore pressure observations. A tiltmeter at the toe showed a lateral displacement of only 1.1 in on the day before failure.

It is important, therefore, to know with which class of material one is dealing. As heavily weathered and fissured overconsolidated materials become stronger on remoulding, but not necessarily on being subjected to large shear strains, their category can best be determined by full scale tests such as the trial embankment at Empingham (see contribution by Vaughan *et al*, p. 617). Both small samples of undisturbed strata, and laboratory compacted samples of fill can give very misleading results with soils of this type.

It has been suggested by the General Reporter that one can obtain an indication of incipient failure from changes in the relationship between increase in pore pressure and increase in total stress. Some support for this view may be obtained from a consideration of the relationship between the incremental A-value and strain for a typical sensitive soil (*Figure 4*). The observed relationships between stress difference $(\sigma_1 - \sigma_3)$, pore pressure u, and axial strain e, suggest that the incremental value of A does in fact undergo a discontinuity at the point of maximum stress difference, since the pore pressure continues to increase as the strength begins to fall.

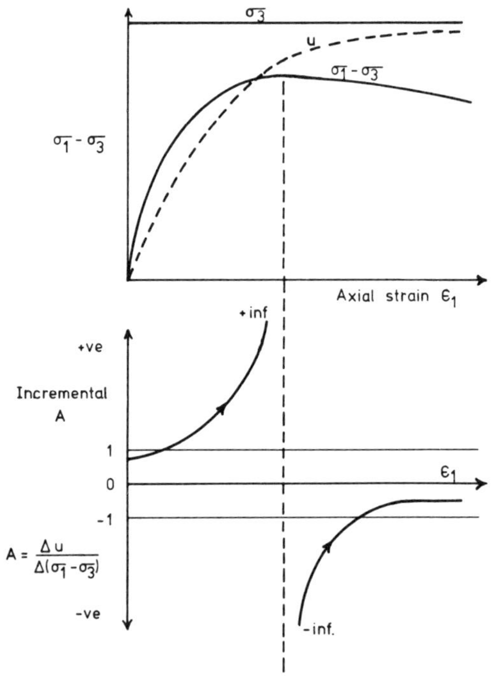

Figure 4 Change in the incremental value of A with strain for a sensitive soil

It might therefore be inferred that a discontinuity in the relation between observed pore pressure and applied stress should be associated with incipient failure in relatively sensitive soils. (The relationships in *Figure 4* are for constant rate of strain tests. Lo (1961) has described the rise in pore pressure at constant stress which occurs during undrained creep in normally consolidated soils. It is important to take account of this effect in drawing inferences about incipient failure.) Whether this is found in practice will clearly depend on the location of the piezometers, since rupture in sensitive or brittle soils tends to be confined to very thin zones (as in the case reported by Ladd, 1972). This may appear to be an incitement to put in more piezometers, since this rise in pore pressure has been reported on only a few occasions (for example, Al-Dhahir *et al*, 1969; Hoeg *et al*, 1970). Possibly the chance of locating a piezometer on the rupture surface in the zone from which progressive failure will propagate (as must happen in a sensitive or brittle soil) is too low to make this a reliable way of anticipating failure.

It is of interest to note that the ultimate negative incremental value of A to which the stress path tends in soils of low plasticity, whether in sensitive soils as *Figure 4* and *Figure 2* (path $X2c$) or in boulder clays as in *Figure 2* (path $X2a$), is a function of ϕ' and is given by the expression (Bishop, 1952):

$$A_{\text{incr}} = -\frac{(1 - \sin\phi')}{2\sin\phi'} \tag{4}$$

In conclusion, it would appear that field observations of pore pressure and displacement are not a very sure guide to the stability of an earth structure without detailed knowlege of the soil types and the stress-strain-pore pressure relationships of the materials involved, as well, of course, as their geological setting. It is with this in mind that I might ask what is to be inferred from the Table of earth dam displacements given by Penman and Charles (1973) in their paper to this Symposium. The engineer may not be concerned overmuch about the lateral displacement of the downstream slope as such. He will be concerned with the margin of safety in relation to the use of the reservoir on a long-term basis and it is how displacements are interpreted with this in mind that is important.

The inclusion in this Table of dams of rockfill of relatively low brittleness index raises the question of how highly rockfill should be stressed for its use to be fully justified. The shear strains corresponding to a reasonably high value of $\phi'_{\text{mobilised}}$ are considerable, and would probably result in larger lateral displacements than were observed at Scammonden, for example, where the design was controlled by other factors. Indeed, the stress ratios involved will probably mean that predictions of performance during construction will have to be based on methods more sophisticated than the simple linear elastic methods of analysis.

REFERENCES

Al-Dhahir, Z. A., Kennard, M. F. and Morgenstern, N. (1969). 'Observations on pore pressures beneath the ash lagoon embankment at Fiddler's Ferry power station.' *Proc. Conf. In Situ Investigations in Soils and Rocks*, pp.175–186.

Banks, J. A. (1948). 'Construction of Muirhead Reservoir, Scotland.' *Proc. 2nd Int. Conf. Soil Mech.*, 2, 24–31.

Bishop, A. W. (1952). 'The stability of earth dams.' PhD Thesis: University of London.

Bishop, A. W. (1971), 'Shear strength parameters for undisturbed and remoulded soil specimens.' From: *Stress-strain behaviour of soils.* Proc. Roscoe Memorial Symp., Cambridge, 3–139.
Bishop, A. W. and Bjerrum, L. (1960). 'The relevance of the triaxial test to the solution of stability problems.' *Proc. A.S.C.E. Res. Conf. Shear strength of the Cohesive Soils,* pp. 437–501.
Bishop, A. W., Webb, D. L. and Skinner, A. E. (1965). 'Triaxial tests on soil at elevated cell pressures.' *Proc. 6th Int. Conf. Soil Mech.*, **1**, 170–174
Bjerrum, L. (1961). 'The effective shear strength parameters of sensitive clays.' *Proc. 5th Int. Conf. Soil Mech.*, **1**, 23–28.
Cooling, L. F. and Golder, H. Q. (1942). 'The analysis of the failure of an earth dam during construction.' *J. Inst. civ. Engrs.*, **19**, 38–55.
Hoeg, K., Andersland, O. B. and Rolfsen, E. N. (1970). 'Undrained behaviour of quick clay under load tests at Asrum.' *Geotechnique*, **19**, 101–115.
Ladd, C. C. (1972). 'Test embankment on sensitive clay.' ASCE *Proc. Speciality Conf. on Performance of Earth and Earth-Supported Structures, Purdue*, **1**, 101–128.
Linell, K. A. and Shea, H. F. (1960). 'Strength and deformation characteristics of various glacial tills in New England.' *Proc. Amer. Soc. civ. Engrs., Res. Conf. on Shear Strength of Cohesive soils*, 275.
Lo, K. Y. (1961). 'Stress-strain relationship and pore water pressure characteristics of a normally consolidated clay.' *Proc. 5th Int. Conf. Soil Mech.*, **1**, 219–224.
Penman, A. D. M. and Charles, J. A. (1973). 'Measuring movements of embankment dams.' *Proc. Symp. on Field Instrumentation in Geotechnical Engineering*, Part 1, 341–358.
Schofield, A. N. and Wroth, C. P. (1968). '*Critical state soil mechanics.*' McGraw-Hill: London.
Skempton, A. W. and Sowa, V. A. (1963). 'The behaviour of saturated clays during sampling and testing.' *Geotechnique*, **8**, 269–290.
Vaughan, P. R. (1971). 'Undrained failure of clay embankments.' From: *Stress-strain behaviour of soils.* Proc. Roscoe Memorial Symp., Cambridge, 683–691.
Ward, W. H., Penman, A. D. M. and Gibson, R. E. 'Stability of a bank on a thin peat layer.' *Proc. Europ. Conf. on Stability of Earth slopes, Stockholm*, **1**, 122–138.

A. D. M. Penman (Building Research Station). The clarifying remarks made by Professor Bishop about obtaining the stability of an earth structure from field measurements are most welcome. Factors of safety calculated from values of pore pressures have been used for a long time – this approach may be regarded as the traditional design method – because there is a reliable method of stability analysis (Bishop 1955) and because it is possible to predict pore pressure values. There are cases where embankment construction has been stopped because measured pore pressures indicated too low a value of factor of safety. There are also cases where construction has been stopped because easily visible movements were excessive. When accurate measurements are made, acceleration towards excessive movement can be observed at an early stage. Clearly field measurements of both pore pressures and deformations should be made, together with many other types of relevant measurements, in making a study of the behaviour of an earth structure. In the past settlement measurements have been made, because there were methods of predicting settlements, but there has been less interest in horizontal deformations, largely because there was no satisfactory method of prediction. Methods of analysis using finite elements (Penman, Burland & Charles 1971) are beginning to give satisfactory predictions which can be of great value to the designer e.g. in relation to the deformations of an upstream membrane.

If the factor of safety falls to one, presumably the movements will be infinite. What will they be if the factor of safety only falls to 1.1? The designer is interested in acceptable movements, and the purpose of the Table referred to by Professor Bishop was to draw attention to the concept of acceptable movement by giving some examples of the movements observed in embankments which would be regarded as satisfactory structures. In order to assess the magnitude of movements considered acceptable much more information is required and it is hoped that as field measurements of movement become more popular, this type of Table will be extended to give the designer guidance to the acceptable movements he can confidently design for.

REFERENCES

Bishop A. W. (1955). 'The use of the slip circle in the stability analysis of slopes.' *Geotechnique* **5**, 1 pp. 7–17.

Penman A. D. M. Burland J. B. and Charles J. A. (1971. 'Observed and predicted deformations in a large embankment dam during construction.' *Proc. I.C.E.* **49**, pp. 1–21.

A. Marsland (Building Research Station). Various speakers have mentioned three possible ways in which field measurements may help us to predict the onset of instability:

(1) By observing changes in the ratio of pore water pressures to applied loads during construction. This method is only of possible use in cases where plastic zones are developed in which the ratio of pore water pressure to load increases appreciably due to the very high shear strains associated with plastic failure.

(2) By observing changes of the ratio of deformation to applied loads.

(3) By analyses estimating the stability of various stages of construction using a selection of pore pressure values and then using the measured pore water pressure to estimate the stability as loading proceeds.

None of these methods are foolproof. The usefulness of methods 1 and 2 depends upon the stress/strain properties and pore water pressure response of the different types of soil as outlined in the contributions to this discussion by Professor Bishop and Dr Wroth. Method 3 requires reliable data on the large-scale behaviour of the ground, the use of the appropriate mechanism in the analyses and in some cases the correct evaluation of the soil/structure interaction.

Banks on soft alluvial soils are often monitored by measuring settlement of the crest and heave near the toes. It is my experience that during construction it is extremely difficult to separate settlement of the crest due to the onset of instability from that due to consolidation. Observations made during a slow failure (*Figure 4* of my paper) show that the heave at the toe is often only about $\frac{1}{5}$ of the vertical settlement at the crest. Measurements of heave at the toe are therefore rather insensitive indicators to provide sufficient warnings of possible instability to enable useful remedial action to be taken during construction. Heave measurements are, however, very useful for measuring movements of 'delayed' failures such as those which sometimes occur on flood banks. Several speakers have shown that lateral movements in the ground below the toe of an embankment are appreciably larger than the vertical movements and provided these can be accurately monitored appear to offer a better alternative.

Another example of the difficulties associated with using settlement as an indicator of the onset of instability can be illustrated by reference to bored piles in stiff fissured clays and weathered rocks. It is well known that the two components of side shear and end bearing have very different stress/strain characteristics as illustrated in *Figures 5a* and *5b* respectively. For a typical large diameter bored pile the two components may be roughly equal. Now consider two piles, one with a good base and one with a poor base, which will have load-settlement characteristics as illustrated in *Figures 5c* and *5d* respectively. When subjected to a working load pile A can still take additional load without catastrophic settlement taking place, whereas pile B is for all practical purposes on the verge of failure. The situation represented by pile B is far from uncommon especially where ground and water conditions prevent adequate cleaning of the base prior to the formation of the piles. This simple example illustrates the need to measure the loads developed at various levels in a foundation in addition to measuring the deformation of the complete foundation.

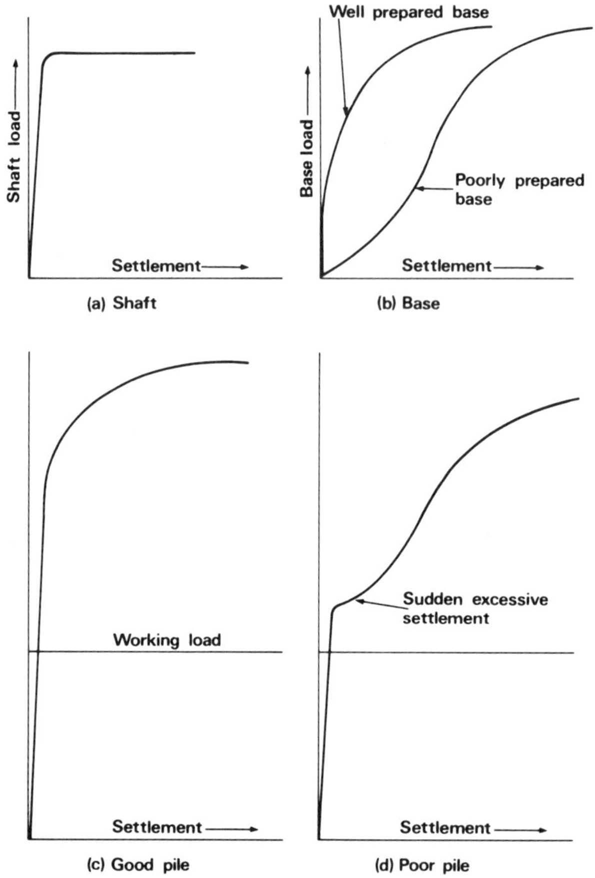

Figure 5 Settlement of large diameter bored pile

I consider that there is a need for many more observations before we can have reasonable confidence in our ability to predict and anticipate the onset of instability especially in earthworks where a large percentage of the deformation takes place in undisturbed natural soils. I consider that reliance on any one of the three methods outlined above can be extremely dangerous and that a flexible approach using as many alternative precedures as practically possible should be used.

With the present state of knowledge our estimates of displacements which are likely to occur are still very crude mainly due to lack of reliable data on the large-scale properties of soils, the effects of constructional procedures including time effects, and of the load transfer mechanism between structures and the adjacent soil. I do not consider that we can rely on measurements of either pore water pressure or deformation to anticipate instability unless we have prior knowledge from similar projects or from adequate trials carried to failure.

D. J. Knight (Sir Alexander Gibb & Partners). This contribution is the necessary sequel to my earlier one on long-term monitoring as given in Session 3 (p. 641), which was restricted to the organisational aspects of instrumentation mainly for remotely sited, overseas embankment dams. The subject of processing of observations was touched upon, but it is necessary to consider the vital question of warning signs from instrumentation – their nature, their numerical values, their interpretation and their treatment, i.e., what action if any is needed in the light of their interpretation.

Much emphasis has rightly been given to factors of safety for overall stability as influenced by pore-water pressures as a type of warning sign, whereby those pressures, when they reach critical levels, serve as indicators of unsatisfactory behaviour. This is an important indicator. Emphasis has also been placed upon deformations serving as warning signs, although the difficulty of determining what is and what is not an acceptable deformation for a given structure is great. One embankment dam for which my firm was responsible was founded on a variable thickness of up to about 20 m of highly expansive clay with liquid limits averaging about 80 %, and as a result of saturation by the reservoir on initial impounding one part of the embankment rose by 14 cm whereas another part some 300 m away settled 17 cm.

For large water-retaining structures, however, such as dams it is highly desirable to monitor not only overall stability by means of pore-water pressures, and deformation, but also the *control* of seepage water; that is, to monitor the general piezometric pattern of the water flow from upstream to downstream via the core and the foundation, and to assess the efficacy of the various internal cut-off and drainage measures provided, by back analysis of the observed piezometric patterns. Examples are the effectiveness of grout curtains and of buried drainage holes installed in the foundation immediately downstream of a core.

A brief mention was made in Little's paper of the importance of seepage measurement, which is generally made by means of V-notches; it would appear that relatively little effort has been made in developing instruments for the measurement of seepage, other than dyes, thermistors and isotopes, (what about, for instance, standard instruments for measuring horizontal seepage velocity at depth in foundations of fill?) Plots of V-notch readings with reservoir level, and of seepage quantities from drainage holes into galleries, are examples of seepage monitoring.

The quantifying of warning signs is difficult, as is illustrated by the following extract from a letter to a dam owner regarding warning levels in the piezometer system: 'The levels quoted are design pore-pressures for the full reservoir condition, Stage 1 . . . which is the one for which warning is required. Should these levels be exceeded in some cases, it does not necessarily mean that a dangerous condition has developed: it is permissible for individual cells to show higher readings provided that other cells in the same cross-section show lower ones. It is impracticable to detail the innumerable combinations of possible maxima.'

Many will have their own examples of instrumentation both wrongly as well as correctly indicating unacceptable structure behaviour. The year-after-year processing and interpretation of long-term monitoring observations is liable, by its very routine, to become a chore, but unless the data are regularly received, assessed and acted upon where necessary, there is no point in the instrumentation being there in the first place. One earth dam, built about half a century ago in the UK and for which my firm is responsible for monitoring and advising upon under the Reservoirs Safety Act was, after a site investigation and initial instrumentation, twelve years ago, judged to require strengthening work, which was later completed. Additional piezometers were also installed to monitor seepage, which was known to take place both beneath and through the porous concrete central core wall. Seepage measurements were also made.

Initially, as piezometer readings were obtained, stability analyses were continued based on highest observed levels to confirm the adequacy of the strengthening works, but later on it was only necessary to ensure that the highest observed levels were not exceeded. The monitoring was thus quite straightforward, being a comparison of readings with acceptable levels. In recent years, however, one part of the embankment, on a flank and thus low in height, exhibited a tendency for downstream shell water levels to rise, and for earlier observed highest levels to be regularly exceeded. A part of the answer was later found to be due to a portion of standpipe having been broken off! This, however, was not sufficient fully to account for the increases, and eventually it became necessary to consider what physical remedy was required. The result is that additional rockfill is due to be placed this year, together with further instrumentation in the affected area.

In conclusion, it is contended that the *sole* object of instrumentation is to monitor a particular structure, and as far as humanly possible to ensure its safety and proper functioning on behalf of its owner.

C. P. Wroth (University of Cambridge). The General Reporter has suggested that field measurements may be used for three main purposes, and my contribution concerns the third purpose that he listed, namely the determination by back analysis of the *in situ* properties of the ground and a proper understanding of its behaviour. In order to make acceptable predictions of the behaviour of a foundation, a model of the soil behaviour is required, and this model should be as simple as possible.

However the behaviour of a particular element of soil depends on its previous consolidation history, and on the stress changes and deformations applied to it. For the same soil it may be necessary to have a family of models, and to select the appropriate one for each stage of the process of computation which leads finally to a prediction. For example, the simple case of one-dimensional reconsolidation of an overconsolidated clay requires two different models for an adequate description of its behaviour; an elastic model will be adequate for applied pressures less than the preconsolidation pressure p_c, whereas a plastic model is necessary for pressures greater than p_c when large irrecoverable volumetric strains occur.

The development of soil models requires a continual process of trial and error. The relevance of a new model must be tested by making a careful comparison of predictions with measured field performance, and then adjusting the model in the light of its deficiencies. This is particularly true for those models that are designed to describe the deformational behaviour of soil and which are not used solely for stability calculations. Dr Atkinson states in his contribution, that the latest evidence indicates an order of magnitude difference between the deformation parameters (i.e. stiffness) for London Clay measured in the field and laboratory. It is essential therefore for progress in our ability to make realistic predictions in foundation engineering, that field measurements are both made and interpreted with great care. Ideally the choice of instruments and their location should be made so as to provide the most searching test possible for a particular model, the method of analysis adopted and the method of computation.

We have heard discussion about the problem of monitoring the safety of embankments and fill on soft ground and whether measurement of pore-pressure or of displacement is a better criterion. The question of a criterion based on allowable displacements is not only complicated by progressive failure, but also by the following factors. Firstly any displacement of a soil element will be made up of both the integrated effects of body distortion and body displacement of the whole soil mass, and (unlike the case of pore-pressure) any prediction will requires a complete solution to a complex boundary value problem. Secondly the shearing behaviour of soil is such that pre-peak the appropriate parameter of distortion is strain (when the soil tends to behave uniformly), whereas post-peak the appropriate parameter is relative displacement. This is because the deformations become concentrated in a narrow failure zone of unknown thickness (the soil behaves non-uniformly) and in such circumstances strain is a meaningless parameter. I believe that a great deal of progress must be made before a displacement criterion can be used with any real confidence.

J. H. Atkinson (Cambridge University). In his General Report for this Session Burland noted the determination, by back analysis, of in situ properties of the ground as a useful purpose of field instrumentation. In this contribution, values for the undrained deformation modulus of London clay obtained from field observations and laboratory tests will be compared; it will be shown that although both sets conform generally to the same basic pattern of behaviour the numerical values differ significantly.

In interpreting field observations due account should be taken of the behaviour of the soil; for example it would seem to be incorrect to use elastic analytical methods for a plastically deforming soil. In addition it may be necessary to account for anisotropy, non-homogeneity and elastic non-linearity otherwise the parameters inferred from the field observations will not correspond exactly to those measured in the laboratory tests. For London clay it is reasonable to regard the material as elastic for small strains although it appears to be anisotropic, non-linear and, in nature, non-homogeneous (Atkinson 1973).

A few case records of the deformations of London clay in situ, both in full scale structures and in instrumented field tests, are available from which values for the undrained Youngs Modulus E_u, may be estimated and the moduli inferred from these records may be compared with the values for the undrained Youngs Modulus measured in laboratory cylindrical compression tests. Values for the field E_u cited were obtained from plate loading tests at Chelsea and Hendon (Marsland 1971), from observations of the heave of a deep excavation (Serota & Jennings 1959), from the deflections of a retaining wall (Cole & Burland 1972) and from the deflections of a tunnel below an excavation (Ward 1961). With the exception of the Hendon plate loading tests these involve sites in or near Central London. The observations and interpretations will not be discussed in detail here but it may be noted that the moduli quoted were calculated from elastic theory assuming isotropy and they are equivalent to secant moduli measured from field stresses to a point on the load deflection curve at working load.

For London clay Wroth (1971) demonstrated that deformation moduli measured in a consistant way are dependent on the mean stress at the start of shear (P'_o) and, to a lesser extent, on the over-consolidation ratio (OCR = P'_m/P'_o where P'_m is the maximum mean stress experienced by the soil) and he suggested a relationship of the form:

$$E/P'_o = F\ (\mathrm{OCR})$$

For London clay in the field and in unconsolidated laboratory tests values for both P'_o and P'_m are rather difficult to estimate. For the present purpose P'_o for both field and laboratory data have been calculated from:

$$P'_o = \tfrac{1}{3}\gamma' h\ (1 + 2k_o)$$

with $\gamma' = 60$ lb/ft^3 and $K_o = 2.5$ at a depth h below exsisting ground surface and, for Central London, P'_m has been taken as 200 lb/in^2 equivalent to a maximum depth of sediment of around 600 ft. with $K_o = 0.64$ for a normally consolidated London clay.

In *Figure 6* values of E_u/P'_o obtained from the field observations have been plotted against OCR. Although the points are rather scattered there is a distinct trend of decreasing E_u/P'_o with decreasing OCR; the overconsolidation ratio has been plotted on a natural scale and no better correlation between E_u/P'_o and OCR would have been obtained if, following Wroth (1971), the OCR had been plotted on a logarithmic scale. It may be noted that the field data were obtained from a wide variety of stress paths involving both loading and unloading and in view of the assumptions made in estimating E_u/P'_o and P'_m the degree of scatter is not unexpected. Nevertheless, the trend of the relationship between E_u/P'_o and OCR is clear and is in agreement with Wroth (1971). Included in *Figure 6* is a line taken fron Atkinson (1973) representing values of E_u/P'_o obtained from laboratory undrained cylindrical compression tests on undisturbed vertical samples of London clay; here the moduli are secants measured from the origin to a point on the stress-strain curve at 50%

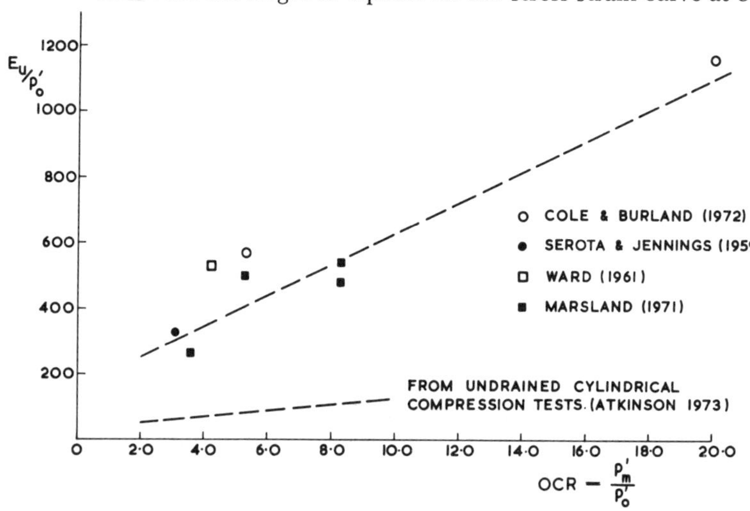

Figure 6 Values of E_u/p'_o plotted against OCR

of the peak deviator stress and are therefore compatible with the field moduli. The laboratory data, which will not be discussed in detail here, were obtained from tube samples taken at Paddington, Victoria and the South Bank (Skempton & Henkel 1957) and from hand cut block samples taken at Ashford Common (Ward, Marsland & Samuels 1965), Wraysbury (Agarwal 1967) and the Barbican (Atkinson 1973). In plotting the original points (Atkinson 1973) the scatter was similar to that found in the field data.

Both sets of moduli in *Figure 6* show the same trend of decreasing E_u/P'_o with decreasing OCR but the two lines drawn differ by a factor of 5.0, the field moduli being larger. This is a significant difference and is apparently more than can be explained by failing to allow for elastic anisotropy in analysing the field data; if anisotropy of the undrained Poissons Ratios is neglected and it is assumed that the apparent field modulus is the mean of the vertical and horizontal undrained Youngs Moduli then, by taking a degree of undrained anisotropy of 0.6 (Atkinson 1973) the field moduli could be reduced by about 30%. Even so, accepting the correctness of the field moduli, this still implies an error of about 4 in the measured laboratory moduli. The reasons for this departure between field and laboratory measured values of the same soil parameter are not entirely clear but may be attributed to sampling disturbance, machine interference and an isotropic starting state of stress in the laboratory tests; nevertheless it is of interest to note the same kind of variation of E_u with P'_o and OCR appearing in both sets of data.

REFERENCES

Agarwal, K. B. (1967), *The influence of size and orientation of sample on the undrained strength of London clay*, Ph.D Thesis, University of London.

Atkinson, J. H. (1973), *The deformation of undisturbed London clay*, Ph.D Thesis, University of London.

Cole, K. W. and Burland, J. B. (1972), 'Observations of retaining wall movements associated with a large excavation', *Proc. 5th European Conference. SM & FE Madrid* 1, pp. 445–453.

Marsland, A. (1971), 'Laboratory and *in situ* measurements of the deformation moduli of London clay.' *Proc. Symp. Interaction of Structure and Foundation. Midland Soc. SM & FE* pp. 7–17.

Serota, S. and Jennings, R. A. J. (1959), 'The Elastic heave of the Bottom of Excavations,' *Geotechique* 9 : 1 pp. 62–70.

Skempton, A. W. and Henkel, D. J. (1957), 'Tests on London clay from deep borings at Paddington, Victoria and S. Bank, *Proc. 4th Int. Conf, SM & FE* 1, pp. 100–105.

Ward, W. H. (1961), 'Displacements and stains in tunnels beneath a large excavation in London,' *Proc. 5th Int. Conf. SM & FE* 2, pp. 749–753.

Ward, W. H., Marsland, A. and Samuels, G. S. (1965), 'Properties of the London clay at the Ashford Common Shaft: *in situ* and undrained strength tests,' *Geotechnique* 15 : 3. n pp. 321–344.

Wroth, C. P. (1971), 'Some aspects of the elastic behaviour of over-consolidated clay,' *Proc. Roscoe Mem. Symp.* pp. 347–361.

M. J. Pender (New Zealand Ministry of Works). I wish to make some comments about the interpretation of results from plate bearing tests in deep boreholes. There are two aspects I shall discuss: the application of a depth connection factor in calculating the apparent elastic modulus and the effect of drilling a hole beneath the centre of the plate to fix deformation gauges in place.

In their paper Marsland and Eason calculate the apparent elastic modulus of the ground by introducing a depth connection factor. This, presumably, makes allowance for the stiffness of the material above the plate assuming that the whole mass behaves as an ideal elastic material. This assumption requires that the soil has some tensile capability because the stiffness of the soil above the plate can only be mobilised by the transmission of tensile stress. If one considers what happens to the soil at the corner of the borehole it seems that this possibility may not be very likely in materials such as soil. Consider the geometry of the 865 mm dia. plate at the bottom of the 900 mm dia borehole given in *Figure 1* of Marsland and Eason's paper (p. 306). With a stiff plate so near the edge of the borehole one might expect a considerable concentration of tensile stress in an ideal elastic material. As the tensile strength of soil is low, a horizontal crack is likely to form at the corner of the borehole when the plate is loaded. Once such a crack was initiated it would spread as the load on the plate was increased. Thus in a deposit of soil the stiffness of the material above the plate would probably not have much effect on the load-deformation behaviour of the plate and I suggest that even in a deep borehole the equivalent modulus should be calculated on the assumption that the plate rests on a simple half-space. In this way the only effect of the material above the plate is to provide a surcharge.

I think that some of the BRS test results may give some support to the point I am making. Tests have been performed at the base of lined and unlined boreholes (reference 4, p. 316) and I understand that there was no significant difference observed in the stiffness of the soil in the two cases. Now the presence of the lining introduces a considerable stiffening effect at the side of the borehole but it would have little effect on the result if a tensile crack spreads from the corner of the borehole.

The second point I want to raise is the effect of drilling a hole beneath the centre of the plate to fix deformation gauges in place as shown in *Figure 2* of the paper (p. 307). Now the diameter of the hole (38 mm) is a mere 4% of the plate diameter and it might be concluded that its effect is negligible. However, when one considers the effect of the hole on displacement pattern it's presence seems to be more important. In the axisymmetric problem of a plate resting on a half space (or at the bottom of a hole) a very important restriction on the deformation is the condition that there is no radial strain along the centre line. The presence of the hole removes this restriction and means that the soil beneath the plate can move inwards, and this for a depth of the first plate diameter or so the system will appear to be more flexible than would be expected from calculations which ignore the presence of the hole.

I suggest that this mechanism may contribute to the rapid increase in secant modulus with depth beneath the plate shown in Table 1 (p. 314). Likewise I feel that it may contribute to explaining the observation that for the first few load increments heave was measured at a depth of 610 mm whilst the plate itself settled. The freedom afforded to the soil immediately beneath the plate by the axial hole may mean that the inward lateral deformation (and hence the settlement of the plate) of the soil in this region is more significant that the contribution due to swelling.

M. B. Jamiolkowski (Politecnico di Torino, Italy) and S. Marchetti (Universita dell' Aquila, Italy). A high precision digitilt inclinometer (Slope Indicator Model 50301) was used during full scale lateral load tests on piles. The rotation of the pile axis along its embedded part was measured. The experimental values of the pile axis rotation were interpolated by means of a regression analysis with polynomials varying from the 5th to the 7th degree.

The use of the so obtained polynomial functions allows one to determine, at each load level, the change, with depth, of:

pile deflection	$y(z)$	$[L]$
bending moment	$M(z)$	$[FL]$
shear	$T(z)$	$[F]$
soil reaction	$P(z)$	$[FL^{-1}]$

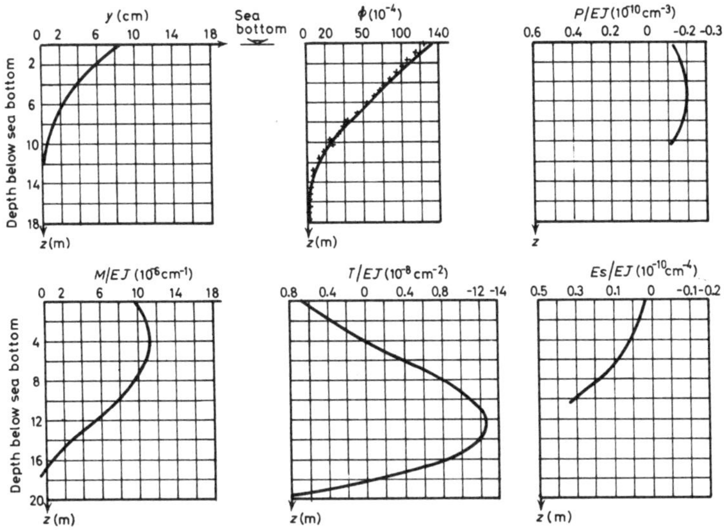

Figure 7 Manfredonia pile De = 122 cm: Pile deflection (Y), bending moment (M), shear (T) and soil reaction (P) as obtained on the basis of measured pile axis relations (ϕ) for H = 20 tons

An example of such interpretation of the experimental inclinometer readings is shown in *Figure 7*. By knowing the values of P and y, it was possible to establish the variation of the soil reaction modulus $E_s = P/Y$ with depth. *Figures 8* and *9* illustrate the results obtained for two steel tubular piles. Also

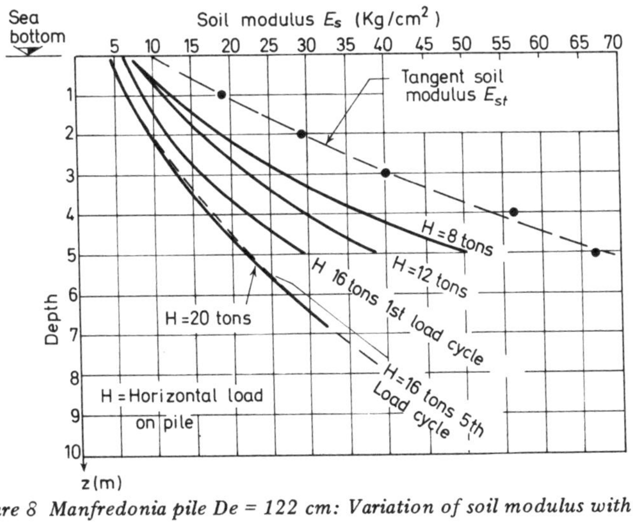

Figure 8 Manfredonia pile De = 122 cm: Variation of soil modulus with depth

reported in the same figures is the value of the initial tangent soil modulus E_{st} as extrapolated on the basis of the P–y curves.

The results reported above refer to the following conditions:

Pile and soil characteristics (Figure 8), Site Manfredonia (South Italy). Open end driven steel tubular pile 14 in o.d. with a thickness of 2.22 cm. The pile penetrates for 7 m into a dense medium sand, passing through a 15 m thick layer of soft NC clay with high plasticity. The point of the lateral load application was 15.4 m above the ground level. The most significant properties of the upper clay layer which are relevant to the laterally loaded pile behaviour are:

Undrained shear strength $C_u = 0.3 \cdot \bar{\sigma}_{vo}$ t/m²
compressibility modulus determined
by means of oedometer tests $E_{ed} = 7 \div 10 \cdot \bar{\sigma}_{vo}$ t/m²
Atterberg limits $70 \leqslant$ L.L. $\leqslant 85$; $50 \leqslant$ P.I. $\leqslant 60$
Natural water content $55 \leqslant W_n \leqslant 70$
Natural bulk density $1.55 \leqslant \gamma_n \leqslant 1.65$ t/m³

where

$\bar{\sigma}_{vo}$ = effective overburden stress.

Pile and soil characteristics (Figure 9), Site Ostiglia (North Italy). Closed end driven steel tubular pile 16 in o.d. with a thickness of 1 cm and filled with

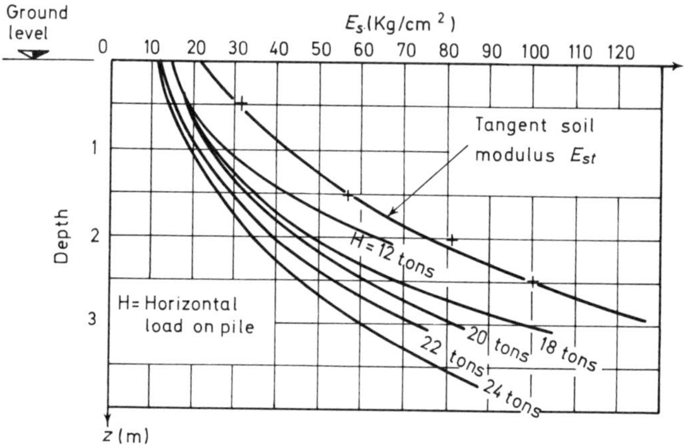

Figure 9 Ostiglia pile De = 40.6 cm: Variation of soil modulus with depth

concrete. The point of the horizontal load application was 1 m above the ground level. The pile penetrates 27.6 m into slightly OC clayey silt with the following significance geotechnical properties:

Undrained shear strength $3 \leqslant C_u \leqslant 6$ t/m²
Compressibility modulus $500 \leqslant E_{ed} \leqslant 800$ t/m²
Atterberg limits $35 \leqslant$ L.L. $\leqslant 55$; $20 \leqslant$ P.I. $\leqslant 35$
Natural water content $25 \leqslant W_n \leqslant 35$
Natural bulk density $1.80 \leqslant \gamma_n \leqslant 1.85$ t/m³

On the basis of the experience gained by the authors in the use of inclinometers for lateral load tests on piles, the following observations may be drawn:

(1) Due to the three times derivation process of the interpolated polynomial functions, which is needed in order to obtain the P values, the derived $E_s(z)$ curves represent only a crude estimate of the reality.
(2) The results illustrated above clearly show that, for the calculation of

laterally loaded piles in soft cohesive soils, non linear soil behaviour and smooth heterogenity of E_s with respect to depth, should be considered even for small loads.

(3) The constructed P–y curves can be well approximated by means of quadratic hyperbola represented by the following equation:

$$P = P_{\text{lim}} \tan\text{h} \frac{E_s\,y}{P_{\text{lim}}} \quad [FL^{-1}]$$

where:

P_{lim} = ultimate lateral soil reaction $[FL^{-2}]$
y = pile deflection $[L]$
P = soil reaction $[PL^{-1}]$

Complete results of the above tests are given in a paper submitted to the XI Italian Conference on SM and FE held in March 1970 in Milan.

G. Price (Building Research Station). The Geotechnics Division of the Building Research Station is currently engaged in field research into the behaviour of pile groups. Instrumented piles are used of 168 mm dia, 5 m long.

In order to compare the behaviour of a single pile with that of the pile when it forms part of a group relevant quantitative information must be obtained from pile tests in both cases. One of the instruments used to obtain this information is described in the paper by Cooke and Price submitted to this Symposium. Other more conventional instruments not described in the paper, but nevertheless vital to an understanding of the problem are: strain gauged load cells built into the pile shaft and movement transducers for measuring pile settlement and soil displacements.

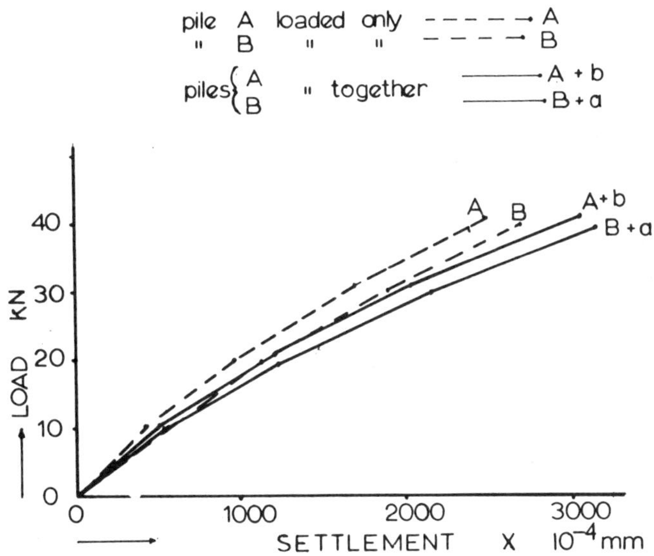

Figure 10 Load settlement graphs of test piles

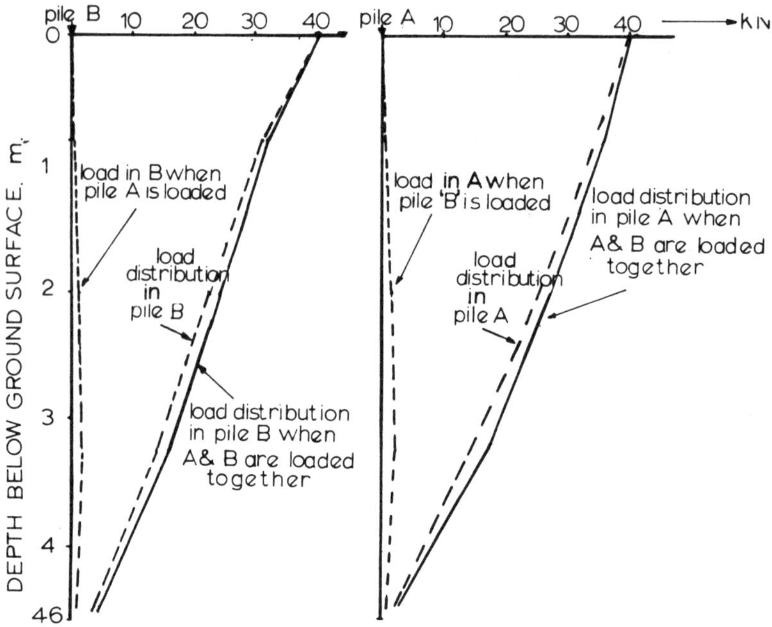

Figure 11 Load distribution with depth of test piles

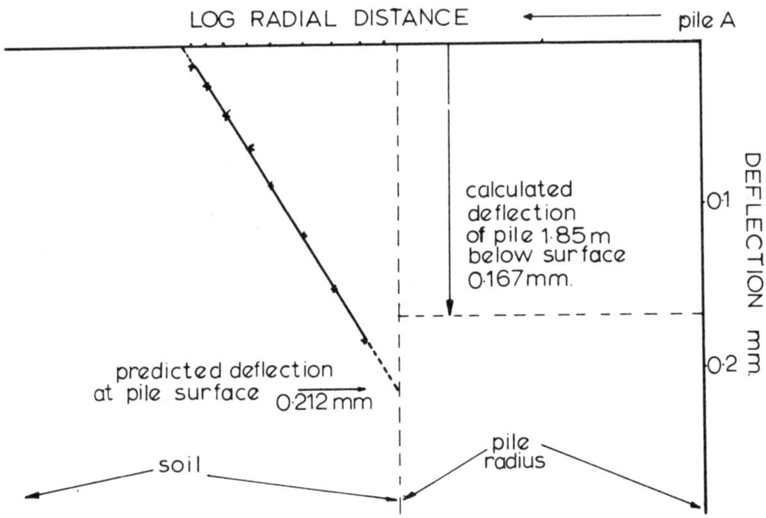

Figure 12 Vertical deflection in soil at 1.85 m below ground surface due to 40 kN load on pile A

To determine the range and sensitivity of the instrumentation needed previous observations on a single pile at the same site were used (Cooke and Price, 1973). Wherever possible the instrumentation was arranged so that:

(a) one system checked another
(b) transducers could be replaced if they malfunctioned
(c) at inaccessible positions gauges were duplicated, i.e. in the pile load cells all electrical circuits were doubled.

The results to be discussed are measurements of the interaction of two piles and of the soil deformation around a single pipe.

Figure 10 shows that the settlement of piles A and B spaced at 3 pile diameters was about 20% greater when loaded together than when loaded individually. It indicates that at this common spacing the piles were interacting. *Figure 11* is a plot of the load distribution in the piles under a 40 kN load. The piles were loaded individually and together. When pile A was loaded it induced a load in pile B and vise versa. The graphs show that the load distribution in a two pile group spaced at three diameters when loaded together is given by the load distribution due to the direct load, plus the induced load.

Figure 12 shows the variation in the vertical soil deflection around pile A as measured by a horizontal inclinometer train 1.85 m below the ground surface. The deflections are plotted against the log of radial distance from the axis of the pile. The significance of the log plot can be seen if one considers:

Stress α strain in an elastic medium: at the test loads used the pile and soil system was found to be elastic.

The transfer of shear stress outward from a unit height of the shaft of a pile in an elastic medium would vary as the increase in area, which increases linearly with radius (r).

$$\text{Since shear stress} = \frac{\text{load}}{\text{area}} \quad \text{(load constant)}$$

shear stress will vary outward from the pile at the rate of $1/r$

$\therefore$ shear strain α $1/r$

but shear strain is equal to the change in slope of the soil relative to the horizontal for an applied load.

The deflection produced in the soil can then be calculated by integrating slope, i.e.

$$\text{Deflection} = \int \text{Slope} = \int \frac{1}{r} = \log_e r$$

$\therefore$ Deflection for a constant E radially from the pile should be related through a log of r plot as shown in *Figure 12*.

To check that the inclinometers are recording the right order of deflection, the curve can be extrapolated to give the predicted pile settlement at 1.85 m below ground level. This can be compared with the calculated settlement given by: Pile head settlement – shortening in top 1.85 m of pile due to load, as shown in *Figure 12*.

The limit of the comparison is governed by the stability of the reference frame which, even at night, when these tests were carried out is of the order of 0.05 mm. From pre-loading observations the inclinometers are considered the more stable.

REFERENCE

Cooke, R. W. and Price, G. (1973), 'Strains and displacements around friction piles'. *Proc. 8th Int. Conf. SM and FE,* Moscow, 2.1. 53–60.

P. D. Morris and M. Roy (S.W. Road Construction Unit). We wish to discuss negative skin friction on piles. In the design of the piled foundations for the structures on the M5 Motorway it was apparent that additional loads would be carried by the piles due to the large settlements in the surrounding soft clay. The soft alluvium extends to 29.0 m average depth and overlies the limestones of the Lower Lias interspersed with weaker layers of clay and mudstone. The down drag force would result in additional settlement, which would be due partly to the compression of the pile and partly to penetration of the pile point into the bearing stratum.

Work by Johannessen and Bjerrum (1965) has shown that the distribution of the adhesion between clay and pile was governed by the distribution of the effective stresses in the surrounding clay and can be expressed as $\Delta v' K \tan \phi'$ where $\Delta v'$ is the effective vertical stress in the soil. The ultimate $K \tan \phi'$ value used for the soft marine clays was 0.20 and for the lower sands and silts 0.25.

Table 1 STEEL PILING FOR OVERBRIDGES

Bridge Number	*Allowable Working Load on Piles (Tonnes)*		
	No Bitumen	*3 mm Bitumen*	*6 mm Bitumen*
08	59	69	76
09	49	66	74
10	58	69	75
13	41	65	73
14	43	65	73

By using a 1 mm thick bitumen slip layer on piles Bjerrum Johannessen and Eide (1969) found that a substantial reduction in the down drag could be achieved. A suitable bitumen for such an application was proposed by Shell International Petroleum Co Ltd to resist the effects of temperature variation and driving stresses. Test piles in the Netherlands where 10 mm thick slip layers were used, confirmed that the amount of negative skin friction shed was related directly to the thickness and viscosity of the bitumen coating and the rate of settlement of the surrounding soil.

From pile tests carried out on the motorway the ultimate load was established from which the working load/pile was calculated as:

$$= \frac{\text{Ultimate load} - \text{Calculated down drag}}{\text{F of S (2)}}$$

The settlement gauges at the overbridge sites indicated settlements of the order of 1 mm/day from which the down drag force was calculated. Table 1 shows the predicted working loads for piles with various thicknesses of bitumen coating, from which it was decided that a 3 mm thick slip layer would be adopted. The additional cost of bitumen coating was approximately £20 per pile $3\frac{1}{3}$ % of the cost of the pile.

REFERENCES

Bjerrum, L. and Johannessen, I. J. (1965), 'Measurement of the compression of a steel pile to rock due to settlement of the surrounding clay. *Proc. of 6th Int. Conf. on Soil Mech and Found Eng.*, 2 pp. 261–264.

Bjerrum, L., Johannessen, I. J. and Eide, O. (1969). 'Reduction of negative skin friction on steel piles to rock. *Proc of 7th Int. Conf. on Soil Mech and Found Eng.*, 2 pp. 27–34. Mexico.

R. H. Bassett (Kings College, London). I would like to discuss the use of centrifugal model testing to predict instrument measurements in the foundations of an embankment. In his introductory notes as the General Reporter to this session Dr Burland has emphasised a number of key points concerning the use and interpretation of instrumentation data. I would like to contribute two of these points with data from a centrifugal model test. First, the use of instruments to check that a soil structural system is behaving adequately in terms of its design expectations and second, the emphasis now being placed on measuring displacements of specific points in a soil mass as a monitoring system.

The proper relevance of both these points can only be realised if a prediction technique of the type described by Professor Lambe in his Rankine Lecture (1973) as a 'grade A prediction', can be made. To be valuable to an instrumentation engineer the prediction must indicate the most suitable locations for instruments and the magnitudes they should be able to measure.

Dr Burland emphasised the power of the finite element technique as a suitable numerical tool in this context; however, I have been disappointed by the lack of its published use for 'grade A' (pre-event) predictions. The present limitation of the finite element technique is the difficulty of describing in a suitable mathematical form the complex, stress level and strain history dependent properties of even the simplest soils. The situation becomes particularly difficult as failure is approached.

The centrifugal model testing technique, I shall describe, is in a sense an analogue type finite element analysis which overcomes the soil property problem more easily than the numerical method. The technique can be criticised in circumstances that lead to significant work softening and progressive failure for reasons given by Bishop (1971) and Palmer and Rice (1973).

The writer's contribution concerns the behaviour of a structure build on soft lightly overconsolidated materials, [Nadarajah (1973)] and is not thought to be particularly susceptible to this mode of failure.

The data to be presented provided a grade A type prediction for the behaviour of the 8 m high prototype road embankment across the flood plane of the Yorkshire Ouse (the investigation was financed by the DOE, North Eastern Road Construction Unit). The results indicated the locations and capacities of the instruments to be inserted into the prototype. First however, I will digress briefly to describe the technique of centrifugal testing; a fuller description can be found in Bassett (1973) or Beasley (1973).

Three research schools are working on centrifugal testing in the UK, two at Manchester and at Cambridge. The latter used a very different monitoring technique and the data obtainable is particularly suited to the observation of unfailed working conditions and consequently is directly applicable to use with field instrumentation. In the writer's tests a 1/60th scale model of the foundation strata below a proposed embankment was made using large undisturbed block samples taken from the prototype location. The foundation model was then subjected to a gravitational field of 60G until an equilibrium state had been reached which approximated closely to the natural field conditions. The proposed embankment structure was formed and the centrifuge run so as to represent construction and the first two years* of the prototypes existence. In the writer's technique (see *Figure 13*), a vertical section through the model and the foundation was observed. This section was instrumented with some 400–600 markers (*Figure 14*) and the movements of these were observed throughout the testing sequence. The analysis of the marker movements provided the displacement data and strain data presented later.

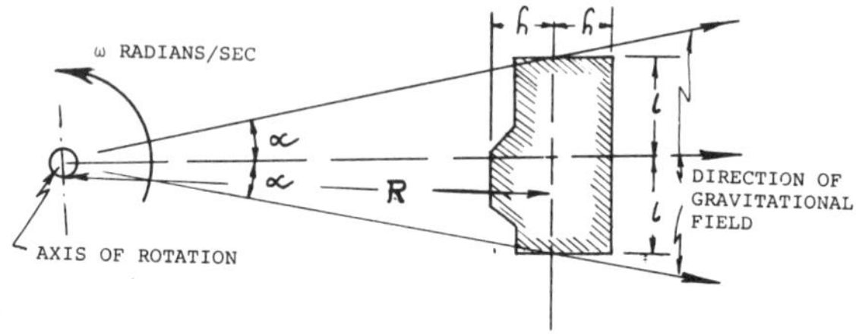

Figure 13 Plan of centrifuge and sample diagrammatic

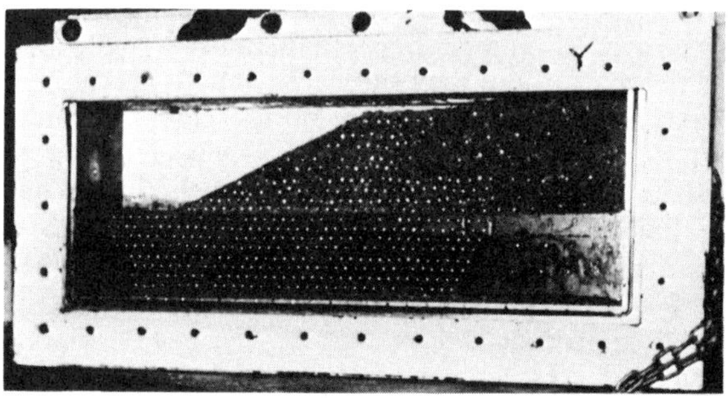

Figure 14 Showing a typical model (1 on 2 side slopes)

A number of models, built on identical foundation strata were tested to give comparative performances. Two are particularly relevant:

Model 1 An 8 m high embankment constructed of compacted fine sand with side slopes of 1 on 2 (*Figure 14*). Slip circle analysis of a prototype to this design would have been expected to suffer large shear deformations in the slopes if not in fact to have failed. A prototype of this section was not expected to be built but the model was run to give an indication of the displacement associated with near failure conditions.

Model 2 An identical embankment to (1) above but with 1 on 3 side slopes. This was the anticipated prototype section.

Figures 15 and *16* show examples of the typical data output for model 1. *Figure 15* is in the form most familiar to instrumentation engineers. It shows the predicted prototype horizontal displacements on four imaginary inclinometer lines (see key diagram) (a) at end of construction and (b) after 23 months.

*(Note that in the centrifuge the time scale is a function of G^2 i.e. 1 second in the centrifuge at 60G equals 1 hour in the prototype and 24 hours at 60G represents 10 years of the prototype life).

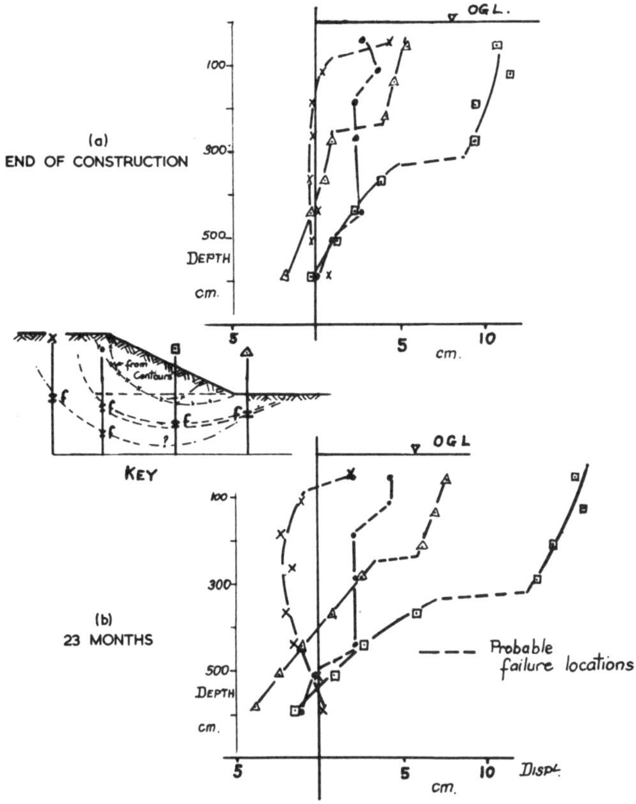

Figure 15 Horizontal displacements on inclinometer lines (1 on 2 slopes)

As mentioned the centrifuge will most probably incorrectly estimate failure displacements when rupture surfaces are involved because of the size effects associated with displacements [Palmer and Rice (1973)], however, the existence and location of failure zones is clearly suggested by the model 23 months after construction. A prototype with 1 on 2 slopes would suffer horizontal displacements at the toe in excess of 15 cm and vertical displacements at the crest in excess of 60 cm. (C. J. F. Jones' contribution was very interesting on this aspect and his prototype shows a similar mechanism, see p. 616).

Figures 17, 18 and *19* present a set of similar data obtained for Model 2 with 1 on 3 side slopes. The data is augmented by the vertical displacements (*Figure 19*). There are now no signs of unmistakeable instabilities although some slight grounds for suspicion might be implied in *Figure 17*. Two significant features are the fact that the horizontal displacements are considerably less than model 1 and that the vertical displacement field is noticeably more uniform. A prototype of this section is therefore probably stable and the displacements anticipated should be close to those indicated by the model.

Further understanding of the behaviour of the embankment can be obtained by examining the data in the form of volumetric strain and shear strain fields. These data are reproduced in *Figure 20a* and *b*. (Note that the situation is assumed to be plane strain and for these conditions the Volumetric strain is

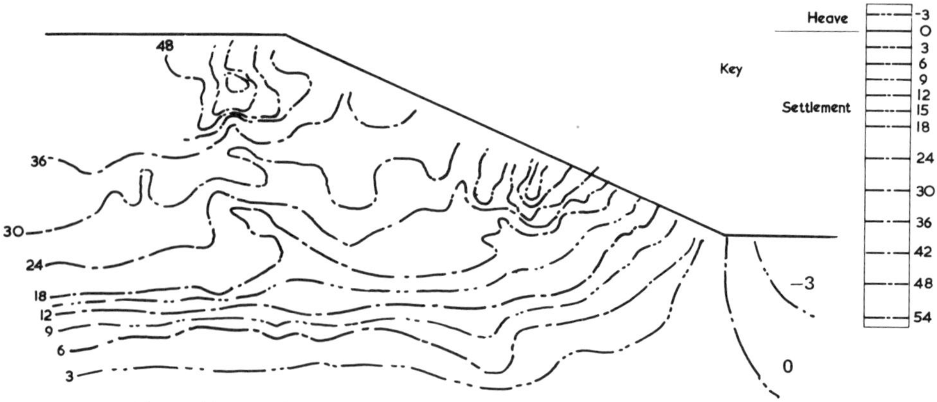

Figure 16a Vertical displacements at end of 23 months (1 on 2 slopes)

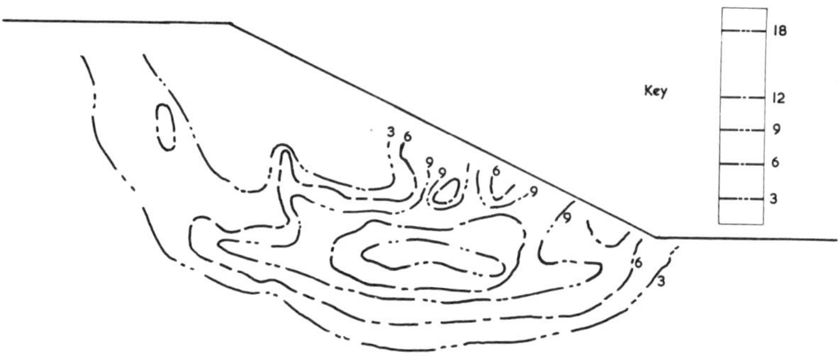

Figure 16b Horizontal displacements at end of 23 months (1 on 2 slopes)

defined as $\Delta V = \Delta\epsilon_1 + \Delta\epsilon_3$ and the shear strain is defined as $\Delta\gamma = \Delta\epsilon_1 - \Delta\epsilon_3$ where $\Delta\epsilon_1$ and $\Delta\epsilon_3$ are the principal incremental strains. Care has to be exercised when examining the shear strain $\Delta\gamma$ fields as one dimensional consolidation commonly associated with vertical settlement involves an equal component of volumetric strain ΔV and shear strain $\Delta\gamma$)

The Volumetric strain plot (*Figure 20a*) shows dilatancy in various parts of the compacted sand of the bank. Particularly noticeable are (i) the area at the toe where shear distortion are probably becoming significant and (ii) the thin zone extending downward from the bank surface at the 0.6 x the full bank height position; this zone may be indicative of a tensile crack forming, associated with a non-circular type of displacement mechanism.

The main areas of consolidation are also clearly shown extending outward from below the full bank height towards the toe. The higher per cent strains being more evident near the upper layers, these areas are near to the surface drainage.

How was this data used in the choice of instrument locations? The model data were not expected to show significant deviation from the pattern anticipated by an experienced instrumentation engineer. The model indicated well

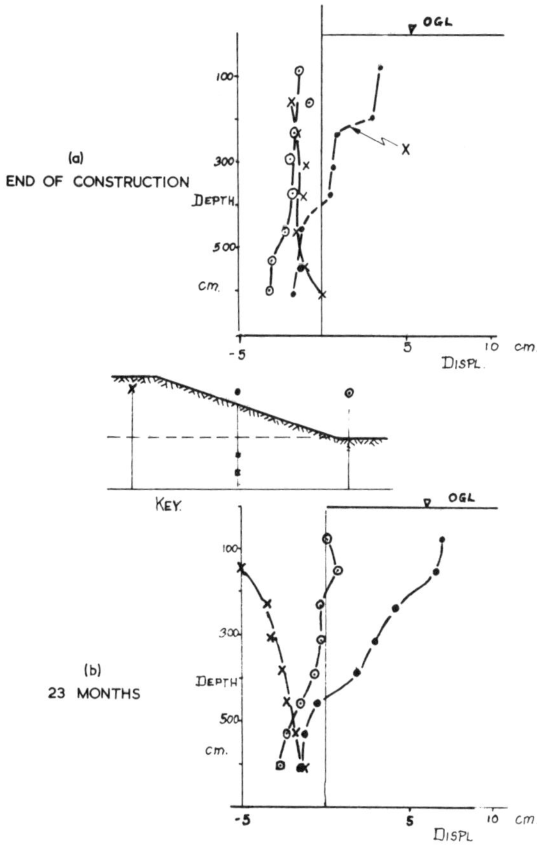

Figure 17 Horizontal displacements on inclinometer lines (1 on 3 slopes)

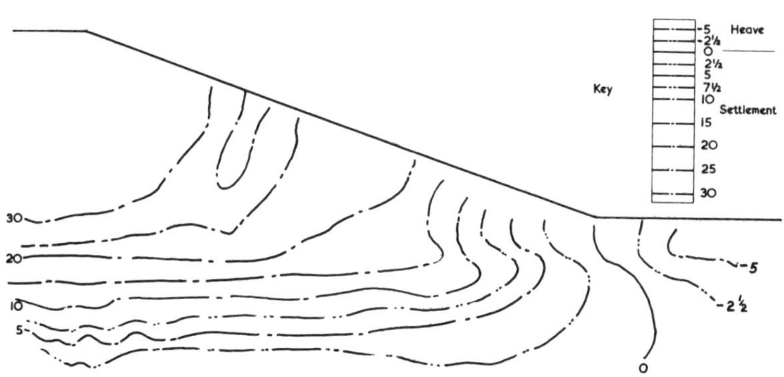

Figure 18a Vertical displacements at end of 23 months (1 on 3 slopes)

Figure 18b Horizontal displacements at end of 23 months (1 on 3 slopes)

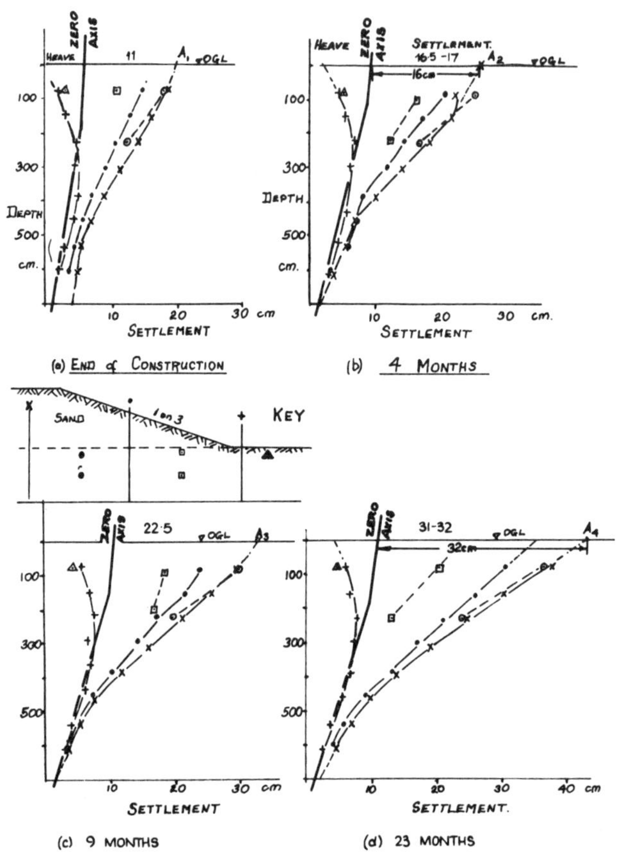

Figure 19 Vertical displacements on inclinometer lines (1 on 3 slopes)

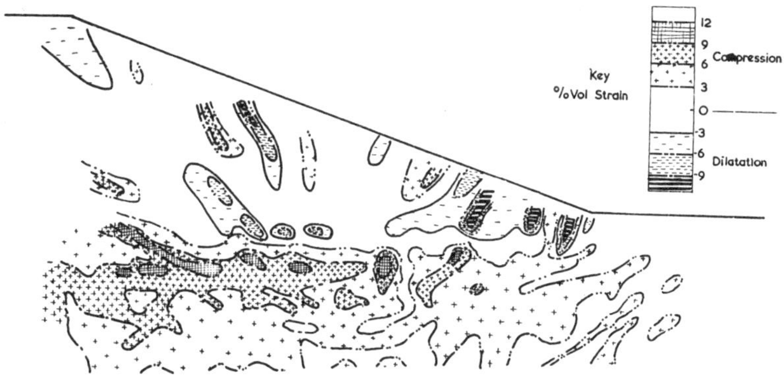

Figure 20a Volumetric strains at end of 23 months (1 on 3 slopes)

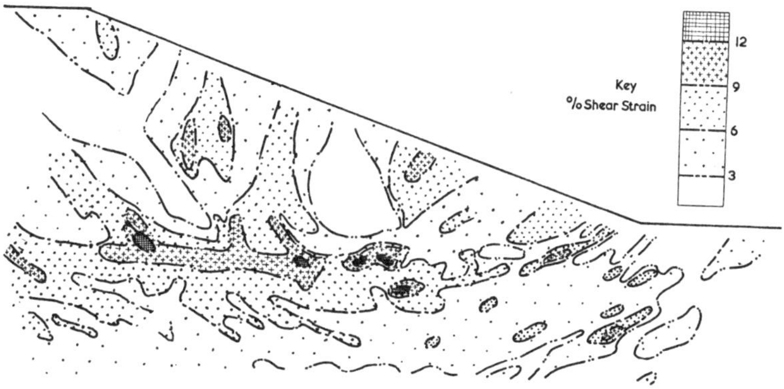

Figure 20b Shear strains at end of 23 months (1 on 3 slopes)

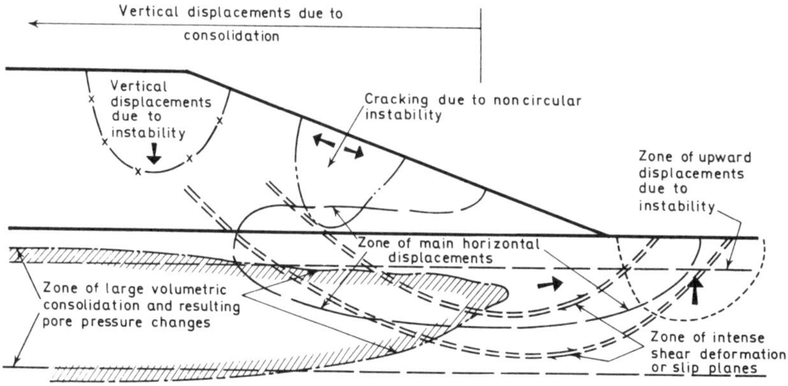

Figure 21 Zones of model showing significant features

known features which were zoned as in *Figure 21*. On this model evidence and in consultation with the BRS the instrumentation at present being installed on the prototype is as shown in *Figure 22a* and *b*. The models did however provide values for the magnitudes of displacements to be expected from a satisfactory embankment section and also gave an idea of the magnitudes which would be associated with possible approaching failure.

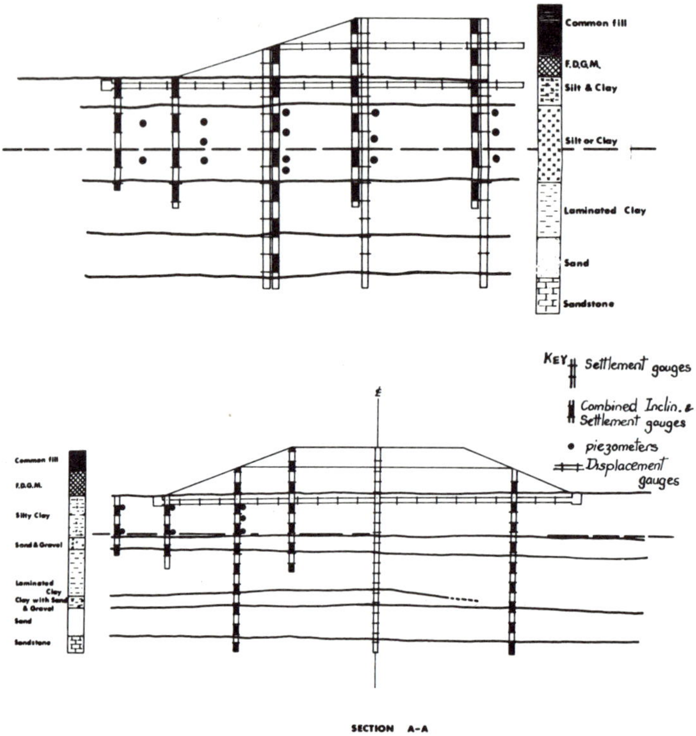

Figure 22 Typical instrumented sections of the prototype

The series of model tests described here give an interesting and I hope, valuable insight into the displacement features associated with a typical embankment on a 'real' soft foundation; they also provided that essential point for instrumentation, a 'grade A prediction' and the writer now waits to see the reliability of the method.

I feel however that though interesting the situations and materials involved in this study were in reality somewhat complex for use in a model technique particularly at such an early stage in the development of centrifugal testing. The important contribution of centrifugal testing is I feel its ability to reproduce small scale models at stress levels which are comparable with those experienced in both the field and those used in reliable laboratory testing.

The writer would like to see the technique used for testing a number of models in which simple loading situations and known boundary conditions are applied to the simplest possible soil types. The model data could then be compared with the predictions obtained using the finite element technique and the soil parameters derived from laboratory data (e.g. Simpson 1973).

Such an approach in which a numerical technique using the same soil data could predict, with acceptable accuracy, different types of loading would give confidence in, and define the limitations for, the use of finite element methods on real situations.

REFERENCES

Lambe, T. W. (1973), 13th Rankine Lecturer. 'Predictions in Soil Engineering' *Geotechnique* 23 No 2, pp. 149–202.

Bishop, A. W. (1971), *Geotechnique* 23 No 2, pp. 168–172.

Palmer, A. C. and Rice, J. (1973), 'The Growth of Slip Surfaces in the Progressive Failure of Overconsolidated Clays', *Royal Society of London* A 332, pp. 527–548.

Nadarajah, V. J. (1973), PhD Thesis, Cambridge.

Bassett, R. H. (1973), 'Centrifugal model tests of embankments on soft alluvial foundations.' *Proc. of VIIth International Conference of Soil Mech & Foundation Eng.*, Moscow.

Beasley, H. (1973), PhD Thesis, Cambridge.

Jones, C. J. F. (1973), Contribution to this conference, p. 616.

P. F. Wilkes (Norfolk County Council). In response to Dr Vaughan's request for more information on the selection of the method of control applied to King's Lynn Southern Bypass, the following may be of interest.

During the filling of the trial embankment the measured pore water pressures were very closely observed and data for selected piezometers plotted as filling proceeded. No rapid rise in pore water pressure was detected as failure approached nor was this indicated on subsequent plots made for all piezometers. Typical pore water pressures for the failed section are shown in *Figure 23*. Thus it was

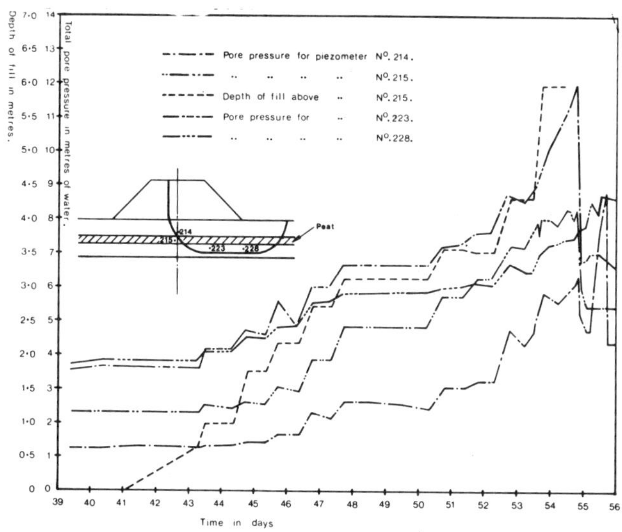

Figure 23 Measured pore pressures in vicinity of slip plane

felt that piezometric readings could not be relied upon to indicate instability.

Total stress stability calculations made for the failed section showed not only variations in the computed factor of safety for different methods of analysis, but more significantly the analysis was very sensitive to the selected strength profile, Wilkes (1972). Attempts to produce design charts based on Margason and Symons (1969) also proved to be difficult due to problems in

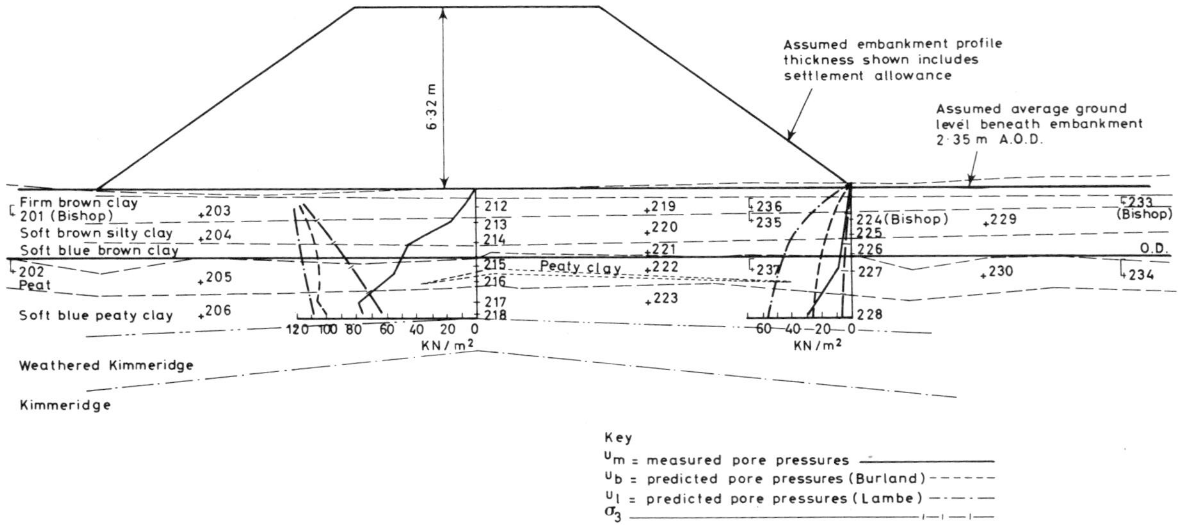

Figure 24 *Distribution of measured and predicted pore pressures beneath the embankment for section 2*

adequately predicting excess pore water pressures under various load conditions, typical results of such predictions are shown in *Figure 24* with measured excess pore water pressures shown for the same loading conditions. There was also a difficulty in deciding the safe computed value for factor of safety.

These problems led to the rejection of stress as the primary control technique and strain methods were then investigated.

Movements of surface pegs for the trial were examined but it was found that greater displacements, under the same load conditions, were being indicated by the inclinometer tubes at about 3 m below ground level. Thus the inclinometer method, outlined in the paper (p. 448), was adopted as being the most sensitive control tool available to the Resident Engineer.

Confidence has been increased in this method on at least one occasion when the $\Delta S/_H$ versus time plot showed an upward deviation from the initial straight line, i.e. instability was indicated. At the same time the pore water pressure shown at a piezometer tip, placed on a probable slip zone based on the trial failure, also showed an abnormal rise, thus indicating compatability between stress and strain measurements.

It is considered that this method is only applicable to similar types of alluvial deposits, i.e. basically very soft, highly compressible, low strength deposits.

REFERENCES

Margason, G. and Symons, I. F. (1969), 'Use of pore-pressure measurements to control embankment construction.' *Proc. 7th ICSMFE*, Mexico 2, pp. 307–315.

Wilkes, P. F. (1972), 'On induced failure at a trial embankment at King's Lynn, Norfolk, England.' *Proc. A.S.C.E., Conf. on Perf. of Earth and Earth Supptd., Structures* 1 part I.

D. C. Snelgrove (Messrs Sandberg). This is intended as a footnote to Peter Wilkes' contribution on the instrumentation of the King's Lynn by-pass.

Prior to fill commencing, readings were taken daily on piezometers to establish initial pore pressures, and as a check that manometer limbs were in balance. Readings were also taken on the inclinometers and vertical and horizontal settlement units to establish datum plots. Piezometers were read three times a day and plotted as pore pressure against time, the fill height was calculated in equivalent metres head of water and plotted on the same graph. Inclinometers were read twice daily and plotted and incremental movements noted. Total deformations (ΔS) were produced as a plot $\Delta S/h$ against time.

With the pore pressure plots it was noted that pore pressures increased almost proportionally with increases in fill height, and potential instability was indicated when pressure increases accelerated out of proportion to fill height (*Figure 25*). The first indication of movement is of course the inclinometer plot, however, if the $\Delta S/h$ plot is produced it will be seen that there is a 'normal' increase in movement proportional to fill height with time. Where there is a steep break away from the normal line, then movement has accelerated out of proportion to the fill added, and again instability is indicated (*Figure 26*).

Filling was in progress over an instrument line, and the fill height had reached approximately 2.5 m when disproportionate pressure increases and inclinometer movements were noted. The $\Delta S/h$ plot showed a sharp upward movement. Further readings were taken which confirmed this trend and the C.R.E. informed, who suspended fill operations in this area. The following day readings showed slight dissipation of pore pressures, and the inclinometer tubes had deformed to such an extent that it was not possible to insert the probe below

5.0 m depth. Lunar tension cracks 25–30 m long in three parallel rows 1 m apart appeared in the fill, the most serious being 1 cm wide. Had filling been allowed to continue there would have undoubtedly been a major slip.

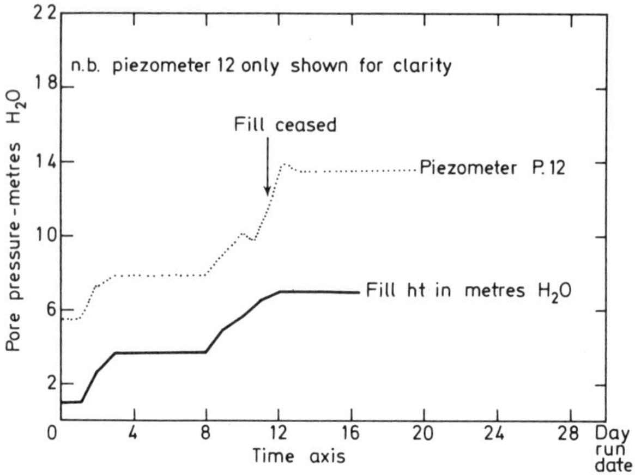

Figure 25

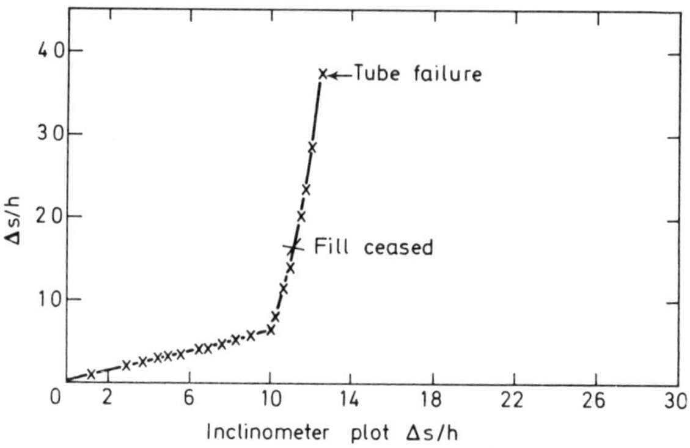

Figure 26

Although this method has been described as 'monitoring by the seat of your pants', it has the great advantage of simplicity and for the conditions which apply at King's Lynn the system works.

The daily monitoring sequence described above has been modified in the light of subsequent experience to twice daily piezometer and inclinometer readings. These are dropped to daily readings for the week after fill, and then further dropped to weekly and monthly readings.

I. F. Symons (Transport and Road Research Laboratory). In Session 3 (p. 623), Cole described the use of porewater pressure measurements for controlling stability during the construction of road embankments on soft foundation soils. The technique is similar in some respects to the method outlined by Margason and Symons (1969) and applied on several road schemes in the UK and abroad. The method has been developed further and there are a number of general points concerning this type of approach to the problem which are relevant to this discussion.

The use of total stress analyses with an adequate factor of safety, as mentioned by Parry (1971), is of value in estimating the safe height for rapid construction. The monitoring of porewater pressures during the construction to this safe height, when little or no control over the rate of construction is necessary, can provide useful information on the vertical and lateral distribution of porewater pressure for inclusion in stability analyses in terms of effective stress. The measured pore pressures can be related to the stress changes occurring in the subsoil using a comparatively simple method of the type described by Burland (1971) and Symons and Murray (1971). Sets of stability charts using a range of assumptions concerning the porewater pressure conditions in the subsoil can then be prepared. These charts should relate the factor of safety to a simple porewater pressure factor for different heights of embankment. Once the assumptions have been checked against the field measurements of porewater pressure and the most suitable chart selected, it can be used by the site staff to obtain a quantitative assessment of stability at any stage during the next phase of the construction. As the work proceeds periodic checks on the validity of porewater pressure assumptions used for the preparation of the charts are then necessary.

Studies of the type described by D'Appolonia *et al* (1971) using a finite element method for stress prediction coupled with an extensive laboratory study of the soil properties in a localised area are of considerable value in providing fundamental information for research purposes. On many road schemes involving embankment construction on alluvial deposits the longitudinal variation in the properties of the subsoil are significant and the sequence of construction cannot be specified easily. In these circumstances a semi-empirical method of the type described in the papers to this Symposium by Cook & Ingold and McKenna and others, such as Margason & Symons (1969), De Lory *et al* (1965), Elais & Storch (1970) and Lobadell (1969) is worthy of consideration since it offers a considerable degree of flexibility in the approach to the problem.

Measurements of deformation can provide useful qualitative information on stability. However it would seem difficult at the present time to specify limits on the amount of movement which can take into account variations in the soil properties and differences in the rate of construction. On a number of sites inclinometer measurements made by the Laboratory have shown lateral movements as large as 35·0 mm without any apparent damage to the embankment. This movement is considerably greater than the values quoted by other contributors.

In his paper Wilkes mentions the use of increases in measured porewater pressure as a means of indicating an incipient failure. Such a change in porewater pressure response reflects the onset of plasticity at a particular point in the subsoil as discussed by Hoeg *et al* (1969) and will occur at different heights of embankment depending upon the location of the measuring point. It would therefore seem difficult to use this approach as a rigorous means of stability control. In this discussion Wilkes has also referred to a high factor of safety from a back analyses of a failure using the measured porewater pressures. One of the difficulties of this type of analysis is in the determination of a reliable value of the cohesion component (c') for lightly overconsolidated deposits. The embankment failure described by Parry (1968) suggests that a low or zero value for the cohesion term may be necessary. The Laboratory has obtained factors of safety as low as 1·2 using effective stress methods of analysis without failure occurring in the embankment.

Finally mention has been made of the use of controlled rates of embankment construction. Where the stability relies on a significant gain in strength from consolidation, construction of the embankment in stages will generally be necessary. In these circumstances the use of a controlled rate of filling is likely to have little effect on the stability and the main advantage is that it provides the necessary time for measurements to be taken and interpreted.

REFERENCES

Margason, G. and Symons, I. F. (1969), 'Use of pore-pressure measurements to control embankment construction on soft foundations.' *Proc. 7th Int. Conf. Soil Mech Foundn Engng, Mexico* (2) pp. 307–315.

Parry, R. G. H. (1971), 'Stability Analyses for low embankments on soft clays.' *Proc. Roscoe Memorial Symposium,* Cambridge pp. 643–668

Burland, J. B. (1971), 'A method of estimating the pore pressures and displacements beneath embankments on soft natural clay deposits.' *Proc. Roscoe Memorial Symposium,* Cambridge pp. 505–536.

Symons, I. F. and Murray, R. T. (1971), 'Contribution to open discussion Session 5. *Proc. Roscoe memorial Symposium,* Cambridge.

D'Appolonia, D. J., Lambe T. W. and Poulos, H. G. (1971), 'Evaluation of pore pressures beneath an embankment.' *Proc. Am. Soc. Civ. Engnrs., Journal Soil Mech Foundns. Div.* SM6 pp. 881–897.

De Lory, F. A., Gass, A. A. and Wong, W. W. (1965), 'Measured pore pressures used for the control of two-stage construction of an embankments on soft foundation soils.' *Proc.* II (3) 216–226.

Elias, V. and Storch, H. (1970), 'Control and performance during construction of a highway embankment on weak soils.' *Highway Research Board,* Washington Record 323.

Lobdell, H. L. (1959), 'Rate of constructing embankment on soft foundation soils.' *Proc. Am. Soc. Civ. Engnrs., Journal Soil Mech, Foundns.* Div. SM5.

Parry, R. G. H. (1968), 'Field and laboratory behaviour of a lightly overconsolidated clay.' *Geotechnique* **18** (2) 151–171.

Hoeg, K., Andersland O. B. and Rolfsen, E. N. (1969), 'Undrained behaviour of quick clay under load tests at Asrum. *Geotechnique* **19** (1) 101–115.

J. N. Hutchinson (Imperial College London). Techniques of measuring small movements of the ground in order to give warning of the approach to conditions of shear failure have featured largely in the Symposium. Implicit in all of these is the assumption that significant pre-failure movements are going to occur. While there is little doubt that this is generally the case for *first-time slides*, there is some evidence to suggest that in the case of *renewals of movement on pre-existing slip surfaces*, movements preceeding failure are relatively small and may not provide a reliable basis for monitoring the approach to failure. Accurate field measurements to check this are needed.

I. W. Johnston (Dames & Moore London). Dr Parry in his contribution to the discussion, emphasised the point that although techniques of instrumentation are fairly sophisticated, quantitative interpretation of such results with regard to stability are still very much open to debate. Often instrumentation is employed to indicate abnormal foundation behaviour in a construction project particularly when there are possibilities of instability. Unfortunately the term 'abnormal foundation behaviour' is rarely defined quantitatively before commencement of construction, but is decided upon and often varies as instruments are read, thereby giving no real warning of impending failure.

However, where trial embankments are built, it should be possible to derive critical parameters from the results so that when full scale construction takes place, the field engineer can continuously plot instrument results and compare

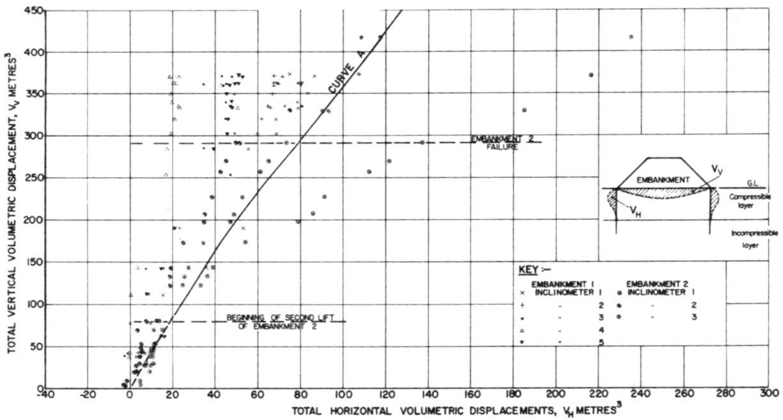

Figure 27 Total volumetric soil displacements in vertical and horizontal directions for the two test embankments at Canvey Island

them with the trial embankment behaviour. He would then be able to adjust loading programmes using a pre-defined critical bound for stability.

The results of the test embankments described in the paper by George and Parry (p. 152), are to be used in such a manner when water testing the instrumented earth supported tanks.

For both the test embankments at Canvey Island, measurements were made of vertical and horizontal displacements. It was then possible to calculate the total vertical volumetric displacement from the results of the mole settlement gauges and also calculate the total horizontal volumetric displacements at various radii around the embankments from the inclinometer results. *Figure 27* shows these results plotted as the total vertical volumetric displacement against total horizontal volumetric displacement, the latter assuming a constant circumferential displacement profile as given by each inclinometer. Embankment 1 was stable throughout, whereas Embankment 2 slipped suddenly as indicated on *Figure 27*. The location of the slip plane appeared to be between inclinometers 1 and 3 on Embankment 2, somewhat nearer to inclinometer 1. As can be seen from *Figure 27*, there is a general increase in the ratio of horizontal to vertical displacement not long after the beginning of the second lift of Embankment 2. This increase in horizontal displacement continues and is most obvious in the results of inclinometer 1 (closest to the slip plane) even before the slip actually occurred. This zone of large horizontal displacement corresponds to the formation of the slip plane, within which soil elements are yielding plastically. It is not until the failure plane is fully developed does the slip occur, with even more horizontal movement. It should be noted that this movement is not fully reflected in the total vertical displacement, as the slip bisects only a portion of the base of the embankment.

It is thus possible to construct a curve, such as Curve A given on *Figure 27*, which may be the critical parameter for the embankments. If a continuous plot of the above volumetric displacements is made, and if this reaches curve A, the building of the embankment should be stopped until sufficient consolidation strengthening of the subsoil can take place, to allow the plot to return to the left hand side of curve A, so loading can be continued. Such a technique may be employed during the water testing of the product tanks to be constructed at the subject site.

In general, by measuring volume deformations and plotting them as above, a warning of impending failure may be obtained in time to apply immediately corrective measures, whereas measurements of pore water pressures alone can only be related to stability via time consuming analyis possibly resulting in a type C prediction (Lambe, T. W. (1973) *Geotechnique* 23, 2, 149–202).

M. Roy (S.W. Road Construction Unit). My contribution deals with horizontal strain under embankments. The line of the motorway severed the existing land drainage system, which was reconnected into large drainage channels (rhynes) adjacent to the motorway. In several places these rhynes were interconnected by culverts.

The culvert units were generally 3.0 x 2.0 x 1.0 m with 125 mm long joints. The units were laid on a hogging curve to allow for the anticipated settlement.

In two cases during the surcharge period, when the ground loading was approximately double the design loading, failure occurred in the end sections of these culverts resulting in the collapse of a headwall. It was apparent that the magnitude of extension exceeded that expected to be induced by vertical displacement. Calculations based on work done by Rutledge and Gould

(Moran *et al* 1960) and reported by Cappelman (1967) gave values of horizontal strain per unit of 125 mm.

Readings taken across the joints subsequently have indicated movements of 100 mm in the tension zone and from the horizontal settlement gauges under similar embankment loadings, strains of up to 100 mm were recorded. It should be emphasised however that the strain readings from this gauge developed by Soil Instruments Ltd (McKenna and Roy 1973) were not reliable because of the temperature effects on the calibrated cable. When this has been overcome this gauge will be a very useful tool in the analysis of the effects of embankment loadings.

References

Moran, Proctor, Mueser and Rutledge (1960), 'Report on investigations of deformations in foundations of earth embankments containing concrete pressure pipe conduits.'

Cappelman, H. L. (1967), 'Movement in pipe conduits under earth drains.' *Proc. ASCE – SM6*, November 1967.

McKenna, J. M. and Roy, M. (1973), 'Performance of the instrumentation used to monitor the M5 motorway embankments built on soft ground between Edithmead and Huntworth.' *Field Instrumentation in Geotechnical Engineering*, 1, 282.

C. J. F. Jones (N.E. Road Construction Unit). In connection with the interprelation of observations Dr Burland in his report for the 4th session shows that there is, at present, an emphasis on deformation measurements as opposed to stress measurements. However, deformation meaurements are not particularly useful to the engineer engaged in the design of earth retaining structures or bridge substructures. The designer is principally interested in the loads and forces acting on a structure, and this explains the West Riding interest in the measurement of stress at discreet points on and near earth retaining structures. The difficulties in making measurements of stress are recognised, but since it is the general stress or stress tendancies that are of interest, as opposed to the exact stresses, it is not particularly important if the instruments used overread or underread by 0.1 % or even 10 %.

The primary object of the research on earth retaining structures that has been carried out in Yorkshire in the last few years is to obtain a better understanding of the soil/structure interaction and hence to produce less expensive structures. The measurements on the M1, M18 and M62 Motorways indicated that the finite element technique could be used to model the action of soil and substructure. In fact, fairly simple elastic analysis using both 2 and 3 dimensional finite elements confirm the West Riding measurements, which indicate that retaining walls represent a 2-dimensional problem and that standard bridge abutments incorporating cantilevered wingwalls, are a 3-dimensional problem.

It is worth considering whether this method can influence design methods. At present the drawing up of a complicated finite element mesh, involving 300 to 400 elements, is not practical for normal problems. However, in the near future higher order 2 dimensional quadrilateral elements will be available for general use in the design office, in which case the design of large walls using the method could be seriously considered.

The abutment problem offers the most scope for material savings in design, due to the large number built annually. The 500th bridge to be built in the West Riding since the beginning of the motorway programme is currently in the preparation stage; other organisations can produce similar figures. The majority of these bridges are standard bread and butter structures similar to those investigated on the M18 Motorway. The majority of design offices will probably not have access to general 3-dimensional finite element programmes for some time, however, the method promises to assist in the solution of one of the current design imponderables of abutments, namely the analysis of the stresses in the wingwall/abutment connection.

B. N. Whittaker (Nottingham University). Dr Burland has raised a very important point in respect of the way in which the surface subsides following undermining. *Figure 28* shows the significance of the point under discussion. The illustration shows a cross-section through a working coalface which is advancing and permitting the overlying beds to collapse. In (a) the ground is shown to behave in the form of sheared blocks which lower in a manner not

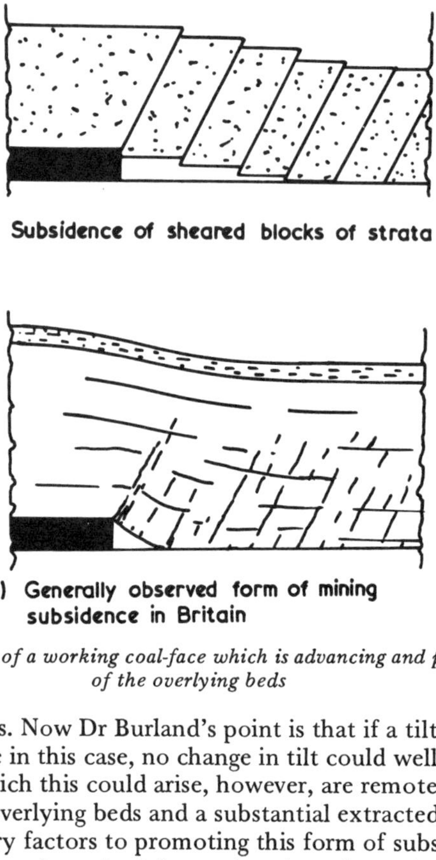

Figure 28 Cross-section of a working coal-face which is advancing and permitting collapse of the overlying beds

unlike a pack of cards. Now Dr Burland's point is that if a tilt meter were located at the surface in this case, no change in tilt could well be observed. The conditions in which this could arise, however, are remote. Shallow conditions, strong brittle overlying beds and a substantial extracted seam height are likely contributory factors to promoting this form of subsidence. However, the more typical picture in coal mining regions is as shown in (b). Shear failure does occur, but at an angle heading over the extracted (and subsequently collapsed) region behind the longwall coalface. These shear planes become dissipated at some distance away from the extraction point. Although collapse of the roof beds is promoted behind the coalface, surface subsidence usually takes the form of a predictable curve. The presence of lateral strata pressures, strata bridging over the working area, the pliability of the overlying coal measure stratified rocks, the invariable presence of a relatively weak surface boundary layer, the uniform thickness of coal removed which is very small compared to the depth, the width/depth ratio of the extraction all contribute to the occurrence of a smooth subsidence profile at the surface. In some metalliferous mines, however, where the deposits are not deep, but very large volumes of ore are removed, shear fractures of the type indicated in (a) can occur and sometimes reach the surface in the manner shown

The instrumentation described in the paper (p. 437) has been employed in normal mining subsidence situations. The tilt measurements all indicate uniformity. Where tilt has been measured on a dynamic subsidence development curve, for example down the central axis of an advancing coalface or within a critical area, the tilt measuring station has returned to its original vertical position after giving sufficient results to indicate the maximum tilt, the subsidence and horizontal strain (since this is a function of curvature). These latter values have been checked against measurements made with the standard surveying equipment used to determine such quantities.

B. O. Skipp (Soil Mechanics Ltd.). My colleague Mr Sherwood has outlined in his paper a telemetered slope monitoring system currently being installed. Mr Wilson also described a sophisticated hillside observation system with data logging and subsequent visual interpretation of recorded data. The proper interpretation of small movements on hillsides as an indicator of potential instability has been discussed by Ter Stepanian (1966). The setting of criteria is in general a very difficult task especially when the instruments sensitivity may well be such as to respond to diurnal and seasonal environmental changes. The problem is one of identifying a trend of movement against a background of 'insignificant' noises. This requires that there be some studies of a composition of the 'noises' and in the example quoted by Mr Sherwood this constitutes the first objective of the exercise.

A computer programme has been written which enables the data on hillside movements from each point of measurement to be statistically examined over a month's period with a 'rolling' aquistion of data and a linear regression fit to a one month's set of data. A prediction is made of the next observation which is accepted if it falls within specified values having regard to the previous trends. It would be interesting to hear from Mr Wilson as to the basis upon which he would set alarm levels not simply upon the magnitude or repeatability of a single reading of any one of his instruments but upon the relationship of that reading to the 'noise' in the system.

Reference

Ter Stepanian, G. (1966), 'Types of deep creep of slopes in rock masses.' *Proc. 1st Cong. Int. Soc. Rock Mech.* Lisbon pp. 157–160.

Sessions 3 and 4. The application of instrumentation in geotechnical engineering and the interpretation of their results

Joint Concluding Remarks: P. R. Vaughan & J. B. Burland

The growth of geotechnical instrumentation can be viewed perhaps as part of the development of a more numerate approach to civil engineering generally. In planning, design, construction, operation and maintenance, throughout the industry, judgement and decision-making have been progressively aided by quantitative methods. Generally, the results of such a process are beneficial. Adverse effects occur if the methods are not applied critically and intelligently, if their sophistication obscures understanding of a problem, if they are based on wrong premises and if they induce a false sense of security or overconfidence.

What have been the effects of this process in geotechnical engineering? There has been a continuous interaction between the development of analytical methods, laboratory testing of materials and observations of field behaviour, with developments in one area stimulating development in another. Field instrumentation and observation has played a vital role. As a result, basic understanding of materials and mechanisms of behaviour has improved and better methods of analysis have been developed. Improved bases for codifying engineering experience have thus been formulated, more difficult problems can be tackled with less uncertainty, the risk of failure and unforseen delays and difficulties has been reduced and construction methods have been aided and made more economic.

While both engineering and construction costs may be increased, an overall saving is probable where safety is improved, cheaper types of construction can be adopted and the overall amount of work reduced. Such a process seems professionally desirable as skill is improved. It is important that such a continuing process should not be assessed by cost-benefit analyses performed on a narrow basis. Failures and disadvantages can usually be identified clearly and costed. Successes and advantages are often more general in nature and are difficult to evaluate in financial terms.

How can this process and, in particular, the role of field instrumentation in geotechnical engineering be made more effective? It must be appreciated that a major professional engineering effort must be made and financed. This effort controls what can be achieved and without it large sums of money can be wasted. A critical attitude must be maintained. Field observations only become useful when they have been evaluated and made use of. Availability of information to others is of critical importance. The aim must be to improve basic understanding and to develop more realistic analyses, rather than the mere accumulation of data or the creation of rigid empirical design rules. Thus there must be a continuous process of assimilation, digestion and formulation of information in usable form.

While both long and short-term benefits are clear, there are short-term disadvantages arising from the engineering effort required for success, the interference of instrumentation with construction and the uncertainties introduced into construction by the flexible methods of control and design modification allowed by instrumentation. These disadvantages must be appreciated and weighed at an early stage of a project. Every effort must be made to minimise them and to allow for them.

It is to be hoped that the papers submitted to this Symposium and the discussions at it will aid the profession in the important area of geotechnical instrumentation and its effective use.

Closing address

Dr L. F. Cooling

I am greatly honoured that you have asked me to give the closing address at this symposium on 'Field Instrumentation' and I must say I have enjoyed the discussions over the past three days. The subject is one with which I have been concerned for a long time and it may be helpful in assessing the present position if I give a brief review of its growth and development as I have seen it over the years.

Some eleven years ago, I gave the second Rankine Lecture to this Society. It was entitled 'Field Measurements in Soil Mechanics', and was based on work done by my colleagues and myself at the Building Research Station over the previous 25 years. The purpose of this paper was to encourage the carrying out of measurements on full-scale structures so as to obtain a better understanding of the sort of deformations and stresses that a normal structure experiences. Already at that time, engineers had come to recognise the importance of making a thorough study of *failures* in civil engineering structures. Such events were openly investigated and detailed tests carrried out to establish the cause of failure and to indicate the appropriate remedial measures. However it was appreciated that such studies were of a 'post-mortem' nature and not always very valuable for establishing precise design criteria. But the importance of building up a body of reliable data by the measurement of the performance of actual structures was not yet fully appreciated and I thought the need for more observational work required to be emphasised and actively promoted. It was beginning to be recognised but not throughout the Construction Industry. Even so, by 1962, substantial progress had already been made, and perhaps I may be permitted to illustrate how far we had come at this stage by an anecdote from the early days of Soil Mechanics in this country.

It occurred during one of Terzaghi's rare visits to the Building Research Station around 1937. It so happened that on that same day the famous American architect Frank Lloyd Wright was also visiting the Station. By chance we all sat down together at one table for lunch. The two famous men had not met each other before, and since they were both accomplished talkers, an animated discussion was soon in progress. Skempton and I were fascinated by their verbal skirmishing and conversation, which soon turned to the subject of structures they had built, etc. The passage I recall, and which is the point of my story, was briefly as follows:

F.L.W. stated 'that the very large hotel he had built in Tokio had been designed so that it should settle just ½ in.

T. came back, as quick as a flash, with the short staccato question, 'And did it?' To which there was no reply.

In all probability the idea of actually measuring the settlement had not even been considered. This illustrates the leeway that had to be made up and the importance of the philosophy Terzaghi preached of never missing the opportunity of finding out by direct observation the difference between forecasts and real developments.

We have travelled some way since that time and have a much clearer insight into what is required and what can be achieved. During my later years at BRS, my colleagues in the Geotechnics Division devoted quite a large proportion of their time to making a special feature of field observations and the instrumentation needed for this work. In 1968, I included the results of a number of these investigations in a paper to the Proceedings of the Fourth Conference of the Australian Road Research Board entitled 'Field measurements for design purposes and for monitoring structural Performance', (Paper No. 504 S, Vol. 4, Pt. 2). The measurements were broadly classified under the following headings:

(1) Measurements for use in preconstruction design,
  (a) field tests to measure properties of soil *in situ*.
  (b) detailed observations on 'trial structures'.
(2) Measurements on full scale structures during construction as part of the 'observational procedure'.
(3) Long-term measurements on full scale structures to develop design methods.

In my experience, the use of field instruments up to this time had been characterized by a number of special features. The instruments were purpose made and were often 'one-off'. They were mainly installed by scientists or people who were specially concerned that they should function properly. Moreover the purpose for which the instruments were intended was clearly recognized.

Since then, progress has been very rapid and the wide increase in the use of instruments has introduced a new situation with new problems. Instruments have been vastly improved and a number of new techniques developed. This is clearly shown in the splendid exhibition of instruments and equipment held as part of this symposium. Moreover the thirty-seven papers published in Part 1 of the Proceedings and the number of participants attending the Symposium has demonstrated how opportune was the time chosen to review the subject. The keeness and quality of the discussions has also shown the great interest in the subject just now.

I thought the Introductory notes produced by the General Reporters were particularly useful in ensuring the smooth running of the four Sessions. I was in full agreement with most of the points made. While I cannot pretend to be able to sum these up even broadly, I may perhaps be permitted to make a few random comments.

In the first session where Dr Arthur set out the fundamental attributes of a satisfactory measuring system, I particularly liked the item he termed 'Environmental calibration.; the relationship between the actual measurement and the value the variable measured would have had, if the presence of the instrument could have been miraculously avoided. Though somewhat involved, this is a very important consideration.

In the second session Dr Green dealt very comprehensively with 'Equipment design, installation and performance of field instruments' I think it is generally agreed that simple, robust instruments are to be preferred when suitable but so often it is necessary to go for a more complicated method to get the sensitivity and accuracy required. Mr Dunnicliff and others stressed the importance of measuring in more than one way. I cannot refrain from quoting the following sentence from my Rankine Lecture – 'It is often an advantage when using specialized techniques to take check readings by an independent set of observations using another form of instrumentation, even if the second method is rather crude. Such results should at least 'give the street you are in' and can lend a good deal of confidence to the more accurate measurements given by the special technique.'

As to the more sophisticated instruments, I can appreciate the advantages of electronic recording equipment, with print-out and direct link to a computer. This gives more and faster results and can obviously be of advantage on occasions. But in the end we need to interpret the results in terms of the practical problem. We should not then take readings just for the sake of taking them, and I feel that the undue proliferation of sophisticated instruments is something to be resisted.

I am indebted to Dr Vaughan for drawing my attention to a recent paper by my friend Professor Peck entitled 'Observation and Instrumentation, some elementary considerations', where in his conclusions he expresses his views on this point. These are put more succinctly and with more authority than I could give:

> 'It concerns me that too many programmes are based on the number of instruments to be used rather than on the questions to be answered. It concerns me that sophistication and automation are substituted for patient proof-testing of equipment under field conditions. To the extent that such practices prevail, they must be discouraged so that the observational approach itself will not be discredited. We need to carry out a vast amount of observational work, but what we do should be done for a purpose and done well.'

If the report of Dr Vaughan on the application of instrumentation and that of Dr Burland on the interpretation of observations are studied and followed, I don't think Professor Peck will have much room for concern.

Finally I think that the British Geotechnical Society has again organised a very successful Symposium, and that the Proceedings will be recognised not only in this country but throughout the world as marking an important stage in the development of Geotechnical Engineering.

I think the Organising Committee, the authors, the General Reporters and those taking part in the discussions should be congratulated on doing a splendid job. It now only remains for me to declare the Symposium closed.

Keyword Index to Papers and Discussion

The Keyword Index consists of three tables, which refer to the papers and discussions in the order in which they appear in the proceedings. The Keywords used are in two groups. The first group is of general Keywords based on the Geotechnical Retrieval System for Geotechnical Abstracts. The second group deals with the type of instrument referred to. The General Reports and the Opening and Closing Addresses have not been indexed.

The papers and the discussions are identified by a contribution number and a page number, as follows

Papers

Discussion. Principles of Measurement

Discussion. Equipment

Discussion. Application

Discussion. Interpretation

Contribution No	Page No	Active earth pressure	Bridge	Calibration	Compaction	Consolidation and compression	Construction control	Contracts and organisation	Corrosion	Costs and benefits	Dam and dam foundation	Deformation and strain	Design	Field performance data	Earthfill and backfill	Earthquake	Elasticity	Equipment design	Equipment installation	Error	Excavations	Failure	Field test	Geophysics	Ground water	Highway	Interpretation and analysis	Lateral yield	Load or plate bearing test	Measurement principles	Mining	Multistorey building	New developments	Operation	Partly saturated soil	Performance
Papers																																				
1	1			x			x					x		x						x							x			x			x			
2	13									x			x	x	x												x									
3	38			x								x						x		x						x				x			x			x
4	52	x					x					x	x	x												x	x	x								x
5	70						x					x	x					x	x		x						x						x	x		x
6	85																	x	x	x												x		x		x
7	100						x			x				x	x												x									
8	112			x								x						x	x	x			x							x			x	x		x
9	126			x								x							x	x								x						x		x
10	136				x								x	x	x							x					x							x		
11	152					x	x	x		x		x							x								x		x							
12	166			x					x			x						x	x	x								x		x			x	x		x
13	180												x					x	x		x							x				x		x		x
14	194			x						x					x			x		x			x										x	x		x
15	207	x	x										x	x					x								x									
16	220						x	x					x		x																					
17	229										x		x						x																	x
18	240										x		x	x					x								x							x		x
19	249				x		x				x	x	x	x	x			x	x								x							x	x	x
20	262					x	x						x		x								x			x	x	x								x
21	275									x					x			x	x							x								x		x
22	287			x		x	x						x		x			x	x						x		x							x		x
23	304					x						x	x	x				x	x				x				x		x				x	x		
24	318		x						x							x									x											x
25	332												x	x	x						x					x	x									
26	341						x				x	x	x						x									x		x				x		x
27	359			x			x					x	x					x	x	x							x	x		x			x	x		x
28	370												x				x		x								x		x					x		x
29	382											x		x	x			x		x														x		x
30	396				x		x						x			x		x	x					x			x			x						x
31	411																	x		x										x				x		x
32	423			x								x							x				x					x					x	x		x
33	437											x						x	x	x										x	x		x	x		x
34	448						x								x				x							x								x		
35	462											x					x			x										x			x			
36	475				x							x	x	x	x			x					x				x		x				x			x
37	488						x	x		x			x		x				x				x											x		x
Discussion - Principles of Measurement																																				
38	527										x	x								x							x			x						
39	528			x				x				x							x											x				x		
40	529												x					x												x				x		
41	530									x									x											x			x			
42	530												x		x			x	x	x										x						x

Contribution No	Permeability	Pile	Pore pressure	Pre-loading	Footing or raft foundation	Repeated loading	Research	Retaining wall	Rock	Safety factor	Seepage	Settlement	Slope stability	State-of-the-art review	Temperature effect	Test procedure	Vibration	Piezometer (hydraulic)	Piezometer (pneumatic)	Piezometer (electric)	Surveying instruments	Surface monuments	Settlement plates & gauges	Settlement (fluid systems)	Settlement (horizontal tubes)	Borehole extensometers	Inclinometers	Tiltmeters	Horizontal displacement gauge	Strain gauge	Failure stakes	Slip plane indicators	Earth pressure cells	Load cells	Flat jacks & stress plugs	Vibration measurements	Seepage measurements
Papers																																					
1							x														x	x															
2							x							x																							
3						x								x																x			x				
4								x																			x						x	x			
5							x	x				x										x				x	x		x								
6							x					x									x	x															
7			x							x								x																			
8		x					x								x	x									x			x									
9																											x										
10			x										x						x			x					x					x					
11				x	x													x						x	x	x	x				x						
12															x												x	x									
13		x			x			x				x														x			x				x	x			
14	x		x									x			x	x		x						x									x				
15							x	x																						x			x				
16																																					
17																		x	x	x		x					x						x				x
18											x							x																			x
19			x				x				x							x				x	x										x				
20		x	x	x						x		x						x				x	x		x		x		x		x						
21			x	x								x						x				x	x		x	x	x		x		x						
22	x		x									x						x	x	x		x				x	x		x								
23						x	x					x			x	x							x			x											
24		x													x																						
25			x							x																											
26												x	x			x						x		x					x								
27		x					x	x							x	x											x										
28									x																										x		
29												x																x								x	
30																	x																				
31			x											x				x	x	x																	
32																x							x				x							x			
33																						x						x	x								
34																		x				x	x	x		x	x		x			x					
35									x																												
36					x																												x				
37			x																																		
Discussion - Principles of Measurement																																					
38																																					
39																																					
40			x																																		
41									x																					x					x		
42												x											x			x											

Contribution No	Page No	Active earth pressure	Bridge	Calibration	Compaction	Consolidation and compression	Construction control	Contracts and organisation	Corrosion	Costs and benefits	Dam and dam foundation	Deformation and strain	Design	Field performance data	Earthfill and backfill	Earthquake	Elasticity	Equipment design	Equipment installation	Error	Excavations	Failure	Field test	Geophysics	Ground water	Highway	Interpretation and analysis	Lateral yield	Load or plate bearing test	Measurement principles	Mining	Multistorey building	New developments	Operation	Partly saturated soil	Performance
43	533											x						x												x			x			
44	533				x									x	x				x								x			x						
45	539																	x												x			x			
46	539											x						x	x	x							x			x			x			
47	543			x															x	x							x			x						x
48	545			x								x							x	x							x			x						
49	546																	x	x														x			
50	549																	x	x				x							x			x			
51	549			x																										x						x
52	551											x																		x						
53	552																	x	x	x										x					x	
Discussion - Equipment																																				
54	556						x					x	x					x	x		x									x			x	x		
55	557											x	x					x	x											x						
56	557											x							x											x						x
57	558											x	x					x												x			x		x	
58	562																	x					x							x			x			
59	563																		x	x															x	
60	565								x									x		x										x			x			
61	566																		x	x										x			x		x	
62	567																		x	x			x							x						
63	568													x					x	x										x			x			x
64	569																	x		x										x				x		
65	570										x	x						x		x										x			x			x
66	572												x	x				x	x	x										x			x			x
67	575								x					x				x		x										x			x			
68	576											x		x				x	x	x										x						x
69	578																	x																		
70	578			x																x										x						x
71	579							x										x		x										x						x
72	580																	x		x										x						x
73	582																													x						x
74	583	x						x		x				x					x	x						x								x		x
75	586							x												x																
76	587																	x	x	x										x			x			
77	589							x	x					x				x																		x
78	594																	x		x										x						x
79	597													x																						
80	597																	x												x						
Discussion - Applications																																				
81	614						x					x															x	x								
82	614									x	x		x		x								x													
83	616						x							x								x				x										
84	617							x		x	x			x	x		x		x	x			x				x	x								

Contribution No	Permeability	Pile	Pore pressure	Pre-loading	Footing or raft foundation	Repeated loading	Research	Retaining wall	Rock	Safety factor	Seepage	Settlement	Slope stability	State-of-the-art review	Temperature effect	Test procedure	Vibration	Piezometer (hydraulic)	Piezometer (pneumatic)	Piezometer (electric)	Surveying instruments	Surface monuments	Settlement plates & gauges	Settlement (fluid systems)	Settlement (horizontal tubes)	Borehole extensometers	Inclinometers	Tiltmeters	Horizontal displacement gauges	Strain gauges	Failure stakes	Slip plane indicators	Earth pressure cells	Load cells	Flat jacks & stress plugs	Vibration measurements	Seepage measurements
43																														x							
44							x								x																		x			x	
45																																	x	x			
46																		x	x	x							x						x				
47																																	x				
48																																	x				
49																				x		x											x				
50																																	x				
51																																					
52																					x	x															
53			x															x																			
Discussion - Equipment																																					
54									x	x			x														x										
55																											x										
56																											x										
57			x													x		x					x				x		x								
58																		x															x				
59			x															x		x																	
60			x																	x																	
61			x																x																		
62							x											x	x																		
63												x											x			x											
64												x											x			x											
65																					x	x															
66		x										x			x						x	x															
67																												x									
68															x													x	x								
69																										x											
70																																	x				
71																																	x				
72																														x			x				
73																																	x				
74																																	x	x			
75																																					
76																		x				x	x			x											
77																			x						x		x										
78															x								x			x							x				
79																																		x			
80																																					
Discussion - Applications																																					
81										x			x																								
82																																					
83													x																								
84			x							x			x																								

Contribution No	Page No	Active earth pressure	Bridge	Calibration	Compaction	Consolidation and compression	Construction control	Contracts and organisation	Corrosion	Costs and benefits	Dam and dam foundation	Deformation and strain	Design	Field performance data	Earthfill and backfill	Earthquake	Elasticity	Equipment design	Equipment installation	Error	Excavations	Failure	Field test	Geophysics	Ground water	Highway	Interpretation and analysis	Lateral yield	Load or plate bearing test	Measurement principles	Mining	Multistorey building	New developments	Operation	Partly saturated soil	Performance
85	623						x	x						x												x	x	x								
86	632							x		x								x																		
87	633					x				x			x	x	x								x			x	x									
88	636						x			x		x														x	x									
89	637						x	x		x				x												x								x		
90	641							x		x																										
91	643							x																												
92	644							x																												
93	645							x		x																						x				
94	646							x		x																						x				
95	646			x										x				x		x												x				x
96	647					x								x					x				x										x			
97	649																															x				
98	649						x					x															x			x						
99	650									x		x								x											x					
100	651											x					x			x			x								x					
101	652						x					x																			x					
102	652											x																			x					
103	653											x		x			x	x	x				x						x	x						
104	665											x																								
Discussion - Interpretation																																				
105	666						x				x	x			x												x	x		x						
106	672						x					x															x	x								
107	673						x					x															x	x								
108	675										x	x													x		x									
109	676											x															x									
110	677											x					x										x		x							
111	679											x					x			x			x				x		x							
112	680											x		x									x				x	x								
113	683											x		x			x						x				x			x						
114	686		x							x																										
115	687											x	x														x	x					x			
116	695						x					x		x												x	x	x								
117	697						x	x				x		x												x								x		
118	698					x	x					x														x	x	x								
119	700						x					x															x	x								
120	700						x					x		x													x	x								
121	702													x												x		x								
122	703	x	x							x			x				x									x	x									
123	704											x								x							x			x	x					
124	705											x								x							x									

Contribution No	Permeability	Pile	Pore pressure	Pre-loading	Footing or raft foundation	Repeated loading	Research	Retaining wall	Rock	Safety factor	Seepage	Settlement	Slope stability	State-of-the-art review	Temperature effect	Test procedure	Vibration	Piezometer (hydraulic)	Piezometer (pneumatic)	Piezometer (electric)	Surveying instruments	Surface monuments	Settlement plates & gauges	Settlement (fluid systems)	Settlement (horizontal tubes)	Borehole extensometers	Inclinometers	Tiltmeters	Horizontal displacement gauges	Strain gauges	Failure stakes	Slip plane indicators	Earth pressure cells	Load cells	Flat jacks & stress plugs	Vibration measurements	Seepage measurements
85			x							x			x																								
86																																					
87												x																									
88			x							x			x																								
89										x			x																								
90																																					
91																																					
92																																					
93							x																														
94							x																														
95		x			x		x								x																		x	x			
96							x																										x				
97							x										x																				
98																					x	x															
99									x																	x											
100							x		x						x											x											
101									x	x																											
102										x			x																								
103									x							x																			x		
104																												x									
Discussion - Interpretation																																					
105			x							x			x																								
106			x							x			x																								
107		x	x							x			x																								
108			x								x																										
109			x				x			x																											
110							x																														
111																																					
112		x					x																				x										
113		x					x					x																									
114		x										x																									
115			x				x					x	x																								
116			x							x			x																								
117			x							x			x																								
118			x							x			x																								
119													x																								
120										x			x																								
121												x												x													
122								x																													
123												x																x									
124										x			x																								

Guarantors and Exhibitors

Guarantors

The British Geotechnical Society gratefully acknowledges the following guarantors of the symposium:—

Allott & Lomax
Ove Arup & Partners
Balfour Beatty & Co. Ltd.
Binnie & Partners
Bullen & Partners
Cementation Ground Engineering Ltd.
Brian Colquhoun & Partners
English Clays Lovering Pochin & Co. Ltd.
Exploration Associates Ltd.
Freeman Fox & Partners
W. & C. French (Construction) Ltd.
Sir Alexander Gibb & Partners
Ground Engineering Ltd.
Sir William Halcrow & Partners
Low and Parsons Brown
G. C. Mander & Partners
Mott, Hay & Anderson
Nuttall Geotechnical Services Ltd.
Messrs. Sandberg
Scott Wilson Kirkpatrick & Partners
Soil Instruments Ltd.
Soil Mechanics Ltd.
Terresearch Ltd.
R. Travers Morgan & Partners
Wimpey Laboratories Ltd.

Exhibitors

AGA (UK) Ltd.
A. Clarkson & Co. Ltd.
Engineering Laboratory Equipment Ltd.
Gage Technique Ltd.
Walter Nold Co.
Nuttall Geotechnical Services Ltd.
Professional & Academic Book Exhibitions.
Robertson Research International Ltd.
Smail, Sons & Co. Ltd.
Soil Instruments Ltd.
Solarton Electronic Group Ltd.
Survey & General Instrument Co. Ltd.
Wild Heerbrugg (UK) Ltd.
Wykeham Farrance Engineering Ltd.

Additionally, there were displays from

Building Research Station
Kent C. C. Sub Unit (South Eastern R.C.U.)
National Physical Laboratory
Transport and Road Research Laboratory.

Printed and bound in Great Britain at The Pitman Press, Bath